West Gondwana: Pre-Cenozoic Correlations Across the South Atlantic Region

Geological Society books refereeing procedures

The Society makes every effort to ensure that the scientific and production quality of its books matches that of its journals. Since 1997, all book proposals have been refereed by specialist reviewers as well as by the Society's Books Editorial Committee. If the referees identify weaknesses in the proposal, these must be addressed before the proposal is accepted.

Once the book is accepted, the Society Book Editors ensure that the volume editors follow strict guidelines on refereeing and quality control. We insist that individual papers can only be accepted after satisfactory review by two independent referees. The questions on the review forms are similar to those for Journal of the Geological Society. The referees' forms and comments must be available to the Society's Book Editors on request.

Although many of the books result from meetings, the editors are expected to commission papers that were not presented at the meeting to ensure that the book provides a balanced coverage of the subject. Being accepted for presentation at the meeting does not guarantee inclusion in the book.

More information about submitting a proposal and producing a book for the Society can be found on its web site: www.geolsoc.org.uk.

It is recommended that reference to all or part of this book should be made in one of the following ways:

PANKHURST, R. J., TROUW, R. A. J., BRITO NEVES, B. B. & DE WIT, M. J. (eds) 2008. *West Gondwana: Pre-Cenozoic Correlations Across the South Atlantic Region*. Geological Society, London, Special Publications, **294.**

MOHRIAK, W., NEMCOK, M. & ENCISO, G. 2008. South Atlantic divergent margin evolution: rift-border uplift and salt tectonics in the basins of SE Brazil. *In*: PANKHURST, R. J., TROUW, R. A. J., BRITO NEVES, B. B. & DE WIT, M. J. (eds) *West Gondwana: Pre-Cenozoic Correlations Across the South Atlantic Region*. Geological Society, London, Special Publications, **294**, 365–398.

GEOLOGICAL SOCIETY SPECIAL PUBLICATION NO. 294

West Gondwana: Pre-Cenozoic Correlations Across the South Atlantic Region

EDITED BY

R. J. PANKHURST
British Geological Survey, UK

R. A. J. TROUW
Universidade Federale do Rio de Janeiro

B. B. DE BRITO NEVES
Universidade de São Paulo, Brazil

and

M. J. DE WIT
University of Cape Town, South Africa

2008
Published by
The Geological Society
London

THE GEOLOGICAL SOCIETY

The Geological Society of London (GSL) was founded in 1807. It is the oldest national geological society in the world and the largest in Europe. It was incorporated under Royal Charter in 1825 and is Registered Charity 210161.

The Society is the UK national learned and professional society for geology with a worldwide Fellowship (FGS) of over 9000. The Society has the power to confer Chartered status on suitably qualified Fellows, and about 2000 of the Fellowship carry the title (CGeol). Chartered Geologists may also obtain the equivalent European title, European Geologist (EurGeol). One fifth of the Society's fellowship resides outside the UK. To find out more about the Society, log on to www.geolsoc.org.uk.

The Geological Society Publishing House (Bath, UK) produces the Society's international journals and books, and acts as European distributor for selected publications of the American Association of Petroleum Geologists (AAPG), the Indonesian Petroleum Association (IPA), the Geological Society of America (GSA), the Society for Sedimentary Geology (SEPM) and the Geologists' Association (GA). Joint marketing agreements ensure that GSL Fellows may purchase these societies' publications at a discount. The Society's online bookshop (accessible from www.geolsoc.org.uk) offers secure book purchasing with your credit or debit card.

To find out about joining the Society and benefiting from substantial discounts on publications of GSL and other societies worldwide, consult www.geolsoc.org.uk, or contact the Fellowship Department at: The Geological Society, Burlington House, Piccadilly, London W1J 0BG: Tel. +44 (0)20 7434 9944; Fax +44 (0)20 7439 8975; E-mail: enquiries@geolsoc.org.uk.

For information about the Society's meetings, consult *Events* on www.geolsoc.org.uk. To find out more about the Society's Corporate Affiliates Scheme, write to enquiries@geolsoc.org.uk.

Published by The Geological Society from:
The Geological Society Publishing House, Unit 7, Brassmill Enterprise Centre, Brassmill Lane, Bath BA1 3JN, UK

(*Orders*: Tel. +44 (0)1225 445046, Fax +44 (0)1225 442836)
Online bookshop: www.geolsoc.org.uk/bookshop

British Library Cataloguing in Publication Data

A catalogue record for this book is available from the British Library.

ISBN 978-1-86239-247-2

Typeset by Techset Composition Ltd, Salisbury, UK

Printed by Cromwell Press, Trowbridge, UK

Distributors

North America
For trade and institutional orders:
The Geological Society, c/o AIDC, 82 Winter Sport Lane, Williston, VT 05495, USA
Orders: Tel +1 800-972-9892
Fax +1 802-864-7626
E-mail: gsl.orders@aidcvt.com

For individual and corporate orders:
AAPG Bookstore, PO Box 979, Tulsa, OK 74101-0979, USA
Orders: Tel +1 918-584-2555
Fax +1 918-560-2652
E-mail: bookstore@aapg.org
Website http://bookstore.aapg.org

India
Affiliated East-West Press Private Ltd, Marketing Division, G-1/16 Ansari Road, Darya Ganj, New Delhi 110 002, India
Orders: Tel +91 11 2327-9113/2326-4180
Fax +91 11 2326-0538
E-mail: affiliat@vsnl.com

Contents

Preface

Although the theme of the construction of Gondwana has been touched on at successive Gondwana Symposia (the first of which was held in 1967, in Mar del Plata, Argentina), at International Geological Congresses, and at the Colloquia on African Geology, no direct focus has yet been dedicated to it in publication. At the Tenth Gondwana Symposium, in Cape Town (1998) it emerged clearly that former links between the Gondwana fragments, such as between Africa and South America, are well defined on a broad scale, yet in detail many correlations remained unresolved.

For this reason, at the Twelfth Gondwana Symposium in Mendoza, Argentina (2005), a special symposium was dedicated to the subject, convened by B. B. Brito Neves and R. A. Armstrong. The idea was to bring together the most prominent geoscientists who have worked in the coastal and flanking areas of the two continents, in order to compare structures and rocks on a scale as detailed as possible, and to establish the present limits and uncertainties of well-known correlation. The choice of the tile for the symposium, West Gondwana, the ties that bind, was meant to be provocative, in that it was recognized that many problems, both of geological and biological nature, continue to hamper the advancement of our detailed understanding of the formation and break up of Gondwana. We are sure that the data and interpretations presented at the symposium, much of which are now collated in this book, will help to forge new strategies for future research on the evolution of Gondwana through refining former geological links across the South Atlantic.

We are especially grateful to the following, who have given their time in reviewing the manuscripts submitted to this volume:

Kodjopa Attoh, Peter Cawood, Umberto Cordani, Silas Dada, Ian Dalziel, Patrick Eriksson, Steve Flint, Hartwig Frimmel, Reinhardt Fuck, Claudio Gaucher, Paulo Gorayeb, David Gray, Chris Hawkesworth, Mônica Heilbron, Patrick Ledru, Nuno Machado, Aroldo Misi, Brendan Murphy, Soenke Neben, Marcio Pimentel, Victor Ramos, Augusto Rapalini, Magdalena Scheck-Wenderoth, Karel Schuman, Alcides Sial, Antonio Pedrosa-Soares, Bryan Storey, Ricardo Trinidade, Alexandre Uhlein, Alan Vaughan, Randy Van Schmus, Andre Steenken, and several anonymous reviewers. RJP also thanks Alan Vaughan for his help with the editing of one paper.

Pre-Cenozoic correlations across the South Atlantic region: 'the ties that bind'

M. J. DE WIT[1], B. B. DE BRITO NEVES[2], R. A. J. TROUW[3] & R. J. PANKHURST[4]

[1]*AEON-Africa Earth Observatory Network and Department of Geological Sciences, University of Cape Town, Rondebosch 7700, South Africa (e-mail: maarten.dewit@uct.ac.za)*

[2]*Instituto de Geociências da Universidade de São Paulo, Rua do Lago 562, CEP 055089-080 São Paulo, Brazil*

[3]*Departamento de Geologia, Universidade Federal do Rio de Janeiro, Ilha do Fundão, CEP 21949-900 Rio de Janeiro, Brazil*

[4]*Visiting Research Associate, British Geological Survey, Keyworth, Nottingham NG12 5GG, UK*

The first to recognize the complementary shapes of Africa and South America and to suggest that these continents were once joined together was Dutch scientist Ortelius in 1596. He was followed in 1620 by Elizabethan philosopher Sir Francis Bacon, who asserted that the similarity of their shapes could not be accidental. Nearly 200 years later, German naturalist von Humboldt described how the two continents may have fitted together, and in 1860 French geographer Antonio Snyder produced the first map that showed South America and Africa in close contact (e.g., Blankett 1965). By 1915 the German meteorologist Alfred Wegener had amassed enough data to publish a comprehensive scientific argument for the past conjunction of these two continents on the basis of similarities in the Palaeozoic–Mesozoic geology on each side of the South Atlantic, and then boldly proposed that 'horizontal displacements of the continents' (Horizontal verschiebungen der Kontinente) caused their subsequent separation (Wegener 1915).

Wegener's original hypothesis of 'continental displacement' (Krause & Thiede 2005) was severely criticized, especially by geophysicists (Oreskes 1999). Nevertheless the concept was successfully transformed into the continental drift hypothesis through the support of, amongst others, two prominent geologists working in South America and Africa, respectively: Argentine Juan Keidel (1916) recognized the geological similarities between the Sierra de La Ventana Fold Belt in Argentina and the Cape Fold Belt in South Africa, whilst South African Alex du Toit (1927), following his extended visit to South America in 1923s, first correlated in detail the litho- and biostratigraphy of the Palaeozoic and Mesozoic Karoo sequences of southern Africa across the Atlantic into Brazil and Argentina, and then summarized these findings in his book *Our Wandering Continents* (1937) (Fig. 1). By the early 1960s, advances in palaeomagnetism and the discovery of apparent polar wander paths finally helped to place Wegener's concept of continental drift on more robust geophysical footing. This period culminated in a well-known Royal Society symposium on continental drift at which the first computer-controlled fit between Africa and South America was presented (Bullard *et al.* 1965). Very shortly thereafter, following the discovery of sea-floor spreading, the emergence of plate tectonic theory rapidly embedded Wegener's continental drift and evolved into a truly new field of solid earth geodynamics (Oreskes 2001). All this stimulated new geological and geochronological research to evaluate and test different South America–Africa reconstructions that had been proposed by then. A comparative survey of ages and structures of the basement rocks on each side of the Atlantic Ocean between Brazil and West Africa was well on the way before the 1970s (e.g., Hurley *et al.* 1967; Almeida & Black 1968).

Similar contributions of this type followed rapidly and a major international programme focussed on cross-Atlantic correlations was initiated with UNESCO support (International Geological Correlation Programme, Projects Nos 108 and 144, 1975–1984). Significant syntheses resulting from this new geological research were published over a period of more than a decade (e.g., Torquato & Cordani 1981; Porada 1989; Trompette 1994). In parallel, geophysical investigations in the southern oceans revealed with increasing detail the magnetic

From: PANKHURST, R. J., TROUW, R. A. J., BRITO NEVES, B. B. & DE WIT, M. J. (eds) *West Gondwana: Pre-Cenozoic Correlations Across the South Atlantic Region.* Geological Society, London, Special Publications, **294**, 1–8. DOI: 10.1144/SP294.1 0305-8719/08/$15.00

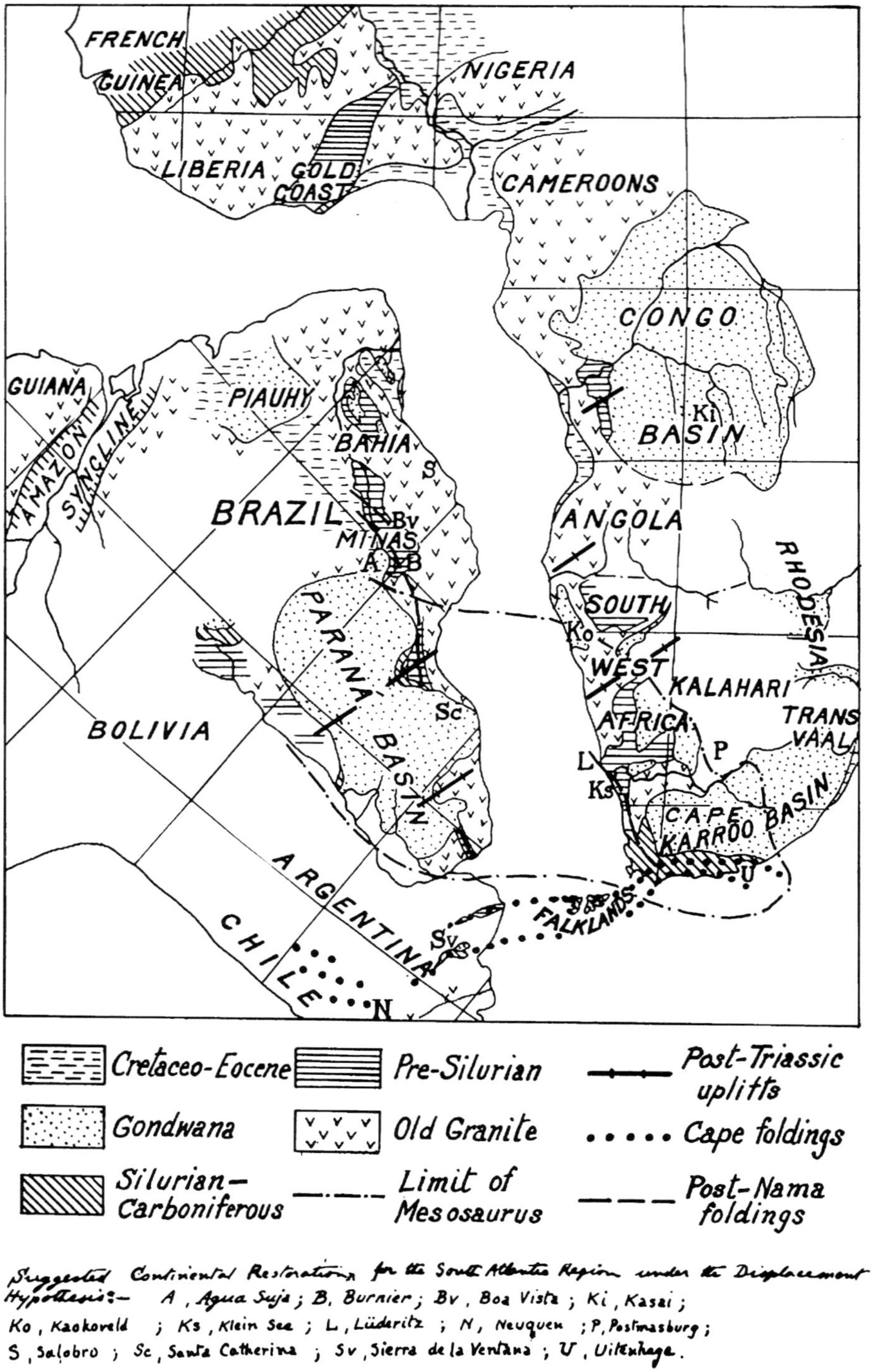

Fig. 1. The first detailed geological comparison between Africa and South America by Alex Logie du Toit. This figure (from the A. du Toit collection, reproduced with permission from the University of Cape Town Library Archives) shows the handwritten proof corrections by du Toit for his book *Our Wandering Continents* published in 1927. This figure was later also published in his presidential address to the Geological Society of South Africa in 1928. Note that du Toit connected the extremities of the Cape Fold Belt and the Sierra de la Ventana Fold Belt directly through the Falkland Islands.

character of the oceanic crust of the South Atlantic: key magnetic anomalies could be correlated on either side of the mid-ocean ridge with great confidence (Rabinowitz & LaBrecque 1979). Using this marine data, new geological reconstruction between these two continents became possible, and by the late 1980s, a new geological map of Gondwana was produced whose reconstruction was based purely on the available marine data (de Wit *et al.* 1988). This map in turn helped stimulate a new phase of geological correlations to further refine the fit between Africa and South America (e.g., Lawver *et al.* 1999). Today, reuniting Gondwana has reached such reliable accuracy that geological features on opposite sides of the South Atlantic can be joined up with a margin of error of less than 100 km (Eagles 2007; de Wit *et al.* this volume).

With this firmer understanding of the relationship between Gondwana continents during the Palaeozoic and early Mesozoic, the geoscience community started to address the question of how Gondwana came to be a supercontinent in the first place; and what might have been the continental precursors to this great landmass. For this, a greater understanding of the building blocks of Gondwana was needed, a requirement that was brought into sharp focus when Canadian geologist Paul Hoffman (1991) suggested that a previous supercontinent, Rodinia, formed at about 1 Ga around the nucleus of Laurentia, the 'Grenvillian' mobile belts representing the associated accretion processes. In this model, Rodinia fragmented during the early Neoproterozoic, the resulting continental blocks drifting away from one another as new ocean basins opened up, and then colliding relatively rapidly again in a complex pattern during the later Neoproterozoic to form the backbone of Gondwana. This new bold step took continental drift much further into the past and nurtured a new concept of supercontinental 'cycles' (e.g., Nance *et al.* 1988; Murphy & Nance 1992; Rogers 1996), almost 100 years after Wegener had introduced the concept of drifting continents. At present the details of Rodinia and its transformation into Gondwana are as controversial as the concept of Gondwana was when it was first formulated (Unrug 1992, 1996; Rogers 1996; Dalziel 1997; Hoffman 1999; Meert 2003; Cordani *et al.* 2003; Mantovani & Brito Neves 2005).

Nomenclature

Differences in the way that geological concepts are used by different geoscientists and on either side of the South Atlantic warrant some discussion. West Gondwana, for example, can be subdivided into cratons, shields, and orogenic or mobile belts (Fig. 2), but there is considerable disagreement about the terms 'craton' and 'shield'. Some of these disagreements stem from the fact that very recent advances in Africa (and Canada), particularly in seismology, tomography, magnetotellurics, geochemistry and mantle petrology, have redefined the shape of cratons more robustly in three and four dimensions: with this, terms such as shield and craton are taking on new meanings. The oldest pristine Archaean terrains are now known to be underlain by unusually thick and depleted mantle lithosphere that stabilized in Archaean times, resulting in a strong lithospheric profile capable of resisting major tectonic and thermal modification for over 3 billion years, except where subsequently rifted apart and broken up below a critical size. Post-Archaean terrains in Africa do not display these unusual lithospheric characteristics. Geoscientists who have focused their studies on these Archaean regions (and their distinct differences with younger continental areas) have suggested that the term craton (or 'tectosphere') should be restricted to these Archaean areas (Jordan 1988; Durheim & Mooney 1994; James *et al.* 2001; Stankiewicz *et al.* 2002; Bell & Moore 2004; Fouch *et al.* 2004; Niu *et al.* 2004; Shirey *et al.* 2005; O'Reilly & Griffith 2006; Chevrot & Zhao 2007). Where cratons have been tectonically fragmented and then reworked by later thermal and tectonic events, they may lose some or all of their cratonic features, especially their thick mantle lithosphere. Such fragments can, in turn, be enlarged through subsequent accretion processes and the addition of new juvenile lithosphere, to form new stabilized regions (within which older cratons, or fragments thereof, may be tectonically embedded), and become covered by undeformed shallow marine and terrestrial sequences. It is suggested that these stable regions should be referred to as shields (de Wit *et al.* this volume). In the present volume, for example, **Pedreira & de Waele** describe Proterozoic (*c.* 1.8 Ga) sedimentary sequences that covered the combined São Francisco–Congo shield prior to Gondwana break-up.

Because in many regions of West Gondwana sufficient seismic/magnetotelluric data and deep mantle petrology/geochemistry are not yet available, the distinction between cratons and shields is not always possible. In this volume therefore the term craton is often used for areas that are composed of both Archaean and Palaeoproterozoic rocks, and that may even include some Mesoproterozoic belts as well, to represent crustal (albeit not necessarily lithospheric) continental units that were essentially unaffected by the late Neoproterozoic to Early Cambrian (650–500 Ma) penecontemporaneous sequence of orogenies traditionally

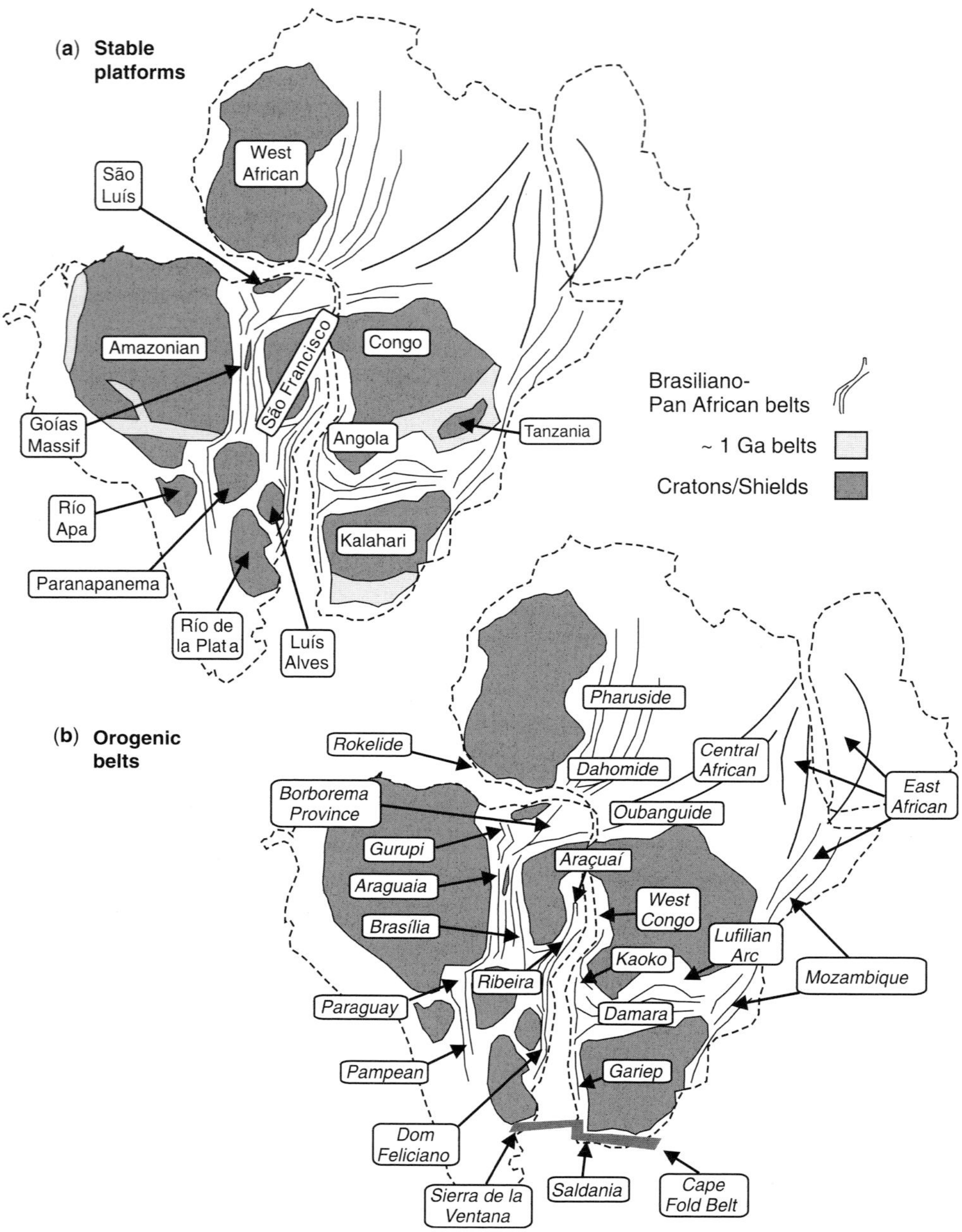

Fig. 2. Modern view of West Gondwana in the mid Palaeozoic with (**a**) the shields and cratonic fragments representing pre-existing continental masses and (**b**) the Pan-African/Brasiliano orogenic belts mainly formed during assembly. N.B. This is a schematic representation, principally to identify the location of named structures dealt with in this volume. Deposition in the Sierra de la Ventana–Cape Fold Belt began in the Early Palaeozoic and continued up to Permian times. After Vaughan & Pankhurst (2008) and Tohver *et al.* (2006).

referred to as *Pan-African* in Africa and *Brasiliano* in South America. These Neoproterozoic 'cratons' represent palaeo-continents (or cores thereof) formed during the Meso–Neoproterozoic break-up of Rodinia, such as the Congo shield in Africa and the São Francisco craton in South America. In cases where theses 'cratons' are relatively small, or their geochronology is poorly defined, they are often referred to as 'blocks' (or crustal fragments), which may have broken off larger palaeocontinents at some earlier stage. One example in South America is the Paranapanema block, which is hidden under the Phanerozoic cover of the Paraná Basin); another, in Africa, is the Latea block of the Hoggar Massif in the Sahara (Caby 2003). Their outlines may be better inferred from gravimetric data, borehole sampling and tectonic inferences (e.g. Mantovani & Brito Neves 2005). In contrast, small fragments of cratonic blocks on one continent may be part of a larger shield on the other continent, for example, the small São Luís fragment in NE Brazil is probably part of the West African shield (see **Klein & Moura** and **de Wit *et al.*** this volume). The Río de la Plata craton (shield) of Uruguay and Argentina is unusual in being predominantly Palaeoproterozoic in age, with only little evidence of Archaean crust: **Pazos *et al.*** review the evidence for Neoproterozoic glaciation of this craton. In Africa, its closest equivalent is the Kalahari shield or the Angola block which, in turn is part of the Congo shield. Clearly then, usage of these different terms for continental lithosphere fragments is confusing. Sorting out these Trans-Atlantic 'geodialects' should be an important quest for future correlation programmes.

Orogenic or mobile belts (also referred to as fold belts, orogens or simply belts) are elongated areas characterized by deformation and/or metamorphism, in the present case mostly related to the Brasiliano/Pan-African orogenies (Fig. 2). They usually contain deformed sedimentary and/or volcanic rocks of Neoproterozoic age, but may contain considerable fragments and slices of older reworked shields or cratons. They may have resulted from collision, transcurrent lithospheric shear zones or progressive accretion of terranes along an active continental margin, but terminal collision is required to explain their position in the interior of West Gondwana. A recent review of the long-term (Neoproterozoic–Palaeozoic) evolution of the accretionary orogenic belts along the proto-Pacific margin of West Gondwana is given by Vaughan & Pankhurst (2007), but the present book is more concerned with the regions within that part of the supercontinent related to initial assembly, which is usually considered to have been completed by mid-Cambrian time. Some of the papers in this book relating to the geology of Brazil represent updated summaries of information presented in the excellent book published for the 31st International Geological Congress in Rio de Janeiro (Cordani *et al.* 2000).

The term 'orogenic cycle', frequently used in the literature on evolution of the Brasiliano/Pan-African belts, meaning to include an initial stage of continental break-up and a final stage of accretion and collision, is largely avoided here for two reasons. First, because in many cases the word 'cycle' is used for the latter part of a full Wilsonian cycle (e.g., that part related to the contractional or orogenic phase), in which case 'orogeny' is preferable. Second, because the concept of the Wilson Cycle with continental break-up followed by collision along the same line of rifting seems not to apply to many orogens under discussion. That is, continents may break up at different times and come together in completely different configurations, possibly on the other side of the Earth. Adherence to the Wilson Cycle concept would appear to be more the exception then the rule. Of course, if the concept of a cycle is understood on a more global scale, as the cycle of formation and destruction of supercontinents (e.g., Nance *et al.* 1988; Murphy & Nance 1992), in this case from Rodinia to Gondwana, then the idea of a supercycle might still be useful.

Instead of 'Brasiliano/Pan-African orogeny' some authors use Brasiliano/Pan-African event or thermo-tectonic event (e.g., de Wit *et al.* 2001, following Kennedy, who first used the expression in Africa in 1964). Since orogenic activity within the whole Gondwana region can now be differentiated using modern geochronology and thermochronology, Pan-African and Brasiliano tectonics are beginning to be recognized as complex and diachronous, and several local orogenies are now identified within the major ones (e.g., the Buzios orogeny within the Ribeira–Araçuai orogenic belt, Schmitt *et al.* 2004; see also Brito Neves *et al.* 1999 and Campos Neto 2000 for syntheses of continental-scale details of the Brasiliano orogeny).

Within the various orogenic belts described in this book many terranes are defined, either exotic or suspect. The precise meaning of this term has been discussed elsewhere in the literature (Coney *et al.* 1980; Howell 1989; Coombs 1997; Vaughan *et al.* 2005) but we should emphasize here that in many cases of contrasting areas of Precambrian rocks, the existing data concerning 'terrane' demarcation and comparison with adjacent areas are relatively scarce and that in several cases these terranes may need to be redefined in the future. Alternatively the term domain may be used for these poorly defined 'possible' terranes.

In summary, much remains to be learned about the details of Brasiliano and Pan-African geology and the various pre-Gondwana basement blocks, before the paleo-geodynamics of Gondwana formation can be fully understood and described. It is therefore perhaps wise that many Gondwana geologists for the moment 'agree to disagree' about the details of their terminology.

Supercontinental origins

The opening of the southern Atlantic Ocean in the Early Cretaceous separated South America from Africa along a line that, south of 12°S, largely follows Neoproterozoic to Cambrian suture belts, but also cuts older cratons and Palaeozoic–Mesozoic sedimentary basins. A best fit of the continents along the 1000 m depth contour shows wide areas where crustal rocks are covered by Mesozoic and Cenozoic shelf sediments, whose disposition has in some cases been disturbed by break-up tectonics (**Mohriak *et al.***), hampering the correlation of older tectonic units across the continents. However, since these older units are mostly cratons, shields and Neoproterozoic mobile belts formed during Gondwana assembly, their detailed comparison and correlation across the present Atlantic Ocean is a crucial step in both the accurate reconstruction of Gondwana and constraining the processes by which it was formed.

The lithospheric nuclei that amalgamated to form Gondwana were essentially fragments of Rodinia. In South America the Brasiliano orogeny records a series of subduction magmatism, accretion and collisional events from 880 to about 530 Ma (Brito Neves *et al.* 1999; Campos Neto 2000). In Africa, the major accretions and collisions of the Pan-African orogeny occurred over a shorter time span, between about 650 and about 530 Ma. Collectively these orogenic events led to the final formation of West Gondwana (Unrug 1996; Brito Neves *et al.* 1999; Meert 2003). The detailed identification, recognition and correlation of tectonic terranes and domains within the various belts and provinces are some of the major issues discussed in this book, together with the ways in which later events that occurred once the supercontinent had achieved stability can be correlated across the Atlantic. Not all these issues are resolved yet in a satisfactory way, and these therefore will need further study in the future.

The assembly of East Gondwana probably resulted from prolonged and/or progressive Pan-African collisions between India, Africa, and East Antarctica–Australia along orogenic belts running from the Arabia–Nubian shield, through the Mozambique belt, to East Antarctica (e.g., Jacobs & Thomas 2004). This process began at 650 Ma or slightly earlier, terminating in some places with a Cambrian-age orogeny at 535–520 Ma (e.g., Meert 2003; Boger & Miller 2004), but this late phase elsewhere may be related to a post-orogenic exhumation history (e.g., de Wit *et al.* 2001). The assembly of the separate fragments that constitute West Gondwana is equally prolonged and, in general, also not well constrained, although some aspects of the puzzle are becoming clearer. Palaeomagnetic data constraining ocean-spreading during the separation of Amazonia, West Africa and Baltica from Laurentia during Rodinia break-up is reviewed by **Pisarevsky *et al.*** They propose that the opening of the main branches of the intervening Iapetus Ocean were probably plume-related, but that a bimodal uncertainty in the database prevents a definitive interpretation, although Tohver *et al.* (2006) consider that some parts of West Gondwana (West Africa–Amazonian shield) may not have been part of Rodinia at all. The time interval between 880 and 650 Ma was marked by the movement of these fragments across Neoproterozoic oceans, generating magmatic arcs (e.g., the Goias magmatic arc in the Brasília Belt, Pimentel *et al.* 2004; **Valeriano *et al.***) and ophiolites (e.g., **Pires Paixão *et al.***; **Pedrosa-Soares *et al.***). The geological evolution of the Borborema Province of NE Brazil up to and including the collisional history recorded in the orogenic belts, and comparisons with evidence from the geological record for these events in West Africa, are reviewed in this volume by **Arthaud *et al.***, **Santos *et al.***, **Van Schmus *et al.*** and **Dada**. To the south of this, the geology and evolution of the Araguaia, Brasília, Araçuaí, and Ribeira belts, together with their probable links to the West Congo region, are treated in this volume by **Pires Paixão *et al.***, **Moura *et al.***, **Valeriano *et al.***, **Pedrosa-Soares *et al.***, **Heilbron *et al.*** and **Schmitt *et al.*** A southern palaeo-ocean, the Adamastor ocean probably existed during much of the Neoproterozoic between the south-central African shields and the south-central South American shields (**Pedrosa-Soares *et al.***, **Gray *et al.***). **Basei *et al.*** present U–Pb data for detrital zircon that elucidate the provenance of sediments deposited on ether margin of this ocean throughout the Neoproterozoic. Collisions between the South American and the African nuclei seem to have culminated at *c.* 520 Ma, essentially at the same time as a terminal event within parts of the East African–Antarctic orogen (Jacobs & Thomas 2004), as demonstrated by the evidence for Cambrian orogeny in the Ribeira Belt of eastern Brazil (**Heilbron *et al.***; **Schmitt *et al.***). **Gray *et al.*** review the history of the orogenic belts on the African side (Damara, Kaoko and Gariep) and deduce that the Adamastor Ocean closed sequentially from north to south, followed by northward thrusting of the Kalahari shield across the Damara Belt.

Between about 520 and 500 Ma, extensive exhumation and erosion led to regional peneplanation, especially across Africa, followed by widespread deposition of siliciclastic sequences such as the Table Mountain Group of southern Africa, the Alto Garças Formation in Brazil, the Caacupé Group in Paraguay and their equivalents in North and West Africa (Burke *et al.* 2003; **Milani & de Wit**).

After the short-lived Ashgill glaciation a gradual transition took place to stable platform conditions, with the development of large intracratonic sedimentary basins, such as the Paraná, Parnaíba, and Amazonas basins in Brazil and the Karoo basin in southern and central Africa, reviewed in this volume by **Milani & de Wit**. During this period of relatively stable internal Gondwana, lasting until Triassic desertification, Palaeozoic accretion continued along its proto-Pacific margin (e.g., Vaughan *et al.* 2005; Vaughan & Pankhurst 2007).

Thus, the formation of Gondwana occurred by the assembly of quite varied fragmented cratonic nuclei from earlier supercontinents, through ocean-spreading, subduction, accretion and collisions over a period of 250–350 million years. In the process, some of the building blocks (shields and cratons) were modified in their form and structure, and even further fragmented. Local and regional orogenic belts developed quasi-simultaneously, often overprinting or cross-cutting earlier belts in a way that could have caused crustal shortening, block rotations and the opening of new basins, even after major stages of assembly were completed. The complexities of these interactions, together with poor exposure or a paucity of good data continue to impede a definitive timetable and exact reconstructions. Continued field and laboratory studies, and in particular aeromagnetic surveys are clearly necessary as called for in the final chapter of this book, in which **de Wit *et al.*** also propose specific geological features that in principle should help to resolve some of the details of how we should envisage West Gondwana in its essentially final form, and constrain parameters in order to model the assembly of Gondwana with greater accuracy and precision.

This is AEON Contribution no. 47.

References

DE ALMEIDA, F. F. M. & BLACK, R. 1968. Geological comparison of northeastern South America and western Africa. *Anais da Academia Brasileira de Ciencias*, **40** (supplement), 317–319.

BELL, D. R. & MOORE, R. O. 2004. Deep chemical structure of the southern African mantle from kimberlite megacrysts. *South African Journal of Geology*, **107**, 59–80.

BLANKETT, P. M. S. 1965. Introduction. *In*: BLANKETT, P. M. S., BULLARD, E. & RUNCORN, S. K. (eds) *A Symposium on Continental Drift*. Philosophical Transactions of the Royal Society, **A258**, vii–x.

BOGER, S. D. & MILLER, J. McL. 2004. Terminal suturing of Gondwana and the onset of the Ross–Delamerian Orogeny: the cause and effect of an Early Cambrian reconfiguration of plate motions. *Earth and Planetary Science Letters*, **219**, 35–48.

BULLARD, E. C., EVERETT, J. E. & SMITH, A. G. 1965. The fit of the continents around the Atlantic. *In*: BLANKETT, P. M. S., BULLARD, E. & RUNCORN, S. K. (eds) *A Symposium on Continental Drift*. Philosophical Transactions of the Royal Society, London, **A258**, 41–51.

BRITO NEVES, B. B., CAMPOS NETO, M. C. & FUCK, R. 1999. From Rodinia to Western Gondwana: an approach to the Brasiliano-Pan African cycle and orogenic collage. *Episodes*, **22**, 155–166.

BURKE, K., MACGREGOR, D. S. & CAMEROON, N. R. 2003. Africa's petroleum systems: four tectonic 'Aces' in the past 600 million years. *In*: ARTHUR, T., MACGREGOR, D. S. & CAMERON, N. R. (eds) *Petroleum Geology of Africa: new themes and developing technologies*. Geological Society, London, Special Publications, **207**, 21–60.

CABY, R. 2003. Terrane assembly and geodynamic evolution of Central-Western Hoggar: a synthesis. *Journal of African Earth Sciences*, **37**, 133–159.

CAMPOS NETO, M. C. 2000. Orogenic Systems from Southwestern Gondwana. *In*: CORDANI, U. G., MILANI, E. J., THOMAZ FILHO, A. & CAMPOS, D. A. (eds) *Tectonic Evolution of South America, 31st International Geological Congress, Rio de Janeiro*, 335–365.

CHEVROT, S. & ZHAO, L. 2007. Multiscale finite-frequency Raleigh wave tomography of the Kaapvaal craton. *Geophyics Journal International*, doi:10.1111/j.1365-246X.2006.03289.x.

CORDANI, U. G., MILANI, E. J., THOMAZ FILHO, A. & CAMPOS, D. A. (eds) 2000. *Tectonic Evolution of South America, International Geological Congress, Rio de Janeiro*. http://www.cprm.gov.br.

CORDANI, U. G., BRITO NEVES, B. B. & D'ÁGRELLA-FILHO, M. S. 2003. From Rodinia to Gondwana: a review of the available evidence from South America. *Gondwana Research*, **6**, 275–283.

DALZIEL, I. W. D. 1997. Neoproterozoic-Paleozoic geography and tectonics: review, hypotheses and environmental speculation. *Geological Society of America Bulletin*, **109**, 16–42.

DE WIT, M. J., JEFFERY, M., BERGH, H. & NICOLAYSEN, L. 1988. *Geological map of sectors of Gondwana, reconstructed to their disposition ~150 Ma*. American Association of Petroleum Geologists, Tulsa, USA.

DE WIT, M. J., BOWRING, S. A., ASHWAL, L. D., RANDRAINASOLO, L. G., MOREL, V. P. I. & RAMBELOSON, R. A. 2001. Age and tectonic evolution of Neoproterozoic ductile shear zones in southwestern Madagascar, with implications for Gondwana studies. *Tectonics*, **20**, 1–45.

DURHEIM, R. J. & MOONEY, W. D. 1994. Evolution of the Precambrian lithosphere: seismological and geochemical constraints. *Journal of Geophysical Research*, **99**, 15359–15374.

DU TOIT, A. L. 1927. *A geological comparison of South America with South Africa*. The Carnegie Institution, Washington, publication **381**.

DU TOIT, A. L. 1937. *Our wandering continents*. Oliver and Boyd, Edinburgh.

EAGLES, G. 2007. New angles on South Atlantic opening. *Geophysical Journal International*, **166**, 353–361.

FOUCH, M. J., JAMES, D. E., VAN DECAR, J. C. & VAN DER LEE, S. 2004. Mantle seismic structure beneath the Kaapvaal and Zimbabwe Cratons. *South African Journal of Geology*, **107**, 33–44.

HOFMANN, P. F. 1991. Did the break-out of Laurentia turn Gondwanaland inside-out? *Science*, **252**, 1409–1412.

HOFMANN, P. F. 1999. The break-up of Rodinia, birth of Gondwana, true polar wander and the snowball Earth. *Journal of African Earth Sciences*, **28**, 17–33.

HURLEY, P. M., ALMEIDA, F. F. *ET AL.* 1967. Test of drift by comparison of radiometric ages. *Science*, **157**, 495–500.

JACOBS, J. & THOMAS, R. J. 2004. Himalayan-type indenter-escape tectonics model for the southern part of the late Neopaleozoic-early Paleozoic East African-Antarctic orogen. *Geology*, **32**, 721–724.

JAMES, D., FOUCH, M., VANDECAR, J. & VAN DER LEE, S. 2001. Tectospheric structure beneath southern Africa. *Geophysical Research Letters*, **28**, 2485–2488.

JORDAN, T. H. 1988. Structure and formation of the continental tectosphere. *Journal of Petrology*, Special Lithosphere Volume, 11–37.

KENNEDY, W. Q. 1964. The structural differentiation of Africa in the Pan African (500 m.y.) tectonic episode. *Annual Reports of the Institute of African Geology*, **8**, session 1962–1963, Leeds University, UK, 48–49.

KEIDEL, J. 1916. La geología de las sierras de la Província de Buenos Aires y sus relaciones con las montañas de Sud Africa y los Andes. *Anales del Ministerio de Agricultura de la Nación, Sección Geología, Mineralogía y Minería, Buenos Aires*, **3**, 1–78.

KRAUSE, R. & THIEDE, J. 2005. *Continental Drift. The original notes and quotations, Alfred Wegener.* Reports on Polar and Marine Research, Alfred-Wegener-Institute fur Polar- und Meereforschung, **516**.

LAWVER, L. A., GAHAGAN, L. M. & DALZIEL, I. W. D. 1999. A tight fit–Early Gondwana, a plate reconstruction perspective. *In*: MOTOYOSHI, Y. & SHIRAISHI, K. (eds) *Origin and evolution of the continents*. Memoirs, National Institute Polar Research, Tokyo, Japan, Special Issue, **53**, 214–229.

MANTOVANI, M. S. M. & BRITO NEVES, B. B. 2005. The Paranapanema Lithospheric block: its importance for the Proterozoic (Rodinia, Gondwana) supercontinents theories. *Gondwana Research*, **8**, 303–315.

MEERT, J. G. 2003. A synopsis of events related to the assembly of eastern Gondwana. *Tectonophysics*, **362**, 1–40.

MURPHY, J. B. & NANCE, R. D. 1992. Mountain Belts and the Supercontinent Cycle. *Scientific American*, 84–91.

NANCE, D., WORSLEY, T. R. & MOODY, J. 1988. The Supercontinent Cycle. *Scientific American*, 44–50.

NIU, F., LEVANDER, A., COOPER, C. M., VAN DER LEE, C. A., LENARDIC, A. & JAMES, D. E. 2004. Seismic constraints on the depth and composition of the mantle keel beneath the Kaapvaal craton. *Earth and Planetary Science Letters*, **224**, 337–346.

ORESKES, N. 1999. *The Rejection of Continental Drift.* Oxford University Press, Oxford.

ORESKES, N. 2001. *Plate tectonics. An insider's history of the modern theory of the earth.* Westview Press, Oxford.

O'REILLY, S. Y. & GRIFFITH, W. L. 2006. Imaging chemical and thermal heterogeneities in the subcontinental lithopshere mantle with garnets and xenoliths: geophysical implications. *Tectonophysics*, **416**, 289–309.

PIMENTEL, M. M., JOST, H. & FUCK, R. A. 2004. O embasamento da Faixa Brailia e o Arco Magmático de Goiás. *In*: MANTESSO NETO, V., BARTORELLI, A., CARNEIRO, C. R. & BRITO NEVES, B. B. (eds) *Geologia do Continente Sul Americano: Evolução da obra de Fernando Flávio Marques de Almeida.* Editora Beca, São Paulo, 355–368.

PORADA, H. 1989. Pan-African rifting and orogenesis in southern to equatorial Africa and Eastern Brazil. *Precambrian Research*, **44**, 103–136.

RABINOWITZ, P. D. & LABRECQUE, J. 1979. The Mesozoic South Atlantic Ocean and evolution of its continental margins, *Journal of Geophysical Research*, **84**, 5973–6002.

ROGERS, J. J. W. 1996. A history of the continents in the past three billions years. *Journal of Geology*, **104**, 217–228

SCHMITT, R., TROUW, R. A. J., VAN SCHMUS, W. R. & PIMENTEL, M. M. 2004. Late amalgamation in the central part of West Gondwana: new geochronological data and the characterization of a Cambrian collisional orogeny in the Ribeira Belt (SE Brazil). *Precambrian Research* **133**, 29–61.

SHIREY, S. B., HARRIS, J. W., RICHARDSON, S. H., FOUCH, M. J., JAMES, D. E., CARTIGNY, P., DEINES, P. & VILJOEN, F. 2005. Diamond genesis, seismic structure, and the evolution of the Kaapvaal-Zimbabwe craton. *Science*, **297**, 1683–1686.

STANKIEWICZ, J., CHEVROT, S., VAN DER HILST, R. D. & DE WIT, M. J. 2002. Crustal thickness, discontinuity depth and upper mantle structure beneath southern Africa: constraints from body wave conversions. *Physics of Earth and Planetary Interiors*, **130**, 235–251.

TOHVER, E., D'AGRELLA-FILHO, M. S. & TRINDADE, R. I. F. 2006. Paleomagnetic record of Africa and South America for the 1200-500 Ma interval, and evaluation of Rodinia and Gondwana assemblies. *Precambrian Research*, **147**, 193–222.

TORQUATO, J. R. & CORDANI, U. C. 1981. Brazil-Africa geological links. *Earth-Science Reviews*, **17**, 155–176.

TROMPETTE, R. 1994. *Geology of Western Gondwana.* A.A. Balkema, Rotterdam.

UNRUG, R. 1992. The supercontinental cycle and Gondwanaland assembly: component cratons and the timing of suturing events. *Journal of Geodynamics*, **16**, 215–240.

UNRUG, R. 1996. The Assembly of Gondwanaland. *Episodes*, **19**, 11–20.

VAUGHAN, A. P. M. & PANKHURST, R. J. 2007. Tectonic overview of the West Gondwana margin. *Gondwana Research*, doi: 10.1016/j.gr.2007.07.004.

VAUGHAN, A. P. M., LEAT, P. T. & PANKHURST, R. J. 2005. Terrane processes at the margins of Gondwana: introduction. *In*: VAUGHAN, A. P. M., LEAT, P. T. & PANKHURST, R. J. (eds) *Terrane Processes at the Margins of Gondwana.* Geological Society, London, Special Publications, **246**, 1–22.

WEGENER, A. 1915 (1920/22/29). *Die Entstehung der Kontinente und Ozeane.* Braunschweig, Germany.

Late Neoproterozoic and Early Cambrian palaeogeography: models and problems

S. A. PISAREVSKY[1,2], J. B. MURPHY[3], P. A. CAWOOD[4] & A. S. COLLINS[4]

[1]*Tectonics Special Research Centre, School of Earth and Geographical Sciences, The University of Western Australia, 35 Stirling Highway, Crawley, 6009, WA, Australia*

[2]*Present address: School of GeoSciences, The University of Edinburgh, Grant Institute, The King's Buildings, West Mains Road, Edinburgh EH9 3JW, UK (e-mail: Sergei.Pisarevsky@ed.ac.uk)*

[3]*Department of Earth Sciences St. Francis Xavier University, Antigonish, Nova Scotia, B2G 2W, Canada*

[4]*Continental Evolution Research Group, School of Earth and Environmental Sciences, The University of Adelaide, Adelaide, SA 5005, Australia*

Abstract: We present two alternative sets of global palaeogeographical reconstructions for the time interval 615–530 Ma using competing high and low-latitude palaeomagnetic data subsets for Laurentia in conjunction with geological data. Both models demonstrate a genetic relationship between the collisional events associated with the assembly of Gondwana and the extensional events related to the opening of the Tornquist Sea, the eastern Iapetus Ocean (600–550 Ma), and the western Iapetus Ocean (after 550 Ma), forming a three-arm rift between Laurentia, Baltica, and Gondwana. The extensional events are probably plume-related, which is indicated in the reconstructions by voluminous mafic magmatism along the margins of palaeo-continents. The low-latitude model requires a single plume event, whereas the high-latitude model needs at least three discrete plumes. Coeval collisions of large continental masses during the assembly of Gondwana, as well as slab pull from subduction zones associated with those collisions, could have caused upper plate extension resulting in the rifted arm that developed into the eastern Iapetus Ocean and Tornquist Sea but retarded development of the western Iapetus Ocean. As a result, the eastern Iapetus Ocean and the Tornquist Sea opened before the western Iapetus Ocean.

The Late Neoproterozoic to Early Cambrian is one of the most enigmatic time intervals in the Earth's history. This interval includes: (i) one or more major global-scale glaciations (e.g., Kirschvink 1992; Hoffman & Schrag 2002; Hoffman 2005), (ii) the explosion of Ediacaran and Cambrian fauna (Knoll 1992; McCall 2006), (iii) the final break-up of the Rodinia supercontinent by opening of the Iapetus ocean, Tornquist Sea and Palaeo-Asian ocean (e.g., Bingen *et al.* 1998; Cawood *et al.* 2001; Cawood & Pisarevsky 2006), and (iv) the assembly of the Gondwana supercontinent by closing the Mozambique, Adamastor and Brasiliano oceans (Pimentel *et al.* 1999; Collins & Pisarevsky 2005). Knowledge of palaeogeography is crucial to understanding these events and possible linkages between them. However, the palaeogeography of this time interval is unresolved (e.g., Hartz & Torsvik 2002; Murphy *et al.* 2004; Collins & Pisarevsky 2005; Cawood & Pisarevsky 2006; McCausland *et al.* 2006; Tohver *et al.* 2006), caused, in part, by controversy over the North American palaeomagnetic data (McCausland & Hodych 1998; Pisarevsky *et al.* 2000, 2001a; Meert & Van der Voo 2001, Cawood & Pisarevsky 2006) and the relatively poor palaeomagnetic database for other continents (Pisarevsky 2005). These controversies have led to hypotheses such as Inertial Interchange True Polar Wander (IITWP, Kirschvink *et al.* 1997; Evans 1998, Maloof *et al.* 2006), or an anomalously large non-dipole component of the Earth's magnetic field (e.g., McCausland *et al.* 2003).

Development of the Iapetus Ocean is one of the key palaeogeographical events during the late Neoproterozoic to early Palaeozoic time interval. Although there is broad agreement on the timing of the rift–drift stages in its development (e.g., Bingen *et al.* 1998; Cawood *et al.* 2001; Williams & Hiscott 1987), understanding of the mechanisms that led its opening remain elusive. In order to gain insights into these mechanisms

From: PANKHURST, R. J., TROUW, R. A. J., BRITO NEVES, B. B. & DE WIT, M. J. (eds) *West Gondwana: Pre-Cenozoic Correlations Across the South Atlantic Region.* Geological Society, London, Special Publications, **294**, 9–31. DOI: 10.1144/SP294.2 0305-8719/08/$15.00

we utilize regional aspects of the latest Neoproterozoic palaeogeography that have already been published (Powell & Pisarevsky 2002; Murphy *et al.* 2004; Collins & Pisarevsky 2005; Cawood & Pisarevsky 2006), with some modifications based on recently published data. We focus on the boundary forces affecting the unified Laurentia–Baltica–Amazonia–West Africa plate immediately prior to the onset of the development of Iapetus opening and on the distribution and orientation of dyke swarms in the heart of that plate.

The assembly of Gondwana is broadly coeval with the development of the Iapetus Ocean (Grunow *et al.* 1996; Cawood 2005; Cawood & Buchan 2007). Models for Gondwana assembly have evolved dramatically through the last decade from a relatively simple situation involving the collision of two large continents, East and West Gondwana (e.g., Dalziel 1992) to a multi-stage assembly of a number of smaller continents and terranes (Collins & Pisarevsky 2005). The latter model includes the accretion of fragments of the future enigmatic Avalonian continent onto the Gondwanan margin: an event that played an important role in Iapetus Ocean history (Murphy *et al.* 2004).

As the emphasis of this paper is on the potential relationship between the development of the Iapetus Ocean and the assembly of Gondwana, we do not consider other continental blocks, such as North and South China, Indochina, Tarim or Omolon. Only a preliminary model of the Siberian drift history is presented in view of ongoing studies on this continent and the controversy regarding its potential Laurentian connections (Pisarevsky & Natapov 2003 and references therein).

Palaeomagnetic constraints for the latest Neoproterozoic–Early Cambrian drift history

Table 1 presents the latest palaeopoles for the interval 615 to 530 Ma with a reliability index $Q \geq 3$ (Van der Voo 1990), dominated by data from three continents: Laurentia, Baltica and Australia. Data from Laurentia are relatively abundant, but also controversial in forming two groups of roughly similar reliability, one that supports a low-latitude position for Laurentia and the other that supports a high-latitude position between *c.* 600 and 560 Ma. Discussion of this controversy is provided by Cawood & Pisarevsky (2006), but we emphasize that if all these late Neoproterozoic palaeopoles are primary and their ages are correctly assigned, then they are difficult to explain through normal plate tectonic mechanisms and processes such as IITPW or a more rapid style of plate-tectonics may need to be invoked (e.g., Evans 2003). An alternative explanation is that some of these data are misleading, being either the result of re-magnetization or incorrect deciphering of the palaeomagnetic signal (e.g., Hodych *et al.* 2004). We think that the question is still open and at present it is impossible to indicate any preference between the two datasets. Accordingly, we evaluate the tectonic significance of two alternative models for the latest Neoproterozoic–Early Cambrian palaeogeography, naming them as 'high-latitude' and 'low-latitude' models depending on the chosen set of Laurentian palaeopoles.

Palaeopoles for Baltica between 650 and 540 Ma (Table 1) are also incompatible with a smooth Apparent Polar Wander Path (APWP), and several alternative APWPs have been debated (Popov *et al.* 2002). This issue is also discussed by Cawood & Pisarevsky (2006) but, importantly, a consistent group of *c.* 550–555 Ma palaeopoles from both northern and southern Baltica (Popov *et al.* 2002, 2005; Nawrocki *et al.* 2004; Iglesia Llanos *et al.* 2005) and the coeval Skinner Cove Laurentian pole (McCausland & Hodych 1998) undoubtedly indicate that Baltica was separated from Laurentia by *c.* 550 Ma (cf. Cawood *et al.* 2001).

In Australia, palaeomagnetic information from a drill hole through strata in the Neoproterozoic Officer basin shows that Australia probably inhabited low latitudes from *c.* 820 Ma until the Early Cambrian (Pisarevsky 2001; Pisarevsky *et al.* 2001*b*; Pisarevsky *et al.* 2007). From *c.* 650 Ma to 550 Ma there is a swathe of palaeopoles, albeit with poor age constraints in many cases. Nevertheless, they form a consistent pattern that places Australia in low latitudes throughout this time, with the southern margin of the continent being near the equator at 600 Ma (Schmidt *et al.* 1991; Schmidt & Williams 1995; Sohl *et al.* 1999). One pole from the Bunyeroo Formation, associated with the Acraman impact structure (Schmidt & Williams 1996), falls outside the main group of palaeopoles. This could be evidence for rapid rotation of Australia, or IITPW, at *c.* 590 Ma. However, impact-related rocks are not well understood palaeomagnetically. Also, no such deflection in palaeolatitude has been found so far by palaeomagnetic studies of Australian sedimentary sections in drill-holes (Pisarevsky 2001; Pisarevsky *et al.* 2001*b*; Pisarevsky *et al.* 2007).

There are no reliable late Neoproterozoic palaeomagnetic poles from the pre-assembly Congo or Kalahari blocks. The post-orogenic 547 Ma Sinyai dolerite intrudes the East African orogen and provides a pole for this part of proto-Gondwana (Meert & Van der Voo 1996; McElhinny *et al.* 2003). In India, palaeomagnetic data from the Bhander and Rewa series are only broadly constrained as Neoproterozoic or Early

Table 1. *Late Neoproterozoic and Early Cambrian palaeomagnetic poles*

Object	Pole		dp/dm	Q	Age	Reference (3)
	(°N)	(°E)	(°)		(Ma)	
Baltica						
Egersund Dykes, Norway	−28	232	15/18	3	616 ± 3	Storetvedt (1966); Bingen *et al.* (1998)
Egersund Dykes, Norway	−22	231	16/21	4	616 ± 3	Poorter (1972); Bingen *et al.* (1998)
Fen Complex, Norway	−56	330	7/10	4	583 ± 15	Meert *et al.* (1998)
Winter Coast sediments, Russia	25	312	2/4	6	555.3 ± 0.3	Popov *et al.* (2002); Martin *et al.* (2000)
Zolotitsa sediments, Russia	32	293	2/3	6	550 ± 5	Popov *et al.* (2005)
Verkhotina sediments, Russia	32	287	2/3	5	550 ± 1	Popov *et al.* (2005)
Zolotitsa sediments, Russia	28	290	4/4	6	550 ± 5	Iglesia Llanos *et al.* (2005)
Volhynia lavas & tuffs*, Russia	34	306	–	4	551 ± 4	Nawrocki *et al.* (2004) Compston *et al.* (1995)
Torneträsk Formation, Sweden	56	296	12/15	4	545–520	Torsvik & Rehnström (2001)
Laurentia						
Long Range Dykes (5 dykes)[†]	19	355	15/21	5	615 ± 2	Murthy *et al.* (1992); Kamo & Gower (1994)
Cloud Mountain Basalt	−5	352	2/4	3	605 ± 10	Deutsch & Rao (1977); Stukas & Reynolds (1974)
Callander Complex	46	301	6/6	5	575 ± 5	Symons & Chiasson (1991)
Catoctin Volcanics A	43	308	9/9	5	564 ± 9	Meert *et al.* (1994); Aleinikoff *et al.* (1995)
Catoctin Volcanics B	4	13	10/10	4	564 ± 9	Meert *et al.* (1994); Aleinikoff *et al.* (1995)
Sept Iles Intrusion A	−20	321	5/9	5	565 ± 4	Tanczyk *et al.* (1987); Higgins & van Breemen (1998)
Sept Iles Dykes B	59	296	10/10	4	<565 ± 4	Tanczyk *et al.* (1987); Higgins & van Breemen (1998)
Buckingham lavas	10	341	7/10	4	573 ± 32	Dankers & Lapointe (1981)
Johnnie Formation	10	342	5/10	4	570 ± 10	Van Alstine & Gillett (1979); Hodych *et al.* (2004)
Skinner Cove Formation	−15	337	9/9	4	550 ± 3	McCausland & Hodych (1998); Cawood *et al.* (2001)
Australia						
Yaltipena Fm., SA	−44	353	8/8	7	620–630	Sohl *et al.*, (1999)
Elatina Fm., SA	−39	6	9/9	7	600–620	Sohl *et al.* (1999)
Elatina Fm., SA	−52	347	11/11	7	600–620	Schmidt & Williams (1995)
Elatina Fm., SA	−54	327	1/1	6	600–620	Schmidt *et al.* (1991)
Elatina Fm., SA	−51	337	2/2	5	600–620	Embleton & Williams (1986)
Brachina Fm., SA	−33	328	16/16	6	590–610	McWilliams & McElhinny (1980)
Bunyeroo Fm., SA	−18	16	7/12	6	550–620	Schmidt & Williams (1996)
Albany Belt Dykes	−38	347	12/12	4	520–600	Harris & Li (1995)
Upper Arumbera Sandstone	−47	333	3/3	7	530–560	Kirschvink (1978)
Todd River Dolomite	−43	340	6/6	7	530–545	Kirschvink (1978)
Hawker Group	−21	15	9/9	5	530–545	Klootwijk (1980)

(*Continued*)

Table 1. *Continued*

Object	Pole		dp/dm	Q	Age	Reference (3)
	(°N)	(°E)	(°)		(Ma)	
Antrim Plateau Volcanics	−9	340	17/17	4	520–570	McElhinny & Luck (1970)
India						
Bhander and Rewa Series	−47	33	6/6	5	530–560	McElhinny *et al.* (1978)
Amazonia						
Puga Cap Carbonate	83	113	6/9	4	580–630	Trinidade *et al.* (2003)
Gondwana						
Mean 540–560 Ma pole	−10	330	7/7		540–560	McElhinny *et al.* (2003)
Mean Early Cambrian pole	23	334	15/15		525–540	McElhinny *et al.* (2003)
Siberia						
Bolshaya Lena redbeds	3	348	6/10	5	542–630	Pisarevsky *et al.* (2000)
Minya Fm	34	217	9/15	3		Kravchinsky *et al.* (2001)
Shaman Fm	32	251	7/14	4	542–630	Kravchinsky *et al.* (2001)
Biryusa dykes	25	301	14/28	4	608–618	Metelkin *et al.* (2005)
Kesyussa Fm, Olenek River	38	345	9/15	5	535–542	Pisarevsky *et al.* (1997)
Lena River sediments	17	245	3/6	5	513–542	Kirschvink & Rozanov (1984)

Q, Quality factor after Van der Voo (1990), which ranges from 0 to 7 with the latter representing the highest quality data.
*Mean of two poles.
†Recalculated by Hodych *et al.* (2004).

Cambrian (McElhinny *et al.* 1978; Evans 2000). Recently Chirananda De (2003) reported the discovery of medusoid fossils of Ediacaran affinity at the base of Bhander Group, apparently below the strata sampled by McElhinny *et al.* (1978). If so, the time range for the Bhander pole might be narrowed to between 560 and 530 Ma.

The Puga cap carbonate palaeopole from Amazonia (Trindade *et al.* 2003) indicates low latitude for Amazonia around 600 Ma. This pole is suspiciously close to the present-day pole, so it may represent a recent re-magnetization. However, the possibility that this pole is correct and that São Francisco/Congo did not collide with Amazonia until after *c.* 635 Ma along the Pampean–Araguaia orogen cannot be discounted (Trindade *et al.* 2006).

The latest Neoproterozoic Siberian palaeomagnetic database is controversial. The poles are derived from sedimentary successions that are poorly dated (Table 1). The only reasonably well dated pole is from the 612 Ma Biryusa dykes (Metelkin *et al.* 2005; age from Gladkochub *et al.* 2006), but has been calculated from only three dykes. Possible reasons for the discordance of Siberian poles have been discussed previously (e.g., Pisarevsky *et al.* 1997; Smethurst *et al.* 1998), including IITWP (Evans 1998). The reconstructions used in those discussions placed the northern margin of Siberia against the northern margin of Laurentia. Recently published Late Mesoproterozoic palaeomagnetic data (Gallet *et al.* 2000; Pavlov *et al.* 2000, 2002) suggest that the only permissible configuration at 1050–950 Ma involved Siberia separated from Laurentia with its southern margin facing towards the northern margin of Laurentia (Pisarevsky & Natapov 2003). On the other hand, the mid-Cambrian position of Siberia is supported by good palaeomagnetic evidence (Smethurst *et al.* 1998), implying that this continent rotated almost 180° during the Neoproterozoic. Possible evidence of a rifting event in southern Siberia (Gladkochub *et al.* 2006) at *c.* 740 Ma implies that this rotation occurred in Late Neoproterozoic times.

Key fragments of the latest Neoproterozoic–Early Cambrian palaeogeography

Iapetian realm

Among various published configurations of Laurentia, Baltica, and Amazonia (e.g., Dalziel 1997; Hartz & Torsvik 2002; Cawood *et al.* 2003; Pisarevsky *et al.* 2003), the reconstruction shown in Figure 1 (after Pisarevsky *et al.* 2003) is the only one that fits both published palaeomagnetic data and geological constraints (Cawood & Pisarevsky 2006). In particular, it is consistent with the existence of the long-lived Meso- to Neoproterozoic passive continental margin along the eastern and northeastern edge of Baltica (Fig. 1, NB unless otherwise stated, all geographic references are in present coordinates), which was converted into an active margin between 600 and 550 Ma (e.g., Nikishin *et al.* 1996; Olovyanishnikov *et al.* 2000; Willner *et al.* 2001, 2003; Maslov & Isherskaya 2002; Roberts & Siedlecka 2002; Puchkov 2003 and references therein). The reconstruction is also supported by the absence of any evidence for the Cambrian rifting and continental break-up along the Uralian margin of Baltica (Maslov *et al.* 1997) required by other models (e.g., Hartz & Torsvik 2002).

At the beginning of the Ediacaran (*c.* 630–600 Ma), Laurentia, Baltica and Amazonia, probably still formed a single continental block: the last fragment of the Rodinia supercontinent (Fig. 1), which had begun to disperse between 800 and 750 Ma (Wingate & Giddings 2000). The *c.* 615 Ma Long Range (Laurentia) and Egersund (Baltica) mafic dyke swarms are both rift-related according to their geochemical characteristics (Bingen *et al.* 1998; Puffer 2002) and are possible indicators for the onset of rifting that eventually resulted in the separation of Baltica from Laurentia and Amazonia and the opening of both the eastern Iapetus Ocean and the Tornquist Sea (Kamo *et al.* 1989; Kamo & Gower 1994; Bingen *et al.* 1998; Puffer 2002). These magmatic events were followed by more widespread rift-related mafic magmatism at *c.* 610–550 Ma along the eastern margin of Laurentia (Halliday *et al.* 1989; Cawood *et al.* 2001; Puffer 2002), the Scandinavian margin of Baltica (e.g., Bingen *et al.* 1998; Roberts *et al.* 2004), and the southwestern margin of Baltica (e.g., Compston *et al.* 1995; Poprawa *et al.* 1999; Shumlyanskyy & Andréasson 2004; Elming *et al.* 2005). Keppie *et al.* (2006) reported a *c.* 546 Ma plume-related mafic dyke swarm in Oaxaquia, Mexico, which in our reconstruction is in close juxtaposition to the western Ukrainian volcanic province of similar age (e.g., Keppie & Ramos 1999; Shumlyanskyy & Andréasson 2004; Elming *et al.* 2005).

The locations of these magmatic provinces and the trends of most of these dyke swarms are shown in Figure 2. This configuration suggests three branches of a plume-related rifting event, with the plume centre at the triple point between Laurentia, Baltica and Amazonia. Two of these branches broadly coincide with the strikes of the Grenville and Sveconorwegian orogenic belts, suggesting a degree of inheritance, whereas the third arm, extending between west Baltica and NE Laurentia, cuts across structural (Cawood *et al.* 2007). In western Scandinavia, the rift-to-drift

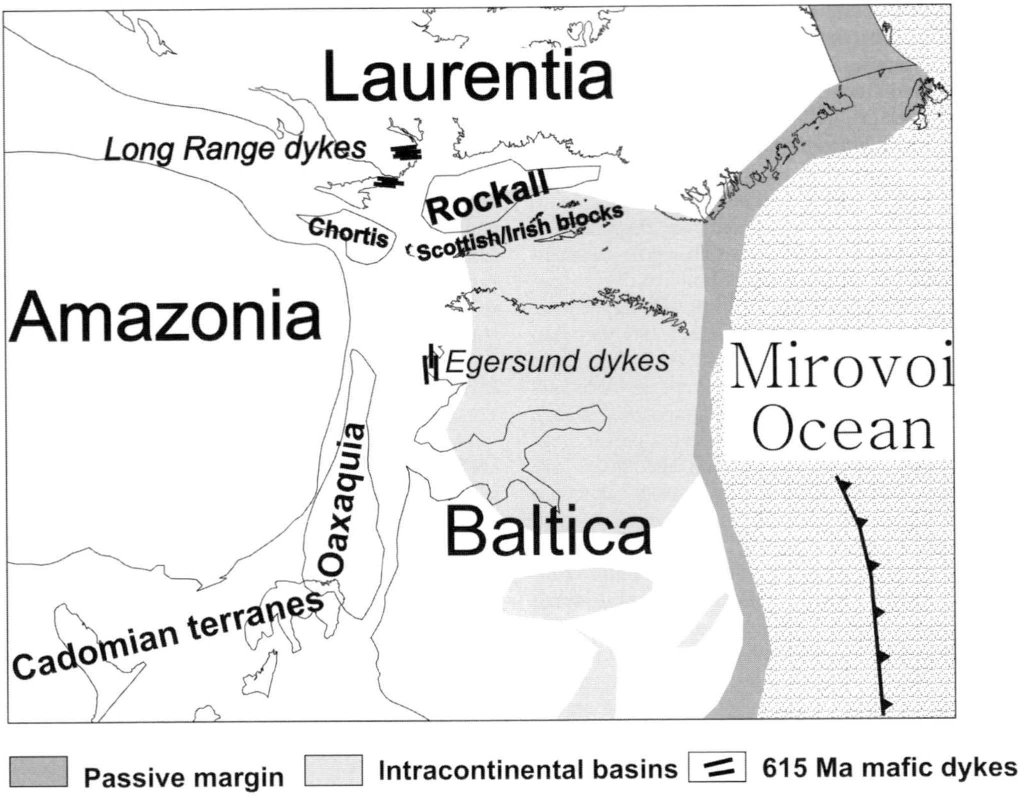

Fig. 1. Pre-Iapetian (*c*. 615 Ma) palaeogeographical reconstruction of Laurentia, Baltica, and Amazonia.

transition occurred around 600–580 Ma, followed by a developing passive continental margin (Bingen *et al.* 1998; Greiling *et al.* 1999; Siedlecka *et al.* 2004). In contrast, Cawood *et al.* (2001) reported that the rift-to-drift transition along the Laurentian mainland (juxtaposed to Amazonia in our reconstruction) occurred in Early Cambrian times, i.e. significantly later than in the western Scandinavia, implying failure of the first attempted rift between Laurentia and Amazonia, possibly indicated by the *c*. 600 Ma Grenville dykes (e.g., Cawood *et al.* 2001). The second attempt in Early Cambrian time, however, was successful. For two other arms, between Baltica and Rockall/Greenland/Scottish blocks and between Baltica and Oaxaquia/Amazonia, the rift-to-drift transition was successfully completed *c*. 600–580 Ma and led to the opening of the eastern Iapetus Ocean and the Tornquist Sea in Ediacaran times (Bingen *et al.* 1998; Greiling *et al.* 1999; Siedlecka *et al.* 2004).

In our scenario opening of the eastern Iapetus Ocean and Tornquist Sea was followed by the opening of the western Iapetus Ocean when Amazonia and Laurentia broke apart (Fig. 2). The configuration of the rifting zone between these two continents has been proposed by Thomas (2005 and references therein). A similar model has been proposed by Bingen *et al.* (1998), but their initial configuration places Baltica against East Greenland. The configuration shown in Figures 1 and 2 is consistent with published palaeomagnetic data (Cawood & Pisarevsky 2006 and references therein). Moreover, Greiling & Smith (2000) noted similarities between the Neoproterozoic sedimentary successions in Scandinavia and Scotland, and proposed a similar Laurentia–Baltica fit. Carbonatite intrusions in Baltica (the 584 Ma Fen and 589 Ma Alnø complexes, Meert *et al.* 1998; Walderhaug *et al.* 2003) and in Laurentia/Greenland at *c*. 574 Ma (St. Honore, Doig & Barton 1968) and *c*. 600 Ma (Sarfartoq, Greenland, Larsen & Rex 1992) also surround the centre of the suggested plume, supporting our reconstruction. Recent palaeomagnetic data showing that Baltica and Laurentia were well separated at *c*. 550 Ma implies that the eastern Iapetus Ocean and the Tornquist Sea were already open by that time. This is in accord with the model shown in Figure 2.

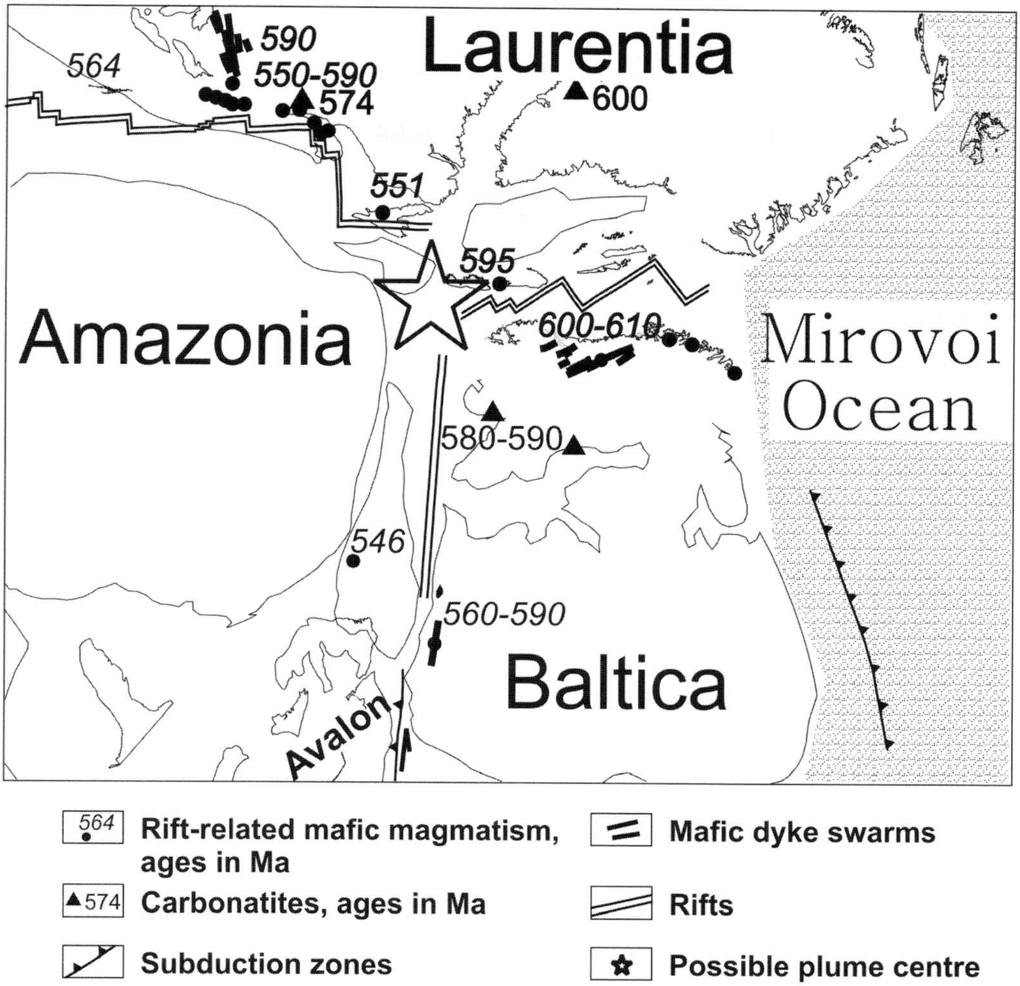

Fig. 2. Distribution of mafic magmatic rocks along incipient rifts between Laurentia, Baltica, and Amazonia at *c*. 600–550 Ma.

Mirovoian realm

Mirovoi is the name given to the ocean that surrounded Rodinia (McMenamin & McMenamin 1990), and we extend that definition here to denote the ocean that surrounded Laurentia–Baltica–Amazonia–West Africa immediately prior to Iapetan rifting. We now examine regional tectonothermal events that are coeval with the development of the Iapetus Ocean and evaluate potential geodynamic linkages between these two oceans.

The failure of the Laurentia–Amazonia rift was coeval with oblique subduction beneath the Avalonian–Cadomian belt (Fig. 2) along the northern margin of Amazonia/West Africa (Murphy *et al.* 2004). We suggest that there may be a geodynamic linkage between these events: subduction directed beneath Amazonia/West Africa could create a counter-force against this rifting of the southern margin of Laurentia. Additionally, the sinistral strike-slip component of Avalonian subduction (Murphy *et al.* 2000) may have provided an additive force for the rifting between Baltica and Amazonia/Oaxaquia (Fig. 2).

Along the northeastern margin of Baltica, Roberts & Siedlecka (2002) proposed that in early Ediacaran time there was a subduction zone outboard of the Timanian part of the Baltican margin, directed ocean-ward (Fig. 2; see also fig. 8 of Roberts & Siedlecka 2002). On the eastern margin of Amazonia, the development of an arc along the margins of the São Francisco Craton (Pimentel *et al.* 1999) indicates another subduction zone outboard of, and directed away from, Amazonia/West

Africa. Taken together, the slab-pull forces associated with these subduction zones would also be consistent with the separation of Baltica from Amazonia/Oaxaquia.

Assembly of Gondwana

Hypotheses about the assembly of Gondwana in the Proterozoic (see Collins & Pisarevsky 2005 for an overview) may be subdivided into three groups: (i) one rigid block, a part of a single supercontinent (e.g., Piper 2000); (ii) two large Neoproterozoic continental masses: East Gondwana (India–Australia–Antarctica) and West Gondwana (Africa–South America) that amalgamated by the end of Neoproterozoic (e.g., McWilliams 1981; Dalziel 1992); (iii) a number of separately drifting continental fragments that assembled by the latest Neoproterozoic–Early Palaeozoic (e.g., Meert *et al.* 1995; Torsvik *et al.* 2001; Powell & Pisarevsky 2002; Pisarevsky *et al.* 2003; Collins & Pisarevsky 2005; Trindade *et al.* 2006).

Trompette (1994, 1997) proposed the existence of a single West Africa–Amazonia--Rio de la Plata mega-craton in the Meso- and Neoproterozoic on the basis of similarities between the Proterozoic sedimentary successions of three blocks. However, he did not exclude the possibility of minor relative movements between its components. Some palaeomagnetic data indicate the possibility of shearing between Amazonia and West Africa (Onstott & Hargraves 1981), but no conclusive evidence has been published.

Rio de la Plata is a poorly known craton, with only a few reliable palaeomagnetic data, that generally support an affinity to Amazonia (Trompette 1994, 1997). Its Precambrian dimensions are similarly uncertain. The Rio de la Plata block, as depicted by Dalziel (1997), for example, included parts of the Pampean terrane as well as the southern extremity of the Guapore block. In contrast, Ramos (1988) envisaged a Pampean–Rio de la Plata collision at 600–570 Ma, although recent studies suggest that this could have happened even later, at 530–520 Ma (e.g., Trindade *et al.* 2006). Trompette (1994) considered the possibility of an Amazonian affinity for the southern Guapore cratonic extension. Pimentel *et al.* (1999) suggested a collision between the São Francisco Craton and the Paraná block between 790 and 750 Ma. In our reconstructions (Fig. 3), we propose three separate blocks: 1) Rio de la Plata *sensu stricto*, which includes basement to the NE and SW of Buenos Aires (Cingolani & Dalla Salda 2000), but does not include the Luis Alves block and the southern extremity of the Guapore Block, (2) the Pampean terrane and (3) the Paraná block. However, we must emphasize that to the best of our knowledge there are no palaeomagnetic data from Pampean terrane and Paraná block, so their inclusion in our reconstructions (Figs 4 and 5) are open to dispute. We have generally followed the tectonic model of Ramos (1988) for the Rio de la Plata and Pampean blocks, keeping them in the vicinity of Laurentia, and the model of Pimentel *et al.* (1999) for Paraná–São Francisco collision. Recent palaeomagnetic data (Sánchez-Bettucci & Rapalini 2002; Rapalini 2006) suggest that Rio de la Plata was part of Gondwana by 550 Ma. These data together with other publications (e.g., Rapela *et al.* 2008) may indicate that our approach to a position and role of Rio de la Plata could be an oversimplification and more sophisticated model for this part of the palaeo-globe should be considered in the future, as was previously suggested by Omarini *et al.* 1999).

The Congo/São Francisco craton, according to palaeomagnetic data (Meert *et al.* 1995; Wingate *et al.* 2004) and geological evidence (Kröner & Cordani 2003; Collins & Pisarevsky 2005), drifted as a separate continent independent of Rodinia. Along its northern margin, peak Neoproterozoic metamorphism accompanied by deformation is reported to have occurred at *c.* 630 Ma both in Uganda (Leggo 1974; Appel *et al.* 2004) and in the Oubanguides Belt of the Central African Republic (Pin & Poidevin 1987). This deformation and metamorphism is interpreted to reflect the collision between this block and the poorly known Saharan metacraton (Abdelsalam *et al.* 2002), which is composed of a number of pre-Neoproterozoic terranes separated by juvenile Neoproterozoic crust (Black *et al.* 1994; Caby 2003; Liégeois *et al.* 2003). This collision is roughly coeval with the *c.* 650–600 Ma collision between the São Francisco Craton and Amazonia along the Brasília and Araguaia belts (Pimentel *et al.* 1991; & Gaudette 1993; Alvarenga *et al.* 2000; Pimentel *et al.* 2000; Valeriano *et al.* 2004). In the Dahomeyide Belt, between the collage of pre-Neoproterozoic terranes that make up Nigeria (Dada 1998) and the West African Craton, $^{40}Ar/^{39}Ar$ hornblende ages of 590–580 Ma in the suture-zone nappes provide a younger age constraint on the collisions of these terranes with the Congo Craton (Attoh *et al.* 1997). Hence we suggest that the collision between Congo/São Francisco and Amazonia/West Africa occurred before 600 Ma: roughly coeval with the opening of the eastern Iapetus Ocean and of the Tornquist Sea, and with oblique subduction beneath Amazonia/West Africa that generated *c.* 630–570 Ma peak magmatism in the Avalonian–Cadomian belt. Smaller continental blocks, such as Afif-Abas and Azania (Collins & Windley 2002; Collins & Pisarevsky 2005), also

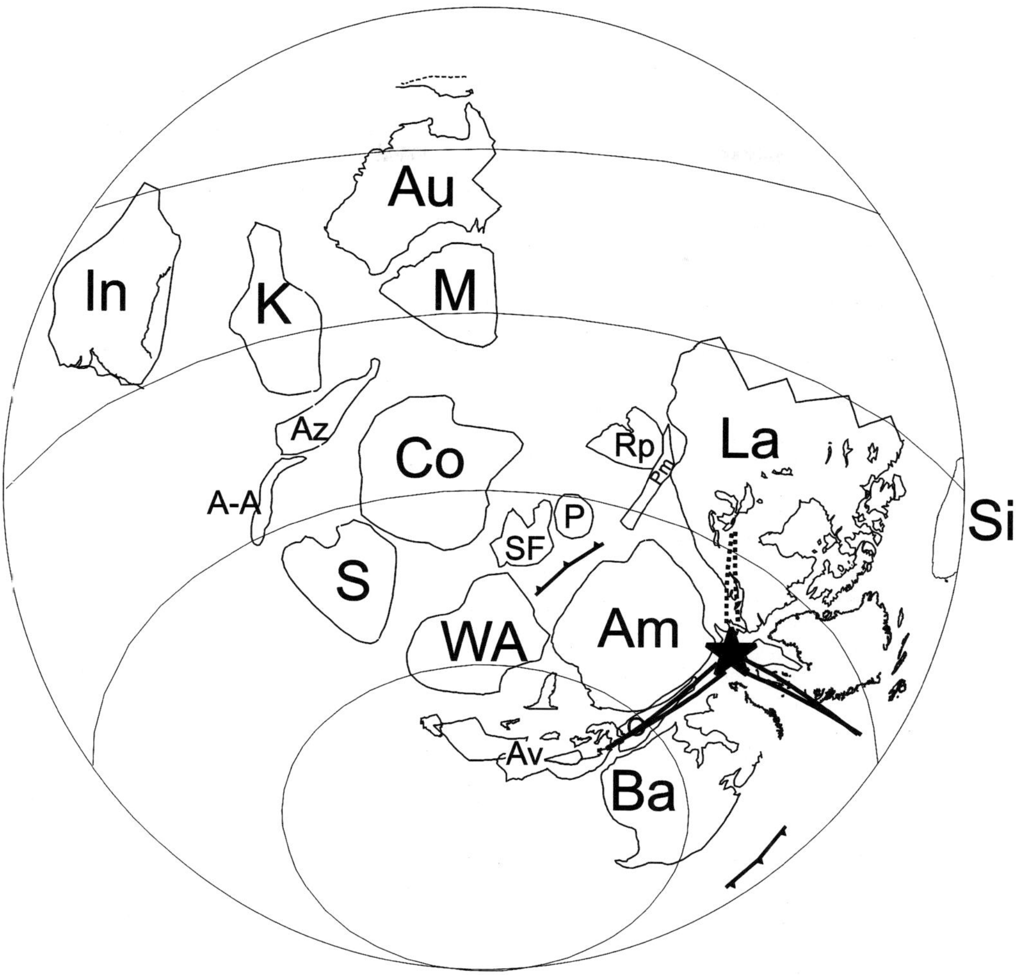

Fig. 3. Palaeogeography at *c.* 600 Ma. Star denotes the centre of a mantle plume. Successive rifts between Baltica and Laurentia, and between Baltica and Amazonia are shown with solid lines; the failed rift between Laurentia and Amazonia is shown with dotted lines. Subduction zones outboard of Baltica and Amazonia are directed in opposite directions, creating an extensional strain by slab-push forces. A-A, Afif-Abas; Am, Amazonia; Au, Australia; Av, Avalonia; Az, Azania; Ba, Baltica; Co, Congo; In, India; K, Kalahri; La, Laurentia; M, Mawson; O, Oaxaquia; P, Paraná; Pm, Pampean; Rp, Rio de La Plata; S, Saharan Metacraton; SF, São Francisco; Si, Siberia.

joined the Congo Craton and the Sahara metacraton roughly at the same time to form the bulk of the East African continent.

Separation between the Kalahari and Congo cratons during most of Neoproterozoic is indicated by Meso- to Neoproterozoic eclogites, arc volcanic rocks and ophiolites in the Zambezi Belt (Oliver *et al.* 1998; Johnson & Oliver 2000, 2004; John *et al.* 2003), which characterize the presence of oceanic crust between them. The timing of the collision between Congo and Kalahari along the Irumide/Zambezi/Damara orogenic system is constrained by the age of the high-pressure metamorphism at 560–505 Ma (Vinyu *et al.* 1999; Jung *et al.* 2000; Hargrove *et al.* 2003; John *et al.* 2003; Goscombe *et al.* 2004; Johnson & Oliver 2004). These and other constraints (Collins & Pisarevsky 2005) suggest that Kalahari collided with Congo significantly later than the collision between Congo/Saõ Francisco and Amazonia (Prave 1996).

The assumption of the integrity of East Gondwana (Australia/East Antarctica/India) was challenged by Fitzsimons (2000), who has shown that three sectors of the assumed Grenville-age orogenic belt in East Antarctica are significantly different in

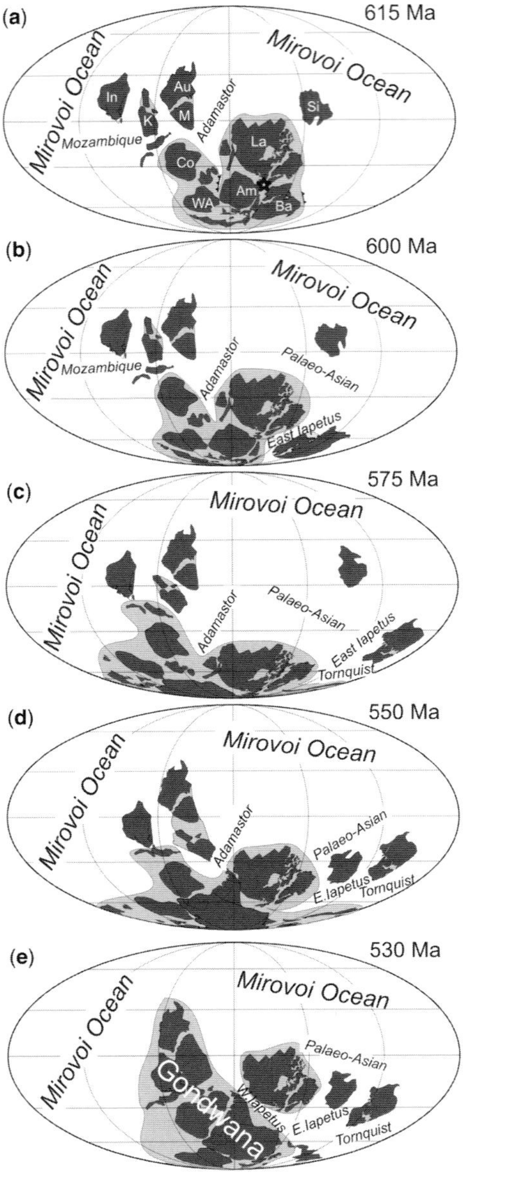

Fig. 4. Global palaeogeography between 615 Ma and 530 Ma, high-latitude model. Euler rotation parameters are in Table 2.

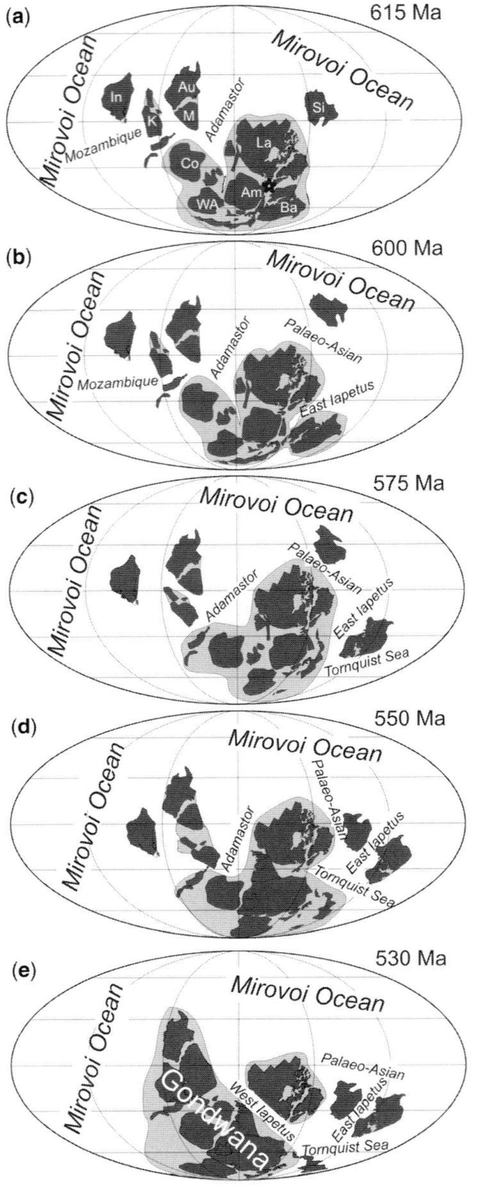

Fig. 5. Global palaeogeography between 615 Ma and 530 Ma, low-latitude model. Euler rotation parameters are in Table 2.

age and history and are separated from each other by Pan-African belts. Subsequent studies (see Collins & Pisarevsky 2005 for an overview), including palaeomagnetic data (Torsvik *et al.* 2001), resulted in the proposal that the Australia/Mawson, India/Rayner and Kalahari/Dronning Maud Land blocks amalgamated shortly before (Powell & Pisarevsky 2002), or during (Boger *et al.* 2001) the final amalgamation of Gondwana. In our reconstructions we suggest that this happened at *c.*530 Ma by the oblique collision of India with Australia/Mawson and East Africa and the coeval docking of Kalahari/Dronning Maud Land to Congo along the Ambezi and Damara belts (Collins & Pisarevsky 2005). At this time East Antarctica became a single continental block.

Table 2. *Euler rotation parameters (to the absolute framework)*

Craton/block/terrane	High-latitude option (Fig. 4)			Low-latitude option (Fig. 5)		
	Pole		Angle	Pole		Angle
	(°N)	(°E)	(°)	(°N)	(°E)	(°)
615 Ma						
E. Antarctica (Mawson)	31.3	90.6	96.4	31.3	90.6	96.4
Dronning Maud Land	24.2	65.5	126.1	24.6	66.0	124.2
Australia	49.0	101.3	83.4	49.0	101.3	83.4
Congo	27.6	115.5	150.7	24.8	116.8	151.5
Kalahari	50.9	81.5	141.9	51.0	83.0	140.7
W. Africa	28.4	120.8	160.1	28.4	120.8	160.1
India	67.4	133.2	80.9	67.4	133.2	80.9
Siberia	62.1	−62.4	−163.9	62.1	−62.4	−163.9
Rockall	25.5	−118.5	−116.0	25.5	−118.5	−116.0
E. Avalon	8.1	115.5	162.5	8.1	115.5	162.5
Baltica	23.8	−92.6	−140.9	23.8	−92.6	−140.9
Rio de La Plata	22.4	88.0	154.2	22.4	88.0	154.2
São Francisco	23.3	86.6	149.0	21.0	88.4	147.0
Pampean	22.4	88.5	153.9	22.4	88.5	153.9
Paraná	23.3	86.6	149.0	21.0	88.4	147.0
Amazonia	23.5	92.5	157.0	23.5	92.5	157.0
Laurentia	15.4	−127.4	−106.8	15.4	−127.4	−106.8
Chortis	2.4	23.6	61.1	2.4	23.6	61.1
Oaxaquia	15.8	96.7	−151.2	15.8	96.7	−151.2
W. Avalon	10.9	85.5	159.1	10.9	85.5	159.1
Baffin Land	16.5	−124.3	−109.7	16.5	−124.3	−109.7
N. Alaska	34.9	−95.4	−159.0	34.9	−95.4	−159.0
Greenland	18.9	−118.7	−116.9	18.9	−118.7	−116.9
600 Ma						
E. Antarctica (Mawson)	30.1	92.4	96.3	30.1	92.4	96.3
Dronning Maud Land	21.3	69.1	126.3	22.9	72.7	120.0
Australia	47.7	103.9	83.9	47.7	103.9	83.9
Congo	35.4	117.6	148.0	25.2	122.0	150.9
Kalahari	47.4	84.2	144.6	47.4	90.9	142.0
W. Africa	35.1	121.6	151.1	24.8	125.2	155.1
India	65.4	136.7	83.5	65.4	136.7	83.5
Siberia	53.4	−93.0	−127.7	68.2	−89.5	−129.1
Rockall	17.9	−122.5	−122.3	24.8	−120.4	−104.7
E. Avalon	14.2	115.2	154.7	3.9	119.0	155.7
Baltica	11.4	−103.2	−136.3	19.2	−103.3	−127.6
Rio de La Plata	31.3	85.7	157.0	22.7	92.1	149.9
São Francisco	31.3	86.4	152.6	22.7	93.2	145.8
Pampean	31.3	86.1	156.6	22.7	92.5	149.7
Paraná	31.3	86.4	152.6	22.7	93.2	145.8
Amazonia	31.3	90.5	154.3	22.3	96.8	148.6
Laurentia	8.6	−131.9	−115.4	13.1	−128.9	−95.9
Chortis	4.4	17.1	75.0	10.4	17.3	54.7
Oaxaquia	21.6	98.7	−155.3	11.1	99.0	−160.3
W. Avalon	18.0	84.2	158.2	9.4	89.2	149.0
Baffin Land	9.5	−128.7	−117.5	14.6	−125.6	−98.5
N. Alaska	27.1	−97.8	−159.5	36.3	−97.4	−148.1
Greenland	11.6	−122.9	−123.4	17.8	−119.9	−105.5
575 Ma						
E. Antarctica (Mawson)	28.1	95.4	96.3	28.1	95.4	96.3
Dronning Maud Land	16.0	86.3	112.9	19.9	88.2	110.4
Australia	45.3	107.8	84.9	45.3	107.8	84.9
Congo	48.6	123.0	136.9	20.0	132.0	146.7
Kalahari	37.2	102.6	145.4	39.4	106.7	144.5

(Continued)

Table 2. *Continued*

Craton/block/terrane	High-latitude option (Fig. 4)			Low-latitude option (Fig. 5)		
	Pole		Angle	Pole		Angle
	(°N)	(°E)	(°)	(°N)	(°E)	(°)
W. Africa	46.7	123.3	137.0	18.2	132.1	147.1
India	62.6	144.9	88.3	62.6	144.9	88.3
Siberia	15.5	−119.2	−98.4	37.6	−126.0	−93.4
Rockall	6.3	−128.1	−134.9	23.5	−124.3	−86.0
E. Avalon	18.1	106.0	135.0	9.6	−60.5	−131.4
Baltica	9.0	59.2	139.8	4.6	−118.6	−109.6
Rio de La Plata	44.3	82.4	154.2	20.8	100.8	135.6
São Francisco	45.8	85.2	152.3	21.6	103.7	136.2
Pampean	44.3	82.6	153.9	20.7	101.0	135.5
Paraná	45.8	85.2	152.3	21.6	103.7	136.2
Amazonia	44.3	86.3	150.9	19.9	104.3	135.0
Laurentia	1.4	41.8	131.7	8.3	−132.0	−78.0
Chortis	6.5	9.8	99.3	26.8	2.9	48.3
Oaxaquia	33.4	100.7	−163.7	6.0	101.4	−177.5
W. Avalon	25.3	73.7	145.4	4.3	90.3	115.0
Baffin Land	1.0	45.0	132.6	10.7	−128.3	−80.2
N. Alaska	14.1	−101.2	−161.4	38.8	−101.4	−130.1
Greenland	0.3	−128.9	−136.2	15.6	−122.1	−86.5
550 Ma						
E. Antarctica (Mawson)	23.8	100.9	91.8	23.8	100.9	91.8
Dronning Maud Land	18.7	100.9	102.2	18.7	100.9	102.2
Australia	40.5	116.2	82.7	40.5	116.2	82.7
Congo	42.0	126.2	137.9	25.1	133.3	142.4
Kalahari	33.4	118.2	144.6	33.4	118.2	144.6
W. Africa	39.3	122.2	136.0	22.5	130.4	138.8
India	58.3	162.0	100.0	58.3	162.0	100.0
Siberia	3.7	47.7	151.5	1.0	−128.3	−118.7
Rockall	13.5	−124.7	−121.0	23.6	−122.8	−90.4
E. Avalon	14.9	104.3	129.8	0.8	−65.6	−122.1
Baltica	10.9	52.7	127.4	8.9	48.0	122.1
Rio de La Plata	38.1	87.7	144.2	25.2	100.6	131.4
São Francisco	41.0	90.6	147.6	27.4	103.2	136.6
Pampean	38.1	87.7	144.2	25.2	100.6	131.4
Paraná	41.0	90.6	147.6	27.4	103.2	136.6
Amazonia	38.1	87.7	144.2	25.2	100.6	131.4
Laurentia	4.0	−134.2	−115.6	9.3	−130.7	−82.3
Chortis	0.5	2.7	77.8	14.9	−11.7	51.1
Oaxaquia	26.8	96.1	−168.3	11.3	98.9	178.5
W. Avalon	23.9	72.0	138.0	14.5	83.6	115.9
Baffin Land	4.9	−130.9	−117.2	11.5	−127.1	−84.6
N. Alaska	23.6	−99.3	−155.0	38.1	−99.9	−134.1
Greenland	7.1	−124.9	−122.2	15.9	−121.1	−91.0
530 Ma						
E. Antarctica	19.4	107.9	91.5	19.4	107.9	91.5
Dronning Maud Land	21.8	107.5	94.1	21.8	107.5	94.1
Australia	35.2	124.1	85.5	35.2	124.1	85.5
Congo	32.3	125.9	140.1	32.3	125.9	140.1
Kalahari	32.3	125.9	140.1	32.3	125.9	140.1
W. Africa	29.5	122.7	137.6	29.5	122.7	137.6
India	51.0	165.2	126.2	51.0	165.2	126.2
Siberia	5.8	−127.5	−135.6	5.8	−127.5	−135.6
Rockall	24.4	−119.1	−106.2	24.4	−119.1	−106.2
E. Avalon	5.0	106.6	128.5	5.0	106.6	128.5

(*Continued*)

Table 2. *Continued*

Craton/block/terrane	High-latitude option (Fig. 4)			Low-latitude option (Fig. 5)		
	Pole		Angle	Pole		Angle
	(°N)	(°E)	(°)	(°N)	(°E)	(°)
Baltica	0.8	67.8	105.1	2.1	58.6	118.1
Rio de La Plata	29.5	91.1	137.4	29.5	91.1	137.4
São Francisco	32.2	93.7	141.5	32.2	93.7	141.5
Pampean	29.5	91.1	137.4	29.5	91.1	137.4
Paraná	32.2	93.7	141.5	32.2	93.7	141.5
Amazonia	29.5	91.1	137.4	29.5	91.1	137.4
Laurentia	12.9	−127.8	−97.6	12.9	−127.8	−97.6
Chortis	5.5	0.9	64.3	5.5	0.9	64.3
Oaxaquia	19.2	96.6	−171.5	19.2	96.6	−171.5
W. Avalon	16.4	75.2	127.3	16.4	75.2	127.3
Baffin Land	14.3	−124.5	−100.3	14.3	−124.5	−100.3
N. Alaska	35.9	−96.2	−149.2	35.9	−96.2	−149.2
Greenland	17.5	−118.8	−107.2	17.5	−118.8	−107.2

Avalonian and related terranes

Avalonia and Cadomian are two of a group of terranes, collectively referred to as peri-Gondwanan. On the basis of faunal, lithostratigraphic, geochemical, and palaeomagnetic data they are traditionally interpreted as remnants of the northern (Amazonian and West African) Gondwanan margin in the Neoproterozoic and Early Palaeozoic (Theokritoff 1979; Van der Voo 1988; Murphy & Nance 1989; Cocks & Fortey 1990; Keppie 1993; McNamara *et al.* 2001; Murphy *et al.* 2002, 2004; Fortey & Cocks 2003; Collins & Buchan 2004), although temporary seaways may have separated Avalonia from Amazonia/West Africa (Landing 1996, 2005). Peri-Gondwanan terranes are characterized by Neoproterozoic magmatism that records a history of subduction beneath the Amazonian/West African margin (e.g., O'Brien *et al.* 1983; Murphy *et al.* 1990; Nance *et al.* 1991; Keppie 1993; Egal *et al.* 1996; Linnemann *et al.* 2000; von Raumer *et al.* 2002). Some peri-Gondwanan terranes, such as Avalonia, and Carolinia (Hibbard 2000; Hibbard *et al.* 2002), were rifted from Amazonia/West Africa by the Early Ordovician and were subsequently involved in Palaeozoic and Mesozoic orogenesis. As a result, they are preserved as a collection of suspect terranes in the younger orogenic belts of Europe and North America. Avalonia stretches from New England to southeastern Newfoundland (O'Brien *et al.* 1983; Murphy & Nance 1989, 1991; Keppie *et al.* 1991) and includes southeastern Ireland (Max & Roddick 1989) and southern Britain (Tucker & Pharoah 1991; Gibbons & Horák 1996). Other peri-Gondwanan terranes occur in the Armorican massif (Cadomia) of northwestern France (Egal *et al.* 1996; Strachan *et al.* 1996), the Iberian peninsula (Quesada 1990; Eguíluz, *et al.* 2000; Fernandez-Suárez *et al.* 2000), isolated inliers in Germany and the Czech Republic (e.g., Bohemian Massif, Zulauf *et al.* 1999; Linnemann *et al.* 2000), and recently recognized vestiges in the Alpine belt (Neubauer 2002; von Raumer *et al.* 2002).

Sm–Nd isotopic data indicate that early arc-related complexes of Avalonia probably formed outboard of the Amazonian/West African margin within the Mirovoi Ocean, whereas their subsequent metamorphism is interpreted as recording their accretion to this margin (Murphy *et al.* 2000, 2004, 2006; Nance *et al.* 2002). In contrast, coeval Cadomian arc magmatism is attributed to melting of the West African Craton.

At about 635 Ma, Avalonian–Cadomian voluminous Andean-style arc-related activity commenced broadly synchronously along the Amazonian/African margin. Arc-related rocks typically include calc-alkaline mafic to felsic volcanic rocks, coeval plutons and pull-apart sedimentary basin deposits which contain detritus derived from the arc. Subduction was oblique and had a sinistral component (Murphy & Nance 1989; Nance & Murphy 1990; Murphy *et al.* 1999; Keppie *et al.* 2003). However, the termination of arc magmatism was diachronous from 590 to 540 Ma (Murphy *et al.* 1999, 2000; Nance *et al.* 2002; Keppie *et al.* 2003), and is marked by the progressive development of an intracontinental strike-slip regime that is interpreted as recording ridge–trench collision, analogous to the Oligocene collision between western North America and the East Pacific Rise and the diachronous initiation of the San Andreas transform margin (Murphy & Nance 1989; Murphy *et al.* 1999; Nance *et al.* 2002).

Middle America terranes

In addition to these terranes, several crustal blocks in Middle America contain assemblages with Early Palaeozoic fauna that suggest an origin along the northern Gondwanan margin (Keppie & Ramos 1999). In the Neoproterozoic–Early Palaeozoic, these terranes are thought to have lain along either the northern or the western margin of Amazonia (Keppie & Ramos 1999; Keppie & Ortega-Gutiérrez 1999; Keppie *et al.* 2003). In our reconstructions we choose the first interpretation, which places Oaxaquia between Amazonia and Baltica, because it is supported by the palaeomagnetic data (Ballard *et al.* 1989) (Figs 1 and 2). The Yucatan block is thought to have been contiguous with the Florida basement (Suwannee terrane, probably part of Avalonia, Heatherington *et al.* 1996). The Grenville-age basement of Oaxaquia and the Chortis block is isotopically transitional between that of the Grenville Belt in Laurentia and basement massifs of Grenville age in Colombia (Ruiz *et al.* 1999) and is attributed to mixing of juvenile Grenville and Archaean sources (Cameron *et al.* 2004).

The basement rocks of Oaxaquia are best known. They consist of: (1) a metavolcanic-metasedimentary juvenile arc sequence of uncertain age; (2) a *c.* 1140 Ma, bimodal, within-plate intrusive suite that was deformed and metamorphosed at *c.* 1100 Ma; (3) a *c.* 1012 Ma anorthosite-gabbro that was deformed and metamorphosed in the granulite facies at *c.* 980–1104 Ma; (4) *c.* 920 Ma post-tectonic calc-alkaline plutonic rocks (Keppie *et al.* 2001; Solari *et al.* 2003; Ortega-Obregón *et al.* 2003). This sequence of magmatic and metamorphic events is broadly consistent with those of the Sveconorwegian orogeny in Baltica (e.g., Gorbatschev & Bogdanova 1993).

615–530 Ma palaeogeography

We choose 615 Ma as a starting point for our reconstructions for the following reasons: (i) the Long Range dykes in Laurentia and Egersund dykes in Baltica are precisely dated at 615 Ma and are indicators of the beginning of rifting events in these areas; (ii) there are several reasonably reliable ($Q \geq 3$, Table 1) published palaeopoles of this age (Storetvedt 1966; Poorter 1972; Embleton & Williams 1986; Schmidt *et al.* 1991; Murthy *et al.* 1992; Kamo & Gower 1994; Schmidt & Williams 1995; Bingen *et al.* 1998; Sohl *et al.* 1999; Metelkin *et al.* 2005), so that the relative locations of Laurentia, Baltica, Australia, India, and Siberia are reasonably well constrained; (iii) our reconstructions for 615 Ma are identical for both high- and low-latitude models for Laurentia (Figs 4a and 5a). These diagrams are slightly simplified versions of the reconstruction shown in Figure 3. We did not use the Amazonian Puga cap carbonate pole (Trindade *et al.* 2003) and the Australian Bunyeroo pole (Schmidt & Williams 1996) for the reasons outlined earlier. The ages of the Siberian poles are poorly constrained and so we urge caution in interpreting the movement of Siberia in Figures 4 and 5. Thus the size and shape of the Palaeo-Asian ocean are schematic.

High-latitude model (Fig. 4)

Laurentian palaeopoles from Table 1 used for this model are: (i) Long Range dykes (data recalculated by Hodych *et al.* 2004), (ii) Callander Complex, (iii) Catoctin Volcanics A, (iv) Sept Iles Dykes B, and (v) Skinner Cove Formation.

Laurentia/Amazonia/West Africa moved gradually to the south, Baltica generally followed them initially, but started to separate at *c.* 600 Ma through the opening of the eastern Iapetus Ocean and the Tornquist Sea (Fig. 4a, b). Australia/Mawson remained at equatorial latitudes. A number of collisions occurred that culminated in the formation of Gondwana. Congo/São Francisco collided with Amazonia, closing the Brasiliano Ocean by *c.* 570 Ma. The final collision of the Sahara metacraton with the Tuareg block and West Africa to form the northern part of Gondwana was completed by *c.* 600 Ma (Attoh *et al.* 1997). Kalahari drifted southeastwards closing the Adamastor Ocean and collided with Congo and Rio de la Plata by 530 Ma. India collided obliquely with Australia/Mawson and East Africa at about the same time. These collisional events could have created southward-directed stresses on the rest of Gondwana, inhibiting it from moving northwards with Laurentia after 550 Ma. This possibly caused Laurentia's separation from Gondwana by opening of the western Iapetus Ocean. Subduction beneath Amazonia/West Africa started at around 600 Ma, creating supportive stresses for the opening of the Tornquist Sea and opposing rifting between Laurentia and Amazonia.

Puffer (2002) suggested that one or two mantle plumes caused the 615–550 Ma mafic magmatism along the east Laurentian margin. The high-latitude model is inconsistent with either of the proposed plumes, if we assume that they were initiated below the asthenosphere. The suggested plume head(s) should occur near 30°S at 615 Ma, near the south pole at 575 Ma and at 60°S at 550 Ma, so at least three discrete plume events are necessary. The migration of Laurentia to high latitudes implied by this model would predict a 615–575 Ma plume track across Laurentia that also would have influenced the Ediacaran-aged passive margin successions deposited along the western margin (modern coordinates). There is no evidence of this plume track and subsidence curves calculated for these successions (Bond *et al.* 1984) do not show evidence for a thermal perturbation.

Low-latitude model (Fig. 5)

Laurentian palaeopoles we used for this model are: (i) Long Range dykes (as above), (ii) Cloud Mountain Basalt, (iii) Catoctin Volcanics B, (iv) Sept Iles Intrusion A, (v) Buckingham Lavas, (vi) Johnnie Formation and (vii) Skinner Cove Formation. In this model, Laurentia remained at equatorial latitudes between 615 and 530 Ma and Laurentia/Amazonia/West Africa rotated about 40° anticlockwise. This rotation was possibly connected to the collision of Congo/São Francisco with Amazonia/West Africa and the closure of the Brasiliano ocean. Baltica rifted away from Laurentia and Amazonia after *c.* 600 Ma, but not as rapidly as in the high-latitude model. Gondwana assembly occurred in a similar way to the high-latitude model, but with less motion of West Gondwana and, correspondingly, slower relative movement of Kalahari and India with respect to West Gondwana. Nevertheless, docking of Kalahari and India probably still created enough force for the southward motion of Gondwana, which in this scenario facilitated the opening of the western Iapetus Ocean by separation of Gondwana from Laurentia.

As Laurentia drifted only a small distance between 615 and 530 Ma in the low-latitude model, it is consistent with the plume hypothesis of Puffer (2002), requiring only a single stationary plume.

The relationship between Gondwanan and Iapetian events

Regardless of the chosen scenario of the palaeogeographical development between 615 and 530 Ma, it is probable that major events are genetically related. The opposing directions of the subduction outboard of the northeastern margin of Baltica (Roberts & Siedlecka 2002) and outboard of the southwestern margin of Amazonia (indicated by the presence of a Neoproterozoic magmatic arc on the northwestern margin of the São Francisco–Congo craton, Pimentel *et al.* 1999) would have necessitated the existence of an extensional regime between these two continents (Figs 4a and 5a). This extensional strain may have supported the rifting along the Tessyer–Tornquist margin of Baltica caused by the coeval mantle plume (Figs 4a and 5a).

The oblique docking of Avalonia to the northern margin of Gondwana at 650 Ma followed by the oblique subduction which produced 630–570 Ma arc magmatism (Murphy *et al.* 2004), together with the final docking of Congo/São Francisco to Amazonia (Collins & Pisarevsky 2005) resulted in a combined force on Amazonia inhibiting its separation from Laurentia (Figs 4b, c and 5b, c). As a result, Baltica drifted away from Laurentia/Amazonia and the eastern Iapetus and the Tornquist Sea opened, but the rifting arm between Laurentia and Amazonia failed (Figs 4b, c and 5b, c). Another plume event near the eastern Laurentian margin (in the high-latitude margin), or the same one (in the low-latitude model), as was suggested by Puffer (2002), is supported by the southward-directed stresses on Gondwana caused by the final docking of Kalahari and India from the north and resulted in the separation between Laurentia and Gondwana and in the opening of the western Iapetus Ocean (Figs 4d, e and 5d, e).

Both sets of palaeogeographical reconstructions (Figs 4 and 5) also show the drift of Siberia, permissible within the age uncertainties of published palaeomagnetic data (Table 1). However, this part of our models should be considered as preliminary due to a number of the ongoing studies in Siberia which may cause significant changes.

Conclusions

Two alternative models for the latest Neoproterozoic–Early Palaeozoic global palaeogeography are based on two subsets of the Laurentian palaeomagnetic poles. These groups of palaeopoles are of similar reliability, and new studies are required to distinguish between these models. Alternatively, a non-uniformitarian approach, such as IITPW or non-dipole field could be invoked to explain the results. Both models suggest a geodynamic relationship between the amalgamation of Gondwana and the opening of the Iapetus Ocean. Although both models suggest plume-related rifting between Laurentia, Baltica and Gondwana, the low-latitude model requires a single plume, whereas the high-latitude model needs more than one plume. These rifting events are indicated by voluminous mafic magmatism on the opposite margins of these palaeo-continents and by the rift-drift transition in the sedimentary successions along these margins. Another difference between the two models is the manner of the opening of the western Iapetus Ocean: through the northward movement of Laurentia in the high-latitude model and through the southward movement of Gondwana in the low-latitude model. The openings of the eastern and western Iapetus Ocean and the Tornquist Sea are genetically related to the subduction processes around Amazonia/Baltica and the collisional dockings of Congo, Kalahari, and India that resulted in the assembly of Gondwana.

Pisarevsky and Collins are grateful to St. Francis Xavier University for the W. F. James Professorship. We thank V. Ramos and A. Rapalini for their reviews. Reconstructions were made using PLATES software from the University of Texas at Austin and the GMT software of Wessel and Smith. This paper forms the Tectonics Special Research Centre publication No. 409.

References

ABDELSALAM, M. G., LIÉGEOIS, J. P. & STERN, R. J. 2002. The Saharan Metacraton. *Journal of African Earth Sciences*, **34**, 119–136.

ALEINIKOFF, J. N., ZARTMAN, R. E., WALTER, M., RANKIN, D. W., LYTTLE, P. T. & BURTON, W. C. 1995. U–Pb ages of metarhyolites of the Catoctin and Mount Rogers formations, central and southern Appalachians: Evidence for two phases of Iapetan rifting. *American Journal of Science*, **295**, 428–454.

ALVARENGA, C. J. S., DE MOURA, C. A. V, GORAYEB, P. S. D. S & DE ABREU, F. D. A. M 2000. Paraguay and Araguaia Belts. *In*: CORDANI, U. G., MILANI, E. J., THOMAZ FILHO, A & CAMPOS, D. A. (eds) *Tectonic evolution of South America.* 31st International Geological Congress, Rio de Janeiro, 183–193.

APPEL, P. W. U., SCHENK, V. & SCHUMANN, A. 2004. P–T conditions and metamorphic ages of pelitic schists at Murchison Falls (NW Uganda): evidence for a Pan-African event within the cratonic crust of Uganda. *In*: *20th Colloquium of African Geology*. BRGM, Orléans, France, 49.

ATTOH, K., DALLMEYER, R. D. & AFFATON, P. 1997. Chronology of nappe assembly in the Pan-African Dahomeyide orogen, West Africa: Evidence from $^{40}Ar/^{39}Ar$ mineral ages. *Precambrian Research*, **82**, 153–171.

BALLARD, M. M., VAN DER VOO, R. & URRUTIA-FUCUGAUCHI, J. 1989. Paleomagnetic results from Grenvillian-aged rocks from Oaxaca, Mexico: evidence for a displaced terrane. *Precambrian Research*, **42**, 343–352.

BINGEN, B., DEMAIFFE, D. & VAN BREEMEN, O. 1998. The 616 Ma old Egersund basaltic dike swarm, SW Norway, and Late Neoproterozoic opening of the Iapetus Ocean. *Journal of Geology*, **106**, 565–574.

BLACK, R., LATOUCHE, L., LIÉGEOIS, J. P., CABY, R. & BERTRAND, J. M. 1994. Pan-African displaced terranes in the Tuareg shield (central Sahara). *Geology*, **22**, 641–644.

BOGER, S. D., WILSON, C. J. L. & FANNING, C. M. 2001. Early Paleozoic tectonism with the East Antarctic craton: the final suture between east and west Gondwana. *Geology*, **29**, 463–466.

BOND, G. C., NICKERSON, P. A. & KOMINZ, M. A. 1984. Breakup of a supercontinent between 625 Ma and 555 Ma: New evidence and implications for continental histories. *Earth and Planetary Science Letters*, **70**, 325–345.

CABY, R. 2003. Terrane assembly and geodynamic evolution of central-western Hoggar: a synthesis. *Journal of African Earth Sciences*, **37**, 133–159.

CAMERON, K. L., LOPEZ, R., ORTEGA-GUTIÉRREZ, F., SOLARI, L. A., KEPPIE, J. D. & SCHULZE, C. 2004. U–Pb constraints and Pb isotopic compositions of leached feldspars: Constraints on the origin and evolution of Grenvillian rocks from eastern and southern Mexico. *In*: TOLLO, R. P., CORRIVEAU, L., MCLELLAND, J. & BARTHOLOMEW, M. J. (eds) *Proterozoic Tectonic Evolution of the Grenville Orogen in North America.* Geological Society of America, Memoirs, **197**, 755–770.

CAWOOD, P. A. 2005. Terra Australis Orogen: Rodinia breakup and development of the Pacific and Iapetus margins of Gondwana during the Neoproterozoic and Paleozoic. *Earth-Science Reviews*, **69**, 249–279.

CAWOOD, P. A. & BUCHAN, C. 2007. Linking accretionary orogenesis with supercontinent assembly. *Earth-Science Reviews*, **82**, 143–258.

CAWOOD, P. A. & PISAREVSKY, S. A. 2006. Was Baltica right way up or upside down in the Neoproterozoic? *Journal of the Geological Society, London*, **163**, 753–759.

CAWOOD, P. A., MCCAUSLAND, P. J. A. & DUNNING, G. R. 2001. Opening Iapetus: Constraints from the Laurentian margin in Newfoundland. *Geological Society of America Bulletin*, **113**, 443–453.

CAWOOD, P. A., NEMCHIN, A. A., SMITH, M. & LOEWY, S. 2003. Source of the Dalradian Supergroup constrained by U/Pb dating of detrital zircon and implications for the East Laurentian margin. *Journal of the Geological Society, London*, **160**, 231–246.

CAWOOD, P. A., NEMCHIN, A. A., STRACHAN, R. A., PRAVE, A. R. & KRABBENDAM, M. 2007. Sedimentary basin and detrital zircon record along East Laurentia and Baltica during assembly and breakup of Rodinia. *Journal of the Geological Society, London*, **164**, 257–275.

CHIRANANDA DE 2003. Possible organisms similar to Ediacaran forms from the Bhander Group, Vindhyan Supergroup, Late Neoproterozoic of India. *Journal of Asian Earth Sciences*, **21**, 387–395.

CINGOLANI, C. & DALLA SADA, L. 2000. Buenos Aires cratonic region. *In*: CORDANI, U. G., MILANI, E. J., THOMAZ FILHO, A. & CAMPOS, D. A. (eds) *Tectonic evolution of South America.* 31st International Geological Congress, Rio de Janeiro, 139–150.

COCKS, L. R. M. & FORTEY, R. A. 1990. Biogeography of Ordovician and Silurian faunas. *In*: MCKERROW, W. S. & SCOTESE, C. R. (eds) *Paleozoic Paleogeography and Biogeography*. Geological Society of London, Memoirs, **12**, 97–104.

COLLINS, A. S. & BUCHAN, C. 2004. Provenance and Age Constraints of the South Stack Group, Anglesey, UK: U–Pb SIMS Detrital Zircon Data. *Journal of the Geological Society, London*, **161**, 743–746.

COLLINS, A. S. & PISAREVSKY, S. A. 2005. Amalgamating Eastern Gondwana: the evolution of the Circum-Indian orogens. *Earth-Science Reviews*, **71**, 229–270.

COLLINS, A. S. & WINDLEY, B. F. 2002. The Tectonic Evolution of central and northern Madagascar and its place in the Final Assembly of Gondwana. *Journal of Geology*, **110**, 325–340.

COMPSTON, W., SAMBRIDGE, M. S., REINFRANK, R. F., MOCZYDLOWSKA, M., VIDAL, G. & CLAESSON, S.

1995. Numerical ages of volcanic rocks and the earliest faunal zone within the late Precambrian of east Poland. *Journal of the Geological Society, London*, **152**, 599–611.

DADA, S. S. 1998. Crust-forming ages and Proterozoic crustal evolution in Nigeria: a reappraisal of current interpretations. *Journal of African Earth Sciences*, **87**, 65–74.

DALZIEL, I. W. D. 1992. On the organization of American plates in the Neoproterozoic and the breakout of Laurentia. *GSA Today*, **2**, 237–241.

DALZIEL, I. W. D. 1997. Neoproterozoic-Paleozoic geography and tectonics: review, hypothesis, environmental speculation. *Geological Society of America Bulletin*, **109**, 16–42.

DANKERS, P. & LAPOINTE, P. 1981. Paleomagnetism of Lower Cambrian volcanics and a cross-cutting Cambro-Ordovician diabase dyke from Buckingham (Quebec). *Canadian Journal of Earth Sciences*, **18**, 1174–1186.

DEUTSCH, E. R. & RAO, K. V. 1977. New palaeomagnetic evidence fails to support rotation of western Newfoundland. *Nature*, **266**, 314–318.

DOIG, R. & BARTON, J. M. 1968. Ages of carbonatites and other alkaline rocks in Quebec. *Canadian Journal of Earth Sciences*, **5**, 1401–1407.

EGAL, E., GUERROT, C., LE GOFF, D., THIÉBLEMONT, D. & CHANTRAINE, J. 1996. The Cadomian orogeny revisited in northern France. *In*: NANCE, R. D. & THOMPSON, M. D. (eds) *Avalonian and Related Peri-Gondwanan Terranes of the Circum-North Atlantic*. Geological Society America, Special Papers, **304**, 281–318.

EGUÍLUZ, L., GIL IBARGUCHI, J. I., ÁBALOS, B. & APRAIZ, A. 2000. Superposed Hercynian and Cadomian orogenic cycles in the Ossa-Morena zone and related areas of the Iberian Massif. *Geological Society of America Bulletin*, **112**, 1398–1413.

ELMING, S.-Å., KRAVCHENKO, S., KHRAMOV, A. N., LAYER, P. & IOSIFIDI, A. G. 2005. Paleomagnetism and $^{40}A/^{39}Ar$ age determinations of the Ediacaran traps from the southwestern margin of the East European Craton, Ukraine: relevance to the Rodinia break-up. *Geophysical Research Abstracts*, **7**, 01467.

EMBLETON, B. J. J. & WILLIAMS, G. E. 1986. Low paleolatitude of deposition for Late Precambrian periglacial varvites in South Australia: Implications for paleoclimatology. *Earth and Planetary Science Letters*, **79**, 419–430.

EVANS, D. A. 1998. True polar wander, a supercontinental legacy. *Earth and Planetary Science Letters*, **157**, 1–8.

EVANS, D. A. D. 2000. Stratigraphic, geochronological, and paleomagnetic constraints upon the Neoproterozoic climatic paradox. *American Journal of Science*, **300**, 347–433.

EVANS, D. A. D. 2003. A fundamental Precambrian–Phanerozoic shift in earth's glacial style? *Tectonophysics*, **375**, 353–385.

FERNÁNDEZ-SUÁREZ, J., GUTIÉRREZ-ALONSO, G., JENNER, G. A. & TUBRETT, M. N. 2000. New ideas on the Proterozoic-early Palaeozoic evolution of NW Iberia; insights from U–Pb detrital zircon ages. *Precambrian Research*, **102**, 185–206.

FITZSIMONS, I. C. W. 2000. Grenville-age basement provinces in East Antarctica: Evidence for three separate collisional orogens. *Geology*, **28**, 879–882.

FORTEY, R. A. & COCKS, L. R. M. 2003. Palaeontological evidence bearing on global Ordovician-Silurian continental reconstructions. *Earth-Science Reviews*, **61**, 245–307.

GALLET, Y., PAVLOV, V. E., SEMIKHATOV, M. A. & PETROV, P.Yu. 2000. Late Mesoproterozoic magnetostratigraphic results from Siberia: paleogeographic implications and magnetic field behaviour. *Journal of Geophysical Research*, **105**, 16481–16499.

GIBBONS, W. & HORÁK, J. M. 1996. The evolution of the Neoproterozoic Avalonian subduction system: Evidence from the British Isles. *In*: CORDANI, U. G., MILANI, E. J., THOMAZ FILHO, A. & CAMPOS, D. A. (eds) *Tectonic evolution of South America*. 31st International Geological Congress, Rio de Janeiro, 269–280.

GLADKOCHUB, D. P., WINGATE, M. T. D., PISAREVSKY, S. A., DONSKAYA, T. V., MAZUKABZOV, A. M., PONOMARCHUK, V. A. & STANEVICH, A. M. 2006. Mafic intrusions in southwestern Siberia and implications for a Neoproterozoic connection with Laurentia. *Precambrian Research*, **147**, 260–278.

GORBATSCHEV, R. & BOGDANOVA, S. 1993. Frontiers in the Baltic Shield. *Precambrian Research*, **64**, 3–21.

GOSCOMBE, B., GRAY, D. R. & HAND, M. 2004. Variation in metamorphic style along the Northern Margin of the Damara Orogen, Namibia. *Journal of Petrology*, **45**, 1261–1295.

GREILING, R. O. & SMITH, A. G. 2000. The Dalradian of Scotland: missing link between the Vendian of northern and southern Scandinavia? *Physics and Chemistry of the Earth*, **A25**, 495–498.

GREILING, R. O., JENSEN, S. & SMITH, A. G. 1999. Vendian-Cambrian subsidence of the passive margin of western Baltica—application of new stratigraphic data from the Scandinavian Caledonian margin. *Norsk Geologisk Tidsskrift*, **79**, 133–144.

GRUNOW, A., HANSON, R. & WILSON, T. 1996. Were aspects of Pan-African deformation linked to Iapetus opening? *Geology*, **24**, 1063–1066.

HALLIDAY, A. N., GRAHAM, C. M., AFTALION, M. & DYMOKE, P. 1989. The depositional age of the Dalradian Supergroup: U–Pb and Sm–Nd isotopic studies of the Tayvallich Volcanics, Scotland. *Journal of the Geological Society, London*, **146**, 3–6.

HARGROVE, U. S., HANSON, R. E., MARTIN, M. W., BLENKINSOP, T. G., BOWRING, S. A., WALKER, N. & MUNYANYIWA, H. 2003. Tectonic evolution of the Zambezi orogenic belt: geochronological, structural, and petrological constraints from northern Zimbabwe. *Precambrian Research*, **123**, 159–186.

HARRIS, L. B. & LI, Z. X. 1995. Palaeomagnetic dating and tectonic significance of dolerite intrusions in the Albany Mobile Belt, Western Australia. *Earth and Planetary Science Letters*, **131**, 143–164.

HARTZ, E. H. & TORSVIK, T. H. 2002. Baltica upside down: a new plate tectonic model for Rodinia and the Iapetus Ocean. *Geology*, **30**, 255–258.

HEATHERINGTON, A. L., MUELLER, P. A. & NUTMAN, A. P. 1996. Neoproterozoic magmatism in the

Suwannee terrane: Implications for terrane correlation. *In*: NANCE, R. D. & THOMPSON, M. D. (eds) *Avalonian and Related Peri-Gondwanan Terranes of the Circum-North Atlantic*. Geological Society America, Special Papers, **304**, 257–268.

HIBBARD, J. 2000. Docking Carolina: Mid-Paleozoic accretion in the southern Appalachians. *Geology*, **28**, 127–130.

HIBBARD, J. P., STODDARD, E. F., STODDARD, E. F., SECOR, D. T. & DENNIS, A. J. 2002. The Carolina Zone: overview of Neoproterozoic to Early Paleozoic peri-Gondwanan terranes along the eastern flank of the southern Appalachians. *Earth-Science Reviews*, **57**, 299–339.

HIGGINS, M. D. & VAN BREEMEN, O. 1998. The Age of the Sept Iles Layered Mafic Intrusion, Canada: Implications for the Late Neoproterozoic/Cambrian History of Southeastern Canada. *Journal of Geology*, **106**, 421–431.

HODYCH, J. P., COX, R. A. & KOŠLER, J. 2004. An equatorial Laurentia at 550 Ma confirmed by Grenvillian inherited zircons dated by LAM ICP-MS in the Skinner Cove volcanics of western Newfoundland: implications for inertial interchange true polar wander. *Precambrian Research*, **129**, 93–113.

HOFFMAN, P. F. 2005. Nature and origin of Neoproterozoic glacials. *In*: WINGATE, M. T. D. & PISAREVSKY, S. A. (eds) *Supercontinents and Earth Evolution Symposium 2005*. Geological Society of Australia Inc. Abstracts, 81, 47–48.

HOFFMAN, P. F. & SCHRAG, D. P. 2002. The snowball Earth hypothesis: testing the limits of global change. *Terra Nova*, **14**, 129–155.

IGLESIA LLANOS, M. P., TAIT, J. A., POPOV, V. & ABLAMASOVA, A. 2005. Palaeomagnetic data from Ediacaran (Vendian) sediments of the Arkhangelsk region, NW Russia: an alternative apparent polar wander path of Baltica for the Late Proterozoic - Early Palaeozoic. *Earth and Planetary Science Letters*, **240**, 732–747.

JOHN, T., SCHENK, V., HAASE, K. & SCHERER & TEMPO, F. 2003. Evidence for a Neoproterozoic ocean in south-central Africa from mid-ocean-ridge-type geochemical signatures and pressure-temperature estimates of Zambian eclogites. *Geology*, **31**, 243–246.

JOHNSON, S. P. & OLIVER, G. J. H. 2000. Mesoproterozoic oceanic subduction, island-arc formation and the initiation of back-arc spreading in the Kibaran Belt of central, Southern Africa; evidence from the ophiolite terrane, Chewore Inliers, northern Zimbabwe. *Precambrian Research*, **103**, 125–146.

JOHNSON, S. P. & OLIVER, G. J. H. 2004. Tectonothermal history of the Kaourera Arc, northern Zimbabwe: implications for the tectonic evolution of the Irumide and Zambezi Belts of south central Africa. *Precambrian Research*, **130**, 71–97.

JUNG, S., HOERNES, S. & MEZGER, K. 2000. Geochronology and petrology of migmatites from the Proterozoic Damara Belt—importance of episodic fluid-present disequilibrium melting and consequences for granite petrology. *Lithos*, **51**, 153–179.

KAMO, S. L. & GOWER, C. F. 1994. Note: U–Pb baddeleyite dating clarifies age of characteristic paleomagnetic remanence of Long Range dykes, southeastern Labrador. *Atlantic Geology*, **30**, 259–262.

KAMO, S. L., GOWER, C. F. & KROCH, T. E. 1989. A birthdate for the Iapetus ocean? A precise U–Pb zircon and baddeleyite age for the Long Range dykes, S. E. Labrador. *Geology*, **17**, 602–605.

KEPPIE, J. D. 1993. Synthesis of Paleozoic deformational events and terrane accretion in the Canadian Appalachians. *Geologische Rundschau*, **82**, 381–431.

KEPPIE, J. D. & ORTEGA-GUTIÉRREZ, F. 1999. Middle American Precambrian basement: A missing piece of the reconstructed 1-Ga orogen. *In*: RAMOS, V. A. & KEPPIE, J. D. (eds) *Laurentia–Gondwana Connections Before Pangea*. Geological Society of America, Special Papers, **336**, 199–210.

KEPPIE, J. D. & RAMOS, V. A. 1999. Odyssey of terranes in the Iapetus and Rheic oceans during the Paleozoic. *In*: RAMOS, V. A. & KEPPIE, J. D. (eds) *Laurentia–Gondwana Connections Before Pangea*. Geological Society of America, Special Papers, **336**, 267–276.

KEPPIE, J. D., NANCE, R. D., MURPHY, J. B. & DOSTAL, J. 1991. Northern Appalachians: Avalon and Meguma Terranes. *In*: DALLMEYER, R. D. & LÉCORCHÉ, J. P. (eds) *The West African Orogens and Circum-Atlantic Correlatives*. Springer-Verlag, New York, 315–334.

KEPPIE, J. D., DOSTAL, J., ORTEGA-GUTIÉRREZ, F. & LÓPEZ, R. 2001. A Grenvillian arc on the margin of Amazonia: evidence from the southern Oaxacan Complex, southern Mexico. *Precambrian Research*, **112**, 165–181.

KEPPIE, J. D., NANCE, R. D., MURPHY, J. B. & DOSTAL, J. 2003. Tethyan, Mediterranean, and Pacific analogues for the Neoproterozoic-Paleozoic birth and development of peri-Gondwanan terranes and their transfer to Laurentia and Laurussia. *Tectonophysics*, **365**, 195–219.

KEPPIE, J. D., DOSTAL, J., NANCE, R. D., MILLER, B. V., ORTEGA-RIVERA, A. & LEE, J. K. W. 2006. Circa 546 Ma plume-related dykes in the *c.*1 Ga Novillo Gneiss (east-central Mexico): evidence for the initial separation of Avalonia. *Precambrian Research*, **147**, 342–353.

KIRSCHVINK, J. L. 1978. The Precambrian-Cambrian boundary problem: Paleomagnetic directions from the Amadeus basin, central Australia. *Earth and Planetary Science Letters*, **40**, 91–100.

KIRSCHVINK, J. L. 1992. Late Proterozoic Low-Latitude Global Glaciation: The Snowball Earth. *In*: SCHOPF, J. W. & KLEIN, C. (eds) *The Proterozoic Biosphere*. Cambridge University Press, Cambridge, 51–52.

KIRSCHVINK, J. L. & ROZANOV, A.Yu. 1984. Magnetostratigraphy of Lower Cambrian strata from the Siberian Platform: a palaeomagnetic pole and preliminary polarity time-scale. *Geological Magazine*, **121**, 189–203.

KIRSCHVINK, J. L., RIPPERDAN, R. L. & EVANS, D. A. 1997. Evidence for large-scale reorganization of Early Cambrian continental masses by inertial interchange true polar wander. *Science*, **277**, 541–545.

KLOOTWIJK, C. T. 1980. Early Palaeozoic palaeomagnetism in Australia. *Tectonophysics*, **64**, 249–332.

KNOLL, A. H. 1992. The early evolution of eukaryotes, a geological perspective. *Science*, **256**, 622–637.

KRAVCHINSKY, V. A., KONSTANTINOV, K. M. & COGNE, J. P. 2001. Paleomagnetic study of Vendian and early Cambrian rocks of South Siberia and

Central Mongolia: was the Siberian platform assembled at this time? *Precambrian Research*, **110**, 61–92.

KRÖNER, A. & CORDANI, U. 2003. African, southern Indian and South American cratons were not part of the Rodinia supercontinent: evidence from field relationships and geochronology. *Tectonophysics*, **375**, 335–365.

LANDING, E. 1996. Avalon: Insular continent by the latest Precambrian. *In*: NANCE, R. D. & THOMPSON, M. D. (eds) *Avalonian and Related Peri-Gondwanan Terranes of the Circum-North Atlantic*. Geological Society of America, Special Papers, **304**, 29–63.

LANDING, E. 2005. Early Paleozoic Avalon-Gondwana unity: An obituary - Response to 'Palaeontological evidence bearing on global Ordovician-Silurian continental reconstructions' by FORTEY, R. A. & COCKS, L. R. M. *Earth-Science Reviews*, **69**, 169–175.

LARSEN, L. M. & REX, D. C. 1992. A review of the 2500 Ma span of alkaline-ultramafic. Potassic and carbonatitic magmatism in West Greenland. *Lithos*, **28**, 367–402.

LEGGO, P. M. 1974. A geochronological study of the basement complex of Uganda. *Journal of the Geological Society, London*, **130**, 263–277.

LIÉGEOIS, J. P., LATOUCHE, L., BOUGHRARA, M., NAVEZ, J. & GUIRAUD, M. 2003. The LATEA metacraton (Central Hoggar, Tuareg shiled, Algeria): behaviour of an old passive margin during the Pan-African orogeny. *Journal of African Earth Sciences*, **37**, 161–190.

LINNEMANN, U., GEHMLICH, M. *ET AL*. 2000. From Cadomian subduction to Early Palaeozoic rifting: The evolution of Saxo-Thuringia at the margin of Gondwana in the light of single zircon geochronology and basin development (Central European Variscides, Germany). *In*: FRANKE, W., ALTHERR, R., HAAK, V., ONCKEN, O. & TANNER, D. (eds) *Orogenic Processes: Quantification and Modelling in the Variscan Belt of Central Europe*. Geological Society, London, Special Publications, **179**, 131–153.

MCCALL, G. J. H. 2006. The Vendian (Ediacaran) in the geological record: Enigmas in geology's prelude to the Cambrian explosion. *Earth-Science Reviews*, **77**, 1–299.

MCCAUSLAND, P. J. A. & HODYCH, J. P. 1998. Palaeomagnetism of the 550 Ma Skinner Cove volcanics of western Newfoundland and opening of the Iapetus Ocean. *Earth and Planetary Science Letters*, **163**, 15–29.

MCCAUSLAND, P. J. A., VAN DER VOO, R. D. & BRANDENBURG, J. P. 2003. Preliminary paleomagnetic results from Late Neoproterozoic intrusions in Quebec, Canada: rapid apparent latitudinal motion of Laurentia. *Geophysical Research Abstracts*, **5**, 09090.

MCCAUSLAND, P. J. A., MURPHY, J. B. & MAC NIOCAILL, C. 2006. Endings and beginnings: Paleogeography of the Neoproterozoic-Cambrian transition. *Precambrian Research*, **147**, 187–192.

MCELHINNY, M. W. & LUCK, G. R. 1970. The palaeomagnetism of the Antrim Plateau Volcanics of Northern Australia. *Geophysical Journal of the Royal Astronomical Society*, **20**, 191–205.

MCELHINNY, M. W., COWLEY, J. A. & EDWARDS, D. J. 1978. Palaeomagnetism of some rocks from Peninsular India and Kashmir. *Tectonophysics*, **50**, 41–54.

MCELHINNY, M. W., POWELL, C. M. & PISAREVSKY, S. A. 2003. Paleozoic terranes of Eastern Australia and the drift history of Gondwana. *Tectonophysics*, **362**, 41–65.

MCMENAMIN, M. A. S. & MCMENAMIN, D. L. 1990. *The Emergence of Animals; The Cambrian Breakthrough*. Columbia University Press, New York.

MCNAMARA, A. K., MAC NIOCAILL, C., VAN DER PLUIJM, B. A. & VAN DER VOO, R. 2001. West African proximity of Avalon in the latest Precambrian. *Geological Society of America Bulletin*, **113**, 1161–1170.

MCWILLIAMS, M. O. 1981. Palaeomagnetism and Precambrian tectonic evolution of Gondwana. *In*: KRÖNER, A. (ed.) *Precambrian Plate Tectonics*. Elsevier, Amsterdam, 649–687.

MCWILLIAMS, M. O. & MCELHINNY, M. W. 1980. Late Precambrian paleomagnetism in Australia: The Adelaide geosyncline. *Journal of Geology*, **88**, 1–26.

MALOOF, A. C., HALVERSON, G. P., KIRSCHVINK, J. L., SCHRAG, D. P., WEISS, B. P. & HOFFMAN, P. F. 2006. Combined palaeomagnetic, isotopic, and stratigraphic evidence for true polar wander from the Neoproterozoic Akademikerbreen Group, Svalbard, Norway. *Geological Society of America Bulletin*, **118**, 1099–1124.

MARTIN, M. W., GRAZHDANKIN, D. V., BOWRING, S. A., EVANS, D. A. D., FEDONKIN, M. A. & KIRSCHVINK, J. L. 2000. Age of Neoproterozoic bilatarian body and trace fossils, White Sea, Russia: implications for metazoan evolution. *Science*, **288**, 841–845.

MASLOV, A. V. & ISHERSKAYA, M. V. 2002. Riphean sedimentary sequences of the eastern and northeastern margins of the Eastern European craton. *Russian Journal of Earth Sciences*, **4**, 271–276.

MASLOV, A. V., ERDTMANN, B. D., IVANOV, K. S., IVANOV, S. N. & KRUPENIN, M. T. 1997. The main tectonic events, depositional history, and the palaeogeography of the southern Urals during the Riphean-early Palaeozoic. *Tectonophysics*, **276**, 313–335.

MAX, M. D. & RODDICK, J. C. 1989. Age of metamorphism in the Rosslare Complex, SE Ireland. *Proceedings of the Geologists' Association*, **100**, 113–121.

MEERT, J. G. & VAN DER VOO, R. 1996. Paleomagnetic and ^{40}Ar/^{39}Ar Study of the Sinyai Dolerite, Kenya: Implications for Gondwana Assembly. *Journal of Geology*, **104**, 131–142.

MEERT, J. G. & VAN DER VOO, R. 2001. Comment on 'New palaeomagnetic result from Vendian red sediments in Cisbaikalia and the problem of the relationship of Siberia and Laurentia in the Vendian' by S. A. Pisarevsky R. A. Komissarova A. N. Khramov. *Geophysical Journal International*, **146**, 867–870.

MEERT, J. G., VAN DER VOO, R. & PAYNE, T. W. 1994. Paleomagnetism of the Catoctin volcanic province: A new Vendian–Cambrian apparent polar wander path for North America. *Journal of Geophysical Research*, **99**, 4625–4641.

MEERT, J. G., VAN DER VOO, R. & AYUB, S. 1995. Paleomagnetic investigation of the Neoproterozoic

Gagwe lavas and Mbozi complex, Tanzania and the assembly of Gondwana. *Precambrian Research*, **74**, 225–244.

MEERT, J. G., TORSVIK, T. H., EIDE, E. A. & DAHLGREN, S. 1998. Tectonic significance of the Fen Province, S. Norway: constraints from geochronology and paleomagnetism. *Journal of Geology*, **106**, 553–564.

METELKIN, D. V., BELONOSOV, I. V., GLADKOCHUB, D. P., DONSKAYA, T. V., MAZUKABZOV, A. M. & STANEVICH, A. M. 2005. Paleomagnetic directions from Nersa intrusions of the Biryusa terrane, Siberian craton, as a reflection of tectonic events in the Neoproterozoic. *Russian Geology and Geophysics*, **46**, 395–410.

MOURA, C. A. V. & GAUDETTE, H. E. 1993. Evidence of Brasiliano/Panafrican deformation in the Araguaia belt: implication for Gondwana evolution. *Revista Brasileira de Geociências*, **23**, 117–123.

MURPHY, J. B. & NANCE, R. D. 1989. Model for the evolution of the Avalonian-Cadomian belt. *Geology*, **17**, 735–738.

MURPHY, J. B. & NANCE, R. D. 1991. Supercontinent model for the contrasting character of Late Proterozoic orogenic belts. *Geology*, **19**, 469–472.

MURPHY, J. B., KEPPIE, J. D., DOSTAL, J. & HYNES, A. J. 1990. The geochemistry and petrology of the late Precambrian Georgeville Group: a volcanic arc–rift succession in the Avalon terrane of Nova Scotia. *In*: D'LEMOS, R. S., STRACHAN, R. A. & TOPLEY, C. G. (eds) *The Cadomian Orogeny*. Geological Society, London, Special Publications, **51**, 383–393.

MURPHY, J. B., KEPPIE, J. D., DOSTAL, J. & NANCE, R. D. 1999. Neoproterozoic-early Paleozoic evolution of Avalonia. *In*: RAMOS, V. A. & KEPPIE, J. D. (eds) *Laurentia–Gondwana Connections Before Pangea*. Geological Society of America, Special Papers, **336**, 253–266.

MURPHY, J. B., STRACHAN, R. A., NANCE, R. D., PARKER, K. D. & FOWLER, M. B. 2000. Proto-Avalonia: a 1. 2–1.0. Ga tectonothermal event and constraints for the evolution of Rodinia. *Geology*, **28**, 1071–1074.

MURPHY, J. B., EGUILUZ, L. & ZULAUF, G. 2002. Cadomian orogens, peri-Gondwanan correlatives and Laurentia-Baltica connections. *Tectonophysics*, **352**, 1–9.

MURPHY, J. B., PISAREVSKY, S. A., NANCE, R. D. & KEPPIE, J. D. 2004. Neoproterozoic–early Paleozoic evolution of peri-Gondwanan terranes: implications for Laurentia–Gondwana connections. *International Journal of Earth Sciences*, **93**, 659–692.

MURPHY, J. B., GUTIÉRREZ-ALONSO, G., NANCE, R. D., FERNÁNDEZ-SUÁREZ, J., KEPPIE, J. D., QUESADA, C., STRACHAN, R. A. & DOSTAL, J. 2006. Origin of the Rheic Ocean: Rifting along a Neoproterozoic suture? *Geology*, **34**, 325–328.

MURTHY, G., GOWER, C., TUBRETT, M. & PÄTZOLD, R. 1992. Paleomagnetism of Eocambrian Long Range dykes and Double Mer Formation from Labrador, Canada. *Canadian Journal of Earth Sciences*, **29**, 1224–1234.

NANCE, R. D. & MURPHY, J. B. 1990. Kinematic history of the Bass River Complex, Nova Scotia: Cadomian tectonostratigraphic relations in the Avalon terrane of the Canadian Appalachians. *In*: D'LEMOS, R. S., STRACHAN, R. A. & TOPLEY, C. G.(eds) *The Cadomian Orogeny*. Geological Society, London, Special Publications, **51**, 395–406.

NANCE, R. D., MURPHY, J. B., STRACHAN, R. A., D'LEMOS, R. S. & TAYLOR, G. K. 1991. Late Proterozoic tectonostratigraphic evolution of the Avalonian and Cadomian terranes. *Precambrian Research*, **53**, 41–78.

NANCE, R. D., MURPHY, J. B. & KEPPIE, J. D. 2002. Cordilleran model for the evolution of Avalonia. *Tectonophysics*, **352**, 11–32.

NAWROCKI, J., BOGUCKIJ, A. & KATINAS, V. 2004. New late Vendian palaeogeography of Baltica and TESZ. *Geological Quarterly*, **48**, 309–316.

NEUBAUER, F. 2002. Evolution of late Neoproterozoic to early Paleozoic tectonic elements in Central and Southeast European Alpine mountain belts: review and synthesis. *Tectonophysics*, **352**, 87–103.

NIKISHIN, A. M., ZIEGLER, P. A. *ET AL*. 1996. Late Precambrian to Triassic history of the East European Craton: dynamics of sedimentary basin evolution. *Tectonophysics*, **268**, 23–63.

O'BRIEN, S. J., WARDLE, R. J. & KING, A. F. 1983. The Avalon zone: a Pan-African terrane in the Appalachian orogen of Canada. *Geological Journal*, **18**, 195–222.

OLIVER, G. J. H., JOHNSON, S. P., WILLIAMS, I. S. & HERD, D. A. 1998. Relic 1.4 Ga oceanic crust in the Zambezi Valley, northern Zimbabwe; evidence for Mesoproterozoic supercontinental fragmentation. *Geology*, **26**, 571–573.

OLOVYANISHNIKOV, V. G., ROBERTS, D. & SIEDLECKA, A. 2000. Tectonics and sedimentation of the Meso- to Neoproterozoic Timan-Varanger belt along the northeastern margin of Baltica. *Polarforschung*, **68**, 267–274.

OMARINI, R. H., SUREDA, R. J., GÖTZE, H.-J., SEILACHER, A. & PFLÜGER, F. 1999. Puncoviscana folded belt in northwestern Argentina: testimony of Late Proterozoic Rodinia fragmentation and pre-Gondwana collisional episodes. *International Journal of Earth Sciences*, **88**, 76–97.

ONSTOTT, T. C. & HARGRAVES, R. B. 1981. Proterozoic transcurrent tectonics: palaeomagnetic evidence from Venezuela and Africa. *Nature*, **289**, 131–136.

ORTEGA-OBREGON, C., KEPPIE, J. D., SOLARI, L. A., ORTEGA-GUTIÉRREZ, F., DOSTAL, J., LOPEZ, R., ORTEGA-RIVERA, A. & LEE, J. W. K. 2003. Geochronology and geochemistry of the *c*.917 Ma, calc-alkaline Etla granitoid pluton (Oaxaca, southern Mexico): evidence of post-Grenvillian subduction along the northern margin of Amazonia. *International Geology Review*, **45**, 596–622.

PAVLOV, V. E., GALLET, Y. & SHATSILLO, A. V. 2000. Palaeomagnetism of the upper Riphean Lakhanda Group of the Uchur-Maya area and the hypothesis of the late Proterozoic supercontinent. *Izvestiya, Physics of the Solid Earth*, **36**, 638–648.

PAVLOV, V. E., GALLET, Y., PETROV, P. Yu., ZHURAVLEV, D. Z. & SHATSILLO, A. V. 2002. Uy series and late Riphean sills of the Uchur-Maya area: isotopic and palaeomagnetic data and the problem of the Rodinia supercontinent. *Geotectonics*, **36**, 278–292.

PIMENTEL, M. M., HEAMAN, L. M. & FUCK, R. A. 1991. U–Pb zircon and sphene geochronology of late Proterozoic volcanic arc rock units from southwestern Goiás, central Brazil. *Journal of South American Earth Sciences*, **4**, 329–339.

PIMENTEL, M. M., FUCK, R. A. & BOTELHO, N. F. 1999. Granites and the geodynamic history of the Neoproterozoic Brasilia belt, Central Brazil: a review. *Lithos*, **46**, 463–483.

PIMENTEL, M. M., FUCK, R. A., JOST, H., FERREIRA FILHO, C. F. & ARAÚJO, S. M. 2000. Geology of the central part of the Tocantins Province: implications for the geodynamic history of the Brasília belt. *In*: CORDANI, U. G., MILANI, E. J., THOMAZ FILHO, A. & CAMPOS, D. A. (eds) *Tectonic evolution of South America*. 31st International Geological Congress, Rio de Janeiro, 195–229.

PIN, C. & POIDEVIN, J. L. 1987. U–Pb zircon evidence for a Pan-African granulite facies metamorphism in the Central African Republic. A new interpretation of the high-grade series of the northern border of the Congo Craton. *Precambrian Research*, **36**, 303–312.

PIPER, J. D. A. 2000. The Neoproterozoic Supercontinent: Rodinia or Palaeopangea? *Earth and Planetary Science Letters*, **176**, 131–146.

PISAREVSKY, S. A. 2001. *Palaeomagnetic study of Vines 1*. Geological Survey of Western Australia, Record **2001/18**, Appendix 4.

PISAREVSKY, S. A. 2005. New edition of the Global Paleomagnetic Database. *EOS Transactions*, **86**, 170.

PISAREVSKY, S. A. & NATAPOV, L. M. 2003. Siberia and Rodinia. *Tectonophysics*, **375**, 221–245.

PISAREVSKY, S. A., GUREVICH, E. L. & KHRAMOV, A. N. 1997. Palaeomagnetism of the Lower Cambrian sediments from the Olenek river section (northern Siberia): paleopoles and a problem of magnetic polarity in Early Cambrian. *Geophysical Journal International*, **130**, 746–756.

PISAREVSKY, S. A., KOMISSAROVA, R. A. & KHRAMOV, A. N. 2000. New palaeomagnetic result from Vendian red sediments in Cisbaikalia and the problem of the relationship of Siberia and Laurentia in the Vendian. *Geophysical Journal International*, **140**, 598–610.

PISAREVSKY, S. A., KOMISSAROVA, R. A. & KHRAMOV, A. N. 2001*a*. Reply to comment by J. G. Meert and R. Van der Voo on 'New palaeomagnetic result from Vendian red sediments in Cisbaikalia and the problem of the relationship of Siberia and Laurentia in the Vendian'. *Geophysical Journal International*, **146**, 871–873.

PISAREVSKY, S. A., LI, Z. X., GREY, K. & STEVENS, M. K. 2001*b*. A palaeomagnetic study of Empress 1A, a stratigraphic drillhole in the Officer Basin: new evidence for the low-latitude position of Australia in the Neoproterozoic. *Precambrian Research*, **110**, 93–108.

PISAREVSKY, S. A., WINGATE, M. T. D., POWELL, C. MCA., JOHNSON, S. & EVANS, D. A. D. 2003. Models of Rodinia assembly and fragmentation. *In*: YOSHIDA, M., WINDLEY, B. & DASGUPTA, S. (eds) *Proterozoic East Gondwana: supercontinent assembly and breakup*. Geological Society, London, Special Publications, **206**, 35–55.

PISAREVSKY, S. A., WINGATE, M. T. D., STEVENS, M. K. & HAINES, P. W. 2007. Paleomagnetic results from the Lancer-1 stratigraphic drill hole, Officer Basin, Western Australia, and implications for Rodinia reconstructions. *Australian Journal of Earth Sciences*, **54**, 561–572.

POORTER, R. P. E. 1972. Palaeomagnetism of the Rogaland Precambrian (southwestern Norway). *Physics of the Earth and Planetary Interiors*, **5**, 167–176.

POPOV, V., IOSIFIDI, A., KHRAMOV, A., TAIT, J. & BACHTADSE, V. 2002. Paleomagnetism of Upper Vendian sediments from the Winter Coast, White Sea region, Russia: implications for the paleogeography of Baltica during Neoproterozoic times. *Journal of the Geophysical Research*, **107** (B11), 2315, doi: 10. 1029/2001JB001607.

POPOV, V., KHRAMOV, A. & BACHTADSE, V. 2005. Palaeomagnetism, magnetic stratigraphy, and petromagnetism of the Upper Vendian sedimentary rocks in the sections of the Zolotitsa River and in the Verkhotina Hole, Winter Coast of the White Sea, Russia. *Russian Journal of Earth Sciences*, **7**, 1–29.

POPRAWA, P., SLIAUPA, S., STEPHENSON, R. & LAZAUSKIENE, J. 1999. Late Vendian–Early Palaeozoic tectonic evolution of the Baltic Basin: regional tectonic implications from subsidence analysis. *Tectonophysics*, **314**, 219–239.

POWELL, C. MCA. & PISAREVSKY, S. A. 2002. Late Neoproterozoic assembly of East Gondwanaland. *Geology*, **30**, 3–6.

PRAVE, A. R. 1996. Tale of three cratons: Tectonostratigraphic anatomy of the Damara orogen in northwestern Namibia and the assembly of Gondwana. *Geology*, **24**, 1115–1118.

PUCHKOV, V. N. (ed.) 2003. *Magmatism on passive margins of the Rodinia. Materials of the international conference*. Ufa, Geological Institute of the USC RAS.

PUFFER, J. H. 2002. A Late Neoproterozoic eastern Laurentian superplume: location, size, chemical composition, and environmental impact. *American Journal of Science*, **302**, 1–27.

QUESADA, C. 1990. Precambrian terranes in the Iberian Variscan foldbelt. *In*: STRACHAN, R. A. & TAYLOR, G. K. (eds) *Avalonian and Cadomian Geology of the North Atlantic*. Blackie, London, 109–133.

RAMOS, V. A. 1988. Late Proterozoic–Early Paleozoic of South America—a collisional history. *Episodes*, **11**, 168–174.

RAPALINI, A. E. 2006. New late Proterozoic paleomagnetic pole for the Rio de la Plata craton: Implications for Gondwana. *Precambrian Research*, **147**, 223–233.

RAPELA, C. W., PANKHURST, R. J., CASQUET, C., FANNING, C. M., BALDO, E. G., GONZÁLEZ-CASADO, J. M., GALINDO, C. & DAHQUIST, J. A. 2008. The Río de la Plata craton and the assembly of SW Gondwana. *Earth-Science Reviews*, **83**, 49–82.

ROBERTS, D. & SIEDLECKA, A. 2002. Timanian orogenic deformation along the northeastern margin of Baltica, northwest Russia and northeast Norway, and Avalonian–Cadomian connections. *Tectonophysics*, **352**, 169–184.

ROBERTS, R. J., CORFU, F., TORSVIK, T. H., RAMSAY, D. M. & ASHWAL, L. D. 2004. Redefining the magmatic evolution of the Kalak Nappe Complex. The

26th Nordic Geological Winter Meeting. *Geologiska Föreningens I Stockholm Förhandlingar*, **126**, 84.

RUIZ, J., TOSDAL, R. M., RESTREPO, P. A. & MURILLO-MUÑETÓN, G. 1999. Pb isotope evidence for Columbia–southern Mexico connections in the Proterozoic. *In*: RAMOS, V. A. & KEPPIE, J. D. (eds) *Laurentia–Gondwana Connections Before Pangea*. Geological Society of America, Special Papers, **336**, 183–197.

SÁNCHEZ-BETTUCCI, L. & RAPALINI, A. E. 2002. Paleomagnetism of the Sierra de Las Animas Complex, southern Uruguay: its implications in the assembly of western Gondwana. *Precambrian Research*, **118**, 243–265.

SCHMIDT, P. W. & WILLIAMS, G. E. 1995. The Neoproterozoic climatic paradox: Equatorial palaeolatitude for Marinoan glaciation near sea level in South Australia. *Earth and Planetary Science Letters*, **134**, 107–124.

SCHMIDT, P. W. & WILLIAMS, G. E. 1996. Palaeomagnetism of the ejecta-bearing Bunyeroo Formation, late Neoproterozoic, Adelaide fold belt, and the age of the Acraman impact. *Earth and Planetary Science Letters*, **144**, 347–357.

SCHMIDT, P. W., WILLIAMS, G. E. & EMBLETON, B. J. J. 1991. Low palaeolatitude of Late Proterozoic glaciation: Early timing of remanence in haematite of the Elatina Formation, South Australia. *Earth and Planetary Science Letters*, **105**, 355–367.

SHUMLYANSKYY, L. & ANDRÉASSON, P. G. 2004. New geochemical and geochronological data from the Volyn Flood Basalt in Ukraine and correlation with large igneous events in Baltoscandia. The 26th Nordic Geological Winter Meeting. *Geologiska Föreningens I Stockholm Förhandlingar*, **126**, 85–86.

SIEDLECKA, A., ROBERTS, D., NYSTUEN, J. P. & OLOVYANISHNIKOV, V. G. 2004. Northeastern and northwestern margins of Baltica in Neoproterozoic time: evidence from the Timanian and Caledonian Orogens. *In*: GEE, D. G. & PEASE, V. L. (eds) *The Neoproterozoic Timanide Orogen of Eastern Baltica*. Geological Society, London, Memoirs, **30**, 169–190.

SMETHURST, M. A., KHRAMOV, A. N. & TORSVIK, T. H. 1998. The Neoproterozoic and Palaeozoic palaeomagnetic data for the Siberian Platform: from Rodinia to Pangea. *Earth-Science Reviews*, **43**, 1–24.

SOHL, L. E., CHRISTIE-BLICK, N. & KENT, D. V. 1999. Paleomagnetic polarity reversals in Marinoan (ca. 600 Ma) glacial deposits of Australia: Implications for the duration of low-latitude glaciation in Neoproterozoic time. *Geological Society of America Bulletin*, **111**, 1120–1139.

SOLARI, L. A., KEPPIE, J. D., ORTEGA-GUTIÉRREZ, F., CAMERON, K. L., LOPEZ, R. & HAMES, W. E. 2003. 990 and 1100 Ma Grenvillian tectonothermal events in the northern Oaxacan Complex, southern Mexico: roots of an orogen. *Tectonophysics*, **365**, 257–282.

STORETVEDT, K. M. 1966. Remanent magnetization of some dolerite intrusions in the Egersund area, southern Norway. *Geophysica Norvegica*, **26**, 1–17.

STRACHAN, R. A., NANCE, R. D., DALLMEYER, R. D., D'LEMOS, R. S., MURPHY, J. B. & WATTS, G. R. 1996. Late Precambrian tectonothermal evolution of the Malverns Complex. *Journal of the Geological Society, London*, **153**, 589–600.

STUKAS, V. & REYNOLDS, P. H. 1974. $^{40}Ar/^{39}Ar$ dating of the Long Range dikes, Newfoundland. *Earth and Planetary Science Letters*, **22**, 256–266.

SYMONS, D. T. A. & CHIASSON, A. D. 1991. Paleomagnetism of the Callander Complex and the Cambrian apparent polar wander path for North America. *Canadian Journal of Earth Sciences*, **28**, 355–363.

TANCZYK, E. I., LAPOINTE, P., MORRIS, W. A. & SCHMIDT, P. W. 1987. A paleomagnetic study of the layered mafic intrusion at Sept-Iles, Quebec. *Canadian Journal of Earth Science*, **24**, 1431–1438.

THEOKRITOFF, G. 1979. Early Cambrian provincialism and biogeographic boundaries in the North Atlantic region. *Lethaia*, **12**, 281–295.

THOMAS, W. A. 2005. Tectonic inheritance at a continental margin. *GSA Today*, **16**, 4–11.

TOHVER, E., D'AGRELLA-FILHO, M. S. & TRINDADE, R. I. F. 2006. Paleomagnetic record of Africa and South America for the 1200–500. Ma interval, and evaluation of Rodinia and Gondwana assemblies. *Precambrian Research*, **147**, 193–222.

TORSVIK, T. H. & REHNSTRÖM, E. F. 2001. Cambrian palaeomagnetic data from Baltica: implications for true polar wander and Cambrian palaeogeography. *Journal of the Geological Society, London*, **158**, 321–329.

TORSVIK, T. H., CARTER, L. M., ASHWAL, L. D., BHUSHAN, S. K., PANDIT, M. K. & JAMTVEIT, B. 2001. Rodinia refined or obscured: palaeomagnetism of the Malani igneous suite (NW India). *Precambrian Research*, **108**, 319–333.

TRINDADE, R. I. F., FONT, E., D'AGRELLA-FILHO, M. S., NOGUEIRA, A. C. R. & RICCOMINI, C. 2003. Low-latitude and multiple geomagnetic reversals in the Neoproterozoic Puga cap carbonate, Amazon craton. *Terra Nova*, **15**, 441–446.

TRINDADE, R. I. F., D'AGRELLA-FILHO, M. S., EPOF, I. & BRITO NEVES, B. B. 2006. Paleomagnetism of Early Cambrian Itabaiana mafic dikes (NE Brazil) and the final assembly of Gondwana. *Earth and Planetary Science Letters*, **244**, 361–367.

TROMPETTE, R. 1994. *Geology of Western Gondwana (2000–500 Ma)*. A. A. Balkema, Rotterdam/Brookfield.

TROMPETTE, R. 1997. Neoproterozoic (*c*. 600 Ma) aggregation of Western Gondwana: a tentative scenario. *Precambrian Research*, **82**, 101–112.

TUCKER, R. D. & PHARAOH, T. C. 1991. U–Pb zircon ages of late Precambrian rocks in southern Britain. *Journal of the Geological Society, London*, **148**, 435–443.

VALERIANO, C. M., MACHADO, N., SIMONETTI, A., VALLADARES, C. S., SEER, H. J. & SIMÕES, L. S. A. 2004. U–Pb geochronology of the southern Brasília belt (SE-Brazil): sedimentary provenance, Neoproterozoic orogeny and assembly of West Gondwana. *Precambrian Research*, **130**, 27–55.

VAN ALSTINE, D. R. & GILLETT, S. L. 1979. Paleomagnetism of Upper Precambrian sedimentary rocks from the Desert Range, Nevada. *Journal of Geophysical Research*, **84**, 4490–4500.

VAN DER VOO, R. 1988. Paleozoic paleogeography of North America, Gondwana, and intervening displaced terranes: comparisons of paleomagnetism with paleoclimatology and biogeographical patterns. *Geological Society of America Bulletin*, **100**, 311–324.

VAN DER VOO, R. 1990. The reliability of paleomagnetic data. *Tectonophysics*, **184**, 1–9.

VINYU, M. L., HANSON, R. E., MARTIN, M. W., BOWRING, S. A., JELSMA, H. A., KROL, M. A. & DIRKS, P. H. G. M. 1999. U/Pb and $^{40}Ar/^{39}Ar$ geochronological constraints on the tectonic evolution of the easternmost part of the Zambezi orogenic belt, northeast Zimbabwe. *Precambrian Research*, **98**, 67–82.

VON RAUMER, J. F., STAMPFLI, G. M., BOREL, G. & BUSSY, F. 2002. Organization of pre-Variscan basement areas at the north-Gondwanan margin. *International Journal of Earth Sciences*, **91**, 35–52.

WALDERHAUG, H. J., TORSVIK, T. H., EIDE, E. A. & MEERT, J. G. 2003. Magnetic properties and age of the Alnø Carbonatite Complex (Sweden). *Geophysical Research Abstracts*, **5**, 10358.

WILLIAMS, H. & HISCOTT, R. N. 1987. Definition of the Iapetus rift-drift transition in western Newfoundland. *Geology*, **15**, 1044–1047.

WILLNER, A. P., ERMOLAEVA, T., STROINK, L., GLASMACHER, U. A., GIESE, U., PUCHKOV, V. N., KOZLOV, V. I. & WALTER, R. 2001. Contrasting provenance signals in Riphean and Vendian sandstones in the SW Urals (Russia): constraints for a change from passive to active continental margin conditions in the Neoproterozoic. *Precambrian Research*, **110**, 215–239.

WILLNER, A. P., SINDERN, S., METZGER, R., ERMOLAEVA, T., KRAMM, U., PUCHKOV, V. & KRONZ, A. 2003. Typology and single grain U/Pb ages of detrital zircons from Proterozoic sandstones in the SW Urals (Russia): early time marks at the eastern margin of Baltica. *Precambrian Research*, **124**, 1–20.

WINGATE, M. T. D. & GIDDINGS, J. W. 2000. Age and palaeomagnetism of the Mundine Well dyke swarm, Western Australia: implications for an Australia-Laurentia connection at 755 Ma. *Precambrian Research*, **100**, 335–357.

WINGATE, M. T. D., PISAREVSKY, S. A. & DE WAELE, B. 2004. Paleomagnetism of the 765 Ma Luakela Volcanics in NW Zambia and Implications for Neoproterozoic Positions of the Congo Craton. *AGU Fall Meeting Supplement*, Abstract U32A-03.

ZULAUF, G., SCHITTER, F., RIEGLER, G., FINGER, F., FIALA, J. & VEJNAR, Z. 1999. Age constraints on the Cadomian evolution of the Teplá-Barrandian unit (Bohemian Massif) through electron microprobe dating of metamorphic monazite. *Zeitschrift der Deutschen Geologischen Gesellschaft*, **150/4**, 627–639.

Contemporaneous evolution of the Palaeoproterozoic–Mesoproterozoic sedimentary basins of the São Francisco–Congo Craton

A. J. PEDREIRA[1] & B. DE WAELE[2,3]

[1]*CPRM–Geological Survey of Brazil, Av. Ulysses Guimarães, 2862, 41213-000 Salvador, Brazil (e-mail: apedreira@sa.cprm.gov.br)*

[2]*Tectonics Special Research Centre, The University of Western Australia, 35 Stirling Highway, WA 6009 Crawley, Australia*

[3]*Present address: British Geological Survey, Keyworth, Nottingham NG12 5GG, UK*

Abstract: Deposition of Palaeo–Mesoproterozoic sedimentary rocks on the São Francisco–Congo craton started during Statherian taphrogenesis (1.8–1.75 Ga), as verified by ages of *c.* 1.7 Ga determined for volcanic rocks of the lower part of the Espinhaço Supergroup in the states of Minas Gerais and Bahia (Brazil). These basins contain volcanic rocks and conglomerates alternating with sandstones, argillites and dolomites, deposited in continental, transitional and marine environments. The rocks in the westernmost sector of the Congo Craton (Central Africa) compose the Chela Group, comprising sandstones, argillites and dolomites. In the easternmost region of the Congo Craton the Kibaran, Akanyaru, Kagera and Muva supergroups occur: the first three in the Kibaran Belt and the last in the Irumide Belt and on the Bangweulu Block. They consist predominantly of pelites and schists, sandstones and, in lesser proportion, conglomerates, deposited in shallow marine, fluvial and lacustrine environments. Their sedimentation ages are constrained through ages on felsic tuff layers as follows: Chela Group 1790 $\pm$ 17 Ma, Kagera Supergroup 1780 $\pm$ 9 Ma, and Muva Supergroup 1879 $\pm$ 13 Ma. These data show that broadly coeval and sedimentologically similar epi-continental sedimentary basins occurred on the São Francisco and Congo cratons, suggesting the possible existence of a long-lived wide epi-continental sea covering large areas of these cratons during Statherian times.

The São Francisco Craton of South America and the Congo Craton of Africa are stable Archaean blocks of a once coherent landmass (Fig. 1) that broke up during the opening of the Atlantic Ocean. These cratonic nuclei are considered to have become stabilized during the Palaeoproterozoic Trans-Amazonian (South America) and Eburnian (Africa) events, and underwent a succession of later events along their margin including the Mesoproterozoic Espinhaço cycle in South America (Brito Neves *et al.* 1980), the Kibaran and Irumide orogens in Africa, often associated with the formation of the Rodinia supercontinent, and the Neoproterozoic Pan-African/Brasiliano orogenic events during the agglutination of Gondwana. The Pan-African/Brasiliano orogenesis reworked the edges of both cratons, giving birth to the Brasília, Araçuaí, Sergipano, Rio Preto and Riacho do Pontal belts in Brazil and the West Congo, Kaoko, Damara, Lufilian, Oubanguides and Zambezi belts and the East African orogen in Africa (Fig. 2).

The initial coherence of the São Francisco and Congo cratons prior to Gondwana was based largely on the occurrence of comparable Precambrian epi-continental sequences on both sides of the Atlantic Ocean (Trompette 1994). In the São Francisco Craton, the Statherian taphrogenesis (1.8–1.75 Ga) opened a series of intra-continental rifts, some of which expanded into sag basins (Brito Neves 2002), into which volcanic and sedimentary rocks were deposited, collectively called the Espinhaço Supergroup. On the Congo Craton, the Palaeo-/Mesoproterozoic successions comprise the Chela Group on the Angola–Kasai Shield (Torquato & Fogaça 1981), the Kibaran and Akanyaru/Kagera supergroups in the central African Kibaran Belt (Royal Museum for Central Africa 1990; Theunissen *et al.* 1991) and the Muva Supergroup on the Bangweulu Block and within the Mesoproterozoic Irumide Belt (Daly & Unrug 1982; De Waele & Mapani 2002; De Waele 2005; Fig. 2).

In Brazil these successions have been studied in detail and have their stratigraphical nomenclature

From: PANKHURST, R. J., TROUW, R. A. J., BRITO NEVES, B. B. & DE WIT, M. J. (eds) *West Gondwana: Pre-Cenozoic Correlations Across the South Atlantic Region.* Geological Society, London, Special Publications, **294**, 33–48. DOI: 10.1144/SP294.3 0305-8719/08/$15.00 © The Geological Society of London 2008.

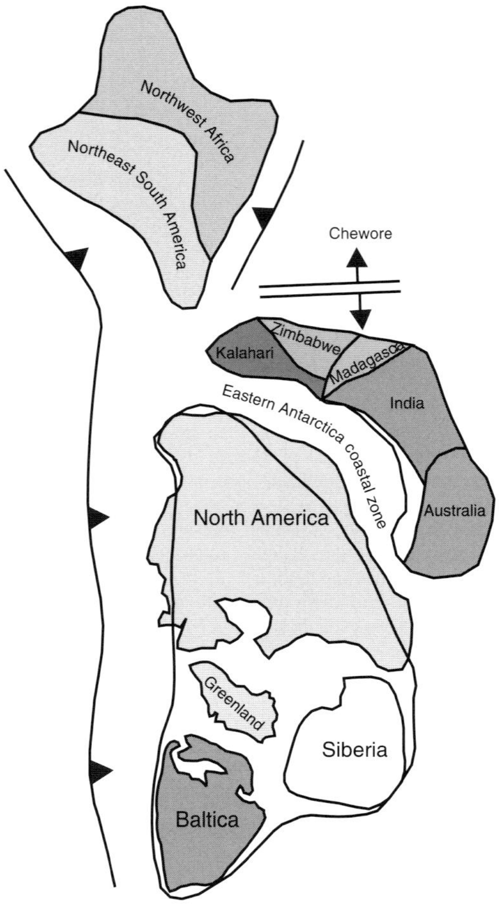

Fig. 1. The Columbia Supercontinent at the beginning of the Mesoproterozoic, according to the configuration of Rogers & Santosh (2002). Modified from Schobbenhaus & Brito Neves (2003).

formalized down to the formation hierarchy and their sedimentary environments described in detail; in Africa, however, most units have only recently been described in any detail, while in most cases stratigraphical nomenclature still needs to be formalized.

In this paper, we will present the known stratigraphy for the successions on the São Francisco Craton based on published papers, while for the African successions we use informal stratigraphies, either published in the literature, or in advanced stage of publication. We will use published age data, basin analysis and, where possible, sequence stratigraphy, to compare these successions. The stratigraphical columns of the sedimentary successions, both in Brazil and Africa, are shown in Figures 3 and 7–9.

Brazilian sedimentary basins: São Francisco Craton

The Palaeo–Mesoproterozoic sedimentary basins of Brazil, started as intra-continental rifts that opened during the Statherian taphrogenesis (1.8–1.75 Ga) and filled with aeolian sands, acid volcanic rocks and conglomerates. The abortion of the rifting process lead to the expansion of the basins into sags, that were filled with conglomerates, sandstones, fine-grained rocks and limestones, and were intruded by basic rocks, collectively named the Espinhaço Supergroup.

Espinhaço Supergroup

The Espinhaço Supergroup crops out in three different domains, all of them related to an orographic system named the Espinhaço Range, developed along a north–south trend between the 10°S and 20°S parallels in eastern Brazil: the southern Espinhaço Range, the northern Espinhaço Range and the Chapada Diamantina (Fig. 3).

In the southern Espinhaço Range, the Espinhaço Supergroup is divided into two groups (the lower Diamantina and the upper Conselheiro Mata), which comprise nine formations composed essentially of sandstones and fine-grained rocks (phyllites, siltstones, quartzites). Conglomerate beds and lenses occur at the base of the succession (Fig. 3a). The Espinhaço Supergroup in the northern Espinhaço Range, was divided by Schobbenhaus (1996) into the Oliveira dos Brejinhos and Santo Onofre groups. The lower, Oliveira dos Brejinhos Group, was deposited in a rift followed by a flexure in the interval comprising the Bom Retiro and Fazendinha formations (Fig. 3b). Recently Danderfer & Dardenne (2002) reviewed the succession from a tectonostratigraphical point of view, describing new formations and groups; the resulting interpretation confirms the one previously presented by Schobbenhaus (1996). In Chapada Diamantina, the Espinhaço Supergroup has been divided into three groups that comprise six formations (Fig. 3c). The succession of Chapada Diamantina and the one of the northern Espinhaço Range were considered to be sub-facies of the same succession (Schobbenhaus 1969).

Deposition of the Espinhaço Supergroup began with a rift phase in all three domains. In the southern Espinhaço Range it encompassed mostly fluvial sandstones and alluvial fan conglomerates that predominate at the base of the Diamantina Group. In the northern Espinhaço Range, a first rift phase occurs during the initial deposition of the Oliveira dos Brejinhos Group, deposited in continental environments: alluvial fan, fluvial–aeolian,

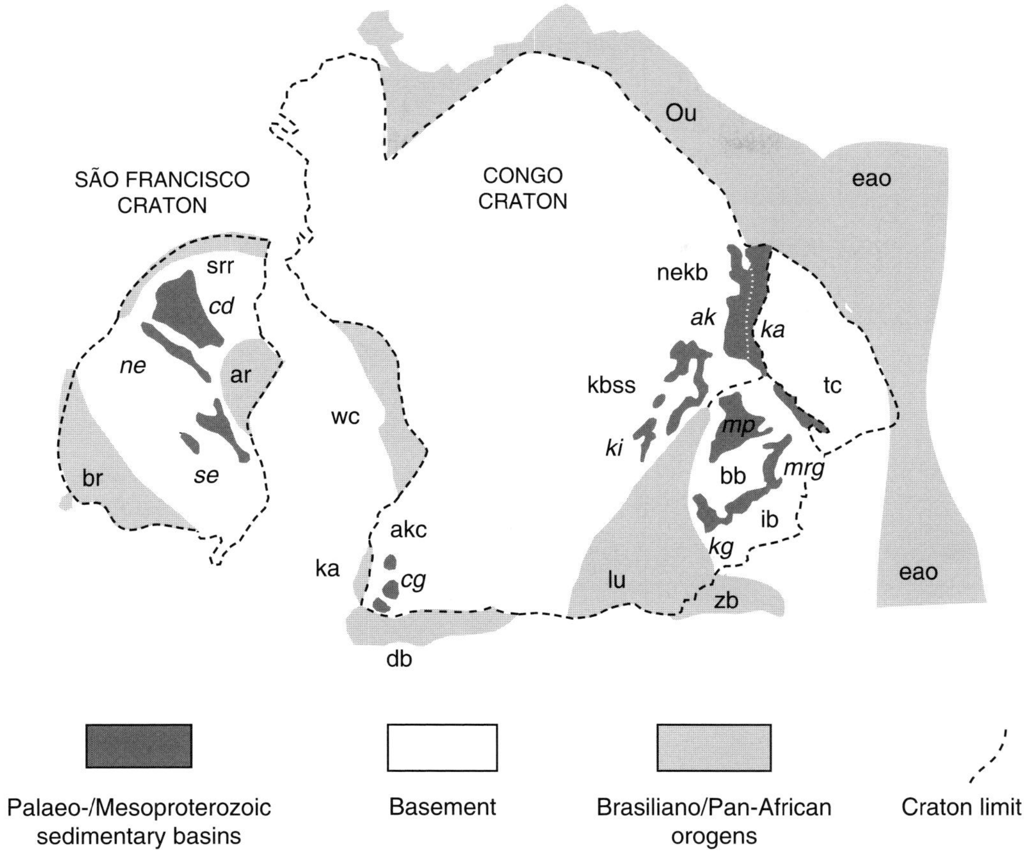

Fig. 2. Approximate Palaeo–Mesoproterozoic positions of the present São Francisco and Congo cratons and Palaeo Mesoproterozoic volcano-sedimentary cover. Key: br, Brasília Belt; srr, Sergipano/Riacho do Pontal/Rio Preto belts; ar, Araçuaí Belt; ka, Kaoko Belt; wc, West Congo Belt; akc, Angola-Kasai Block; db, Damara Belt; lu, Lufilian Belt; kbss, Kibaran Belt *sensu stricto*; ou, Oubanguides; nekb, Northeastern Kibaran Belt; bb, Bangweulu Block; ib, Irumide Belt; zb, Zambezi Belt; tc, Tanzania craton; eao, East African Orogen. Sedimentary successions discussed in the text are indicated in italics as follows: *ne*, northern Espinhaço Supergroup; *se*, southern Espinhaço Supergroup; *cd*, Chapada Diamantina; *cg*, Chela Group; *ki*, Kibaran Supergroup; *ak*, Akanyaru Supergroup; *ka*, Kagera Supergroup; *mp*, Mporokoso Group; *mrg*, Manshya River Group (including the Kasama Formation); and *kg*, Kanona Group. Modified from Brito Neves (2004).

deltaic and lacustrine. The Rio dos Remédios Group of the Chapada Diamantina is composed of acid volcanic rocks, aeolian sandstones and polymict conglomerates interpreted as alluvial fans; it was also deposited during the rift phase.

The rift phase of the Espinhaço Supergroup is followed in the three domains by a transitional phase well characterized in the southern Espinhaço Range (Martins-Neto 2000). This transitional phase is represented in the top of the Diamantina Group by the Galho do Miguel Formation, a thick sandstone formation deposited in aeolian and shallow-marine environments (Dossin *et al.* 1987). Danderfer & Dardenne (2002) correlated this formation with the Bom Retiro and Mangabeira formations respectively of the northern Espinhaço Range (Fig. 3b) and Chapada Diamantina (Fig. 3c).

The transitional phase is followed in the southern Espinhaço and Chapada Diamantina domains by the sag phase. In the southern Espinhaço Range, this sag phase is represented by the Conselheiro Mata Group, composed of interstratified sandstones and shales/argillites that represent shallow-marine deposits with occasional incursions of fluvial and aeolian sediments (Almeida Abreu & Renger 2002) and terminated with some scarce limestone intercalations. In the northern Espinhaço Range, the sag phase comprises only the Bom

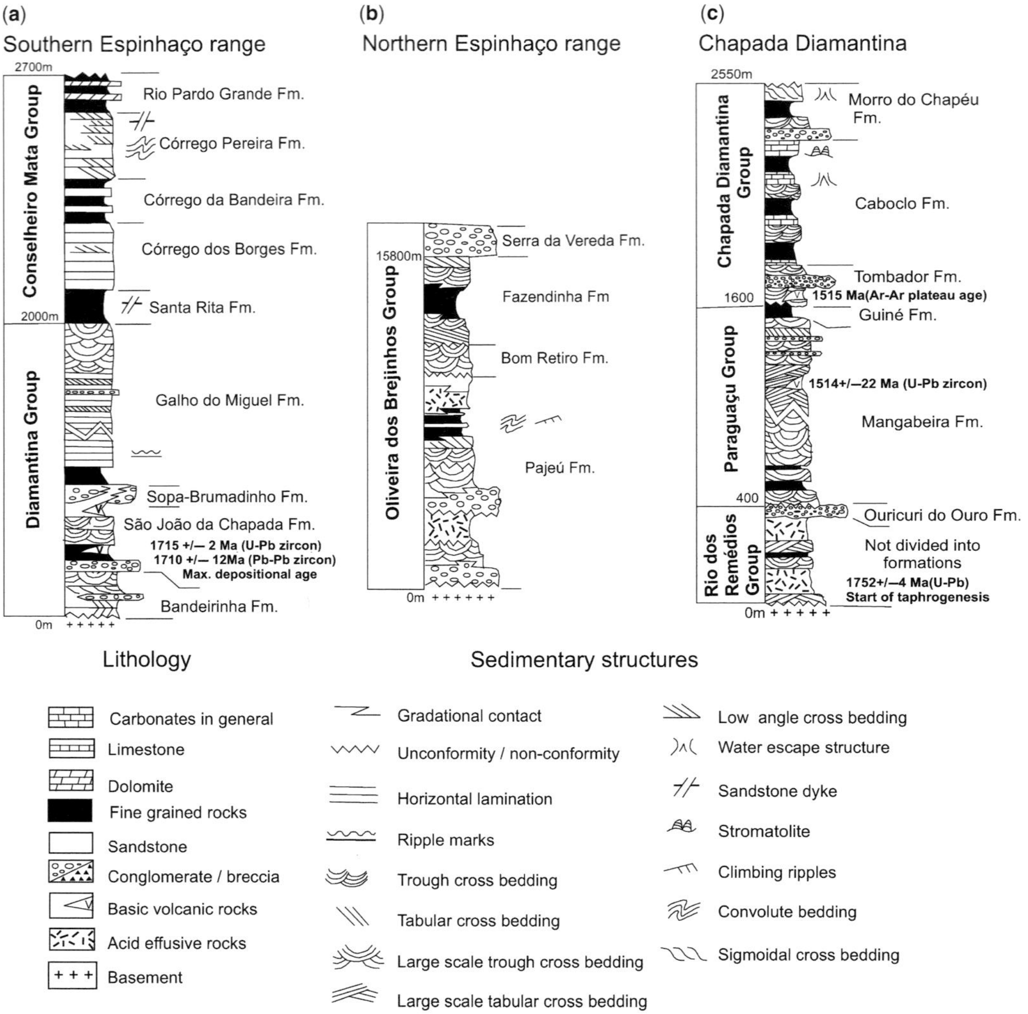

Fig. 3. Stratigraphical columns of the Espinhaço Supergroup: (**a**) Diamantina and Conselheiro Mata groups, after Martins-Neto (2000), Schöll & Fogaça (1979) and Silva 1999; (**b**) Oliveira dos Brejinhos and Santo Onofre groups, modified from Danderfer & Dardenne (2002) and Schobbenhaus (1993); (**c**) Rio dos Remédios, Paraguaçu and Chapada Diamantina groups, modified from Pedreira (1994) and Pedreira & Rocha (2004).

Retiro and Fazendinha formations, composed of interstratified sandstones and shales/argillites (Schobbenhaus 1996). The Chapada Diamantina Group represents the sag phase in the Chapada Diamantina. In this group the basal and upper formations were deposited both in continental and transitional/shallow-marine environments; the intermediate formation was entirely deposited in shallow-marine environment, containing at least four intercalations of stromatolitic carbonate.

In the southern Espinhaço Range, the supergroup crops out in a rift-sag type basin, whose stratigraphic evolution was controlled by its subsidence history (Martins-Neto 2000). The rift phase of the basin comprises four depositional sequences (Silva 1993): Olaria, Natureza, São João da Chapada and Sopa–Brumadinho. Each of these comprises one or more depositional systems and they are approximately correlated with lithostratigraphic units as shown in Figure 4, corresponding to one pre-rift and three rift stages.

The Olaria and Natureza depositional sequences correspond to the Bandeirinha Formation (Fig. 4). The first is bound at the base by a thrust fault and at the top by an erosional unconformity. The Natureza depositional sequence is composed of four depositional systems: alluvial fan, fluvial braided, aeolian and transitional, and is bound at the top by

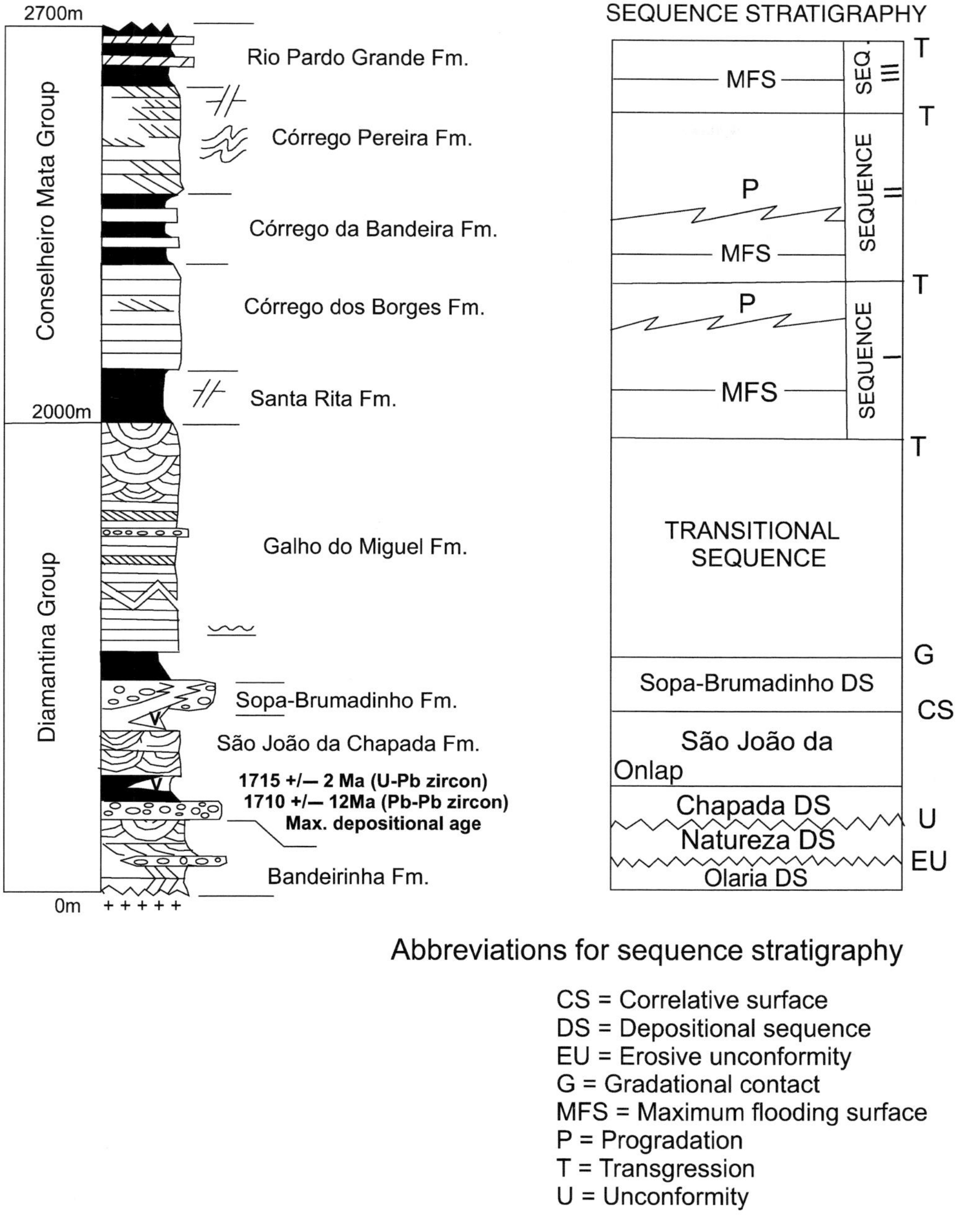

Fig. 4. Lithostratigraphy and sequence stratigraphy of the Espinhaço Supergroup in the southern Espinhaço Range. Lithology: see Figure 3a. Abbreviations: CS, correlative surface; DS, depositional sequence; EU, erosive unconformity; G, gradational contact; MFS, maximum flooding surface; P, progradation; T, transgression; U, unconformity. Modified from Martins-Neto (2000), Dupont (1995) and Silva (1993).

an angular unconformity. The São João da Chapada sequence begins with a talus deposit, with shallow-marine or lacustrine sedimentary rocks that onlap over it, indicating a transgression of the depositional base level. The Sopa-Brumadinho sequence is limited at the base by an angular unconformity, correlated with an erosional unconformity that crops out elsewhere. Lacustrine pelites cover the unconformity and grade vertically into polymict diamond-bearing conglomerates interpreted as a deltaic system. The top of the Sopa–Brumadinho depositional sequence consists of deltaic fans that grade into the aeolian sandstones of the Galho do Miguel Formation. In this formation, the subsidence rates are relatively low and balanced by basin fill. A marine transgression at the top of the formation is followed by higher subsidence rates and the beginning of the sag phase.

The formations that comprise the Conselheiro Mata Group (Fig. 3a) were grouped by Dupont (1995) into three sequences (Fig. 4). The lower one (Sequence I) comprises the Santa Rita and Córrego dos Borges formations; the intermediate (Sequence II) corresponds to the Córrego da Bandeira and Córrego Pereira formations; and the uppermost (Sequence III) coincides with the Rio Pardo Grande Formation. The first two begin with a transgression, reach the maximum flooding surface (MFS), and finish with progradation. The uppermost sequence (Sequence III) begins and ends with transgressions.

In the northern Espinhaço Range, Dominguez & Rocha (1993) divided the Espinhaço stratigraphic column into three depositional sequences (Fig. 5).

According to these authors, the lowermost unit, named Borda Leste, begins with banded

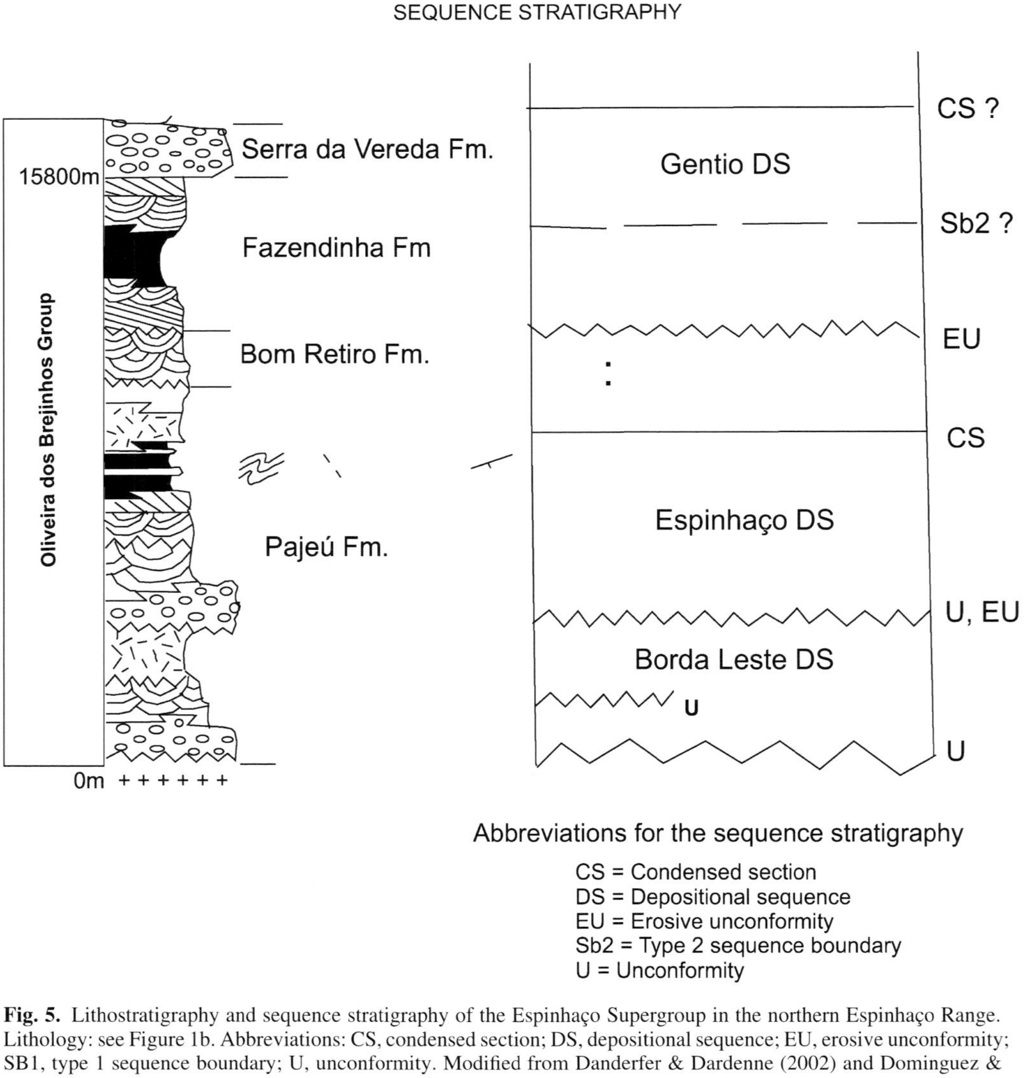

Fig. 5. Lithostratigraphy and sequence stratigraphy of the Espinhaço Supergroup in the northern Espinhaço Range. Lithology: see Figure 1b. Abbreviations: CS, condensed section; DS, depositional sequence; EU, erosive unconformity; SB1, type 1 sequence boundary; U, unconformity. Modified from Danderfer & Dardenne (2002) and Dominguez & Rocha (1993).

Fe–Mg formations overlain by conglomerates and fluvial sediments, whose deposition was followed by an episode of uplift and then a major episode of subsidence. The unconformity that separates the two episodes may be comparable to the break-up unconformity that characterizes the evolution of rift basins. The subsidence episode allowed sedimentation of the Espinhaço depositional sequence. This sequence consists of fluvial sediments that were rapidly covered by aeolian sands and then by coarse-grained shallow-water arenaceous sediments; these were deposited under the influence of waves and currents and were followed by fine-grained graphitic sediments.

A major drop in the sea level led to incision of the Espinhaço depositional sequence by rivers and deposition of a third depositional sequence, the Gentio depositional sequence, composed of coarse grained turbiditic sediments, possibly deposited on a sequence boundary without sub-aerial exposure (Fig. 5). These relationships led Dominguez & Rocha (1993) to interpret the Espinhaço Supergroup in the northern Espinhaço Range as a rift-sag basin.

The Espinhaço succession in the Chapada Diamantina consists of four depositional sequences (Pedreira 1994; Fig. 6): the basal Rio dos Remédios sequence is composed of aeolian sands, effusive rocks and polymict conglomerates, the later named Ouricuri do Ouro Formation.

The succession that follows, the Paraguaçu depositional sequence, begins with a transgression of the coast line (Souza 1986) followed by deposition of fluvial and thick aeolian sands of the Mangabeira Formation. At the top of the formation, closely-spaced argillaceous levels herald a new sea-level elevation (see Shaney & McCabe 1994; Pedreira 2003), an event that began with deposition of the Guiné Formation and closed with deltaic deposits that characterize a regression (Pedreira 1995). The Tombador–Caboclo depositional sequence that overlies it has a lower section (Tombador Formation) that comprises both continental and transitional deposits (Castro 2003); the upper section (Caboclo Formation) is essentially marine. Along its stratigraphic column there are two drops in sea level, represented by incised valleys (Pedreira & Rocha 2004), first filled with fluvial sediments and then by marine deposits representing transgressions of the coast line. A last drop in sea level started the deposition of the Morro do Chapéu Formation, whose base consists of fluvial diamond-bearing conglomerates. After the deposition of the fluvial conglomerates, a new elevation of the sea level deposited a sandy tidal flat followed by regressive deltaic sediments. The top of the Morro do Chapéu depositional sequence is truncated by the unconformity below the glacial Neoproterozoic deposits.

The depositional age of the Espinhaço basins has been determined by dating zircons from volcanic units in the lower parts of the Espinhaço Supergroup in the southern Espinhaço Range and Chapada Diamantina only (Fig. 3a and c). In the former area they comprise *c.* 1711 Ma (zircon U–Pb, Schobbenhaus 1993), 1710 ± 12 Ma (Pb–Pb on carbonate, Dussin & Dussin 1995), and 1715 ± 12 Ma (zircon U–Pb, Machado *et al.* 1989). Similar ages were found in rhyolites, rhyodacites and dacites of the Rio dos Remédios Group at the base of the Chapada Diamantina (Fig. 3c): 1752 ± 4 Ma (zircon U–Pb, Schobbenhaus *et al.* 1994). Higher up in the stratigraphical column, at the base of the Chapada Diamantina Group, intrusive dykes have been dated as *c.* 1515 Ma (Ar/Ar plateau ages, Battilani *et al.* 2005). In the northern Espinhaço Range there are no age data for any acid volcanic rocks of similar stratigraphical position to the Rio dos Remédios Group.

African sedimentary basins: Congo Craton

Several sedimentary successions occur on the Congo Craton (Fig. 2), from west to east, the Chela Group in Angola, the Kibaran Supergroup in the Mesoproterozoic Kibaran Belt of the CDR, the Akanyaru and Kagera supergroups of the CDR, Rwanda and Burundi (formerly known as the Rwanda or Burundi supergroups), and the Muva Supergroup on the Bangweulu Block and in the Mesoproterozoic Irumide Belt of Zambia and southeast CDR. Of those, only the Chela Group and parts of the Kagera and Muva supergroups are relatively undeformed, allowing reconstruction of a stratigraphic column, while for the more deformed and metamorphosed successions of the Kibaran and Irumide belts only a synthetic stratigraphic column can be presented. Basin analysis and sequence stratigraphy is only attempted on the undeformed Mporokoso Group.

Chela Group

The Chela Group crops out in the neighbourhood of the town of Sá da Bandeira on the Umpata Plateau, Angola. It is the westernmost succession on the Congo Craton (Fig. 2) and comprises five formations (Torquato & Fogaça 1981; Fig. 7). From the base to top these are the Tundavala Formation (begins with a basal conglomerate followed by sandstones with pyroclastic intercalations), the Humpata Formation (acid volcaniclastic rocks, products of explosive volcanism, with sandstone

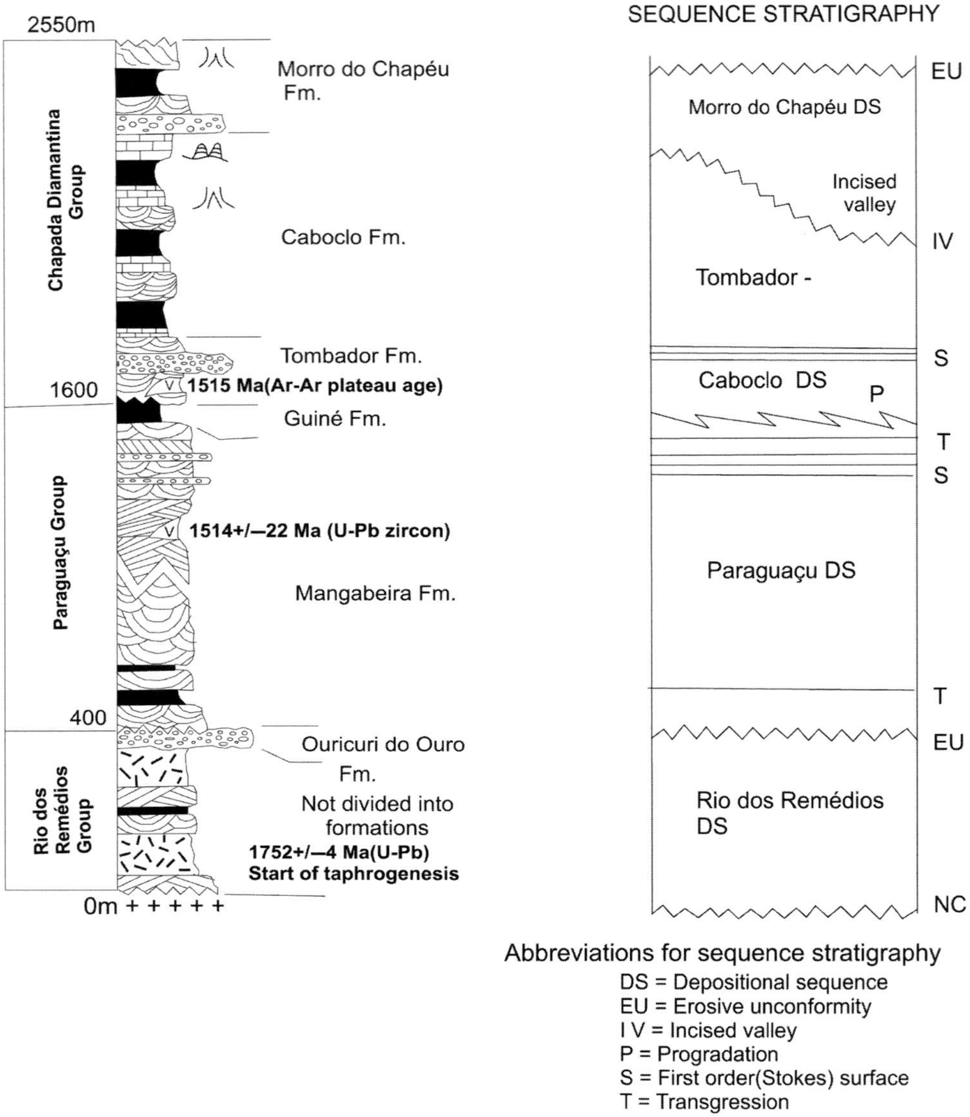

Fig. 6. Lithostratigraphy and sequence stratigraphy of the Espinhaço Supergroup in Chapada Diamantina. Lithology: see Figure 3c. Abbreviations: DS, depositional sequence; EU, erosive unconformity; I, incised valley; P, progradation; S, first-order (Stokes) surface; T, transgression. Modified from Pedreira (1994, 1995, 2003).

intercalations), the Bruco Formation (basal volcanogenic conglomerate followed by interbedded sandstones and siltstones with volcanic and conglomeratic levels), the Cangalongue Formation interbeds (argillite, limestone and arkosic sandstone), and the Leba Formation (cherts, argillite and stromatolitic dolomites). With the exception of the Bruco Formation, partially deposited in a fluvial environment, and the Cangalongue Formation, interpreted as continental red beds, the Chela Group was deposited in shallow-marine environments.

The depositional age of the Chela Group was determined at 1790 ± 17 Ma by the U–Pb SHRIMP method on magmatic zircons from an ignimbrite of the Humpata Formation (McCourt *et al.* 2004), similar to the Rio dos Remédios effusive rocks in the São Francisco Craton.

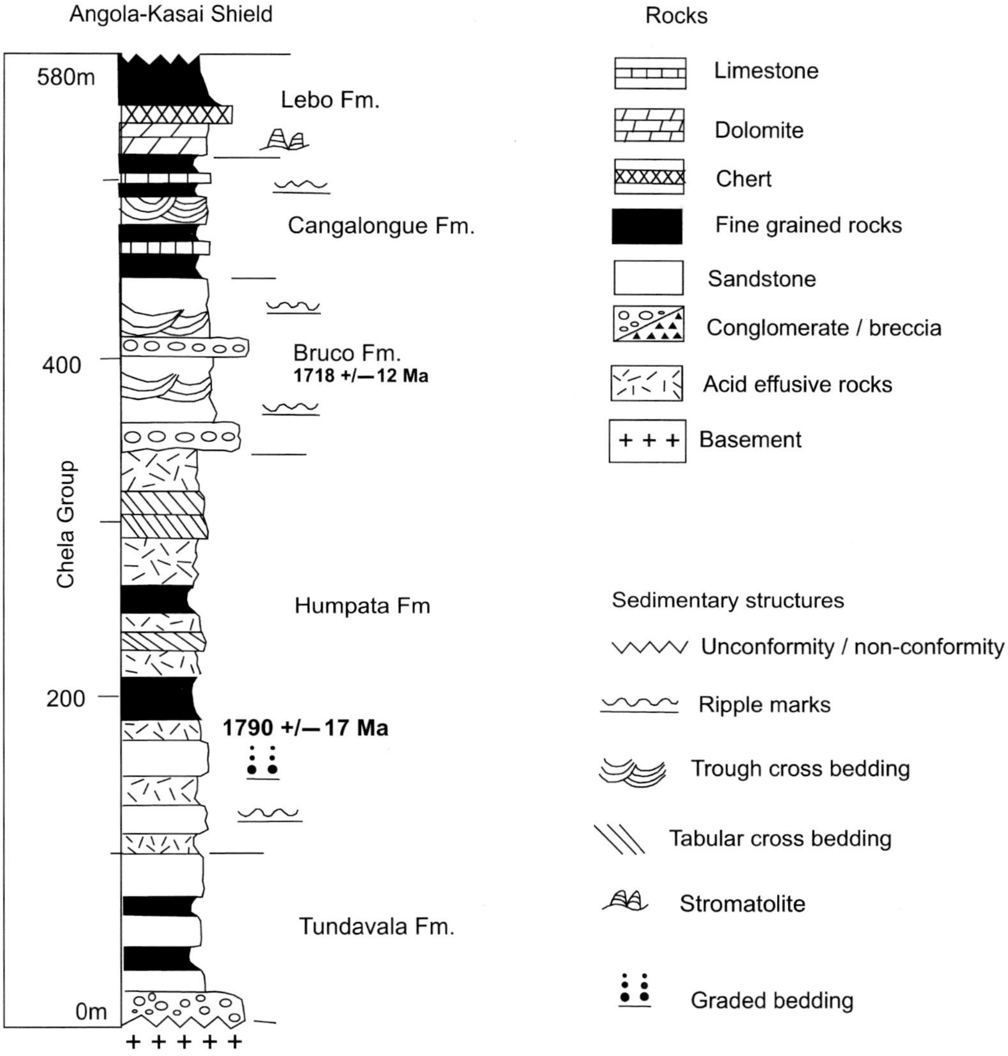

Fig. 7. Stratigraphical column of the Chela Group in the region of the Humpata Plateau, Angola (modified from Torquato & Fogaça 1981 and McCourt *et al.* 2004).

Kibaran and Akanyaru/Kagera supergroups

The Kibaran and Akanyaru/Kagera supergroups occur in the central eastern sector of the Congo Craton, in the Kibaran Belt. Their outcrops are distributed in two belts: the Kibaran Belt *sensu stricto* and the Northeastern Kibaran Belt (Tack *et al.* 1994) (Fig. 8). The Northeastern Kibaran Belt can be further divided into a Western Internal Domain and an Eastern External Domain, separated by a basement rise of Palaeoproterozoic gneisses and schists referred to as the Rusizian Rise (Lavreau 1985; Tack *et al.* 1994).

Burundi Group and Rwanda Group were the names of the succession in Burundi and Rwanda respectively (Royal Museum for Central Africa 1990; Theunissen *et al.* 1991), but similar successions extend into northern Tanzania, southern Uganda and eastern Congo, where they were sometimes referred to under the same name, but more often than not, were attributed local and informal stratigraphical names. The successions of the Northeastern Kibaran Belt have recently been redefined into the Akanyaru Supergroup, which occurs in the Western Internal Domain, and the Kagera Supergroup, occurring in the Eastern

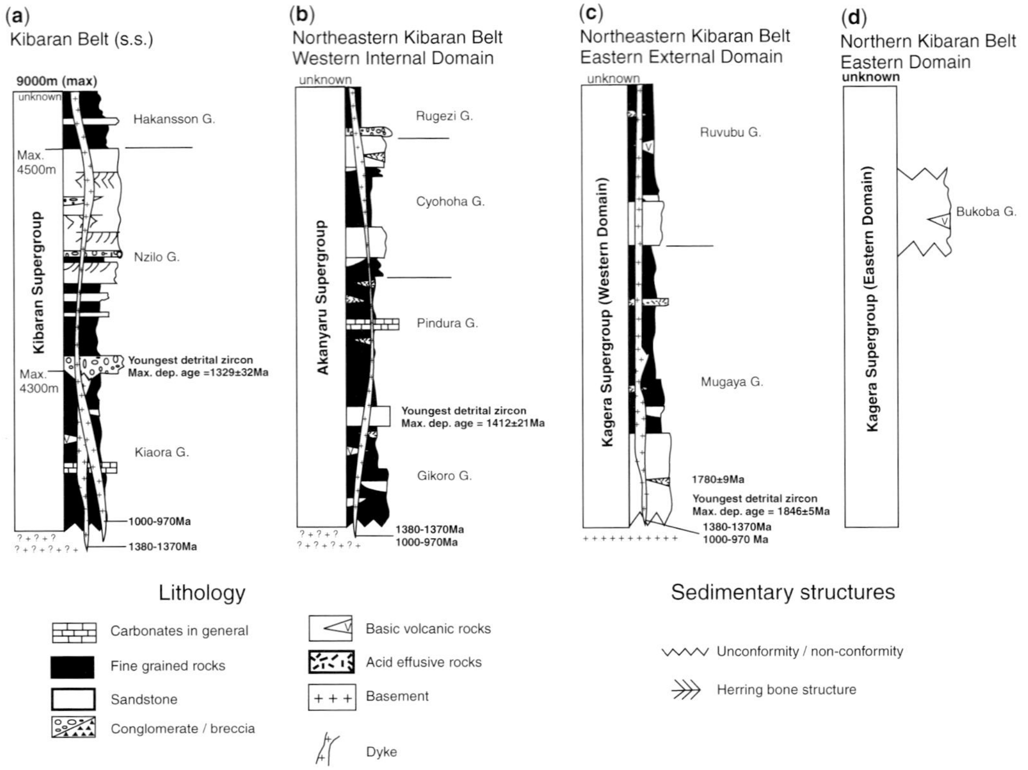

Fig. 8. Stratigraphical columns of (**a**) Kibaran, (**b**) Akanyaru and (**c**, **d**) Kagera supergroups (see Fig. 2 for location within the Kibaran Belt).

External Domain. The sedimentary succession in the Kibaran Belt *s.s.* was formalized as the Kibaran Supergroup, comprising the Kiaora, Nzilo and Hakansson groups (Kokonyangi *et al.* 2001, 2006).

In the Kibaran Belt *s.s.*, the Kibaran Supergroup consists of three groups: the basal Kiaora Group dominated by pelites/schists, the middle Nzilo Group dominated by quartzites, and the upper Hakansson Group, dominated by pelitic units (Fig. 8; Kokonyangi *et al.* 2001, 2006). The two lower successions are separated by the regionally defined Kataba Conglomerate. The only age data available for these units are from the Nzilo Group, which unconformably overlies 1.38 Ga granitoids, and is intruded by *c.* 1.00 Ga Sn-bearing granitoids, providing its maximum and minimum ages, respectively (Kokonyangi *et al.* 2006). The older Kiaora Group is intruded by the 1.38 Ga granitoids (i.e., its minimum age is 1.38 Ga), while no age constraints are available for the Hakansson Group.

In the Western Internal Domain of the Northeastern Kibaran Belt, the Akanyaru Supergroup comprises four groups (Fig. 8), which include pelites/schists and sandstones/quartzites and at least one marble bed. In the basal Gikoro and Pindura groups schisto-arkosic successions with volcanic rocks and basic intrusions are reported. In the westernmost part of the Eastern External Domain, the Kagera Supergroup is subdivided into the Mugaya and Ruvubu groups, whereas in the eastern part only one, the Bukoba Group, is recognized. In the eastern domain the Kagera Supergroup unconformably overlies the Archaean Tanzania Craton, but for the western domain of the Northeastern Kibaran Belt and the Kibaran Belt *s.s.* the presence of the basement is unproved.

The depositional environment of the Kibaran and Akanyaru/Kagera supergroups is interpreted as shallow-marine, but in the Northeastern Kibaran Belt a turbiditic facies has been reported (Baudet *et al.* 1988), attesting to the presence of deeper environments. The depositional age of the Kagera Supergroup was determined by a zircon U–Pb crystallisation age of 1780 $\pm$ 9 Ma in the Murore Tuff at the base of the Mugaya Group (Cutten *et al.* 2004). The age of the Akanyaru Supergroup is constrained by the youngest detrital

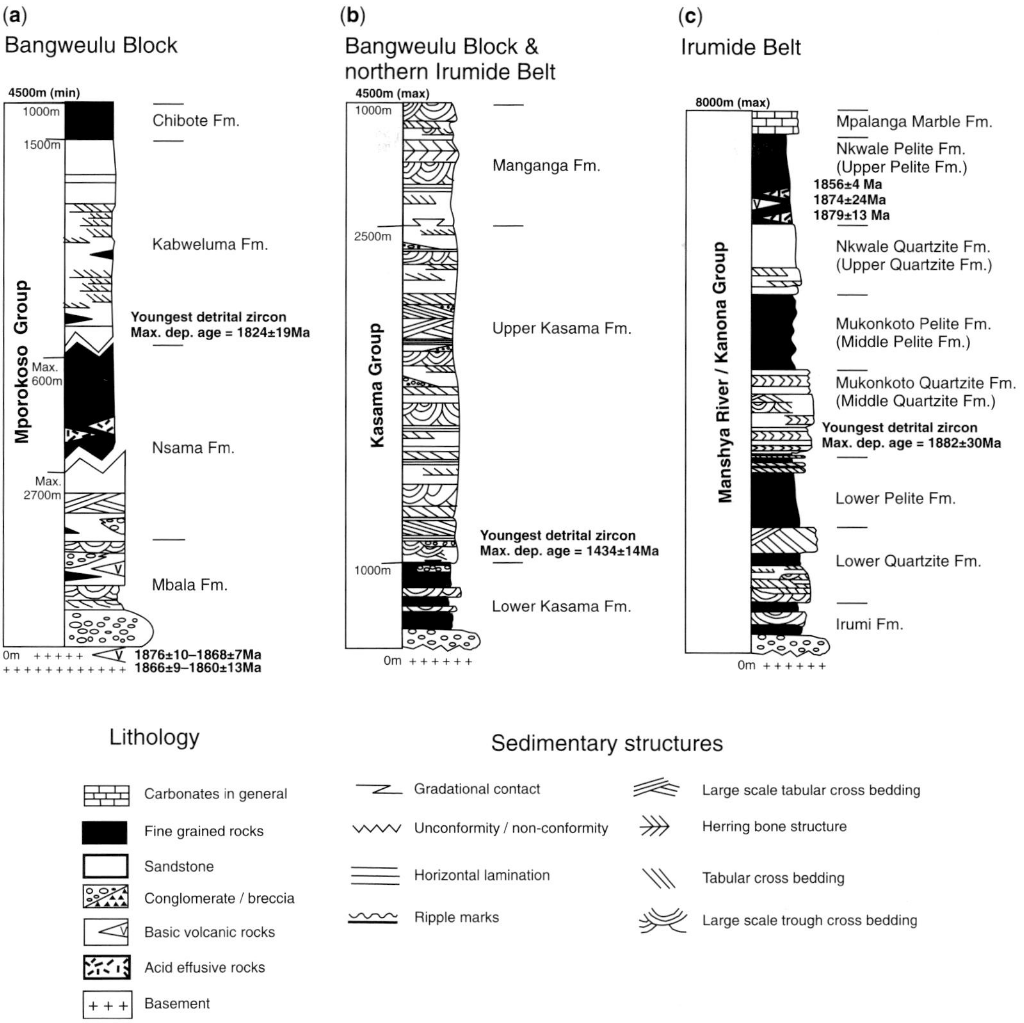

Fig. 9. Stratigraphical columns of the (**a**) Mporokoso, (**b**) Kasama and (**c**) Manshya River/Kanona groups in the Bangweulu Block and Irumide Belt. Formation names in brackets refer to equivalent formations in the southern part of the Irumide Belt (Kanona Group of De Waele & Mapani 2002). Modified from Daly & Unrug (1982) and De Waele & Mapani (2002).

zircon identified in the Gikoro Group at 1412 ± 21 Ma, indicating its maximum age of deposition, and granitoid intrusions intruding the sequence dated at *c.* 1.38 Ga (Tack *et al.* 2002; Kokonyangi *et al.* 2004).

Muva Supergroup

The Muva Supergroup comprises two groups that occur in the Bangweulu Block and within the Irumide Belt, with the names of Mporokoso Group (Andersen & Unrug 1984) and Manshya River Group (De Waele & Mapani 2002; De Waele *et al.* 2006), respectively (Fig. 9). According to Andrews-Speed (1989), the Mporokoso Group consists of both immature and mature sandstones, conglomerates, ordered or not, cherts, tuffs, and volcanic rocks. It was deposited in both fluvial and shallow-marine environments: the former is represented by the immature sandstones and the conglomerates, the latter by the mature sandstones. The Manshya River Group was involved in the folding of the Irumide Belt and consists of meta-siltstones, phyllites, slates and quartzites, with sporadic calc-silicate rock and marble at the top (Daly & Unrug 1982; De Waele & Mapani 2002;

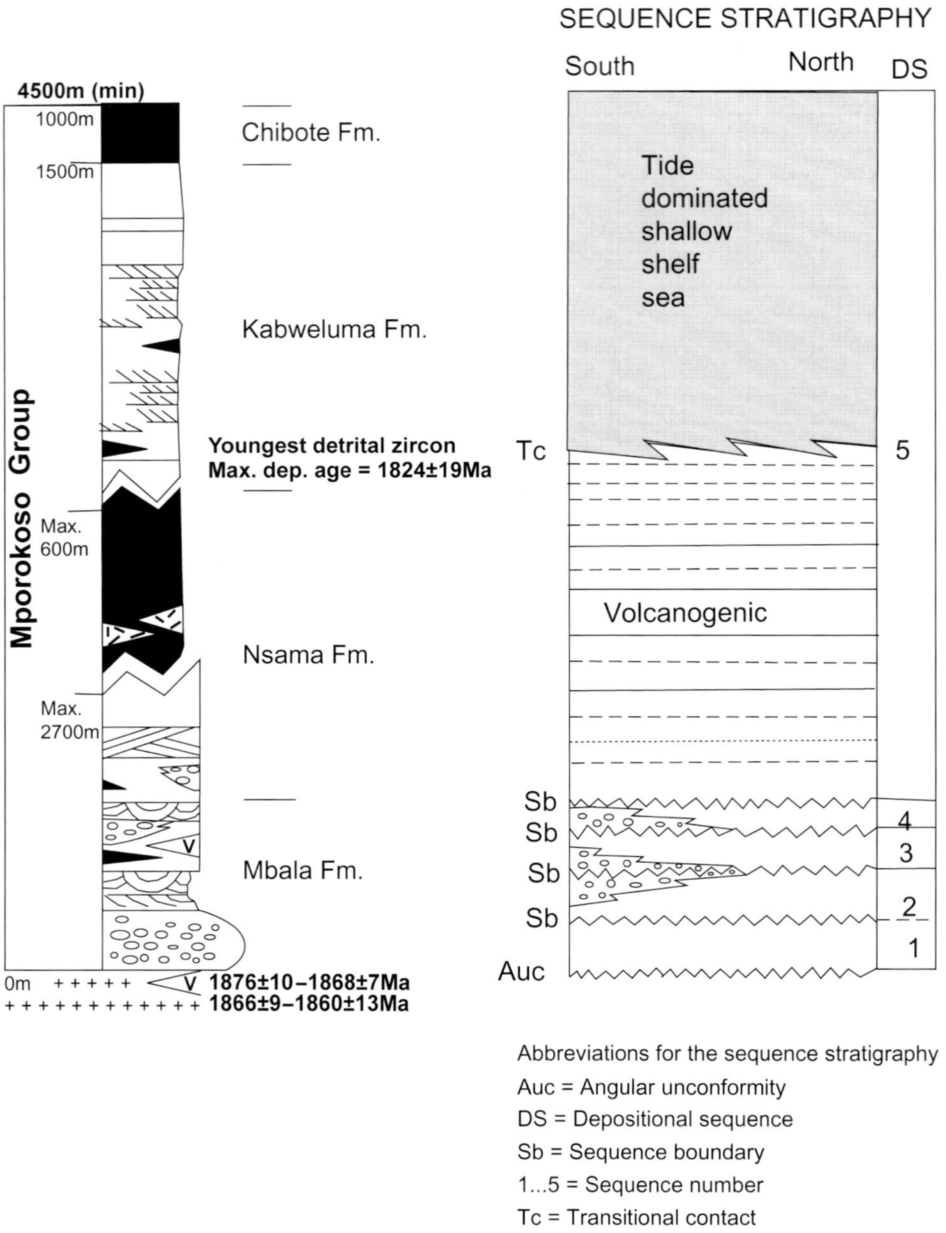

Fig. 10. Lithostratigraphy and sequence stratigraphy of the Mporokoso Group in the Bangweulu Block. Lithology: see Figure 9a. Abbreviations: AU, angular unconformity; DS, depositional sequence; SB, sequence boundary; TC, transitional contact. Modified from Andrews-Speed (1989).

De Waele *et al.* 2006). Its depositional environment is interpreted as shallow-marine (Daly & Unrug 1982; De Waele & Mapani 2002), although fluvial units have been recognized in the northeastern Irumide Belt (Daly & Unrug 1982).

The Mporokoso Group was classified by Andrews-Speed (1989) as a megasequence bound at the base by a regional angular unconformity and at the top by the present erosional surface; no angular unconformities were recognized within the group (Fig. 10). It is divided into five depositional sequences: the lower four correspond to the Mbala Formation, the upper one comprises the Nsama and Kabweluma formations (Fig. 10).

The first and lowermost sequence consists of planar cross-bedded fluvial sandstones interpreted as braided-river deposits. Sequence 2 (Fig. 10) contains a coarsening-upward succession of immature trough cross-bedded sandstone interpreted as the deposits of sandy to pebbly sheet-braided rivers, capped by conglomerates deposited in wide shallow channels. The base of Sequence 3 is formed by an influx of debris flow or grain flows; the coarse, poorly sorted conglomerates include intra-basin pebbles of reworked sandstone. These conglomerates are overlain by a fining-up sequence of fluvial sandstones followed by marine ones, which represent a transgressive sequence. A final flux of fluvial sandstones marks the start of Sequence 4 of the Mbala Formation. These fluvial sediments pass upwards into shallow-marine sandstones. Sequence 5 comprises both the Nsama and Kabweluma formations. The former is essentially composed by mudstone tuff and chert with minor cross-bedded sandstones; the latter comprises essentially cross-bedded and rippled sandstones.

The Mporokoso Group of the Bangweulu Block unconformably overlies a plutono-volcanic basement dated using zircon U–Pb at 1.87–1.86 Ga (De Waele *et al.* 2004; De Waele 2005). Tuff layers, associated with this basement occur within the basal parts of the Mporokoso Group, strongly suggesting a depositional age of *c.* 1.86 Ga. Similar tuffs and lavas also occur within the Manshya River succession of the Irumide Belt, and yielded zircon U–Pb ages of 1879 ± 13, 1871 ± 24 and 1856 ± 4 Ma, directly constraining ages of deposition for that group to be broadly coeval with deposition of the Mporokoso Group (De Waele 2005).

Discussion

A correlation between the Espinhaço Supergroup in the Brazilian states of Minas Gerais and Bahia (Fig. 3a, c) and the Chela Group (Fig. 7) was made by Torquato & Fogaça (1981) on a bed-to-bed basis. These authors extended this correlation to the Neoproterozoic Nosib and Khoabendus formations of Namibia, but because of the 1.78 Ga age of a felsic tuff in the Chela Group, this latter correlation is unsustainable (McCourt *et al.* 2004). As correlation from one continental landmass to another is not advisable, we will limit the discussion to a comparison of the Palaeo–Mesoproterozoic basins of the São Francisco and Congo cratons.

Comparisons between the Espinhaço Supergroup in the southern Espinhaço Range and the Chapada Diamantina can be made on several counts. There were similar depositional environments: continental in the Diamantina and Rio dos Remédios, and most of the Paraguaçu groups, and shallow-marine with continental incursions in the Conselheiro Mata and Chapada Diamantina groups. Tectonic settings are also similar: extensional in the Diamantina and Rio dos Remédios/Paraguaçu groups and flexural/thermal in Conselheiro Mata and Chapada Diamantina groups. Moreover, as indicated by the geochronological determinations already mentioned (Schobbenhaus 1993; Dussin & Dussin 1995; Schobbenhaus *et al.* 1994), these successions were deposited simultaneously at *c.* 1.75 Ga.

For the northern domain of the Espinhaço Supergroup, the Pajeú Formation was also deposited in an intra-continental rift and the Bom Retiro Formation is transitional; the Fazendinha Formation was deposited in the sag phase. Geochronological dating of felsic tuffs in the successions of the northern Espinhaço Range would determine whether deposition was coeval with that in the southern Espinhaço Range and Chapada Diamantina.

The Espinhaço Supergroup and the Chela Group have different depositional ages: 1790 ± 17 Ma for the Chela Group and 1710 ± 12 Ma for the Diamantina Group. On a lithological basis, the Humpata Formation of the Chela Group and the Rio dos Remédios Group of the Chapada Diamantina, both composed of acid effusive rocks (Figs 7 and 3c), are comparable, but the age determinations clearly indicate that volcanism was strongly diachronous. Within the Congo Craton, the Chela Group and the Kibaran, Ankanyaru/Kagera groups appear to be coeval with depositional ages of 1790 ± 17 Ma for the Chela Group and 1780 ± 9 Ma for the Akanyaru Supergroup. Additionally, both successions were deposited in similar shallow-marine environments, making a direct correlation between these sequences plausible. It must, however, be noted that geochronological data for the Kibaran successions are very sparse at present, and that multiple sedimentation cycles may be present in these successions, spanning the Palaeoproterozoic through to the end of the Mesoproterozoic.

Finally, comparison of the Kibaran, Akanyaru and Kagera supergroups with the Muva Supergroup is somewhat problematic, because the latter is about 100 Ma older (*c.* 1.86 Ga) and one of its components (the Mporokoso Group on the Bangweulu Block) was deposited in a continental rather than shallow-marine environment. However, the depositional age of the Manshya River Group (*c.* 1.8 Ga, De Waele & Mapani 2002; De Waele 2005), which does represent a shallow-marine succession, places its deposition in the Palaeoproterozoic and makes it coeval with the Mporokoso Group.

The intracratonic basins (Espinhaço and Mporokoso) may be compared in terms of sequence stratigraphy, especially in the southern Espinhaço Range. The lower parts of both were divided into several depositional sequences with conglomerates, probably characterizing the retreat of escarpments. The uppermost depositional sequences also appear to be similar (sequences I, II and III of Fig. 4 and the tide-dominated shallow-marine sections of Fig. 10). To a lesser degree, similar features may be seen in the Chapada Diamantina. Thus it is suggested that in the São Francisco–Congo craton, the rift shoulders at the base of the Espinhaço Supergroup and the Bangweulu Block represented topographic highs, dominated by alluvial fans, lacustrine, aeolian and fluvial environments, while the Conselheiro Mata and Chapada Diamantina groups, and the Kibaran and Irumide belts, were the loci of shallow seas within or at the margin of the Congo Craton.

Conclusions

The sedimentological, stratigraphic, tectonic and geochronological data presented here show that broad coeval epi-continental sedimentary basins were forming on the São Francisco–Congo craton during the Palaeo–Mesoproterozoic and over a relatively long period of time (1880–1700 Ma, at least). The tectonic regime that governed the deposition of these basins (mostly rift-sag) was similar over the entire São Francisco–Congo continent, as were the sedimentary environments. This suggests the possible existence of a wide shallow-water epi-continental sea covering large extents of the São Francisco–Congo craton during Statherian times.

The senior author (AJP) thanks B.B. de Brito Neves and his fellow editors for the invitation to present this paper at the Gondwana 12 Conference, and CPRM–Geological Survey of Brazil for logistical support. The authors also thank Alexandre Uhlein, B.B. de Brito Neves and an anonymous reviewer for suggestions that considerably improved an earlier version of this paper. This is TSRC publication 380.

References

Almeida Abreu, P. & Renger, F. E. 2002. Serra do Espinhaço meridional: um orógeno de colisão do Mesoproterozóico. *Revista Brasileira de Geociências*, **32**, 1–14.

Andersen, L. S. & Unrug, R. 1984. Geodynamic evolution of the Bangweulu block, northern Zambia. *Precambrian Research*, **25**, 187–212.

Andrews-Speed, C. P. 1989. The Mid-Proterozoic Mporokoso Basin, Northern Zambia: Sequence Stratigraphy, Tectonic Setting and Potential for Gold and Uranium Mineralisation. *Precambrian Research*, **44**, 1–17.

Battilani, G. B., Vasconcelos, P. M., Gomes, N. S. & Guerra, W. J. 2005. Geochronological data of dykes and sills intruding Proterozoic sequences of the Tombador Formation, Bahia-Brazil. *In*: Rosa, M. L. et al. (eds) *Anais do 3° Simpósio sobre o Craton do São Francisco, CBPM/UFBA/SBG, Salvador*, 139–142.

Baudet, D., Hanon, M., Lemonne, E. & Theunissen, K. 1988. Lithostratigraphie du domaine sédimentaire de la chaine Kibarienne au Rwanda. *Annales de la Société Géologique Belge*, **112**, 225–246.

Brito Neves, B. B. 2002. Main Stages of the development of the sedimentary basins of South America with the tectonics of supercontinents. *Gondwana Research*, **5**, 175–196.

Brito Neves, B. B. 2004. A História dos Continentes–Trajetórias e tramas tectônicas. *In*: Mantesso-Neto, V., Bartorelli, A ., Carneiro, C. D. R. & Brito Neves, B. B. (eds) *Geologia do Continente Sul-Americano; Evolução da Obra de Fernando Flávio Marques de Almeida*. Beca, São Paulo, 123–149.

Brito Neves, B. B., Cordani, U. G. & Torquato, J. R. 1980. Evolução geológica do Precambriano no estado da Bahia. *In*: Inda, H. V. & Duarte, J. B. (eds) *Geologia e Recursos Minerais do Estado da Bahia*, **3**, 1–101.

Castro, M. R. 2003. *Estratigrafia de Seqüências na Formação Tombador, Grupo Chapada Diamantina, Bahia*. PhD Thesis, Instituto de Geociências, Universidade de São Paulo.

Cutten, H., de Waele, B., Fernandez-Alonso, M. & Tack, L. 2004. Age constraints for basin evolution and sedimentation in the 'Northeastern Kibaran Belt.' *In*: *20th Colloquium of African Geology, Orléans, France*, **123**.

Daly, M. C. & Unrug, R. 1982. The Muva Supergroup, northern Zambia. *Transactions of the Geological Society of South Africa*, **85**, 155–165.

Danderfer, A. & Dardenne, M. A. 2002. Tectonoestratigrafia da bacia Espinhaço na porção norte do Craton do São Francisco: registro de uma evolução poliistórica descontínua. *Revista Brasileira de Geociências*, **32**, 449–460.

De Waele, B. 2005. *The Proterozoic geological history of the Irumide Belt, Zambia.*, PhD Thesis, Curtin University of Technology, Australia.

De Waele, B. & Mapani, B. 2002. Geology and correlation of the central Irumide Belt. *Journal of African Earth Sciences*, **35**, 385–397.

De Waele, B., Fitzsimons, I. C. W., Wingate, M. T. D., Mapani, B. & Tembo, F. 2004. A U–Pb SHRIMP geochronological database for the Irumide Belt of Zambia: from Palaeoproterozoic sedimentation to late Mesoproterozoic magmatism. *In*: *32nd International Geological Congress, Florence, Italy*, 870–871.

De Waele, B., Kampunzu, A. B., Mapani, B. S. E. & Tembo, F. 2006. The Irumide Belt of Zambia. *Journal of African Earth Sciences*, **46**, 36–70.

Dominguez, J. M. L & Rocha, G. M. F. 1993. Middle Proterozoic paleogeography of the São

Francisco Craton in the state of Bahia. *In*: *Anais do II Simpósio do Cráton do São Francisco, Salvador*, 158–159.

DOSSIN, I. A., GARCIA, A. J. V., UHLEIN, A., DARDENNE, M. A. & DOSSIN, T. M. 1987. Fácies Eólico na Formação Galho do Miguel, Supergrupo Espinhaço (MG). *In*: *Sociedade Brasileira de Geologia, Núcleo de Minas Gerais, Boletim*, **6**, 85–96.

DUPONT, H. 1995. O Grupo Conselheiro Mata no seu quadro paleogeográfico e estratigráfico. *Sociedade Brasileira de Geologia, Núcleo de Minas Gerais, Boletim* **13**, 9–10.

DUSSIN, I. A. & DUSSIN, T. M. 1995. Supergrupo Espinhaço: Modelo de evolução geodinâmica. *Geonomos*, **III**, 19–26.

KOKONYANGI, J., OKUDIAIRA, T., KAMPUNZU, A. B. & YOSHIDA, M. 2001. Geological evolution of the Kibarides Belt, Mitwaba, Democratic Republic of Congo, central Africa. *Gondwana Research*, **4**, 663–664.

KOKONYANGI, J., ARMSTRONG, R. A., KAMPUNZU, A. B., YOSHIDA, M. & OKUDAIRA, T. 2004. U–Pb zircon geochronology and petrology of granitoids from Mitwaba (Katanga, Congo): implications for the evolution of the Mesoproterozoic Kibaran belt. *Precambrian Research*, **132**, 79–106.

KOKONYANGI, J., KAMPUNZU, A. B., ARMSTRONG, R., YOSHIDA, M., OKUDAIRA, T. & ARIMA, M. 2007. The Mesoproterozoic Kibaride belt (Katanga, SE D. R. Congo). *Journal of African Earth Sciences*, **46**, 1–35.

LAVREAU, J. 1985. Le Groupe de la Rusizi (Rusizien du Zaïre, Rwanda et Burundi) à la lumière des connaissances actuelles. *Annual Report 1983–84, Department of Geology and Mining, Royal Museum for Central Africa, Tervueren, Belgium*, 111–119.

MACHADO, N., SCHRANK, A., ABREU, F. R., KNAUER, L. G. & ALMEIDA ABREU, P. 1989. Resultados preliminares da geocronologia U/Pb na Serra do Espinhaço Meridional. *Sociedade Brasileira de Geologia, Núcleo Minas Gerais, Boletim*, **10**, 171–174.

MARTINS-NETO, M. A. 2000. Tectonics and sedimentation in a Paleo/Mesozoic rift-sag basin (Espinhaço basin, southeastern Brasil). *Precambrian Research*, **103**, 147–173.

MCCOURT, S., ARMSTRONG, R. A., KAMPUNZU, A. B., MAPEO, R. B. M. & MORAIS, E. 2004. New U-Pb SHRIMP ages for zircons from the Lubango region, SW Angola: insights into the Proterozoic evolution of South-Western Africa. *In*: *Geoscience Africa 2004*. Geological Society of South Africa, 438–439.

PEDREIRA, A. J. 1994. *O Supergrupo Espinhaço na Chapada Diamantina Centro-oriental, Bahia: Sedimentologia, Estratigrafia e Tectônica*. PhD Thesis, Instituto de Geociências, Universidade de São Paulo.

PEDREIRA, A. J. 1995. Estratigrafia de seqüências e modelo deposicional da Formação Guiné (Mesoproterozoico) na Chapada Diamantina, Bahia. *Sociedade Brasileira de Geologia, Núcleo de Minas Gerais, Boletim*, **13**, 28–29.

PEDREIRA, A. J. 2003. Characterization of marine flooding surfaces in Proterozoic strata of the Chapada Diamantina, Brazil. *In*: ROSSETTI, D. F., GÓES, A. M. & TRUCKENBRODT, W. (eds) *3rd Latinamerican Congress of Sedimentology, Abstracts, Belém, Session 14*, 232–234.

PEDREIRA, A. J. & ROCHA, A. J. D. 2004. Revisão estratigráfica do Grupo Chapada Diamantina, Bahia. *In*: *Congresso Brasileiro de Geologia 42, Araxá, Anais*. Sociedade Brasileira de Geologia. (CD-ROM).

ROYAL MUSEUM FOR CENTRAL AFRICA 1990. *Carte Géologique du Burund*, 1 : 1,000,000 scale. Tervueren, Royal Museum for Central Africa.

ROGERS, J. J. W. & SANTOSH, M. 2002. Configuration of Columbia, a Mesoproterozoic supercontinent. *Gondwana Research*, **5**, 5–22.

SCHOBBENHAUS, C. 1969. Mapa geológico preliminar da região setentrional da serra do Espinhaço. *In*: *Congresso Brasileiro de Geologia 23, Salvador, Anais*. Sociedade Brasileira de Geologia. 74–86.

SCHOBBENHAUS, C. 1993. *Das Mittlere Proterozoikum Brasiliens mit besonderer Berücksichtung des zentralen Osten: Eine Revision*. PhD Thesis, Albert-Ludwiges Universität, Freiburg im Bresgau, Germany.

SCHOBBENHAUS, C. 1996. As tafrogêneses superpostas Espinhaço e Santo Onofre, Estado da Bahia: Revisão e novas propostas. *Revista Brasileira de Geociências*, **26**, 265–276.

SCHOBBENHAUS, C. & BRITO NEVES, B. B. 2003. A geologia do Brasil no contexto da Plataforma Sul-Americana. *In*: BIZZI, L. A., SCHOBBENHAUS, C., VIDOTTI, R. M. & GONÇALVES, J. H. (eds) *Geologia, Tectônica e Recursos Minerais do Brasil*. Companhia de Pesquisa de Recursos Minerais – Geological Survey of Brazil, Brasília, 5–54 (Explanatory text for the Geological Map of Brazil, 1:2,500,000 scale).

SCHOBBENHAUS, C., HOPPE, A., BAUMANN, A. & LORK, A. 1994. Idade U/Pb do Vulcanismo Rio dos Remédios Chapada Diamantina, Bahia. *In*: *Congresso Brasileiro de Geologia, Balneário Camboriú 38, Resumos Expandidos*. Sociedade Brasileira de Geologia, **2**, 397–398.

SCHÖLL, W. U. & FOGAÇA, A. C. 1979. Estratigrafia da serra do Espinhaço na região de Diamantina (M. G.). *Sociedade Brasileira de Geologia, Núcleo de Minas Gerais, Boletim*, **1**, 55–73.

SHANEY, K. W. & MCCABE, P. J. 1994. Perspectives on sequence stratigraphy of continental strata. *American Association of Petroleum Geologists, Bulletin*, **78**, 544–568.

SILVA, R. R. 1999. Evolução tectônica durante o Proterozóico das bacias do Espinhaço e São Francisco em Minas Gerais. Anais do VII Simpósio Nacional de Estudos Tectônicos/Simpósio Internacional de Tectônica da SBG. Lençóis, Brazil, Session 2, 12–14.

SOUZA, W. S. 1986. Contribuição aos estudos de revisão estratigráfica do Pré-Cambriano Brasileiro a partir da utilização do conceito de unidade deposicional. *In*: *Congresso Brasileiro de Geologia 34, Goiânia, Anais*, Sociedade Brasileira de Geologia, **1**, 391–401.

TACK, L., LIÉGEOIS, J. P., DEBLOND, A. & DUCHESNE, J. C. 1994. Kibaran A-type granitoids and mafic rocks generated by two mantle sources in a late orogenic setting (Burundi). *Precambrian Research*, **68**, 323–356.

TACK, L., FERNANDEZ-ALONSO, M., TAHON, M., WINGATE, M. T. D. & BARRITT, S. 2002. The 'northeastern Kibaran belt' (NKB) and its mineralisations reconsidered: new constraints from a revised lithostratigraphy, a GIS-compilation of existing geological maps and a review of recently published as well as unpublished igneous emplacement ages in Burundi. *In*: *11th IAGOD Quadrennial Symposium and Geocongress*. Geological Survey of Namibia, Windhoek, Namibia, 6.

THEUNISSEN, K., HANON, M. & FERNANDEZ-ALONSO, M. 1991. *Carte Géologique du Rwand*, 1:1,000,000 scale. Royal Museum for Central Africa, Tervueren.

TORQUATO, J. R. & FOGAÇA, A. C. 1981. Correlação entre o Supergrupo Espinhaço no Brasil, o Grupo Chela em Angola e as formações Nosib e Khoabendus da Namíbia. *In*: *Anais do Simpósio sobre o Craton do São Francisco e suas Faixas Marginais*, Sociedade Brasileira de Geologia/Coordenação da Produção Mineral, Salvador, 87–98.

TROMPETTE, R. 1994. *Geology of western Gondwana (2000-500 Ma). Pan-African-Brasiliano aggregation of South America and Africa.* A. A. Balkema, Rotterdam.

Geology of the northern Borborema Province, NE Brazil and its correlation with Nigeria, NW Africa

M. H. ARTHAUD[1], R. CABY[2], R. A. FUCK[3], E. L. DANTAS[3] & C. V. PARENTE[1]

[1]*Departamento de Geologia, Universidade Federal do Ceará, CE, Brazil (e-mail: michel.h@terra.com.br)*

[2]*Laboratoire de Tectonophysique, Université des Sciences et Techniques du Languedoc, Montpellier, France*

[3]*Instituto de Geociências, Universidade de Brasilia – DF, Brazil*

Abstract: The Borborema and Benin–Nigeria provinces of NE Brazil and NW Africa, respectively, are key areas in the amalgamation of West Gondwana by continental collision during the Brasiliano/Pan-African orogenies. Both are underlain by complex basement: Nigeria has *c*. 3.05 Ga Archaean crust but no known Palaeoproterozoic rocks >2.0 Ga; in NE Brazil, 2.6–3.5 Ga Archaean rocks form small cores within Palaeoproterozoic gneiss terrains affected by plutonism at *c*. 2.17 Ga. Both regions exhibit Late Palaeoproterozoic (*c*. 1.8 Ga) rift-related magmatism and metasedimentary sequences overlying the basement. The Seridó Group of NE Brazil (<0.65 Ga) is similar to the Igarra Sequence in SW Nigeria. The Ceará Group, which may date back to *c*. 0.85 Ga, is a passive margin deposit on crust thinned during initiation of an oceanic domain. In both provinces, basement and sedimentary cover were involved in tangential tectonics that resulted in crust-thickening by nappe-stacking associated with closure of this ocean. Frontal collision between *c*. 0.66 and 0.60 Ga later evolved to an oblique collision, generating north–south continental strike-slip shear zones at *c*. 0.59 Ga. In NE Brazil, the main Pan-African suture is probably buried beneath the Parnaíba Basin. The Transbrasiliano Lineament, interpreted as the prolongation of the Kandi–4°50 Lineament in Hoggar, may represent a cryptic suture.

Opening of the Atlantic Ocean in the Mesozoic led to the break-up of Pangaea (and West Gondwana) and the consequent individualization of the South America and Africa continents, each one containing part of the Brasiliano/Pan-African fold belt.

The Borborema Province (Almeida *et al.* 1981) in northeastern Brazil (Fig. 1) was built during the Brasiliano/Pan-African orogeny, as the result of convergence and final collision of the São Luis–West Africa and São Francisco–Congo cratons, in the context of West Gondwana amalgamation. The present structural framework of the province (Fig. 2) dates from the end of the Brasiliano/Pan-African orogeny, forming a mosaic of independent domains juxtaposed along large crustal-scale shear zones (Vauchez *et al.* 1995) in a continental-scale collage (Van Schmus *et al.* 1998).

Since the first attempts to reconstruct West Gondwana, the similarity of geological features between NE Brazil and NW Africa were used to argue for the juxtaposition of the two continents. However, there is still much uncertainty, and correlation between the two domains remains imprecise (Brito Neves *et al.* 2002). The northern part of the Borborema Province, north of the Patos Lineament, and the Benin–Nigeria province are the key for correlation between the continents (Fig. 2). Integration of recent geological and geochronological data allows re-evaluation of the geological framework of the two provinces and provides new insights for the correlation between the two provinces.

Summary of the Precambrian geology of the northern Borborema Province

The northern part of the Borborema Province is limited by the Atlantic Ocean to the north and east, the Parnaíba Basin to the west, and by the Patos Lineament to the south (Fig. 3). In this summary we do not deal with the NW Ceará Domain, which is discussed in detail by Santos *et al.* (2008). More than 80% of the northern Borborema Province comprises Precambrian metamorphic rocks, with ages ranging from Palaeoarchaean to Neoproterozoic.

Archaean record

Archaean rocks have been identified in four areas. *The São José do Campestre Massif*, in the eastern part of the province, close to Natal (Fig. 3), is the oldest continental crust segment in South America. Calc-alkaline rocks recording 3.45 Ga

From: PANKHURST, R. J., TROUW, R. A. J., BRITO NEVES, B. B. & DE WIT, M. J. (eds) *West Gondwana: Pre-Cenozoic Correlations Across the South Atlantic Region.* Geological Society, London, Special Publications, **294**, 49–67. DOI: 10.1144/SP294.4 0305-8719/08/$15.00

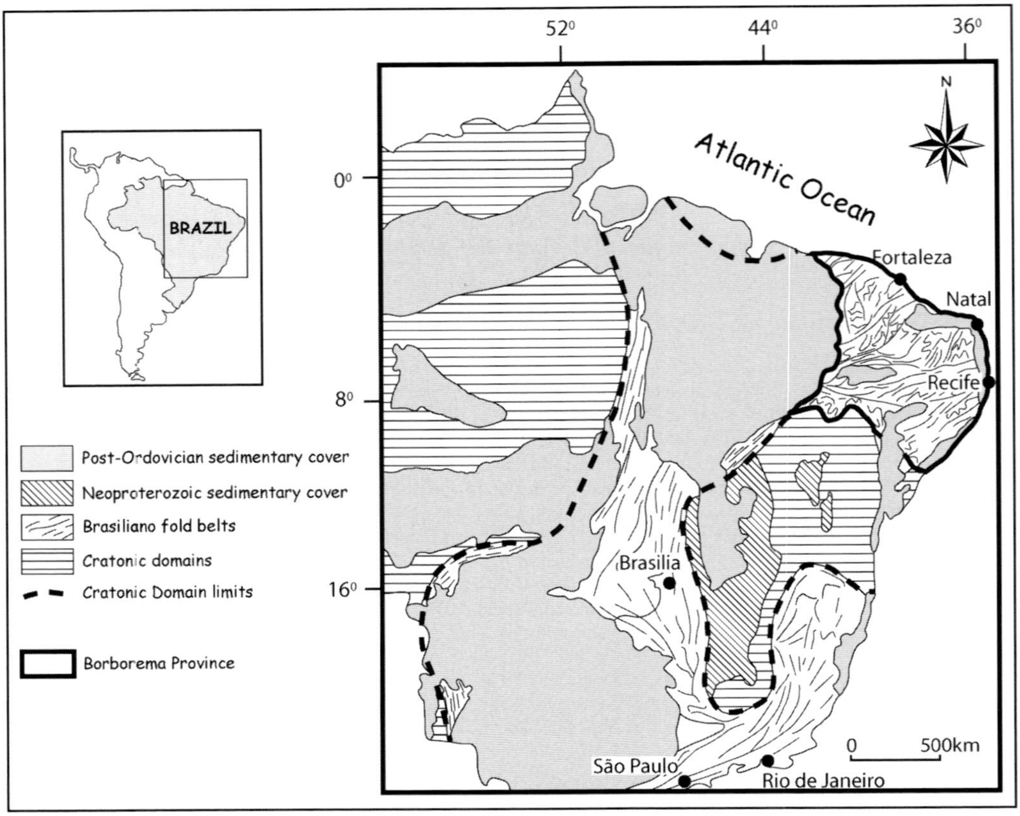

Fig. 1. The Borborema Province in NE Brazil (Schobbenhaus & Campos 1984).

U–Pb zircon ages are derived from an older sialic crust (T_{DM} model ages of 3.77 Ga, Dantas *et al.* 2004), but U–Pb and Sm–Nd data also reveal the existence of a juvenile crustal segment dated at 3.3 Ga (Dantas *et al.* 2004). Different events of trondhjemite magmatism at 3.25 and 3.18 Ga characterize repeated recycling/reworking events and the growth of juvenile crust. Archaean metamorphic events in amphibolite facies with migmatization are recorded by U–Pb zircon and monazite ages in different rock units of the complex. Syenogranite dated at 2.7 Ga represents the youngest and most evolved plutonic unit of this Archaean nucleus. The São José do Campestre massif is surrounded by Palaeoproterozoic gneisses and was affected by Palaeoproterozoic magmatism and tectonism. The complex and protracted evolution of this massif suggests that it may represent a detached fragment of a larger Archaean cratonic mass.

The Granjeiro Unit is in southern Ceará, limited to the north by the Farias Brito Shear Zone and to the south by the Patos Lineament (Fig. 3). It comprises plutonic rocks of tonalite to granodiorite composition cross-cutting mafic metavolcanic rocks of tholeiitic affinity and associated metasedimentary rocks (Arthaud *et al.* 1998). Plutonic rocks have U–Pb zircon ages of 2.55 Ga (Silva *et al.* 2002). Rocks of this domain display medium- to high-temperature amphibolite-facies paragenesis and were strongly deformed during the Brasiliano orogeny (Monié *et al.* 1997). Foliation tends to vertical near the limiting shear zones.

The Mombaça and Cruzeta complexes are located in central Ceará and are mostly made of migmatitic gneiss, displaying complex compositional layering, predominant granodiorite bands alternating with tonalite and granite bands. Mafic and ultramafic boudins are common, representing fragmented layered bodies, some with chromite mineralization. Quartz–feldspar veins and pegmatites are commonly observed along the foliation planes and locally represent more than 50% of the rock volume. Metasedimentary rocks are scarce, being mostly represented by banded iron formation associated with amphibolite and tourmalinite. Relationships between the supracrustal and metaplutonic rocks are not unequivocal due to the intensity of deformation, during which the main

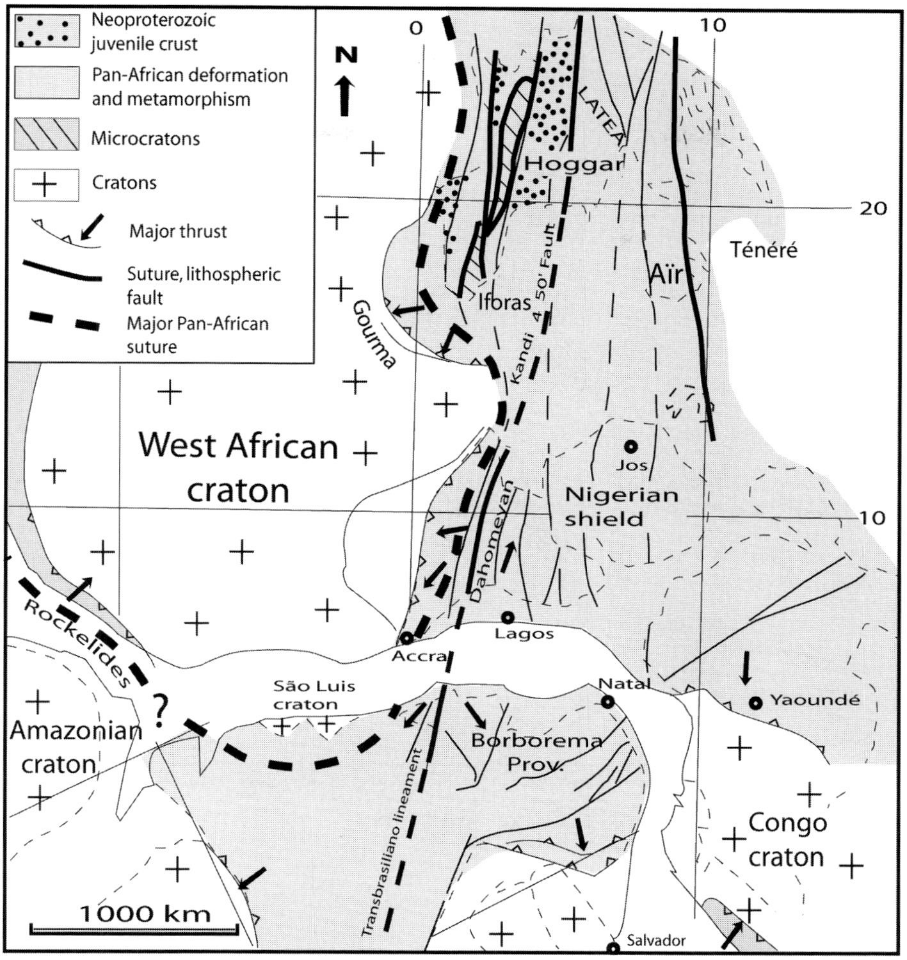

Fig. 2. Pre-drift reconstruction of NE Brazil and NW Africa in late Neoproterozoic and early Paleozoic times (adapted from Caby 1989).

structural trends of all rock types became parallel; syn-metamorphic magmatic injections of Neoproterozoic age also occurred. Foliation is generally low-dipping, usually less than 30° to the SE, becoming vertical along the Sabonete–Inharé and Senador Pompeu shear zones, as well as along the western contact with Palaeoproterozoic units. Metamorphism is of high-temperature amphibolite facies, usually associated with partial melting and migmatite generation.

Although these two complexes are separated by the Sabonete–Inharé shear zone, there is little lithological difference between them. Discrimination is due to contrasting signatures based on airborne gamma-spectrometry and slightly different U–Pb zircon ages (Fetter 1999); the Cruzeta Complex was dated at *c.* 2.7 Ga and the Mombaça Complex at *c.* 2.8 Ga. Recently, a SHRIMP U–Pb age of 3.27 Ga, interpreted as crystallization age, was determined by Silva *et al.* (2002) for zircon grains from a meta-tonalite of the Cruzeta Complex. The discordant zircon analyses may be interpreted as inherited, indicating that older Archaean rocks may have been cut by the Neoarchaean plutons. T_{DM} model ages of the Archaean rocks from the Cruzeta Complex are between 2.7 and 2.8 Ga and from the Mombaça Complex between 2.9 and 3.0 Ga. The former have been considered as juvenile, whereas the latter appear to correspond to rocks with a contribution from older Archaean crustal material (Fetter *et al.* 2000). However, Palaeoproterozoic model ages have also been recorded in the Cruzeta Complex (Arthaud unpublished results), indicating reworking of the Archaean crust.

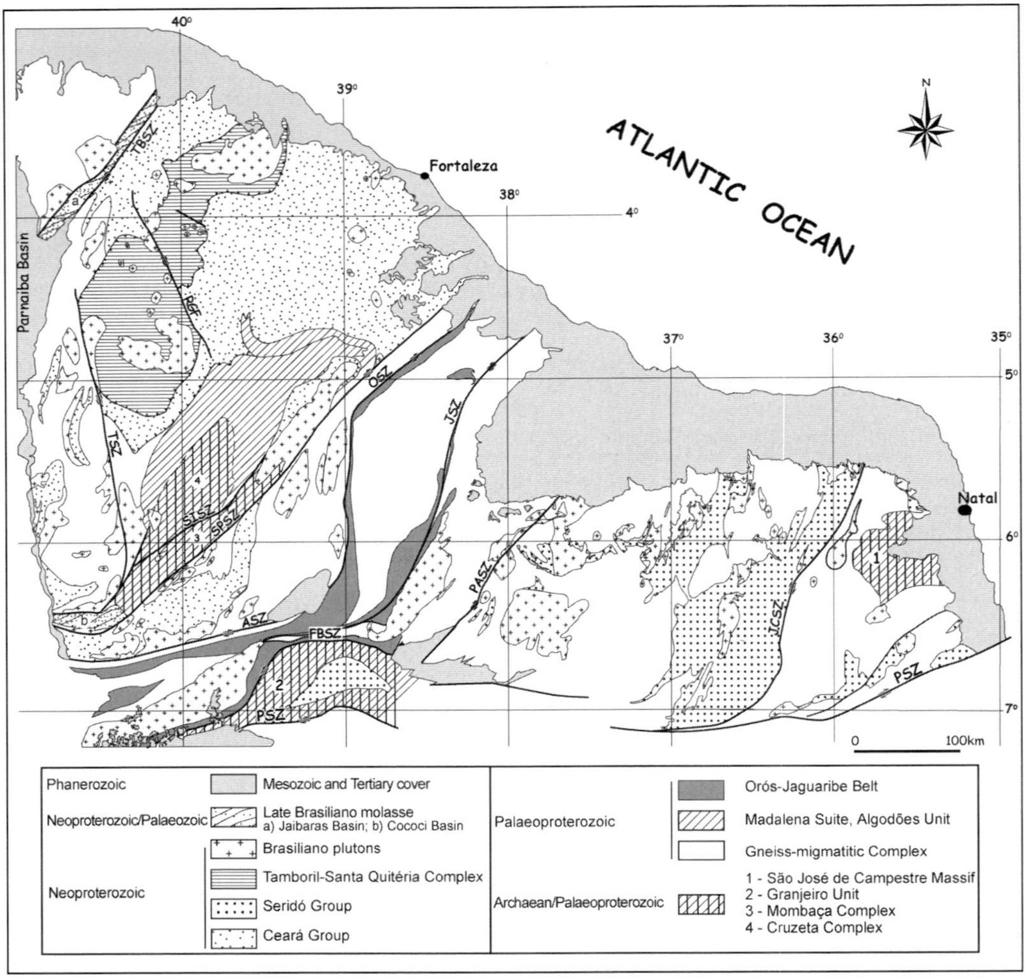

Fig. 3. Simplified geological map of the northern part of the Borborema Province. Main shear zones: TBSZ, Transbrasiliano (Sobral–Pedro II); TSZ, Tauá; SISZ, Sabonete–Inharé; SPSZ, Senador Pompeu; OSZ, Orós; JSZ, Jaguaribe; PASZ, Portalegre; JCSZ, João Câmara; ASZ, Aiuába; FBSZ, Farias Brito; PSZ, Patos. RGF, Rio Groaíras fault. Modified from Mont'Alverne *et al.* (1998), Cavalcante (1999), Cavalcante *et al.* (2003), Van Schmus *et al.* (2003), Dantas *et al.* (2004) and Arthaud (2005).

Palaeoproterozoic record

Four large rock assemblages are of Palaeoproterozoic age, namely the Gneiss–Migmatite complex, the Madalena Suite, the Algodões–Choró Unit, and the Orós–Jaguaribe Belt (Fig. 3).

Gneiss–Migmatite associations underlie large areas of the basement of the northern Borborema Province. They mostly comprise tonalite to granodiorite orthogneiss, generally metamorphosed under high-temperature amphibolite-facies conditions with variable degrees of migmatization. Al-rich metasedimentary rocks are not common, except for the domain between the Orós and Senador Pompeu shear zones, where they are dominant. Marble and quartzite are absent. Based on geochronological data, Fetter *et al.* (2000) argue for differences between the rocks east (Rio Piranhas and São José do Campestre massifs) and west of the Senador Pompeu Shear Zone (Central Ceará domain). Although U–Pb zircon ages between *c.* 2.11 and 2.19 Ga are similar for both areas (Hackspacher *et al.* 1990; Van Schmus *et al.* 1995; Martins *et al.* 1998; Fetter *et al.* 2000; Castro *et al.* 2003; Castro 2004), there is a difference with respect to model ages. T_{DM} model ages are between 2.5 and 2.6 Ga in the Rio Grande do

Norte domain, suggesting reworked older crust, whereas in the Central Ceará domain they are between 2.42 and 2.48 Ga with positive εNd(t) values, suggesting accretion of juvenile Palaeoproterozoic crust.

The Madalena Suite is an association of quartz diorite and syn-plutonic micro-diorite dykes intruding the Cruzeta Complex. These rocks display little deformation and were not migmatized, suggesting that they were emplaced after the Cruzeta Complex had been deformed and metamorphosed. U–Pb zircon ages of the Madalena Suite range between 2.15 and 2.2 Ga (Castro 2004; M. Arthaud unpublished results). Several other similar intrusions, lacking age determinations but with similar T_{DM} model ages of *c.* 2.35 Ga (M. Arthaud unpublished results), are probably part of the same magmatic event.

The Algodões Unit (not shown in Fig. 3) surrounds the Cruzeta Complex. It probably represents a Palaeoproterozoic cover of the complex or, alternatively, constitutes an allochthonous unit resulting from Neoproterozoic tectonics. It comprises supracrustal rocks displaying sub-horizontal foliation and lacking migmatization features. Dominant rock types are amphibolite derived from basalt, fine-grained leucocratic gneiss (meta-tuff), meta-greywacke, meta-arkose, rare metapelite (bearing graphite and kyanite in some localities), pure or micaceous quartzite, meta-conglomerate with carbonate clasts and cement, and calc-silicate rocks; a narrow strip of banded Mn-rich formation is also recognized. A suite of intrusive sheets of actinolite/tremolite and talc-rich meta-ultramafic rocks includes tholeiite and metabasalts that are picritic and komatiitic according to their major and REE element contents. Meta-komatiites display εNd values (at 2.0 Ga) of +7.6 to 7.9 and define a Sm–Nd whole rock isochron of 2.06 ± 0.1 Ga (J. P. Liégeois, pers. com.). Felsic metavolcanic rocks are found close to the top of the sequence. A >500 m thick package of pinkish fine-grained and porphyritic orthogneiss, including fine-grained porphyritic meta-dacite and micro-diorite bearing biotite-rich enclaves is interpreted as a sill complex. The SHRIMP U–Pb zircon age of a meta-rhyolite sample is *c.* 2.13 Ga (Castro 2004), confirming that this unit is younger than the Madalena Suite.

The Orós-Jaguaribe Belt, striking NNE–SSW and bending to east–west in its southern sector, comprises two supracrustal rocks units (Orós to the west and Jaguaribe to the east), separated by a Palaeoproterozoic gneissic basement. The belt is limited eastwards by the Portalegre Shear Zone, and westwards by the Orós Shear Zone; the Jaguaribe Shear Zone cuts through the eastern belt. The main rock types belong to a metavolcanic–sedimentary sequence, associated with orthogneiss (Caby & Arthaud 1986; Mendonça & Braga 1987; Sá & Bertrand 1992; Sá *et al.* 1995; Parente 1995; Parente & Arthaud 1995). The metasedimentary rocks are mainly Al-rich schist, intercalated with a narrow strip of quartzite, as well as lenses of Ca- or Mg-rich marble, calc-silicate rocks, carbonaceous schist and quartzite. The occurrence of magnesite layers and gypsum pseudomorphs indicates a former evaporitic environment (Parente *et al.* 2004*a*). Metavolcanic rocks comprise dominant porphyritic meta-rhyolite, along with meta-rhyodacite and meta-dacite; the meta-rhyolite and felsic tuff usually bear bluish quartz phenocrysts. At the Orós dam the main quartzite layer is limited, at its base, by channels of pebbly meta-arenite and rests on top of the porphyritic meta-rhyolite. The granite orthogneiss is subalkaline in composition, usually porphyritic and displays relict Rapakivi textures.

This rock association is interpreted as formed in a continental rift, the sediments having been deposited on a thinned crust, initially stable, whereas the volcanic and plutonic rocks are the result of the active phase of rifting (Parente & Arthaud 1995).

U–Pb zircon ages of the metavolcanic rocks are in the interval 1.75–1.8 Ga; an orthogneiss sample has been dated at *c.* 1.69 Ga (Sá 1991). These ages allow correlation with crustal thinning and rifting events known in other areas in central-eastern Brazil (the Espinhaço Event, Alkmin *et al.* 1993) as well as in the Pan-African belt in West Africa (Caby & Andreopoulos-Renaud 1983). Foliation within the belt is sub-vertical. Metavolcanic rocks were mylonitized in a dextral transpressive regime. Metamorphism is locally of greenschist facies (white mica + chloritoid) in the central part, grading to low-temperature amphibolite facies (staurolite + andalusite + biotite + garnet), increasing northwards to high-temperature amphibolite and granulite facies in the northern sector of the Orós–Jaguaribe Belt.

The Brasiliano structural and metamorphic evolution of the belt and its geochronological characteristics differ completely from the gneisses and migmatites exposed to the east of the Portalegre Shear Zone and to the west of the Orós Shear Zone. These rocks do not represent the original basement affected by the Statherian rift, and were brought to their present position due to displacement along large dextral strike-slip shear zones, probably nucleated in the precursor normal faults that limited the rift.

Neoproterozoic record

The Neoproterozoic record is represented by the Ceará and Seridó groups, the Tamboril–Santa

Quitéria Complex and numerous syn- to late- and post Brasiliano granitoids.

The Ceará Group is a thick sequence of terrigenous metasedimentary rocks dominated by metapelites. Associated rock types are thin quartzite beds that form important ridges in the regional relief and locally bear Al-minerals, and lens-shaped marble and calc-silicate rocks, frequently associated with amphibolite that may represent basaltic flows or mafic tuffs. We interpret the Ceará Group as a passive margin-type sedimentary unit. In Central Ceará, this unit tectonically overlies the less deformed and metamorphosed Palaeoproterozoic rocks of the Algodões Unit, with a retrograde mylonite sole.

Foliation in the Ceará Group is relatively simple, with shallow dips, easterly to the west of the Tamboril–Santa Quitéria Complex and westerly to the east of the Complex. Emplacement of the nappes was accompanied initially by the development of recumbent to isoclinal folds well preserved in the quartzite layers and, after emplacement, by a later phase of upright folds. Numerous internal low-angle shear zones add to the complexity of the structural framework. Sillimanite or kyanite mineral lineations striking WNW–ESE indicate the main direction of tectonic mass transport, in contrast with NNE–SSW directions observed in the underlying units (Cruzeta Complex, Madalena Suite, Algodões Unit).

Migmatization of metapelite is the rule in this unit. The presence of the mylonite sole, the marked differences in structure and the metamorphic gap between the Ceará Group and the underlying Palaeoproterozoic units indicate that the contact between these rock units is a fundamental low-angle thrust of the Brasiliano chain, characterizing nappe tectonics similar to that observed in Himalayan-type collisional mountain chains (Caby & Arthaud 1986).

Metamorphism in Central Ceará is typically inverted (Caby & Arthaud 1986). Above the sole, the typical Al-silicate in metapelites is kyanite, and partial melting is poorly developed. Kyanite + white mica + garnet + rutile, with kyanite included in garnet and muscovite, is the typical paragenesis in the metapelites. The occurrence of clinopyroxene- and garnet-bearing amphibolites is evidence of high-pressure metamorphism, possibly under eclogite-facies conditions (Castro 2004; Garcia & Arthaud 2004) and the widespread occurrence of amphibole and Na-poor (<2 wt%) clinopyroxene–plagioclase symplectites relates to high-temperature decompression toward the transition between granulite and high-temperature amphibolite-facies conditions (Castro 2004; Garcia & Arthaud 2004). Sheets of kyanite-bearing charnockitic granites indicate high-pressure anhydrous partial melting of pelites ($P \geq 14$ kbar, $T \geq 750$ °C). Most metapelites were partly recrystallized in the stability field of prismatic sillimanite and fibrolite. In the middle portion of the nappe kyanite disappears progressively, and both Al-silicates frequently coexist. Sillimanite is formed either by kyanite break-down or as a result of the reaction muscovite + quartz + $H_2O \rightarrow$ sillimanite + melt. Higher up in the sequence sillimanite is the only Al-silicate present. In the upper part of the nappe, metapelites are thoroughly migmatized, and progressively grade to metatexite and diatexite with frequent intercalations of aluminous cordierite-bearing anatectic granites. The age of metamorphism is poorly defined; the oldest $^{40}Ar/^{39}Ar$ mineral ages (amphibole, phlogopite, biotite, muscovite) obtained by Monié *et al.* (1997) range around 0.575–0.56 Ga in this domain.

For many years the Ceará Group has been considered as Palaeoproterozoic in age (see Cavalcante *et al.* 2003). However, Sm–Nd T_{DM} model ages range between 2.4 Ga and 1.09 Ga (Fetter 1999; Castro *et al.* 2003; Santos *et al.* 2003, 2004; Castro 2004; M. Arthaud unpublished results), suggesting a Mesoproterozoic upper limit for the age of deposition, with an important detrital contribution from Palaeoproterozoic and possibly Archaean sources. A meta-rhyolite layer intercalated in metapelite with U–Pb zircon age of 0.77 ± 0.03 Ga was interpreted by Fetter (1999) as possibly representing the precursor rift formed prior to the opening of the ocean basin subsequently closed during the Brasiliano orogeny. Recent SHRIMP U–Pb zircon ages (M. Arthaud unpublished results) provide constraint on the depositional history of the Ceará Group. Among the detrital zircons extracted from a metapelite sample (biotite gneiss with white mica, garnet and kyanite), two populations yield Neoproterozoic ages, clustering at *c.* 0.63 and 0.8 Ga. The younger value is interpreted as corresponding to the age of metamorphism, whereas the older may be associated with a rifting episode and subsequent opening of an ocean.

The Seridó Group is a thick metasedimentary sequence exposed in eastern Rio Grande do Norte (Fig. 3). It comprises three formations known as, from base to top, the Jucurutu, Equador and Seridó formations. The Jucurutu Formation comprises biotite gneiss, calc-silicate rocks, marble and subordinate quartzite; concordant intercalations of mafic and felsic metavolcanic rocks are common (Caby *et al.* 1995). The Equador Formation, observed in unconformity over the basement (Caby *et al.* 1995), is represented by quartzite with conglomeratic layers, whereas metapelites are the main rock types of the Seridó Formation, including rare impure quartzite, quartz schist and carbonate schist. Preserved primary structures in low strain areas suggest that the biotite schists represent turbidite deposits. Internal unconformities appear to separate the constituent formations of the Seridó Group.

Structurally, the Seridó Group is characterized by upright folds with NNE–SSW–trending axes; the regional slaty cleavage, which grades eastwards to a two-mica foliation, represents the main axial planar surface to these folds (Archanjo & Bouchez 1991; Caby *et al.* 1995). Jardim de Sá (1984, 1994) reported structural complexity interpreted as symptomatic of polyphase deformation. Polyphase deformation is indeed observed within the thermal aureoles of the Acari Pluton (Archanjo 1993). However, syn-sedimentary recumbent folds are also likely. Similarly, the metamorphism displays striking polarity, increasing from greenschist facies in the west to high-temperature amphibolite-facies conditions in the east, with the presence of sillimanite, cordierite (Lima 1992) and, locally, kyanite (Sá & Legrand 1983; Archanjo & Bouchez 1991). Southwards, close to the Patos Lineament, the rocks display low-pressure mineral assemblages representing garnet–cordierite–spinel ± corundum diatexite closely associated with stocks of cordierite granite (Corsini *et al.* 1998).

In the eastern sector, the relationship between the Seridó Group and gneiss–migmatite rocks is obliterated by the truncating João Câmara Shear Zone. In the NW, rocks of the Seridó Formation overlie the gneissic basement through a stratigraphic unconformity, starting with a polymictic conglomerate layer including clasts of the basement units in a sandy matrix (Caby *et al.* 1995). The conglomerate suggests that in this area the Seridó Group is autochthonous relative to the basement.

In previous models the Seridó Group has been considered to be of Palaeoproterozoic age with a polycyclic tectonic and metamorphic history (Jardim de Sá 1984, 1994; Bertrand & Jardim de Sá 1990; Jardim de Sá *et al.* 1997). Field evidence and geochronological data obtained by several authors show that this unit was in fact deposited and deformed in the Neoproterozoic (Caby 1989; Caby *et al.* 1990; Van Schmus *et al.* 1995, 1997, 2000; Archanjo & Legrand 1997). SHRIMP U–Pb ages show that the youngest fraction of detrital zircon from both the Seridó and Jucurutu formations is 0.65 ± 0.05 Ga, establishing, therefore, the maximum depositional age (Van Schmus *et al.* 2003). Sedimentation must have lasted for 50 million years, at the most, since the Brasiliano collision took place at *c.* 0.6 Ga. Both formations present Sm–Nd model ages of 1.2–1.6 Ga, indicating mixture of detrital components derived from Archaean to Neoproterozoic terrains (Van Schmus *et al.* 2003).

The Lavras da Mangabeira Sequence is a small metasedimentary sequence with meta-conglomerate and quartzite at the base, followed by metapelite unconformably overlying deformed meta-tonalites of the Granjeiro domain (Caby *et al.* 1995). This monocyclic cover, that probably correlates with the Seridó Group, was metamorphosed under high-temperature greenschist-facies conditions (andalusite + biotite + garnet).

Neoproterozoic granitoids

The Borborema Province is characterized by widespread Neoproterozoic granitic plutonism. In the northern part of the province, this magmatism is still under scrutiny. However some basic observations can be put forward.

(i) According to available U–Pb age determinations, most of the dated intrusions crystallized within the 0.62–0.63 Ga interval (Brito Neves *et al.* 2003).

(ii) The syn-Brasiliano magmatism comprises two granite associations. Firstly, S-type granites result from partial melting of metasedimentary sequences coeval with the thickening of the crust: in general, they form small intrusions, more or less directly associated with the anatectic sources; only two S-type granites, intruded between the Senador Pompeu and Orós shear zones, display batholith dimensions. These bodies have not been dated properly, but they were affected by solid-state deformation along the shear zones. The second association is of granites emplaced in transtensional domains of the main ductile strike-slip shear zones. U–Pb zircon age determinations of granites intruded along the Senador Pompeu Shear Zone indicate that they crystallized around 0.58–0.59 Ga (J. F. Nogueira 2004).

(iii) A large number of late- to post-Brasiliano granite intrusions with ages of 0.58–0.53 Ga are recognized. They form small stocks or large batholiths and their emplacement seems to be partially controlled by the large strike-slip shear zones.

The Tamboril–Santa Quitéria Complex is a large anatectic/igneous complex with a thin, low-angle, basal metasedimentary belt including high-temperature mylonites. The plutonic rocks display a ubiquitous syn- to late-magmatic deformation that was coeval with the injection of younger, less deformed magmas. Large volumes of magma were intruded in the form of veins, layers, sheets and plutons. They range in composition from mafic diorite to tonalite to granodiorite and granite. This complex plutonic association intruded supracrustal rocks that are preserved only as large restitic pendants and enclaves of calc-silicate rocks and amphibolite, probably derived from former basalt and lesser sillimanite-bearing metapelite. Initially interpreted as an allochthonous unit of possible Archaean age (Caby & Arthaud 1986), the Tamboril–Santa Quitéria Complex is in fact of Neoproterozoic age, as shown by several Sm–Nd model ages and U–Pb zircon ages (Fetter *et al.* 2003; Castro 2004). Fetter *et al.* (2003) interpreted the complex as a continental-margin magmatic arc emplaced during the Brasiliano collision. It is therefore a complex Neoproterozoic, pre- to early Brasiliano batholith, characterized by numerous magmatic pulses, the last of which was synchronous with reheating and remelting of previous tonalite–granodiorite plutons, providing the thermal energy for partial melting in an environment of intermediate hot crust. Although the allochthonous character of the complex is likely, since the basal contacts observed on all sides of the complex dip inwards with the same features, such an interpretation is still challenged by one of us (RC) and requires further field and petro-structural observations.

Neoproterozoic–Palaeozoic record

Late Brasiliano molasses. Several small sedimentary basins in the northern sector of the Borborema Province record the transition from Neoproterozoic to Phanerozoic times. These basins developed along large-scale ductile strike-slip shear zones, and were partially preserved in pull-apart structures controlled by reactivation of the shear zones.

In general, the basins contain two continental sedimentary sequences separated by an erosional unconformity. The lower sequence, Ediacaran–Cambrian in age (Parente *et al.* 2004*b*), comprises polymict clast-supported conglomerate and breccia. Clasts are mostly of gneiss, quartz vein, amphibolite and arkosic sandstones. The sandy matrix is purplish-grey and fine-grained. The upper sequence (Cambrian–Ordovician) is also characterized by clast-supported conglomerate with a sandstone matrix; most of the clasts here are of volcanic origin, including basalt and rhyolite, but fragments from the underlying sedimentary rocks are also common.

Recurring volcanism is recorded within and outside the basin limits. The first phase is of dyke swarms of basalt, quartz diorite, dacite, rhyodacite and porphyritic rhyolite, which occur at the NW border of the Jaibaras Basin (Almeida 1998). Porphyry rhyolite dykes also occur close to the northern border of Cococi Basin (Parente *et al.* 2004*b*). Rb–Sr age determinations on whole-rock samples have yielded ages of *c.* 580 Ma (Novais *et al.* 1979), 0.56 $\pm$ 0.02 Ga (Sial & Long 1987), and 0.562 $\pm$ 0.01 Ga (Tavares *et al.* 1990). Within the basins, magmatism is mostly extrusive, bimodal, associated with or cutting through sedimentary units of different stratigraphic positions (Parapuí Formation, Costa *et al.* 1979). Volcanic rocks comprise mainly continental tholeiite, and alkaline and andesitic basalt, which may be amygdaloidal. The rocks are hydrothermally altered, with sodic and propyllitic alterations being the most common. Granite stocks and batholiths, such as the Mucambo granite (0.53 Ga, Fetter 1999, U–Pb zircon age), intrude the basal sequences (Parente *et al.* 2004*b*).

A wide variety of models have been proposed to explain the origin and evolution of the basins: graben filled with molasse sediments (Kegel *et al.* 1958; Brito Neves 1975; Mello 1978; Costa *et al.* 1979; Nascimento & Gava 1979; Cavalcante *et al.* 1983), intermontane basins filled with molasse sediments (Almeida 1967, 1969; Mabesoone *et al.* 1971; Danni 1972), graben filled with volcano-sedimentary sequences (Parente & Fuck 1987; Quadros *et al.* 1994; Quadros & Abreu 1995), pull-apart basins (Gorayeb *et al.* 1988; Parente *et al.* 1990; Abreu *et al.* 1993; Vasconcelos *et al.* 1998), extrusion basins (Brito Neves 1998, 2002), and rift-activated basins (Oliveira 2000, 2001; Oliveira & Mohriak 2003).

Most of the large shear zones that control these basins are Neoproterozoic in age, and were active between 0.58 and 0.50 Ga, according to $^{40}Ar/^{39}Ar$ data (Corsini *et al.* 1998). At least two transtensional phases are recognized, each responsible for the formation or modification of the basins, in part accompanied by bimodal volcanism, allowing them to be classified as rift basins associated with transtension, or strike-slip-type basins with oblique lateral movement.

The geochronology of these basins is based essentially on Rb–Sr ages for the volcanic rocks and is therefore not well constrained. Parente *et al.* (2004*b*) suggest intervals of 0.56–0.53 Ga for deposition of the basal sequence and 0.53–0.44 Ga for the upper sequence.

Post-Brasiliano granites. Recent U–Pb zircon age determinations indicate that several granite intrusions, including ring complexes, are Ordovician in age (0.47 $\pm$ 0.02 Ga, Pajé suite, Teixeira 2005; 0.478 $\pm$ 0.03 Ga, Quintas Ring Complex, Castro 2004), and therefore post-Brasiliano. Progress of geochronological knowledge indicates that other granite intrusions are of similar age, as suggested by work in progress in Rio Grande do Norte.

Major transcurrent shear zones

One of the most outstanding features of the northern Borborema Province is the presence of a number of

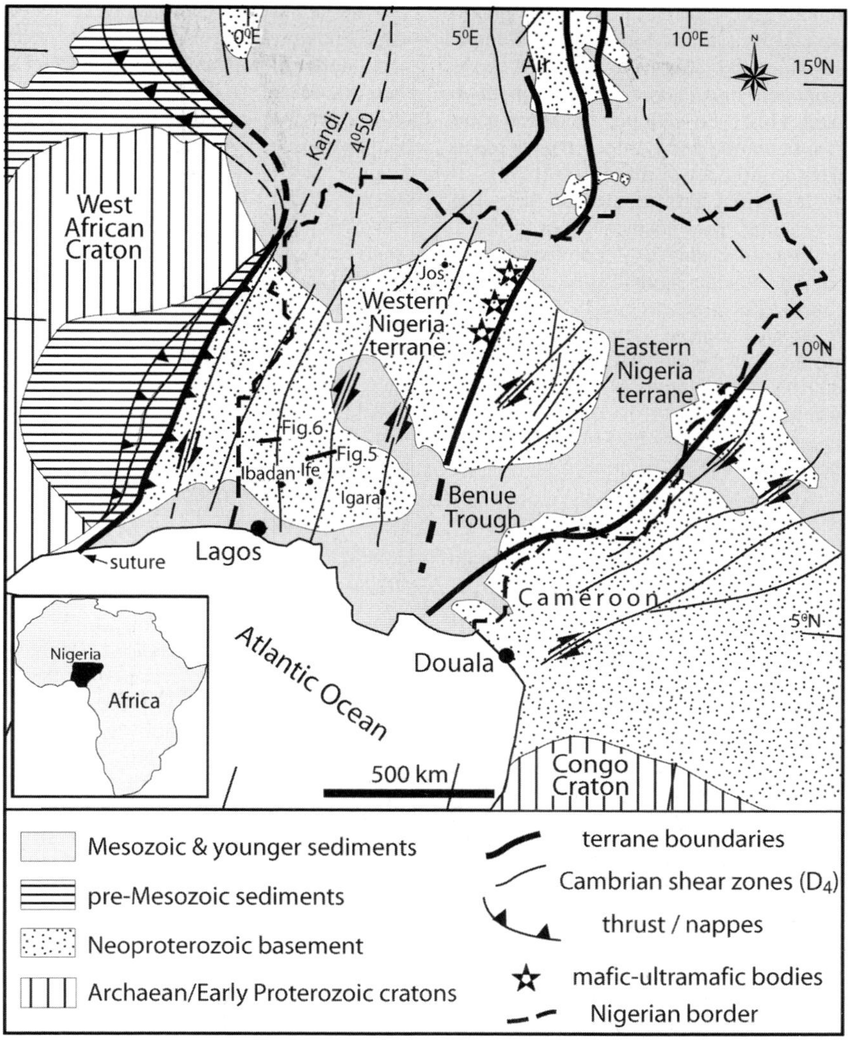

Fig. 4. Simplified map of the southern part of Nigeria and adjacent areas (Ferré & Caby 2006).

continental-scale strike-slip-type shear zones (Fig. 3). The lineaments are delineated by mylonite and ultramylonite belts, tens to hundreds of metres wide, with nearly horizontal stretching lineations. Consistent shear criteria indicate that, except for the Tauá Shear Zone, these belts formed as a result of dextral motion. Besides the greenschist-facies mylonites observed in the core of the shear zones that record the late movement at shallow crustal level, most of the shear zones (e.g., the Patos Lineament) are also marked by gneissic mylonite up to 20 km wide, generally developed under high-temperature and low-pressure conditions, accompanied by anatexis (Vauchez *et al.* 1995; Corsini *et al.* 1998).

The Patos Lineament, which is the southern limit of the northern sector of the Borborema Province, strikes east–west. The other dextral shear zones strike NNE–SSW, parallel to large transcurrent shear zones in the Pan-African chain, particularly the Kandi–4°50 lineament (Fig. 4), which represents a suture. In West Africa, the NNE–SSW shear zones are interpreted as the product of trans-pressure, resulting from oblique collision between the São Francisco/Congo and West Africa cratons (Castaing *et al.* 1994), an interpretation that can also be put forward for the strike-slip shear zones in northern Borborema Province. The Transbrasiliano Lineament (Sobral-Pedro II Shear Zone) is taken as the continuation of the Pan-African Kandi–4°50 lineament.

Geochronological data from syn-shear granite intrusions (*c.* 0.59 Ga in the Senador Pompeu Shear Zone, Nogueira 2004) indicate that movement is younger than nappe emplacement in Central Ceará. However, some of the shear zones such as Orós, Jaguaribe and Portalegre have reactivated older structures, reworking the normal faults limiting the Statherian Orós–Jaguaribe Belt. Tectonic activity along the lineaments took place in shallow depths until Cambrian–Ordovician times, and even until Devonian times, in the case of the Transbrasiliano Lineament, as can be seen in outcrops of tilted and sheared Devonian sandstones near Acaraú. It is possible that the Patos Lineament, which produces drag structures on other shear zones, bending their strike to the east–west direction, is younger than the NNE–SSW shear zones and the result of change in the continental convergence to a less oblique geometry.

Geology and geodynamic evolution of the Pan-African belt in Nigeria (with special reference to SW Nigeria)

The Nigerian shield is exposed east of the Kandi fault, one of the major Pan-African steep strike-slip shear zones that matches the 4°50 shear zone of Hoggar (Caby 2003) and the Sobral-Pedro II Shear Zone (Transbrasiliano Lineament (Caby 1989; Figs 3 & 4). This major shear zone is nearly parallel, but it cuts across the gently east-dipping high-temperature units exposed to the east of the Pan-African suture shown in Figure 4. The Nigerian shield comprises a polycyclic basement and remnants of metamorphic cover sequences preserved in large synforms and schist belts with variable metamorphic mineral assemblages ranging from greenschist to amphibolite facies (Rahaman 1976; Turner 1983; Fitches *et al.* 1985; Ajibade *et al.* 1987). Basement–cover relationships identified in different regions (Grant 1978; Mullan 1979; Caby 1989) suggest that a great part of the shield represents a former ensialic domain that was buried underneath peri-cratonic sediments and subsequently thoroughly reworked during the multistage Pan-African orogeny.

The pattern of large-scale shear zones (Caby 1989) indicates that the Nigerian shield may represent, in part, the southern extension of the central/eastern Tuareg terranes identified by Black *et al.* (1994). Available geochronological data for syn- to late-kinematic plutons suggest that two distinct tectonic and magmatic events can be recognized in eastern Nigeria (Tubosun *et al.* 1984; Rahaman *et al.* 1991; Ferré *et al.* 1998), the older dated at around 0.64–0.62 Ga and the younger at around 0.59 Ga.

Contrary to the complex three- or even four-stage polyorogenic development proposed by some authors (Ajibade *et al.* 1988 and references therein), recent U–Pb zircon ages and a critical reinterpretation of the geochronological data coupled with new petro-structural and lithostratigraphic data have led Caby (1989) and Caby & Boessé (2001) to propose a simpler scenario. The Pan-African event appears not only as a simple 'rejuvenation' of an older polyorogenic domain, as previously advocated by several authors (Rahaman 1988 and references therein), but it represents the major tectono-metamorphic episode, as shown in central Hoggar (Caby 2003). Ferré & Caby (2006) have shown that the Pan-African continental collision in northeastern Nigeria produced a low-angle foliation coeval with high-grade metamorphism (up to granulite facies) and pervasive migmatization in monocyclic supracrustal units of Proterozoic age.

Archaean basement

The northwestern part of the shield is floored by Archaean TTG type granodiorite–tonalite gneisses of early Archaean age (Dada *et al.* 1993), whereas the cover, of assumed Proterozoic age, has been preserved in north–south trending schist belts (Turner 1983). Zircon grains from orthogneisses in the Kaduna area have given U–Pb ages of 3.04–3.05 Ga for the emplacement of Archaean granodioritic magmas (Bruguier *et al.* 1994). These authors pointed out the lack of Palaeo- or Mesoproterozoic tectono-metamorphic events. A U–Pb monazite age of 0.62 Ga from leucosome of Archaean migmatitic gneiss indicates that the basement of this area underwent high-grade metamorphism and anatexis during the Pan-African orogeny (Dada *et al.* 1993).

In SW Nigeria, monocyclic metasedimentary rocks represent the cover deposited on the ≥ 2.5 Ga polycyclic Archaean basement, both being involved in nappe tectonics (Caby 1989; Caby & Boessé 2001). Grey gneisses from the Ife–Ilesha area (SW Nigeria) display a monotonous mineralogy dominated by plagioclase, quartz, minor K-feldspar, amphibole, biotite, ilmenite, titanite, allanite and late epidote. Lenses of orthogneisses, occasionally porphyritic, display preserved igneous layering and rare highly stretched enclaves, suggesting that large parts of the grey gneisses were derived from strongly deformed and layered igneous protoliths ranging in composition from tonalite to granodiorite with minor layered trondhjemite (TTG-type). Subsequent intense deformation has generally converted the protoliths into banded grey gneiss. Shearing and refolding of this tectonometamorphic banding was coeval with partial melting of some thinly layered protoliths of appropriate composition, producing up to 20% of plagioclase-rich leucosomes with minor amounts of amphibole and/or biotite, titanite and ilmenite. Anatexis of rocks of such leucotonalitic to trondhjemitic compositions requires regional temperatures of *c.* 700 °C. Lenses and boudins of amphibolites, sometimes fragmented and cemented by plagioclase-rich leucocratic material, are also observed within the grey gneiss complex. Injection of undeformed Pan-African granitoid sheets is common.

In the Ife University campus, metatexites have been cut by undeformed syn-metamorphic, syn-anatectic magmatic veins ranging in composition from granodiorite to biotite-rich tonalite and diorite that are slightly oblique to foliation. These latter are considered to be Pan-African intrusive rocks. U–Pb zircon ages of *c.* 2.6 Ga (upper intercepts, conventional method) obtained by Rahaman (1988) constrain an Archaean age of the grey gneisses. Lower intercepts with Concordia at *c.* 0.6 Ga testify to the major influence of high-temperature Pan-African metamorphism and anatexis.

Proterozoic units from southern Nigeria

Caby & Boessé (2001) have recognized the following Proterozoic units in southwest Nigeria.

The Late Palaeoproterozoic Supergroup comprises ortho-quartzites and aluminous metapelites such as those described by Grant (1978) from the Ibadan region and from the Ife–Ilesha area (Oke–Mesi Formation, Rahaman 1988). The basal stratigraphic unconformity of ortho-quartzites with ghost oblique bedding has been locally observed along the contact with the grey gneisses near Ibadan (Caby 1989). In southern Benin, the Badagba sillimanite-bearing quartzites overlain by marbles and calc-silicate layers (Caby 1989) may represent the same unit, resting directly on the gneissic basement. Marbles, calc-silicate gneisses and amphibolites apparently overlie the quartzites, but a major part of the carbonates has been removed by erosion and reworked in conglomerates interlayered with a turbiditic greywacke–semipelite association of the much younger Igara Sequence described below.

In the Ife–Ilesha area, the monotonous sequence of metapelites and quartz-schists displaying a recumbent foliation tectonically overlies the Archaean grey gneisses and pink orthogneisses of the Ife dome (Fig. 5). Micaschists and quartz-schists frequently display preserved sedimentary bedding which has been involved in recumbent to isoclinal folds. The schists alternate with quartzites and thin (a few centimetres) layers richer in mica or mica + sillimanite. Aluminous quartzites comprise a prominent unit cropping out in ridges east of the Ifewara Shear Zone where they are associated with the pink orthogneisses. The mineral assemblage of micaschists is dominated by quartz–muscovite–biotite with local staurolite, garnet and rarely preserved fibrolite.

Late Palaeoproterozoic K-rich pink orthogneisses are porphyritic red granitoids and metaporphyries representing meta-intrusive rocks emplaced in the quartzite sequence. NW of Ife, pink porphyritic orthogneisses form the core of a dome surrounded by the Archaean grey gneisses (Fig. 5). Pink augen-gneisses represent former K-rich sub-alkaline porphyritic granites. Dark green, Fe-rich biotite and dark-green amphibole of the hastingsite group represent polycrystalline pseudomorphs after coarse-grained ($\geq$1 cm) igneous phases. Fluorite, allanite and zircon are the common accessory minerals. Preserved igneous layering defined by variable abundances of deformed K-feldspar phenocrysts, biotite and

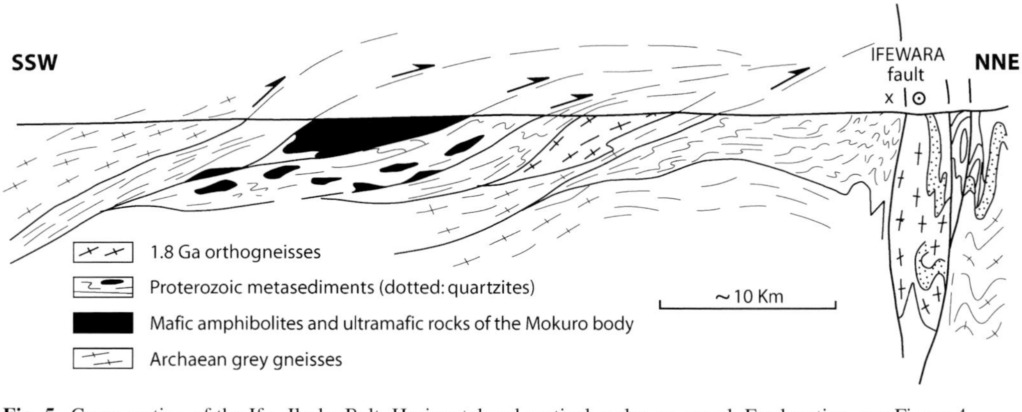

Fig. 5. Cross-section of the Ife–Ilesha Belt. Horizontal and vertical scales are equal. For location, see Figure 4.

amphibole is obliquely cut by the regional foliation, as well as some meta-pegmatite veins and meta-aplite sheets. Because of the bulk rheology and mineralogy, severe shearing is uncommon and only incipient partial melting is observed in these orthogneisses, in contrast with the adjacent grey gneisses. Contacts with surrounding Archaean grey gneisses are abrupt and almost parallel to the regional foliation. The pink subalkaline orthogneisses in the university campus and in the city of Ibadan have both yielded zircon U–Pb ages of about 1.85 Ga (Rahaman 1988).

The Mokuro body (Fig. 5) comprises mainly meta-pyroxenites, gabbro-norites and melanocratic amphibolites, alternating at the scale of some metres to tens of metres (Ige *et al.* 1998). Plagioclase-rich meta-gabbros are also a major constituent of the northern part of the body. Ultramafic rocks occur as lenses and boudins contain tremolite, talc and Mg chlorite, relict olivine (Fo_{90}), and anthophyllite. Some thin strips of quartzite, biotite–staurolite and garnet–quartz schist occur in the central part of the body. Ige *et al.* (1998) have clearly pointed out the cumulate character of most rocks of the complex, using discriminant diagrams on trace elements and REE patterns, as well as identifying preserved petrographic and mineralogical features of typical igneous origin. They interpret the massif as a large basaltic sill with MORB-like characteristics, affected by crystal settling that produced the ultramafic cumulates. The close association of syenites with mafic rocks, however, suggests that part of the magmas was extracted from an enriched mantle, thus suggesting the same extensional setting that produced the 1.85 Ga sub-alkaline granitoids. The occurrence of boudins of mafic amphibolite and garnet-bearing meta-anorthosite within the micaschists implies that the body is a strongly boudinaged sill originally emplaced within the schists and quartzites prior to regional metamorphism, in agreement with the conclusions of Ige *et al.* (1998). However, detailed mapping and structural observations show that all contacts are tectonic and marked by sheared rocks (Fig. 5).

The Late Neoproterozoic Supergroup is characterized, as in the Tuareg shield, by turbiditic meta-greywackes and schists. Ubiquitous greywacke deposits derived from arc terrains in northern Mali are younger than 0.69 Ga, as indicated by the U–Pb zircon age of a pre-tectonic tonalite reworked as pebbles in the greywackes (Caby & Andreopoulos-Renaud 1989).

The Igarra formation exposed in southern Nigeria (Fig. 6) was described by Odeyemi (1982) and Omitogun *et al.* (1991). Most of the unit is a rhythmic formation of assumed turbiditic character formed by alternating impure quartzites, greywackes and semipelites. Its lower section includes quartz-schists capped by carbonates and calc-silicate layers. Its unconformable upper part encompasses polymict conglomerates containing

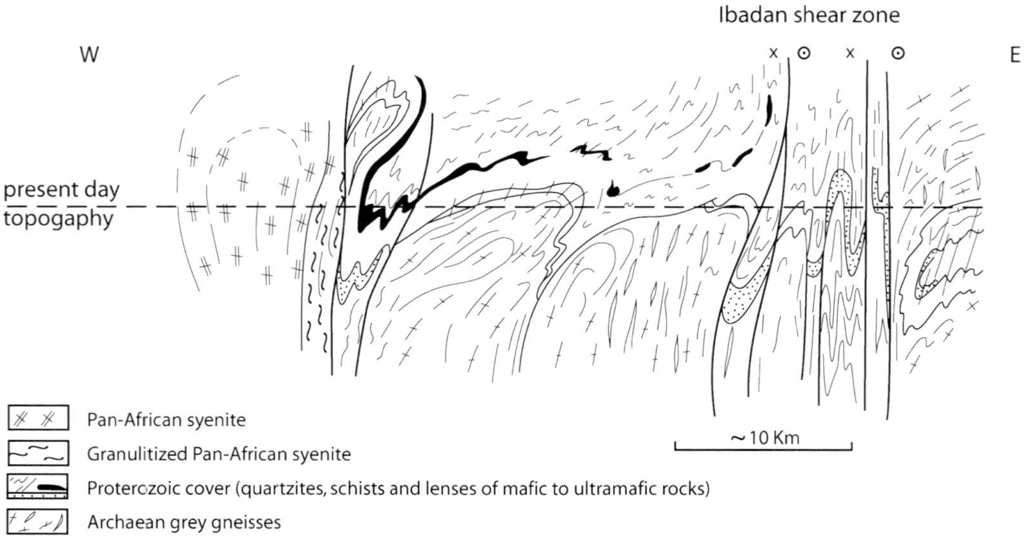

Fig. 6. Interpretative section showing geometrical relationships between Archaean basement and cover, 70 km NNW of Iabadan. Pan-African syenite, locally granulitized, cuts the Proterozoic cover. Archaean grey gneisses have the geometry of domes east of the Ibadan shear zone, which is a second-order fault. Horizontal and vertical scales are equal. For location, see Figure 4.

reworked angular clasts up to 1 m in size, derived from the gneissic basement, granites, quartzites and the underlying carbonates of assumed late Palaeoproterozoic age.

Although regional correlations with units identified in the Tuareg shield are difficult to establish due to the paucity of available robust geochronological and structural data in Nigeria, the proposed two-fold division is similar to that well established in the western part of the Trans-Saharan Belt (Caby 2003) and in NE Brazil, where turbiditic sediments including greywackes are younger than 0.65 Ga (Van Schmus *et al.* 2003). A small basin of non-metamorphic red molassic rocks in southern Benin was described by Boussari & Rollet (1974); the red arenites were derived from rhyolitic volcanic rocks and an intercalation of pillowed basalts of tholeiitic character occurs in the series.

Overview of the Pan-African regional metamorphism

The Pan-African metamorphism in southern Nigeria is generally of high-temperature, medium- to low-pressure type. The lower temperature assemblages are preserved in simple synformal schist belts (Onyeagocha & Ekwueme 1990), whereas lower amphibolite- to granulite-facies conditions (Omitogun *et al.* 1991) were reached in adjacent antiforms. Rahaman & Ocan (1988) considered the granulite-facies metamorphism to be pre-Pan African in age. Our reconnaissance work in southern/central Nigeria shows that large domains include low-pressure migmatites and granulite-facies rocks, displaying a recumbent foliation that affected monocyclic metasedimentary sequences. Progressive charnockitization is observed in adjacent granite orthogneisses and Archaean grey gneisses in which partial melting generated mesoperthite–orthopyroxene-bearing leucosomes that merge to form intrusive veins. As in NE Nigeria (the Jos-Bauchi area, Ferré & Caby 2006), such high-grade metamorphic domains are spatially associated with the emplacement of syn-kinematic massifs (granite, monzonite, granodiorite, charnockite). This has been pointed out by Sacchi (1968) who reported the occurrence of dumortierite in peraluminous metapelitic diatexites closely associated with charnockite plutons emplaced within a schist belt in northern Nigeria. All U–Pb zircon ages from Nigerian charnockites are bracketed between 0.64 and 0.58 Ga (Tubosun *et al.* 1984; Dada 1998 and references therein). Hot anhydrous (charnockite) magmas of monzodioritic affinity generated through a high degree of melting of the lower crust (Dada 1998) are characteristic of the entire Nigerian province (Rahaman 1988; Ferré *et al.* 1998). The close association of gabbroic rocks and potassic syenites in SW Nigeria (Rahaman *et al.* 1991) also implies magma extraction from an enriched mantle roughly synchronous with high-temperature metamorphism.

Structural styles in SW Nigeria

In most schist belts the structural style is portrayed by north–south upright folds with low-plunging stretching lineations (Turner 1983), suggesting a general transpressive regime similar to that described in many parts of the Tuareg shield (Caby 1987; Boullier 1982; Black *et al.* 1994). The anatectic basement in such domains was exhumed through elongate domes frequently flanked by north–south-trending syn- to late-metamorphic faults. There is no metamorphic gap or structural break between high-grade gneisses close to granulite facies and adjacent meta-turbidites of the Igarra Schist Belt, but a sharp metamorphic field gradient is observed (Omitogun *et al.* 1991). Other domains, in contrast, are characterized by a recumbent foliation over large areas, both in the southern (Ajibade & Wright 1989; Caby 1989) and northern parts of the shield (Ferré *et al.* 1998; Ferré & Caby 2006). However, the real significance of the recumbent foliation is not clear, since these flat-lying structures formed under low-pressure metamorphism and may have been related to an extensional setting of post-collision gravitational collapse coeval with magma emplacement. Crystalline nappes are observed at mid-crustal levels in the Ife–Ilesha area, where upright folding is recorded (Fig. 5).

Link with the frontal units of the Dahomeyan belt and with the Pan-African suture

The suture zone that can be traced from the south Saharan desert to the Guinean Gulf (Caby 1989; Caby *et al.* in press) is delineated on its eastern side in Benin, Togo and Ghana by meta-gabbroic massifs affected by granulite facies metamorphism of Pan-African age (Attoh 1998; Affaton *et al.* 2000). Based on petrological and geochemical studies, the larger massif represents the granulitized root of a Neoproterozoic magmatic arc (Duclaux *et al.* 2006). It rests with a flat tectonic contact above the Atakora nappe (Caby 1989), which includes eclogite slices (Agbossoumonde *et al.* 2001). Above this, the poorly studied 'Dahomeyan gneisses' were affected by high-pressure, high-temperature Pan-African metamorphism and anatexis; this unit includes both polycyclic grey gneisses of assumed Archaean age similar to those of Ibadan–Ife, and a monocyclic unit that consists of sillimanite quartzite, kyanite-bearing kinzigite and garnet–pyroxene amphibolite possibly derived from high-temperature eclogite. This monometamorphic unit is petrographically similar to the Granja granulites (Caby 1989; see Santos *et al.* 2008). As in Central Ceará, an inverted metamorphism is thus clearly evidenced in an east–west section, with temperature increasing towards the top of the nappe pile (Caby 1989). The Kandi fault is a *c.* 400 m thick steep band of ultramylonites with shallow-plunging stretching lineations, as in southern Hoggar. However, in southern Benin the fault is also delineated by a belt of two-pyroxene charnockitic/mangeritic gneiss. Petrological reconnaissance on these rocks suggests they were derived from syn-kinematic sheeted charnockite plutons emplaced during the early stage of movement of the shear zone.

Discussion and conclusions

Northern Borborema Province and Nigeria are partially underlain by Archaean crust formed mainly by grey gneisses derived from TTG-type igneous protoliths ranging in age from 2.7 to 3.5 Ga. It is not certain if Palaeoproterozoic igneous or tectono-thermal events affected the Archaean crust in Nigeria. Indeed, only a few Palaeoproterozoic ages have been obtained by the Rb–Sr isochron method in Nigeria and Benin, and some of these dates have been proved to be inaccurate by U–Pb zircon ages. In NE Brazil, a significant plutonic event is well documented at *c.* 2.17 Ga (Castro 2004; Fetter *et al.* 2000; Martins *et al.* 1998).

The late Palaeoproterozoic (Statherian) period is marked in NE Brazil and in Nigeria by the emplacement of rift-related anorogenic sub-alkaline magmatism, the ages being bracketed between 1.85 and 1.73 Ga in both continents. This 'Espinhaço'-age magmatism is well developed in the Orós–Jaguaribe palaeo-rift (Cavalcante 1999; Fetter 1999; Sá 1991; Sá *et al.* 1995).

Rock associations of the Igarra Sequence in SW Nigeria are very reminiscent of the Seridó Group, both units being composed mainly of impure sandstones, meta-greywackes, black metapelites and semi-pelites, and some carbonates, the base of both sequences being marked by polygenetic conglomerates. The depositional age of the Seridó Formation is younger than 0.65 Ga (Van Schmus *et al.* 2003). With respect to the tectonic style, many similarities are observed between Africa and South America. Flat-lying foliations formed as a result of piling of large-scale crystalline nappes above low-angle ductile thrusts that were active until the late stages of the collision. The nappe vergence seems to be opposite (northeastward in Nigeria, southeastward in central Ceará). Linear domains of schist belts associated with steep strike-slip continental-scale shear zones occur in both continents, whereas the structure of some domains is characterized by syn-metamorphic upright folding synchronous with emplacement of syn- to late-kinematic plutons preferentially aligned along transpressive belts.

Earlier regional correlations presented by Caby (1989) consider that the aluminous São Joaquim quartzite nappe (see Santos *et al.* 2008) that was transported southwestward and characterized by kyanite–rutile assemblages correlates very well with the Atacora nappe in Benin–Togo. For this reason and according to gravimetric anomalies (Lesquer *et al.* 1981), the prolongation of the major Pan-African suture zone may be buried under the Parnaíba Basin west of the Granja–Médio Coreaú domain (Caby 1989; Monié *et al.* 1997). It is thus speculated that the suture bends parallel to the Brazilian portion of the West African Craton (São Luis Craton). Its northward prolongation can be also inferred west of the Rockelides, where Pan-African granulites similar to those of Granja and Benin ('Dahomeyan') have been recognized (Delor *et al.* 2001).

The main query is that the amount of lateral displacements along the Kandi/Sobral –Pedro II shear zone is impossible to estimate. The lithospheric nature of this first order fault in Hoggar has been discussed by Caby (2003) who proposed that the early Pan-African eclogites from the Latea terrane necessarily root to the west, i.e. within the 4°50 shear zone. This author also proposed that the two shear zones that delimit the granulite microcraton in Hoggar (the In Ouzzal and Iforas microcontinents) also represent two cryptic sutures, since they have controlled the exhumation at 0.61 Ga of slices of blueschist and kyanite-bearing eclogite, the latter equilibrated at about 18 kbar (Caby & Monié 2003). Note that the *c.* 400 km wide domain of active margin assemblages of western Hoggar, which includes large volumes of juvenile crust, disappears to the south of the Saharan region, speculatively due to their subduction, in agreement with the terrane model of Black *et al.* (1994). Therefore it is proposed that the Kandi/Sobral Pedro II shear zone should be also considered a cryptic suture along which large amounts of lithosphere might have been consumed before the onset of oblique collision. In this context, the allochthoneity of the Tamboril–Santa Quiteria arc complex requires further confirmation.

The geological background of the Northern Borborema and Nigerian provinces displays some structural and lithological similarities that should help to elaborate more accurate pre-drift evolution of West Gondwana. Some of the apparent differences may relate to the different crustal levels observed on both sides of the Atlantic Ocean, related in the first instance to variable rates of exhumation of the mountain roots at the time of the Cambrian molasse stage.

This research is supported by CAPES/PROCAD grant nº 0015/05-9, and CNPq/Institutos do Milênio grant 420222/05-7. We would like to thank Hartwig Frimmel and two anonymous referees for their helpful reviews and Bob Pankhurst for improvement of the English.

References

ABREU, F. A. M., HASUI, Y. & GORAYEB, P. S. S. 1993. Grábens eopaleozóicos do oeste cearense. Considerações sobre as seqüências litoestratigráficas. *In*: *15 Simposio Geologico, NE, Natal, Anais*, 29–31.

AFFATON, P., KRÖNER, A. & SEDDOH, K. F. 2000. Pan-African granulite formation in the Kabyé Massif of northern Togo (West Africa): Pb-Pb zircon ages. *International Journal Earth Sciences*, **88**, 778–790.

AGBOSSOUMONDÉ, Y., MÉNOT, R. P. & GUILLOT, S. 2001. Metamorphic evolution of Neoproterozoic eclogites from south-Togo (West Africa). Geodynamic implications. *Journal of African Earth Sciences*, **33**, 227–244.

AJIBADE, A. C. & WRIGHT, J. B. 1989. The Togo Benin Nigeria Shield: evidence of crustal aggregation in the Pan-African Belt. *Tectonophysics*, **165**, 125–129.

AJIBADE, A. C., WOAKES, M. & RAHAMAN, M. A. 1987. Proterozoic crustal development in the Pan-African regime of Nigeria. *In*: KRÖNER, A. (ed.) *Proterozoic Lithospheric Evolution*. American Geophysical Union, Geodynamic Series, **17**, 259–271.

AJIBADE, A. C., RAHAMAN, M. A. & OGEZI, A. E. O. 1988. The Precambrian of Nigeria: a geochronological summary. *In*: OLUYIDE, P. O. *ET AL*. (eds) *Precambrian Geology of Nigeria*. Geological Survey of Nigeria. 313–324.

ALKMIN, F. F., BRITO NEVES, B. B. & CASTRO ALVES, J. A. 1993. Arcabouço tectônico do Cráton do São Francisco. Uma revisão. *In*: DOMINGUEZ, J. M. L. & MISI, A. (eds) *O Cráton do São Francisco*. Sociedade Brasileira de Geologia/Superintendência de Geologia e Mineração, Salvador, 45–62.

ALMEIDA, A. R. 1998. *O magmatismo Parapuí e a evolução geológica da Bacia de Jaibaras*. Titular Professor Thesis, DEGEO, Universidade Federal do Ceará.

ALMEIDA, F. F. M. 1967. *Origem e evolução da plataforma brasileira. Rio de Janeiro*. Departamento Nacional da Produção Mineral/Divisão de Minas e Geologia, Boletim **241**.

ALMEIDA, F. F. M. 1969. Diferenciação tectônica da plataforma brasileira. *In*: *Congresso Brásileiro Geologia 23, Salvador, Anais*. Sociedade Brasileira de Geologia, 29–40.

ALMEIDA, F. F. M., HASUI, Y., BRITO NEVES, B. B. & FUCK, R. A. 1981. Brasilian Structural Provinces: an introduction. *Earth-Science Reviews*, **17**, 1–29.

ARCHANJO, C. J. 1993. *Fabrique des plutons granitiques et déformatrion crustale du Nord-Est du Brésil*. Doctoral thesis, Université de Toulouse III, France.

ARCHANJO, C. J. & BOUCHEZ, J. L. 1991. Le Seridó, une chaine transpressive dextre au Proterozoique Supérieur du nord-est du Brésil. *Bulletin de la Societé Géologique de France*, **4**, 637–647.

ARCHANJO, C. J. & LEGRAND, J. M. 1997. Pre-brasiliano orogenic evolution in the Seridó Belt, NE Brazil: conflicting geochronological and structural data (comentários). *Revista Brasileira do Geociências*, **27**, 309–310.

ARTHAUD, M. 2005. *Tectônica de nappes e espessamento crustal brasiliano na região de Madalena? Boa Viagem, Ceará Central*. Exame de Qualificação, Instituto de Geociências, Universidade de Brasilia.

ARTHAUD, M. H., VASCONCELOS, A. M. *ET AL*. 1998. Main Structural Features of Precambrian Domains from Ceará (NE Brazil). *In*: *International Conference on Precambrian and Cratons Tectonics, Ouro Preto, Abstracts*, 84–85.

ATTOH, K. 1998. High-pressure granulite facies metamorphism in the Pan-African Dahomeyide orogen, West Africa. *Journal of Geology*, **106**, 236–246.

BERTRAND, J. M. & JARDIM DE SÁ, E. F. 1990. Where are the Eburnian–Transamazonian collisional belts? *Canadian Journal of Earth Sciences*, **27**, 1382–1393.

BLACK, R., LATOUCHE, L., LIÉGEOIS, J. P., CABY, R. & BERTRAND, J. M. 1994. Pan African displaced terranes in the Tuareg shield (central Sahara). *Geology*, **22**, 641–644.

BOULLIER, A. M. 1982. *Etude structurale du centre de l'Adrar de Iforas (Mali). Mylonites et tectogénèse*. Thesis, Institut National Polytechnique de Loraine, University of Nancy, France.

BOUSSARI, M. W. & ROLLET, M. 1974. Découverte d'un bassin volcano-sedimentaire dans la région Centre-Ouest du Dahomey. *Comptes Rendus de l' Académie des Sciences, Paris*, **279**, 29–32.

BRITO NEVES, B. B. 1975. *Regionalização Geotectônica do Pré-Cambriano Nordestino*. Doctoral thesis, Instituto de Geociências, Universidade de São Paulo.

BRITO NEVES, B. B. 1998. The Cambro-Ordovician of the Borborema Province. *Boletim Instituto de Geociências, Universidade de São Paulo, Brazil, Série Científica*, **29**, 175–193.

BRITO NEVES, B. B. 2002. Main stages of the development of the sedimentary basins of South América and their relationship with tectonics of supercontinents. *Gondwana Research*, **5**, 175–196.

BRITO NEVES, B.B., VAN SCHMUS, W. R. & FETTER, A. 2002. North-western Africa-North-eastern Brazil. Major tectonic links and correlation problems. *Journal of African Earth Sciences*, **34**, 275–278.

BRITO NEVES, B. B., PASSARELI, C. R., BASEI, A. A. S. & SANTOS, E. J. 2003. U–Pb age of some classic granites of the Borborema Province. *In*: *IV South American Symposium of Isotope Geology, Salvador, Brazil, Short Papers*, 158–159.

BRUGUIER, O., DADA, S. & LANCELOT, J. R. 1994. Early Archean component (>3.5 Ga) within a 3.05 Ga orthogneiss from northern Nigeria: U–Pb zircon evidence. *Earth and Planetary Science Letters*, **125**, 89–103.

CABY, R. 1987. The Pan-African belt of West Africa from the Sahara Desert to the Gulf of Benin. *In*: SCHAER, J. P. & RODGER, J. (eds) *Anatomy of Mountain*

Ranges. Princeton University Press, Princeton, 129–170.

CABY, R. 1989. Precambrian terranes of Benin–Nigeria and northeast Brazil and the Late Proterozoic south Atlantic fit. *In*: DALLMEYER, R. D. (ed.) *Terranes in the Circum-Atlantic Paleozoic Orogens*. Geological Society of America, Special Papers, **230**, 145–158.

CABY, R. 2003. Terrane assembly and geodynamic evolution of Central-Western Hoggar: a synthesis. *Journal of African Earth Sciences*, **37**, 133–159.

CABY, R. & ANDREOPOULOS-RENAUD, U. 1983. Age à 1800 Ma du magmatisme subalcalin associé aux méta-sédiments monocycliques dans la chaine Pan-Africaine du Nord du Sahara Central. *Journal of African Earth Sciences*, **1**, 193–197.

CABY, R. & ANDREOPOULOS-RENAUD, U. 1989. Age U–Pb à 620 Ma d'un pluton synorogénique de l'Adrar des Iforas (Mali). Conséquences pour l'age de la phase majeure de l'orogène pan-africain. *Comptes Rendus de l' Academie des Sciences, Paris*, **308**, 307–314.

CABY, R. & ARTHAUD, M. H. 1986. Major Precambrian nappes of the Brazilian Belt. Ceará, Northeast Brazil. *Geology*, **14**, 871–874.

CABY, R. & BOESSÉ, J. M. 2001. Pan-African nappe system in southwest Nigeria: the Ife-Ilesha schist belt. *Journal of African Earth Sciences*, **33**, 211–225.

CABY, R. & MONIÉ, P. 2003. Neoproterozoic subductions and differential exhumation of western Hoggar (southwest Algeria): new structural, petrological and geochronological evidence. *Journal of African Earth Sciences*, **37**, 269–293.

CABY, R., SIAL, A., ARTHAUD, M. H. & VAUCHEZ, A. 1990. Crustal evolution and the Brasiliano Orogeny in Northeast Brazil. *In*: DALLMEYER, R. D. & LECORCHÉ, J. P. (eds) *The West African Orogens and Circum-Atlantic Correlatives*. Springer Verlag, New York, 373–397.

CABY, R., ARTHAUD, M. H. & ARCHANJO, C. J. 1995. Lithostratigraphy and petrostructural characterization of supracrustal units in the Brasiliano Belt of Northeast Brazil: geodynamic implications. *Journal of South American Earth Sciences*, **8**, 235–246.

CABY, R., BUSCAIL, F., DEMBÉLÉ, D., DIAKITÉ, S., SACKO, S. & BAL, M. 2008. Neoproterozoic garnet-glaucophanites and eclogites: new insights for subduction metamorphism of the Gourma fold- and- thrust belt (eastern Mali). *In*: ENNIH, N. & LIÉGEOIS, J. P. (eds) *Boundaries of the West African Craton*. Geological Society, London, Special Publications, in press.

CASTAING, C., FEYBESSE, D., THIÉBLEMONT, D, TRIBOULET, C. & CHÉVREMONT, P. 1994. Paleogeographical reconstructions of the Pan-African/Brasiliano orogen: closure of an oceanic domain or intracontinental convergence between major blocks? *Precambrian Research*, **69**, 327–344.

CASTRO, N. A. 2004. *Evolução geológica proterozóica da região entre Madalena e Taperuaba, domínio tectônico Ceará Central (Província Borborema)*. Doctoral thesis, Universidade de São Paulo.

CASTRO, N. A., BASEI, M. A. S. & SATO, K. 2003. Nd data from Central Ceara Domain: Archean relicts, Paleoproterozoic record and Neoproterozoic crustal reworking. *In*: *IV South American Symposium on Isotope Geology, Salvador, Brazil, Short Papers*, 161–163.

CAVALCANTE, J. C. 1999. *Limites e evolução do Sistema Jaguaribeano, Província Borborema, Nordeste do Brasil*. Masters dissertation, Universidade Federal do Rio Grande do Norte, Brazil.

CAVALCANTE, J. C., FERREIRA, C. A. *ET AL*. 1983. *Mapa Geológico do Estado do Ceará – Escala 1:500.000*. Ministério das Minas e Energia/Companhia de Pesquisa de Recursos Minerais, Fortaleza.

CAVALCANTE, J. C., VASCONCELOS, A. M. *ET AL*. 2003. *Mapa Geológico do Estado do Ceará – Escala 1:500.000*. Ministério das Minas e Energia/Companhia de Pesquisa de Recursos Minerais, Fortaleza.

CORSINI, M., LAMBERT DE FIGUEIREDO, L., CABY, R., FÉRAUD, G., RUFFET, G. & VAUCHEZ, A. 1998. Thermal history of the Pan African/Brasiliano Borborema Province of northeast Brazil deduced from $^{40}Ar/^{39}Ar$ analysis. *Tectonophysics*, **285**, 103–117.

COSTA, M. J., FRANÇA, J. B., LINS, C. A. C., BACCHIEGA, I. F., HABEKOST, C. R. & CRUZ, W. B. 1979. *Geologia da Bacia Jaibaras: Ceará, Piauí e Maranhão – Projeto Jaibaras*. Departamento Nacional da Produção Mineral/Ministério das Minas e Energia, Geologia 14, Seção Geologia Básica 11.

DADA, S. S. 1998. Crust-forming ages and Proterozoic crustal evolution in Nigeria: a reappraisal of current interpretations. *Precambrian Research*, **87**, 65–74.

DADA, S. S., TUBOSUN, I. A., LANCELOT, J. R. & LAR, A. U. 1993. Late Archean U–Pb age for the reactivated basement of Northeastern Nigeria. *Journal of African Earth Sciences*, **16**, 405–412.

DANNI, J. C. M. 1972. Geologia da porção sul do Grupo Jaibaras-Ceará. *Revista Brásileira de Geociências*, **2**, 85–87.

DANTAS, E. L., VAN SCHMUS, W. R. *ET AL*. 2004. The 3.4–3.5 Ga São José do Campestre massif, NE Brazil: remnants of the oldest crust in South America. *Precambrian Research*, **130**, 113–137.

DELOR, C., LAFON, J. M., MILÉSI, J. P. & FANNING, C. M. 2001. First evidence of 560–575 Ma granulites and syn-tectonic magmatism in the Rockelides belt: geology, geochronology and geodynamic implications. *In*: *19th Colloquium of African Geology, El Jadida, Morocco, Abstracts*, 58–59.

DUCLAUX, D., MÉNOT, P., AGBOSSOUMONDÉ, Y. & HILAIRET, N. 2006. The mafic layered complex of the Kabyé massif (north Togo and north Benin): evidence of a Pan-African granulitic continental root. *Precambrian Research*, **151**, 101–118.

FERRÉ, E. & CABY, R. 2006. Granulite facies metamorphism and charnockite plutonism: examples from the Neoproterozoic Belt of northern Nigeria. *Proceedings of the Geologists' Association*, **118**, 1–8.

FERRÉ, E. C., CABY, C., MONIÉ, P. & PEUCAT, J. J. 1998. Pan-African, post-collisional, A2 type alkaline granite and monzonite plutons of Eastern Nigeria. *Lithos*, **45**, 225–279.

FETTER, A. H. 1999. *U–Pb and Sm–Nd geochronological constraints on the crustal framework and geologic history of Ceará State, NW Borborema Province, NE Brazil: implications for the assembly of Gondwana*. PhD thesis, Kansas University, USA.

FETTER, A. H., VAN SCHMUS, W. R., SANTOS, T. J. S., ARTHAUD, M. & NOGUEIRA NETO, J. A. 2000. U–Pb and Sm–Nd geochronological constrains on the crustal evolution and basement architecture of Ceará State, NW Borborema Province, NE Brazil: implications for the existence of the paleoproterozóico supercontinent Atlantica. *Revista Brasileira de Geociências*, **30**, 102–106.

FETTER, A. H., SANTOS, T. J. S. *ET AL.* 2003. Evidence for Neoproterozoic continental arc magmatism in the Santa Quitéria Batholith of Ceará State, NW Borborema Province, NE Brazil: implications for the assembly of West Gondwana. *Gondwana Research*, **6**, 265–273.

FITCHES, W. R., AJIBADE, A. C., EGBUNIWE, I. G., HOLT, R. W. & WRIGHT, J. B. 1985. Late Proterozoic schist belts and plutonism in Northwest Nigeria. *Journal of the Geological Society, London*, **142**, 319–337.

GARCIA, M. G. M. & ARTHAUD, M. H. 2004. Caracterização de trajetória P–T em nappes brasilianas: região de Boa Viagem/Madalena – Ceará Central (NE Brasil). *Revista de Geologia, Universidade Federal do Ceará*, **17**, 173–191.

GORAYEB, P. S. S., ABREU, F. A. M., CORREA, J. A. M. & MOURA, C. A. V. 1988. Relações Estratigráficas entre o Granito Meruoca e a Sequência Ubajara-Jaibaras. *In*: *Congresso Brasileiro de Geologia 35, Belém, Anais* **6**. Sociedade Brasileira de Geologia, 2678–2688.

GRANT, N. K. 1978. Structural distinction between a metasedimentary cover and underlying basement in the 600 m.y. -old Pan-African domain of northwestern Nigeria, West Africa. *Geological Society of America, Bulletin*, **89**, 50–58.

HACKSPACHER, P. C., VAN SCHMUS, W. R. & DANTAS, E. L. 1990. Um embasamento Transamazônico na Província Borborema. *In*: *Congresso Brasileiro de Geologia 36, Natal, Anais* **6**. Sociedade Brasileira de Geologia, 2683–2696.

IGE, O. A., OKRUSCH, M., SCHÜSSLER, U., SCHMÄDICKE, E. & COOK, N. J. 1998. The metamorphosed mafic-ultramafic complex of Mokuro, Ilesha schist belt, southwestern Nigeria. *Journal of African Earth Sciences*, **26**, 593–618.

JARDIM DE SÁ, E. F. 1984. A evolução proterozóica da Província Borborema. *In*: *12 Simposio de Geologia do Nordeste, Anais*, 33–45.

JARDIM DE SÁ, E. F. 1994. *A Faixa Seridó (Província Borborema, NE do Brasil) e o seu significado geodinâmico na Cadeia Brasiliana/Pan Africana*. Doctoral thesis, Instituto de Geociências, Universidade de Brasília.

JARDIM DE SÁ, E. F., FUCK, R. A. *ET AL.* 1997. Pré-brasiliano orogenic evolution in the Seridó Belt, NE Brazil: conflicting geochronological and structural data (reply). *Revista Brasileira de Geociências*, **25**, 307–314.

KEGEL, W., SCORZA, E. T. & COELHO, F. C. T. 1958. *Estudos Geológicos do Norte do Ceará*. Ministério das Minas e Energia/Departamento Nacional da Produção Mineral-Divisão de Geologia e Minas, Rio de Janeiro, Boletim, **134**.

LESQUER, A., BELTRÃO, J. F. & DE ABREU, F. A. M. 1981. Proterozoic links between northeastern Brazil and West Africa: a plate tectonic model based on gravity data. *Tectonophysics*, **110**, 9–26.

LIMA, E. A. M. 1992. Metamorphic conditions in the Seridó region of North-eastern Brazil during the Brasilian Cycle (Late Proterozoic). *Journal of African Earth Sciences*, **5**, 265–273.

MABESOONE, J. M., BEURLEN, K. *ET AL.* 1971. Geologia da Bacia de Jaibaras, Ceará. *Estudos e Pesquisas, Universidade Federal de Pernambuco/Instituto de Geociências, Recife*, **2**, 1–116.

MARTINS, G., OLIVEIRA, E. P., SOUZA FILHO, C. R. & LAFON, J. M. 1998. Geochemistry and geochronology of the Algodões sequence, Ceará, NE Brazil: a paleoproterozoic magmatic arc in the Central Ceará domain of the Borborema Province? *In*: *Congresso Brasileiro de Geologia 40, Anais*. Sociedade Brasileira de Geologia, 28.

MELLO, Z. F. 1978. Evoluções finais do Ciclo Geotectônico Brasiliano no Nordeste Oriental. *In*: *Congresso Brasileiro de Geologia 30*, Recife, *Anais*, **6**. Sociedade Brasileira de Geologia, 2438–2450.

MENDONÇA, J. C. G. S. & BRAGA, A. P. G. 1987. As faixas vulcano-sedimentares de Orós e Jaguaribe: um greenstone belt? *Revista Brasileira de Geociências*, **17**, 225–341.

MONT'ALVERNE, A. A. F., JARDIM DE SÁ, E. F., DERZE, G. R., DANTAS, J. R. A. & VENTURA, P. E. O. 1998. *Mapa Geológico do Rio Grande do Norte – 1:500.000*. Departamento Nacional da Produção Mineral/Universidade Federal do Rio Grande do Norte/PETROBRÁS/CRM.

MONIÉ, P., CABY, R. & ARTHAUD, M. 1997. The Neoproterozoic orogeny in northeast Brasil: 40Ar/39Ar ages and petrostructural data from Ceará. *Precambrian Research*, **81**, 241–264.

MULLAN, H. S. 1979. Structural distinction between a metasedimentary cover and a underlying basement in the 600 m.y. old Pan-African domain of Northwestern Nigeria, West Africa. *Geological Society of America, Bulletin*, **90**, 983–984.

NASCIMENTO, D. A. & GAVA, A. 1979. Novas Considerações sobre a Estratigrafia da Bacia Jaibaras. *In*: *9 Simposio de Geologia do Nordeste, Natal, Atas*, 9–29.

NOGUEIRA, J. F. 2004. *Estrutura, geocronologia e alojamento dos batólitos de Quixadá, Quixeramobim e Senador Pompeu – Ceará Central*. Doctoral thesis, Instituto de Geociências e Ciências exatas -Universidade Estadual Paulista, Rio Claro.

NOVAIS, F. R. G., BRITO NEVES, B. B. & KAWASHITA, K. 1979. Reconhecimento cronoestratigráfico da região nordeste do Estado do Ceará. *In*: *7 Simposio de Geologia do Nordeste, Natal, Atas*, 93–110.

ODEYEMI, I. B. 1982. Lithostratigraphy and structural relationships of the Upper Precambrian metasediments in Igarra area, southwest Nigeria. *In*: OLUYIDE, P. O. *ET AL.* (eds) *Precambrian Geology of Nigeria*, Geological Survey of Nigeria, 111–125.

OLIVEIRA, D. C. 2000. Stratigraphic interplays between igneous and sedimentary events in the early Paleozoic Jaibaras Trough (Northeast Brazil). *Revista Brasileira de Geociências*, **30**, 423–427.

OLIVEIRA, D. C. 2001. Reavaliação da EvoluçãoTectono-Magmática do Graben de Jaibaras (Nordeste do Brasil). *Acta Geológica Hispanica*, **36**, 53–95.

OLIVEIRA, D. C. & MOHRIAK, W. U. 2003. Jaibaras trough: an important element in the early tectonic evolution of the Parnaiba interior sag basin, Northern Brazil. *Marine and Petroleum Geology*, **20**, 351–383.

OMITOGUN, A. O., CABY, R., DEBAT, P. & MERCIER, A. 1991. Le métamorphisme pan-africain de la partie centrale de la zone mobile du Nigéria (Ceinture schisteuse d'Igarra s.l.). *Comptes Rendus de l'Academie des Sciences, Paris*, **313**, 1417–1423.

ONYEAGOCHA, A. C. & EKWUEME, B. N. 1990. Temperature-pressure distribution patterns in metamorphosed rocks of the Nigerian Basement Complex - a preliminary analysis. *Journal of African Earth Sciences*, **11**, 83–93.

PARENTE, C. V. 1995. *Géologie et pétrographie d'une plate-forme à évaporites et magnésite d'âge protérozoique (2 Ga): le cadre géotectonique initial de la ceinture mobile Orós dans la région d'Alencar (Ceará-Brésil)*. Doctoral thesis, Université de Nantes, France.

PARENTE, C. V. & ARTHAUD, M. 1995. O Sistema Orós-Jaguaribe no Ceará, NE do Brasil. *Revista Brasileira de Geociências*, **25**, 297–306.

PARENTE, C. V. & FUCK, R. A. 1987. Geologia da ocorrência de cobre de Mandacaru (PI). *Revista Brasileira Geociências*, **17**, 21–32.

PARENTE, C. V., ARTHAUD, M. H. & OLIVEIRA, F. V. C. 1990. Geologia da bacia eo-paleozóica de Iara, CE. *Revista de Geologia – Universidade Federal do Ceará*, **3**, 29–39.

PARENTE, C. V., RONCHI, L. H. *ET AL*. 2004*a*. Geology and geochemistry of paleoproterozoic magnesite deposits (~1.8 Ga), State of Ceará, Northeastern Brazil. *Carbonates and Evaporites*, **19**, 28–50.

PARENTE, C. V., SILVA FILHO, W. F. & ALMEIDA, A. R. 2004*b*. *Bacias do Estágio da Transição do Domínio Setentrional da Província Borborema*. *In*: MANTESSO-NETO, V., BARTORELLI, A., CARNEIRO, C. D. R. & BRITO NEVES, B. B. (eds) *Geologia do Continente Sul-Americano: Evolução da obra de Fernando Flávio Marques de Almeida*. Editora Beca, São Paulo, 525–536.

QUADROS, M. L. E. S. & ABREU, F. A. M. 1995. Bacia de Jaibaras–NW do Ceará: Considerações sobre a evolução tectono-sedimentar. *In*: *5 Simposio Nacional de Estudos Tectônicos Gramado, Brazil, Boletim de Resumos Expandidos*. Sociedade Brasileira de Geologia, 307–309.

QUADROS, M. L. E. S., ABREU, F. A. M. & GORAYEB, P. S. S. 1994. Considerações sobre os ambientes deposicionais das formações Pacujá e Aprazível, Bacia de Jaibaras–NW do Ceará. *In*: *Congresso Brasileiro de Geologia 38, Camboriú, Anais*, **3**. Sociedade Brasileira de Geologia, 240–242.

RAHAMAN, M. A. 1976. Progressive polyphase métamorphisme in pelitic schists around Aiyetoro, Oyo State, Nigeria. *Journal of Mining Geology*, **12** (2), 33–44.

RAHAMAN, M. A. 1988. Recent advances in the study of the basement complex of Nigeria. *In*: OLUYIDE, P. O. *ET AL*. (eds) *Precambrian Geology of Nigeria*. Geological Survey of Nigeria, 11–43.

RAHAMAN, M. A. & OCAN, O. 1988. The nature of granulite facies metamorphism in Ikare area, Southwestern Nigeria. *In*: OLUYIDE, P. O. *ET AL*. (eds) *Precambrian Geology of Nigeria*. Geological Survey of Nigeria, 157–163.

RAHAMAN, M. A., TUBOSUN, I. A. & LANCELOT, J. R. 1991. U–Pb geochronology of potassic syenites from SW Nigeria and the timing of deformation events during the Pan-African orogeny. *Journal of African Earth Sciences*, **13**, 387–395.

SÁ, J. M. 1991. *Évolution géodynamique de la ceinture protérozoique d'Orós, Nord-Est du Brésil*. Doctoral thesis, Université de Nancy, France.

SÁ, J. M. & BERTRAND, J. M. 1992: Transpressão dextral no sudeste do Estado do Ceará. Província Borborema. *In*: *Congresso Brasileiro de Geologia 37, São Paulo, Boletim de Resumenes Expandido*. Sociedade Brasileira de Geologia, 368–370.

SÁ, J. M. & LEGRAND, J. M. 1983. Superposição de fases metamórficas na região da Serra do Chico, Lages, RN. *Revista Ciências da Terra, Sociedade Brasileira de Geologia*, **7**, 12–15.

SÁ, J. M., MCREATH, I. & LETERRIER, J. 1995. Petrology, geochemistry and geodynamic setting of Proterozoic igneous suites of Orós fold belt (Borborema Province, Northeast Brazil). *Journal of South American Earth Sciences*, **8**, 299–314.

SACCHI, R. 1968. The geology of the region around Bena in Northern Nigeria. *Memorie degli Istituti di Geologia e Mineralogia dell'Università di Padova, Italy*, **26**, 1–47.

SANTOS, T. J. S., SANTOS, A. A., DANTAS, E. L., FUCK, R. A. & PARENTE, C. V. 2003. Nd isotopes and the provenance of metassediments of the Itataia Group, Northwest Borborema Province, NE Brazil. *In*: *IV South American Symposium of Isotope Geology, Salvador, Brazil, Short Papers*, 286–289.

SANTOS, T. J. S., DANTAS, E. L., ARTHAUD, M. H., FUCK, R. A., PIMENTEL, M. M. & FETTER, A. H. 2004. Evidências de crosta juvenil neoproterozóica no Ceará. *In*: *Congresso Brasileiro de Geologia 42, Araxá, Anais Digitais*. Sociedade Brasileira de Geologia, 1175.

SANTOS, T. J. S., FETTER, A. H. & NOGUEIRA NETO, J. A. 2008. Comparisons between the northwestern Borborema Province, NE Brazil, and the southwestern Pharusian Dahomey Belt, SW Central Africa. *In*: PANKHURST, R. J., TROUW, R. A. J., BRITO NEVES, B. B. & DE WIT, M. J. (eds) *West Gondwana: pre-Cenozoic Correlations Across the South Atlantic Region*. Geological Society, London, Special Publication, **294**, 101–119.

SCHOBBENHAUS, C. & CAMPOS, D. A. 1984. A evolução da Plata-Forma Sul-Americana no Brasil e suas principais concentrações minerais. *In*: SCHOBBENHAUS, C., CAMPOS, D. A., DERZE, G. R. & ASMUS, H. E. (eds) *Geologia do Brasil*. Ministério das Minas e Energia/Departamento Nacional da Produção Mineral, Brasília, 9–53.

SIAL, A. N. & LONG, L. E. 1987. Rb–Sr and oxygen isotope study of the Meruoca and Mucambo Granites, Northeastern Brazil. *In*: *Fourth International Colloquium of Geochronology, Cosmochemistry and Isotope Geology (USGS Report 78–701)*, 398–400.

SILVA, L. C., ARMSTRONG, R. *ET AL.* 2002. Reavaliação da evolução geológica em terrenos pré-cambrianos brasileiros com base em novos dados U-Pb SHRIMP. Parte III: província Borborema, Mantiqueira Meridional e Rio Negro-Jurena. *Revista Brasileira de Geociências*, **32**, 529–544.

TAVARES, S. S. JR., GORAYEB, P. S. S. & LAFON, J. M. 1990. Petrografia e geocronologia Rb/Sr do feixe de diques da borda oeste do Granito de Meruoca (CE). *In*: *Congresso Brasileiro de Geologia 36, Natal, Anais*. Sociedade Brasileira de Geologia, 337–338.

TEIXEIRA, M. L. A. 2005. *Integração de dados aerogeofísicos, geológicos e isotópicos do limite norte do Complexo Tamboril-Santa Quitéria – CE (Província Borborema)*. MSc thesis, Instituto de Geociências, Brasília.

TUBOSUN, I. A., LANCELOT, J. R., RAHAMAN, M. A. & OCAN, O. O. 1984. U–Pb Pan-African ages of two charnockite–granite associations from SW Nigeria. *Contributions to Mineralogy and Petrology*, **88**, 188–195.

TURNER, D. C. 1983. Upper Proterozoic Schist Belts in the Nigerian sector of the Panafrican province of West Africa. *Precambrian Research*, **21**, 55–79.

VAN SCHMUS, W. R., BRITO NEVES, B. B., HACKSPACHER, P. & BABINSKI, M. 1995. U/Pb and Sm/Nd geochronologic studies of Eastern Borborema Province, northeast Brazil: initial conclusions. *Journal of South American Earth Sciences*, **8**, 267–288.

VAN SCHMUS, W. R., BRITO NEVES, B. B., HACKSPACHER, P., BABINSKI, M., FETTER, A. H. & DANTAS, E. L. 1997. Application of U–Pb and Sm–Nd geochronology to understanding the geotectonic history of the Borborema Province, NE Brazil and its implications for the evolution of West Gondwana. *In*: *South American Symposium on Isotope Geology, Extended Abstracts*, 27–29.

VAN SCHMUS, W. R., BRITO NEVES, B. B. *ET AL.* 1998. The Borborema Province: a collage of polycyclic domains in northeast Brazil. *In*: *International Conference on Precambrian Craton Tectonics, Ouro Preto, Brazil, Abstracts*, 80–83.

VAN SCHMUS, W. R., BRITO NEVES, B. B., HACKSPACHER, P. C., WILLIAMS, L. S. & FETTER, A. H. 2000. The Seridó Group, NE Brazil: a late Neoproterozóic (650 ma), pre-collisional, Brasiliano flysch basin? *In*: *31st International Geological Congress, Rio de Janeiro, Abstract volume (CD-ROM)*. Sociedade Brasileira de Geologia.

VAN SCHMUS, W. R., BRITO NEVES, B. B. *ET AL.* 2003. The Seridó Group of NE Brazil, a late pre- to syn-collisional basin in West Gondwana: insights from SHRIMP U–Pb detrital zircon ages and Sm-Nd crustal residence (T_{DM}) ages. *Precambrian Research*, **127**, 287–327.

VASCONCELOS, A. M., PRADO, F. D. *ET AL.* 1998. *Folha Iguatu (Folha SB.24-Y-B) Estado do Ceará. Escala 1:250.000.* Companhia de Pesquisa de Recursos Minerais/Divisão de Editoração Geral/Departamento de Apoio, Brasília.

VAUCHEZ, A., NEVES, S. *ET AL.* 1995. The Borborema shear zone system, NE Brazil. *Journal of South American Earth Sciences*, **8**, 247–266.

Proterozoic links between the Borborema Province, NE Brazil, and the Central African Fold Belt

W. R. VAN SCHMUS[1], E. P. OLIVEIRA[2], A. F. DA SILVA FILHO[3], S. F. TOTEU[4], J. PENAYE[4] & I. P. GUIMARÃES[3]

[1]*Department of Geology, University of Kansas, Lawrence, Kansas, 66045 USA (e-mail: rvschmus@ku.edu)*

[2]*Instituto Geociências, UNICAMP, Campinas, 13083-970 Brazil*

[3]*Departamento de Geologia, Universidade Federal de Pernambuco, 50739 Recife, Brazil*

[4]*Centre de Recherches Géologiques et Minières, BP 333, Garoua, Cameroon*

Abstract: The Congo (CC) and the São Francisco (SFC) cratons were joined at about 2.05 Ga; northern parts of Palaeoproterozoic basement subsequently underwent extension at about 1 Ga, forming intracratonic basins. Neoproterozoic metasedimentary rocks in these basins yield detrital zircons as young as 630 Ma. The Brasiliano and Pan-African (*c.* 620–580 Ma) assembly of West Gondwana extensively altered this system. The Sergipano domain occurs north of the SFC, and the comparable Yaoundé domain occurs north of the CC. Crust north of the Sergipano domain comprises the Pernambuco–Alagoas (PEAL) domain. The NE–SW-striking Tchollíré–Banyo fault in Cameroon may extend southwestwards between the PEAL and Sergipano domains, defining northern limits of abundant SFC/CC basement. The Adamawa–Yadé domain in Africa does not appear to extend into Brazil. The Transverse domain of Brazil is a collage of Palaeoproterozoic crustal blocks, the 1.0 Ga Cariris Velhos orogen (CVO), late Neoproterozoic basins, and Brasiliano granites. The CVO extends ENE for more than 700 km in Brazil, but eastern continuation into Africa has not been identified. North of the Transverse domain contiguous *c.* 2.15 Ga gneisses comprise basement of Rio Grande do Norte and Ceará domains, which continue eastwards into western Nigeria and western Sahara.

This report presents a summary of continuing studies in Brasiliano and Pan-African domains of the Borborema Province in NE Brazil and of the Central African Fold Belt in west-central Africa (Fig. 1). It has long been recognized that there is strong geological correlation between NE Brazil and west-central Africa (e.g., Caby 1989; Castaing *et al.* 1994; Trompette 1997; Neves 2003), but detailed understanding of the Proterozoic history of the association is far from complete (see Brito Neves *et al.* 2002). We believe that the overall relationships are consistent with a model in which late Mesoproterozoic to early Neoproterozoic break-up of a Palaeoproterozoic supercontinent (e.g., Atlantica, Rogers 1996) created a large region between the Congo–São Francisco and West African–Amazonian cratons, consisting of extensional basins floored by Palaeoproterozoic crust, local basins approaching small oceans, and a larger ocean between the northern edge of extensional crust and the West African–Amazonian craton. At present, there is no clear link between this tectonism and formation or break-up of Rodinia (see Kröner & Cordani 2003). During the middle to late Neoproterozoic the major cratons converged, forming juvenile oceanic terranes and major collisional belts. The convergent phase culminated with final assembly of West Gondwana, followed by post-collisional tectonic adjustments, mainly post-tectonic magmatism, and transcurrent faulting. *Note*: In this paper we will use 'craton' to refer to relatively rigid, large continental masses against which major Pan-African and Brasiliano orogenic domains formed about 600 Ma and which were not significantly affected internally by that tectonism and metamorphism (in this case, since 2.0 Ga).

Several types of data are crucial for developing a stratigraphic and tectonic history for the regions concerned. Traditional K–Ar and Rb–Sr results have been available for four decades, and these quickly established the major craton–orogenic belt architecture of Africa and South America. Over the past 15 years U–Pb geochronology of zircon and other minerals, Sm–Nd model crust-formation ages (T_{DM}), and Ar/Ar thermochronology have provided major refinements and, in some cases, significant reinterpretations. Most of the recent and current U–Pb data for primary ages or major secondary crystallization events are based

From: PANKHURST, R. J., TROUW, R. A. J., BRITO NEVES, B. B. & DE WIT, M. J. (eds) *West Gondwana: Pre-Cenozoic Correlations Across the South Atlantic Region.* Geological Society, London, Special Publications, **294**, 69–99. DOI: 10.1144/SP294.5 0305-8719/08/$15.00

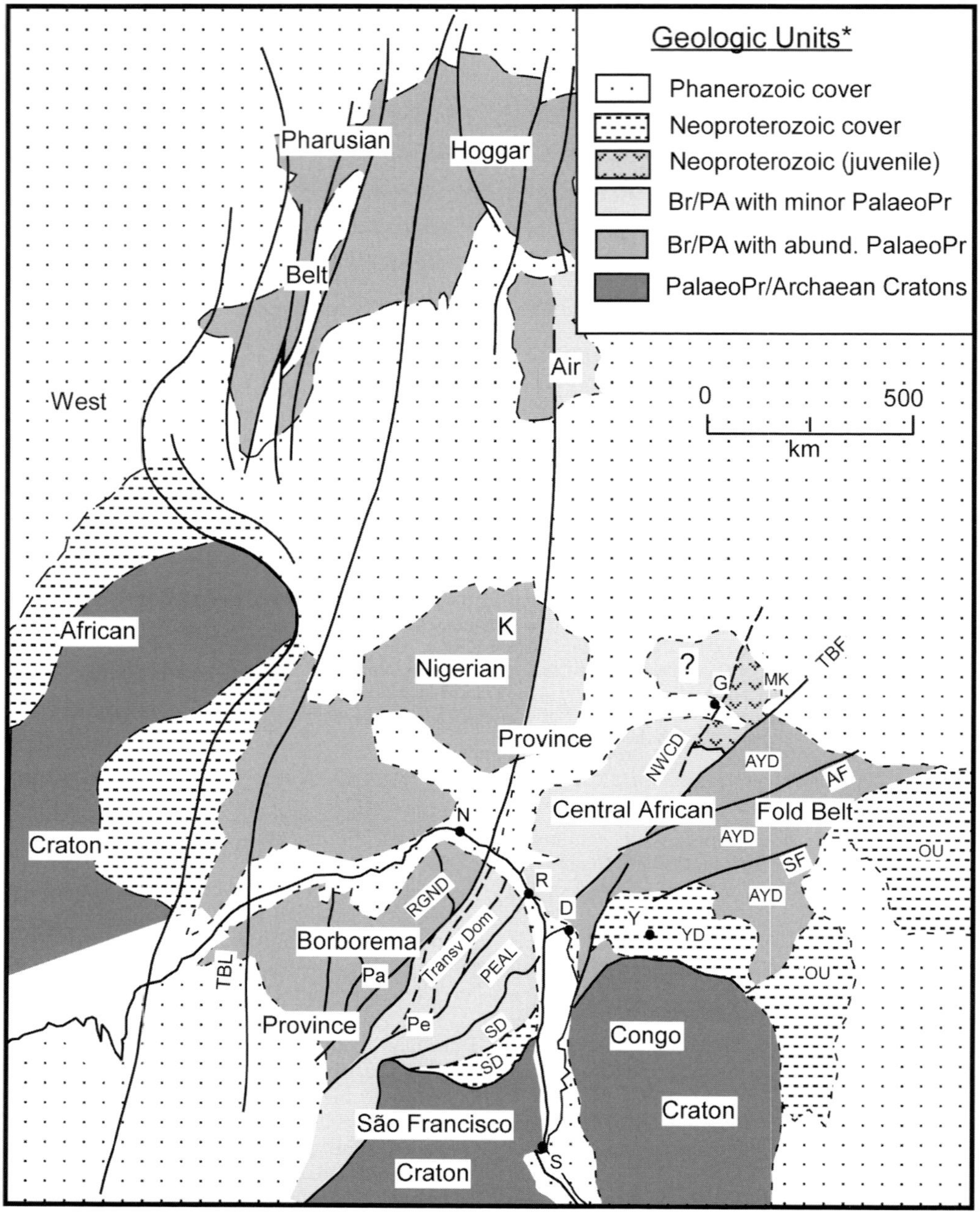

Fig. 1. A portion of West Gondwana about 500 Ma showing inferred geological provinces and potential correlations from Brazil to west-central Africa. Legend: Br/PA, Brasiliano/Pan-African orogenic domains; PalaeoPr, Palaeoproterozoic crust. *Major domains and regions used in this paper*: AYD, Adamawa–Yadé domain; MK, Mayo Kebi terrane; NWCD, NW Cameroon domain; OU, Oubanguide fold belt; PEAL, Pernambuco–Alagoas domain; RGND, Rio Grande do Norte domain; SD, Sergipano domain; Transv. Dom., transverse domain; YD, Yaoundé domain. *Faults and shear zones*: AF, Adamawa fault; Pa, Patos fault zone; Pe, Pernambuco fault zone; SF. Sanaga fault; TBF, Tchollіré–Banyo fault; TBL, Transbrasiliano Lineament. *Cities*: D. Douala; G, Garoua; K, Kaduna area of Nigeria; R, Recife; N, Natal; S, Salvador; Y, Yaoundé.

on work from several laboratories using standard thermal ionization mass spectrometry (TIMS) methods (single and multi-grain approaches) and Pb-evaporation $^{207}Pb/^{206}Pb$ ages. Determination of ages for individual grains in large suites of detrital zircon from metasedimentary rocks to examine provenance and to set limits on depositional ages is an important and growing avenue of research in Precambrian terranes and is based largely on secondary-ion mass spectrometry (SIMS, e.g., SHRIMP) methods and laser-ablation ICP (LA-ICP) methods. Sm–Nd whole-rock analyses yield depleted-mantle crustal formation ages (e.g., T_{DM}, DePaolo 1981), which provide very important constraints for crustal provenance, not only for igneous and orthogneiss suites, but also for metasedimentary and paragneiss sequences. Although such data are not precise, they still allow determination of maximum ages of crystallization or deposition that can be crucial to defining crustal provinces.

Detrital zircon ages are able to set maximum (but not actual or minimum) ages of sedimentation, although maximum ages can often bracket depositional ages narrowly in the context of other information. Detrital zircons can also give an indication of major provenance, which in some cases can be important for interpretation of assembly histories of crustal collages. T_{DM} ages are, similarly, maximum ages for mantle extraction and can often be used to determine whether a crustal terrane is Archaean (or Palaeoproterozoic) crust, derived from such crust, or must be younger. For example, although an orthogneiss having a T_{DM} age of 1.50 Ga must be Mesoproterozoic or younger; the exact age cannot be determined from this data; the T_{DM} age could be a mixture of older crustal material (up to and including Archaean) with juvenile material of an unknown younger age. Regional geological relationships in conjunction with other dating methods (e.g., U–Pb ages of zircons) must be used to constrain the age of a crustal block or terrane further. Nonetheless, Sm–Nd model ages are essential for working in large, often poorly defined, regions such as NE Brazil and west-central Africa and are used extensively in this report.

Northern São Francisco Craton, Brazil

The São Francisco Craton underlies the southern boundary of the Borborema Province of NE Brazil (Fig. 2), and it is also the structural buttress against which terranes to the north were accreted. Its main geological features are outlined by Teixeira & Figueiredo (1991), Teixeira *et al.* (2000), and Bizzi *et al.* (2001). In general, the São Francisco Craton consists of Archaean to Palaeoproterozoic high-grade (migmatite, granulite) gneisses and granite–greenstone supracrustal terranes overlain by middle to late Proterozoic platform-type cover. In the northern part of the craton there are several greenstone belts ranging in age from *c.* 3.3 Ga to *c.* 2.1 Ga.

The high-grade metamorphic terranes in the northern part of the craton are traditionally separated into an Archaean block (the Jequié migmatite–granulite complex) and a Palaeoproterozoic mobile belt known as the Itabuna–Salvador–Curaçá orogen (Barbosa & Sabaté 2002; Oliveira *et al.* 2004). Geochronological data of Marinho *et al.* (1994), Ledru *et al.* (1994), and Barbosa & Sabaté (2004) on the southern segment of this orogen suggest that plate collision occurred after 2.4 Ga and that metamorphism reached its peak at about 2.05 Ga.

The northern segment of the Itabuna–Salvador–Curaçá orogen originated by collision between two Archaean blocks (Barbosa & Sabaté 2002, 2004), namely the Gavião block to the west, and the Serrinha block to the east. The Serrinha block, of major interest as potential sediment sources for the Sergipano domain, contains an Archaean basement with U–Pb ages between 3120 Ma and 2980 Ma covered by volcanic rocks of the Rio Itapicuru greenstone belt (2200–2100 Ma, Silva 1992), both of which were intruded by granites with U–Pb ages in the range 2160–2080 Ma (Rios 2002; Oliveira *et al.* 2004). This basement crops out along the southern edge of the Sergipano domain (Fig. 2), where it forms basement to platform deposits. The northern edge of contiguous São Francisco Craton basement is marked by the São Miguel do Aleixo shear zone, which also delineates southernmost occurrences of Brasiliano granites and first appearance (going northward) of Mesoproterozoic T_{DM} ages (Van Schmus *et al.* 1995). Palaeoproterozoic to Archaean rocks that are probably correlative with São Francisco Craton basement also occur as smaller uplifted, possibly isolated, blocks to the north.

Borborema Province, NE Brazil

The Borborema Province of NE Brazil (e.g., Brito Neves *et al.* 2000) can be divided into several geotectonic fold belts, domains, massifs, or terranes. In this paper we will group them into six major regions (Fig. 2) under the terminology of *domains*. Northward from the São Francisco Craton they are: (1) the Sergipano domain; (2) the Pernambuco–Alagoas (PEAL) domain; (3) the Riacho do Pontal domain, to the west of the PEAL domain; (4) the Transverse domain (Ebert 1970; Jardim de Sa 1994); (5) the Rio Grande do Norte and Ceará domains in Rio Grande do Norte state and central

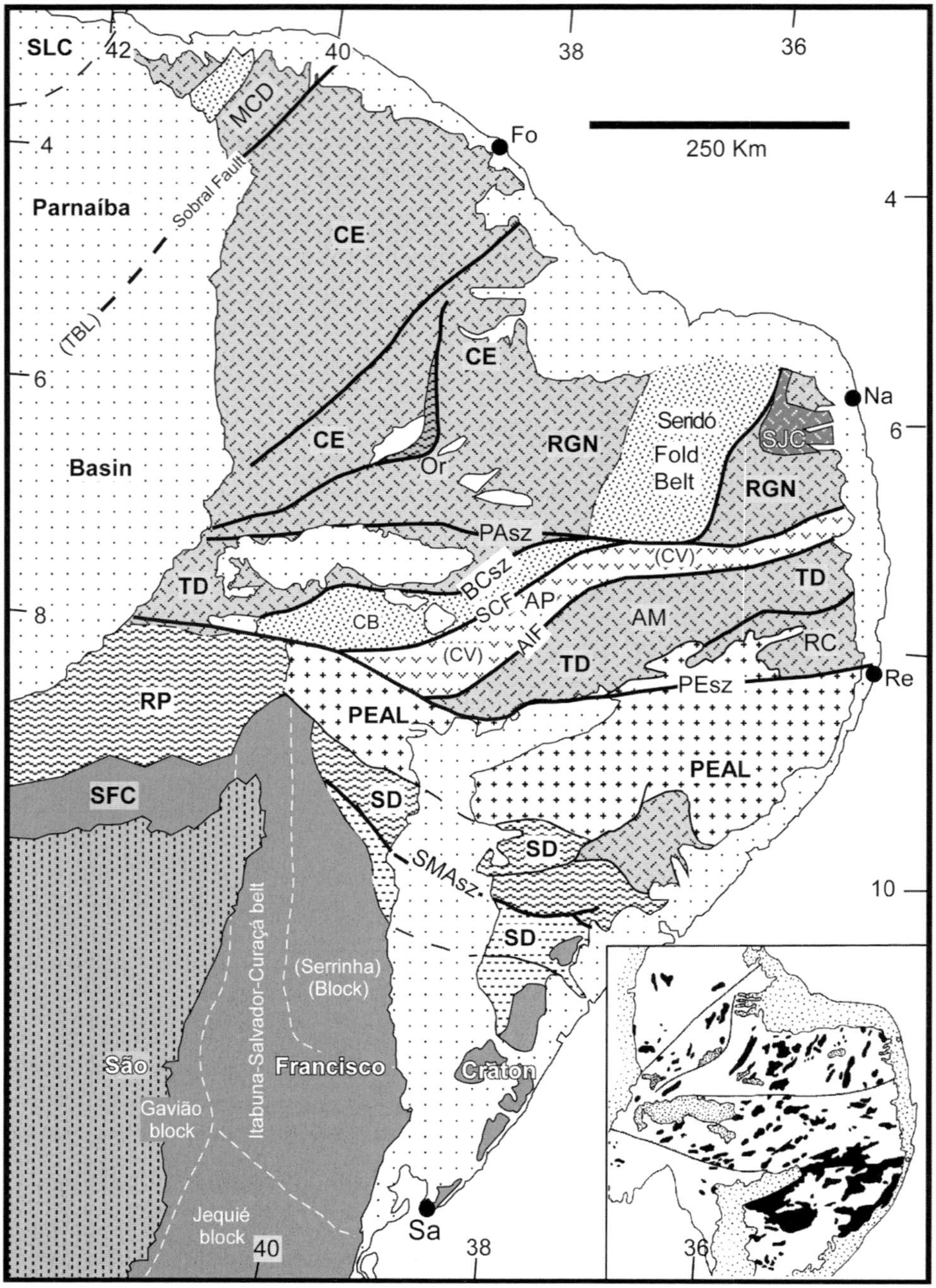

Fig. 2. Borborema Province. *Major domains and terranes*: CE, Ceará domain (Or, 1.8 Ga Orós fold belt); MCD, Médio Coreaú domain; PEAL, Pernambuco Alagoas domain; RGN, Rio Grande do Norte domain (SJC, São José do Campestre Archaean nucleus); RP, Riacho do Pontal domain; SD, Sergipano domain; SFC, São Francisco Craton; SLC, São Luís Craton; TD, Transverse domain (AP, Alto Pajeú terrane; AM, Alto Moxoto terrane; CB, Cachoerinha belt; CV, Cariris Velhos orogenic belt; RC, Rio Capibaribe terrane). *Faults and shear zones*: AIF, Afagados do Ingrazeira fault; BCsz, Boqueirão dos Conchos shear zone; PAsz, Patos shear zone; PEsz, Pernambuco shear zone; SCF, Serra do Caboclo fault; SMAsz, São Miguel do Aleixo shear zone; TBL, Transbrasiliano Lineament. *Cities and towns*: Fo, Fortaleza; Na, Natal; Re, Recife; Sa, Salvador. Neoproterozoic plutons or supracrustal units in Ceará domain are not shown. Inset: general distribution of Brasiliano granites.

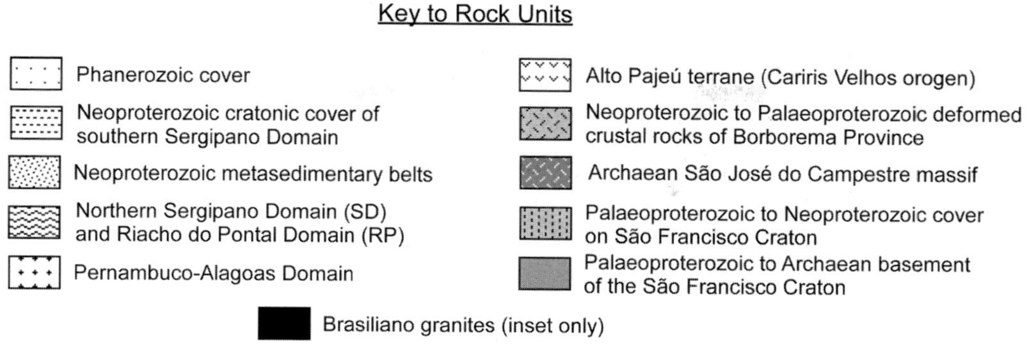

Fig. 2. (*Continued.*)

to eastern Ceará state; (6) the Médio Coreaú domain, west of the Sobral fault in NW Ceará state. The PEAL domain is not a distinct lithostratigraphic terrane, but a region of higher grade derivatives of rocks similar to those in the Transverse and Sergipano domains; therefore it will be discussed after these two regions. The Riacho do Pontal, Rio Grande do Norte–Ceará and Médio Coreú domains are important in the overall synthesis, but will not be covered in as much detail, in part because others (Santos *et al.* 2008) will discuss the latter two areas, and the first is not well known in detail at present.

Brasiliano plutonism, deformation and metamorphism

Brasiliano granites comprise a significant portion of outcrop in the Borborema Province (Fig. 2, inset). These granites are important because their genesis can tell us much about composition of deeper levels of crust in the province (e.g., Sial 1986; Ferreira *et al.* 1998) and their isotopic characteristics, especially Sm–Nd T_{DM} model ages, can tell us much about the age of the rocks from which the magmas were derived, as discussed below. In addition, age relationships among various types of granites, in conjunction with their modes of occurrence, can help substantially to define the tectonic history of the Brasiliano orogeny in the province.

The duration of deformation is best controlled by U–Pb ages of pre-, syn- and post-tectonic Brasiliano plutons, and over the past 15 years many new U–Pb ages on zircon, monazite, and titanite have been reported for the Borborema Province (e.g., Jardim de Sá 1994; Van Schmus *et al.* 1995; Fetter 1999; Dantas 1997; Kozuch 2003; Guimarães *et al.* 2004; Neves *et al.* 2006). Several pre-tectonic plutons, now often gneissic granites, have ages of 640 to 620 Ma, indicating that deformation began after 620 Ma; plutons in the 620–640 Ma range are generally found south of the Patos shear zone, in the central and southern regions. U–Pb crystallization ages of syn-deformational igneous rocks or U–Pb ages on metamorphic zircon, monazite, or titanite suggest that thermal activity peaked at about 600 Ma. Post-tectonic plutons in all regions commonly have ages of 580 to 570 Ma, indicating that compressive ductile deformation had essentially ceased by 580 Ma.

The Borborema Province also contains many Brasiliano shear zones (Brito Neves *et al.* 1982; Jardim de Sá 1994; Vauchez *et al.* 1995). Some, such as the Patos and Pernambuco shear zones (Fig. 2), can be traced into comparable shear zones in Africa (e.g., Toteu *et al.* 2004, and as discussed below). In many cases the shear zones represent major faults within former continental blocks; others, however, may represent major terrane boundaries (Brito Neves *et al.* 2000). For example, the eastern part of the Patos shear zone probably represents a major terrane boundary, formed by convergence between the Rio Grande do Norte domain and the Alto Pajeú terrane of the Transverse domain prior to the Brasiliano orogeny (Van Schmus *et al.* 2003; Brito Neves *et al.* 2001*a*). On the other hand, the Pernambuco shear zone is mainly intracontinental (e.g., Neves & Mariano 1999). A major shear couple formed between the Patos and Pernambuco shear zones during later stages of Brasiliano deformation, resulting in the creation of many transverse faults with block rotation and internal deformation within the Transverse domain (Jardim de Sá 1994), which complicate internal tectonic reconstructions.

Sergipano domain

The Sergipano domain (formerly 'belt' or 'fold belt') is one of the most important Precambrian orogenic regions of NE Brazil, not only because it was early considered as evidence for continental drift (Allard & Hurst 1969), but also because it contains several structural and lithological subdomains that allow it to be compared with Phanerozoic orogens. This domain has been interpreted previously as a typical geosyncline (Humphrey & Allard 1968; Silva Filho & Brito Neves 1979), as a collage of lithostratigraphic domains (Davison & Santos 1989; Silva Filho 1998), or as a Neoproterozoic fold-and-thrust belt produced by inversion of a passive margin basin located at the northeastern edge of the ancient São Francisco plate (D'el-Rey Silva 1999). Much of the Sergipano domain was formed by compression between the São Francisco Craton and the Borborema Province during the Brasiliano orogeny (Brito Neves *et al.* 1977); during this convergence the PEAL domain acted as a major crustal block, or 'massif', compressing units of the Sergipano domain against the São Francisco Craton and thrusting many units southward over it.

According to Santos & Souza (1988), Davison & Santos (1989), and Silva Filho (1998) the Sergipano domain consists of six lithostratigraphic subdomains which are, from south to north: Estância, Vaza Barris, Macururé, Marancó, Poço Redondo and Canindé, each separated from the other by major shear zones (Fig. 3). The first three are mostly composed of metasedimentary rocks, with metamorphic grade increasing from weakly or nonmetamorphic in the Estância subdomain through greenschist grade in the Vaza Barris subdomain to amphibolite facies in the Macururé subdomain. Higher-grade equivalents of the Macururé subdomain (granulite retrograded to amphibolite) occur within the PEAL domain (Silva Filho & Torres 2002; Silva Filho *et al.* 2003). Brasiliano granitoid plutons occur in all regions north of the São Miguel do Aleixo fault, but they are absent in the southernmost Vaza Barris and Estância subdomains. These last two areas are underlain by Palaeoproterozoic to Archaean basement contiguous with the Sao Francisco Craton and are relatively undeformed; geotectonically they could be regarded as part of that craton.

The *Estância subdomain* comprises, from base to top, sandstones and argillites of the Jueté Formation, dolomites and limestones of the Acauã Formation, sandstones and argillites of the Lagarto Formation, and sandstone and minor conglomerate lenses of the Palmares Formation (Silva Filho & Brito Neves 1979); primary sedimentary structures are ubiquitous. The *Vaza Barris subdomain* mostly consists of greenschist facies clastic and chemical sedimentary rocks. D'el-Rey Silva & McClay (1995) subdivided rocks of this area into the Miaba Group (quartzite–conglomerate, at the base, followed upwards by phyllites,

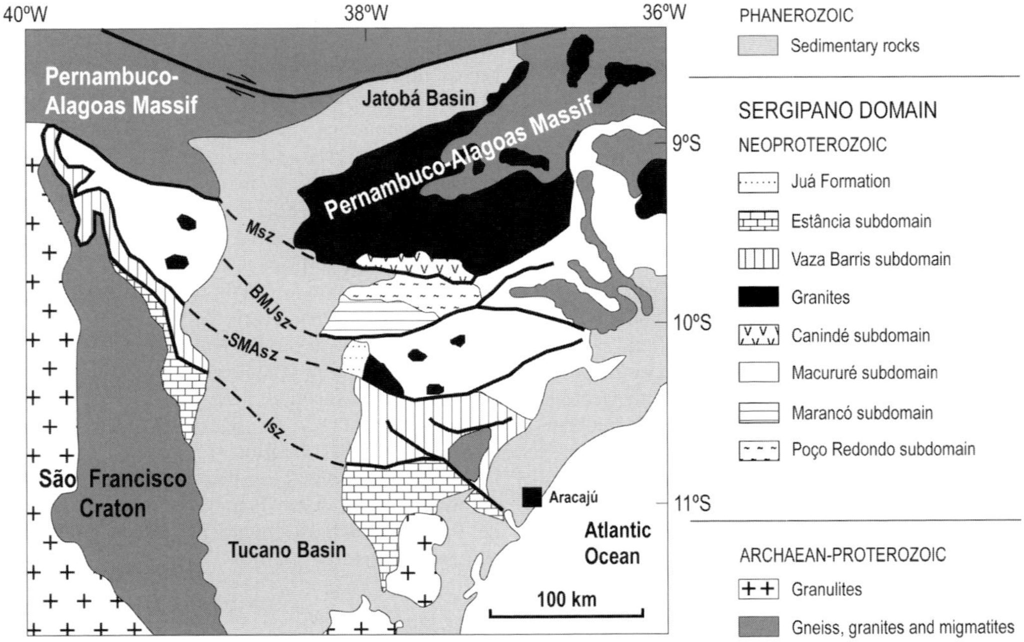

Fig. 3. Sergipano domain (modified from Oliveira *et al.* 2006). BMJsz, Belo Monte–Jeremoabo shear zone; Isz, Itaporanga shear zone; Msz, Macururé shear zone; SMAsz, São Miguel do Aleixo shear zone.

meta-greywackes, chlorite-schist, and meta-limestone), Simão Dias Group (sandstones, mudstones, meta-siltites, meta-greywackes, phyllites, and meta-rhythmites), and the Vaza–Barris Group (meta-diamictites, phyllites, and meta-limestone).

D'el-Rey Silva (1999) interpreted sediment deposition in the Estância and Vaza Barris subdomains as recording two cycles of sedimentation onto a passive continental margin of the ancient São Francisco plate. In this model, the São Francisco Craton should be the source of detritus. However, on the basis of detrital zircon populations with ages between 570 Ma and 657 Ma (Fig. 4b, c), Oliveira *et al.* (2005*a*, *b*) proposed that the uppermost clastic sedimentary rocks of the Vaza Barris and Estância subdomains were deposited in foreland basins, with detritus sources from the Sergipano fold belt and other sources farther north in the Borborema Province. Depleted-mantle Sm–Nd T_{DM} model ages of 1.8 to 1.2 Ga for fine-grained clastic metasedimentary rocks from the Macururé and Estância subdomains (Fig. 5a) are also

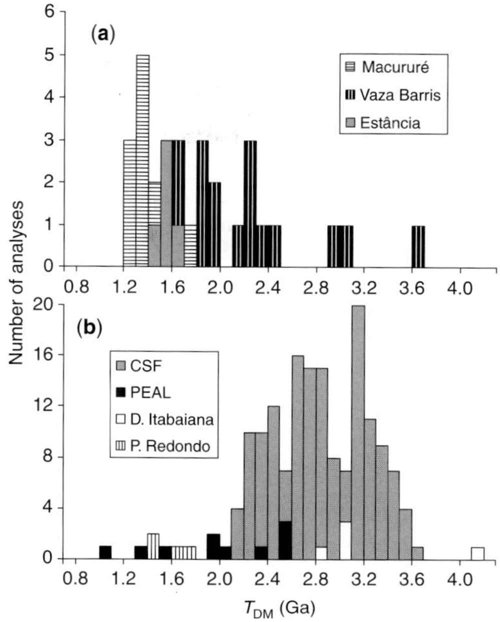

Fig. 5. Sm–Nd T_{DM} model ages from Sergipano domain (modified from Oliveira *et al.* 2006). CSF, São Francisco Craton; D. Itabaiana, Itabaiana Dome; P. Redondo, Poço Redondo subdomain; PEAL, Pernambuco–Alagoas domain.

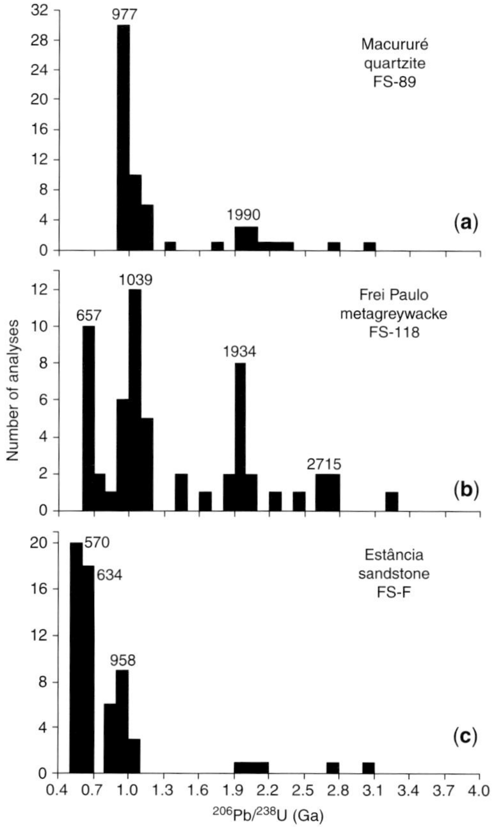

Fig. 4. Detrital zircon populations from Sergipano domain (modified from Oliveira *et al.* 2006).

consistent with provenance from sources other than the São Francisco Craton, although a minor contribution from the craton cannot be firmly ruled out. U–Pb SHRIMP ages on detrital zircon grains from the Frei Paulo meta-greywacke in the Simão Dias Group cluster about 657 Ma, 1039 Ma, 1934 Ma and less often about 2715 Ma (Fig. 4b); the youngest zircon grain is 615 Ma, thus constraining deposition of the original sediment to less than this age.

The *Macururé subdomain* lies to the north of the Vaza Barris subdomain and extends the width of the belt. The São Miguel do Aleixo shear zone separates these two domains and is a major crustal boundary. Brasiliano plutons do not occur south of the shear zone, and all Brasiliano plutons north of it have Mesoproterozoic Sm–Nd T_{DM} ages suggesting that São Francisco Craton crust, if present, must lie below the depth of magma genesis (Van Schmus *et al.* 1995; Oliveira *et al.* 2006). The Macururé subdomain mostly consists of garnet–mica schists and phyllites with minor quartzite and marble, all intruded by granitic plutons (628–625 Ma: Bueno *et al.* 2005; Long *et al.* 2003) and a few mafic to ultramafic sheets. Detrital zircon grains cluster mostly in the intervals 980–1100 Ma and 1800–2100 Ma (Fig. 4a); there are no zircons younger than

900 Ma (Oliveira *et al.* 2006), resulting in a relatively unconstrained depositional age between 900 Ma and 625 Ma.

The *Marancó subdomain* contains a metavolcanic–metasedimentary sequence (quartzite, conglomerate, micaschists, phyllites, and lenses of andesite, dacite and quartz-porphyry) with peridotite and amphibolite lenses. Andesite and dacite lenses occur conformably interleaved with phyllites and yield U–Pb SHRIMP ages between 602 and 603 Ma (Carvalho *et al.* 2005). Detrital zircon grains from four samples of clastic metasedimentary rocks have U–Pb SHRIMP ages clustering mostly between 950 and 1100 Ma, with the youngest zircon at 914 Ma, indicating deposition after this age and with detritus provenance dominantly from late Mesoproterozoic and early Neoproterozoic sources similar in age to the Poço Redondo gneisses (Carvalho *et al.* 2005). Intercalated lenses of 603–602 Ma andesites and dacites and stocks of granodiorite dated at 595 $\pm$ 11 Ma (Rb–Sr isochron, Silva Filho *et al.* 1997*a*) that cross-cut the sedimentary sequence indicate that deposition probably occurred in the late Neoproterozoic.

The *Poço Redondo subdomain* is composed of migmatites, biotite gneisses and several granitic intrusions, such as the Serra Negra augen gneiss (952 $\pm$ 2 Ma; U–Pb SHRIMP, Carvalho *et al.* 2005), and leucogranite sheets similar to ones in the Canindé subdomain (below). Grey gneisses from the palaeosome of migmatites yield U–Pb SHRIMP ages from 960 to 980 Ma (Carvalho *et al.* 2005), equivalent to orthogneisses and metavolcanic rocks of the Cariris Velhos orogen in the Transverse domain (discussed below). Sm–Nd T_{DM} model ages are typically between 1.7 and 1.3 Ga (Fig. 5b) showing that provenance includes sources that are Mesoproterozoic or younger. This subdomain is particularly important since it is the primary evidence for *c.* 1.0 Ga igneous rocks south of the PEAL domain.

The *Canindé subdomain* is composed of (i) an elongated pink granite sheet (Garrote unit) dated at about 715 Ma by Van Schmus (unpublished data in Santos *et al.* 1998); (ii) a metavolcanic–sedimentary sequence (Novo Gosto unit) represented by fine-grained amphibolite, marble, graphite schist, mica-schist, and meta-greywacke containing younger detrital zircon grains dated between 625 and 629 Ma (U–Pb SHRIMP; Nascimento *et al.* 2005); (iii) a sub-volcanic microgabbro – quartz-monzodiorite complex (Gentileza unit) with microgabbro, porphyritic quartz-monzodiorite (688 $\pm$ 15 Ma, U–Pb SHRIMP, Nascimento *et al.* 2005), Rapakivi granite (684 $\pm$ 7 Ma U–Pb TIMS, Nascimento *et al.* 2005) and quartz-porphyry; (iv) the Canindé gabbro–leucogabbro complex with gabbro, gabbro-norite, leucogabbro, peridotite, and pegmatitic gabbro (690 $\pm$ 16 Ma, U–Pb SHRIMP, Nascimento *et al.* 2005); (v) several granitoid bodies such as tonalite (634 $\pm$ 10 Ma, U–Pb SHRIMP, Nascimento *et al.* 2005), granodiorite, quartz-syenite (617 $\pm$ 7 Ma, Rb–Sr isochron, Silva Filho *et al.* 1997*a*) and leucogranite sheets (609 $\pm$ 11 Ma, Rb–Sr isochron, Silva Filho *et al.* 1997*a*). On the basis of major and trace elements, and Nd-isotope geochemistry, Nascimento *et al.* (2005) suggested that the Canindé domain is the root of an inverted continental rift sequence.

Structural evolution of Sergipano domain. The Sergipano domain underwent three main deformation episodes related to the Brasiliano collisional event (D_1–D_3, Jardim de Sá *et al.* 1986; Campos Neto & Brito Neves 1987; D'el-Rey Silva 1995; Araújo *et al.* 2003). These deformations are best recognized in supracrustal sequences of the Vaza Barris and Macururé subdomains, as well as in the basement rocks exposed in domal uplifts. The collisional event reworked older gneiss–migmatitic fabrics (D_n) that can be either remnants of a pre-Brasiliano deformation event or an early structure related to the very beginning of the collision. D_1 is characterized by south-verging nappes and thrust zones that transported the metasedimentary rocks of the Macururé, Vaza Barris, and Estância subdomains large distances southwards over the São Francisco Craton. D_2 deformation is marked by extensive reactivation of the D_1 episode and is associated with a transpressive regime that affected the entire belt. Some 625 Ma granites in the Macurure subdomain were intruded between D_1 and D_2, since they are affected by the D_2 phase of deformation. D_3 probably took place after 600 Ma when the domain experienced a large amount of uplift during which the rock units had a brittle to ductile–brittle behaviour.

Transverse domain

The Transverse domain is a complex, heterogeneous collage of several crustal units, ranging from Palaeoproterozoic (Transamazonian, *c.* 2.15 Ga) terranes and isolated basement blocks to late-Brasiliano, post-collisional (*c.* 540–580 Ma) granite plutons, all cut by late transcurrent faults within the shear couple formed by the Patos and Pernambuco shear zones (Fig. 6; Jardim de Sá 1994). In spite of this complexity, several specific geotectonic units have been recognized on the basis of geochronology, isotopic studies, and field studies over the past 30 years (e.g., Brito Neves 1978; Brito Neves *et al.* 2000; Santos & Brito Neves 1984; Santos *et al.* 1997). Terminology has

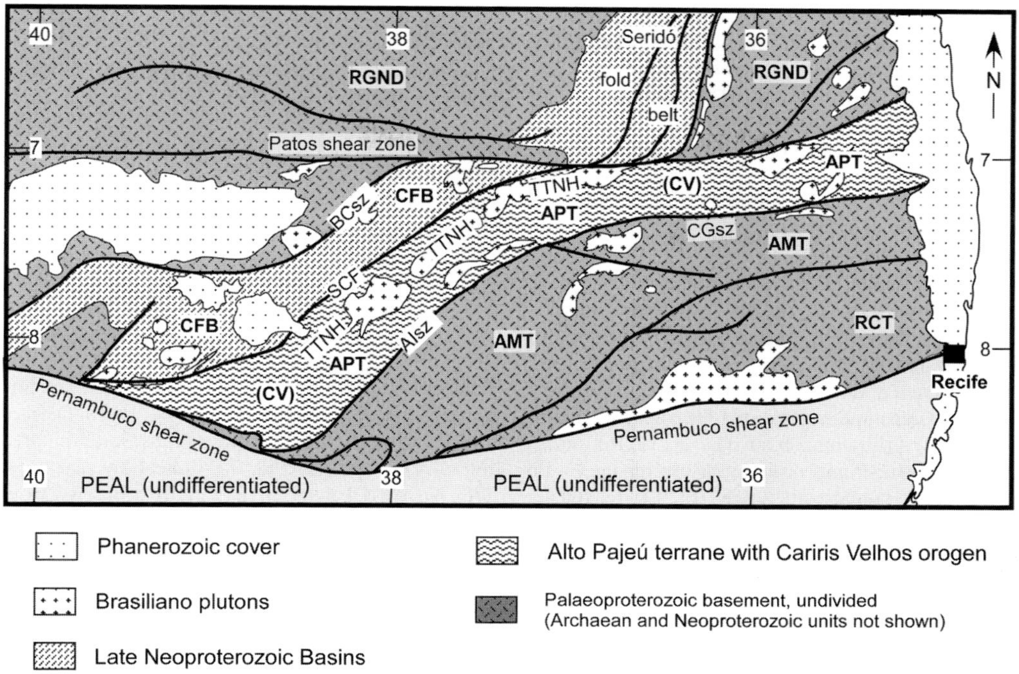

Fig. 6. General geology in the Transverse domain. TTNH, Teixeira–Terra Nova structural high, which coincides with trend Brasiliano syenitic plutons (Syenitoid Line). Major fault zones in the DZT that are pertinent to this discussion include AIsz (Afogados da Ingazeira shear zone), BCsz (Boqueirão dos Conchos shear zone), CGsz (Campina Grande shear zone), SCF (Serra do Caboclo fault). Boundaries of Alto Pajeú terrane (APT), Alto Moxotó terrane (AMT), and Rio Capibaribe terranes (RCT) are from Santos *et al.* (1997). CFB, Cachoerinha fold belt; CV, Cariris Velhos orogen; PEAL, Pernambuco–Alagoas domain; RGND, Rio Grande do Norte domain. Map modified from Kozuch (2003), Bittar (1998) and Brito Neves *et al.* (2005).

also changed significantly, and current conventions will be followed, with annotations to former terminology as needed.

The eastern part of the area, the former 'Pajeú–Paraíba fold belt' of Brito Neves (1978) was divided into several fault-bounded tectonic 'terranes' by Santos *et al.* (1997): from ESE to NNW they are the Rio Capibaribe, Alto Moxotó, and Alto Pajeú terranes (Figs 2 & 6). The *Rio Capibaribe terrane* in the ESE part of the Transverse domain is underlain mainly by Palaeoproterozoic basement, but it includes areas of undated metasedimentary rocks which, at least in some cases, represent sequences correlative with late Neoproterozoic units found to the west and north (Neves *et al.* 2005, 2006). The *Alto Moxotó terrane* (Brito Neves *et al.* 2001*b*) lies north of the Rio Capibaribe terrane and is dominated by large areas of Palaeoproterozoic basement with relatively few Brasiliano granites. Several metasedimentary sequences are known to be Palaeoproterozoic, but some areas mapped as Palaeoproterozoic may include late Mesoproterozoic to late Neoproterozoic sequences.

Recent and current studies (Van Schmus *et al.* 1995; Brito Neves *et al.* 1995, 2000, 2001*a*; Kozuch 2003; Guimarães *et al.* 2004) show that the *Alto Pajeú terrane* is dominated by the 990–940 Ma Cariris Velhos orogen (CV in Fig. 6; see following section) and intrusive Brasiliano plutonic rocks, with some remnants of Palaeoproterozoic gneiss. An important aspect of this terrane is that Brasiliano plutons along its western part are mainly high-K syenitic rocks that yield Palaeoproterozoic Sm–Nd T_{DM} ages and form the core of a major topographic ridge called the 'Teixeira–Terra Nova High', or 'Syenitoid Line' (Sial 1986; Brito Neves *et al.* 2005; TTNH in Fig. 6). These rocks have been difficult to date precisely because of lack of zircon and/or complication by inherited Palaeoproterozoic zircons. The western part of the Alto Pajeú terrane consists of metasedimentary and metavolcanic rocks of the early Neoproterozoic Riacho Gravatá sequence (Bittar 1998), which occurs on the west side of the TTNH and is bounded to the west by the Serra do Caboclo fault.

Guimarães *et al.* (2004) reported U–Pb zircon ages for several plutons in the eastern part of the

Transverse domain, with samples coming from all three terranes mentioned above. The ages ranged from 512 to 640 Ma and were divided into four main groups: 640–610 Ma, 590–580 Ma, *c.* 570 Ma, and 545–512 Ma. The oldest group is syntectonic, whereas the other three post-date the peak of compressive, ductile deformation.

Cariris Velhos orogen. Bimodal (but mostly felsic) volcanism and granitic plutonism 1.0 to 0.9 Ga is widespread in the Alto Pajeú terrane. Such ages had been postulated on the basis of Rb–Sr data a decade prior to the first zircon work (Brito Neves *et al.* 1984, 1995), but many workers attributed ages in the range of 1 Ga to partial degradation of Transamazonian (*c.* 2.1 Ga) systems during the Brasiliano orogeny. More recent U–Pb zircon ages confirm that ages of 1.0 to 0.9 Ga reflect a distinct event and lithostratigraphic assemblage (Van Schmus *et al.* 1995; Kozuch 2003). Campos Neto *et al.* (1994) proposed the term Cariris Velhos orogeny for 1 Ga events in the central part of the Transverse domain (Figs 2 & 3). Based on data currently available, the Cariris Velhos orogen is a 50–100 km wide belt that extends for more than 700 km, from the northeastern part of the Transverse domain west-southwestwards to the Riacho Pontal fold belt (Fig. 6).

The core of the orogen is comprised of augen gneisses representing granitic plutons that were intruded into bimodal (but mostly felsic) volcanic successions, which crop out on both sides of the gneissic core. The age of the orogen is currently constrained between 990 and 940 Ma based on detrital zircons in metasedimentary rocks (Van Schmus *et al.* 1999) and U–Pb ages of Cariris Velhos plutons and volcanic rocks (Kozuch 2003). Sm–Nd model ages of 1.2 to 1.9 Ga for igneous rocks yielding *c.* 0.94 to 0.99 Ga crystallization ages (Fig. 7) indicate that some older, probably Palaeoproterozoic, crust was involved in magma genesis, yielding hybrid Sm–Nd T_{DM} ages. Detrital zircons from metasedimentary units are dominated by *c.* 1 Ga grains (Fig. 8) indicating that their Sm–Nd model ages are inherited primarily from igneous rocks of the orogen that provided most of the detritus. The Sm–Nd T_{DM} ages of the igneous rocks are, in turn, presumably hybrids of *c.* 1 Ga juvenile magma contaminated by Palaeoproterozoic crust.

The northern boundary of the Cariris Velhos orogen (and, hence, the Alto Pajeú terrane), occurs along the eastern part of the Patos shear zone, and then it swings southwest along the Serra

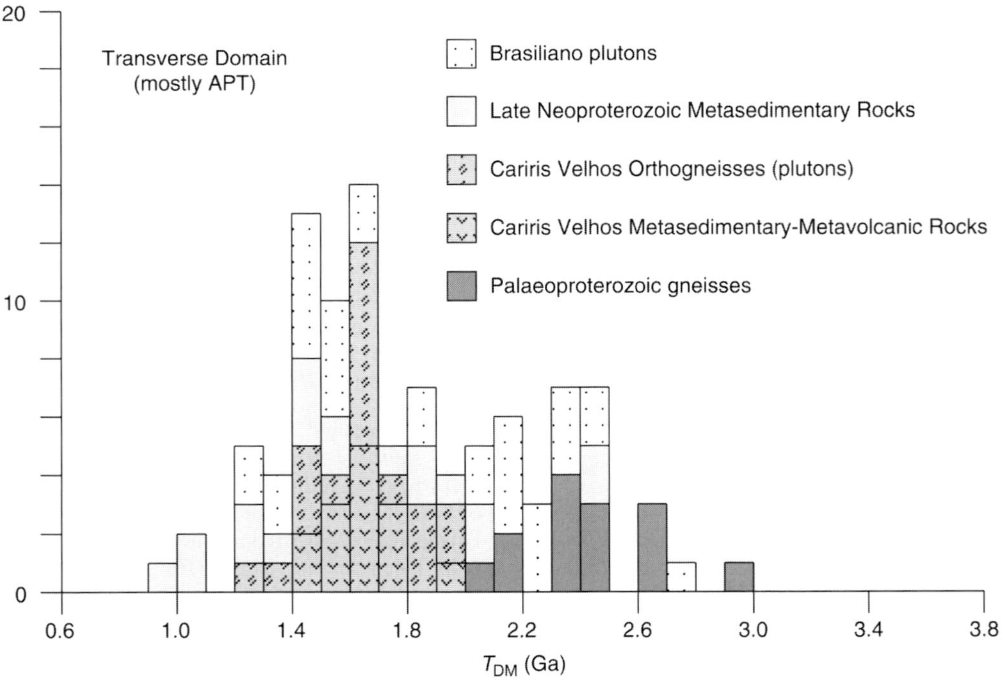

Fig. 7. Sm–Nd T_{DM} model ages within the Transverse domain (mostly from the Alto Pajeú terrane and Cachoerinha fold belt).

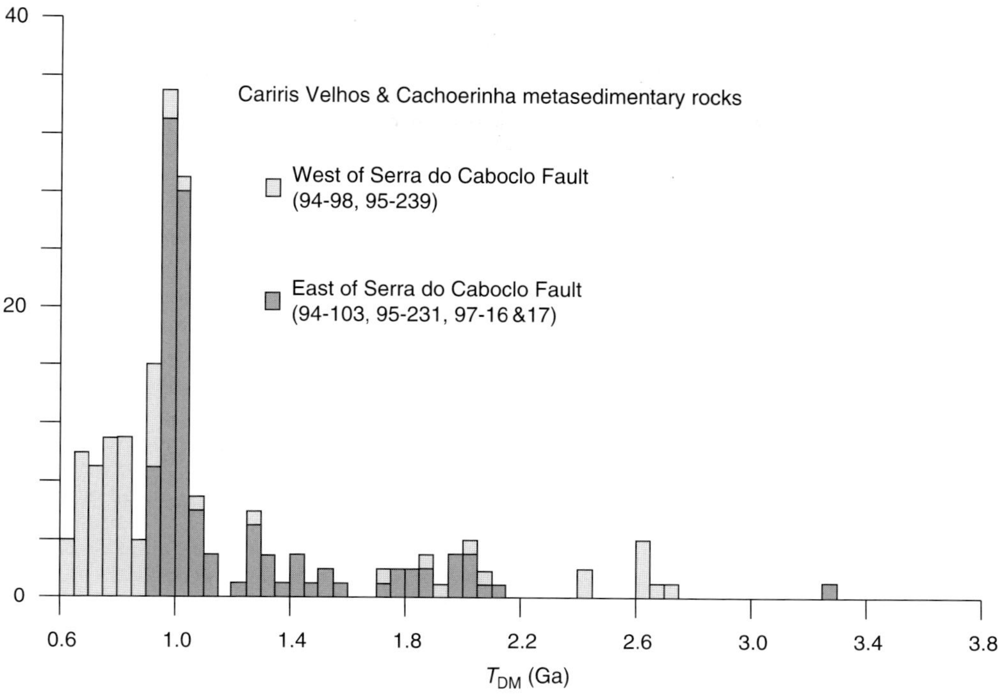

Fig. 8. Detrital zircons from Transverse domain (mostly from the Alto Pajeú terrane and Cachoerinha fold belt, Van Schmus *et al.* 1999 unpublished).

do Caboclo fault. There is a major contrast in Sm–Nd T_{DM} ages on either side of the eastern part of the shear zone (Archaean model ages to the north and Mesoproterozoic model ages to the south), suggesting that it represents a major terrane boundary. It is presently unclear, however, whether the Serra do Caboclo fault, on the west side of the Riacho Gravatá sequence, is also a major terrane boundary; it may include significant vertical offset (east side up) between the Cariris Velhos orogen and the Cachoerinha Belt to the west.

Two distinctly different models have been proposed for the Cariris Velhos orogen: (a) that it represents a continental margin magmatic arc (overlying Palaeoproterozoic crust: Van Schmus *et al.* 1995; Brito Neves *et al.* 1995; Kozuch *et al.* 2003) or (b) that it represents a major extensional belt that formed about 1 Ga (Guimarães & Brito Neves 2005). Up to now geochemical data on the orthogneisses are equivocal, but overall consideration of the geochemistry, petrology, isotopic data and regional geology tends to favour an extensional environment for the Cariris Velhos orogen (Guimarães & Brito Neves 2005).

Cachoerinha Belt. The former 'Piancó–Alta Brígida fold belt' of Brito Neves, (1978; Brito Neves *et al.* 1984) or Salgueiro–Cachoerinha fold belt of Sial (1986) included both higher-grade metasedimentary rocks in the east and lower-grade metasedimentary rocks in the west. The eastern units (Riacho Gravatá sequence, Bittar 1998) are now known to be early Neoproterozoic and related to the Cariris Velhos orogen, so that they are now included within the Alto Pajeú terrane. They are separated from lower grade rocks to the west by a major fault system (Serra do Caboclo fault). The lower grade rocks to the west yield detrital zircon populations $\geq$625 Ma (Van Schmus *et al.* 1999), similar to those from the Seridó Group to the north (cf. Van Schmus *et al.* 2003). The metasedimentary sequences also include intercalated *c.* 625 Ma metavolcanic rocks (Kozuch 2003), showing that they were deposited just prior to the peak of the Brasiliano orogeny. These units comprised the 'Cachoerinha' part of the Salgueiro–Cachoerinha fold belt of Sial (1986), and that terminology will be retained here. Metasedimentary rocks of the Cachoerinha Belt are intruded by a variety of Brasiliano granites (Sial 1986, Ferreira *et al.* 1998) ranging in age from 640 to 580 Ma (Brito Neves *et al.* 2003; Kozuch 2003). In its northern part the Cachoerinha Belt is bordered on the west by the Boqueirão dos

Conchos shear zone; the western boundary in the south is not well defined at present. To the west of the belt the geology is relatively poorly known, although it is clear that much of it includes Palaeoproterozoic basement.

A major characteristic of this belt is the presence of numerous Brasiliano granites having Mesoproterozoic Sm–Nd T_{DM} ages that are similar to or slightly older than those from the metasedimentary rocks themselves (Fig. 7). Thus, it is possible to conclude that Palaeoproterozoic basement played a limited role in magma genesis. The age and nature of basement to the Cachoerinha Belt is unknown, although Sm–Nd results for the granites suggest that Palaeoproterozoic crust, if present, must be very deep. Since the western side of this belt appears to be faulted against blocks of Palaeoproterozoic basement, the Cachoerinha Belt could be a deep intracratonic basin.

Riacho do Pontal domain

The Riacho do Pontal domain occurs west of the Sergipano and PEAL domains, along the northwestern part of the São Francisco Craton (Fig. 2). This region is not known in as much detail as domains to the east, but the data that are available (e.g., Van Schmus *et al.* 1995; Brito Neves *et al.* 2000) show that it contains units potentially correlative with Palaeoproterozoic basement, with the Cariris Velhos orogen of the Transverse domain, with younger (late Neoproterozoic) metasedimentary sequences similar to those in the Cachoerinha Belt and the Sergipano domain, and with various Brasiliano granite suites. In any case, this domain is peripheral to our discussions about Brazil–Africa correlations and will not be discussed further.

Pernambuco–Alagoas domain

The Pernambuco–Alagoas (PEAL) domain is bordered on the north and south by inward dipping thrust faults and is a large region of high-grade gneisses, migmatites and Brasiliano granites that acted as a large structural massif during late Brasiliano deformation. This domain was originally identified in the Borborema Province (Brito Neves *et al.* 1982) as the 'Pernambuco–Alagoas Massif'. At that time it was thought to consist primarily of Archaean to Palaeoproterozoic (Trans-Amazonian) basement gneisses with intrusive Brasiliano granites. It was also interpreted as a crystalline core ('massif') within the Borborema Province. Subsequent work (Van Schmus *et al.* 1995; Silva Filho *et al.* 2002; Oliveira *et al.* 2006), now shows that the PEAL complex is a collage of many units of diverse ages (Fig. 9a), and Sm–Nd model ages of 1.0 to 1.5 Ga require that large parts of the protolith (including sources for many Brasiliano plutons) must be Mesoproterozoic or younger (Silva Filho *et al.* 2002, 2005*a*, *b*), although many gneisses also show Archaean to late Palaeoproterozoic origin (Figs 9b & 10). Thus, the PEAL domain is not a distinct lithostratigraphic terrane, but instead, it is comprised of higher-grade derivatives of rocks similar to those in the Transverse domain. In order to reflect this reality, we will not use 'massif', but will refer to the region as a domain.

Brasiliano plutonic rocks in the PEAL domain. Brasiliano plutonic rocks are abundant in the PEAL domain and can range from highly deformed, pre- to syn-tectonic units to late- to post-tectonic plutons that are largely undeformed except for late transcurrent faulting. The pre-to syn-tectonic Brasiliano plutons are commonly deformed or migmatized and thus often very difficult to distinguish in the field from older (Mesoproterozoic to Palaeoproterozoic) crustal rock, and it is necessary to rely on isotopic results for identification. Silva Filho *et al.* (1996, 1997*b*, 2002) identified various late-tectonic granitic intrusions and suites in the eastern part of the PEAL domain, with compositions ranging from high-K, calcalkaline, shoshonitic, mildly alkaline, rocks to various peraluminous ($\pm$ garnet bearing) granites. Based on locations, petrography, and geochemistry, these intrusions may be grouped into four major suites. The *Buique–Paulo Afonso suite* (PAB, Fig. 9a) is comprised of several granitic plutons of wide compositional range, which were intruded into tonalitic orthogneisses to the north-northeast of Paulo Afonso. The *Águas Belas–Canindé suite* (ABC, Fig. 9a) is bordered on the south by the Sergipano domain and lies between the towns of Paulo Afonso and Palmeira dos Indios. This suite contains small plutons of biotite monzogranite, amphibole granodiorite, amphibole tonalite to diorite, and shoshonitic composition intruded into metatexites and diatexites of tonalitic bulk composition. The granitic intrusions are both metaluminous and peraluminous, syn- to late-tectonic, with compositions ranging from medium to high-K calc-alkaline. The *Ipojuca–Atalaia suite* (IA, Fig. 9a) runs parallel to the present coast southwest of Recife and is bordered on the west by migmatites and orthogneisses. It consists of leucocratic peraluminous granitic plutons intruded into diatexites. The *Marimbondo–Correntes suite* (MC, Fig. 9a) is a cluster of plutons northeast of Palmeira dos Indios; it contains calc-alkaline peraluminous and metaluminous plutons which intrude older gneisses and migmatites. A fifth group of plutons called the Garanhuns batholith by Silva Filho *et al.* (2002) is not utilized here. There are relatively few published U–Pb

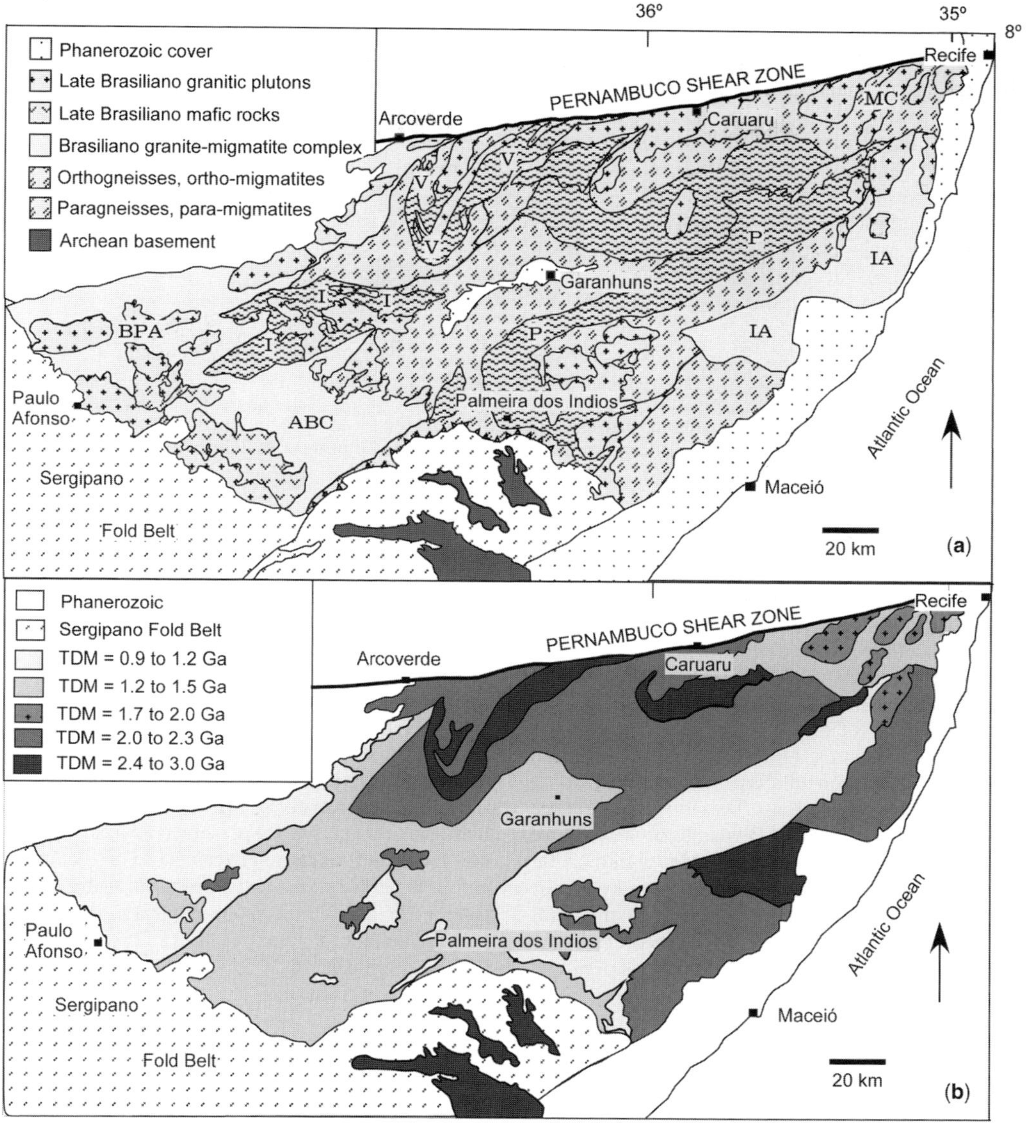

Fig. 9. (**a**) Eastern part of Pernambuco–Alagoas domain showing geological units being defined through on-going field, petrological, and isotopic studies (modified from Silva Filho *et al.* 2002). ABC, Aguas Belas–Canindé magmatic suite; BPA, Buique–Paulo Afonso magmatic suite; IA, Ipojuca–Atalaia magmatic suite; MC, Marimbondo–Correntes magmatic suite; I, Inhapi metamorphic suite; P, Palmares metamorphic suite; V, Venturosa metamorphic suite. (**b**) Distribution of Sm–Nd T_{DM} ages in the Pernambuco–Alagoas domain.

crystallization ages for plutons of these suites, but results that are available (Silva Filho & Guimarães 2000) show that they typically range from 625 Ma to 590 Ma, with some post-tectonic plutons as young as 520 Ma.

Gneiss and migmatite complexes. Santos (1995) and Medeiros & Santos (1998) recognized two major subdivisions of metamorphic rocks in the PEAL domain. Rocks assigned to the *Cabrobó Complex* are dominantly derived from metasedimentary to metavolcanic protoliths and include biotite–garnet gneiss, orthogneiss, and migmatitic orthogneisses, plus other migmatitic varieties with intercalation of quartzite, quartz-schist, calc-silicate rocks and amphibolite. Rocks assigned to the *Belém do São*

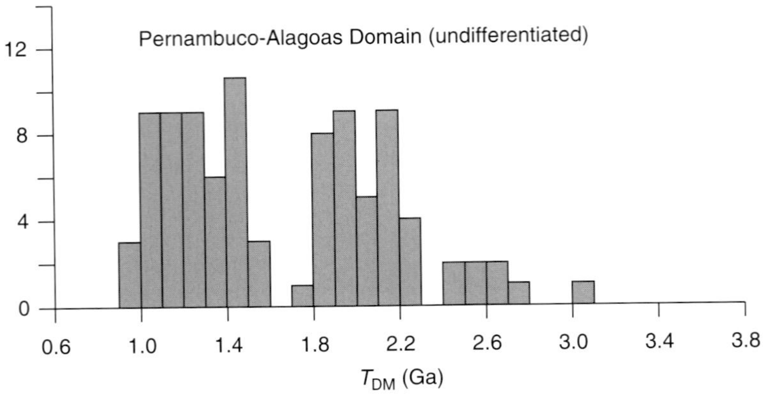

Fig. 10. Sm–Nd model ages for a comprehensive suite of rocks from the PEAL domain. The bimodal distribution shows distinctly different amounts of older lithosphere in various samples. Compare with Figures 5 and 7. From Silva Filho *et al.* (2002) and unpublished new data.

Francisco Complex mainly represent deeper crustal (mainly igneous) rocks and include migmatite, biotite gneiss, tonalitic orthogneiss, amphibolitic gneiss and leuco-granodiorite to leuco-monzogranite, with some included remnants of the Cabrobó-type supracrustal rocks.

The Cabrobó Complex was further divided by Medeiros & Santos (1998) into three different suites based on their lithotypes: *Venturosa*, *Inhapi* and *Palmares* successions. The Venturosa succession occurs south of the Pernambuco shear zone, southeast of the town of Arcoverde. Osako (2005) showed that this succession is composed of quartzites and a series of meta-igneous and metasedimentary migmatites overlying Palaeoproterozoic basement. This region also includes amphibole-bearing migmatites which are considered to be part of the Belém do São Francisco Complex (Medeiros & Santos 1998) and which show sharp contact with the metasedimentary rocks. The Inhapi succession crops out in a wedge-shaped area about *c.* 100 km long and *c.* 30 km wide between the Águas Belas–Canindé suite to the south and the Buique–Paulo Afonso suite to the north. This succession shows pre- to syn-tectonic, syn-tectonic and late- to post-tectonic garnet-bearing S-type syenogranites intruded into flat-lying, foliated metasedimentary units. The metasedimentary rocks are represented mainly by two associations: (a) locally migmatized sillimanite–garnet–muscovite–biotite gneisses, with small carbonate lenses and common amphibolite lenses and (b) biotite–muscovite–garnet gneisses, sometimes migmatized, with amphibolite lenses and calc-silicate lenses. The Palmares succession is about 30 km wide and extends about 200 km NE from Palmeira dos Índios. It is comprised of garnet gneisses, greywackes, and amphibolites intruded by syn-tectonic tonalitic to amphibole–granodioritic gneisses and gabbros. This complex shows vergence to the NW (Medeiros & Santos 1998).

The ages of these units are still poorly known, and it is probable that grouping based primarily on rock-type has lumped together units having primary depositional or plutonic ages ranging from Palaeoproterozoic (or older) to Brasiliano. Brito Neves *et al.* (1995) presented Rb–Sr geochronological ages ranging between 1.13 Ga (diatexites) and 0.96 Ga for rocks assigned to the Cabrobó and Belém do São Francisco complexes in western PEAL domain. Pessoa *et al.* (1978) reported a Rb–Sr isochron of 1.53 ± 73 Ma for tonalitic orthogneisses from the Ibirajuba area. Very few U–Pb ages have been reported for the PEAL domain. Van Schmus *et al.* (1995) reported a zircon upper-intercept apparent age of 1.58 ± 73 Ma for garnet-bearing migmatite west of Palmeira dos Indios. A single U–Pb zircon age of 2.0 Ga was obtained by LA-ICPMS for migmatized tonalitic orthogneisses in the Jupi area, NE of Garanhuns (Neves *et al.* 2005). Thus, in spite of early interpretations of a Palaeoproterozoic to Archaean age for most of the PEAL domain, few direct ages confirm this. Lithostratigraphic terminology and correlations will have to be revised extensively in the future as primary crystallization or depositional ages are defined for individual local units.

Santos (1995) and Van Schmus *et al.* (1995) reported Sm–Nd T_{DM} model ages for rocks of the PEAL domain that range between 1.2 and 1.6 Ga; a larger data set reported by Silva Filho *et al.* (2002, 2005*b*) expanded this range to 0.9–2.8 Ga

(Fig. 10). These data have a bimodal distribution, indicating that most of the primary ages are either Brasiliano or Trans-Amazonian. The low frequency of samples with T_{DM} from 1.5 to 1.8 Ga may indicate that Cariris Velhos igneous rocks, which commonly have T_{DM} ages in this range (see Fig. 7), are scarce or not found in the PEAL domain.

Isotopic subdomains. Silva Filho *et al.* (2002) evaluated PEAL crustal evolution based on Sm–Nd isotopic data from the Neoproterozoic granitoids, and they defined two smaller crustal subdomains. New Sm–Nd isotopic data from the metamorphic complexes, added to the original data, still show a major two-fold grouping of model ages (Fig. 10). However, when these data are coupled with geological mapping, the results indicate that the two-fold subdivision of Silva Filho *et al.* (2002) is an oversimplification. Five distinct model age subdivisions can be recognized (Fig. 9b). They are (a) T_{DM} older than 2.40 Ga, represented by several local occurrences of gneiss and migmatite, (b) T_{DM} between 2.00 and 2.20 Ga, represented by large areas in the northeastern half of the PEAL domain, (c) T_{DM} between 1.70 and 2.00 Ga, represented by several Brasiliano plutons in the NE corner of the domain, (d) T_{DM} between 1.20 and 1.50 Ga, represented by large parts of the southwestern half of the PEAL domain and (e) T_{DM} between 0.90 and 1.20 Ga, represented mainly by the Buique–Paulo Afonso batholith in the west and the Palmares succession and adjacent orthogneisses in the east. Of these, (a) corresponds to a small cluster in Figure 10, (b) and (c) represent the older large cluster in Figure 10, and (d) and (e) represent the younger large cluster in Figure 10. None of the areas defined here are 'pure'; that is, each contains occurrences of younger or older T_{DM} depending on geological complexities within each region. As more data become available, both from the field and the laboratory, we expect that this general picture will change in detail, in part because of complexities due to tectonic imbrication.

Rio Grande do Norte and Ceará domains

The basement complex in the Rio Grande do Norte domain is composed mostly of the 2.15 Ga Rio Piranhas massif, with a 2.6–3.6 Ga Archaean remnant, the São José do Campestre massif, in the east (Brito Neves *et al.* 2000; Dantas *et al.* 1998, 2004) and smaller remnants of Archaean crust dispersed irregularly elsewhere. Rocks with Transamazonian crystallization or metamorphic ages (*c.* 2.15 Ga) typically have Sm–Nd crustal residence ages (T_{DM}) of 2.4 to 3.0 Ga (Van Schmus *et al.* 1995; Dantas 1997), indicating that these units were not wholly juvenile when they formed (Fig. 11). Younger Sm–Nd model ages in the Ceará domain indicate greater amounts of juvenile Transamazonian contribution in the west (Fetter 1999).

Several metasedimentary and metavolcanic–metasedimentary basins are present in the RGN/CE domain. The oldest ones are the 1.8–1.7 Ga intracratonic Orós and Jaguaribeano fold belts in the eastern part of Ceará state (e.g., Sá *et al.* 1995), but few other sequences of similar age

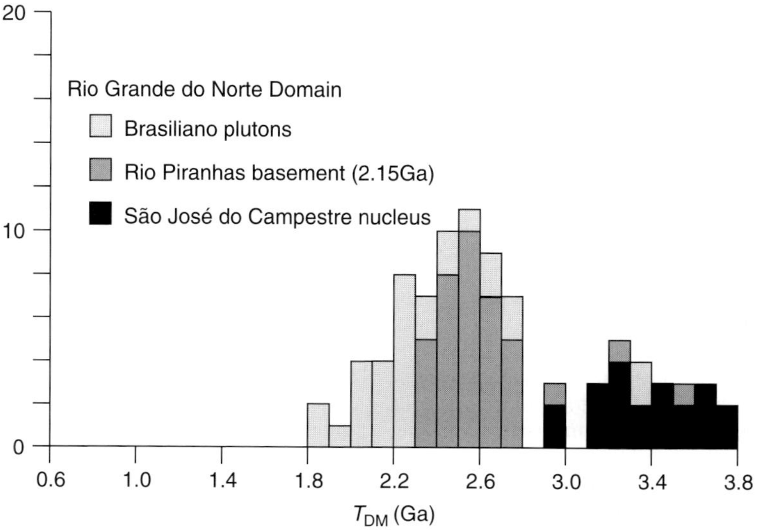

Fig. 11. Sm–Nd T_{DM} model ages from the Rio Grande do Norte domain. Mostly from Van Schmus *et al.* (1995) and Dantas (1997).

have been documented in NE Brazil. The *Seridó fold belt* is the largest Neoproterozoic fold belt in the Rio Grande do Norte domain (Fig. 2). SHRIMP U–Pb dating of detrital zircons in metagreywackes of the Seridó fold belt (Van Schmus *et al.* 2003) shows the presence of a large, late-Neoproterozoic (*c.* 640–650 Ma) population, indicating that the basin developed just prior to or during the early phases of the Brasiliano orogeny which caused its deformation and metamorphism at *c.* 620–580 Ma. C- and Sr-chemostratigraphic data for marbles from the Jucurutu Formation in the lower Seridó Group (Nascimento *et al.* 2004) also support late Neoproterozoic (640–570 Ma) deposition. Sm–Nd T_{DM} ages of 1.1 to 1.5 Ga for metasedimentary rocks of the Seridó fold belt show that most of it was probably derived from distal, younger sources rather than proximal, underlying Palaeoproterozoic basement (Van Schmus *et al.* 2003). On the other hand, Palaeoproterozoic Sm–Nd T_{DM} ages for Brasiliano plutons which intrude the Seridó Group (Dantas 1997; Hollanda *et al.* 2003) show that it must overlie continuous basement of the Rio Grande do Norte domain.

Médio Coreaú domain

The Médio Coreaú domain in the northwest corner of the Borborema Province consists of Palaeoproterozoic basement with Neoproterozoic supracrustal rocks and Brasiliano plutons. This domain separated from the Ceará domain by a major fault, the Sobral fault, which is part of the Transbrasiliano Lineament. Santos *et al.* (2008) discuss this domain in detail and its correlation with the SE part of the West African Craton and SW part of the Pan-African fold belt. This domain is not central to our discussion and will not be presented in further detail.

Northern Congo Craton

The northern part of the Congo Craton is composed of an Archaean core, the Ntem Complex, and peripheral Palaeoproterozoic rocks of the Nyong Complex along the northwest margin of the Ntem Complex. U–Pb zircon geochronology allows definition of three main stages of crustal evolution in the Ntem Complex: (a) formation of greenstone belt (ortho-amphibolites and metasedimentary rocks) formed about 3.1 Ga, (b) a major phase of crustal formation with emplacement of charnockite and tonalite (TTG) about 2.9 Ga and (c) a final stage corresponding to melting of greenstone belts and TTG at deeper levels to form K-rich granitoids between 2.7 and 2.5 Ga (Nédélec *et al.* 1990; Toteu *et al.* 1994; Tchameni *et al.* 2000; Shang *et al.* 2004). Most Sm–Nd T_{DM} ages are similar to U–Pb zircon ages indicating that this Archaean core is essentially juvenile.

The Nyong Complex is dominated by metasedimentary rocks, including quartzite, paragneiss, schist and migmatite. This complex is largely at granulite grade as a result of the *c.* 2050 Ma (Eburnian–Transamazonian) fusion of the Congo and São Francisco cratons (Ledru *et al.* 1994; Penaye *et al.* 2004; Lerouge *et al.* 2006) during assembly of a middle to late Palaeoproterozoic continent (e.g., 'Atlantica' of Rogers 1996). Lerouge *et al.* (2006) summarized detrital zircon ages from Nyong Complex metasedimentary rocks and found a range of 2400 to 3100 Ma, similar to that for the core of the Congo Craton. These ages and compositions should be reflected in any material in the Central African Fold Belt that was derived from the Congo Craton.

Central African Fold Belt

The late Neoproterozoic (Pan-African) Central African Fold Belt north of the Congo Craton was recognized in the early 1960s by the widespread occurrence of *c.* 500–600 Ma Rb–Sr whole rock and mineral ages (Lasserre 1967). This fold belt underlies Cameroon, Chad, and the Central African Republic, between the Congo Craton to the south and the Western Nigerian Shield to the north (Figs 1 & 12) and corresponds to the southern part of the Saharan metacraton (Abdelsalam *et al.* 2002; Toteu *et al.* 2004). Interactions between this mobile domain and the major cratons are only partially understood. Along the eastern border of the West African Craton evolution seems to be well constrained between 630 and 580 Ma, with eastward subduction followed by the collision between the craton and the Tuareg–Nigerian shield (Caby 1989; Santos *et al.* 2008). On the northern edge of the Congo Craton, the tectonic evolution is still enigmatic since no clear evidence of oceanic rocks has yet been found, despite many features that characterize a collisional belt: long-lived (800–600 Ma) arc-type magmatism, external nappes of regional extent, granulitic metamorphism, intensive plutonism associated with crustal melting, regional strike-slip faults (some of which extend to NE Brazil), and the possible presence of molasse basins. Proposed tectonic models broadly correspond to collision between the Congo Craton and the mobile belt (Abdelsalam *et al.* 2002) or collision among different blocks within the mobile belt (Toteu *et al.* 2004).

The following domains can be distinguished in the Central African Fold Belt north of the Congo Craton (Toteu *et al.* 2004). The *Yaoundé domain* extends east–west north of the Congo Craton and

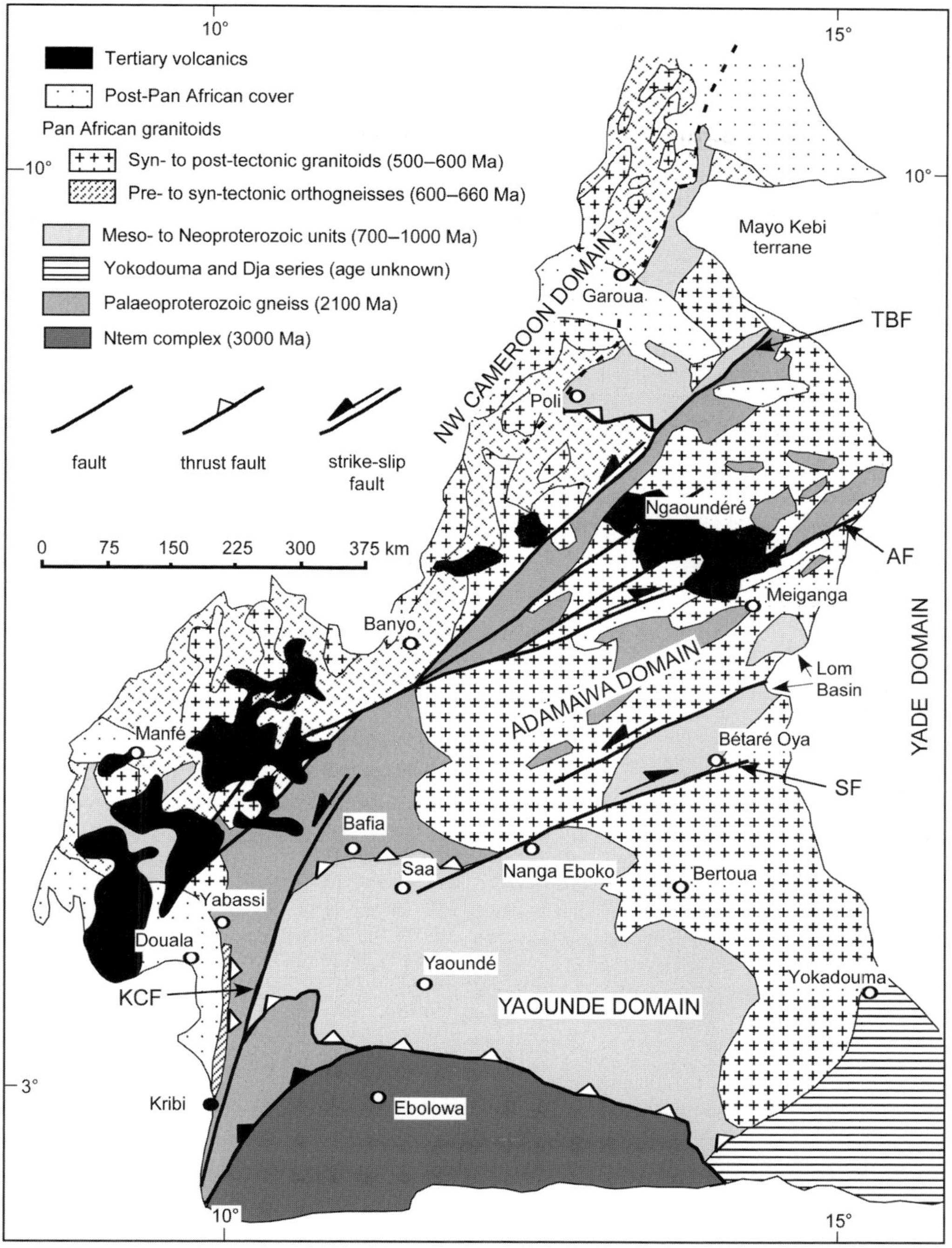

Fig. 12. Geological map of Cameroon showing major lithotectonic suites and domains. Modified from Toteu *et al.* (2001). AF, Adamawa fault; KCF, Kribi–Campo fault; SF, Sanaga fault; TBF, Tcholliré–Banyo fault.

in large part consists of units thrust southwards over the craton; it continues eastwards as the Oubanguide Belt in the Central African Republic. The *Adamawa–Yadé domain* extends eastwards from central Cameroon and is bordered on the north by the Tcholliré–Banyo fault and on the south by the Yaoundé domain; it is a complex and heterogeneous domain of Palaeo- to Neoproterozoic high-grade

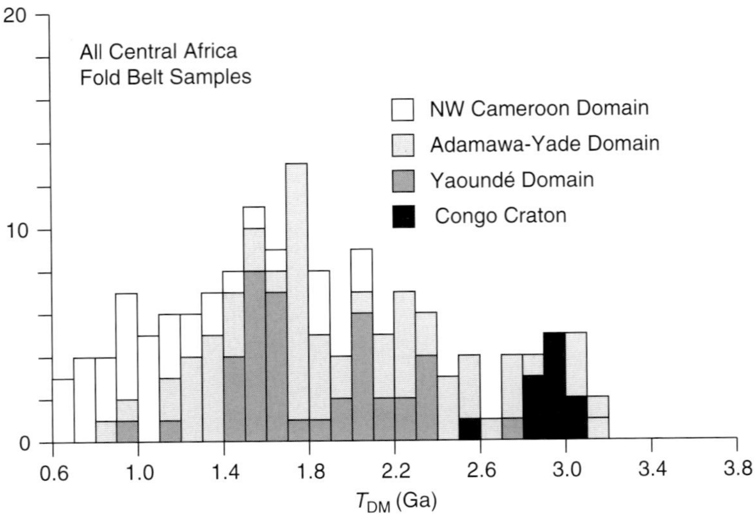

Fig. 13. Sm–Nd T_{DM} model ages from the Central African Fold Belt.

rocks extensively intruded by Pan-African granitoids and cut by late transcurrent faults. The *NW Cameroon domain* occurs NW of the Tchollíré–Banyo fault and is characterized by lesser contributions of Palaeoproterozoic crust in the Pan-African plutonic rocks, suggesting that Palaeoproterozoic basement is discontinuous to absent. Its westward continuation is ambiguous since isotopic data indicate that Palaeoproterozoic inheritance is more important in the Eastern Nigeria terrane. Geophysical information across the Benue Trough will be necessary in the future to determine whether or not the Eastern Nigeria terrane belongs to a different block and is separated from the NW Cameroon domain by a major crustal boundary.

Yaoundé domain of Cameroon and Central African Republic

The Yaoundé domain (Fig. 12) contains three major units: (a) low to medium-grade schists of the *Mbalmayo group*, passing in continuity northwards to (b) a unit of garnetiferous micaschists and gneisses, granulites and migmatitic gneisses of the *Yaoundé Group* (Nédélec *et al.* 1986; Nzenti *et al.* 1988); (c) a *Bafia Group* of assumed Palaeoproterozoic age that consists of high-grade gneisses and orthogneisses (Noizet 1982; Tchakounté 1999).

The Mbamayo and Yaoundé groups are dominated by metasedimentary rocks (pelites, meta-greywackes, quartzites, amphibolites, talc-schists) and meta-plutonic rocks (meta-diorites, meta-gabbros, meta-syenites, meta-granites and meta-peridotites), all of which were involved in Pan-African nappe formation (Toteu *et al.* 2006*b*). Figure 13 illustrates Sm–Nd T_{DM} ages for rocks of the Yaoundé domain and their potential sources. The bulk of the detritus in Yaoundé domain metasedimentary rocks may have come from sources to the north, rather than from the Congo Craton (Penaye *et al.* 1993; Toteu *et al.* 1994, 2001), which is comparable to the case for the Sergipano domain in Brazil (Oliveira *et al.* 2006). The presence of *c.* 626 Ma detrital zircon in a Yaoundé mica-schist suggests it was deposited after this age (Toteu *et al.* 2006*a*), making sedimentary units of the Yaoundé domain essentially coeval with units from the Sergipano domain. The Yaoundé domain metasedimentary rocks are interpreted as the products of reworking of Neoproterozoic Pan-African arc rocks developed in the southern part of the Adamawa–Yadé domain (Toteu *et al.* 2006*a*).

The Bafia Group is poorly known. From the tectonic point of view it is part of the Yaoundé domain as it is involved in the nappe tectonics. However, available Sm–Nd T_{DM} ages (all Palaeoproterozoic) suggest that this group may be part of the Adamawa–Yadé domain. It may comprise a tectonic intercalation of Neoproterozoic, Mesoproterozoic (?) and Palaeoproterozoic units (Toteu *et al.* 2001, 2006*a*). The main difference relative to the Yaoundé Group is the scarcity of metapelitic rocks and abundance of meta-greywackes (biotite and amphibole gneisses) and amphibolites. Both groups were intruded by Pan-African granitoids prior to nappe formation.

Deformation history in the Yaoundé domain involved two successive episodes, D_1 and D_2, with peak conditions of the Pan-African granulite facies metamorphism occurring between D_1 and D_2. D_2 regional flat-laying foliation is the result of southward-directed thrusting (Toteu *et al.* 2004, 2006*b*) or extension (Mvondo *et al.* 2003). The low-grade rocks of the Mbalmayo Group are interpreted as the sole of the nappes (Nédélec *et al.* 1986). The high-grade metamorphism in the Yaoundé region is constrained between 620 and 610 Ma (Penaye *et al.* 1993; Toteu *et al.* 1994, 2006*b*; Stendal *et al.* 2006). The Yaoundé Group underwent a rapid evolution (deposition, burial and metamorphism to granulite facies, followed by exhumation and thrusting onto the Congo Craton) in a time span of about 20 Ma (Toteu *et al.* 2006*b*).

Adamawa–Yadé domain

The Adamawa–Yadé domain (Toteu *et al.* 2004) is characterized by Palaeoproterozoic upper intercept ages for zircons and by Palaeoproterozoic to Archaean Sm–Nd T_{DM} ages recorded in both metasedimentary rocks and orthogneisses. This indicates that the domain is underlain by Archaean–Palaeoproterozoic basement. However, although 2.1 Ga granulitic metamorphism has been proposed (Toteu *et al.* 2001), no clear Palaeoproterozoic relict metamorphic age (garnet or titanite) has so far been recorded. The domain includes: (a) remnants of metasedimentary rocks and orthogneisses showing retrogressed granulitic metamorphism; (b) the Lom schist belt, which is composed of low- to medium-grade metasedimentary rocks and felsic volcaniclastic rocks with Pan-African amphibolite facies metamorphism; (c) the poorly known 'Yadé Massif' in the Central African Republic; (d) widespread syn- to late-tectonic granitoids of transitional composition and crustal-derived origin.

Although Palaeoproterozoic juvenile material exists in some areas of this domain, Sm–Nd T_{DM} ages (Fig. 13) and inherited zircons show that most basement rocks were derived from recycling (melting or erosion and sedimentation) of Archaean crust similar in age to the Congo Craton. This is also the case for the Nyong Series in SW Cameroon, which is regarded as southern extension of basement rocks of the Adamawa region (Toteu *et al.* 2001) but which has remained linked with the Congo Craton during Neoproterozoic fragmentation.

Toteu *et al.* (2006*a*) reported U–Pb ion microprobe ages for zircons from metasedimentary and metavolcanic rocks of the Lom Basin of the Adamawa–Yadé domain in eastern Cameroon. These data show that depositional ages for the supracrustal rocks are late Neoproterozoic, with detrital zircons up to 2800 Ma and Sm–Nd T_{DM} ages of 1.4 to 2.2 Ga. This basin is probably underlain by Palaeoproterozoic crust, but the younger Sm–Nd model ages indicate that significant amounts of detritus must have come by lateral transport from presently unknown, but younger and probably Neoproterozoic, sources.

The Yadé massif in the Central African Republic is a poorly surveyed complex of granitic gneiss that has been assumed to be Archaean (Poidevin 1991). However, its proximity with the Adamawa domain in Cameroon and the continuity of structures suggest that the Yadé massif is probably an eastern continuation of that domain, consistent with current field work and Sm–Nd isotopic studies indicating that the Yadé massif is underlain by Palaeoproterozoic crust.

NW Cameroon domain

The NW Cameroon domain lies west of the Tchollíré–Banyo shear zone and continues into eastern Nigeria and southwestern Chad (Figs 12 & 14). It consists of (a) Neoproterozoic medium- to high-grade schists and gneisses of volcanic and volcano-sedimentary origin (Poli Group) that were formed *c.* 700 Ma on, or in the vicinity of, young magmatic arcs (Toteu *et al.* 2006*a*), (b) Pan-African pre-, syn-, to late-tectonic calk-alkaline granitoids emplaced between 660 and 580 Ma (Toteu *et al.* 1987, 2001; Penaye *et al.* 2006), (c) post-tectonic alkaline granitoids which comprise mafic and felsic dykes cross-cut by sub-circular granites and syenites and (d) numerous basins with low-grade metasedimentary and metavolcanic rocks that may correspond to molasse deposits of the Pan-African orogeny (Montes-Lauar *et al.* 1997). Rocks of the NW Cameroon domain commonly yield Mesoproterozoic to Neoproterozoic Sm–Nd model ages, without the abundant Palaeoproterozoic ages found in the Adamawa–Yadé domain (Fig. 13). This, along with Rb–Sr and U–Pb data, indicate that most of the gneissic and granitic rocks of this domain are Neoproterozoic with relatively low contribution from 2.1 Ga crust. No Archaean inheritance has yet been recognized (Toteu *et al.* 2001), but sampling is still relatively sparse.

Toteu *et al.* (2006*a*) reported new U–Pb ages and Sm–Nd T_{DM} ages for various units in the Poli region in NW Cameroon. They documented *c.* 920–730 Ma metavolcanic rocks and detrital magmatic zircons in the supracrustal suite. Sm–Nd T_{DM} ages of 0.8 to 1.1 Ga for these rocks indicate a substantial juvenile component. To the NE of Poli, in the Mayo Kebi region of SW Chad, Penaye *et al.* (2006) reported U–Pb zircon ages of *c.* 740 Ma with Sm–Nd T_{DM} ages of 0.6 to 0.8 Ga for gabbroic to dioritic rocks that have been interpreted as part of a juvenile arc that lies parallel to and on the NW side of the Tchollíré–Banyo shear

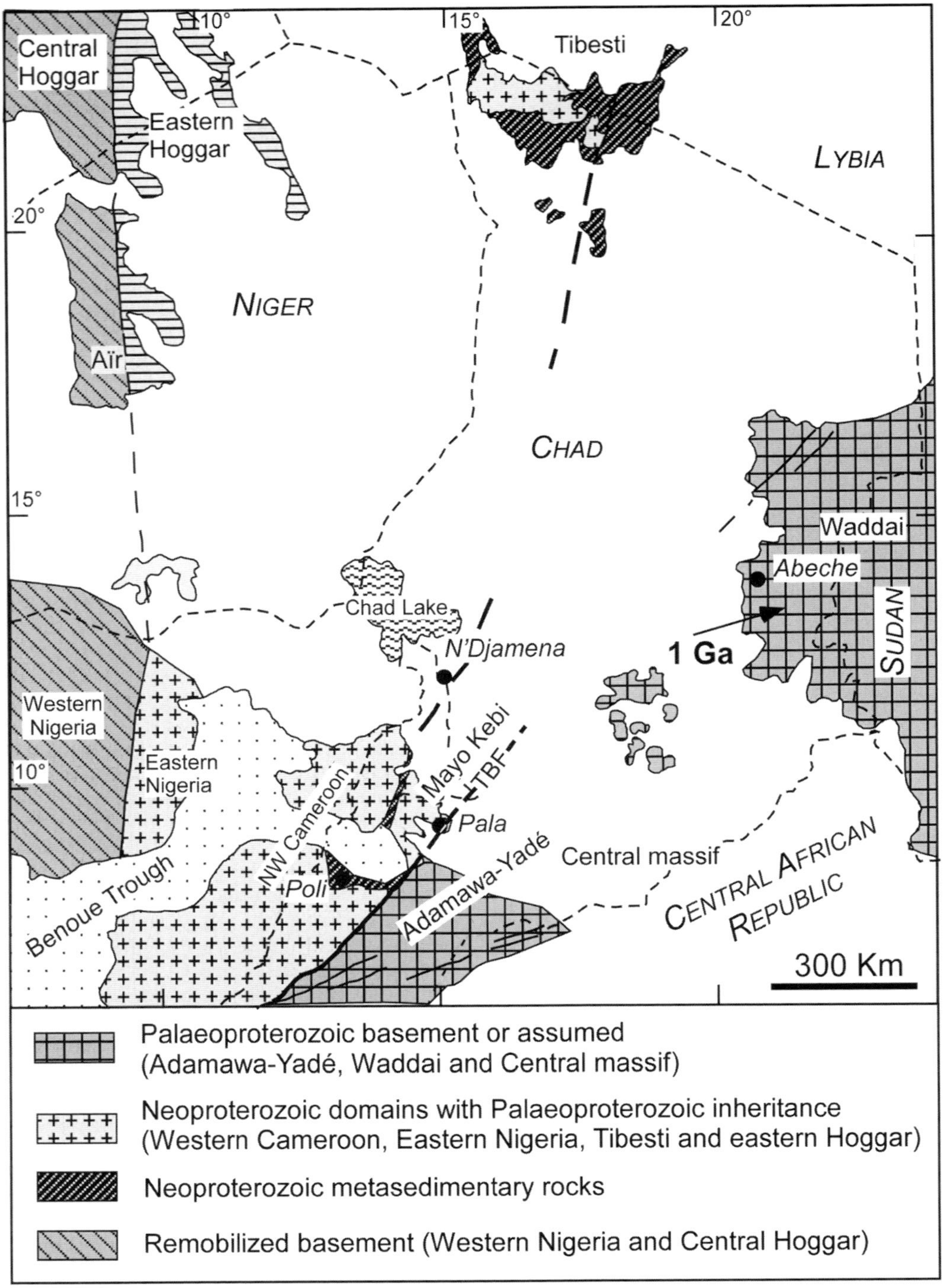

Fig. 14. Geological map of northern Cameroon, adjacent parts of Nigeria, Chad, and the Central African Republic, and areas in the Saharan region to the north. Modified from Figure 7 of Penaye *et al.* (2006). '1 Ga' denotes region from which de Wit *et al.* (2005) reported *c.* 1000 Ma gneisses. TBF, Tchollíré–Banyo fault.

zone. Other plutonic units in the Mayo Kebi area yield U–Pb ages of 640 to 665 Ma, with one late pluton at *c.* 567 Ma. This region is particularly notable because it clearly documents the existence of Pan-African juvenile rocks in the Central African Fold Belt (Penaye *et al.* 2006; Pouclet *et al.* 2006). It also suggests that the Tcholliré–Banyo shear zone is a major terrane boundary in Cameroon, with dominantly juvenile Neoproterozoic upper crust on the NW side and Palaeoproterozoic crust on the SE side.

The westward extent of this juvenile belt is not well constrained. Geological control in the region (Penaye *et al.* 2006) and Sm–Nd T_{DM} ages (Toteu *et al.* 2001) suggest that it is a narrow juvenile belt with rocks to the west, especially in eastern Nigeria, containing a greater Palaeoproterozoic component (see following).

Nigeria and Sahara

The eastern part of the Nigerian shield appears to be westward continuation of the Central African Fold Belt. U–Pb results and Sm–Nd T_{DM} ages reported by reported by Dada (1998), Ekwueme & Kröner (2006) and Ferré *et al.* (1996, 1998, 2002) show that, while Pan-African metamorphism and magmatism dominate the tectonic history of the region from 660 to 580 Ma, Palaeoproterozoic basement, presumably *c.* 2.2–2.0 Ga (Birimian/Eburnian-age) crust, is ubiquitous and dominates the Sm–Nd model ages. This is also true for Pan-African granites, which have T_{DM} ages of *c.* 2.0 Ga; Mesoproterozoic or younger T_{DM} ages seem to be lacking in eastern Nigeria. This is consistent with the suggestion above that the Poli–Mayo Kebi terrane is a narrow juvenile belt, with western parts of the NW Cameroon domain more similar to eastern Nigeria or to the Adamawa–Yadé domain.

Ferré *et al.* (1996, 2002) argued that there is a major north–south boundary in Nigeria which separates the Eastern Nigeria terrane from a dominantly Palaeoproterozoic–Archaean Western Nigeria terrane. Ekwueme & Kröner (1993) and Bruguier *et al.* (1994) reported Archaean U–Pb ages of *c.* 3.5 Ga in the Kaduna region in the northern part of the Western Nigeria terrane, enhancing the contrast with the Eastern Nigeria terrane. Liégeois *et al.* (1994) presented a tectonic history for the Aïr Massif in the southeastern Tuareg Shield, to the north of the Nigerian terranes. They argued for a major terrane boundary between the eastern and western part of the massif, and this boundary can be continued southwards to match up with the terrane boundary between the Western Nigeria and Eastern Nigeria terranes (Fig. 14). Gneissic basement in eastern Chad (western Darfur region) yields U–Pb zircon ages of *c.* 1.0 Ga with Sm–Nd T_{DM} ages of 1.70 to 3.0 Ga (de Wit *et al.* 2005). This suggests that the Darfur region may be northeastward continuation of the Adamawa–Yadé domain.

Geological links between the Borborema Province and the Central African Fold Belt

Sm–Nd model ages, in conjunction with U–Pb zircon ages and other geological data, show that Palaeoproterozoic (and locally Archaean) basement gneisses occur discontinuously throughout most of the Borborema Province and the Central African Fold Belt, either as discrete basement blocks and terranes, lower crust inferred from isotopic studies of younger plutons, or as protoliths to Brasiliano and Pan-African metamorphic complexes. These same data show that Neoproterozoic supracrustal sequences also occur throughout the region, as thin units on Palaeoproterozoic basement, as thick sequences in deep basins with no known basement, or as clearly juvenile terranes between major structural boundaries. The presence or absence of Palaeoproterozoic basement, either at the surface or at depth, and the distribution of Neoproterozoic supracrustal sequences are an important aspect of correlation between Brazil and Africa. In this section we will examine these correlations, beginning in the south with the stable cratons.

Correlation between Congo Craton and São Francisco Craton

It is widely accepted that before the Mesozoic break-up of Gondwana the São Francisco and the Congo cratons were part of a larger continent, and these cratons are deformed on their margins by Pan-African/Brasiliano orogens (e.g., Alkmim *et al.* 2001). Both cratonic areas contain Archaean nuclei such as the Jequié, Gavião, and Serrinha blocks in Brazil and the Ntem, Equatorial Guinea, Gabon, and Congo blocks in western Central Africa. These nuclei have ages ranging from 3.4 to 2.6 Ga (Toteu *et al.* 1994; Barbosa & Sabaté 2004; Caen-Vachette *et al.* 1988; Campos *et al.* 2003; Kosin *et al.* 2003; Oliveira *et al.* 2004; Shang *et al.* 2004; Milesi *et al.* 2006 and were amalgamated during 2.2–2.0 Ga orogenesis. This amalgamation is represented in Brazil by the Itabuna–Salvador–Curaçá orogen (e.g., Teixeira & Figueiredo 1991; Barbosa & Sabaté 2004; Oliveira *et al.* 2004) and in Africa by the West Central African Belt, which is a Palaeoproterozoic structure that extends south from Cameroon, along the western side of the Congo Craton (Feybesse *et al.* 1998, Penaye *et al.* 2004). These two Palaeoproterozoic collisional belts match each other very well on a pre-drift reconstruction of South America–Africa and are mainly composed of high-grade gneisses, syn- to late tectonic granitoids and terrigenous molasse (Jacobina Basin and upper units of Francevillian Basin).

The Palaeoproterozoic collisional belt between the northeastern part of the São Francisco Craton and the northwestern part of the Congo Craton is continuous on pre-drift reconstructions, without evidence of any Brasiliano/Pan-African suture. To the south, however, there is evidence of opening of a rift in the southern part of the Palaeoproterozoic craton, forming a south-facing oceanic basin that was closed at the north end (Pedrosa-Soares *et al.* 1992). Correlative late Mesoproterozoic to early Neoproterozoic (1100–910 Ma) magmatic and sedimentary sequences on the eastern edge of the São Francisco Craton and western edge of the Congo Craton include the Ilhéus–Salvador mafic dyke swarms (Correa-Gomes & Oliveira 2000) and the West Congo Supergroup (Tack *et al.* 2001). Closure of this basin during assembly of West Gondwana produced the Araçuaí–West Congo orogen. Alkmim *et al.* (2006) have presented a 'nutcracker' model in which they argue that the northern end remained closed because of rotation between the São Francisco and Congo cratons, with an early Neoproterozoic opening phase followed by a late Neoproterozoic (Brasiliano/Pan-African) closing phase and a pivot point at the north end of the orogen.

Correlation between the Sergipano and Yaoundé domains

The major feature of the Sergipano and Yaoundé domains is their formation as a result of the continental collision between the Pernambuco–Alagoas (PEAL) and Adamawa–Yadé massifs to the north and the São Francisco and Congo cratons to the south. The resulting structure in both regions is the nappe stacking of tectono-stratigraphic units with different characteristics.

Dating of detrital zircons in the Sergipano domain shows that lithostratigraphic domains are broadly young (<650 Ma) and probably formed between 630 and 600 Ma apart from the Estância subdomain foreland basin (Fig. 3). The Macururé subdomain, which does not yield detrital zircons younger than 850 Ma, could be older; the 628–625 Ma intrusive granitoids indicate that they may have been deposited at any time between 850 and 625 Ma. Except for the Estância subdomain, for which an equivalent has not yet been found in Cameroon, ages obtained for the other domains of the Sergipano domain can be found in Cameroon. For example, the Yaoundé mica-schists yield a detrital zircon age of 626 Ma comparable to that of Vaza Barris and Canindé subdomains. On the other hand, detrital zircon from the Mahan amphibolite in Cameroon yielded ages of 1072 and 820 Ma (Toteu *et al.* 2006*a*), comparable to the ages recorded in the Macururé and Marancó subdomains in Brazil.

New SHRIMP geochronology, Sm–Nd model ages, and major and trace element data for units of the Sergipano domain indicate that the Canindé subdomain is the root of a 715–680 Ma rift-related volcano-plutonic-sedimentary sequence (Nascimento *et al.* 2005). The Poço Redondo gneiss migmatite subdomain is possibly the root of a 980–960 Ma Andean-type batholith intruded by a post-collision A-type batholith (*c.* 950 Ma) and intensively reworked during the Late Neoproterozoic (Carvalho *et al.* 2005). Sedimentary provenance studies on the Marancó and Macururé metasedimentary subdomains indicate detritus provenance mostly from 980–1020 Ma old terranes, possibly the Poço Redondo subdomain or Cariris Velhos-age terranes farther north (Oliveira *et al.* 2006); no detrital zircons younger than 900 Ma were found in these metasedimentary rocks. On the other hand, the southernmost Vaza Barris and Estância subdomains contain Brasiliano/Pan-African age detrital zircons (680–570 Ma) probably derived from Brasiliano/Pan-African granitoids, in addition to older Meso- to Neoproterozoic (920–1110 Ma) and Archaean zircon grains. The uppermost sedimentary units of these two subdomains are interpreted as having been deposited on peripheral foreland basins, with detritus coming mostly from terranes to the north (Oliveira *et al.* 2006). Zircon provenance studies in Cameroon are scarce: Toteu *et al.* (2006*a*) and Toteu *et al.* (2001) suggested that metasedimentary rocks in the Yaoundé domain are mainly the product of detritus from Neoproterozoic magmatic arcs and Palaeoproterozoic igneous basement of the Adamawa–Yadé domain, although a few Archaean xenocrysts are also present. Sm–Nd model ages (Toteu *et al.* 2001) are compatible with sediment transport from the north with insignificant or no contribution from the Congo Craton. Geological units in the age range 980–950 Ma, comparable to rocks from the Cariris Velhos orogen (Brito Neves *et al.* 1995), have not yet been found in Cameroon, but there is a strong possibility that they might be found in the future as zircons of this age are already reported elsewhere in Central Africa (de Wit *et al.* 2005; Toteu *et al.* 2006*a*).

Both domains are dominated by south-verging Brasiliano/Pan-African thrusting that lead to nappe stacking upon the Congo–São Francisco cratons. Boundaries between tectono-stratigraphic units are well defined in the Sergipano domain. In the Yaoundé domain they are assumed, since detailed geological mapping is lacking in much of the region. All plutonic rocks of both belts are involved in this thrusting and nappe formation. To the north of both belts, e.g., in the PEAL domain of the Borborema Province and Adamawa–Yadé domain of the Central African Fold Belt, there is extensive development of late- to post-tectonic granitoids and strike-slip shear zones. This suggests that plutonism in both regions continued after nappe emplacement.

Both regions are characterized by a decrease in metamorphic grade towards the cratons. In the Sergipano domain there is no direct estimate of the age of the metamorphism. However, the minimum age of Pan-African/Brasiliano collision is constrained by a pre-D_2 tonalite intrusion in the Sergipano domain dated at 628 ± 12 Ma (Bueno *et al.* 2005). Consistent Sm–Nd whole rock–garnet pair and Pb–Pb step-leaching ages on garnet constrain metamorphism in the Yaoundé domain between 616 and 611 Ma, which broadly corresponds to the onset of nappe formation that continued until about 600 Ma (Toteu *et al.* 2006b; Stendal *et al.* 2006). The presence of 620 ± 10 Ma, pré-D_2 granulite facies meta-diorite in the Yaoundé series (Toteu *et al.* 1994) suggests that the collisional event is broadly coeval in both belts.

Correlation of PEAL and Transverse domains of NE Brazil with the Central African Fold Belt

A common feature within the central domains of both the Borborema Province (Pernambuco–Alagoas and Transverse domains) and the Central African Fold Belt (Adamawa–Yadé and NW Cameroon domains, Eastern Nigeria terrane) is the prevalence of metasedimentary, metavolcanic, and meta-plutonic rock units with Sm–Nd model ages (T_{DM}) between 1.0 and 1.6 Ga, substantially younger than in domains to the south (São Francisco and Congo cratons) and north (Rio Grande do Norte domain, Western Nigeria terrane). U–Pb ages of detrital zircons in metasedimentary rocks or of igneous zircons in plutons show that most of the rock units with Mesoproterozoic T_{DM} ages probably formed between 1.0 and 0.6 Ga, with the apparent Mesoproterozoic model ages being due to mixing of variable amounts of Palaeoproterozoic to Archaean crustal material with 1.0 to 0.6 Ga juvenile material. Brasiliano and Pan-African plutons emplaced into Palaeoproterozoic crust normally show Palaeoproterozoic Sm–Nd model ages, reflecting the sources of the magmas. Young plutons emplaced into Neoproterozoic metasedimentary rocks overlying older basement may also show Palaeoproterozoic Sm–Nd model ages, but in this case they reflect magma generation from older crust underlying the younger supracrustal rocks.

In many Neoproterozoic metasedimentary sequences in NE Brazil (e.g., Cariris Velhos, Cachoerinha, Sergipano, parts of the PEAL domain) the Sm–Nd model ages in Brasiliano plutons are Mesoproterozoic (1.0–1.6 Ga), often virtually the same as the T_{DM} ages of the host metasedimentary rocks. In these cases it is difficult to argue for significant contribution of underlying Palaeoproterozoic crust to generation of the magma. Although Sm–Nd model ages alone can usually delineate regions of thick post-Palaeoproterozoic (probable Neoproterozoic) crust, zircon ages are required to determine accurate and precise crystallization or maximum depositional ages within the Neoproterozoic. For these supracrustal sequences we must conclude that either the basins were very deep (>20 km, with or without Palaeoproterozoic basal crust) so that they could be melted to form the granite magmas, or that juvenile, late Neoproterozoic mantle-derived magma fortuitously mixed with Palaeoproterozoic crust to yield Mesoproterozoic T_{DM} ages.

Neoproterozoic rocks with Mesoproterozoic model ages also occur in the Central African Fold Belt, notably in the Yaoundé Group (presented above), the Lom basins in the Adamawa domain, and the Poli Basin in the NW Cameroon domain. Pan-African granites in the region also commonly show Mesoproterozoic T_{DM} ages, although they are more dominant in the NW Cameroon domain than in the Adamawa–Yadé domain (Fig. 13). In general, however, the abundance and distribution of such plutons are not as well known in the Central African Fold Belt, making precise correlations with the Borborema Province of Brazil difficult.

There are two major structures in Cameroon that project westwards toward Brazil: the Tcholliré–Banyo fault and the cross-cutting Adamawa fault. As argued above, the Tcholliré–Banyo fault appears to be a significant terrane boundary, with lesser contributions from Palaeoproterozoic crust in the NW Cameroon domain than in the Adamawa–Yadé domain. A potential counterpart in Brazil, based on geology (Fig. 9) and spatial relationships (Fig. 1) could be the northern margin of the Sergipano domain, which continues westwards as the Macururé shear zone (Fig. 3), separating the Sergipano domain from the central part of the PEAL domain. A consequence of this correlation is that the *c.* 750–640 Ma Poli–Mayo Kebi terrane in Cameroon–Chad (Figs 12 & 14) and the *c.* 720–630 Ma Canindé subdomain in Brazil (Fig. 3) would both lie on the north side of the boundary, perhaps being remnants of juvenile complexes along a major rift and/or suture.

The Adamawa fault is a younger transcurrent feature within the Adamawa domain and has commonly been extrapolated to join with the Pernambuco shear zone, which is similarly a late-tectonic transcurrent feature (Neves & Mariano 1999). Neither appears to represent a significant terrane boundary, so that although the correlation is permissible, it is not well constrained by geological relationships. The extrapolation of these two structures towards each other is not well aligned in the de Wit *et al.* (1988) reconstruction used (Fig. 1), suggesting either that they are not correlated or that the fit between Cameroon and NE Brazil needs to be modified.

There are also some major structures in the Transverse domain of Brazil which project eastwards toward Africa. One is the boundary between the Alto Moxotó terrane and the Rio Capibaribe terrane (Fig. 6). The Rio Capibaribe terrane consists of Palaeoproterozoic basement, late Neoproterozoic supracrustal rocks (e.g., Suburim schists, Neves *et al.* 2005, 2006), and Brasiliano granites. This terrane bears superficial resemblance to western parts of the NW Cameroon domain, NW of the Poli–Mayo Kebi terrane, although an exact comparison is not possible with the limited data currently available.

The Alto Moxotó terrane is somewhat different from the two terranes which flank it in the Transverse domain. Unlike the Rio Capibaribe terrane, the Alto Moxotó terrane contains relatively few Brasiliano plutons or supracrustal rocks and is dominated by *c.* 2.1 Ga Palaeoproterozoic crust (Brito Neves *et al.* 2001*b*). In this respect it is more like the Eastern Nigeria terrane, which is dominated by Palaeoproterozoic crustal parentage (T_{DM} *c.* 2.0 Ga). Although the Eastern Nigeria terrane has abundant Pan-African plutons (see Ferré *et al.* 2002), this could be a local manifestation in an otherwise dominantly Palaeoproterozoic terrane. If this correlation is valid, then the boundary between the Alto Moxotó and Rio Capibaribe terranes could continue from eastern Brazil (Fig. 6) into the covered region (Benoue Trough) between eastern Nigeria and NW Cameroon (Figs 1 & 14).

The Alto Pajeú terrane, with its included Cariris Velhos orogen, is quite different from all other terranes in both Brazil and Africa. Although it includes some Palaeoproterozoic remnants, it is dominated by early Neoproterozoic crustal rocks formed at 990–940 Ma. There are major contrasts with terranes on either side: the Alto Moxotó terrane on the south is dominantly Palaeoproterozoic gneiss, whereas the Palaeoproterozoic basement of the Rio Grande do Norte domain on the north includes a significant Archaean component as well as a discrete 3.5 Ga Archaean block (Fig. 2). So far no counterpart to the Alto Pajeú terrane is known in Africa, and it is shown in Figure 1 as pinching out eastward. Thus, its northern and southern boundaries merge to become a major terrane boundary in Nigeria and to the north (see below).

One consequence of the correlations proposed above is that, whereas the Pernambuco–Alagoas and Transverse domains in Brazil correlate with the NW Cameroon and Eastern Nigeria domains in Africa (Fig. 1), there is no major counterpart on the Brazilian side to the Adamawa–Yadé domain of Africa. Furthermore, because of relatively poor geological control, there is no clearly defined boundary in Cameroon between the Yaoundé domain and the Adamawa domain. These are problems that will have to await future resolution.

Western Nigeria terrane–Tuareg Shield–Rio Grande do Norte/Ceará domains

The eastern part of the Patos shear zone in Brazil represents a major terrane boundary between Archaean and Palaeoproterozoic cratonic rocks of the Rio Grande do Norte domain and dominantly early Neoproterozoic (Cariris Velhos) to Brasiliano rocks of the Alto Pajeú terrane. Recent work (e.g., Brito Neves *et al.* 2001*a*) shows that this terrane boundary follows a fault that swings northeast before it leaves the continent north of João Pessoa (Fig. 2). In Gondwana reconstructions, this boundary trends into Nigeria and can easily be aligned with the major terrane boundary (Fig. 14) that separates the Archaean–Palaeoproterozoic Western Nigeria terrane from the Eastern Nigeria terrane (Ferré *et al.* 1996). This boundary can, in turn, be aligned northward with a major terrane boundary in the Aïr massif and which separates Central Hoggar from Eastern Hoggar in the Tuareg Shield (Liégeois *et al.* 1994; Caby 2003; Fig. 14). This proposed alignment separates older cratonic basement on the north and west (Rio Grande do Norte domain, Western Nigeria terrane, Central Hoggar) from terranes which consist of a collage of isolated Palaeoproterozoic blocks and Neoproterozoic basins (Transverse domain in Brazil, NW Cameroon domain, Eastern Nigeria terrane, and western parts of the Saharan metacraton) (Figs 1, 2 & 14). Continuation of this boundary west and SW of the city of Patos in Brazil is not well constrained, although it probably trends SW along the west side of the Cachoerinha Belt (Boqueirão dos Conchos shear zone, Fig. 2).

Médio Coreaú domain and West African Craton

The northwestern part of the Borborema Province contains a major fault where the Médio Coreaú domain is joined to the Ceará domain. This Sobral fault is part of the Transbrasiliano lineament (Fig. 2), and it has commonly been correlated with the Kandi fault east of the West African Craton (Caby 1989; Castaing *et al.* 1993, 1994; Brito Neves *et al.* 2002; Fetter *et al.* 2003; Santos *et al.* 2008).

São Francisco/Congo craton–Boborema Province/Central African Fold Belt/Sahara–West Africa/Angola craton history

We believe that the São Francisco and Congo cratons are relatively undeformed remnants of a large Palaeoproterozoic continent that extended northwards throughout the Borborema Province and central and northern Africa. At some time the northern part of this continent underwent major crustal extension, creating a large expanse of faulted Palaeoproterozoic crust with major depositional basins created over down-dropped blocks or where extension may have created small oceanic rifts. Because the oldest metasedimentary and metavolcanic rocks associated with these basins are earliest Neoproterozoic (about 1 Ga; Cariris Velhos in Brazil), we believe that this is a likely time for onset of extensional tectonics. Because extension was also occurring to the south in the subsequent Araçuaí–West Congo orogen, extension in the Borborema Province–West African Fold Belt may have resulted largely from northward movement of crustal blocks farther north, perhaps during separation of the Amazonian–West African craton from the São Francisco Craton–Congo Craton–Saharan metacraton to form a major Neoproterozoic ocean (e.g., Alkmim *et al.* 2001). Kröner & Cordani (2003) argued that, while the West African Craton and the Amazon Craton were part of Rodinia, there was a major Neoproterozoic ocean between them and the São Francisco–Borborema–Central African Fold Belt–Sahara–Congo continental complex.

Evidence for *c.* 700–800 Ma sequences and igneous rocks in both Brazil and Africa may testify to a second pulse of extension in the middle Neoproterozoic, and some of the oceanic rifts may have closed about 640 Ma, preserving remnants of juvenile terranes (e.g., Poli–Mayo Kebi terrane in the Central African Fold Belt, Canindé subdomain in Brazil). Subsequent late Neoproterozoic convergence resulted in assembly of West Gondwana, with substantial terranes of juvenile crust accreted between the Amazon and São Francisco cratons (Pimentel *et al.* 2000) and between the West African Craton and the Saharan metacraton. The assembly of all these terranes, domains and cratons resulted in extensive deformation, metamorphism, and granitic magmatism from about 640 to 540 Ma, with the major peak of activity between 620 and 580 Ma. The proposed 'nutcracker' cycle proposed for the Araçuaí–West Congo orogen (Alkmim *et al.* 2006) further indicates that motions between and among cratons during assembly of this part of West Gondwana must have been very complex.

Van Schmus acknowledges support from several NSF grants from 1983 to 2002 in support of his studies in Brazil and Cameroon, with additional support from the Department of Geology, University of Kansas. He also gratefully acknowledges the collaborative research with many students and colleagues during this time, including but not limited to Marly Babinski, B. Bley de Brito Neves, Umberto Cordani, Elton Dantas, Allen Fetter, Peter C. Hackspacher, Emanuel F. Jardim de Sá, and Marianne Kozuch.

E. P. Oliveira acknowledges the financial support of the Brazilian agencies CNPq (301025/2005-3) and FAPESP (02/03085-2, 02/07536-9). A.F. da Silva Filho acknowledges the financial support of the Brazilian agencies CAPES (AEX0753/99-8), CNPq (521.031/95-8, 520.012/96-8) and FINEP (88.98.0745.00). He also acknowledges the important contribution of research students M. F. Lyra de Brito, L. Osako, C. Carmona, E. B. Luna and D. V. Siqueira, and the colleagues B. Bley de Brito Neves, E. J. dos Santos and J. M. Rangel da Silva. The isotopic work was carried out in the Isotope Geochemistry Laboratory (IGL), University of Kansas. He is very grateful to M. Kozuch and to Allen Fetter for their technical support in the IGL. Cameroon portions of this work were supported in early phases by the Institute for Geological and Mining Research (IRGM) in Cameroon, which supported field work in northern Cameroon and isotopic analyses at CRGP, Nancy (France). In 1993, S. F. Toteu benefited from special funding from Cimentéries du Cameroun (CIMENCAM) to organize a field trip for sample collection in north-central Cameroon. He also wishes to pay special tribute to the United States Information Agency, which awarded him a Fulbright Grant in 1993 for a six-month visit at the IGL, University of Kansas.

Much research that forms the basis for this report was done in conjunction with IGCP-426, *Granite Systems and Proterozoic Lithospheric Processes*, and IGCP-470, *The 600 Ma Pan-African belt of Central Africa*. Support of UNESCO and IUGS for these projects is gratefully acknowledged. Thoughtful reviews by Alcides Sial and Maarten de Wit substantially improved this paper.

References

ABDELSALAM, M. G., LIÉGEOIS, J. P. & STERN, R. J. 2002. The Saharan metacraton. *Journal of African Earth Sciences*, **31**, 119–136.

ALLARD, G. O. & HURST, V. J. 1969. Brazil-Gabon geologic link supports continental drift. *Science*, **163**, 528–532.

ALKMIM, F. F., MARSHAK, S. & FONSECA, M. A. 2001. Assembling West Gondwana in the

Neoproterozoic: Clues from the São Francisco craton region, Brazil. *Geology*, **29**, 319–322.

ALKMIM, F. F., MARSHAK, S., PEDROSA-SOARES, A. C., PERES, G. G., CRUZ, S. C. & WHITTINGTON, A. 2006. Kinematic evolution of the Araçuaí–West Congo orogen in Brazil and Africa: Nutcracker tectonics during the Neoproterozoic assembly of Gondwana. *Precambrian Research*, **149**, 43–64.

ARAÚJO, M. N. C., OLIVEIRA, E. P. & CARVALHO, M. J. 2003. Tectônica de endentação na Faixa Sergipana, NE do Brasil: compatibilização entre os elementos estruturais e cinemáticos. *In*: *IX Simpósio Nacional de Estudos Tectônicos, Búzios, Boletim de Resumos*, 115–117.

BARBOSA, J. S. F. & SABATÉ, P. 2002. Geological features and the Paleoproterozoic collision of four Archean crustal segments of the São Francisco Craton, Bahia, Brazil. A synthesis. *Anais da Academia Brasileira de Ciências*, **74**, 343–359.

BARBOSA, J. S. F. & SABATÉ, P. 2004. Archean and Paleoproterozoic crust of the São Francisco Craton, Bahia, Brazil: geodynamic features. *Precambrian Research*, **133**, 1–27.

BITTAR, S. M. B. 1998. *Faixa Piancó-Alto Brígida: Terrenos tectonoestratigráficos sob regimes metamórficos e deformacionais contrastantes*. PhD thesis, Instituto de Geociências, Universidade de São Paulo.

BIZZI, L. A., SCHOBBENHAUS, C. *ET AL*. 2001. Geologia, Tectônica e Recursos Minerais do Brasil:Sistema de Informações Geográficas-SIG. Companhia de Pesquisa de Recursos Minerais, Brasília, 2001.4 CD-ROM, Scale 1:2.500.000 (http://wwwa.geoambiente.com.br/website/cprm1/viewer.htm).

BRITO NEVES, B. B. 1978. A propósito da evolução litoestratigráfica do Precambriano Superior do Nordeste. Volume Djalma Guimarães, *Jornal de Mineralogia, Recife*, **7**, 19–27.

BRITO NEVES, B. B., SIAL, A. N. & ALBUQUERQUE, J. P. T. 1977. Vergência centrífuga residual no Sistema de Dobramentos Sergipano. *Revista Brasileira de Geociências*, **7**, 102–114.

BRITO NEVES, B. B., SIAL, A. N., RAND, H. M. & MANSO, V. V. 1982. The Pernambuco–Alagoas Massif, northeastern Brazil. *Revista Brasileira de Geociências*, **12**, 240–250.

BRITO NEVES, B. B., PESSOA, D. A. R., PESSOA, R. J. R. & CORTÊS, P. L. 1984, Estudo geocronológico das rochas do embasamento da Folha Salgueiro, Pernambuco. *In*: *Congresso Brasileiro de Geologia 33, Rio de Janeiro, Anais*. Sociedade Brasileira de Geologia, 2473–2491.

BRITO NEVES, B. B., VAN SCHMUS, W. R., SANTOS, E. J., CAMPOS NETO, M. & KOZUCH, M. 1995. O evento Cariris Velhos na Província Borborema: integração de dados, implicações e perspectivas. *Revista Brasileira de Geociências*, **25**, 279–296.

BRITO NEVES, B. B., DOS SANTOS, E. J. & VAN SCHMUS, W. R. 2000, Tectonic history of the Borborema Province, Northeastern Brazil. *In:* CORDANI, U. G., MILANI, E. J., THOMAZ FILHO, A. & CAMPOS, D. A. (eds) *Tectonic evolution of South America, 31st International Geological Congress, Rio de Janeiro, Brazil*, 151–182.

BRITO NEVES, B. B., CAMPOS NETO, M. C., VAN SCHMUS, W. R. & SANTOS, E. J. 2001*a*. O "Sistema Pajeú-Paraíba" e o "maciço" São José do Campestre no leste da Borborema. *Revista Brasileira de Geociências*, **31**, 173–184.

BRITO NEVES, B. B., CAMPOS NETO, M. C., VAN SCHMUS, W. R., FERNANDEZ, T. M. G. & SOUZA, S. L. 2001*b*. O terreno Alto Moxotó no leste da Paraiba ("Maciço Caldas Brandão"). *Revista Brasileira de Geociências*, **31**, 185–194.

BRITO NEVES, B. B., VAN SCHMUS, W. R. & FETTER, A. H. 2002. North-western Africa–North-eastern Brazil. Major tectonic links and correlation problems. *Journal of African Earth Sciences*, **34**, 275–278.

BRITO NEVES, B. B., PASSARELLI, C. R., BASEI, M. A. S. & SANTOS, E. J. 2003. Idades U–Pb em zircão de alguns granitos clássicos da Província Borborema. *Geologia USP, Série Científica, São Paulo*, **3**, 25–38.

BRITO NEVES, B. B., VAN SCHMUS, W. R., KOZUCH, M., DOS SANTOS, E. J. & PETRONILHO, L. 2005. A zona tectônica Teixeira Terra Nova-ZTTTN-fundamentos da geologia regional e isotópica. *Geologia USP, Série Científica, São Paulo*, **5**, 57–80.

BRUGUIER, O., DADA, S. & LANCELOT, J. R. 1994. Early Archean component (>3.5 Ga) within a 3.05 Ga orthogneiss from northern Nigeria: U–Pb zircon evidence. *Earth and Planetary Science Letters*, **125**, 89–103.

BUENO, J. F., OLIVEIRA, E. P., ARAÚJO, M. N. C., CARVALHO, M. J. & MCNAUGHTON, N. 2005. Granitos e a deformação na Faixa Sergipana: o início da colisão entre o Cráton São Francisco e o Maciço Pernambuco-Alagoas. *In*: *III Simpósio sobre o Cráton do São Francisco, Salvador, Bahia, Anais*, 192–195.

CABY, R. 1989. Precambrian terranes of Benin-Nigeria and northeast Brazil and the Late Proterozoic south Atlantic fit. *In:* DALLMEYER, R. D. (ed.) *Terranes in the circum-Atlantic Paleozoic orogens.* Geological Society of America, Special Papers, **230**, 145–158.

CABY, R. 2003. Terrane assembly and geodynamic evolution of central-western Hoggar: a synthesis. *Journal African Earth Sciences*, **37**, 133–159.

CAEN-VACHETTE, M., VIALETTE, Y., BASSOT, J. P. & VIDAL, P. 1988. Apport de la géochronologie isotopique à la connaissance de la géologie gabonaise. *Chronique de Recherche Minière*, **49**, 35–54.

CAMPOS, J. C. S., CARNEIRO, M. A. & BASEI, M. A. S. 2003. U–Pb evidence for late Neoarchean crustal reworking in the southern São Francisco craton (Minas Gerais, Brazil). *Annals of the Brazilian Academy of Sciences*, **75**, 497–511.

CAMPOS NETO, M. C. & BRITO NEVES, B. B. 1987. Considerações sobre a organização e geometria do Sistema de Dobramentos Sergipano. *In*: *I Simposio Nacional Estudos Tectonicos, Salvador, Bahia, Abstract volume*, 90–93.

CAMPOS NETO, M. C., BITTAR, S. M. B. & BRITO NEVES, B. B. 1994. Domínio tectônico Rio Pajeú-Província Borborema: orogêneses superpostas no ciclo Brasiliano/Pan-Africano. *In*: *Congresso Brasileiro de Geologia 38, Balneário*

Camboriú, Anais, **1**. Sociedade Brasileira de Geologia, 221–222.

CARVALHO, M. J., OLIVEIRA, E. P., DANTAS, E. L. & MCNAUGHTON, N. 2005. Evolução tectônica do Domínio Marancó - Poço Redondo: registro das orogêneses Cariris Velhos e Brasiliana na margem norte da Faixa Sergipana. *In*: *III Simpósio sobre o Cráton do São Francisco, Salvador, Bahia, Brazil, Anais*, 204–207.

CASTAING, C., TRIBOULET, C., FEYBESSE, J. L. & CHEVREMONT, P. 1993. Tectonometamorphic evolution of Ghana, Togo and Benin in the light of the Pan-African/Brasiliano orogeny. *Tectonophysics*, **218**, 323–342.

CASTAING, C., FEYBESSE, J. L., THIÉBLEMONT, TRIBOULET, C. & CHEVREMONT, P. 1994. Paleogeographical reconstructions of the Pan-African/Brasiliano orogen: closure of an oceanic domain or intracontinental convergence between major blocks. *Precambrian Research*, **69**, 327–344.

CORRÊA GOMES, L. C. & OLIVEIRA, E. P. 2000. Radiating 1.0 Ga Mafic Dyke Swarms of Eastern Brazil and Western Africa: Evidence of Post-Assembly Extension in the Rodinia Supercontinent? *Gondwana Research*, **3**, 325–332.

D'EL-REY SILVA, L. J. H. 1995. Tectonic evolution of the Sergipano Belt, NE Brazil. *Revista Brasileira de Geociências*, **25**, 315–332.

D'EL-REY SILVA, L. J. H. 1999. Basin infilling in the southern-central part of the Sergipano Belt (NE Brazil) and implications for the evolution of Pan-African/Brasiliano cratons and Neoproterozoic sedimentary cover. *Journal South American Earth Sciences*, **12**, 453–470.

D'EL-REY SILVA, L. J. H. & MCCLAY, K. R. 1995. Stratigraphy of the southern part of the Sergipano belt, NE Brazil: tectonic implications. *Revista Brasileira de Geociências*, **25**, 185–202.

DADA, S. S. 1998. Crust-forming ages and Proterozoic crustal evolution in Nigeria: a reappraisal of current interpretations. *Precambrian Research*, **87**, 65–74.

DANTAS, E. L. 1997. *Geocronologia U/Pb e Sm/Nd de terrenos Arqueanos e Paleoproterozóicos do Maciço Caldas Brandão, NE Brasil.* PhD thesis, Instituto de Geociências, Universidade Estadual Paulista, Brazil.

DANTAS, E. L., HACKSPACHER, P. C., VAN SCHMUS, W. R. & BRITO NEVES, B. B. 1998. Archean accretion in the São José do Campestre massif, Borborema Province, northeast Brazil. *Revista Brasileira de Geociências*, **28**, 221–228.

DANTAS, E. L., VAN SCHMUS, W. R. *ET AL*. 2004. The 3.4–3.5 Ga São José do Campestre massif, NE Brazil: remnants of the oldest crust in South America. *Precambrian Research*, **130**, 113–137.

DAVISON, I. & SANTOS, R. A. 1989. Tectonic Evolution of the Sergipano Fold Belt, NE Brazil, during the Brasiliano Orogeny. *Precambrian Research*, **45**, 319–342.

DEPAOLO, D. J. 1981. A neodymium and strontium isotopic study of the Mesozoic calc-alkaline granitic batholiths of the Sierra Nevada and Peninsular Ranges, California: *Journal of Geophysical Research*, **86**, 10 470–10 488.

DE WIT, M., JEFFERY, M., BERGH, H. & NICOLAYSEN, L. 1988. *Geological map of sectors of Gondwana reconstructed to their disposition c. 150 Ma*. American Association of Petroleum Geologists, Tulsa, Oklahoma.

DE WIT, M. J., BOWRING, S., DUDAS, F. & TAGNE-KAMGA, G. 2005. Saharan Africa and the tectonic assembly of the northern margin of Gondwana. *In*: PANKHURST, R. J. & VEIGA, G. D. (eds) *Gondwana 12: Geological and Biological Heritage of Gondwana, Abstracts*. Academia Nacional de Ciéncias, Cordoba, Argentina, 133.

EBERT, H. 1970. The Precambrian geology of the Borborema Belt (states of Paraíba and Rio Grande do Norte, northeastern Brazil), and the origin of its mineral resources. *Geologisches Rundschau*, **59**, 1299–1326.

EKWUEME, B. N. & KRÖNER, A. 1993. Preliminary zircon evaporation ages from migmatitic gneisses in Kaduna, N. Nigeria: evidence for Early Archean (Pre-Leonian) event in the Nigeria basement complex. *In*: *NMGS 29th Annual Conference Abstracts*. Nigerian Mining and Geosciences Society, 61.

EKWUEME, B. N. & KRÖNER, A. 2006. Single zircon ages of migmatitic gneisses and granulites in the Obudu Plateau: Timing of granulite-facies metamorphism in southeastern Nigeria. *Journal of African Earth Sciences*, **44**, 459–469.

FERREIRA, V., SIAL, A. N. & JARDIM DE SÁ, E. F. 1998. Geochemical and isotopic signatures of the Proterozoic granitoids in terranes of the Borborema Province, northeastern Brazil. *Journal of South American Earth Sciences*, **11**, 439–455.

FERRÉ, E., DÉLÉRIS, J., BOUCHEZ, J. L., LAR, A. U. & PEUCAT, J. J. 1996. The Pan-African reactivation of contrast Eburnéen and Archean provinces in Nigeria: stuctural and isotopic data. *Journal of Geological Society, London*, **153**, 719–728.

FERRÉ, E. C., CABY, R., PEUCAT, J., CAPDEVILA, R. & MONIÉ, P. 1998. Pan-African post-collisional, ferro-potassic granite and quartz-monzonite plutons of Eastern Nigeria. *Lithos*, **45**, 255–279.

FERRÉ, E., GLEIZES, G. & CABY, R. 2002. Obliquely convergent tectonics and granite emplacement in the Trans-Saharan belt of Eastern Nigeria: a synthesis. *Precambrian Research*, **114**, 199–219.

FETTER, A. H. 1999. *U–Pb and Sm–Nd constraints on the crustal framework and geologic history of Ceará State, NW Borborema Province, NE Brazil: Implications for the assembly of Gondwana*. PhD thesis, University of Kansas, USA.

FETTER, A. H., SARAIVA DOS SANTOS, T. J. *ET AL*. 2003. Evidence for Neoproterozoic Continental Arc Magmatism in the Santa Quitéria batholith of Ceará State, NW Bornorema Province, NE Brazil: Implications for the Assembly of West Gondwana. *Gondwana Research*, **6**, 265–273.

FEYBESSE, J. L., JOHAN, V. *ET AL*. 1998. The West Central African Belt: a model of 2.5–2.0 Ga accretion and two-phase orogenic evolution. *Precambrian Research*, **87**, 161–216.

GUIMARÃES, I. P. & BRITO NEVES, B. N. 2005. Geoquímica e significado tectonico do plutonismo Eo-Neoproterozóico no limite norte do domínio estructural central da Província Borborema. *In*: *XXI*

Simpósio de Geologia do Nordeste, Recife, Brazil, Boletim, **19**, 68–70.

GUIMARÃES, I. P., SILVA FILHO, A. F. *ET AL.* 2004. Brasiliano (Pan-African) granitic magmatism in the Pajeú-Paraíba belt, Northeast Brazil: an isotopic and geochronological approach. *Precambrian Research*, **135**, 23–53.

HOLLANDA, M. H. B. M., PIMENTEL, M. M. & JARDIM DE SA, E. F. 2003. Paleoproterozoic subduction-related metasomatic signatures in the lithospheric mantle beneath NE Brazil: inferences from trace element and Sm–Nd–Pb isotopic compositions of Neoproterozoic high-K igneous rocks. *Journal of South American Earth Sciences*, **15**, 885–900.

HUMPHREY, F. L. & ALLARD, G. O. 1968. The Propriá Geosyncline, a newly recognized Precambrian tectonic province in the Brazilian shield. *In*: *23rd International Geological Congress*, **4**, 123–139.

JARDIM DE SÁ, E. F. 1994. *A Faixa Seridó (Província Borborema, NE do Brasil) e o seu significado geodynâmico na cadeia Brasiliana/Pan-Africana*. PhD thesis, Instituto de Geociências, Universidade de Brasilia, Brazil.

JARDIM DE SÁ, E. F., MORAES, J. A. C. & D'EL-REY SILVA, L. J. H. 1986. Tectônica tangencial na Faixa Sergipana. *In*: *Congresso Brasileiro de Geologia 34, Goiania, Anais*, **3**. Sociedade Brasileira de Geologia, 1246–1259.

KOSIN, M., DE MELLO, R. C., DE ZOUZA, J. D., DE OLIVEIRA, A. P., CARVALHO, M. J. & LEITE, C. M. 2003. Geology of the northern segment of the Itabuna-Salvador-Curaçá orogen and field trip guide. *Revista Brasileira de Geociências*, **33**, 15–26.

KOZUCH, M. 2003. *Isotopic and trace element geochemistry of early Neoproterozoic gneissic and metavolcanic rocks in the Cariris Velhos orogen of the Borborema Province, Brazil, and their bearing on tectonic setting*. PhD thesis, University of Kansas, USA.

KRÖNER, A. & CORDANI, U. 2003. African, southern Indian and South American cratons were not part of the Rodinia supercontinent: evidence from field relationships and geochronology. *Tectonophysics*, **375**, 325–352.

LASSERRE, M. 1967. Données géochronologiques nouvelles acquises au 1er Janvier 1967 par la méthode au strontium appliquées aux roches cristallophylliennes du Cameroun. *Annales de la Faculté des Sciences, Université Clermont Ferrand*, **36**, 109–144.

LEDRU, P., COCHERIE, A., BARBOSA, J., JOHAN, V. & ONSTOTT, T. 1994. Ages du métamorphisme granulitique dans le craton du São Francisco (Brésil). Implications sur la nature de l'orogène transamazonien. *Comptes Rendus de l'Academie des Sciences, Paris*, Série II, **318**, 251–257.

LEROUGE, C., COCHERIE, A. *ET AL.* 2006. Shrimp U–Pb zircon age evidence for Paleoproterozoic sedimentation and 2.05 Ga syntectonic plutonism in the Nyong Group, South-Western Cameroon: consequences for the Eburnean–Transamazonian belt of NE Brazil and Central Africa. *Journal of African Earth Sciences*, **44**, 413–427.

LIÉGEOIS, J. P., BLACK, R. & NAVEZ, J. 1994. Early and late Pan-African orogenies in the Air assembly of terranes (Tuareg shield, Niger). *Precambrian Research*, **67**, 59–88.

LONG, L. E., CASTELLANA, C. H. & SIAL, A. N. 2003. Cooling history of the Coronel João Sá Pluton, Bahia. Brazil. *In*: *IV South American Symposium on Isotope Geology, Salvador, Brazil, Short Papers*, **1**, 92–94.

MARINHO, M. M., VIDAL, P., ALIBERT, C., BARBOSA, J. S. F. & SABATÉ, P. 1994. Geochronology of the Jequié-Itabuna granulitic belt and of the Contendas-Mirante volcano-sedimentary belt. *Boletim IG-USP, Publicação Especial*, **17**, 73–96.

MEDEIROS, V. C. & SANTOS, E. J. 1998. *Folha Garanhuns (SC.24-X-B, escala 1:250.000)*. Integração Geológica (Relatório Interno), CPRM. Recife-PE, Brazil.

MILESI, J. P., TOTEU, S. F. *ET AL.* 2006. An overview of the geology and major ore deposits of Central Africa: Explanatory note for the 1:4,000,000 map "Geology and major ore deposits of Central Africa". *Journal of African Earth Sciences*, **44**, 571–595.

MONTES-LAUAR, C. R., TROMPETTE, R. *ET AL.* 1997. Pan-African Rb-Sr isochron of magmatic rocks from northern Cameroon. Preliminary results. *In*: *1st South American Symposium on Isotope Geology, Brazil*, 204–205.

MVONDO, H., DEN BROK, S. W. J. & ONDOA, J. M. 2003. Evidence for symmetric extension and exhumation of the Yaoundé nappe (Pan-African fold belt, Cameroon). *Journal of African Earth Sciences*, **36**, 215–231.

NASCIMENTO, R. S., OLIVEIRA, E. P., CARVALHO, M. J. & MCNAUGHTON, N. 2005. Evolução Tectônica do Domínio Canindé, Faixa Sergipana, NE do Brasil. *In*: *III Simpósio sobre o Cráton do São Francisco, Salvador, Bahia, Anais*, 239–242.

NASCIMENTO, R. S. C., SIAL, A. N. & PIMENTEL., M. M. 2004. Chemostratigraphy of medium-grade marbles of the Late Neoproterozoic Seridó Group, Seridó Fold Belt, Northeastern Brazil. *Gondwana Research*, **7**, 731–744.

NÉDÉLEC, A., MACAUDIÈRE, J., NZENTI, J. P. & BARBEY, P. 1986. Evolution métamorphique et structurale des schistes de Mbalmayo (Cameroun). Implications pour la structure de la zone mobile panafricaine d'Afrique centrale, au contact du craton du Congo. *Comptes Rendus de l'Académie des Sciences, Paris*, **303**, 75–80.

NÉDÉLEC, A., NSIFA, E. N. & MARTIN, H. 1990. Major and trace element geochemistry of the Archaean Ntem plutonic complex (South Cameroon): petrogenesis and crustal evolution. *Precambrian Research*, **47**, 35–50.

NEVES, S. P. 2003. Proterozoic history of the Borborema province (NE Brazil): Correlations with neighboring cratons and Pan-African belts and implications for the evolution of western Gondwana. *Tectonics*, **22**, 1031–1044.

NEVES, S. P. & MARIANO, G. 1999. Assessing the tectonic significance of a large-scale transcurrent shear zone system: the Pernambuco lineament, northeastern Brazil. *Journal of Structural Geology*, **21**, 1369–1383.

NEVES, S. P., BRUGUEIR, O., SILVA, J. M. R. & MARIANO, G. 2005. Age, provenance, and metamorphism of the Suburim Complex (eastern Borborema Province, NE Brazil): A LA-ICP-MS study. *In*: *X Congresso Brasileiro de Geoquímica e II Simpósio de Geoquímica dos Países do Mercosul, Porto de Galinhas, Pernambuco, Brazil, Abstract CD*, 13-512.

NEVES, S. P., BRUGUEIR, O., VAUCHEZ, A., BOSCH, D., SILVA, J. M. R. & MARIANO, G. 2006. Timing of crust formation, deposition of supracrustal sequences, and Transamazonian and Brasiliano metamorphism in the East Pernambuco belt (Borborema Province, NE Brazil): Implications for western Gondwana assembly. *Precambrian Research*, **149**, 197–216.

NOIZET, G. 1982. Disposition géologique des régions de Yaoundé et Bafia. *Annales de la Faculté des Sciences, Université de Yaoundé, Cameroun*, **1**, 95–103.

NZENTI, J. P., BARBEY, P., MACAUDIÉRE, J. & SOBA, D. 1988. Origin and evolution of the late Precambrian high-grade Yaoundé gneisses (Cameroon). *Precambrian Research*, **38**, 91–103.

OLIVEIRA, E. P., WINDLEY, B. F., MCNAUGHTON, N., PIMENTEL, M. & FLETCHER, I. R. 2004. Contrasting copper and chromium metallogenic evolution of terranes in the Palaeoproterozoic Itabuna–Salvador–Curaçá Orogen, São Francisco Craton, Brazil: new zircon (SHRIMP) and Sm-Nd (model) ages and their significance for orogen-parallel escape tectonics. *Precambrian Research*, **128**, 143–165.

OLIVEIRA, E. P., ARAÚJO, M. N. C. *ET AL*. 2005*a*. Timing and Duration of Collision in the Neoproterozoic Sergipano Belt, NE Brazil: Age Constraints from Major Shear Zones, Orogenic Granites and Foreland Basin Filling. *In*: *Atas do XXI Simpósio Geologia do Nordeste, Brazil*, 95–98.

OLIVEIRA, E. P., CARVALHO, M. J. *ET AL*. 2005*b*. Evidence from detrital zircon geochronology and whole-rock Sm-Nd isotopes for off-craton provenance of clastic metasedimentary units of the Sergipano belt, NE Brazil. *In*: *X Simpósio Nacional de Estudos Tectônicos, Curitiba, Boletim de Resumos Expandidos*, 308–311.

OLIVEIRA, E. P., TOTEU, S. F. *ET AL*. 2006. Geologic correlation between the Neoproterozoic Sergipano belt (NE Brazil) and the Yaoundé schist belt (Cameroon, Africa). *Journal of African Earth Sciences*, **44**, 470–478.

OSAKO, L. 2005. *Caracterização geológica da região entre as localidades de Paranatama e Curral novo, PE, porção centro-norte do Complexo Pernambuco-Alagoas, Província Borborema*. PhD thesis, Universidade Federal de Pernambuco, Brazil.

PEDROSA-SOARES, A. C., NOCE, C. M., VIDAL, P., MONTEIRO, R. L. B. P. & LEONARDOS, O. H. 1992. Toward a new tectonic model for the Late Proterozoic Aracuai (SE Brazil)-West Congolian (SW Africa) Belt. *Journal of South American Earth Sciences*, **6**, 33–47.

PENAYE, J., TOTEU, S. F., VAN SCHMUS, W. R. & NZENTI, J. P. 1993. U–Pb and Sm–Nd preliminary geochronologic data on the Yaoundé series, Cameroon: re-interpretation of the granulitic rocks as the suture of a collision in the "Centrafrican belt". *Comptes Rendus de l'Académie des Sciences, Paris*, **317**, 789–794.

PENAYE, J., TOTEU, S. F. *ET AL*. 2004. The 2.1 Ga West Central African Belt in Cameroon: extension and evolution. *Journal of African Earth Sciences*, **39**, 159–164.

PENAYE, J., KRÖNER, A., TOTEU, S. F., VAN SCHMUS, W. R. & DOUMNANG, J.-C. 2006. Evolution of the Mayo Kebbi region as revealed by zircon dating: An early (*c*. 740 Ma) Pan-African magmatic arc in southwestern Chad. *Journal of African Earth Sciences*, **44**, 530–542.

PESSOA, D. R., PESSOA, R. R., BRITO NEVES, B. B. & KAWASHITA, K. 1978. Magmatismo tardi-tectônico brasiliano no Maciço PE-AL: o quartzo-sienito de Cachoeirinha-PE. *In*: *Congresso Brasileiro de Geologia 30, Recife, Anais*. Sociedade Brasileira de Geologia, 1279–1287.

PIMENTEL, M. M., FUCK, F. A., JOST, H., FERREIRA FILHO, C. F. & ARAÚJO, S. M. 2000. The basement of the Brasília fold belt and the Goiás magmatic arc. *In*: CORDANI, U., MILANI, E. J., THOMAZ FILHO, A. & CAMPOS, D. A. (eds) *Tectonic evolution of South America. 31st International Geological Congress, Rio de Janeiro, Brazil*, 195–229.

POIDEVIN, J. L. 1991. *Les ceintures de roches vertes de la République Centrafricaine. Contribution à la connaissance du précambrien du nord du craton du Congo*. Thèse de Doctorat d'Etat, Université Blaise Pascal, Clermont-Ferrand, France.

POUCLET, A, VIDAL, M., DOUMNANG, J.-C., VICAT, J.-P. & TCHAMENI, R. 2006. Neoproterozoic crustal evolution in Southern Chad: Pan-African ocean basin closing, arc accretion and late- to post-orogenic granitic intrusion. *Journal of African Earth Sciences*, **44**, 543–560.

RIOS, D. C. 2002. *Granitogênese no Núcleo Serrinha, Bahia, Brasil: Geocronologia e litogeoquímica*. PhD thesis, Instituto de Geociências, Universidade Federal da Bahia, Brazil.

ROGERS, J. J. W. 1996. A history of continents in the past three billion years. *Journal of Geology*, **104**, 91–107.

SÁ, J. M., MCREATH, I. & LETERRIER, J. 1995. Petrology, geochemistry, and geodynamic setting of Proterozoic igneous suites of the Orós fold belt (Borborema Province, Northeast Brazil). *Journal of South American Earth Sciences*, **8**, 299–314.

SANTOS, E. J. 1995. *O Complexo Granítico Lagoa das Pedras:Acresção e colisão na região de Floresta (Pernambuco), Província Borborema*. PhD thesis, Instituto de Geociências, Universidade de São Paulo, Brazil.

SANTOS, E. J. & BRITO NEVES, B. B. 1984. Província Borborema. *In*: ALMEIDA, F. F. M. & YOCITERU HASUI, Y. (eds) *O Pré-Cambriano do Brasil*. Edgar Blucher Ltd., São Paulo, 123–186.

SANTOS, E. J., VAN SCHMUS, W. R., BRITO NEVES, B. B., OLIVEIRA, R. G. & MEDEIROS, V. C. 1997. Terranes and their boundaries in the Proterozoic Borborema Province, northeast Brazil. *In*: *VII Simpósio Nacional Estudos Tectônicos, Bahia, Brazil, Extended Abstracts*, 120–124.

SANTOS, R. A. & SOUZA, J. D. 1988. *Programa Levantamentos Geológicos Básicos do Brasil:* Piranhus, Folha SC.24-X-C-VI, Estados de Sergipe, Alagoas e Bahia. DNPM/CPRM, Brasilia.

SANTOS, R. A., MARTINS, A. A. M., NEVES, J. P. & LEAL, R. A. 1998. *Geologia e Recursos Minerais do*

Estado de Sergipe. Companhia de Pesquisa de Recursos Minerais/Codise.

SANTOS, T. J. S., FETTER, A. H. & NOGUEIRA NETO, J. A. 2008. Comparisons between the northwestern Borborema Province, NE Brazil, and the southwestern Pharusian Dahomey Belt, SW Central Africa. *In*: PANKHURST, R. J., TROUW, R. A. J., BRITO NEVES, B. B. & DE WIT, M. J. (eds) *West Gondwana: Pre-Cenozoic correlations Across the South Atlantic Region. Geo logical Society*, London, Special Publications, **294**, 101–119.

SHANG, C. K., SATIR, M., SIEBEL, W., NSIFA, N. E., TAUBALD, H., LIEGEOIS, J. P. & TCHOUA, F. M. 2004. TTG magmatism in the Congo craton: a view from major and trace elements geochemistry, Rb-Sr and Sm-Nd systematics: case of the Sangmelima region, Ntem Complex, southern Cameroon. *Journal of African Earth Sciences*, **40**, 61–70.

SIAL, A. N. 1986. Granite-types in northeast Brazil: Current knowledge. *Revista Brasileira de Geociências*, **16**, 54–72.

SILVA, M. G. 1992. Evidências isotópicas e geocronológicas de um fenômeno de acrescimento crustal transamazônico no Cráton do São Francisco, Estado da Bahia. *In*: *Congresso Brasileiro de Geologia 37, São Paulo, Anais*, **2**. Sociedade Brasileira de Geologia, 181–182.

SILVA FILHO, A. F. & GUIMARAES, I. P. 2000. Sm/Nd isotopic data and U/Pb geochronology of collisional to post-collisional high-K shoshonitic granitoids from the Pernambuco-Alagoas terrane, Borborema Province, NE. Brazil. *In*: *31st International Geological Congress, Rio de Janeiro, Brazil, Abstracts Volume CD-ROM*.

SILVA FILHO, A. F., GUIMARÃES, I. P., SAMPAIO, M.A & LUNA, E. B. A. 1996. A super suite de granitóides ricos em K Neoproterozóicos tardi a pós - tectônicos da parte sul do Maciço PE-AL; magmatismo intraplaca? *In*: *Congresso Brasileiro de Geologia 39, Salvador, Resumos Expandidos*, **6**. Sociedade Brasileira de Geologia, 318–320.

SILVA FILHO, AF., VAN SCHMUS, W. R., GUIMARÃES, I. P. & LUNA, E. B. A. 1997*a*. Nd signature of PE-AL massif late tectonic granitic rocks, NE Brazil: evidence of sucessive crustal accretion during the Proterozoic. *In*: *1st South American Symposium on Isotope Geology*, 304–306.

SILVA FILHO, A. F., GUIMARÃES, I. P., LYRA DE BRITO, M. F. & PIMENTEL, M. M. 1997*b*. Geochemical signatures of the main Neoproterozoic late tectonic granitoids from the Proterozoic Sergipano fold belt, NE Brazil and its significance for the Brasiliano orogeny. *International Geology Review*, **39**, 639–659.

SILVA FILHO, A. F., GUIMARÃES, I. P. & VAN SCHMUS, W. R. 2002. Crustal evolution of the Pernambuco-Alagoas complex, Borborema Province, NE Brazil: Nd isotopic data from Neoproterozoic granitoids. *Gondwana Research*, **5**, 409–422.

SILVA FILHO, AF, GUIMARÃES, I. P. *ET AL*. 2005*a*. Caracterização geológica e geoquímica dos granitóides e ortognaisses Proterozóicos cálcio-alcalinos de alto-K do Domínio Crustal Garanhuns, Terreno Pernambuco-Alagoas, e seu significado tectônico. *In*: *Atas do XXI Simpósio Geologia do Nordeste, Brazil*, 119–123.

SILVA FILHO, A. F., VAN SCHMUS, W. R., BRITO NEVES, B. B., GUIMARÃES, I. P., TOTEU, S. F. & OSAKO, L. S. 2005*b*. Geological fit between the Pernambuco-Alagoas terrane of NE Brazil and Central African Fold Belt in Cameroon, based on Proterozoic structures and magmatism. *In*: PANKHURST, R. J. & VEIGA, G. D. (eds) *Gondwana 12: Geological and Biological Heritage of Gondwana, Abstracts*. Academia Nacional de Ciéncias, Cordoba, Argentina, 120.

SILVA FILHO, M. A. 1998. Arco vulcânico Canindé-Marancó e a Faixa Sul-Alagoana: seqüências orogênicas mesoproterozóicas. *In*: *Congresso Brasileiro de Geologia 40, Belo Horizonte, Anais*. Sociedade Brasileira de Geologia, 16.

SILVA FILHO, M. A. & BRITO NEVES, B. B. 1979. O Sistema de dobramentos Sergipano no Nordeste da Bahia. *Geologia Recursos Minerais do Estado da Bahia, Textos Básicos*, **1**, 203–217.

SILVA FILHO, M. A. & TORRES, H. H. F. 2002. A new interpretation on the Sergipano belt domain. *Anais Academia Brasileira de Ciências*, **74**, 556–557.

SILVA FILHO, M. A., ACIOLY, A. C. A., TORRES, H. H. F. & ARAÚJO, R. V. 2003. O Complexo Jaramataia no contexto do Sistema Sergipano. *Revista de Geologia, Fortaleza*, **16**, 99–110.

STENDAL, H., TOTEU, S. F. *ET AL*. 2006. Derivation of detrital rutile in the Yaoundé region from the Neoproterozoic Pan-African belt in southern Cameroon (Central Africa). *Journal of African Earth Sciences*, **44**, 443–458.

TACK, L., WINGATE, M. T. D., LIÉGEOIS, J.-P., FERNADEZ-ALONSO, M. & DEBLOND, A. 2001. Early Neoproterozoic magmatism (1000-910 Ma) of the Zadinian and Mayumbian Groups (Bas-Congo): onset of Rodinia rifting at the western edge of the Congo Craton. *Precambrian Research*, **110**, 277–306.

TCHAKOUNTÉ, J. 1999. *Etude géologique de la région d'Etoundou-Bayomen dans la série de Bafia (province du Centre) : tectonique, géochimie-métamorphisme*. Thèse de Doctorat 3è Cycle, Université de Yaoundé I, Cameroun.

TCHAMENI, R., MEZGER, K., NSIFA, N. E. & POUCLET, A. 2000. Neoarchean crustal evolution in the Congo craton: evidence from K-rich granitoids of the Ntem complex, southern Cameroon. *Journal of African Earth Sciences*, **30**, 133–147.

TEIXEIRA, W. & FIGUEIREDO, M. C. H. 1991. An outline of Early Proterozoic crustal evolution in the São Francisco Craton, Brazil: a review. *Precambrian Research*, **53**, 1–22.

TEIXEIRA, W., SABATÉ, P., BARBOSA, J., NOCE, C. M. & CARNEIRO, M. A. 2000. Archean and Paleoproterozoic tectonic evolution of the São Francisco Craton. *In*: CORDANI, U. G., MILANI, E. J., THOMAZ FILHO, A. & CAMPOS, D. A. (eds) *Tectonic Evolution of South America. 31st International Geological Congress*, Rio de Janeiro, 101–137.

TOTEU, S. F., MICHARD, A., BERTRAND, J. M. & ROCCI, G. 1987. U–Pb dating of Precambrian rocks from northern Cameroon, orogenic evolution and chronology of the Pan-African belt of central Africa. *Precambrian Research*, **37**, 71–87.

TOTEU, S. F., VAN SCHMUS, W. R., PENAYE, J. & NYOBÉ, J. B. 1994. U–Pb and Sm–Nd evidence for Eburnian and Pan-African high-grade metamorphism in cratonic rocks of southern Cameroon. *Precambrian Research*, **67**, 321–347.

TOTEU, S. F., VAN SCHMUS, W. R., PENAYE, J. & MICHARD, A. 2001. New U–Pb and Sm–Nd data from north-central Camroon and its bearing on pre-Pan African history of central Africa. *Precambrian Research*, **108**, 45–73.

TOTEU, S. F., PENAYE, J. & POUDJOM DJOMANI, Y. 2004. Geodynamic evolution of the Pan-African belt of Central Africa with special reference to Cameroon. *Canadian Journal of Earth Sciences*, **41**, 73–85.

TOTEU, S. F., PENAYE, J., DELOULE, E., VAN SCHMUS, W. R. & TCHAMENI, R. 2006*a*. Diachronous evolution of volcano-sedimentary basins north of the Congo craton: insights from U–Pb ion microprobe dating of zircons from the Poli, Lom and Yaounde Series (Cameroon). *Journal of African Earth Sciences*, **44**, 428–442.

TOTEU, S. F., FOUATEU, R. Y. *ET AL*. 2006*b*. U–Pb dating of plutonic rocks involved in the nappe tectonic in southern Cameroon: consequence for the Pan-African orogenic evolution of the central African fold belt. *Journal of African Earth Sciences*, **44**, 479–493.

TROMPETTE, R. 1997. Neoproterozic (~600 Ma) aggregation of Western Gondwana: a tentative scenario. *Precambrian Research*, **82**, 101–112.

VAN SCHMUS, W. R., BRITO NEVES, B. B., HACKSPACHER, P. & BABINSKI, M. 1995. U/Pb and Sm/Nd geochronolgic studies of eastern Borborema Province, northeastern Brazil: initial conclusions. *Journal of South American Earth Sciences*, **8**, 267–288.

VAN SCHMUS, W. R., FETTER, A. H., BRITO NEVES, B. B. & WILLIAMS, I. S. 1999. Ages of detrital zircon populations from Neoproterozoic supracrustal units in NE Brazil: Implications for assembly of West Gondwanaland. *Geological Society of America Abstracts with Programs*, **31**, A-299.

VAN SCHMUS, W. R., BRITO NEVES, B. B. *ET AL*. 2003. Seridó Group of NE Brazil, a Late Neoproterozoic pre- to syn-collisional flysch basin in West Gondwanaland? : insights from SHRIMP U-Pb detrital zircon ages. *Precambrian Research*, **127**, 287–327.

VAUCHEZ, A., NEVES, S. *ET AL*. 1995. The Borborema shear zone systems, NE Brazil. *Journal of South American Earth Sciences*, **8**, 247–266.

Comparisons between the northwestern Borborema Province, NE Brazil, and the southwestern Pharusian Dahomey Belt, SW Central Africa

T. J. S. DOS SANTOS[1], A. H. FETTER[2] & J. A. N. NETO[3]

[1]*Instituto de Geociências, Universidade Estadual de Campinas (UNICAMP), PO Box 6152, CEP 13081-970, Campinas, SP, Brazil (e-mail: ticiano@ige.unicamp.br)*

[2]*U.S. Nuclear Regulatory Commission, 11555 Rockville Pike, Rockville, MD 20852, USA*

[3]*Universidade Federal do Ceará, Campus do Pici, CEP 60455-760 Fortaleza, CE, Brazil*

Abstract: Geological and geochronological data for the northwestern part of the Brasiliano Borborema Province are described and compared with their counterparts in the Pan-African Dahomey (Pharusian) belt that flanks the southeastern margin of the West African Craton, where outcrops are sufficiently continuous to discern the nature of the collision during West Gondwana assembly. In the Médio Coreaú domain, NW Borborema Province, U–Pb and Sm–Nd data have revealed unusual basement rocks representing 2.35–2.30 Ga juvenile crust, along with large tracts of 2.15–2.10 Ga juvenile gneisses in the Ceará Central domain. These basement blocks were affected by two pulses of intracratonic extension at 1785 and 775 Ma. Prior to West Gondwana collision, a continental arc (the Santa Quitéria batholith) developed between 665 Ma and 620 Ma. The presence of this arc strengthens the hypothesis that convergence between the Borborema Province and the São Luis craton involved closure of an oceanic basin. New geochronological data are presented showing that Palaeoproterozoic orthogneisses (U–Pb upper intercept 2288 $\pm$ 2 Ma) were affected by a major late Neoproterozoic event (554 $\pm$ 4 Ma U–Pb lower intercept, 558 $\pm$ 3 Ma Sm–Nd whole-rock and mineral isochron). Exhumation and cooling of granulite rocks between 568 and 550 Ma in the Médio Coreaú domain and between *c.* 587 and 576 in the West African Dahomey Belt indicate that the final tectonic phase was not simultaneous along this front of the orogen.

A correlation between Africa and South America was originally postulated by Benjamin Franklin based on the apparent fit between their two margins. Since then, the advent of plate tectonic theory, along with the accumulation of geophysical, geological and geochronological data, especially in the regions of NE Brazil and central Western Africa, have led to the development of a more sophisticated understanding of the geological connections, as well as apparent discontinuities between the two continents (e.g., Almeida & Black 1968; Lesquer *et al.* 1984; Caby 1989; Trompette 1994; Brito Neves *et al.* 2002). Geological investigations on both landmasses have revealed that vast areas of South America and West Africa were tectonized during the assembly of Gondwana *c.* 600 Ma, referred to as Brasiliano and Pan-African thermotectonic event, respectively. One of the main Brasiliano domains that is key for understanding the evolution of West Gondwana is the Borborema Province of NE Brazil. This province developed as a result of convergence of the West African–São Luís and São Francisco Cratons during the amalgamation of West Gondwana (Brito Neves *et al.* 2000). Recent U–Pb zircon and Sm–Nd whole-rock studies of the Borborema Province (Van Schmus *et al.* 1995; Fetter *et al.* 1997, 2000; Dantas *et al.* 1998) have yielded important constraints on the structural and tectonic evolution of the province. This information is complemented by a number of similar investigations of the corresponding regions of West Africa (Bruguier *et al.* 1994; Attoh *et al.* 1997; Toteu *et al.* 2001; Gasquet *et al.* 2003; Caby 2003), which also provide a better overview of the nature of West Gondwana assembly.

In this paper we deal with geological and structural evolution of northwestern part of Borborema Province (Médio Coreaú or NW Ceará), NE Brazil, and compare it with the Pharusian Belt region along the southeastern margin of the West African Craton (from the Dahomey Belt to the Hoggar). We show that complementary geological information from these continents contributes to

From: PANKHURST, R. J., TROUW, R. A. J., BRITO NEVES, B. B. & DE WIT, M. J. (eds) *West Gondwana: Pre-Cenozoic Correlations Across the South Atlantic Region*. Geological Society, London, Special Publications, **294**, 101–119. DOI: 10.1144/SP294.6 0305-8719/08/$15.00

our understanding of the pre-collisional development and assembly history of this part of West Gondwana.

First-order geological correlations

The general crustal architecture of the Brasiliano/Pan-African provinces is characterized by large tracts of Palaeoproterozoic gneisses and migmatites surrounding several smaller nuclei of Archaean crust. These basement gneisses are overlain by remnants of Palaeoproterozoic to Neoproterozoic supracrustal rocks, all of which are intruded by ubiquitous Neoproterozoic granitoids and dissected by a network of anastomosing shear zones. Some of these shear zones represent local structural adjustments within individual crustal blocks, whereas others are major tectonic lineaments that bound crustal domains and extend across the Gondwana provinces (Fig. 1). These lineaments are particularly useful in matching up corresponding geological domains in Africa and South America, although variable exposure of different crustal levels along the coastal areas on both sides of the Atlantic makes exact geological correlations challenging.

Based on pre-drift reconstructions of Africa and South America, the Médio Coreaú and Ceará Central domains of NE Brazil correspond to the allochthonous Accra Plains Migmatite and Dzodze Gneiss in the Dahomey Belt, respectively. In Brazil, the boundary between the Médio Coreaú and Ceará Central domains is marked by the

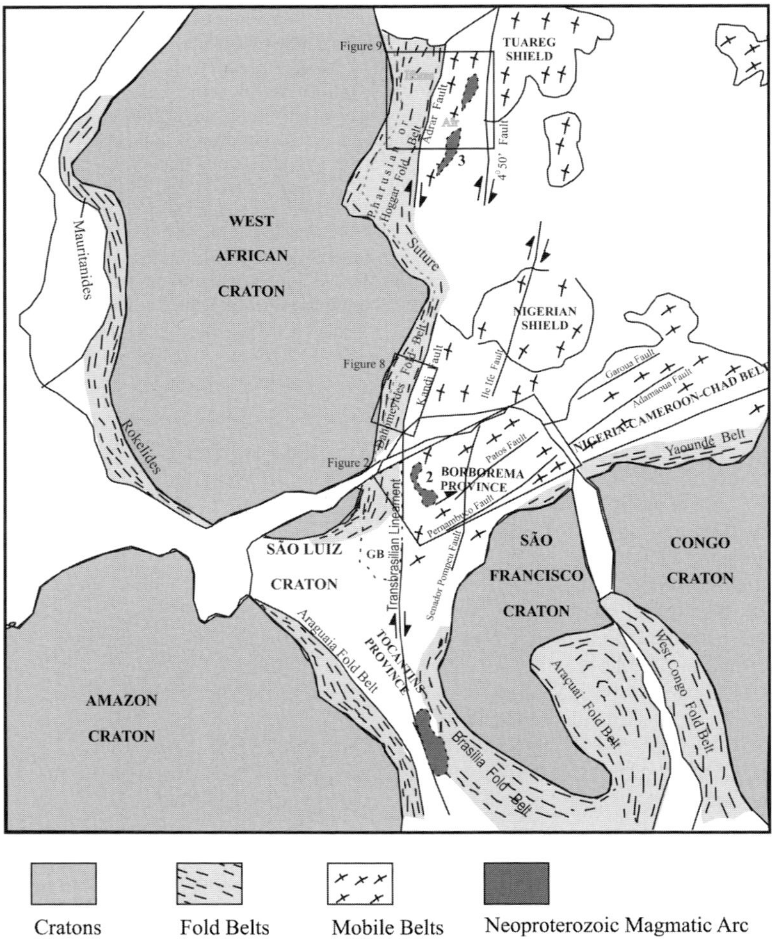

Fig. 1. Pre-Mesozoic drift reconstruction of parts of Africa and South America showing the major cratonic and Pan-African/Brasiliano provinces (modified from Trompette 1994). Neoproterozoic magmatic arc: 1, Goiás; 2, Santa Quitéria; 3, arc batholiths of the Pharusian terrane; GB, (Granja Block).

Transbrasiliano Lineament, which may be correlated with the 4°50′–Kandi fault in Africa (Fig. 1). The southeastern limit of the Ceará Central domain, the Senador Pompeu Lineament, corresponds to the Ile–Ife shear zone in Nigeria. Whereas in Africa the transition between the Accra Plains Migmatite and the West African Craton is relatively well exposed and characterized as representing a probable oceanic suture zone, the contact between the Médio Coreaú belt and the São Luís Craton to the NW is obscured beneath the Phanerozoic sedimentary rocks of the Parnaíba Basin. Nonetheless, other lines of geological and isotopic evidence from Brazil strongly suggest that the Pharusian oceanic domain was continuous into South America prior to final Gondwana assembly (Fetter *et al.* 2003). Conversely, though distinctive differences have been identified with respect to the U–Pb crystallization ages and Nd isotopic signatures of the Médio Coreaú and Ceará Central domains (Fetter *et al.* 2000), a similar study has not been carried out between the corresponding Accra Plains Migmatite and Dzodze Gneiss in Africa. In spite of these limitations, the available information and data from both sides of the Atlantic are complementary in nature; hence we are able to understand the dynamics of the wider tectonic setting of West Gondwana better by looking at the geology of both continents than by looking at each continent separately.

Crustal architecture and lithostratigraphy of the northwestern Borborema Province

The northern tectonic domain of the Borborema Province is located north of the Patos Lineament and corresponds to the largest and most diverse and complex crustal architecture in this region. Sm–Nd isotopic studies and U–Pb geochronology of basement rocks in the northern tectonic domain have been used to identify three distinct crustal blocks: the Rio Grande do Norte, the Ceará Central, and the Médio Coreaú (or NW Ceará belt) blocks, the limits of which are marked by Senador Pompeu and Transbrasiliano lineaments, respectively (Fig. 2). These data show that the Médio Coreaú and Ceará Central domains have distinct and unrelated crustal histories, and it is clear that the Transbrasiliano Lineament represents a major crustal-block boundary. Because significant strike-slip movement has occurred along the lineament, the original timing and nature of the tectonic contact between the two domains remains unclear. Specifically, it is unknown whether these two domains were juxtaposed much earlier than the assembly of West Gondwana.

Médio Coreaú domain

The main basement unit of the Médio Coreaú domain is the Granja Complex, a high-grade metamorphic complex (granulite–upper amphibolite facies) composed mainly of orthogneisses with tonalite–trondhjemite–granodiorite (TTG) affinities, as well as amphibolite gneisses, amphibolites, leucogranites, mafic granulites, and enderbites, kondalites, kinzigites and migmatites (Fig. 3; Santos *et al.* 2001). U–Pb and Sm–Nd studies of these rocks (Fetter *et al.* 1997, 2000) show that samples from all main gneisses are early Palaeoproterozoic in age (2.36–2.29 Ga). This age range is reinforced by a new U–Pb zircon upper intercept age of 2288 ± 2 Ma for zircon and titanile from a Granja Complex tonalite gneiss (see below). Granja Complex gneisses yield Nd crustal residence (T_{DM}) ages between 2.61 and 2.38 Ga (Fetter *et al.* 2000). As most of the εNd(t) values (at $t =$ crystallization age) of the basement gneisses are positive (+0.5 to +1.9, Fig. 4), they are interpreted as representing juvenile crustal growth and are thought to have been generated in an arc-type setting. This age range is unique, not only because it does not belong to the classical 'Transamazonian–Eburnian' event (*c.* 2.2–2.0 Ga), but also because it represents a period in which very little crustal growth has been recognized worldwide.

The 'Saquinho Volcanic Sequence' comprises felsic to intermediate volcanic rocks, including trachyandesite, rhyodacite, rhyolite and volcanic breccias and tuffs (Santos *et al.* 2002). At present, it is unclear how this sequence is related to the surrounding units of the Ubajara Group. U–Pb zircon data from a meta-rhyolite have yielded a concordant age of 1785 ± 2 Ma, interpreted as the crystallization age of the volcanic rock (Santos *et al.* 2002). This age corresponds to a widespread 1.8–1.7 Ga extensional event recognized throughout South America (Statherian taphrogenesis, Brito Neves *et al.* 1995) that is also recorded by the Orós and Jaguaribeano volcano-sedimentary sequences in the western part of Rio Grande do Norte domain (Sá *et al.* 1995). Consequently, the Saquinho Volcanic Sequence may be a manifestation of this Palaeoproterozoic extensional event in the Médio Coreaú domain. Sm–Nd data from the metasedimentary rocks of the Ubajara Group and Saquinho Volcanic Sequence show a wide range of T_{DM} ages, from 2.87 Ga to 1.62 Ga, but mostly around 2.4 Ga (Santos *et al.* 2002). These Nd data indicate that the principal age of the sedimentary source material is early Palaeoproterozoic, which is consistent with the observed basement rocks.

Two main supracrustal sequences unconformably overlie the basement complex of the Granja Complex: the Ubajara and Martinópole groups.

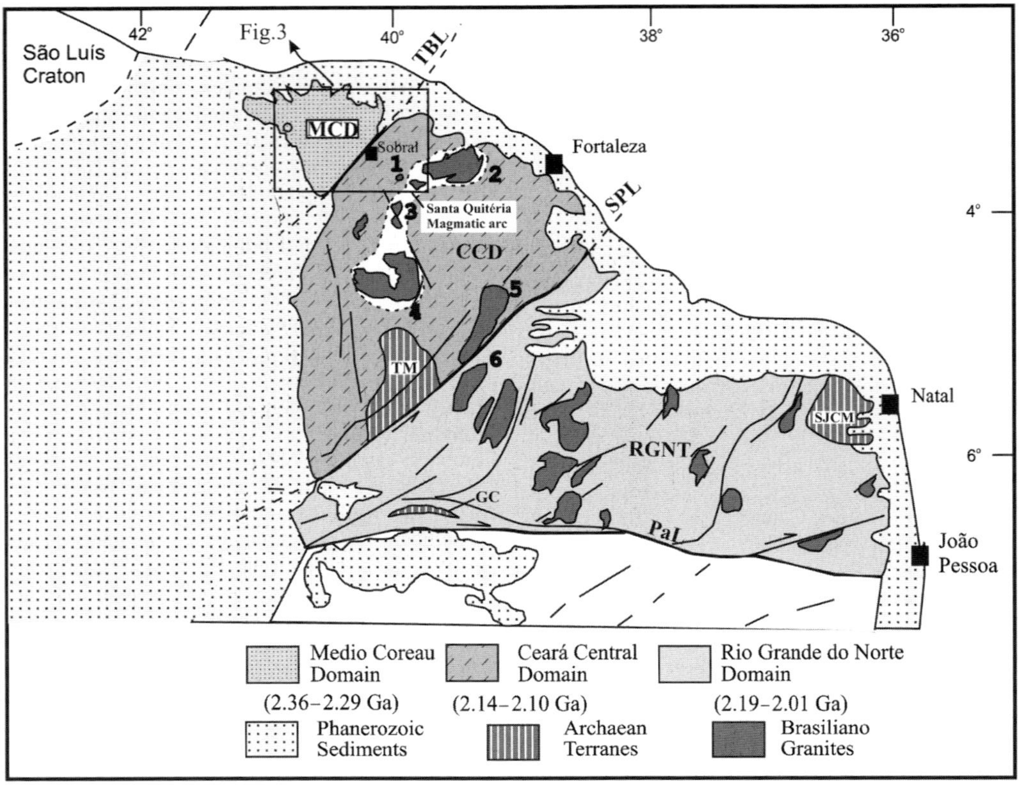

Fig. 2. Geological map of the Northern domain of the Borborema Province, showing the main domains and Brasiliano granitic plutons situated north of the Patos Lineament (PaL): TM, Tróia Massif (Archaean inlier); SJCM, São José de Campestre massif (Archaean inlier); TBL, Transbrasiliano Lineament; SPL, Senador Pompeu Lineament; RGNT, Rio Grande do, Norte Terrane; CCD, Ceará Central domain; MCD, Médio Coreaú domain; GC, Granjeiro Complex. Main Brasiliano plutons: 1, Serra da Barriga; 2, Itapajé; 3, Pajé; 4, Monsenhor Tabosa; 5, Quixeramobim; 6, Quixadá; Dashed area corresponds to the Santa Quitéria continental magmatic arc.

The Martinópole Group (Fig. 3) is subdivided, from base to top, into four formations: Goiabeira, São Joaquim, Covão and Santa Terezinha. These are comprised mainly of fine-grained sedimentary rocks: metapelites, meta-greywackes and quartzites with minor intercalations of calc-silicates, meta-carbonates and metavolcanic rocks, which are suggestive of sedimentation in a low-energy environment, while the tectonic environment probably corresponds to an intracontinental rift evolving to marine conditions. Regardless of the tectonic setting, the two lower formations of the Martinópole Group were metamorphosed to greenschist facies during Brasiliano orogenesis, while the two upper formations reached amphibolite facies.

The lowermost Goiabeira Formation is in tectonic contact (thrust-fault) with the Granja Complex and consists of garnet–chlorite schist, staurolite schist, muscovite–chlorite schist, biotite schist, kyanite schist, garnet–biotite schist and quartz–feldspar paragneiss, representing metamorphosed pelitic protoliths.

The São Joaquim Formation is comprised primarily of quartzites, with minor intercalations of schist, calc-silicates and meta-rhyolite. Metamorphic minerals present in this formation include kyanite, sillimanite, muscovite and staurolite, indicating intermediate-pressure amphibolite-facies metamorphism below the K-feldspar stability field. Retrograde metamorphism to greenschist-facies conditions during the final stage of Neoproterozoic deformation is constrained by an overprinting assemblage of chlorite and muscovite. Meta-rhyolites occur interlayered with quartzites, and both are strongly deformed and show mylonitic textures.

The Covão Formation is a sequence consisting of muscovite–quartz–sericite–chlorite schists and minor quartzite layers. These clastic metasedimentary rocks show typical Bouma sequence

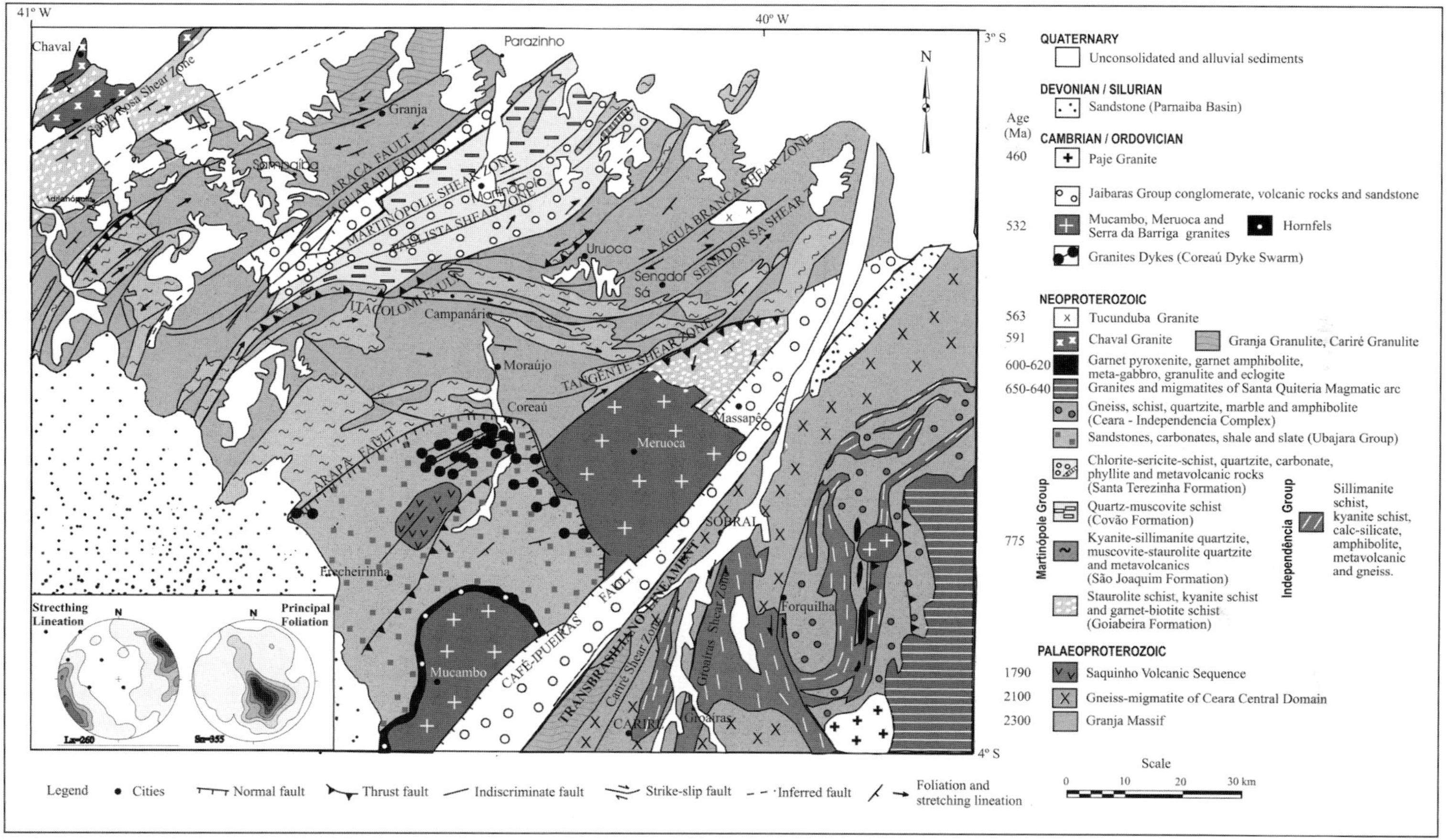

Fig. 3. Simplified geological map of the Médio Coreaú domain, in the NW portion of the Borborema Province.

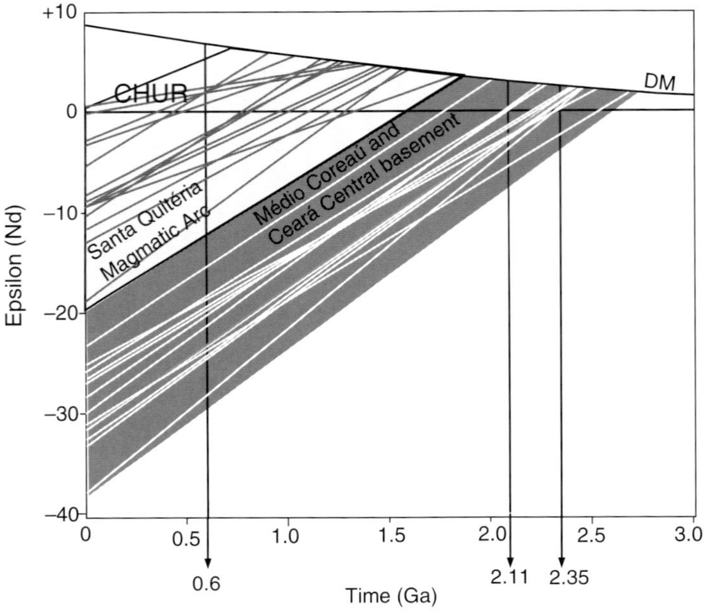

Fig. 4. Nd evolution diagram for rocks from the Palaeoproterozoic basement of the Médio Coreaú domain and Ceará Central, and rocks of the Neoproterozoic Santa Quitéria magmatic arc (Fetter 1999; Fetter *et al.* 2003).

features and are interpreted as marine, continental slope turbidite deposits. The mineral assemblage (chlorite–quartz–muscovite) and micro-structural features of quartz in quartzite (lamellae and deformation bands, sub-grain boundary) suggest lower greenschist-facies metamorphism.

The uppermost Santa Terezinha Formation consists of quartz-rich schists, metapelites, and

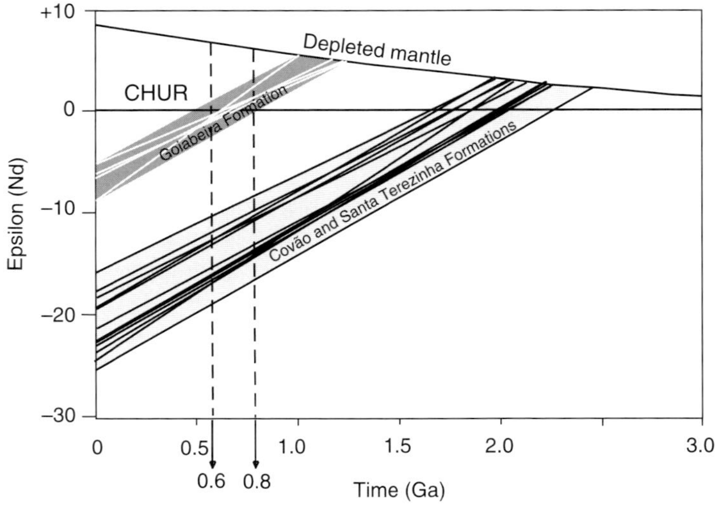

Fig. 5. Nd evolution diagram for rocks from the Martinópole Group. Note that the data fall into two populations, having T_{DM} ages of *c.* 1.24 and 1.32 Ga (Goiabeira Formation) and 1.61 and 2.69 Ga (Santa Terezinha and Covão formations), with ε_{Nd} ($t = 800$ Ma) of +2 to +0.5 and −8 to −17, respectively. The rocks of the Santa Terezinha Formation at the top of the Martinópole Group and Covão Formation show variable mixture between Neoproterozoic juvenile sources and Palaeoproterozoic basement gneiss (Fetter 1999; Fetter *et al.* 2003).

meta-carbonates with meta-greywacke associations, meta-rhythmites, quartzite and interlayered meta-rhyolites. These units crop out between the Maravilha fault and Uruoca-Tucunduba fault (Fig. 3). The meta-rhyolite units contain phenocrysts of blue quartz and crop out as narrow (1–2 m), semi-continuous beds near the Paulista fault. They are interlayered with quartzites and muscovite schists and are locally boudinaged.

To place precise constraints on the timing of volcanism and sedimentation, felsic volcanic units from Santa Terezinha formation were analysed using U–Pb zircon methods to yield an upper intercept age of 777 $\pm$ 11 Ma. This age was interpreted as the crystallization age of the rhyolite and the probable age of active sedimentation in this formation of the Martinópole Group (Fetter *et al.* 2003).

Neodymium isotopic data from the Martinópole Group indicate that clastic contributions involved two distinctively different source areas (Fig. 5). The base of the Martinópole Group, consisting of schists of the Goiabeira Formation, yield T_{DM} values between 1.24 and 1.32 Ga and εNd values ($t = 777$ Ma) from +2.4 to −0.3. In contrast, the overlying Santa Terezinha and Covão formations display more enriched Nd signatures; their Nd T_{DM} values vary from 1.61 to 2.69, with εNd values ($t = 777$ Ma) below −15. The likely dominant source of the more enriched clastic component is the surrounding Paleoproterozoic basement as the more negative εNd values are quite similar. Slightly less negative εNd values are interpreted to represent mixtures of basement clastic rocks and some Neoproterozoic juvenile material. As for the basal juvenile formation, the only likely source for the younger juvenile material would be the rocks that developed early in the formation of the rift. The only other regional source of Neoproterozoic juvenile material is the Santa Quitéria complex (Fig. 2) (Fetter *et al.* 2003), but its youngest age so far identified (665 Ma) post-dates the Martinópole Group, and hence it cannot be a candidate.

The Ubajara Group is a proximal, stable-shelf sedimentary sequence composed of three formations (metamorphosed), from base to top: the Caiçaras (slate, siltstone and sandstone), Trapiá–Frecheirinha (sandstone and carbonate, respectively), and Coreaú (sandstone and greywacke) formations (Fig. 3).

The Médio Coreaú domain contains four granite plutons that intrude the basement rocks and the Martinópole, Ubajara, and Jaibaras groups. Two of these granites (Chaval and Tucunduba) were affected by (and are related to) the Brasiliano deformation. The Chaval granite crops out in the NW part of the Médio Coreaú domain and is covered to the NW by sediments of the Parnaíba Basin. This pluton has a porphyritic texture, and the central portion of the body displays southeast-dipping low-angle foliation. U–Pb monazite data yield a crystallization age of 591 $\pm$ 10 Ma (Fetter *et al.* 2000). The Tucunduba porphyritic granite-granodiorite shows a predominantly transcurrent deformational pattern with a general NE–SW trend, parallel to foliation and lineation developed within the basement wall-rocks. The emplacement of this pluton was contemporaneous with the late transcurrent tectonic phase of the Brasiliano orogeny (discussed below). Two sets of strike-slip faults (Água Branca and Senador Sá faults) bound the pluton, which has yielded a U–Pb zircon age of 563 $\pm$ 17 Ma (Santos *et al.* 2007). This age is coincident with uplift and cooling of the granulite of Granja Complex, which is here constrained by a precise a Sm–Nd whole-rock mineral isochron (plagioclase, whole-rock, and garnet) of 558 $\pm$ 3 Ma (Fig. 6a, Table 1, Appendix 1). The uplift of this granulite is associated with late-stage transcurrent tectonic movements in the domain

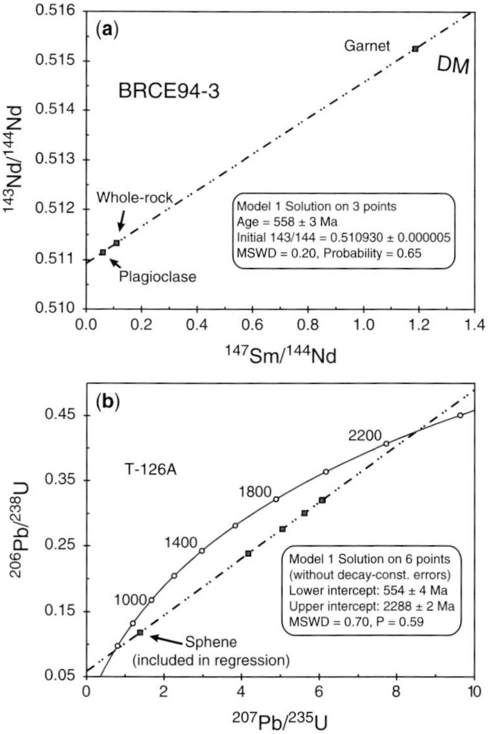

Fig. 6. (**a**) Sm–Nd isochron of plagioclase (pg), whole-rock (wr) and garnet (gt) analyses from a kinzingite in the Granja Complex; (**b**) Concordia plot of zircon and titanite analyses from a tonalitic gneiss in the Granja Complex. The isochron age and the lower intercept age from the Concordia plot are interpreted as the cooling age of minerals in the Médio Coreaú domain.

Table 1. *Sm–Nd whole-rock data sample BRCE94-3 (kinzigite) from the Granja Complex*

Fraction	Nd (ppm)	Sm (ppm)	^{147}Sm/^{144}Nd*	^{143}Nd/^{144}Nd*	ε_{Nd} (600 Ma)†	(t)‡ Ga	ε_{Nd} (t)	T_{DM} Ga§
Whole-rock	26.96	4.93	0.11052	0.511334	−18.9	2.33	0.4	2.54
Plagioclase	6.51	0.65	0.06067	0.511150	−18.6			
Garnet	5.98	11.72	1.18579	0.515260	−24.8			

Notes: *2σ analytical uncertainty = 0.5% on ^{147}Sm/^{144}Nd and 0.0015% on ^{143}Nd/^{144}Nd.
†ε_{Nd} (600) = ((^{143}Nd/^{144}Nd [sample 600 Ma]/^{143}Nd/^{144}[CHUR 600 Ma]) − 1) × 10^4, where CHUR has present-day ^{147}Sm/^{144}Nd = and ^{143}Nd/^{144}Nd = 0.512368.
‡U–Pb zircon age of Granja Complex (Fetter *et al.* 2000).
§Calculated according to the single-stage depleted-mantle model of DePaolo (1981).

thought to be responsible for the exhumation of much of the massif. Additional evidence for general cooling at this time comes from a U–Pb discordia line regressed through zircon and titanite fractions from a tonalitic gneiss in the region. The regression through six mineral fractions (five fractions of zircon and one of titanite) yields an upper intercept age of 2288 ± 2 Ma (Fig. 6b, Table 2), which is consistent with the protolith age of the Granja Complex as discussed above, as is its Sm–Nd T_{DM} age of 2.54 Ga (Table 1). Although lower intercept ages must be interpreted with caution, in this case the age obtained of 554 ± 4 Ma is consistent with two other ages associated with the final transcurrent motions in the domain (Monié *et al.* 1997; Santos *et al.* 2004).

The undeformed Meruoca and Mucambo granites (Fig. 3) represent post-orogenic plutonism in the Médio Coreaú domain. These granites are intimately linked to the development of the Jaibaras graben, and are hence the result of a subsequent rift-related event as opposed to late-tectonic effects of the Brasiliano orogeny. U–Pb zircon data from the Mucambo granite yield a crystallization age of 532 ± 6 Ma (Fetter *et al.* 2003).

The Jaibaras Group corresponds to an important and extensive exposure of continental immature siliciclastic sedimentary deposits (conglomerates, sandstones, and reddish shale) and volcaniclastic rocks of Early Palaeozoic age. This group has traditionally been considered as representing an early molasse-type deposit post-dating the Brasiliano collage. An updated overview of the Jaibaras Basin is summarized by Oliveira & Mohriak (2003); Oliveira (2000) and Oliveira & Mohriak (2003) show that this basin was generated by reactivation of the Transbrasiliano Lineament that took place in a stage of rifting, with a minor amount of lateral displacement.

Ceará Central domain

The Ceará Central domain is a composite crustal block composed of (1) vast tracts of juvenile middle Palaeoproterozoic high-grade felsic orthogneisses and migmatites (predominantly 2.14–2.10 Ga, Fetter *et al.* 2000; Martins *et al.* 1998; Santos *et al.* 2003), (2) an inlier of Archaean crust known as the Tróia-Pedra Branca massif and (3) the roots of a Neoproterozoic continental arc called the Santa Quitéria batholith (Fetter *et al.* 2003; Arthaud *et al.* this volume) (Fig. 2).

The Santa Quitéria batholith is a NE–SW trending complex of Brasiliano granitoid plutons and migmatites, covering some 40,000 km^2. It is located *c.* 200 km SE of high-density gravity anomalies beneath the Parnaíba Basin, which probably represent the extension of the suture of southern West Africa (Fig. 2). This batholithic suite comprises a large variety of dioritic to granitic rocks and is flanked by classical fore-arc and back-arc sediments (volcanic, volcaniclastic, and calc-silicate rocks along with turbidites derived mainly from the volcanic arc carapace). U–Pb ages so far obtained from the complex range from 665 to 622 Ma, although outlying remnants of supracrustal rocks suggest that development of this arc could have begun as early as 775 Ma. The Nd isotopic signatures are consistent with variable mixtures of juvenile Neoproterozoic magmas and the surrounding Palaeoproterozoic gneisses. This fact, combined with our other observations of the Santa Quitéria batholith, leads us to believe that it probably represents the remnants of a Brasiliano continental arc. As the construction of a continental arc requires the subduction of oceanic crust, this implies that an oceanic domain existed to the NW of the Borborema Province prior to the amalgamation of West Gondwana, presumably a southernmost branch of the Pharusian Ocean.

The western portion of the Santa Quitéria batholith encloses a sequence of biotite gneiss, sillimanite–garnet gneiss, muscovite–sillimanite quartzite, quartz–muscovite schist, amphibole gneiss, augen gneiss granitic, garnet-bearing leucogranite, and meta-mafic or gabbroic rocks. These rocks display a predominant north–south low-angle

Table 2. *U–Pb zircon and titanite data for sample T-216 (migmatitic tonalite gneiss) from the Granja Complex*

Fraction*	Size (mg)	U ppm	Pb ppm	^{206}Pb/^{204}Pb (obs.)	^{207}Pb/^{235}U†	2σ%	^{206}Pb/^{238}U†	2σ%	(rho)	^{207}Pb/^{206}Pb†	2σ%	^{206}Pb/^{238}U age (Ma)†	2σ	^{207}Pb/^{235}U age (Ma)†	2σ	^{207}Pb/^{206}Pb† age (Ma)†	2σ
T-126A																	
NM(−2) zircon	0.054	217	79	4695	6.0857	0.53	0.32090	0.51	0.975	0.13754	0.12	1794	9	1988	11	2197	2.0
M(−2) zircon	0.059	169	58	3517	5.6163	0.55	0.30090	0.53	0.979	0.13537	0.11	1794	10	1919	11	2169	1.9
M(−1) zircon	0.052	216	66	4435	5.0491	0.54	0.27622	0.52	0.976	0.13258	0.12	1794	9	1828	10	2132	2.0
M(0) zircon	0.028	194	71	1381	6.0613	0.66	0.32034	0.64	0.975	0.13723	0.15	1794	11	1985	13	2192	2.6
M(1) zircon	0.011	309	82	886	4.1732	1.02	0.23842	0.99	0.977	0.12695	0.22	1794	18	1669	17	2056	3.8
M(1.5A) titanite	0.302	114	15	688	1.3842	0.50	0.11817	0.49	0.972	0.08496	0.12	720	4	882	4	1315	2.3

*NM, non-magnetic at 1.5 amps; M, magnetic at 1.5 amps; single-digit numbers in parentheses indicate side tilt on Frantz separator.
†Total U and Pb concentrations corrected for analytical blank; ^{206}Pb/^{204}Pb not corrected for blank or non-radiogenic Pb; Radiogenic Pb corrected for blank and initial Pb; U corrected for blank.
Ages given in Ma using decay constants recommended by Steiger & Jäger (1977).
Intercept ages (titanite not included): upper intercept age = 2286 ± 6 Ma; lower intercept age = 549 ± 17 Ma; MSWD = 0.7.
Intercept ages (all mineral fractions regressed): upper intercept age = 2288 ± 2 Ma; lower intercept age = 554 ± 4 Ma; MSWD = 0.8.

east-dipping foliation and low–medium (~30°) east- or SE-plunging stretching lineation defined by sillimanite, quartz, biotite and muscovite. Westward, the thrusting event evolved progressively to a strike-slip regime whose main manifestation is a pervasive NE–SW to north–south trending and steeply dipping foliation with NE–SW to north–south orientated stretching lineation.

The meta-mafic rocks (garnet-clinopyroxenites, garnet-bearing amphibolites and incipient metagabbro-norite rocks) display medium- to coarse-grained textures and may be massive or foliated. The mineral assemblage is composed of garnet, clinopyroxene (jadeite-rich), amphibole, plagioclase and quartz. Rutile, ilmenite and titanite occur as accessory. The main textural features common to these rocks are symplectites (clinopyroxene + plagioclase ± amphibole intergrowths), or aggregates formed after omphacite pseudomorphs and reaction coronas around garnet (Fig. 7a, b). These mineral assemblages clearly indicate retrograde changes from eclogite- to granulite- to amphibolite-facies conditions; preliminary thermobarometric data for these rocks have given pressures of up to 17 kbar and temperature around 820 °C. These rocks

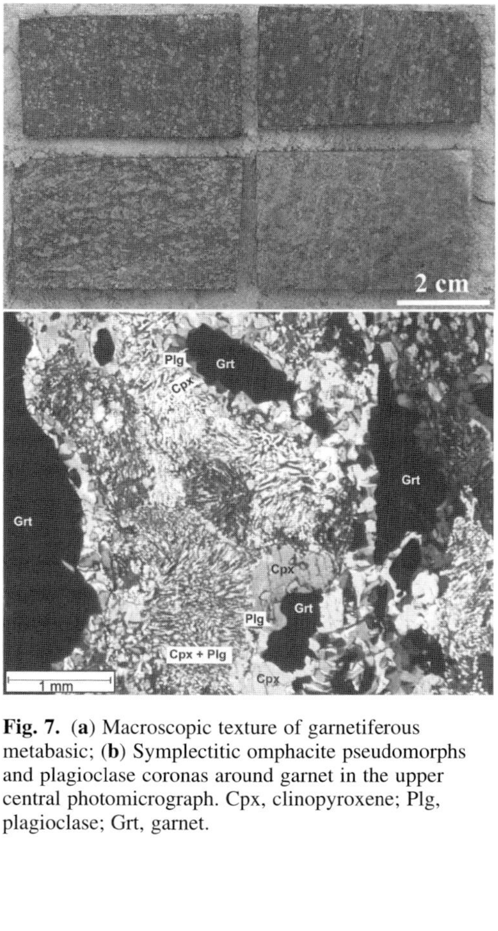

Fig. 7. (**a**) Macroscopic texture of garnetiferous metabasic; (**b**) Symplectitic omphacite pseudomorphs and plagioclase coronas around garnet in the upper central photomicrograph. Cpx, clinopyroxene; Plg, plagioclase; Grt, garnet.

display N-MORB type geochemical signatures. The tectonic implication of these small relicts of eclogites is currently unclear, but they constitute some of the unsolved questions with regard to final stages of collision between the Borborema Province and the São Luís Craton. Specifically, convergence could have involved closure of more than one small ocean basin or the collage of several smaller terrains during the collisional process.

Tectonic evolution of northwestern Borborema Province

Structural Evolution of the Médio Coreaú domain

One of the principal structural studies regarding the tectonic evolution of northwestern NE Brazil is that of Santos *et al.* (2004) in the Médio Coreaú domain. This domain records evidence of polycyclic deformation, inferred to have affected only the basement rocks during Palaeoproterozoic orogenesis (2.2–2.0 Ga), and three further phases affecting most rocks in the domain during Brasiliano–Pan-African cycle (*c.* 620–557 Ma).

In addition to being the oldest part of the domain, the Granja Complex gneisses also represents the base of the structural pile in the Médio Coreaú domain and highest metamorphic grade: middle amphibolite to granulite facies. High-grade metamorphic conditions are defined by orthopyroxene and clinopyroxene, which commonly occur in these rocks; however, during uplift and exhumation, some retrograde mineral assemblages developed, characterized by amphibole–quartz–plagioclase and garnet–biotite associations.

The first Neoproterozoic deformation is characterized by medium- to low-angle SE-dipping foliation and gently plunging NE- or SE-stretching lineation, recorded by sillimanite, kyanite, staurolite and muscovite. This fabric developed under amphibolite facies in the Martinópole Group and greenschist facies in the Ubajara Group, with crystallization of chlorite and sericite. In the Granja Complex, the continuous SE–NW movement was responsible for the uplift and cooling of granulite rocks during the final stage of Brasiliano orogenesis.

The tangential compressional event evolved progressively to transcurrent deformation, which is characterized by strongly sub-vertical foliation striking NE–SW to east–west and defines large dextral shear zones; the gently plunging stretching mineral lineations are defined by amphibole, sillimanite and kyanite. The orientation of the foliations and lineation associated with tangential and transcurrent events indicate an evolutionary history with northwestward thrusting progressively changing to a predominant northeastward transcurrent tectonic regime.

The available U–Pb zircon and monazite ages from variably deformed igneous bodies in the Médio Coreaú and Ceará Central domains indicate that tangential collision happened there after 620 Ma, and the transition to transcurrent tectonics occurred between 614 and 591 Ma. During the transcurrent tectonic regime, granitoid plutons and other intrusive bodies were emplaced in transtensional zones. The youngest of these intrusions dated during this study indicates that transcurrent tectonics were active at least until 563 Ma in the NW Borborema Province (Santos *et al.* 2004). Sm–Nd mineral (garnet–plagioclase–whole rock), monazite and titanite cooling ages obtained by Fetter (1999) and in this study indicate that cooling and uplift occurred in the region between *c.* 568 and 557 Ma. Mineral cooling ages obtained by Monié *et al.* (1997), however, suggest that transcurrent motions may have continued until about 540 Ma in the NW Borborema Province. Movement along shear zones probably ceased before 532 Ma, as evidenced by the emplacement of post-tectonic extensional granitoids adjacent to the Transbrasiliano Lineament, but cooling of the more deeply buried parts of the orogen in Ceará continued until about 522 Ma (Monié *et al.* 1997).

Crustal architecture and lithostratigraphy of the Pharusian Belt

Dahomeyide Belt

The Pan-African Dahomeyide Belt extends from the southeast of the West African Craton to the western part of the Nigerian Province, throughout eastern Ghana, Togo, Benin, Nigeria and Cameroon (Fig. 1). This orogen seems to represent a collisional event between the passive margin of the West African Craton and many small plates and terranes of the Nigerian–Beninian domain (Caby 1989; Attoh *et al.* 1997).

In the Dahomeyide orogen deep-level crustal rocks are tectonically juxtaposed with the foreland fold and thrust belt in *c.* 100 km wide transect. The Dahomeyide orogen consists of three tectonic zones: (1) western external zone (Buem, Kande and Atacora structural units), (2) suture zone and (3) extensive internal zone comprising the Benin Plains and Accra Plains gneisses that underlie much of the Benin–Nigerian shield (Fig. 8) (Attoh *et al.* 1997).

The Buem structural unit to the west is composed of non-metamorphic rocks, arkoses, shales

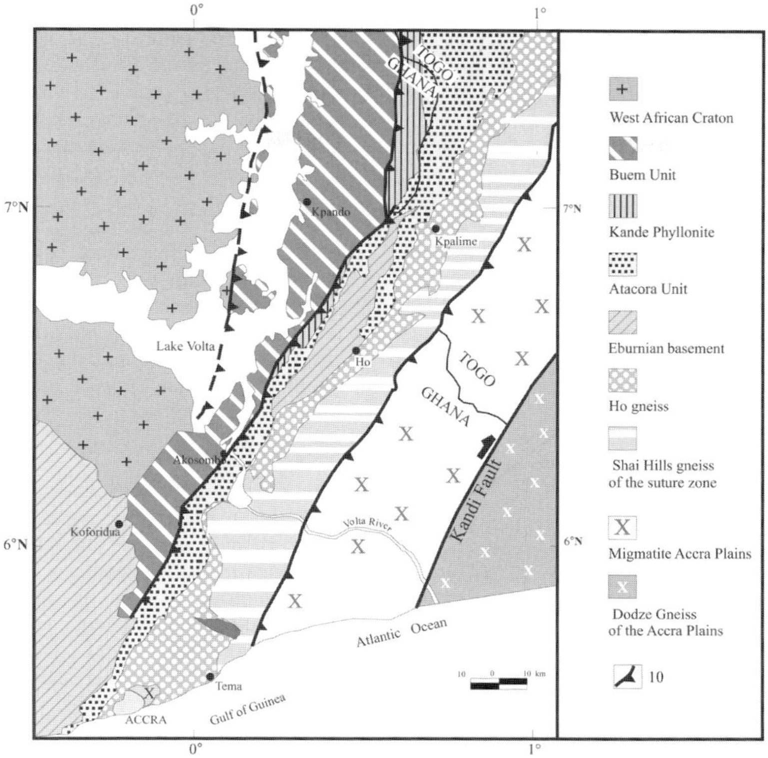

Fig. 8. Geological map of the Dahomeyide orogenic belt. 1, Cratonic platform; 2, Buem Unit; 3, Kande Phyllonite; 4, Atacora Unit; 5, Eburnian basement; 6, Ho Gneiss; 7, Migmatite Accra Plains; 8, Dodze Gneiss of the Accra Plains; 9, Shai Hills gneiss of the suture zone; 10, Thrust fault. (Modified from Attoh *et al.* 1997.)

and mudstones interbedded with cherts and limestones, and volcanic assemblages such as basalt pillow-lava, diabase and peridotitic cumulates (Jones 1990; Castaing *et al.* 1994; Attoh *et al.* 1997).

The Atacora structural unit lies on the crystalline basement series with an angular unconformity (Affaton *et al.* 1991). This unit is composed of aluminous quartzite, albitic micaschist, leucocratic gneiss, and meta-mafic and ultramafic rocks (Castaing *et al.* 1994). The Atacora quartzite shows sheath folds and NE–SW stretching lineation defined by kyanite and white mica, which define a southwestward nappe transport (Castaing *et al.* 1993).

Muscovite in the Atacora unit records an Ar–Ar plateau age of 579.4 ± 0.8 Ma, which correspond to the youngest ages of orogen-parallel nappe transport (Attoh *et al.* 1997). The Kande schist is composed of serpentinite bodies rich in chromite, which when sheared show a phyllonitic texture of chloritic and graphitic schist. This schist zone separates the Buem unit from the Atacora nappes (Attoh *et al.* 1997).

The suture zone corresponds to a narrow area that is marked by positive gravity and magnetic anomalies (El-Hadj Tidjani *et al.* 1997). It contains mafic and ultramafic rocks, with granulite, eclogite and amphibolite facies mineral assemblages (Attoh *et al.* 1997; Agbossoumondé *et al.* 2001, 2004; Attoh & Morgan 2004; Duclaux *et al.* 2006) (Fig. 8). This area represents the boundary between the autochthonous West African Craton and the exotic terranes that comprise the gneiss complexes to the east. Structural features indicate early northwestward thrusting followed by southeasterly nappe transport in the southern Dahomeyides (Castaing *et al.* 1993).

Geochronological data of Shai Hills gneiss from the suture zone yields a U–Pb zircon age of 610 ± 2 Ma (Attoh *et al.* 1991), interpreted as the peak metamorphic age. Ar–Ar analyses of muscovite and hornblende from the high-pressure granulite facies garnet–hornblende gneiss show ages of 587 ± 4.3 Ma and 581.9 ± 2.4 Ma, respectively, interpreted as the date of exhumation of suture zone nappes (Attoh *et al.* 1997). U–Pb zircon data from garnet–hornblende gneiss record peak regional granulite facies at 603 ± 5 Ma, and a rutile age of 576 ± 2 Ma suggests the time of regional cooling through about 400 °C (Hirdes & Davis 2002).

The internal zone corresponds to the western part of the granitic gneiss–migmatite Nigerian province and the eastern portion of the suture zone. The internal zone enclosed high-grade gneiss that is subdivided into two units: (1) biotite migmatites and (2) the Dodze orthogneiss (Fig. 8) derived from a dioritic protolith (Attoh *et al.* 1997). Granitoids of calc-alkaline geochemical affinity increase in abundance towards the east. Southeast and northeast of the Kandi fault, two basins, Dahou Mahou and Kandi, respectively, represent Pan-African molasse deposited in a graben associated with post-tectonic uplift of the high-grade terrains (Affaton *et al.* 1991).

Western Hoggar

A complete synthesis of the geodynamic evolution of central-western Hoggar was presented by Caby (2003) and Caby & Monié (2003). Based mostly upon these recent papers, we present a summary of the geological features of the terrains between the 4°50′ fault and the West African Craton.

The Pan-African belt of NW Africa corresponds to a segment of the Tran-Saharan belt that comprises the Hoggar, the Adrar des Iforas and the Aïr regions and is interpreted as a collision belt formed during a Wilson cycle (Trompette 1994) (Fig. 1). The Trans-Saharan orogeny resulted from closure of a Pan-African ocean between the West African Craton and the Tuareg palaeo-continent along an easterly-dipping subduction zone (Trompette 1994). The suture is marked by remnants of oceanic crust, magmatic arc assemblages, high- and ultrahigh-pressure metamorphic rocks and a positive gravity anomaly that may be traced from northern Mali to the Guinean Gulf (Caby 1989; Dostal *et al.* 1994; Caby 1994) (Figs 1 & 9).

Twenty-three terranes of variable nature, age (Archaean to Neoproterozoic), and evolution were assembled in a very complex way and juxtaposed by north–south orientated strike-slip mega-shear zones during the Pan-African orogeny (750–550 Ma). Two main collision phases are registered (Liégeois *et al.* 1994): A first intense collision occurred between the East Saharan craton and the easternmost Tuareg terranes at 750–660 Ma. The second more oblique collision with the West African Craton began at approximately 630 Ma, and its post-collisional movements and magmatism lasted until approximately 525 Ma (Acef *et al.* 2003). These collisions induced large displacements, which were responsible for calc-alkaline magmatism and intrusion of high-level plutons, dyke swarms, and plateau lavas of alkaline composition or of transitional affinity (Azzouni-Sekkal & Boissonnas 1993). Terrane boundaries are often marked by mafic–ultramafic, potentially ophiolitic bodies, and by late-stage elongated molasse basins.

The Archaean–Palaeoproterozoic rocks of the western Hoggar are characterized by the In Ouzzal granulite, the Iforas granulite, the Kidal terrane and the Tassendjanet basement (Fig. 9). These units are composed of granitic gneisses, charnockites, ultramafic and mafic rocks, and granulitic metasedimentary rocks (Caby 2003). Caby & Monié (2003) consider that the tectonic contacts of the In Ouzzal granulite with the Kidal and Tassendjane terranes represent possible sutures, along which eclogites and blue-schist facies rocks are identified.

The Kidal terrane constitutes an old basement overlain by metamorphic cover sequences that may be correlated with the Neoproterozoic platform stromatolites in the Tassendjanet terrane (Fig. 9). Also of Neoproterozoic age, the Kidal terrane contains volcano-sedimentary sequences, meta-tonalite and meta-diorite plutons, dyke swarms, ultramafic to gabbroic rocks, and anorthosites. The Tassendjanet terrane is Palaeoproterozoic but was strongly remobilized at the end of the Neoproterozoic by major thrust sheets, high T–high P metamorphism, abundant magmatism and volcanism (Caby 2003).

The Tilemsi belt (Fig. 9) is composed of mafic volcanic rocks of tholeiitic affinities, mafic to felsic plutons, volcanic sequences of tholeiitic or calc-alkali affinity, volcaniclastic rocks and several pre-, syn- and post-tectonic intrusions of predominantly mafic composition (Dostal *et al.* 1994). A minimum age of 726 Ma for the insular Tilemsi arc is given by the U–Pb zircon age of a cross-cutting pre-Pan African quartz diorite sheet (Caby *et al.* 1989). This arc may have been generated by intra-oceanic subduction around 730 Ma.

Closure of the western Pharusian Ocean is recorded by calc-alkaline suites that range from 685 to 633 Ma along the active continental margin of the Tuareg palaeo-continent (Caby 2003). The continental collision between the West African Craton and Tuareg palaeo-continent, recorded by high- and ultrahigh-pressure metamorphism with preserved coesite (Caby 1994), documents eastward subduction of the West African palaeo-margin. The time of the eclogitization, based on Rb–Sr and Sm–Nd isotope analyses on omphacite–kyanite micaschist and a mafic coesite-bearing eclogite in the nappes of the Gourma area, is constrained by concordant ages of about 620 Ma (Jahn *et al.* 2001).

Caby (2003) suggested that the major east-dipping subduction of the large Pan-African ocean (680–620) post-dates an older west-dipping

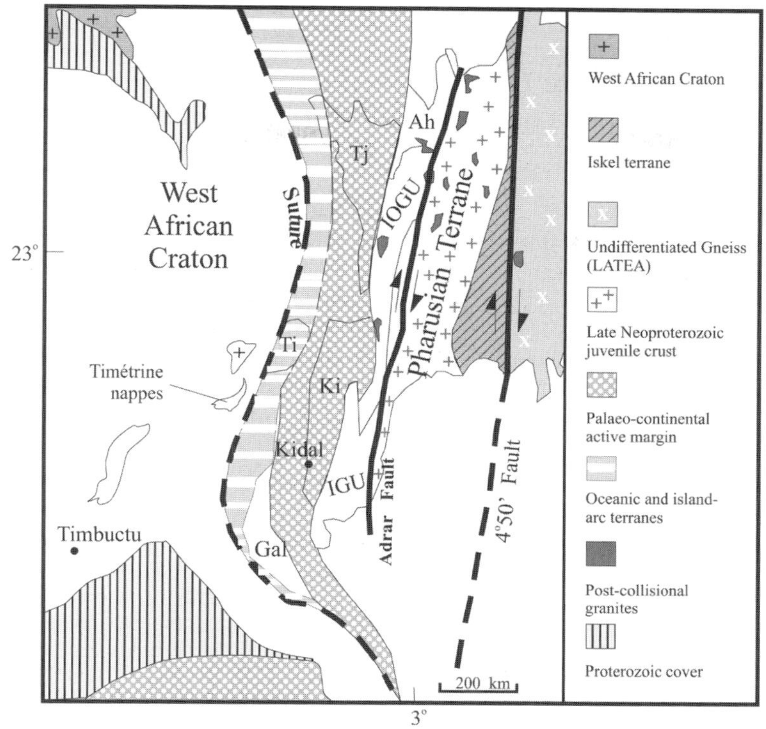

Fig. 9. Schematic map of the western terrane of the Trans-Sahara belt. Iogu, In Ouzzal granulite;Igu, Iforas granulite; Ah, Ahnet terrane; Tas, Tassendjanet terrane; Ki, Kidal terrane; Ti, Tilemsi arc. (Modified from Caby 2003 and Caby & Monié 2003.)

subduction to the east (Pharusian subduction, 690–650 Ma), corresponding to the main phase of juvenile crust generation. The boundary west and east of the In Ouzzal granulite terrane is characterized by two shear zones, which probably correspond to cryptic sutures, with Late Neoproterozoic palaeosubduction in two opposite directions and the 4°50′ shear zone corresponding to the eastern suture (Caby & Monié 2003).

The final evolution of the Pharusian Belt is characterized by ductile uplift and molasse deposition along the Adrar fault and the western portion of the 4°50′ fault. A solid-state dextral shear zone affected the margins of the late-kinematic plutons around 530–523 Ma (Paquette *et al.* 1998).

Summary of NW Borborema Province–NW Africa comparisons

A first-order observation of the current state of knowledge between NE Brazil and SW Central Africa is that, overall, geological work in Africa has focused mainly on the Neoproterozoic evolution of the region, with less emphasis on the older crustal evolution. In contrast, more recent geochronological and isotopic studies in NE Brazil focused first on the older crustal architecture, then on the Neoproterozoic tectonic history of the region. Consequently, a weak link in the correlation between the two continents is the limited geological, U–Pb zircon and Nd data for basement gneisses along the southwestern coast of Africa.

Relative to correlation between basement rocks of two continents, the question remains open as to whether the rare 2.35–2.29 Ga juvenile crust identified in the Médio Coreaú domain is limited to NE Brazil, or if it continues farther north into Central Africa or further west to the São Luís Craton. Based on geophysical (seismic, gravimetric and airborne magnetometric) and geological data, Nunes (1993) showed that the Granja Complex constitutes a large NE–SW elongate block beneath the sedimentary rocks of the Parnaíba Basin, isolated from the São Luís Craton and Ceará Central domain (Fig. 1). In the West African Craton the presence of pre-Birimian crust is shown by Gasquet *et al.* (2003), who obtained a U–Pb age of 2.31 Ga from a zircon core in the Dabakala tonalite.

Based on the current data, however, it is probable that the Médio Coreaú basement, i.e., the Granja Complex, constitutes a single crustal block with no clear genetic relationship or affinity with either the West African Craton or Ceará Central domain, which represents mainly juvenile 2.15 Ga crust. The current geochronological database of the African basement precludes that the Granja Complex corresponds to the eastern extension of the São Luís–West Africa craton, which was strongly reworked during Brasiliano orogenesis. This data gap presently hampers the establishment of a definitive genetic crustal connection between the Médio Coreaú domain and the corresponding Accra Plains Migmatite of the Dahomey Belt. A large part of the Dahomey Belt, however, appears to comprise Birimian rocks remobilized during the Pan-African orogeny. Some reconnaissance U–Pb geochronology from Accra Plains Migmatite would be useful for establishing a more definitive correlation with Médio Coreaú domain.

Regarding the Neoproterozoic evolution in these two continents, it is clear that in Africa the Western Hoggar appears to have a longer and more complex accretionary and collisional history than that observed farther south in the Dahomey Belt. In Brazil, the geological record is more deeply buried or obscured (Fig. 10). In this context, what is well-characterized in Africa provides a better basis for interpreting the palaeogeography of northwestern NE Brazil than can be observed in South America. Specifically, it seems that the oceanic suture zone rocks associated with the closure of the Pharusian Ocean do not terminate in the southern Dahomey Belt. This suture zone is represented by mafic and ultramafic rocks (e.g., Shai Hills gneiss in SE Ghana, Fig. 8) that display granulite-, eclogite-, and amphibolite–facies mineral assemblages (Attoh *et al.* 1997; Agbossoumondé *et al.* 2004; Attoh & Morgan 2004; Duclaux *et al.* 2006). Whether these suture zone rocks continue into South America is unclear, as the corresponding rocks in Brazil lie beneath the sedimentary fill of the Parnaíba Basin.

There are two lines of evidence to support the continuation of a suture zone into South America: (1) the presence of high-density gravity anomalies beneath the Parnaíba Basin interpreted by Lesquer *et al.* (1984) as representing buried mafic and ultramafic rocks of a suture zone and (2) the Santa Quitéria batholith, located approximately 200 km SE of these gravity anomalies, displays both lithologies and isotopic signatures consistent with a continental arc complex (Fetter *et al.* 2003). As a continental arc requires subduction of oceanic crust to develop, it appears likely that the Pharusian Ocean continued into South America, albeit on a smaller scale than can be inferred in Africa.

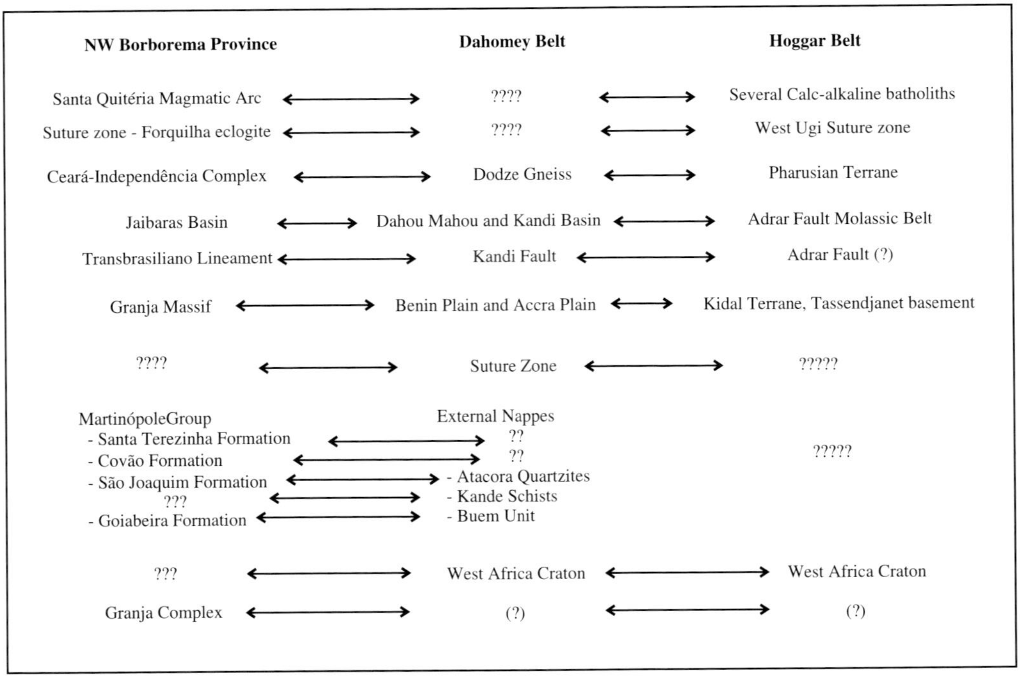

Fig. 10. Comparative tectonic evolution of the NW Borborema Province, Dahomey Belt, and Hoggar Belt.

The mafic and ultramafic rocks that crop out between the Transbrasiliano Lineament and the Santa Quitéria continental arc have eclogitic metamorphic assemblages (the Forquilha eclogite), for which preliminary estimates of $P-T$ conditions indicate a baric peak of up to 17 kbar for a temperature of 823 ± 20 °C (Santos *et al.* 2007) These rocks were subjected to retrograde events marked by transitions from eclogite to granulite facies and granulite to amphibolite facies, and the last stage marked by greenschist-facies assemblages. Correlation of these rocks with those of the suture zone in the Dahomey Belt is not tenable, because in Brazil these rocks are smaller bodies and crop out east of the Transbrasiliano Lineament, while in Africa large bodies crop out to the west.

As mentioned earlier, buried remnants of the main suture in Brazil are inferred to lie beneath the sediments of the Parnaíba Basin. Nonetheless, the presence of these small eclogite bodies between the Santa Quitéria complex and the Transbrasiliano Lineament are tantalizing discoveries that may have a bearing for the refining of our tectonic models for this region as more data become available. A definitive correlation between the basement rocks of the Médio Coreaú and those of the southern Dahomey Belt is hampered by sparse U–Pb and Sm–Nd data from that part of Africa. For example, in the internal zone of the Dahomey Belt, represented by biotite migmatites and the Dodze Gneiss (Fig. 8) (Attoh *et al.* 1997), the geochronology is of limited value as the K–Ar and Ar–Ar ages represent the final Neoproterozoic metamorphic history.

Structural features and lithological similarities, however, such as shear zones and analogous Neoproterozoic supracrustal sequences support a general correlation (Fig. 10). An analogue to the Santa Quitéria magmatic arc has not yet been recognized in Africa, but granitoids of calc-alkaline geochemical affinity are frequent east of the Kandi fault. We find it unlikely that the Santa Quitéria continental arc would be truncated in Africa; hence the concentration of calc-alkaline plutons east of the Kandi fault may represent the arc's continuation. This concentration of plutons in the southern Dahomey Belt is another area that begs systematic studies using Sm–Nd and U–Pb geochronology.

Farther north into the West African Hoggar belt there is a considerable departure from the geology observed in the Médio Coreaú and Ceará Central domains. This change in the geology is not surprising given the significant distance of thousands of kilometres. The Hoggar Belt comprises several terranes recording variable tectonic settings, several palaeo-subduction zones, abundant arc-related precollisional calc-alkaline batholiths, syn-collisional granitoids, and high-pressure suture-zone rocks. A structural continuation of the 4°50′ fault with the Transbrasiliano Lineament is uncertain, and the latter could be correlated with the Adrar fault.

Based on the geodynamic evolution of western-central Hoggar proposed by Caby (2003, Fig. 11), four phases of subduction have been recognized between 900 Ma and 520 Ma. The phase in the Hoggar that is most analogous with the tectonic evolution of the northwestern Borborema Province is the east-dipping subduction that occurred along the western margin of the Iforas granulite terrane between 680 and 610 Ma (Caby 2003). These zones show similar east-dipping subduction of Early Neoproterozoic oceanic lithosphere under the Palaeoproterozoic continental crust of the Ceará Central domain and the western terranes in Hoggar, respectively. In the same way, both systems display northwestward thrusting with medium- to low-angle SE- or east-dipping foliation and gently plunging NE- or SE-stretching lineations. The sequence of chronological events is well constrained in the Hoggar Belt (Caby 2003).

In Brazil, both geophysical data and the presence of a continental arc complex, the Santa Quitéria batholith, and high-pressure belts strongly suggest that Pharusian Ocean continued into South America prior to West Gondwana assembly.

With respect to the structural evolution of the two continents, they share broadly common collisional histories, though they are somewhat diachronous in terms of late-stage uplift and cooling. In southwestern Central Africa, granulite cooling ages range between *c.* 587 and 576 Ma, whereas exhumation of granulites in NE Brazil occurred between *c.* 568 and 550 Ma.

The authors are grateful to the organizers of Gondwana 12 (editorial board), and to W. R. Van Schmus and an anonymous reviewer for their comments on the manuscript. Ticiano J.S. Santos thanks the Conselho Nacional de Desenvolvimento Científico e Tecnológico (CNPq) and FAPESP (Fundação de Amparo a Pesquisa do Estado de São Paulo), which provided a research fellowship during his doctoral thesis. This work benefited from financial support from FAPESP, grant 03/07663-3.

Appendix

Analytical methods for analyses of U and Pb isotopes

Analytical data for this study were obtained at the Isotope Geochemistry Laboratory at the University of Kansas (USA). Sample T-126A is a migmatitic tonalite gneiss from the Granja massif within the Médio Coreaú domain (3°10.25′ N, 40°52.55′ W, 'Riacho de Gangorra', adjacent to a low concrete bridge on the road between Granja and

Terezinha). The sample was crushed, pulverized and heavy minerals were concentrated and passed through a Frantz magnetic separator to concentrate zircon (non-magnetic at 1.5A). The concentrate was washed in hot (*c.* 100 °C) 7 N HNO_3 for 2 hours to remove any iron oxide from the mineral grains prior to final separation of multiple zircon fractions on the Frantz separator at 1.5A, with side tilts of $-2°$ to $+1°$. Mineral fractions were prepared, dissolved, and Pb and U were purified using procedures modified after Krogh (1973, 1982) and Parrish (1987) and were spiked with a $^{205}Pb-^{235}U$ tracer solution prior to dissolution. Isotopic ratios were measured using a VG sector multi-collector mass spectrometer for samples with strong signals (>200 mV for ^{206}Pb) and in single-collector mode using the Daly detector for weaker signals. Both Pb and U isotopic compositions were analysed on the same single Re filament using silica gel and phosphoric acid and corrected for average mass discrimination of 0.12 ± 0.05% per mass unit for multi-collector mode and 0.18 ± 0.05% per mass unit for single-collector mode (based on replicate analyses of common Pb standard SRM 981). Uranium fractionation was monitored by replicate analyses of SRM U-500. Uncertainties in U/Pb ratios were ±0.5%; in some instances weak signals caused uncertainties up to 2%. Radiogenic Pb isotopes were calculated by correcting for modern blank Pb and for original non-radiogenic Pb using the Stacey & Kramers (1975) model for the approximate age of the sample. Uncertainties in radiogenic Pb ratios are typically ±0.1% except where low $^{206}Pb/^{204}Pb$ ratios indicate larger common Pb corrections. Total procedure blanks over the course of analyses ranged from 2 to 25 pg for Pb and 0.5 to 4 pg for U. Data were regressed using the ISOPLOT program of Ludwig (1993).

Analytical methods for analyses of Sm and Nd isotopes

Sample BCRE94-3 (weighing 15 kg) is a migmatitic garnet granulite from the Granja massif in the Medio Coreaú domain (3°3.78′N, 78°49.84′W, *c.* 6 km north of Granja 1–2 km east of the BR-071). It was crushed to granule-sized fragments and split systematically in order to obtain a smaller representative whole-rock sample, which was then reduced to powder in an alumina ceramic ball mill. Mineral separates were obtained by additional crushing of the granules, washing and sieving to 150–300 μm, before using a Frantz Isodynamic magnetic separator and heavy-liquids. All samples were loaded in 23 ml PTFE Parr bombs and spiked with a mixed $^{149}Sm-^{150}Nd$ tracer. Dissolution in $HF-HNO_3$ included a first step at 125 °C on a hotplate and 5–7 days at 180 °C in an oven (Patchett & Ruiz 1987). Rare-earth elements were concentrated using ion-exchange columns containing 100–200 mesh AG50W-X8 cation resin in an HCl media; Sm and Nd were separated with a second HCl chromatigraphic procedure using 100–200 mesh EiChrom LN-SPEC resin (Richard *et al.* 1976; White & Patchett 1984; Patchett & Ruiz 1987).

Sm samples were loaded on single Ta filaments with 0.25 M phosphoric acid and analysed as Sm^+ in static multi-collector or single collector mode. Nd samples were loaded on single Re filaments with AG50W-X8 cation resin and 0.25 M phosphoric acid and analysed as Nd^+ in a dynamic multi-collector mode, collecting 50–100 ratios with *c.* 1V ^{144}Nd peak intensity. This generally yields an internal precision for $^{143}Nd/^{144}Nd$ of 18 to 30 ppm (2σ) for individual runs; external precision based on repeated analyses of internal standard KU NNd-1A varied from 40 to 70 ppm (2σ). All analyses were adjusted for instrumental bias using the La Jolla Nd standard $^{143}Nd/^{144}Nd$ ratio of 0.511860, after normalization to $^{146}Nd/^{144}Nd = 0.7219$. During the course of these analyses Nd blanks ranged from 500 to <100 pg, with corresponding Sm blanks of 100 to <50 pg, insignificant for the determined Sm and Nd concentrations and Nd isotopic compositions. $\varepsilon Nd(t)$ values calculated for replicate samples from the same rock powder were generally reproducible within ±0.5. Depleted-mantle model ages T_{DM} were calculated using the equation $\varepsilon Nd(T) = 0.25T^2 - 3T + 8.5$ for the mantle evolution curve (DePaolo 1981), where T is the model age.

References

Acef, K., Liégeois, J. P., Ouabadi, A. & Latouche, L. 2003. The Anfeg post-collisional Pan-African high-K calc-alkaline batholith (Hoggar, Algeria), emplaced within the LATEA metacraton. *Journal of African Earth Sciences*, **37**, 295–311.

Affaton, P., Rahaman, M. A., Trompette, R. & Sougy, J. 1991. The Dahomeyide orogen: Tectonothermal evolution and relationship with the Volta basin. *In*: Dallmeyer, R. D. & Lecorche, J. P. (eds) *The West African orogens and circum-Atlantic correlatives*. Springer, New York, 95–111.

Agbossoumondé, Y., Menot, R.-P. & Guillot, S. 2001. Metamorphic evolution of Neoproterozoic eclogite from south Togo (West Africa). *Journal of African Earth Sciences*, **33**, 227–244.

Agbossoumondé, Y., Guillot, S. & Menot, R. 2004. Pan-African subduction–collision event evidenced by high-*P* coronas in metanorites from the Agou massif (southern Togo). *Precambrian Research*, **135**, 1–21.

Almeida, F. F. M. & Black, R. 1968. Geological comparison of northeastern South America and Western Africa. *Anais de Academia Brasileira de Ciências*, **40**, 317–319.

Arthaud, M. H., Caby, R., Fuck, R. A., Dantas, E. L. & Parente, C. V. 2008. Geology of the northern Borborema Province, NE Brazil and its correlation with Nigeria, NW Africa. *In*: Pankhurst, R. J., Trouw, R. A. J., Brito Neves, B. B. & De Wit, M. J. (eds) *West Gondwana: Pre-Cenozoic Correlations Across the South Atlantic Region*. Geological Society, London, Special Publications, **294**, 49–67.

Attoh, K. & Morgan, J. 2004. Geochemistry of high-pressure granulites from the Pan-African Dahomeyide orogen, West Africa: constraints on the origin and composition of the lower crust. *Journal of African Earth Sciences*, **39**, 201–208.

ATTOH, K., HAWKINS, D., BOWRING, S. & ALLEN, B. 1991. U–Pb zircon ages of gneisses from the Pan-African Dahomeyide orogen, West Africa. *EOS Transactions of the American Geophysical Union*, **72**, 299

ATTOH, K., DALLMEYER, R. D. & AFFATON, P. 1997. Chronology of nappe assembly in the Pan-African Dahomeyide orogen, West Africa: evidence from $^{40}Ar-^{39}Ar$ mineral ages. *Precambrian Research*, **82**, 153–171.

AZZOUNI-SEKKAL, A. & BOISSONNAS, J. 1993. Une province magmatïque de transition du calco-alcalin à l'alcalin: les granitoïdes Pan-africains à structure annulaire de la chaîne pharusienne du Hoggar (Algérie). *Bulletin de la Société Géologique de France*, **164**, 597–608.

BRITO NEVES, B. B., SÁ, J. M., NILSON, R. R. & BOTELHO, N. F. 1995. A tafrogênese Estateriana nos blocos paleoproterozóicos da América do Sul e processos subseqüentes. *Geonomos*, **3**, 1–21.

BRITO NEVES, B. B., DOS SANTOS, E. J. & VAN SCHMUS, W. R. 2000. Tectonic history of the Borborema Province, northeastern Brazil. *In*: CORDANI, U., MILANI, E. J., THOMAZ FILHO, A. & CAMPOS, D. A. (eds) *Tectonic Evolution of South America. 31st International Geological Congress, Rio de Janeiro, Brazil*, 151–182.

BRITO NEVES, B. B, VAN SCHMUS, W. R. & FETTER, A. H. 2002. Northwestern Africa–Northeastern Brazil. Major tectonic links and correlation problems. *Journal of African Earth Sciences*, **34**, 275–278.

BRUGUIER, O., DADA, S. & LANCELOT, J. R. 1994. Early Archean component (>3.5 Ga) within a 3.05 Ga orthogneiss from northern Nigeria: U/Pb zircon evidence. *Earth and Planetary and Science Letters*, **125**, 89–103.

CABY, R. 1989. Precambrian terranes of Benin Nigeria and Northeast Brazil and the Late Proterozoic South Atlantic. *In*: DALLMEYER, R. D. (ed.) *Terranes in the Circum-Atlantic Paleozoic Orogens*. Geological Society of America, Special Papers, **230**, 145–158.

CABY, R. 1994. Precambrian coesite from northern Mali: first record and implications for plate tectonics in the trans-Saharan segment of the Pan-African belt. *European Journal of Mineralogy*, **6**, 235–244.

CABY, R. 2003. Terrane assembly and geodynamic evolution of central –western Hoggar: a synthesis. *Journal of African Earth Sciences*, **37**, 133–159

CABY, R. & MONIÉ, P. 2003. Neoproterozoic subduction and differential exhumation of western Hoggar (southwest Algeria): new structural, petrological and geochronological evidence. *Journal of African Earth Sciences*, **37**, 269–293.

CABY, R., ANDREOPOULOS-RENAUD, U. & PIN, C. 1989. Late Proterozoic arc-continent and continent collision in the Pan-African Trans-Saharan belt of Mali. *Canadian Journal Earth Sciences*, **26**, 1136–1146.

CASTAING, C., TRIBOULET, C., FEYBESSE, J.-L. & CHEVREMENT, P. 1993. Tectonometamorphic evolution of Ghana, Togo and Benin in the light of the Pan-African Brasiliano orogeny. *Tectonophysics*, **28**, 423–432.

CASTAING, C., FEYBESSE, J. L., THIÉBLEMONT, C., TRIBOULET, C. & CHEVREMONT, P. 1994. Paleogeographical reconstructions of the Pan-Africano/ Brasiliano Orogen: closure of an oceanic domain or intracontinental convergence between major blocks? *Precambrian Research*, **69**, 327–344.

DANTAS, E. L., HACKSPACHER, P. C., VAN SCHMUS, W. R. & BRITO NEVES, B. B. 1998. Archean accretion in the São José do Campestre massif, Borborema Province, northeast Brazil. *Revista Brasileira de Geociências*, **28**, 221–228.

DEPAOLO, D. J. 1981. Neodymium isotopes of the Colorado Front Range and crust-mantle evolution in the Proterozoic. *Nature*, **291**, 193–196.

DOSTAL, J., DUPUY, C. & CABY, R. 1994. Geochemistry of the Neoproterozoic Tilemsi belt of the Iforas (Mali, Sahara): a crustal section of an oceanic island arc. *Precambrian Research*, **65**, 55–69.

DUCLAUX, G., MÉNOT, R. P., GUILLOT, S., AGBOSSOU-MONDÉ, Y. & HILAIRET, N. 2006. The mafic layered complex of the Kabý e massif (north Togo and north Benin): Evidence of a Pan-African granulitic continental arc root. *Precambrian Research*, **151**, 101–118.

EL-HADJ TIDJANI, AFFATON, P., LOUIS, P. & SOCOHOU, A. 1997. Gravity characteristics of the Pan-African Orogen in Ghana, Togo and Benin (West Africa). *Journal of African Earth Sciences*, **24**, 241–258

FETTER, A. H. 1999. *U–Pb and Sm–Nd Geochronological constraints on the crustal framework and geological history of Ceará State, NW Borborema Province, NE Brazil: implications for the assembly of Gondwana*. PhD thesis, University of Kansas.

FETTER, A. H., VAN SCHMUS, W. R., DOS SANTOS, T. J. S., ARTHAUD, M. H. & NOGUEIRA NETO, J. A. 1997. Geologic history and framework of Ceará State: northwest Borborema Province, NE Brazil. *In*: *Actas I South American Symposium on Isotope Geology, Brazil, Extended Abstracts*, 112–114.

FETTER, A. H., VAN SCHMUS, W. R., DOS SANTOS, T. J. S., ARTHAUD, M. & NOGUEIRA NETO, J. A. 2000. U–Pb and Sm–Nd geochronological constraints on the crustal evolution and basement architecture of Ceará State, NW Borborema Province, NE Brazil: implications for the existence of the Paleoproterozoic supercontinent "Atlantica". *Revista Brasileira de Geociências*, **30**, 102–106.

FETTER, A. H., DOS SANTOS, T. J. S., VAN SCHMUS, W. R., HACKSPACHER, P. C., BRITO NEVES, B. B., ARTHAUD, M. H., NOGUEIRA NETO, J. A. & WERNICK, E. 2003. Evidence for Neoproterozoic continental arc magmatism in the Santa Quitéria batholith of Ceará State, NW Borborema Province, NE Brazil: implications for assembly of West Gondwana. *Gondwana Research*, **6**, 265–273.

GASQUET, D., BARBEY, P., DOU, M. & PAQUETTE, J. L. 2003. Structure, Sr–Nd isotope geochemistry and zircon U–Pb geochronology of the granitoids of the Dabakala area (Côte d'Ivoire): evidence for a 2.3 Ga crustal growth event in the Palaeoproterozoic of West Africa? *Precambrian Research*, **127**, 329–354.

HIRDES, W. & DAVIS, D. W. 2002. U–Pb zircon and rutile metamorphic ages of Dahomeyan garnet-hornblende gneiss in southeastern Ghana, West Africa. *Journal of African Earth Sciences*, **35**, 445–449.

JAHN, B. M., CABY, R. & MONIÉ, P. 2001. The oldest UHP eclogites of the world: age of UHP metamorphism, nature of protoliths and tectonic implications. *Chemical Geology*, **178**, 143–158.

JONES, W. B. 1990. The Buem volcanic and associated sedimentary rocks, Ghana: a field and geochemical investigation. *Journal of African Earth Sciences*, **11**, 373–383.

KROGH, T. E. 1973. A low contamination method for hydrothermal decomposition of zircon and extraction of U and Pb for isotopic age determinations. *Geochimica et Cosmochimica Acta*, **37**, 485–494.

KROGH, T. E. 1982. Improved accuracy of U–Pb zircon ages by the creation of more concordant systems using an air abrasion technique: *Geochimica et Cosmochimica. Acta*, **46**, 637–649.

LESQUER, A., BELTRÃO, J. F. & DE ABREU, F. A. M. 1984. Proterozoic links between northeastern Brazil and West Africa: a plate tectonic model based on gravity data. *Tectonophysics*, **110**, 9–26.

LIÉGEOIS, J. P., BLACK, R. & NAVEZ, J. 1994. Early and late Pan-African orogenies in the Aïr assembly of terranes (Tuareg shield, Niger). *Precambrian Research*, **67**, 59–88.

LUDWIG, K. R. 1993. *ISOPLOT — A plotting and regression program for radiogenic-isotope data: version 2.70*. United States Geological Survey Open-File Report **91–445**.

MARTINS, G., OLIVEIRA, E. P., SOUZA FILHO, C. R. & LAFON, J. M. 1998. Geochemistry, geochronology of the Algodões sequence, Ceará, NE Brazil: A Paleoproterozoic magmatic arc in the Central Ceará domain of the Borborema Province? *In*: *Congresso Brasileiro de Geologia 40, Belo Horizonte, Anais*. Sociedade Brasileira de Geologia, 28.

MONIÉ, P., CABY, R. & ARTHAUD, M. H. 1997. The Neoproterozoic Brasiliano Orogeny in Northeast Brazil: 40Ar/39Ar and petrostructural data from Ceará. *Precambrian Research*, **81**, 241–264.

NUNES, K. C. 1993. Interpretação integrada da Bacia do Parnaíba com ênfase nos dados aeromagnéticos. *In*: *3rd International Congress of the Brazilian Geophysical Society, Expanded Abstracts*, **1**, 152–175.

OLIVEIRA, D. C. 2000. Stratigraphic interplays between igneous and sedimentary events in the Early Paleozoic Jaibaras trough (Northeast Brazil). *Revista Brasileira de Geociências*, **30**, 423–427.

OLIVEIRA, D. C. & MOHRIAK, W. U. 2003. Jaibaras trough: an important element in the early tectonic evolution of the Parnaíba interior sag basin, Northern Brazil) *Marine and Petroleum Geology*, **20**, 351–383.

PAQUETTE, J. L., CABY, R., DJOUADI, M. T. & BOUCHEZ, J. L. 1998. U–Pb dating of the Pan-African orogeny in the Tuareg shield: the post-collisional syn-shear Tioueine pluton (western Hoggar, Algeria). *Lithos*, **45**, 245–253.

PARRISH, R. R. 1987. An improved micro-capsule for zircon dissolution in U–Pb geochronology. *Isotope Geoscience*, **66**, 99–102.

PATCHETT, P. J. & RUIZ, J. 1987. Nd isotopic ages of crust formation and metamorphism in the Precambrian of eastern and southern Mexico. *Contributions to Mineralogy and Petrology*, **96**, 523–528.

RICHARD, P. N., SHIMIZU, N. & ALLEGRE, C. J. 1976. 143Nd/146Nd, a natural tracer, an application to oceanic basalts. *Earth and Planetary Science Letters*, **31**, 269–278.

SÁ, J. M., MCREATH, I. & LETERRIER, J. 1995. Petrology, geochemistry, and geodynamic setting of Proterozoic igneous suites of the Orós fold belt (Borborema Province, northeast Brazil). *Journal of South American Earth Sciences*, **8**, 299–314.

SANTOS, T. J. S., NOGUEIRA NETO, J. A., FETTER, A. H. & HACKSPACHER, P. C. 2001. Petrografia e Litogeoquímica das Rochas do Embasamento Cristalino da Região de Granja - CE. *Revista de Geologia*, **14**, 33–48.

SANTOS, T. J. S., SOUZA, G. M., QUEIROZ, H. B., NOGUEIRA NETO, J. A. & PARENTE, C. V. 2002. Tafrogênese estateriana no embasamento paleoproterozóico do NW da Província Borborema: Uma abordagem petrográfica, geoquímica e geocronológica *In*: *Congresso Brasileiro de Geologia 41, João Pessoa, Actas*. Sociedade Brasileira de Geologia, 337.

SANTOS, T. J. S., SANTOS, A. A., DANTAS, E. L., FUCK, R. A. & PARENTE, C. V. 2003. Nd Isotopes and the provenance of metasediments of the Itataia Group, northwest Borborema Province, NE Brazil. *In*: *IV Symposium on South American Isotope Geology, Salvador, Short Papers*, 286–289.

SANTOS, T. J. S., FETTER, A., HACKSPACHER, P., VAN SCHMUS, W. R. & NOGUEIRA NETO, J. A. 2004. Structural and geochronological studies of the Médio Coreaú Domain, NE Brazil: Constraints on Brasiliano/Pan-African tectonic evolution in the NW part of the Borborema Province. *In*: WEINBERG, R., TROUW, R., FUCK, R. & HACKSPACHER, P. (eds) *The 750–550 Ma Brasiliano Event of South America*. Journal of the Virtual Explorer, Electronic Edition, 17, Paper 9.

SANTOS, T. J. S., DANTAS, E. L., FUCK, R. A., ARAÚJO, C. E. G., ROSA, F. F. & AMARAL, W. S. 2007. The geology and U–Pb and Sm–Nd geochronology from the northern portion of the Santa Quitéria Batholith, NE Brazil. *In XI Simpósio nacional de estudos tectônicos – V International symposium on tectonics of Brazil Actas, Extended Abstracts*. Sociedade Brasileira de Geologia, 142–144.

SANTOS, T. J. S., FETTER, A. H., HACKSPACHER, P. C., VAN SCHMUS, W. R. & NOGUEIRA NETO, J. A. 2007. Neoproterozoic tectonic and magmatic episodes in the NW segment of the Borborema Province, NE Brazil, during the assembly of the western Gondwana. *Journal of South American Earth Science*, doi: 10.1016/j.jsames.2007.05.006.

STACEY, J. S. & KRAMERS, J. D. 1975. Approximation of terrestrial lead isotope evolution by a two-stage model. *Earth and Planetary Science Letters*, **26**, 207–221.

STEIGER, R. H. & JÄGER, E. 1977. Subcommision on Geochronology: convention on the use of decay constants in geo- and cosmochronology. *Earth and Planetary Science Letters*, **36**, 359–362.

TOTEU, S. F., VAN SCHMUS, W. R., PENAYE, J. & MICHARD, A. 2001. New U–Pb and Sm–Nd data from north-central Camaroon and its bearing on pre-Pan African history of central Africa. *Precambrian Research*, **108**, 45–73.

TROMPETTE, R. 1994. *Geology of western Gondwana, Pan-African/Brasiliano aggregation of South America and Africa.* A.A. Balkema, Rotterdam.

VAN SCHMUS, W. R., BRITO NEVES, B. B., HACKSPACHER, P. C. & BABINSKI, M. 1995. U/Pb and Sm/Nd geochronolgic studies of eastern Borborema Province, northeastern Brazil: initial conclusions. *Journal of South American Earth Sciences*, **8**, 267–288.

WHITE, W. M. & PATCHETT, P. J. 1984. Hf–Nd–Sr isotopes and incompatible element abundances in island arcs: implications for magma origins and crust-mantle evolution. *Earth and Planetary Science Letters*, **64**, 167–185.

Proterozoic evolution of the Nigeria–Boborema province

S. S. DADA

Department of Earth Sciences, Ajayi Crowther University, Oyo, Oyo State, Nigeria (e-mail: ssdada47@yahoo.com)

Abstract: Structural, geochronological, geochemical and mineralization patterns in the Nigeria–Borborema province of western Africa and NE Brazil reflect a complex Proterozoic evolution culminating in the Neoproterozoic Pan-African/Brasiliano orogenesis (*c.* 600 Ma). Reworking of the Archaean–early Proterozoic crust produced heterogeneous deformation exemplified by prevalent shears, migmatization, granitization and intrusion of large volumes of granitoids typical of a Himalayan-type thickened crust resulting from continent–continent collision. Dominant north–south to east–west structures, with prominent penetrative fabric and mylonitised wrench faults, refolded, transpressed, or even obliterated older structural trends, which are preserved in nappes of the central Sahara region (NW Africa to Nigeria) and in NE Brazil. Anatexis and recrystallisation were coeval with emplacement of Pan-African granitoids throughout this mobile belt. Bulk chemical modification, especially affecting magmatophile elements and REE patterns, attest to chemical exchange between Archaean basement and Pan-African/Brasiliano rocks. Older crust is present in both regions, including early (3.6–3.5 Ga), mid (3.1 Ga) and late (2.7–2.5 Ga) Archaean, as well as large areas of Palaeoproterozoic rocks reworked by the *c.* 600 Ma tectonothermal events. The extent and interpretation of Eburnian/Transamazonian (2.1–2.0 Ga) events have not yet been resolved due to inadequate structural and isotopic data. Litho-structural control of Au, Sn, Nb and Ta mineralization relates to main or late-stage Pan-African deformation.

The continents of Africa and South America occupy a strategic place in global tectonic understanding and have attracted geoscientific attention since the beginning of the continental drift hypothesis (Hurley 1968; Torquato & Cordani 1981). The Nigerian Proterozoic province provides a link between the Hoggar Massif to the north and the Borborema Province to the south; both of which are assemblages of contrasted terrains of metasedimentary and exhumed crystalline basement rocks. Multidisciplinary studies carried out by various workers in both regions over the years now provide a fairly coherent picture regarding unequivocal similarities in the evolution of the two provinces (Almeida 1968; Brito Neves 1975; Caby & Arthaud 1986; Caby 1989; Dada 1989, 1998; Caby *et al.* 1990; Van Schmus *et al.* 2003; Dantas *et al.* 2004; Guimeraes *et al.* 2004). Structural, geophysical and geochronological data in the last three decades have reinforced earlier evidence and conclusions (Almeida 1968) that the geodynamic evolution of the Nigeria–Borborema Proterozoic is related to continent–continent collision at about 600 Ma (Burke & Dewey 1972; Black *et al.* 1979; Caby *et al.* 1981; Brito Neves 1982). Throughout this region, Neoproterozoic intrusions (belonging to the Pan African/Brasiliano sequence) include lower crustal granitoids associated with dioritic, gabbroic and charnockitic rocks (Dada *et al.* 1989, 1995). The Pan African/Brasiliano tectonic events caused heterogeneous reworking of pre-existing terranes through extensive deformation, migmatization, granitization and intrusion of a whole range of granitoids at elevated temperatures (T $\geq$ 600 °C), together with the development of dominant north–south, NE–SW and east–west shears defining the main structural fabric (folds, foliations, schistosity, lineations, etc.) of the entire region. This complex tectonic history has given rise to great difficulties in accurate interpretation of radiometric ages and isotopic characteristics, as well as in the structural analysis of older trends. Paradoxically, the problems of provenance of the metasedimentary rocks have in the last decade become amenable to combined structural and isotopic analysis (Caby & Arthaud 1986; Caby 1987, 1989; Annor 1995, 1998; Dada & Rahaman 1995; Caby & Boesse 2001). The basement of the Nigeria–Borborema shield (Fig. 1), which is overlain by inland and marginal Phanerozoic sediments fringing the Atlantic Ocean, consists of three major rock assemblages: (i) an Archaean migmatite gneiss complex; (ii) Proterozoic schist belts (metasedimentary and metavolcanic rocks); (iii) Pan-African/Brasiliano granitoids.

Archaean migmatite-gneiss complex

Long regarded as basement (*s.s.*), extensively but variably migmatized Archaean gneisses are well exposed in Nigeria (McCurry 1976; Rahaman

From: PANKHURST, R. J., TROUW, R. A. J., BRITO NEVES, B. B. & DE WIT, M. J. (eds) *West Gondwana: Pre-Cenozoic Correlations Across the South Atlantic Region.* Geological Society, London, Special Publications, **294**, 121–136. DOI: 10.1144/SP294.7 0305-8719/08/$15.00

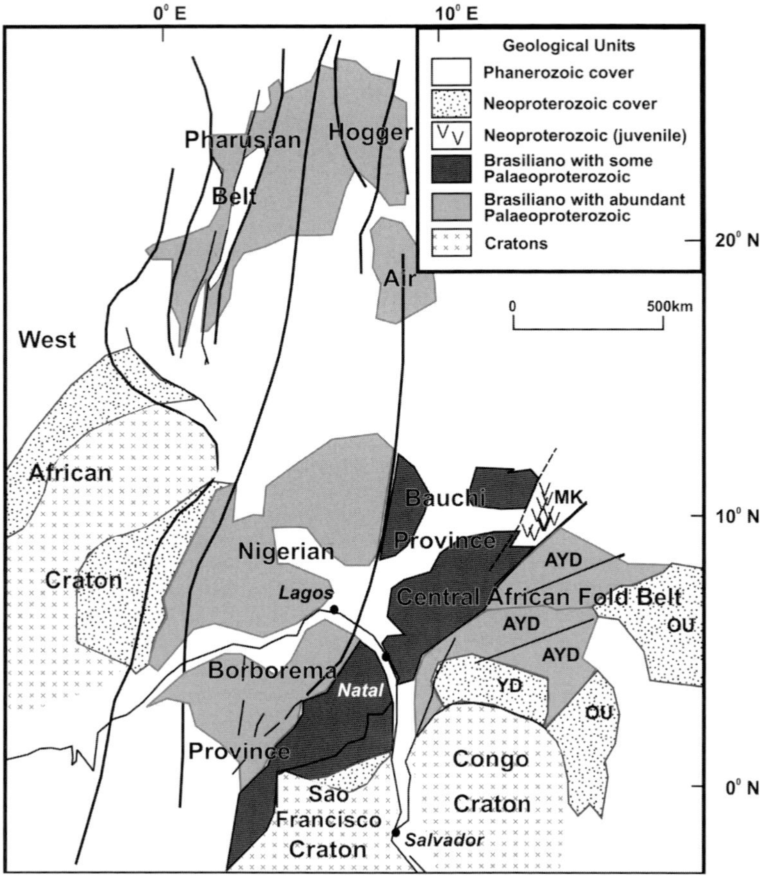

Fig. 1. The Trans-Saharan and Nigeria–Borborema Neoproterozoic belt of NW Africa and NE Brazil in a pre-Mesozoic drift reconstruction, after Van Schmus *et al.* (2008). AYD, Adamawa–Yadé domain; MK, Mayo Kebi terrane; OU, Oubanguide fold belt; YD, Yaoundé domain.

1976; Dada 1989; Ekwueme 1991) as well as in Ceará and Rio Grande do Norte, in the northern part of Borborema Province of Brazil (Brito Neves *et al.* 1975; Caby & Arthaud 1986). It is a heterogeneous assemblage including migmatized gneisses, orthogneisses, paragneisses and a series of metamorphosed basic and ultrabasic rocks. Petrographic evidence indicates that Pan-African/Braziliano reworking led to recrystallization of many of the constituent minerals of the migmatite-gneiss complex during partial melting, and most display medium to upper amphibolite-facies-metamorphism.

In both Nigeria and Ceará, the gneisses of the migmatite-gneiss complex are interleaved with amphibolites that may be derived from Mg-rich rocks such as continental basalts (Caby *et al.* 1990; Dada 1999*a*). However, there are no conclusive age and isotopic data to elucidate their origin. Gneisses and amphibolites in Nigeria form a bimodal association whose petrological and geochemical characteristics indicate a primary igneous origin (Dada 1989, 1999*a*). The Archaean migmatite-gneiss complex represents a reworked TTG terrain of migmatite gneisses, including plagioclase-rich leucosomes and potassic augen, cross-cut by quartz veins, aplites and pegmatites of late Proterozoic age as determined from lower intercept ages on U–Pb Concordia. However, a great proportion of the gneisses and migmatites in Ceará and Rio do Grande do Norte have a sedimentary origin. Both multiple and single zircon U–Pb, as well as Rb–Sr studies, have confirmed metamorphic events at 3.1–3.0, 2.7 and 0.6 Ga (Santos & Brito Neves 1984; Pessoa *et al.* 1986; Bruguier *et al.* 1994; Dada & Rahaman 1995; see Table 1), showing that the migmatite-gneiss complex is a relict component within the mobile belt.

Table 1. *Geological (U–Pb, Rb–Sr) and model (Nd, Sr) ages for rocks of the Nigerian Basement and the Jurassic ring complexes*

Lithology	Geological age	TNd (Ga)	Nd(t)	TSr	T= t-TNd	Reference(s)
Kaduna granodiorite gneiss	3.46 Ga (U–Pb)	3.57	–	3.49 Ga	10 Ma	Bruguier *et al.* (1994)
	3.46 Ga (U–Pb)	–	–	–	–	Ekwueme & Kröner (1992)
Kaduna early gneiss	3.1 Ga (U–Pb, Rb–Sr)	3.54	–	3.18 Ga	440	Dada (1989); Bruguier *et al.* (1994)
Kaduna late gneiss	2.75 Ga (Rb–Sr)		3.51	–	2.73 Ga 760	Dada *et al.* (1998)
Ibadan Aplite	2.75 Ga (Pb-Pb)	–	–	–	–	Oversby (1975)
Odo Ogun Gneiss	2.5 Ga (U–Pb)	–	–	–	–	Pidgeon *et al.* (1976)
Ile–Ife grey gneiss	2.3 Ga/439 Ma (U–Pb)	–	–	–	–	Rahaman (1988)
Ile–Ife granite gneiss	1.85 Ga/550 Ma (U–Pb)	–	–	–	–	Rahaman (1988)
Igbetti augen gneiss	1.9 Ga (Rb–Sr)	–	–	–	–	Rahaman (1988)
Egbe gneiss/ Kabba–Okene gneiss		2.56	–	2.36 Ga	–	Dada & Rahaman (1995)
Tiden Fulani migmatite	2.5 Ga/500 Ma (U–Pb)	1.80	−1.3	680 Ma	1300 – Ma	Dada *et al.* (1993*a*, *b*)
Badiko granite gneiss	2.5 Ga/500 Ma (U–Pb)	2.10	−14.6	740 Ma	1600 Ma	Dada *et al.* (1993*a*, *b*)
Okene Granodiorite Gn	2.1 Ga (U–Pb)	2.10	–	2.78 Ga	–	Annor (1995), Dada & Rahaman (1995)
Sarkin Pawa syntectonic	635 Ma (U–Pb)	1.50	−3.4	710 Ma	865 Ma	Dada (1999*b*)
Migmatite Badiko syntectonic diorite	623 Ma (U–Pb)	1.90	−12.3	780 Ma	1297 Ma	Dada (1999*b*)
Ikerre massive charnockite	620 Ma (U–Pb)	–	–	–	–	Tubosun *et al.* (1984)
Akure gneissic charnockite	634 Ma (U–Pb)	–	–	–	–	Tubosun *et al.* (1984)
Akure porphyritic granite	621 Ma (U–Pb)	–	–	–	–	Tubosun *et al.* (1984)
Idanre gneissic charnockite	580 Ma (U–Pb)					Tubosun *et al.* (1984)
Idanre massive charnockite	593 Ma (U–Pb)	–	–	–	–	Tubosun *et al.* (1984)
Idanre porphyritic granite	587 Ma (U–Pb)	–	–	–	–	Tubosun *et al.* (1984)
Toro Biot-Hbd granite	607 Ma (U–Pb)	2.10	−12.6	670 Ma	1493	Dada *et al.* (1989)
Toro charnockitic diorite	638 Ma (U–Pb)	2.50	−15.6	1.07 Ma	1915	Dada *et al.* (1989)
Bauchi quartz fayalite monzonite(bauchite)	638 Ma (U–Pb)	1.70	−3.9	1.80 Ga	1062	Dada & Respaut (1989)
Toro migmatite	581 Ma	1.75	−1.86	–	1280	Ferre *et al.* (1996)
Toro anatectic granite	616 Ma	1.88	−14	–	1380	Ferre *et al.* (1996)
Toro migmatite granite	715 Ma	2.71	−12.4	–	1960	Ferre *et al.* (1996)
Ring complex 473	170 Ma	1.46	−3.2	–	1290	Dickin *et al.* (1991)
Ring Complex 412	170 Ma	1.92	−5.6	–	1750	Dickin *et al.* (1991)

Populations of zircon with and without inherited cores occur in the same rock: a common situation in complex reworked terrains.

The heterogeneous nature of the Pan-African remobilization is evident in places where Palaeoproterozoic rocks have survived 600 Ma resetting, e.g., the Kabba–Okene gneisses (Annor 1995; Table 1). On the other hand, in some areas the isotopic record of accessory minerals such as zircon, monazite, titanite and apatite in pre-Pan African rocks has been completely reset during the Neoproterozoic (Dada 1999*b*). The latter rocks show fractionated REE patterns with negative Eu anomalies, although less pronounced than in the Neoproterozoic granitoids. This implies some degree of fractionation and retention of residual plagioclase during partial melting (Dada *et al.* 1993*a*; Dada 1999*b*). Radiogenic isotope data (Nd, Sr, Pb) confirm the above observation, indicating extensive reworking and remelting of older crust during the Neoproterozoic (see Figs 4b and 7a, b, c).

The Proterozoic schist belts

Schist belts constitute one of the most remarkable structural features in the Nigeria–Borborema shield. Various workers cited above have recognized and described the major north–south and east–west elongate belts that define the structural grains of the Nigeria–Borborema basement. These belts belong to two groups: (1) the older metasediments, which include quartzite, marble, micaschist and metavolcanic rocks and (2) the younger psammitic to pelitic metasediments with varying amounts of mafic rocks (amphibolites). The first group is well developed in SW Nigeria (Figs 2 & 4a) and the second is widespread in NW Nigeria

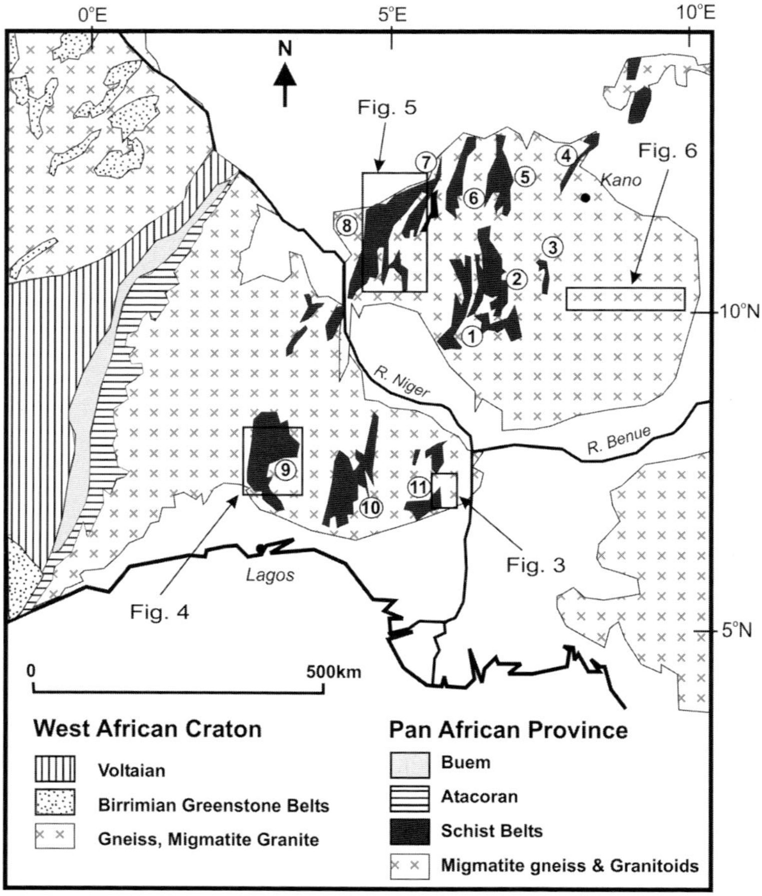

Fig. 2. Location of Nigerian schist belts on the eastern margin of the West African craton, after Turner (1983). 1, Zungeru–Birnin Gwari; 2, Kushaka; 3, Karaukarau; 4, Kazaure; 5, Wonaka; 6, Maru; 7, Anka; 8, Zuru; 9, Iseyin–Oyan River; 10, Ilesha; 11, Igarra.

(Figs 2 & 5a) and the Igarra schist belts (Oyawoye 1972; McCurry 1976; Rahaman 1976).

In NW Nigeria, there is a dominant series of schists of greywacke origin that range from metapelites to quartzites; in detail they are made up of phyllites, schists (*s.s.*), quartzo-feldspathic schists, paragneiss, Fe–Mn-bearing (ferruginous) quartzites and garnet amphibolites. Acid and intermediate volcanic rocks are interbedded with the metamorphosed pelitic to semi-pelitic rocks in the Anka, Birnin Gwari and Zungeru schist belts (Figs 2 & 5a), which are recognizable discrete belts with distinct and contrasted lithologies, separated by the Archaean migmatite-gneiss complex or by Pan-African granitoids. This has led to the suggestion of several palaeo-depocentres. While the problem of their possible co-sanguinity remains unresolved, Ajibade *et al.* (1987) have suggested that the prevalence of inter-belt schist relics in the intervening deformed granite terrains is strong evidence that the schist belts were not confined to their presently mapped areas. The consequence of such a suggestion is that the metasedimentary sequence was dismembered during Pan-African deformation and the schist belts are now rafted segments or relics of a single supracrustal cover.

In SW Nigeria, three major schist belts have been recognized (Turner 1983). They are the Iseyin–Oyan River, the Ilesha and the Igarra–Kabba–Lokoja schist belts (Figs 3 & 4a). The Iseyin–Oyan River belt, which continues into the Ibadan area, appears to form part of the late Archaean to Palaeoproterozoic banded gneiss–quartzite–schist sequence of Jones & Hockey (1964) and Burke *et al.* (1976). The Ife–Ilesha belt consists of two contrasting rock assemblages separated by the NNE-trending Ifewara fault. To the west, the belt includes massive amphibolite, amphibole schist, talc–tremolite schist and pelitic rocks, whereas the eastern unit is made up of quartzite, quartz schist, ferruginous quartzite and schist with minor amphibolite (Rahaman 1976). The work of Bafor (1988) shows close similarities between the Egbe–Isanlu and Ilesha schist belts. The Igarra–Kabba–Lokoja belt rocks are essentially metapelites with inter-layered quartzite and marble. Structural evidence suggests that rocks of the western belt are older than those of the eastern belt (Ajibade *et al.* 1987).

In the Borborema Province, the metasediments are similar to those described above in Nigeria. They are also preserved within elongated fault-bounded structures (Brito Neves *et al.* 1984), as metavolcano-sedimentary fold belts composed essentially of mica schist, phyllites, quartzites, marbles and calc-silicate rocks, ranging in metamorphic grade from upper greenschist to almandine–amphibolite facies (Arthaud *et al.* 2008; Santos *et al.* 2008). However, the most striking tectonic structures are the east–west trending Pernambuco and Patos shears (Braun 1982; Brito Neves 1983; Caby & Arthaud 1986; Jardim de Sa *et al.* 1987; Caby 1989). In detail, the Serido and north Ceará regions are easily correlated with the Nigerian schist belt, petrologically, in the degree of metamorphism and, particularly, in structural style. In the Borborema Province, the rocks display a pervasive flat-lying metamorphic foliation parallel to lithological boundaries (Caby *et al.* 1990; Arthaud *et al.* 2008).

Rb–Sr and K–Ar ages of 700–450 Ma give minima for metamorphic cooling, but there are no reliable ages of formation for the Nigerian metasediments. Indirect evidence for a Palaeoproterozoic age is the 2.1 Ga U–Pb zircon date of the Kabba–Okene gneiss (Annor 1995, table 1). The gneiss hosts metasedimentary xenoliths and shows the same early tectono-metamorphic fabric exhibited by the Okene–Igarra schist (Annor 1998; fig. 3). A similar interpretation has been suggested for the Jucurutu Group which has yielded a whole-rock Rb–Sr isochron age of 2.1–2.0 Ga (Jardim de Sa *et al.* 1987). Extensive application of the Rb–Sr method on metasediments by several workers from different laboratories (Holt 1982; Caen-Vachette & Umeji 1983; Fitches *et al.* 1985; Caen-Vachette & Ekwueme 1988; Ogezi 1988) suggests that these rocks suffered extensive reworking during Pan-African orogenies. In most cases, the rehomogenisation results in errorchrons of between 1400 and 450 Ma (Fig. 5b) suggesting mixture between the pre-Pan African basement and *c.* 600 Ma events; these ages are often wrongly interpreted as Kibaran (1300–900 Ma), e.g, Ogezi (1988), Holt (1982), Fitches *et al.* (1985), Caen-Vachette & Umeji (1983), Caen-Vachette & Ekwueme (1988). More recent studies have suggested a Neoproterozoic age for the deposition of the Jucurutu and Serido supracrustal rocks in northeastern Brazil, based on the presence of Neoproterozoic detrital zircons in the metasediments (e.g., Van Schmus *et al.* 2003).

Pan-African/Brasiliano granitoids

Migmatization of the older basement and generation of Pan-African granitoids constitute the most widespread manifestations of the 600 Ma orogenies in the Nigeria–Borborema shield. The Neoproterozoic granitoids are composed of several contemporaneous petrological groups. They vary from granites (*s.s.*) and their associated charnockitic, dioritic, monzonitic, syenitic rocks, to gabbros, serpentinites and anorthosites. Felsic and mafic dykes in the form of pegmatites and dolerites, as well as

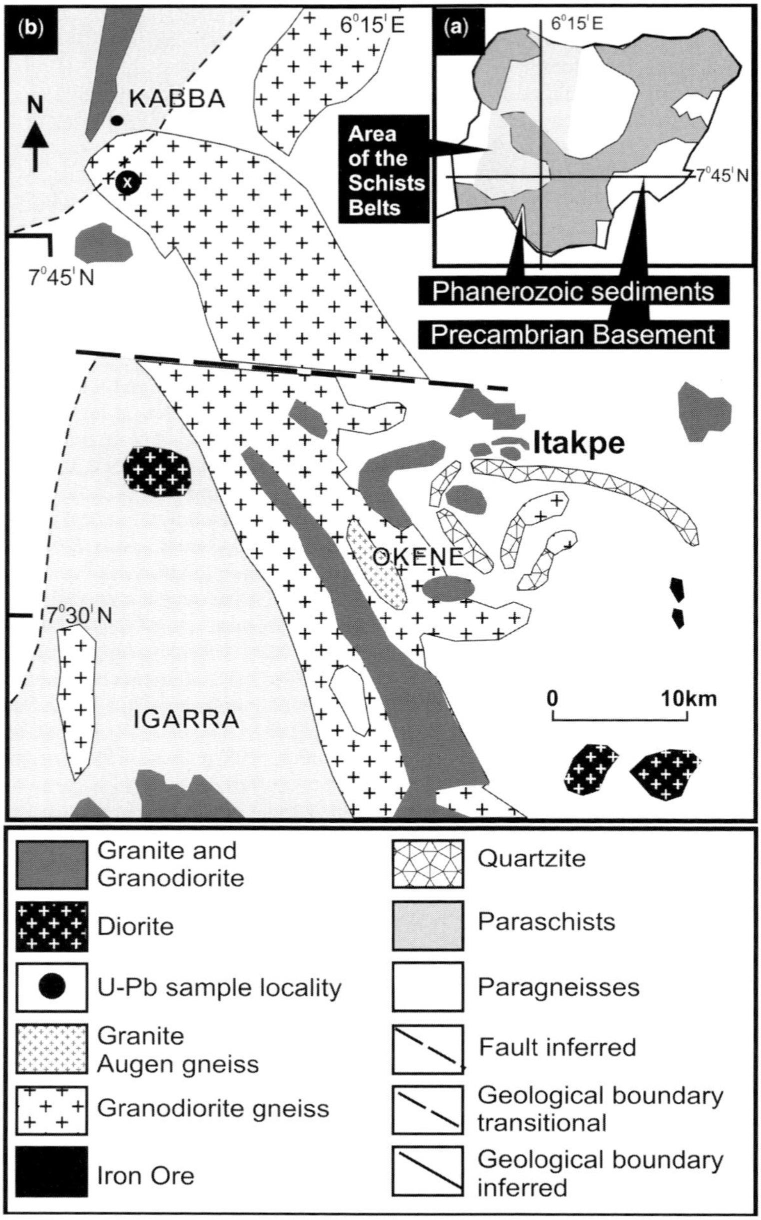

Fig. 3. Geological map of Kabba–Okene with banded iron ore north of Igarra schist belt, southwestern Nigeria, after Annor (1995).

extensive migmatized and granitized pre-Pan-African basement, are well exposed. The structural trends formed during this widespread event subsequently controlled the emplacement of the Jurassic alkaline to super-alkaline ring complexes to a large extent (Rahaman *et al.* 1984; Dickin *et al.* 1991).

Major and trace element geochemistry combined with U–Pb geochronology and Pb-, Sr- and Nd-isotope geochemistry in a large segment of northern Nigeria, from Kaduna in the west to Bauchi in the east (Fig. 6), favours a mixing model between juvenile Pan-African material and the Archaean basement, with a predominant

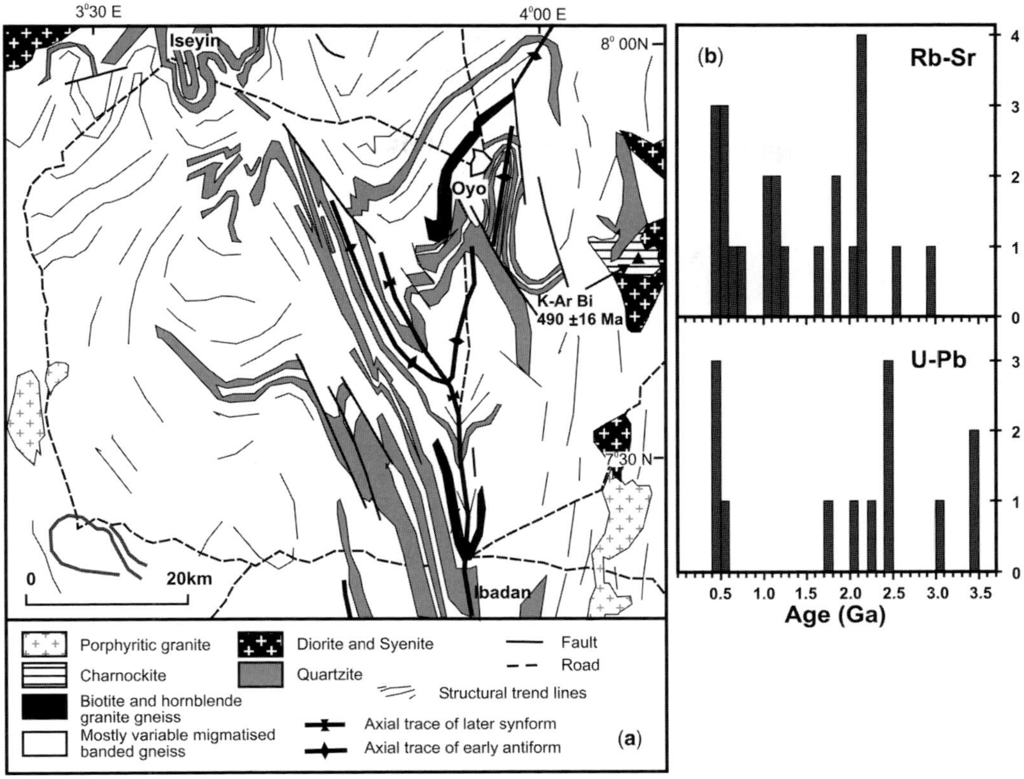

Fig. 4. (**a**) The regional geology of Iseyin–Oyo–Ibadan schist belt, southwestern Nigeria, showing the mode of occurrence of the gneisses and the refolded quartzites, after Grant (1970); (**b**) Histograms of Rb–Sr and U–Pb ages of Nigerian migmatite-gneisses.

involvement of the latter component, in the genesis of the Pan-African granitoids (Dada *et al.* 1995; Dada 1998; Fig. 7).

Trace element studies indicate high LREE abundances in the granitoids with prominent negative Eu anomalies (Olarewaju 1988; Dada *et al.* 1995), due to intra-crustal melting (Taylor & McLennan 1981) during the Pan African event. Modification of the bulk chemistry by chemical exchange between the Archaean and late Proterozoic rocks is evident in the high concentration of magmatophile elements (K, Rb, Ba, Sr, La, Ce), in agreement with isotope geochemical data (Sr, Nd, Pb) on these rocks. It has been suggested that partial melting in the mantle resulted from Pan-African plate collision, giving rise to juvenile magma which, together with the inherent heat, led to large-scale reworking with concomitant assimilation of older material (Dada *et al.* 1995). The resulting contamination produced Pan-African initial $^{87}Sr/^{86}Sr$ (0.70617–0.71015) and $^{143}Nd/^{144}Nd$ (0.511071–0.511599) ratios that are closer to crustal than mantle values ($\varepsilon Sr_i = +30$ to $+86$; $\varepsilon Nd_i = -15.5$ to -4.0). This is true for a large part of the Nigerian basement and shows that each of the granitic episodes represents mixture of mantle and assimilated older crustal components during Pan-African continent–continent collisional geodynamic evolution *c.* 600 Ma (e.g., Burke & Dewey 1972; Black *et al.* 1979; Caby 1989).

Neoproterozoic (*c.* 600 Ma) U–Pb, Rb–Sr, K–Ar ages have been reported from granitoids within the Nigerian basement (Grant 1978; Matheis & Caen-Vachette 1983; Tubosun *et al.* 1984; Fitches *et al.* 1985; Ogezi 1988; Rahaman 1988; Rahaman *et al.* 1991). In particular, U–Pb data on zircons confirm Pan-African ages of emplacement for the charnockitic rocks that were previously thought to be Kibaran (*c.* 1100 Ma) or even Archaean (Jones & Hockey 1964; Cooray 1977; Hubbard 1975). Combination of the available structural data and U–Pb ages suggest the following sequence of events in the reworked Nigerian Pan-African orogen: (i) early deformational phase D_1 with migmatization and local anatexis at 640–620 Ma; (ii) main deformational phase D_2 with

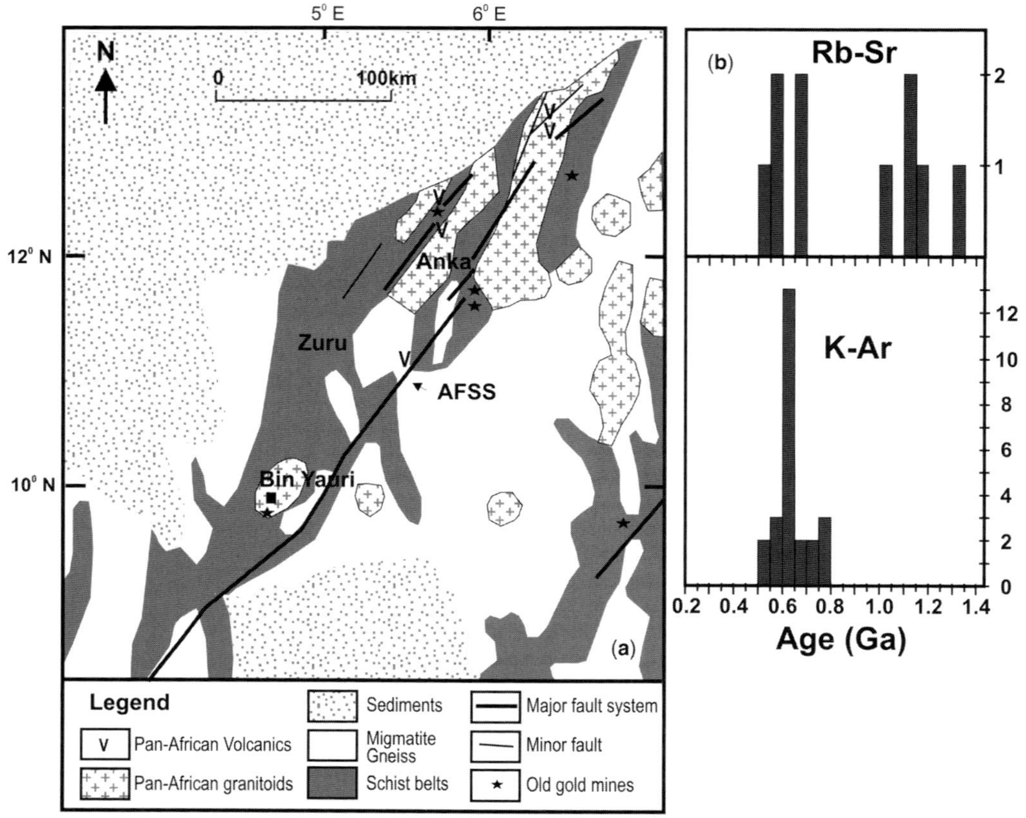

Fig. 5. (**a**) Regional geology showing the major NNE–SSW Anka–Yauri fault associated with gold mineralization, after Garba (2000); (**b**) Histograms of Rb–Sr and K–Ar ages of Nigerian metasedimentary rocks.

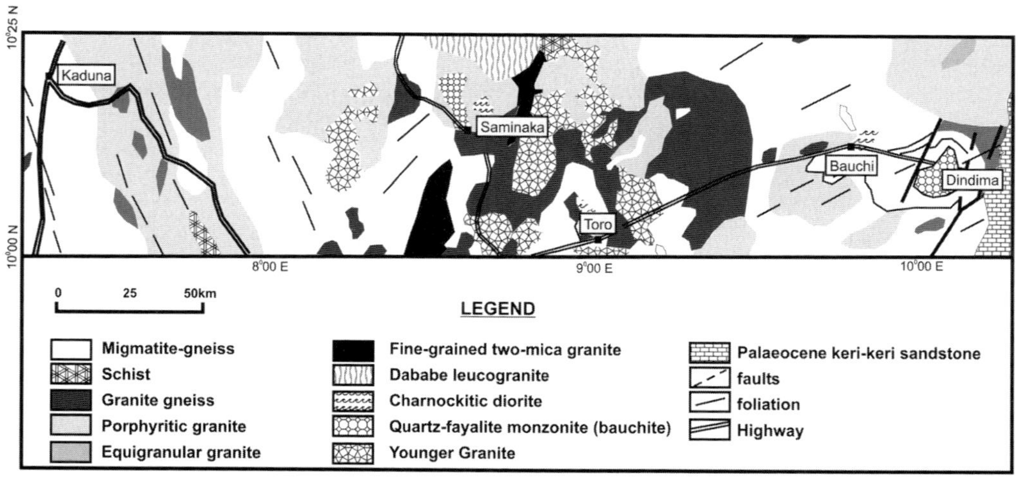

Fig. 6. Geological map of Kaduna–Toro–Bauchi region in north-central Nigeria, after Dada *et al.* (1995).

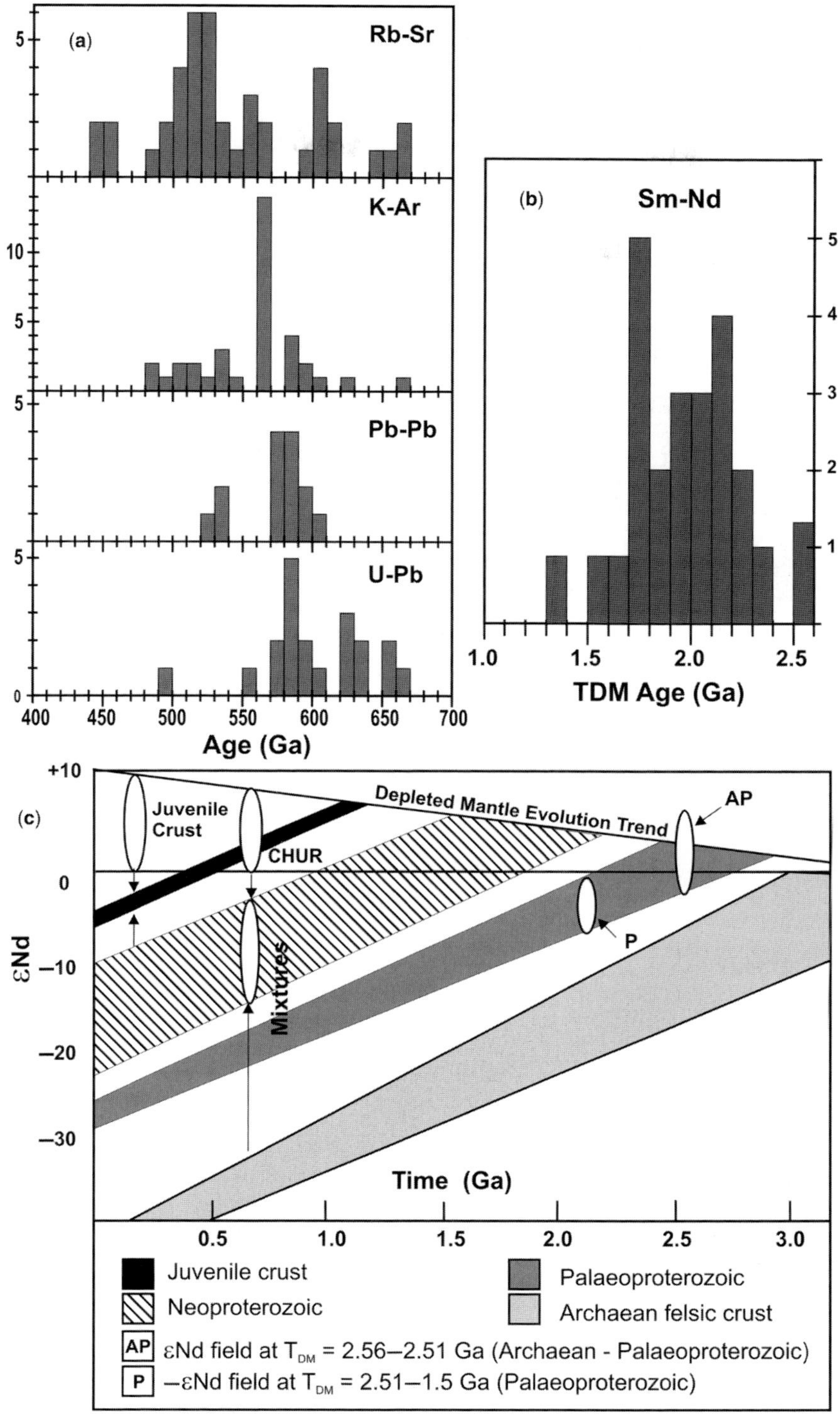

Fig. 7. (**a**) Histograms of Rb–Sr, K–Ar, Pb–Pb, and U–Pb ages of Nigerian granitoids; (**b**) Histograms of Nd model ages (T_{DM}) of Nigerian granitoids; (**c**) Nd isotope evolution diagram showing possible mixing between proposed Palaeoproterozoic, Neoproterozoic and juvenile crusts and the Archaean (>2.5 Ga) felsic component of the Nigerian migmatite-gneiss basement. Depleted mantle evolution trend assumes a linear growth from a DM source with present-day εNd = +10 (Jahn *et al.* 1988), after Dada *et al.* (1995).

the formation of shear zones and emplacement of syntectonic granitoids at the climax of Pan-African magmatism (620–600 Ma); (iii) emplacement of late to post-tectonic granitoids during the late second phase (D_2) deformation (600–580 Ma).

Further studies will probably modify or refine this scheme in detail, particularly because Pan-African tectono-metamorphism was heterogeneous in style, degree and grade (Annor & Freeth 1985). In addition, increasing evidence suggests that deformation may not have been synchronous with magmatism (Grant 1978; Rahaman *et al.* 1991). A similar sequence has been described in the Borborema Province (Guimaraes *et al.* 2004; Van Schmus *et al.* 2003, 2008). The cooling ages obtained from Rb–Sr on whole rocks are similar for both regions, clustering around 500 Ma.

The age of the felsite dykes has been established as between 580 and 535 Ma (Rb–Sr whole-rock ages, Van Breemen *et al.* 1977; Matheis & Caen-Vachette 1983), whereas the basic dykes seem to be considerably younger, with ages of *c.* 500 Ma (478 ± 19 Ma, Grant 1970). The structural and geochronological importance of this suite of rocks is often overlooked in consideration of the Nigeria–Borborema shield although their emplacement ages are instrumental for the establishment of the chronostratigraphic and structural history of the region. Whereas the basic dykes are interpreted as representing early Brasiliano magmatic activity in NE Brazil (Bernasconi 1987), there seems to be general structural and geochronological evidence that they constitute the post-tectonic units in the Nigerian basement (Rahaman 1976).

Structural geology

The structural similarities between the Precambrian terrains exposed in Nigeria and the Borborema Province of NE Brazil have long been recognized (Torquato & Cordani 1981, and see several other contributions to this volume). The dominant structural features of the Nigeria–Borborema basement are apparent from studies in the schist belts and conclusively show that such structures were developed during the Pan-African/Brasiliano sequence of orogenies; pre-existing structures were overprinted or obliterated. While there is a gross similarity in tectonic style, the observed patterns vary in detail due to the variable degree of rock exposure and differences in lithological distribution. Earlier workers described the Nigerian basement as a combination of well developed metasedimentary cover to the west and a largely vestigial crystalline terrain to the east (McCurry 1971; Oyawoye 1972; Rahaman 1976, 1988; Grant 1978; Turner 1983; Fitches *et al.* 1985; Ajibade *et al.* 1987; Ferre *et al.* 1996). Recent isotopic data, gravity evidence and structural analysis (Lesquer *et al.* 1984; Caby 1989; Caby *et al.* 1990; Black *et al.* 1994; Dada 1998; Ferre *et al.* 1998) have confirmed the allochthonous nature of the supracrustal terrains that were welded together, presumably in contiguity with the Hoggar Massif to the north and the Borborema Province to the south. Sutures have been proposed along the two transcurrent fault zones, and in particular within the Ife–Ilesha schist belt, which has been interpreted as a back-arc marginal basin (Rahaman *et al.* 1988), and east-verging nappes (Caby & Boesse 2001).

In Nigeria, structural studies of the metasedimentary belts have led to the proposition of two major phases of Pan-African deformation (McCurry 1971, 1976; Rahaman 1976, 1988). Phase 1 is characterized by isoclinal folds (F_1) with subhorizontal S_1 axial schistosity planes and a dominantly east–west mineral lineation (L_1) parallel to the fold axis. Phase 2 is characterized by regional isoclinal folds (F_2) with subvertical axial planes and subhorizontal axes and an S_2 axial plane of schistosity. The micro-folds associated with the major folds define a crenulation lineation (L_c) parallel to the L_2 mineral lineation, in conformity with the general north–south F_2 fold axis (i.e., NNW to NNE, Figs 1–5). Discrete brittle major faults have N20 to NE–SW trend within the schist belts and on a regional scale (Figs 4, 5, 6); both McCurry (1976) and Rahaman (1976) have described sinistral N130 to north–south conjugate faults.

However, it is the second phase of deformation that is regionally most pervasive; it has left a most dominant submeridianal (*c.* north–south) imprint, not only in the schist belts but all over the Nigerian basement. This is roughly parallel to the outcrops of the syntectonic Pan-African granitoids, which were preferentially emplaced within north–south shear zones (McCurry 1976; Cahen *et al.* 1984) and on which late Pan-African deformation (Rahaman 1976) was super-imposed as NNE–SSW to north–south trending mylonites (McCurry 1971; Ajibade *et al.* 1979). Detailed description and analysis of the structural patterns for the Borborema Province are given by Caby & Arthaud (1986) and Caby *et al.* (1995).

Metamorphism

Variations in the metamorphic imprint on the rocks of the Nigeria–Borborema province are observed in the mineral assemblages associated with penetrative fabrics in the older rock units and, to a lesser extent, in the granitoids, and reflect the heterogeneity of the metamorphism. The relationship between phases of deformation and prograde metamorphism shows that the Pan-African/Brasiliano deformation took place under medium to high amphibolite-facies conditions. In general, there is a contrast between greenschist to almandine–amphibolite facies in the metasediments and upper amphibolite to granulite facies in the gneisses.

Greenschist facies is indicated in the metasediments by the presence of chlorite, while biotite, garnet, plagioclase ($\pm$staurolite) define the almandine–amphibolite facies. Muscovite after pre-existing chlorite is common in the phyllites. In the metasediments of northwestern Nigeria, McCurry (1976) identified two periods of syntectonic progressive metamorphism, separated and followed by periods of static metamorphism. In southwestern Nigeria on the other hand, Rahaman (1976) recognized three metamorphic episodes, on both macroscopic and microscopic scales. While there seems to be agreement in the progressive nature of the metamorphism by these two authors, Annor *et al.* (1996) and Annor (1998) have recorded retrograde metamorphism in the Egbe–Isanlu and the Okene–Igarra schist belts.

The Archaean migmatite-gneiss complex, on the other hand, displays higher metamorphic grade with mineral associations including sillimanite and kyanite (McCurry 1976). Most assemblages reflect staurolite–almandine sub-facies conditions of the amphibolite facies (Rahaman 1976). Rahaman *et al.* (1991) suggested that Pan-African magmatism was the main heat source for the metamorphism, and that it took place in the interval between 630 and 600 Ma, whereas deformation was diachronous from west to east in tandem with the prograde metamorphic gradient (Rahaman & Ocan 1978), until granulite-facies conditions were locally attained in the Ikare area (Rahaman & Ocan 1988). The major thrusts recognized in the Nigerian schist belts (Rahaman 1976; Odeyemi 1988; Odeyemi & Rahaman 1992; Ajibade *et al.* 1979; Annor & Freeth 1985; Caby 1989; Annor *et al.* 1996; Annor 1998) must have continued at lower crustal levels, merging with each other in the layered granulitic lower crust that may underlie most of these areas (Caby & Boesse 2001). Identical relationships have been established in Ceará, NE Brazil, by Pessoa & Archanjo (1984) and Caby & Arthaud (1986).

Mineralization

Pre-drift reconstruction of the structural patterns and other geological features of the Nigeria–Borborema province also shows overwhelming correspondence in the control of mineralization by deformation processes during the Pan-African/Brasiliano orogenies (Torquato & Cordani 1981). Among these are: (i) well-defined pegmatitic provinces with Sn, Nb, W, Au mineralization and gemstones; (ii) Fe–Mn mineralization of the schist belts; (iii) the late Gondwana fragmentation with associated marginal basins of high potential for mineral fuels such as petroleum, coal, bituminous schist and uranium (Beurlen & Cassedanne 1981) as well as Pb, Zn and evaporates in inter-continental basins.

While the Nigerian schist belts can be regarded as a metallogenetic province (Woakes *et al.* 1987) on the basis of general association of particular minerals, its assignment to the Pan-African is fraught with many ambiguities due to the polycyclic nature of the basement. In particular, the relationship of the Archaean migmatite-gneiss complex with the banded iron formation and mineralization in Pan-African quartz veins and pegmatites make the proposition of Pan-African metallogeny tenuous, especially in the light of compelling structural (Fig. 3) and isotopic data for the Okene–Igarra schist belt indicating that ore deposits may be inherited from earlier metallogenic processes (Annor 1995, 1998).

Pan-African redistribution and concentration of minerals can be discussed in broad terms and in relation with rock associations and structural controls. For example, the two regional NNE–SSW wrench faults (Anka–Yauri–Iseyin and Kalangai–Zungeru–Ifewara, Figs 4 & 5) have long been recognized as possible Pan-African crustal sutures (Wright 1976; McCurry & Wright 1977; Ajibade & Wright 1988), and as loci of economic mineralization. Several geological and mineral exploration programmes have been carried out in well-defined schist belts (Maru, Anka, Yauri, Igbetti–Shaki, Malumfashi, Birnin Gwari, Minna–Izom, Egbe–Isanlu–Kabba, Ijero, and Ilesha, e.g., Garba 2000; fig. 5a). Many of these areas host gold, talc, anthophyllitic asbestos, Sn–Nb–Ta and Fe–Mn deposits. Iron-ore deposits in the Okene–Kabba and Muro are the most prominent of the several deposits and prospects of Palaeoproterozoic age that bear the imprint of Pan-African structural styles. To what extent the *c.* 600 Ma events have concentrated or dispersed earlier mineralization is unknown; suffice it to say that the main-phase granitoids in the Nigeria–Borborema province are themselves markedly poor in mineralization. The late- to post-orogenic granitoids such as the pegmatites, quartz veins, microgranites and the basic and ultrabasic intrusive rocks deserve further studies, especially the latter as possible sources of sulphides, chrome, nickel (magmatic) ores and kimberlite. Added to marbles, dolomites and graphitic schists in gneisses in Jakura, Ubo, Osara, Burum, Muro, Igbetti and several other localities, metasediments and metavolcanic rocks of the Nigerian schist belts hold promise not only for iron ore but also the much-needed refractory, fluxes and foundry materials needed for iron and steel industries (Dada 1988).

Conclusions

Positive initial εNd values combined with U–Pb zircon crystallisation ages for Archaean orthogneisses suggest juvenile crustal addition during the Archaean and at the Archaean–Proterozoic boundary (Dada & Rahaman 1995; Dada 1998). U–Pb zircons from orthogneisses in northern Nigeria do not show the imprint of the Eburnian orogeny, but exhibit very strong Pan-African influence, with precise definition of lower or upper intercepts around 600 Ma. Nevertheless, the work of Annor (1995) in the SW and the recent single zircon ages of Ekwueme & Kröner (2006) in south-eastern Nigeria are in agreement with well recognized Palaeoproterozoic ages comparable to those that occur throughout the West African and Sao Francisco cratons and in the Borborema Province.

Rocks in SW Nigeria rocks in SW Nigeria have Nd model ages of 2.56–2.51 Ga (Dada & Rahaman 1995) and negative εNd values at 2.1 Ga, suggesting that some Archaean crustal components were incorporated into the original Palaeoproterozoic granitoid magmas. These results are in good agreement with available evidence in the basement of the Hoggar and the Nigerian–Borborema regions (Caby 1987, 1989; Caby & Arthaud 1986), both of which have Archaean enclaves (Macambira 1992; Dantas *et al.* 2004) in essentially reworked Proterozoic terrains. Rocks from the Nigerian–Borborema mobile belt for which Eburnian/Transamazonian ages (2.1–2.0 Ga) have been reported, but without the Nd and Sr isotopic characteristics of juvenile additions found on the cratons (e.g., Abouchami *et al.* 1990; Boher *et al.* 1991), may similarly be interpreted as derived from Neoproterozoic magmas with significant Archaean crustal component. Alternatively, they could represent post-Archaean/Early Proterozoic crust-stabilization processes, much like the anorogenic 1.9–1.8 Ga magmatism within the NW African shield. In most cases, there are no unequivocal Palaeoproterozoic structural fabrics, as these would have been largely obliterated during the Pan African/Brasiliano tectono-thermal events.

Many workers have long recognized the Nigeria–Borborema province as an assemblage of contrasted metasedimentary and crystalline terranes representing a continuation of the geology of the Hoggar to the north. The widespread U–Pb Pan-African/Brasiliano ages of the granitoids, along with the pervasive deformation and metamorphism, emphasize its early recognition as an orogenic belt (e.g., as Pan-African by Kennedy 1964). Abundant lithostructural, trace element and isotopic evidence for Proterozoic rocks in an intracratonic setting (Caby & Arthaud 1986; Caby 1989; Macambira 1992; Dada *et al.* 1995; Caby & Boesse 2001) strongly support a model with significant involvement of Archaean components in the formation of Pan-African/Brasiliano granitoids, in contrast to the largely juvenile Birrimian (2.2–2.0 Ga) rocks on the cratons. Therefore, post-Archaean ages obtained on rocks in the Nigeria–Borborema province cannot be interpreted as representing purely juvenile additions, particularly when such rocks give Nd crustal residence ages which do not agree with U–Pb zircon ages or well established structural evidence.

The author wishes to thank two other reviewers, Maarten de Wit and Bob Pankhurst for their constructive contributions to the manuscript.

References

ABOUCHAMI, W., BOHER, M., MICHARD, A. & ALBAREDE, F. 1990. A major 2.1 Ga old event of mafic magmatism in West Africa: an early stage of crustal addition. *Journal of Geophysical Research*, **95**, 17605–17629

AJIBADE, A. C. & WRIGHT, J. B. 1988. Structural relationships in the schist belts of Northwestern Nigeria. *In*: OLUYIDE, P. O. *ET AL.* (eds) *Precambrian Geology of Nigeria*. Geological Survey of Nigeria Publication, Kaduna, 103–109.

AJIBADE, A. C., FITCHES, W. R. & WRIGHT, J. B. 1979. The Zungeru mylonites, Nigeria: recognition of a major tectonic unit. *Revue de Geologie et de Geographie et Physique*, **21**, 359–363.

AJIBADE, A. C., WOAKES, M. & RAHAMAN, M. A. 1987. Proterozoic crustal development in the Pan African regime of Nigeria. *In*: KRÖNER, A. (ed.) *Proterozoic crustal evolution*. American Geophysical Union, Geodynamic Series, **17**, 259–271.

ALMEIDA, F. F. M. 1968. Precambrian geology of northeastern Brazil and western Africa and the theory of continental drift. *In*: *UNESCO, Proceedings of the Symposium on the granites of West Africa, Natural Resources Research*, **4**, 151–162.

ANNOR, A. E. 1995. U–Pb zircon age for Kabba-Okene granodiorite gneiss: implication for Nigeria's basement chronology. *African Geoscience Reviews*, **2**, 101–105.

ANNOR, A. E. 1998. Structural and chronological relationship between the low grade Igarra schist and its adjoining Okene migmatite-gneiss terrain in the Precambrian exposure of southwestern Nigeria. *Journal of Mining & Geology*, **34**, 187–196.

ANNOR, A. E. & FREETH, S. J. 1985. Thermo-tectonic evolution of the Basement Complex around Okene,

Nigeria, with special reference to deformation mechanism. *Precambrian Research*, **28**, 73–77.

ANNOR, E. A., OLOBANIYI, S. B. & MUCKE, A. 1996. A note on the geology of the Isanlu area, in the Egbe-Isanlu schist belt, S.W. Nigeria. *Journal of Mining Geology*, **32**, 47–52.

ARTHAUD, M. A., CABY, R., FUCK, R. A., DANTAS, E. L. & PARENTE, C. V. 2008. Geology of the northern Borborema Province, NE Brazil and its correlation with Nigeria, NW Africa. *In*: PANKHURST, R. J., TROUW, R. A. J., BRITO NEVES, B. B. & DE WIT, M. J. (eds) *West Gondwana: Pre-Cenozoic Correlations Across the South Atlantic Region*. Geological Society, London, Special Publications, **294**, 49–67.

BAFOR, B. E. 1988. Some geochemical considerations in the evolution of the Nigerian basement in the Egbe area of southwestern Nigeria. *In*: OLUYIDE, P. O. ET AL. (eds) *Precambrian Geology of Nigeria*. Geological Survey of Nigeria Publication, Kaduna, 277–288.

BERNASCONI, A. 1987. The major Precambrian terranes of eastern South America: A Study of their regional and chronological evolution. *Precambrian Research*, **37**, 107–174.

BEURLEN, H. & CASSEDANNE, J. P. 1981. The Brazilian Mineral Resources. *Earth-Science Reviews*, **17**, 177–206.

BLACK, R., CABY, R., MOUSSINE-POUCHKINE, A., BAYER, R., BERTRAND, J. M., BOULLIER, M. M., FABRE, J. & LESQUER, A. 1979. Evidence for late Precambrian plate tectonics in Western Africa. *Nature*, **278**, 223–227.

BLACK, R., LATOUCHE, L., LIEGEOIS, J. P., CABY, R. & BERTRAND, J. M. 1994. Pan-African displaced terranes in the Tuareg shield (central Sahara). *Geology*, **22**, 641–644.

BOHER, M., ABOUCHAMI, W., MICHARD, A., ALBAREDE, F. & ARNDT, N. T. 1991. Crustal growth in West Africa at 2.1 Ga. *Journal of Geophysical Research*, **97**, 345–369.

BRAUN, O. P. G. 1982. A structural synthesis of Brazil, based on the study of major lineaments derived from remote sensing imagery interpretation. *Photogrametria*, **37**, 77–108.

BRITO NEVES, B. B. 1975. *Regionalização Geotectônica do Precambriano Nordestino.*, Doctoral thesis, Instituto de Geociencias-USP, São Paulo, 198 p.

BRITO NEVES, B. B. 1982. Síntese da Geocronologia. *In*: DANTAS, J. R. A. (ed.) *Mapa Geológico do Estado da Paraíba, Texto Explicativo*. Companhia de Desenvolvimento de Recursos Minerais do Estado da Paraíba, Campina Grande-PB, 55–69.

BRITO NEVES, B. B. 1983. *O Mapa Geológico do Nordeste Oriental do Brasil, Escala 1/ 1 000 000*. Tese de Livre-Docência, Universidade de São Paulo.

BRITO NEVES, B. B., KAWASHITA, K. & PESSOA, D. A. R. 1975. A posição Estratigráfica do Complexo Caicó. *In*: *Atas do VII Simpósio de Geologia do Nordeste*. Sociedade Brasileira de Geologia-Núcleo Nordeste, 289–297.

BRITO NEVES, B. B., FUCK, R. A., CORDANI, U. G. & THOMAS, A. F. 1984. Influence of basement structures in the evolution of the major sedimentary basins of Brazil; a case of tectonic heritage. *Journal of Geodynamics*, **1**, 495–510.

BRUGUIER, O., DADA, S. & LANCELOT, J. R. 1994. Early Archaean component (>3.5 Ga) within a 3.05 Ga orthogneiss from northern Nigeria: U-Pb zircon evidence. *Earth and Planetary Science Letters*, **125**, 89–103.

BURKE, K. C. & DEWEY, J. F. 1972. Orogeny in Africa. *In*: DESSAUAGIE, T. F. J. & WHITEMAN, A. J. (eds) *African Geology*, Ibadan University Press, Ibadan, 583–608.

BURKE, K., FREETH, S. J. & GRANT, N. K. 1976. The structure and sequence of geological events in the basement complex of the Ibadan area western Nigeria. *Precambrian Research*, **3**, 537–545.

CABY, R. 1987. The pan-african belt of West Africa from the Sahara Desert to the Gulf of Benin. *In*: SCHAER, J. P. & RODGERS, J. (eds) *Anatomy of mountain ranges*. Princeton University Press, Princeton, USA, 129–170.

CABY, R. 1989. Precambrian terranes of Benin-Nigeria and Northeast Brazil and the Late Proterozoic South Atlantic fit. *In*: DALLMEYER, R. D. (ed.) *Terranes in the Circum-Atlantic Paleozoic Orogens*. Geological Society America, Special Papers, **230**, 145–158.

CABY, R. & ARTHAUD, M. 1986. Major Precambrian nappes of the Brazilian belt, Ceará, northeast Brazil. *Geology*, **14**, 571–874.

CABY, R. & BOESSE, M. 2001. Pan African nappe system in southwest Nigeria: the Ife -Ilesha schist belt. *Journal of African Earth Sciences*, **33**, 211–225.

CABY, R., BERTRAND, J. M. L. & BLACK, R. 1981. Pan African ocean closure and continental collision in the Hogger-Iforas segment, central Sahara. *In*: KRÖNER, A.(ed.) *Precambrian Plate Tectonics*. Elsevier, Amsterdam, 407–434.

CABY, R., SIAL, A., ARTHAUD, M. H. & VAUCHEZ, A. 1990. Crustal evolution and the Brasiliano orogeny in northeast Brazil. *In:* DALLMEYER, R. D. & LECORCHE, J. P. (eds) *The West African orogens and circum-Atlantic correlatives*. Springer Verlag, New York, 373–397.

CABY, R., ARTHAUD, M. H. & ARCHANJO, C. J. 1995. Lithostratigraphy and petrostructural characterization of supracrustal units in the Brasiliano belt of northeastern Brazil: geodynamic implications. *Journal of South American Earth Sciences*, **8**, 235–246.

CAEN-VACHETTE, M. & UMEJI, A. C. 1983. Whole-rock Rb-Sr dating of two monzogranites in southeast Lokoja and their implications for the Precambrian evolution in central Nigeria. *Journal of African Earth Sciences*, **7**, 121–126.

CAEN-VACHETTE, M. & EKWUEME, B. N. 1988. Rb-Sr ages of schists in the metasedimentary belts in southeast Lokoja and their implications for the Precambrian evolution of central Nigeria. *Journal of African Earth Sciences*, **7**, 127–131.

CAHEN, L., SNELLING, N. J., DELHAL, J. & VAIL, J. R. 1984. The trans-Saharan mobile belt. *In*: CAHEN, L. & SNELLING, N. J. (eds) *The Geochronology and Evolution of Africa*. Oxford Science Publication, 349–374.

COORAY, P. G. 1977. Classification of charnockitic rocks of Nigeria. *Journal of Mining & Geology*, **14**, 1–6.

DADA, O. 1988. The Schist Belt of the Nigerian Precambrian and its potentials for raw material development for Steel Plants in Nigeria. *In*: OLUYIDE, P. O. *ET AL*. (eds) *Precambrian Geology of Nigeria*. Geological Survey of Nigeria Publication, Kaduna, 211–218.

DADA, S. S. 1989. *Evolution de la croute continental au Nord Nigeria: apport de la geochimie, de la geochronologie U–Pb et des traceurs isotopiques Sr Nd et Pb*. Doctorate Thesis, Université des Sciences et Techniques du Languedoc, Montpellier, France.

DADA, S. S. 1998. Crust-forming ages and Proterozoic crustal evolution in Nigeria: a reappraisal of current interpretations. *Precambrian Research*, **87**, 65–74.

DADA, S. S. 1999*a*. Geochemistry and petrogenesis of the reworked Archaean gneiss complex of northcentral Nigeria: major and trace element studies on Kaduna amphibolites and migmatitic gneisses. *Global Journal of Pure & Applied Sciences*, **5**, 535–543.

DADA, S. S. 1999*b*. Geochemistry, U–Pb, Rb–Sr and accessory mineralogical characteristics of the Sarkin Pawa Migmatite Gneiss, North Central Nigeria. *Spectrum Journal*, **6**, 126–136.

DADA, S. S. & RAHAMAN, M. A. 1995. Archaean–Lower Proterozoic crustal evolution in Nigeria. *African Geoscience Reviews*, **2**, 219–225.

DADA, S. S. & RESPAUT, J. P. 1989. La monzonite a fayalite de Bauchi (bauchite): Nouveau emoin d'ún magmatisme syntectonique pan-africain au nord du Nigeria. *Comptes Rendus de l'Academie des Sciences de Paris*, **309**, 887–892.

DADA, S. S., LANCELOT, J. R. & BRIQUEU, L. 1989. Age and origin of the annular charnockitic complex at Toro, Northern Nigeria: U-Pb and Rb-Sr evidence. *Journal of African Earth Sciences*, **9**, 227–234.

DADA, S., BIRCK, J. L., LANCELOT, J. R. & RAHAMAN, M. A. 1993*a*. Archean migmatite-gnesis complex of North Central Nigeria: its geochemistry, petrogenesis and crustal evolution. *In*: *16th International Colloquium on African Geology, Mbabane, Swaziland, Extended Abstracts*, **1**, 97–102.

DADA, S. S., TUBOSUN, I. A., LANCELOT, J. R. & LAR, A. U. 1993*b*. Late Archaean U–Pb age for the reactivated basement of northeastern Nigeria. *Journal of African Earth Sciences*, **16**, 405–412.

DADA, S. S., BRIQUEU, L., HARMS, U., LANCELOT, J. R. & MATHEIS, G. 1995. Charnockitic and monzonitic Pan-African series from north central Nigeria: Trace-element and Nd, Sr, Pb isotope constraints on their petrogenesis. *Chemical Geology (Isotope Geoscience)*, **124**, 233–252.

DADA, S. S., BRIQUEU, L. & BIRCK, J. L. 1998. Primordial crustal growth in northern Nigeria: Preliminary Rb-Sr and Sm-Nd constraints from Kaduna migmatite-gneiss complex. *Journal of Mining & Geology*, **34**, 1–6.

DANTAS, E. L., VAN SCHMUS, W. R., HACKSPACHER, P. C., FETTER, A. H., BRITO NEVES, B. B., CORDANI, U., NUTMAN, A. P. & WILLIAMS, I. S. 2004. The 3.4–3.5 Ga Sao Jose do Campestre massif, NE Brazil: remnants of the oldest crust in South America. *Precambrian Research*, **130**, 113–137.

DICKIN, A. P., HALLIDAY, A. N. & BOWDEN, P. 1991. A Pb, Sr, Nd isotope study of the basement and Mesozoic ring complexes of the Jos Plateau, Nigeria. *Chemical Geology (Isotope Geoscience)*, **94**, 23–32.

EKWUEME, B. N. 1991. The geology of the area around Obudu Cattle Ranch, SE Nigeria. *Journal of Mining & Geology*, **27**, 129–134.

EKWUEME, B. N. & KRÖNER, A. 1992. Preliminary zircon evaporation ages from migmatitic gneisses .in Kaduna, northern Nigeria: evidence for early Archaean (Leonian) event in the Nigerian basement complex. *In*: *Nigerian Mining & Geosciences Society 29th Annual Conference Abstracts*, 61.

EKWUEME, B. N. & KRÖNER, A. 2006. Single zircon ages of migmatite gneiss and granulite in the Obudu Plateau: Timing of granulite facies metamorphism in SE Nigeria. *Journal of African Earth Sciences*, **44**, 459–469.

FERRE, E., DELERIS, J., BOUCHEZ, J. L., LAR, A.U. & PEUCAT, J.-J. 1996. The Pan-African reactivation of contrasted Eburean and Archaean provinces in Nigeria: structural and isotopic data. *Journal of the Geological Society, London*, **153**, 719–728.

FERRE, E., CABY, R., PEUCAT, J. J., CAPDEVILA, R. & MONIE, P. 1998. Pan-African, post-collisional, ferro-potassic granite and quartz-monzonite plutons of Eastern Nigeria. *Lithos*, **45**, 25–279.

FITCHES, W. R., AJIBADE, A. C., EGBUNIWE, I. G., HOLT, R. W. & WRIGHT, J. B. 1985. Late Proterozoic schist belts and plutonism in Northwest Nigeria. *Journal of the Geological Society, London*, **142**, 319–337.

GARBA, I. 2000. Gold prospect of the Nigerian Pan-African terrain of West Africa. *Journal of Mining & Geology*, **36**, 123–156.

GRANT, N. K. 1970. Geochemistry of Precambrian basement rocks from Ibadan, southwestern Nigeria. *Earth & Planetary. Science Letters*, **10**, 29–38.

GRANT, N. K. 1978. Structural distinction between a metasedimentary cover and an underlying basement in the 600 m.y. old Pan African domain of Northwestern Nigeria, West Africa. *Geological Society of America, Bulletin*, **89**, 50–58.

GUIMARAES, I. P., DA SILVA FILHO, A. F. *ET AL*. 2004. Brasiliano (Pan African) granitic magmatism in the Pajeu-Paraiba belt, Northeast Brazil: an isotopic and geochronological approach. *Precambrian Research*, **135**, 23–53.

HOLT, R. 1982. *Geotectonic evolution of the Anka belt in the Precambrian basement complex of Northwest Nigeria*. PhD Thesis, The Open University, UK.

HUBBARD, F. H. 1975. Precambrian crustal development in Western Nigeria: Indications from the Iwo region. *Geology Society of America, Bulletin*, **86**, 548–554.

HURLEY, P. 1968. The confirmation of continental drift. *Scientific American*, **218**, 52–64.

JAHN, B. M., AUREY, B., SHEN, Q. H., LIU, D. Y., ZHANG, Z. Q., DONG, Y. J., YE, X. J., ZHANG, Q. Z., CORNICHET, J. & MACE, J. 1988. Archaean crustal evolution in China: The Taishan complex and evidence for juvenile crustal addition from long-term depleted mantle. *Precambrian Research*, **38**, 381–403.

JARDIM DE SA, E. F., WACEDO, W. H., LEGRAND, J. M., MCREATH, L., GALINDO, A. & SA, J. M. 1987. Proterozoic granitoids in a polycyclic setting; the Serido region, northeast Brazil. *In*: *International Symposium on Granites and Associated Mineralization, Salvador, Brazil, January 1987, Extended Abstracts*, 103–110.

JONES, H. A. & HOCKEY, R. D. 1964. The geology of part of Southwestern Nigeria. *Geological Survey of Nigeria, Bulletin*, **31**, 1–101.

KENNEDY, W. Q. 1964. The structural differentiation of Africa in the Pan African ±500 Ma tectonic episode. *In*: *8th Annual Report of the Research Institute of African Geology, Leeds University*, UK, 128.

LESQUER, A., BELTRAO, J. F. & DE ABREAU, F. A. M. 1984. Proterozoic links between Northeastern Brazil and West Africa: A plate tectonic model based on gravity data. *Tectonophysics*, **110**, 9–10.

MACAMBIRA, M. 1992. *Chronologie U-Pb, Rb-Sr K-Ar et croissance de la croute continental dans l'Amazonie du Sud-Est: exemple de la region de Rio Maria, Province de Carajas, Bresil*. Doctorate Thesis, Université des Sciences et Techniques du Languedoc, Montpellier, France.

MATHEIS, G. & CAEN-VACHETTE, M. 1983. Rb-Sr isotopic study of rare metal-bearing and barren pegmatites in the Pan African reactivation zone of Nigeria. *Journal of African Earth Sciences*, **1**, 35–40.

MCCURRY, P. 1971. Pan-African orogeny in northern Nigeria. *Geological Society of America Bulletin*, **82**, 3251–3263.

MCCURRY, P. 1976. The geology of the Precambrian to Lower Paleozoic rocks of Northern Nigeria. *In*: KOGBE, C. A. (ed.) *Geology of Nigeria*, Elizabethan Publishing Company, Lagos, 67–99.

MCCURRY, P. & WRIGHT, J. B. 1977. Geochemistry of calc-alkaline volcanics in Northwestern Nigeria, and a possible Pan-African suture zone. *Earth and Planetary Science Letters*, **37**, 90–96.

ODEYEMI, I. B. 1988. Lithostratigraphy and structural relationships of the upper Precambrian metasediments in Igarra area, Southwestern Nigeria. *In*: OLUYIDE, P. O. ET AL. (eds) *Precambrian Geology of Nigeria*. Geological Survey of Nigeria Publication, Kaduna, 111–125.

ODEYEMI, I. B. & RAHAMAN, M. A. 1992. The petrology of a composite syenite dyke in Igarra, Southwestern Nigeria. *Journal of Mining & Geology*, **28**, 255–263.

OGEZI, O. A. E. 1988. Origin and evolution of the Basement Complex of Northwestern Nigeria in the light of new geochemical and geochronological data. *In*: OLUYIDE, P. O. ET AL. (eds) *Precambrian Geology of Nigeria*. Geological Survey of Nigeria Publication, Kaduna, 301–312.

OLAREWAJU, V. O. 1988. REE in the charnockitic and associated granitic rocks of Ado-Ekiti Akure, Southwest Nigeria. *In*: OLUYIDE, P. O. ET AL. (eds) *Precambrian Geology of Nigeria*, Geological Survey of Nigeria Publication, Kaduna, 231–239.

OVERSBY, V. M. 1975. Lead isotopic study of aplites from the Precambrian basement complex rocks near Ibadan, Southwest Nigeria. *Earth and Planetary Science Letters*, **27**, 177–180.

OYAWOYE, M. O. 1972. The basement complex of Nigeria. *In*: DESSAUVAGIE, T. F. J. & WHITEMAN, A. J. (eds) *African Geology*. Ibadan University Press, Ibadan, 66–102.

PESSOA, R. R. & ARCHANJO, C. J. 1984. Tectonica de empurroes na região de Troia-Ce: *In*: *33rd Congress of Brazilian Geology, Rio de Janeiro*, Sociedade Brasileira de Geologia, 1721–1728.

PESSOA, R. R., BRITO NEVES, B. B., KAWASHITA, K., PESSAU, D. R. & FACK, A. R. 1986. Contribuical as esuado da evolucao geocronologica do macico de Troia-Ce. *In*: *12 Simposio de Geologia do Nordeste, Joao Pessoa, Brazil*, 75–93.

PIDGEON, R. T., VAN BREEMEN, O. & OYAWOYE, M. O. 1976. Pan African and earlier events in the basement complex of Nigeria. *In*: *25th International Geological Congress, Sydney, Australia*, **3**, 667.

RAHAMAN, M. A. 1976. Review of the basement geology of southwestern Nigeria. *In*: KOGBE, C. A. (ed.) *Geology of Nigeria*. Elizabethan Publishing Co., Lagos, 41–58.

RAHAMAN, M. A. 1988. Recent advances in the study of the basement complex of Nigeria. *In*: OLUYIDE, P. O. ET AL. (eds) *Precambrian Geology of Nigeria*. Geological Survey of Nigeria Publication, Kaduna, 11–43.

RAHAMAN, M. A. & OCAN, O. 1978. On relationships in the Precambrian migmatitic gneiss of Nigeria. *Journal of Mining & Geology*, **15**, 23–32.

RAHAMAN, M. A. & OCAN, O. 1988. The nature of granulite facies metamorphism in Ikare area, southwestern Nigeria. *In*: OLUYIDE, P. O. ET AL. (eds) *Precambrian Geology of Nigeria*, Geological Survey of Nigeria Publication, Kaduna, 157–163.

RAHAMAN, M. A., VAN BREEMEN, O., BOWDEN, P. & BENNETT, J. N. 1984. Age migrations of anorogenic ring complexes in Northern Nigeria. *Journal of Geology*, **92**, 173–184.

RAHAMAN, M. A., AJAYI, T. R., OSHIN, I. O. & ASUBIOJO, F. O. I. 1988. Trace element geochemistry and geotectonic setting of Ile-Ife schist belt. *In*: OLUYIDE, P. O. ET AL. (eds) *Precambrian Geology of Nigeria*. Geological Survey of Nigeria Publication, Kaduna, 241–257.

RAHAMAN, M. A., TUBOSUN, I. A. & LANCELOT, J. R. 1991. U-Pb geochronology of potassic syenites from southwestern Nigeria and the timing of deformational events during the Pan-African orogeny. *Journal of African Earth Sciences*, **13**, 387–395.

SANTOS, E. J. & BRITO NEVES, B. B. 1984. Provincia Borborema. *In*: ALMEIDA, F. F. M. & HAUI, Y. (eds) *O Pre-Cambriano do Brazil*. Edgar Blucher, 123–186.

SANTOS, T. J. S., FETTER, A. H. & NOGUEIRA NETO, J. A. 2008. Comparisons between the northwestern Borborema Province, NE Brazil, and the southwestern Pharusian Dahomey Belt, SW Central Africa. *In*: PANKHURST, R. J., TROUW, R. A. J., BRITO NEVES, B. B. & DE WIT, M. J. (eds) *West Gondwana: Pre-Cenozoic Correlations Across the South Atlantic Region*. Geological Society, London, Special Publications, **204**, 101–119.

TAYLOR, S. R. & MCLENNAN, S. M. 1981. The composition and evolution of the continental crust: REE evidence from sedimentary rocks. *Philosophical Transactions of the Royal Society, London*, **103**, 381–399.

TORQUOTO, J. R. & CORDANI, U. G. 1981. Brazil – Africa Geological Links. *Earth-Science Reviews*, **17**, 155–176.

TUBOSUN, I. A., LANCELOT, J. R., RAHAMAN, M. A. & OCAN, O. 1984. U-Pb Panafrican ages of two charnockite-granite association from southwestern Nigerian. *Contributions to Mineralogy and Petrology*, **88**, 188–195.

TURNER, D. C. 1983. Upper Proterozoic Schist Belts in the Nigerian sector of the Pan African Province of West Africa. *Precambrian Research*, **21**, 55–79.

VAN BREEMEN, O., PIDGEON, R. T. & BOWDEN, P. 1977. Age and isotopic studies of some Pan African granites from Northcentral Nigeria. *Precambrian Research*, **4**, 307–319.

VAN SCHMUS, W. R., BRITO NEVES, B. B. *ET AL*. 2003. The Serido Group of NE Brazil, a late Neoproterozoic pre- to syn-collisional basin in West Gondwana: insights from SHRIMP U-Pb detrital zircon ages and Sm-Nd crustal residence (TDM) ages. *Precambrian Research*, **127**, 287–327.

VAN SCHMUS, W. R., TOTEU, F., DA SILVA FILHO, A. F., GUIMARAES, I. P. & OSAKO, L. S. 2008 Geological framework of the West Central and Northeast Brazil. *In*: PANKHURST, R. J., TROUW, R. A. J., BRITO NEVES, B. B. & DE WIT, M. J. (eds) *West Gondwana: Pre-Cenozoic Correlations Across the South Atlantic Region*. Geological Society, London, Special Publications, **294**, 69–99.

WOAKES, M., RAHAMAN, M. A. & AJIBADE, A. C. 1987. Some metallogenic features of the Nigerian basement. *Journal of African Earth Sciences*, **6**, 655–664.

WRIGHT, J. S. 1976. Fracture systems in Nigeria and initiation of fracture zones in the South Atlantic. *Tectonophysics*, **34**, 43–47.

São Luís Craton and Gurupi Belt (Brazil): possible links with the West African Craton and surrounding Pan-African belts

E. L. KLEIN[1,2] & C. A. V. MOURA[3]

[1]*CPRM (Companhia de Pesquisa de Recursos Minerais)/Geological Survey of Brazil, Av. Dr. Freitas, 3645, Belém-PA, CEP 66095-110, Brazil (e-mail: eklein@be.cprm.gov.br)*

[2]*Researcher at CNPq (Conselho Nacional de Desenvolvimento Científico e Tecnológico)*

[3]*Laboratório de Geologia Isotópica/Pará-Iso, Universidade Federal do Pará, Centro de Geociências, CP 1611, Belém-PA, Brazil, CEP 66075-900*

Abstract: The São Luís Craton and the Palaeoproterozoic basement rocks of the Neoproterozoic Gurupi Belt in northern Brazil are part of an orogen having an early accretionary phase at 2240–2150 Ma and a late collisional phase at 2080 ± 20 Ma. Geological, geochronological and isotopic evidence, along with palaeogeographic reconstructions, strongly suggest that these Brazilian terrains were contiguous with the West African Craton in Palaeoproterozoic times, and that this landmass apparently survived subsequent continental break-up until its incorporation in Rodinia.

The Gurupi Belt is an orogen developed in the southern margin of the West African–São Luís Craton at *c.* 750–550 Ma, after the break up of Rodinia. Factors such as present-day and possible past geographical positions, the timing of a few well-characterized events, the structural polarity and internal structure of the belt, in addition to other indirect evidence, all favour correlation between the Gurupi Belt and other Brasiliano/Pan-African belts, especially the Médio Coreaú domain of the Borborema Province and the Trans-Saharan Belt of Africa, despite the lack of proven physical links between them. These Neoproterozoic belts are part of the branched system of orogens associated with amalgamation of the Amazonian, West Africa–São Luís, São Francisco and other cratons and minor continental blocks into the West Gondwana supercontinent.

The São Luís Craton and the Neoproterozoic Gurupi Belt in northern Brazil crop out through Phanerozoic sedimentary cover in response to Cretaceous tectonic uplift and doming that preceded the rifting stage and opening of the Atlantic Ocean and subsequent erosive removal of more than 6 km of Mesozoic and Palaeozoic sediments (Rezende & Pamplona 1970). The widespread remains of the sedimentary cover and the absence of palaeomagnetic information hinder a better understanding of the relationships, if any, of these two Precambrian terrains to the surrounding Precambrian units of the present-day South American continent, such as the Amazonian and São Francisco cratons, and the Neoproterozoic Borborema and Araguaia belts (Fig. 1).

Nevertheless, some attempts have been made to correlate the Precambrian units of northern and northeastern Brazil with those of northwestern Africa in pre-Pangaea palaeogeographic reconstructions. Hurley *et al.* (1967, 1968) and Torquato & Cordani (1981) presented Rb–Sr and K–Ar geochronological data and, following the pre-drift continental fit for the Atlantic Ocean (Bullard *et al.* 1965), observed geochronological similarities between Brazil and Africa in that litho-structural units having a Neoproterozoic imprint surrounded units having a Palaeoproterozoic signature. Lesquer *et al.* (1984) used regional scale structural and geophysical information to discuss the correlation.

These studies were important in establishing the broad limits between major geotectonic units, such as cratons, mobile belts and sedimentary basins. However, the internal geological framework of each of these units remained uncertain. Despite unavoidable and continuous debate, knowledge of the geology and tectonic evolution of the northwestern African terrains has experienced significant advances in the last 15 years (Abouchami *et al.* 1990; Feybesse & Milési 1994; Trompette 1997; Egal *et al.* 2002; Caby 2003, and many others). In Brazil, only recently the geological evolution of the São Luís Craton and Gurupi Belt became better understood, on the basis of regional mapping programmes (Pastana 1995; Costa 2000), more robust geochronological information and reinterpretation of tectonic settings (Klein & Moura 2001, 2003; Moura *et al.* 2003; Klein *et al.* 2005*a*, *b*).

Taking into account these new advances and problems, this paper intends to reassess possible

From: PANKHURST, R. J., TROUW, R. A. J., BRITO NEVES, B. B. & DE WIT, M. J. (eds) *West Gondwana: Pre-Cenozoic Correlations Across the South Atlantic Region.* Geological Society, London, Special Publications, **294**, 137–151. DOI: 10.1144/SP294.8 0305-8719/08/$15.00

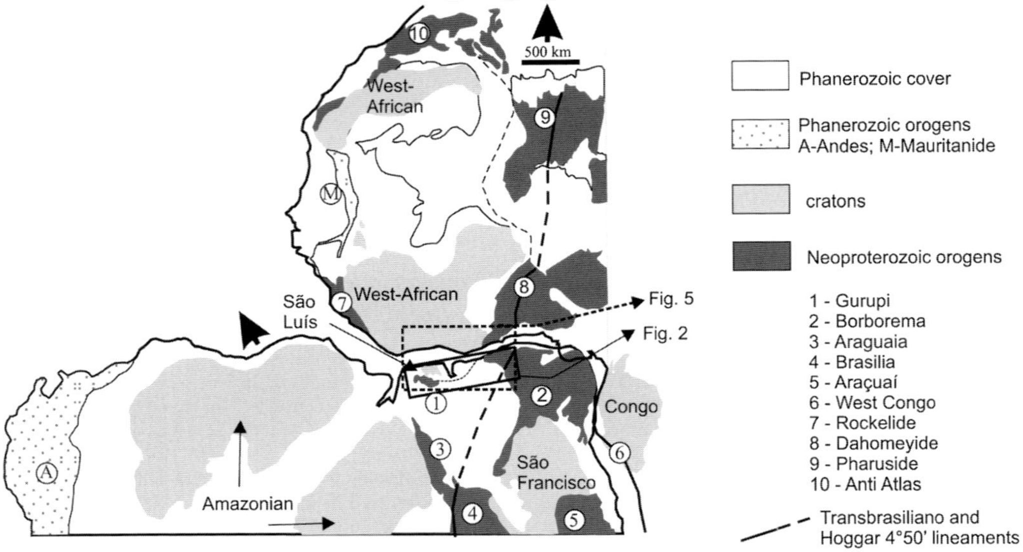

Fig. 1. Location of the São Luís craton and Gurupi Belt in relation to the main tectonic units of northern South America and northwestern Africa.

correlations between the São Luís Craton and the West African Craton (Fig. 1), and between the Gurupi Belt and the Brasiliano/Pan-African belts of northern Brazil and northwestern Africa (Figs 1 and 2). This will be done by comparing similarities (and differences) between rock associations and the tectonic settings in which these associations formed, and by the investigation of the timing of relatively well characterised geological events, such as magmatism, metamorphism, sedimentation and tectonism. Correlation problems will also be discussed, as well as their implications for the assembly and dispersal of supercontinents.

Geological overview

The São Luís Craton

The São Luís Craton is composed of a metavolcano-sedimentary sequence and a few generations of granitoids (Fig. 3). The former consists of schists of variable composition, metavolcanic and metapyroclastic rocks, quartzite, metachert, and meta-mafic–ultramafic rocks. The metamorphic conditions are predominantly greenschist facies, but locally reached lower amphibolite facies. This supracrustal sequence is considered to have formed in island-are setting (Klein *et al.* 2005*a*). Zircon geochronology of this sequence is still limited. A single sample of metapyroclastic rock yielded an age of 2240 ± 5 Ma (single zircon Pb evaporation), and Sm–Nd T_{DM} model ages vary from 2.21 to 2.48 Ga, with εNd(t) values of +0.8 to +3.5 (Klein & Moura 2001; Klein *et al.* 2005*a*). A similar metavolcano-sedimentary sequence that occurs in the basement sequence of the Gurupi Belt (Fig. 3) has zircon Pb–Pb ages of 2148–2160 Ma (Klein & Moura 2001) and it is possible that the supracrustal sequence of the São Luís Craton also continued to develop until this time.

Granitoids make up the major part of the cratonic area, forming batholiths and stocks. The Tromaí Suite is composed of equigranular and massive to weakly foliated quartz-diorite, tonalite, diorite, granodiorite and minor trondhjemite. These rocks are calc-alkaline, metaluminous and sodic, with low to moderate K_2O contents. Zircon crystallisation ages vary between 2168 Ma and 2147 Ma, and Sm–Nd T_{DM} model ages are 2.22 to 2.26 Ga, with εNd(t) values of +1.9 to +2.6 (Klein & Moura 2001; Klein *et al.* 2005*a*). The Areal granite (2149 ± 4 Ma) is weakly peraluminous, K_2O-enriched and has similar Sm–Nd patterns to the Tromaí Suite (Klein & Moura 2003; Klein *et al.* 2005*a*); it is interpreted as having formed from parental magmas similar to those that produced the Tromaí calc-alkaline granitoids, along with reworked products of the island arcs in which they formed (Klein *et al.* 2005*a*).

A third suite of granitoids (Tracuateua, Fig. 3) consists of strongly peraluminous, S-type two-mica granites, derived from the partial melting of crustal rocks (Lowell & Villas 1983). These granites have zircon Pb–Pb crystallization ages of 2086–2091 Ma and Sm–Nd model ages

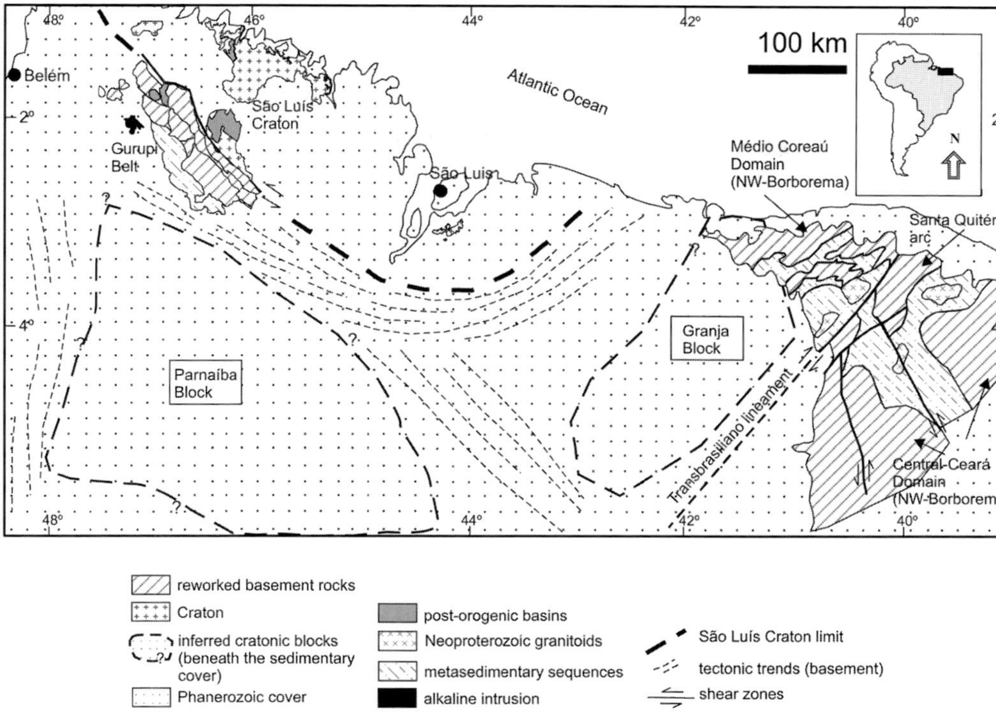

Fig. 2. Simplified map showing the location of the São Luís craton, Gurupi Belt and Médio Coreaú and Central-Ceará domains (NW Borborema Province) in relation to the interpreted tectonic trends of the basement rocks of the Phanerozoic sedimentary cover. These trends indicate the possible continuation, beneath the sedimentary cover, of the São Luís craton, Gurupi and Araguaia belts, and of the Transbrasiliano lineament. Also shown are the Parnaíba and Granja blocks inferred to underlie the Phanerozoic basins (adapted from Brito Neves *et al.* 1984; Nunes 1993; Fetter *et al.* 2003).

(T_{DM}) varying between 2.31 and 2.50 Ga, with εNd(t) values ranging from −1.3 to +1.1 (Moura *et al.* 2003).

The geodynamic evolution of the São Luís Craton has been discussed on the basis of rock assemblages and affinities, structural features, and limited geochemical information, combined with zircon geochronology and Nd isotope data (Klein *et al.* 2005*a*). At least three periods of rock generation have been recognized, occurring at about 2240 ± 5 Ma (supracrustal rocks), 2168–2147 Ma (calc-alkaline granitoids + supracrustal rocks), and 2086–2090 Ma (S-type granites), with almost all sequences showing positive εNd(t) values and therefore being derived from juvenile protoliths. The large association of juvenile calc-alkaline granitoids and volcano-sedimentary rocks and the lack of voluminous mafic rocks have been interpreted by Klein *et al.* (2005*a*) as indicating an intra-oceanic, are-related subduction setting for these sequences.

Only scarce relicts of a reworked Archaean crust have been indicated by slightly negative εNd(t) values and Sm–Nd model ages of the younger S-type granites (Moura *et al.* 2003). However, Sm–Nd T_{DM} model ages of 2.48 Ga and 2.42 Ga found in meta-dacite and dacite suggest that protoliths older than 2.24 Ga might have been involved at least in part of the cratonic evolution. As such, the model ages of about 2.4 Ga may record the age of the mafic protoliths (ocean crust?) that could have formed at that time and that were subsequently melted.

The time interval of 2240–2150 Ma records an accretionary phase of the Palaeoproterozoic orogen, whereas the collisional phase is represented by the S-type granitoids of 2086–2090 Ma, produced by melting of pre-existing crustal material. This phase is better represented in the basement of the Gurupi Belt (see next section).

The Gurupi Belt

The Gurupi Belt is a Neoproterozoic mobile belt located along the southern margin of the São Luís Craton (Fig. 3); it shows metamorphosed

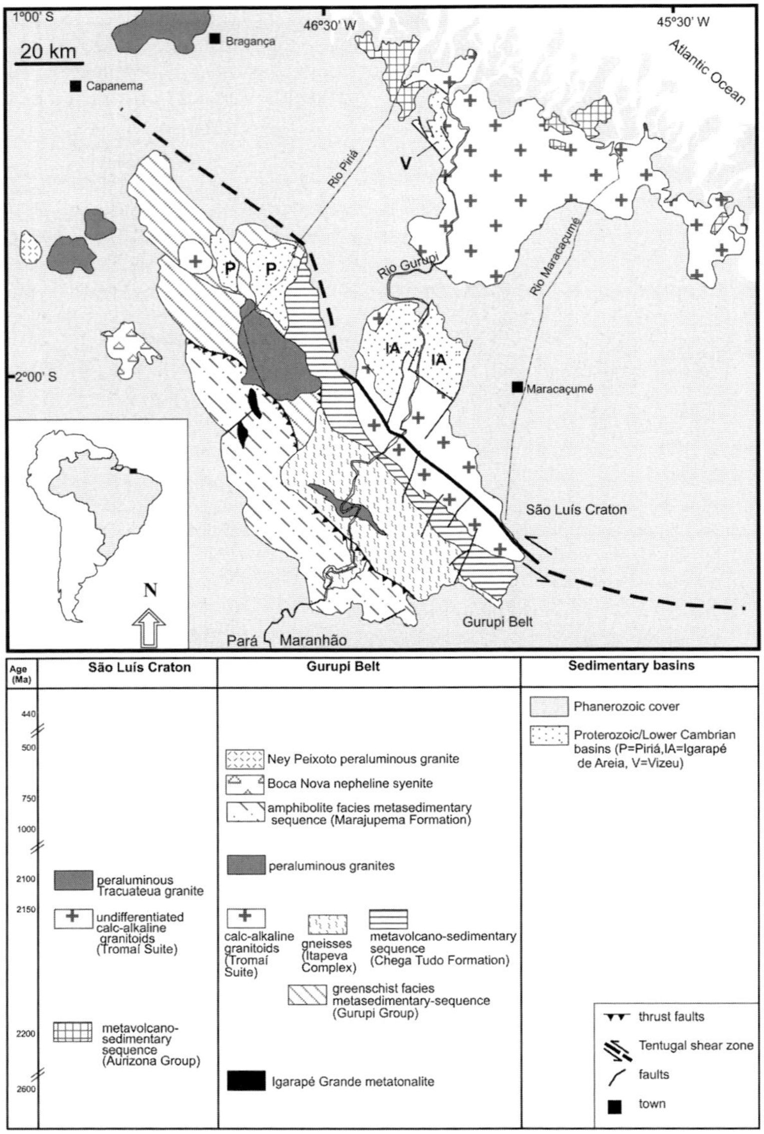

Fig. 3. Geological map of the São Luís craton and Gurupi Belt (adapted from Pastana 1995 and Klein *et al.* 2005*b*).

supracrustal and plutonic units originally formed in Archaean, Palaeoproterozoic and Neoproterozoic times (Klein & Moura 2001; Moura *et al.* 2003; Klein *et al.* 2005*b*). With the exception of some rounded granite stocks, most of the sequences of the Gurupi Belt form elongated bodies parallel to NW–SE cross-cutting structures. Structures dipping at low to moderate angles to SSW, with down-dip or oblique lineations, are mostly confined to the northwestern portion of the belt. These structures record tectonic transport from SW to NE, toward the São Luís Craton, and it is possible that they resulted from the convergence between this craton and an inferred landmass existing to the south. This block may correspond to the concealed Parnaíba block (Fig. 2), a cratonic nucleus having distinct age and structural trends in relation to the surrounding terrains, which has been proposed on the basis of geophysical evidence in addition to petrography and Rb–Sr and K–Ar geochronology of the basement rocks of the Phanerozoic basins (Brito Neves *et al.* 1984; Nunes 1993).

Steeply dipping strike-slip shear zones are concentrated in the central and southeastern portions of the belt, being associated with the Tentugal Shear Zone, which represents the litho-structural and geochronological (Rb–Sr, K–Ar) boundary between the São Luís Craton and the Gurupi Belt (Klein *et al.* 2005*b*). The Tentugal Shear Zone was active during the Neoproterozoic orogeny, but probably resulted from the reactivation of older structures related to the Palaeoproterozoic evolution of the São Luís Craton.

Most of the exposed rocks of the Gurupi Belt are Palaeoproterozoic units that show physical continuity and broadly exhibit the same age and Nd isotope patterns displayed by the Palaeoproterozoic rocks of the São Luís Craton. For instance, bodies correlated with the cratonic calc-alkaline granitoids (2168–2147 Ma) occur within the belt, where they show variable effects of deformation. In addition, a tonalite gneiss of 2167 $\pm$ 2.5 Ma (protolith crystallization age), has T_{DM} model ages between 2.22 and 2.31 Ga, and εNd(t) values ranging from +1.4 to +1.6 (Klein *et al.* 2005*b*). These gneisses could represent the same calc-alkaline granitoids as those of the São Luís Craton that underwent more severe metamorphic and deformational conditions, but additional studies are needed to solve this issue.

A metavolcano-sedimentary succession (Chega Tudo Formation) contains felsic volcanic rocks of 2148–2160 Ma with a juvenile Nd isotope signature, and was probably formed within are systems (Klein & Moura 2001; Klein *et al.* 2005*b*). Several plutons of peraluminous, biotite- and muscovite-bearing granite intruded the Palaeoproterozoic supracrustal and gneissic units between 2100 and 2060 Ma, and at least one of these plutons is clearly syntectonic. Inherited zircon crystals and Nd isotopes indicate that the peraluminous rocks were formed by variable degrees of reworking of Palaeoproterozoic and minor Archaean crust (Moura *et al.* 2003; Klein *et al.* 2005*b*). A sub-greenschist to greenschist facies metasedimentary sequence of unknown age (the Gurupi Group) is tentatively considered to be older than 2159 $\pm$ 13 Ma, based on supposed intrusion relationships (Costa 2000).

All these Palaeoproterozoic units appear to have been originally related to the evolution of the present-day São Luís Craton sequences, broadly representing an accretionary (São Luís) and a collisional phase (Gurupi) of a Palaeoproterozoic orogen. However, the units located in the Gurupi Belt show variable but widespread evidence of resetting of Rb–Sr and K–Ar isotopic systems by Neoproterozoic events (Klein *et al.* 2005*b* and references therein). Furthermore, they show a distinct structural pattern in relation to the present-day São Luís Craton. These Palaeoproterozoic rocks, along with subordinate lenses of an Archaean metatonalite (Igarapé Grande, Fig. 3) represent the continental basement on which the Neoproterozoic Gurupi orogen developed, and now survive as the external portion of the orogen.

Apart from the generalized resetting of isotopic systems, only two magmatic events after the Palaeoproterozoic orogeny have so far been characterized in the Gurupi Belt, both occurring in Neoproterozoic times. The first event was the intrusion of the Boca Nova nepheline syenite pluton 732 $\pm$ 7 Ma ago (Klein *et al.* 2005*b*), probably recording the rifting of the pre-existing crust amalgamated in the Palaeoproterozoic era. This intrusion (Figs 2 and 3) was subsequently deformed and metamorphosed under amphibolite-facies conditions (Lowell & Villas 1983); gneissic banding strikes NW–SE and dips at low angles to the WSW. The second event was the intrusion of a peraluminous, muscovite-bearing granite at 549 $\pm$ 4 Ma (Ney Peixoto Granite, Moura *et al.* 2003). This granite was only moderately affected by the widespread NW–SE strike-slip shearing, and is interpreted as a late- to post-tectonic intrusion.

Further evidence of younger (Neoproterozoic?) activity is the presence of detrital zircon crystals as young as 1100 Ma in the amphibolite-facies metasedimentary Marajupema Formation (Fig. 3), which has Sm–Nd T_{DM} model age of 1.41 Ga (Klein *et al.* 2005*b*). These authors suggested that the sedimentation of detritus from Archaean, Palaeoproterozoic, and Mesoproterozoic/Neoproterozoic sources could have occurred in the rift in which the nepheline syenite intruded, or on a continental margin. A possibility (not unique) is that this basin was subsequently closed at the end of the Ediacaran period (580–550 Ma), after a period of inferred subduction, arc construction, and collision.

Small sedimentary basins formed over sequences of the São Luís Craton (Vizeu and Igarapé de Areia basins) and Gurupi Belt (Piriá Basin) and show similar lithological, metamorphic and structural aspects (Fig. 3). The basins comprise variable proportions of arkose, sandstone, pelite and conglomerate that record continental semi-arid conditions to shallow lake or marine waters (Pastana 1995). A large proportion of detrital zircons (>80%) found in an arkose from one of these basins shows Pb–Pb ages between 700 and 500 Ma (Pinheiro *et al.* 2003), indicating that this basin at least formed late in the Neoproterozoic era or even in the Early Cambrian epoch and leading most authors to relate the basins to the post-orogenic development of the Gurupi Belt (Pinheiro *et al.* 2003; Klein *et al.* 2005*b*).

The Médio Coreaú Domain of the Borborema Province

The Borborema Province in northeastern Brazil (location in Fig. 1) is a large branching system of orogenic belts that shows a long-lived and polycyclic geological history. Its basement rocks record orogenic and anorogenic processes from 3500 Ma to 940 Ma. The Neoproterozoic (Brasiliano) orogeny took place diachronously throughout the province, but involved basically the same sequence of events, which can be summarized as follows (Brito Neves *et al.* 2000; Fetter *et al.* 2000, 2003): (1) 850–700 Ma, continental rift phase, with small volcano-sedimentary and plutonic complexes; (2) 650–620 Ma, extensive subduction-related calc-alkaline magmatism, and pre- to syn-collisional metavolcano-sedimentary sequences; (3) 620–570 Ma, collisional plutonism; (4) 580–510 Ma, post-tectonic to anorogenic plutonism, uplift and extrusion tectonics.

The northwestern portion of the Borborema Province is represented by the Médio Coreaú domain (Figs 1 and 2). This domain, which occurs to the west of the Transbrasiliano lineament, comprises sedimentary and volcano-sedimentary sequences deformed and metamorphosed in the Neoproterozoic era, during the Brasiliano cycle of orogenies. These sequences were deposited over a middle- to high-grade basement composed of gneiss and migmatite, along with subordinate enderbite, charnockite, and kinzigite, with magmatic ages varying from 2356 Ma to 2176 Ma and Sm–Nd model ages clustering between 2.42 and 2.48 Ga, indicating a juvenile character for most of these rocks (Brito Neves *et al.* 2000; Fetter *et al.* 2000). Neoproterozoic orogenic magmatism (777 ± 11 to 591 ± 8 Ma) falls in approximately the age intervals reported in the previous paragraph for the Borborema province as a whole and the tectonic setting of the orogenic granites is interpreted as related to back-arc and fore-arc basins (Fetter *et al.* 2003).

The West African Craton

The West African Craton is represented by two Precambrian shields, Reguibat to the north and Man to the south, separated by Neoproterozoic to Palaeozoic cover (Fig. 4a). The Man shield, which is of more interest for the present work, is subdivided into a western domain, predominantly of Archaean age (Kénema–Man domain), and a central–eastern domain, composed of Palaeoproterozoic rocks (Baoulé–Mossi domain). These domains are separated by the strike-slip Sassandra Shear Zone and by an Archaean-Palaeoproterozoic transitional domain (Feybesse & Milési 1994; Caby *et al.* 2000; Egal *et al.* 2002).

The Kénema–Man domain is composed of granulite gneiss, migmatite and charnockite of Archaean age, along with subordinate Palaeoproterozoic granitoid, volcanic and sedimentary rocks. Crust formation events, at least in part juvenile, have been identified at 3542 ± 13 Ma, 3300–3200 Ma (pre-Leonian and Leonian orogeny), whereas an extensive period of granite magmatism and granulite-facies metamorphism, reworking of older crust and apparent absence of deposition of supracrustal rocks occurred between 2910 Ma and 2800 Ma (Liberian orogeny). Perturbation of isotopic systems occurred between 2250 Ma and 2020 Ma (Thiéblemont *et al.* 2004 and references therein), as result of the Palaeoproterozoic Eburnian orogeny that is widespread in the Baoulé–Mossi domain.

The Baoulé–Mossi domain (Fig. 4a) consists of several NNE–SSW-orientated belts of metavolcanic, metasedimentary and metavolcano-sedimentary rocks, with voluminous batholiths of granitoid rocks having variable ages and sources, as well as variable petrographic, geochemical and structural characteristics, and subordinate mafic–ultramafic rocks. These rocks formed mostly during, and were affected by, the widespread Palaeoproterozoic Eburnian orogeny. The metavolcanic rocks comprise tholeiitic to subordinate komatiitic basalts, along with calc-alkaline rhyolite and rhyodacite, whereas the sedimentary and metavolcano-sedimentary belts are composed of clastic sedimentary rocks intercalated with metavolcaniclastic and felsic to intermediate metavolcanic and pyroclastic rocks (Sylvester & Attoh 1992; Feybesse & Milési 1994; Caby *et al.* 2000; Hein *et al.* 2004). Greenschist-facies metamorphism is largely predominant, but lower amphibolite conditions occur locally (Vidal *et al.* 1996). Gabbro–diorite–pyroxenite bodies intrude the supracrustal sequences and appear to predate the granitic magmatism (Hein *et al.* 2004).

Two main phases of granitic magmatism have been characterised in the Baoulé–Mossi domain (Doumbia *et al.* 1998). The dominant type of granitoids comprises sodic–calcic and metaluminous calc-alkaline tonalites and granodiorites. These rocks derived from juvenile sources and intruded supracrustal rocks chiefly at about 2155 ± 15 Ma. The other type is composed of more potassic and peraluminous granitoids, including crust-derived two-mica granites that intruded the supracrustal sequences at 2100 ± 10 Ma.

There is a general consensus about the above description regarding the Baoulé–Mossi domain. However, at least three main subjects of debate still remain. Firstly, the stratigraphic position of

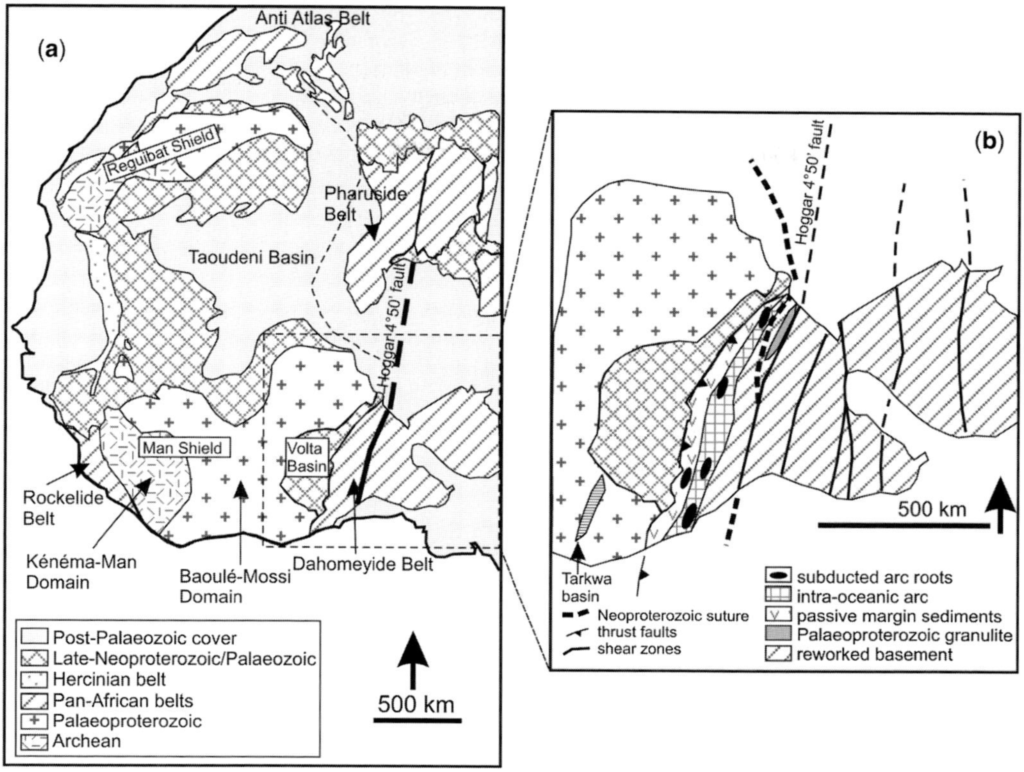

Fig. 4. (**a**) Simplified tectonic map of northwestern Africa (adapted from Feybesse & Milési 1994); (**b**) Detail of the Dahomeyide Belt in the Ghana–Nigeria region (adapted from Caby 1998), showing the tectonic elements of the Neoproterozoic orogen. The location of the Tarkwa basin is also shown.

the metasedimentary units, which are placed either below (Feybesse & Milési 1994) or above (Hirdes *et al.* 1996; Pouclet *et al.* 1996) the metavolcanic sequences.

Secondly, the tectonic setting in which the granitoid and (meta-)volcanic rocks formed: continental rift in a pre-Birimian basement older than 2200 Ma that would have evolved to an oceanic basin (Vidal & Alric 1994); oceanic within-plate setting (Abouchami *et al.* 1990); immature island arcs constructed over an oceanic crust (Sylvester & Attoh 1992); marginal marine setting adjacent to a volcanic centre or island arc (Hein *et al.* 2004). Because of the large extension of the Palaeoproterozoic domain, it is reasonable to expect some variability in terms of tectonic settings. Thus, since different authors have often worked in distinct portions of the domain, it is likely that most of the interpretations are to some extent feasible.

The third subject of discussion concerns the geotectonic evolution of the Palaeoproterozoic domain, with one group of researchers accepting modern plate tectonics (e.g., Abouchami *et al.* 1990; Ledru *et al.* 1994; Egal *et al.* 2002), and other researchers invoking vertical, plume-related, 'Archaean type' dynamics associated with strike-slip tectonics (e.g., Vidal *et al.* 1996; Caby *et al.* 2000). Irrespective of the models, the two schools describe basically the same sequence of events that can be grouped as follows, at least for the post-2200 Ma period: (1) early oceanic stage producing tholeiitic basic rocks; (2) juvenile, metaluminous, calc-alkaline, sodic plutonism and volcanism; (3) clastic and volcaniclastic sedimentation derived from the precedent stages; (4) another generation of calc-alkaline magmatism and extensive production of crust derived leucogranites; (5) metamorphism and deformation.

Compilation of more than one hundred zircon U–Pb and Pb–Pb age determinations and Nd information shows that this evolution is nearly continuous between at least 2200 Ma and 2060 Ma, with peaks of activity around 2155 Ma and 2095 Ma, and that the Palaeoproterozoic Baoulé-Mossi domain is essentially juvenile. Notwithstanding, the existence of a not yet understood pre-2.2 Ga (early-Birimian?) crustal growth episode is becoming evident, as suggested by geochronological (Lahondère *et al.* 2002; Gasquet *et al.* 2003) and stratigraphic–structural evidence (Hein *et al.* 2004). As such, recycling of pre-2.2 Ga crust might also have occurred in the genesis of the Birimian rocks of the Baoulé–Mossi domain.

The Tarkwa sedimentary sequence

The Tarkwa sedimentary mega-sequence shows very low metamorphism and well-preserved sedimentary structures. From the bottom to the top it consists of: (1) polymict conglomerate with pebbles of phyllite, granite, felsic and mafic lavas, pyroclastic rocks, quartz, and chert; (2) sandstone and auriferous monomict conglomerate with quartz pebbles; (3) phyllite and sandstone. These rocks were deposited in elongated graben formed over the Birimian belts and record continental to mature fluviatile and deltaic environmental conditions, and they are locally associated with pyroclastic flows and rhyolites, and with dolerite and gabbro sills (Davis *et al.* 1994; Feybesse & Milési 1994; Ledru *et al.* 1994). Most of the detrital zircons of this sequence have ages in the range of 2185–2155 Ma, with subordinate ages of 2245 Ma, 2132 Ma and 2124 Ma (Davis *et al.* 1994; Bossière *et al.* 1996), and the maximum age of sedimentation would be bracketed by intrusion relationships as between 2080 and 1960 Ma (Ledru *et al.* 1994; Bossière *et al.* 1996). There is a certain agreement in that both the Birimian and Tarkwa successions have been deformed together during the final stages of the Palaeoproterozoic Eburnian orogeny (Davis *et al.* 1994; Ledru *et al.* 1994).

The Pan-African belts

The West African Craton is nearly entirely surrounded (Fig. 4a) by orogenic belts of Neoproterozoic (Dahomeyide, Pharuside, Anti-Atlas, Bassaride, Rockelide) and Hercynian (Mauritanide) age. In general, the Neoproterozoic belts have been interpreted as the result of a protracted and diachronous (1000–500 Ma) succession of subduction and collisional events that involved disparate tectonic settings, including passive and active continental margins, platform covers, fragments of older continents, ocean basins and crust and magmatic arcs (Trompette 1997; Caby 2003).

The Pan-African Rockelide–Bassaride and Dahomeyide belts are located in the southern margins of the West African Craton (Fig. 4a). These belts were initiated as the margins of the West African Craton rifted, forming passive margins with clastic and carbonate sedimentation, followed by opening of an oceanic basin. Both belts are characterized by extensive reworking of older sequences and high-grade metamorphism (Villeneuve & Cornée 1994; Trompette 1997; Caby 1998).

The Rockelide Belt seems to represent an intracontinental orogeny, or passive margin, with diachronous evolution from north to south. The reworked basement is composed of high-grade gneisses, deep crustal granitoids, and Mesoproterozoic to Neoproterozoic (>700 Ma) cratonic covers, whereas the rift sequences comprise volcanic and volcaniclastic rocks. Intracontinental rifting occurred before about 550 Ma, and collision with the oriental portion of the Guyana Shield occurred at 550 Ma, provoking the thrusting of the Rockelide successions onto the West African Craton (Villeneuve & Cornée 1994).

The Dahomeyide Belt is the southern portion of the larger Trans-Saharan Belt that includes the Pharuside Belt to the north (Fig. 4a). It comprises a pre-orogenic rift phase that evolved to an active margin, with subduction and calc-alkaline magmatism occurring between 700 Ma and 600 Ma, and final collision against the eastern margin of the West African Craton at 610–600 Ma, producing granulite-facies metamorphism. Post-collisional plutonism occurred until 500 Ma (Villeneuve & Cornée 1994; Trompette 1997). In more detail, Caby (1998) recognized several tectonic elements that make up the Dahomeyide Belt in the southeastern portion of the West African Craton. These include (Fig 4b): (1) basement rocks composed of Eburnian (2.2 to 2.1 Ga) and allochthonous polycyclic gneisses that represent subducted fragments of Palaeoproterozoic continental crust; (2) passive margin sediments (Volta Basin); (3) subducted passive palaeo-margin sediments (mainly quartzites) of the West African Craton; (4) an intra-oceanic island arc (volcano-volcaniclastic sequence) and subducted arc roots (mafic–ultramafic to tonalitic rocks); (5) Pan-African suture zone; (6) thrust faults; (7) major ductile shear zones (e.g., the Hoggar 4°50' lineament). Caby (1998) describes voluminous calc-alkaline magmatism only in the northern continuation of the Dahomeyide Belt, the Pharusian Belt.

Sedimentary covers of the West African Craton

The West African Craton is widely covered by sedimentary sequences whose infilling occurred mostly in Neoproterozoic and Cambrian times and extended, in some places, until the Carboniferous period, being more or less affected by tectonism during the Pan-African cycle of orogenies. These sequences include those of the Tindouf Basin to the north, the Volta Basin to the SE, and several sub-basins in the central portion of the craton separating the Reguibat and Man shields (Fig. 4a) that may be broadly grouped together in the Taoudeni Basin. The thickness of these basins attained 8 km in some places, with glacial, marine and continental sediments recording a complex tectono-sedimentary evolution (Villeneuve & Cornée 1994).

Attempting to correlate northern Brazil and Western Africa

In this section we reassess possible links between the Precambrian terrains of northern Brazil and Western Africa, based on similarities (and differences) of rock association and geochemistry, zircon geochronology and isotope data. These elements can help in the discussion of the assembly and break up events through the identification of possible tectonic environments in which the associated rocks formed and, in consequence, of processes such as rifting, formation of oceanic basins and island arcs, and collision.

The Palaeoproterozoic domains

Palaeomagnetic data are lacking for the São Luís Craton. There is recent information for the Amazonian and West African cratons (Nomade *et al.* 2003; Tohver *et al.* 2006), but the two studies show different reconstructions, and one (Nomade *et al.* 2003) does not take the São Luís Craton into account. An alternative starting point comes from the study of the evolution of the central Atlantic transform faults based on sea floor topographic and gravity evidence (Sandwell & Smith 1995), and of the sedimentary and tectonic evolution of Phanerozoic sedimentary basins on both sides of the equatorial Atlantic (Matos 2000; Pletsch *et al.* 2001). Palaeogeographic reconstructions based on these studies consistently put the São Luís Craton and the Gurupi Belt opposite the present-day coastline of Ivory Coast (Fig. 5).

The three periods of magmatic activity defined in the São Luís Craton and in the Palaeoproterozoic portion of the Gurupi Belt (2240 ± 5 Ma, 2160 ± 10 Ma, 2080 ± 20 Ma) are also found in the West African Craton, having the same lithological and tectonic characteristics (Fig. 6). Furthermore, the geotectonic evolution of São Luís and

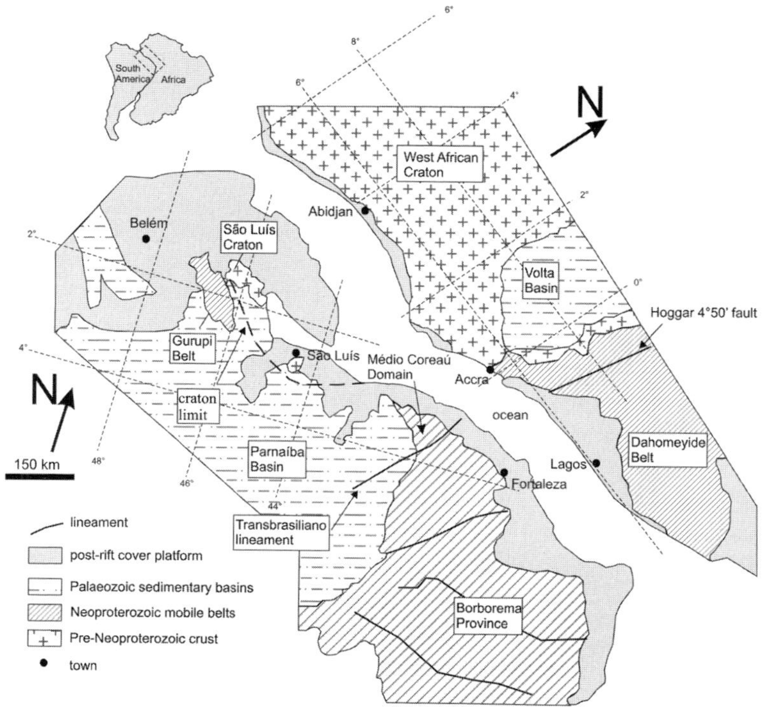

Fig. 5. Pre-drift palaeogeographic reconstruction of north-northeastern Brazil and northwestern Africa for the Aptian period, based on the similarity of sedimentary sequences on both sides of the Atlantic ocean (modified from Matos 2000), showing the main tectonic elements discussed in the text.

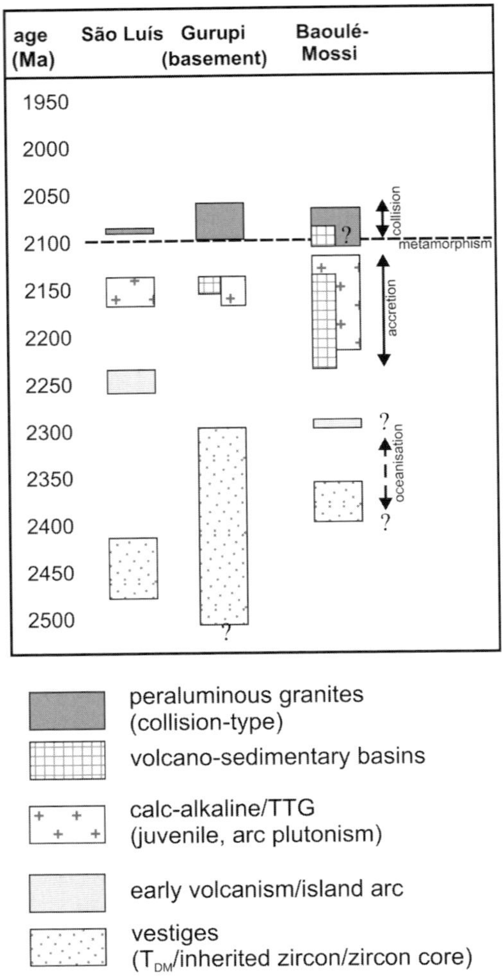

Fig. 6. Tectonic correlation chart for the Paleoproterozoic domains. See text for references and discussion.

West Africa in the Palaeoproterozoic era is interpreted in a similar way. This supports the correlation between the Brazilian and African terrains and the view that the São Luís Craton represents a fragment of the southern portion of the West African Craton (Baoulé–Mossi domain).

Several differences must be highlighted, because the West African Craton records a much more complex and nearly continuous evolution between *c*. 2220 Ma and 2060 Ma. For instance, in the São Luís Craton/Gurupi Belt there is no known geological record in the 2220–2170 Ma and 2140–2100 Ma time intervals, both of which are characterized by plutonic and volcanic activity in the West African Craton. Also, the granitoids of the West African Craton show a wider diversity in terms of petrology, geochemistry, age and association. The initial oceanic stage in the West African Craton, marked by extensive mafic–ultramafic volcanism, including abundant komatiite, has not yet been recognized in the São Luís Craton/Gurupi Belt, where basic magmatism is subordinate. The Tarkwa sedimentary sequence is a very important component of the West African Craton stratigraphy but no such sequence is known in the São Luís Craton so far. Despite the fact that the age and tectonic meaning of the Viseu Basin (Fig. 3) is far from being understood, this basin discordantly overlies the Palaeoproterozoic rocks of the São Luís Craton.

These differences may be real or due to distinct levels of knowledge in Brazilian and African counterparts, but could also be related to the fact that the Brazilian area is comparatively smaller than the African one.

The Neoproterozoic domains

The correlation of the Gurupi Belt with any other Brasiliano/Pan-African belt is not straightforward (see a comparison between the main events described for the Neoproterozoic marginal belts of the São Luís–West African craton in Fig. 7) because of fundamental problems that come from the Phanerozoic tectonic (uplift), sedimentary and erosive history, the widespread Phanerozoic cover and the relatively limited exposure of the Gurupi Belt with consequent difficulties in understanding the limits and the internal architecture of the belt.

The problem of the concealed limits of the belt was at least partially resolved by the geophysical evidence and the study of the basement rocks of the Phanerozoic basins (Lesquer *et al.* 1984; Nunes 1993). These data indicate that the Gurupi Belt extends 60–80 km to the south below the basins, where it is limited by the inferred Parnaíba block (Fig. 2).

To the east, the geophysical information (Lesquer *et al.* 1984; Nunes 1993) highlights the curvilinear shape of the Gurupi Belt, probably outlining the margin of the São Luís Craton (Fig. 2). Furthermore, the basement rocks a few tens of kilometres east of the town of São Luís have been affected by Neoproterozoic events. As such, the Gurupi Belt is probably connected with the Médio Coreaú domain (Fig. 2), despite differences in the age, metamorphism and lithological content of the basement rocks (older, higher-grade rocks occur in the Médio Coreaú domain).

To the west, the continuation of the Gurupi Belt is still obscure. An interesting solution has been proposed by Villeneuve & Cornée (1994), in which the Rockelide–Araguaia–Gurupi belts formed a triple junction by the convergence of the Amazonian and São Luís–West African cratons,

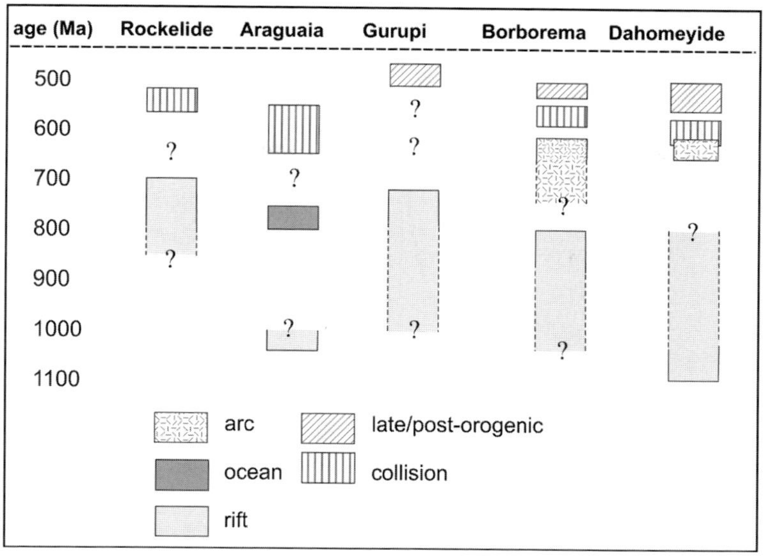

Fig. 7. Tectonic correlation chart for Brasiliano/Pan-African belts. See text for references and discussion.

and a third block located between these two cratons, possibly the Parnaíba block. In fact, convergence between the São Luís–West African craton and the Parnaíba block is a possible hypothesis to explain origin of the Gurupi Belt (Klein *et al.* 2005*b*). This is, however, a working hypothesis, since not even those works dealing with the basement of the Phanerozoic cover have provided any explanation for the continuation of the Gurupi Belt to the west (Fig. 2).

The closest and apparently least debatable link between north-northeastern Brazil and western Africa appears to be between the Médio Coreaú domain and the Dahomeyide Belt along the Hoggar 4°50' –Transbrasiliano lineaments (Figs 1 and 2). Brito Neves *et al.* (2002), however, stated that many features of the Dahomeyide Belt described by Caby (1998) are not found in the Médio Coreaú domain. This includes the tectonosedimentary record and the absence on the Brazilian side of collisional linearity and of the Neoproterozoic suture. Brito Neves *et al.* (2002) argued that this could result from the drag effect related to the Hoggar 4°50' –Transbrasiliano lineaments, and to the shape of the Neoproterozoic ocean that developed off the eastern margin of the West African Craton. We understand that this comment is also valid for the whole southeastern margin of the West African–São Luís craton, i.e., we also include the area that encompasses the Gurupi Belt.

Positive gravity anomalies have been interpreted as reflecting the concealed suture zone between the Neoproterozoic belts (Gurupi, Borborema) and the West African–São Luís craton (Lesquer *et al.* 1984). This is a possibility since mafic–ultramafic and granulite-facies rocks have been documented in the eastern margin of the West African Craton. However, the lithological record of this suture has been found neither in the Gurupi Belt nor in the Médio Coreaú domain, and metamorphic conditions attained in the Gurupi Belt are only of amphibolite facies. Moreover, the Tentugal strike-slip shear zone (Fig. 3) represents only the geochronological (Rb–Sr, K–Ar) boundary between the São Luís Craton and the Gurupi Belt and not a suture, since the same rock sequences are found on both sides (Klein *et al.* 2005*b*). An alternative location for the suture could be south of the Tentugal shear zone, approximately near the gneissose nepheline syenite (Fig. 3), since deformed alkaline rocks are good indicators of the proximity of suture zones (see Burke *et al.* 2003). Therefore, the gravity contrasts described by Lesquer *et al.* (1984) may still indicate the presence of denser rocks (granulite facies?), but these would underlie the exposed Gurupi Belt.

A major problem with any model for the Neoproterozoic tectonic evolution (Klein *et al.* 2005*b*) and the discussion of the internal architecture of the Gurupi Belt is its incomplete lithological record. Most of the rocks that crop out in the belt are Palaeoproterozoic rocks of the São Luís Craton that have been reworked during the Neoproterozoic orogenic events, i.e., they represent the cratonic margin and the external domain of the orogen.

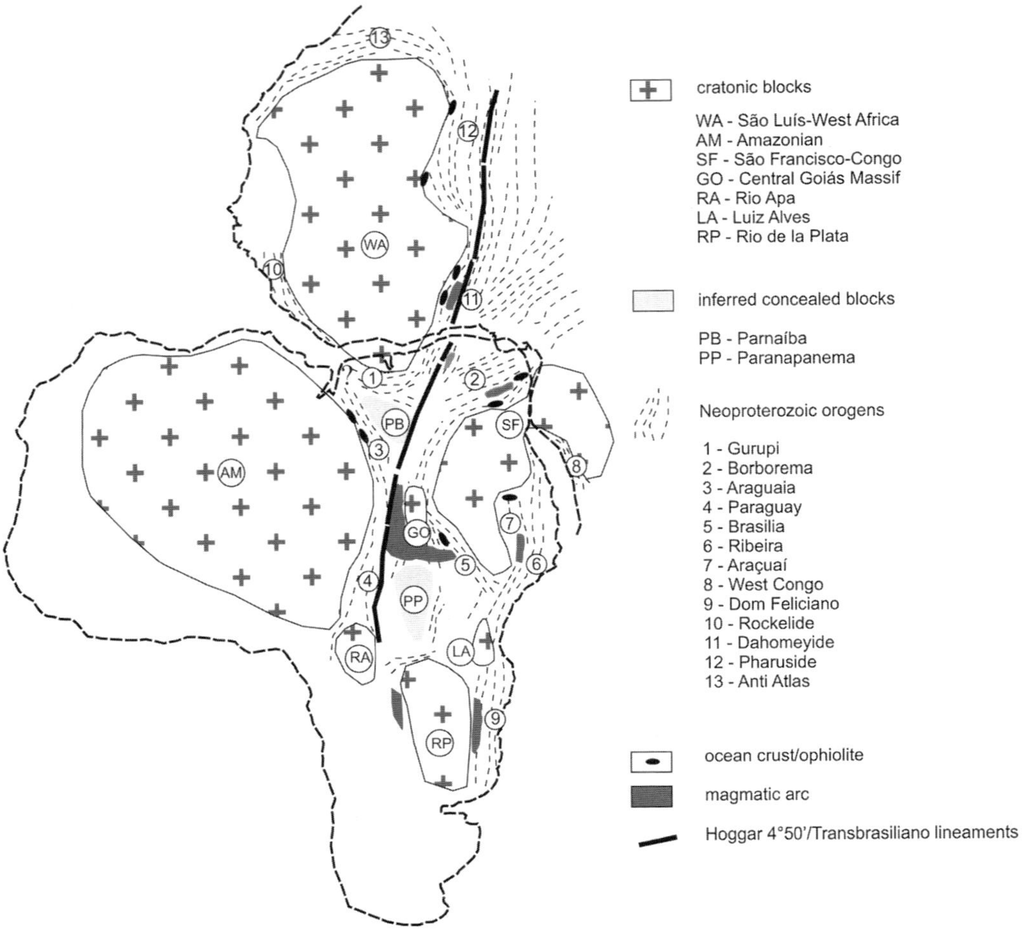

Fig. 8. Cartoon (not to scale) showing the tectonic elements of South America and northwestern Africa that participated of the West Gondwana assembly (references in the text). Although based primarily on geological comparisons and correlations, this reconstruction is essentially compatible with that of Tohver *et al.* (2006), which is based on palaeomagnetic control (albeit not for the São Luís craton). Both reconstructions differ from that of Nomade *et al.* (2003), which is also based on palaeomagnetic data, but without taking into account the position of the São Luís craton.

Otherwise, only the pre-orogenic (or syn-rift) and late/post-orogenic magmatic stages have been recognised at *c.* 732 and 550 Ma, respectively (Fig. 7) and we can only speculate as to what happened between these events. For instance, the existence of more felsic to intermediate rocks (arc-related?) within, or in the proximity of, the Gurupi Belt has been inferred (Pinheiro *et al.* 2003; Klein *et al.* 2005*b*) from the presence of abundant 550 Ma old detrital zircons in an immature (proximal) arkose of the Igarapé de Areia Formation and from a Sm–Nd T_{DM} model age of 1.4 Ga found in the Neoproterozoic amphibolite-facies metasedimentary sequence of the Marajupema Formation (Klein *et al.* 2005*b*).

The tectonic setting of this amphibolite-facies sequence is not clear. However the abundance of quartz-rich rocks (quartzite, feldspar-rich quartzite and coarse-grained quartz–mica schist) and the absence of associated igneous rocks suggest a passive margin environment, without carbonate deposition. This passive margin probably developed from the continental rift that contained the nepheline syenite intrusion and then evolved to an active margin.

The age of metamorphism is still unconstrained. There are a number of K–Ar and Rb–Sr mineral ages in the 466–618 Ma interval (see Klein *et al.* 2005*b* for a review and primary references). It is uncertain if these ages represent metamorphism, cooling/uplift, or late extrusion tectonics. Since the Neoproterozoic peraluminous granite of 549 ± 4 Ma has not been affected by the low-angle deformation imparted to the supracrustal rocks and the nepheline syenite in the north-western portion of the belt, metamorphism and tangential deformation are interpreted as older than the emplacement of this granite (Klein *et al.* 2005*b*).

Implications for the assembly and break-up of supercontinents

São Luís Craton and Gurupi Belt in West Gondwana

There is reasonable geological, geochronological and isotopic evidence that in Neoproterozoic times a large branching system of ocean basins extended from western Africa to central Brazil, passing through north-western Brazil. The evidence is provided by (Fig. 8): (1) mafic–ultramafic rocks and calc-alkaline granitoids in the Pharusian–Dahomeyide belts (Caby 1998); (2) continental magmatic arc rocks of 650–620 Ma in the northwestern (Médio Coreaú domain) and eastern portions of the Borborema Belt (Brito Neves *et al.* 2000; Fetter *et al.* 2003); (3) mafic–ultramafic rocks of 757 ± 49 Ma that represent obducted oceanic crust in the Araguaia Belt (Paixão *et al.* 2002); (4) juvenile magmatic arc rocks of 890–600 Ma in the Brasília Belt (Pimentel *et al.* 2005).

The closure of these ocean basins via subduction and the attendant convergence of several continental blocks in Neoproterozoic/Early Cambrian times resulted in orogenic belts, including the Gurupi Belt, that amalgamated these blocks forming the western part of the Gondwana supercontinent. The ocean closure was not linear but it probably formed a branching system and the main suture is approximately outlined by the Hoggar 4°50'–Transbrasiliano lineaments (Fig. 8). The overall evolution of these orogenic belts involved the formation of continental rifts, marginal and oceanic basins, island arcs, subduction and collision, with the events occurring diachronously (Fig. 7) in different geographic areas of West Gondwana. In the study area, the continental blocks that participated in West Gondwana assembly include, among others, the Amazonian and West African–São Luís cratons, which are the best-preserved cratonic blocks, being affected only in their margins, and the concealed Parnaíba block (Fig. 8).

São Luís Craton in Rodinia

Geological, tectonic and geochronological aspects of the São Luís and West African cratons (and of the Palaeoproterozoic portion of the Gurupi Belt), along with previous palaeogeographic and geochronological reconstructions, strongly support the interpretation that these terrains were contiguous in Palaeoproterozoic times.

Most of this Palaeoproterozoic landmass remained relatively stable until its incorporation into Rodinia (Condie 2002; Rogers & Santosh 2002). This is almost certainly true with respect to the São Luís Craton, where there is no known geological activity in the time interval between about 2000 and 750 Ma. Rifting of the (present-day) southern margin of the São Luís Craton occurred slightly before the intrusion of the nepheline syenite pluton (732 ± 7 Ma). This event marks an early stage of the Gurupi orogen and the beginning of Rodinia break-up in this region. At this time, and even before, ocean realms had already been developed in other parts of future West Gondwana.

The editors of the special issue are thanked for the invitation to write this paper. B.B. Brito Neves is acknowledged for the review of an early draft of the manuscript, which also benefited from the comments of two anonymous reviewers, and from the editorial handling of Robert Pankhurst.

References

ABOUCHAMI, W., BOHER, M., MICHARD, A. & ALBAREDE, F. 1990. A major 2.1 Ga event of mafic magmatism in West Africa: an early stage of crustal accretion. *Journal of Geophysical Research*, **95**(B11), 17605–17629.

BOSSIÈRE, G., BONKOUGOU, I., PEUCAT, J. J. & PUPIN, J. P. 1996. Origin and age of Paleoproterozoic conglomerates and sandstones of the Tarkwaian Group in Burkina Faso, West Africa. *Precambrian Research*, **80**, 153–172.

BRITO NEVES, B. B., FUCK, R. A., CORDANI, U. G. & THOMAZ FILHO, A. 1984. Influence of basement structures on the evolution of the major sedimentary basins of Brazil: a case of tectonic heritage. *Journal of Geodynamics*, **1**, 495–510.

BRITO NEVES, B. B., SANTOS, E. J. & VAN SCHMUS, W. R. 2000. Tectonic history of the Borborema Province. *In*: CORDANI, U. G., MILANI, E. J., THOMAZ FILHO, A. & CAMPOS, D. A. (eds) *Tectonic Evolution of South America, 31st International Geological Congress, Rio de Janeiro*, 151–182.

BRITO NEVES, B. B., VAN SCHMUS, W. R. & EGAL, A. H. 2002. North-western Africa–North-eastern Brazil. Major tectonic links and correlation problems. *Journal of African Earth Sciences*, **34**, 275–278.

BULLARD, E., EVERETT, J. E. & SMITH, A. G. 1965. The fit of the continents around the Atlantic: Symposium on Continental Drift. *Philosophical Transactions of the Royal Society of London*, **258**, 41–51.

BURKE, K., ASHWAL, L. D. & WEBB, S. J. 2003. New way to map old sutures using deformed alkaline rocks and carbonatites. *Geology*, **31**, 391–394.

CABY, R. 1998. Tectonic history and geodynamic evolution of northern Africa during the Neoproterozoic. *In*: *14 International Conference on Basement Tectonics, Ouro Preto, Abstracts*, 72–75.

CABY, R. 2003. Terrane assembly and geodynamic evolution of central–western Hoggar: a synthesis. *Journal of African Earth Sciences*, **37**, 133–159.

CABY, R., DELOR, C. & AGOH, O. 2000. Lithologie, structure et métamorphisme des formations birimiennes dans la région d'Odienné (Côte d'Ivoire): rôle majeur du diapirisme des plutons et des décrochements en bordure du craton de Man. *Journal of African Earth Sciences*, **30**, 351–374.

CONDIE, K. C. 2002. Breakup of a Paleoproterozoic supercontinent. *Gondwana Research*, **5**, 41–43.

COSTA, J. L. 2000. *Programa Levantamentos Geológicos Básicos do Brasil. Programa Grande Carajás. Castanhal, Folha SA.23-V-C. Estado do Pará*. Companhia de Pesquisa de Recursos Minerais Belém.

DAVIS, D. W., HIRDES, W., SCHALTEGGER, U. & NUNOO, E. A., 1994. U–Pb age constraints on deposition and provenance of Birimian and gold-bearing Tarkwaian sediments in Ghana, West Africa. *Precambrian Research*, **67**, 89–107.

DOUMBIA, S., POUCLET, A., KOUAMELAN, A., PECAUT, J. J., VIDAL, M. & DELOR, C. 1998. Petrogenesis of juvenile-type Birimian (Paleoproterozoic) granitoids in Central Côte-d'Ivoire, West Africa: geochemistry and geochronology. *Precambrian Research*, **87**, 33–63.

EGAL, E., THIÉBLEMONT, D. *ET AL*. 2002. Late Eburnian granitization and tectonics along the western and northwestern margin of the Archaean Kenema–Man domain (Guinea, West African craton). *Precambrian Research*, **117**, 57–84.

FETTER, A. H., VAN SCHMUS, W. R., SANTOS, T. J. S., NOGUEIRA NETO, J. A. & ARTHAUD, M. H. 2000. U–Pb and Sm–Nd geochronological constraints on the crustal evolution and basement architecture of Ceará State, NW Borborema province, NE Brazil: implications for the existence of the Paleoproterozoic supercontinent 'Atlantica'. *Revista Brasileira de Geociências*, **30**, 102–106.

FETTER, A. H., SANTOS, T. J. S. *ET AL*. 2003. Evidence for Neoproterozoic continental arc Magmatism in the Santa Quitéria Batholith of Ceará State, NW Borborema Province, NE Brazil: Implications for the assembly of West Gondwana. *Gondwana Research*, **6**, 265–273.

FEYBESSE, J. L. & MILÉSI, J. P. 1994. The Archaean/Proterozoic contact zone in West Africa: a mountain belt of décollement thrusting and folding on a continental margin related to 2.1 Ga convergence of Archaean cratons? *Precambrian Research*, **69**, 199–227.

GASQUET, D., BARBEY, P., ADOU, M. & PAQUETTE, J. L. 2003. Structure, Sr–Nd isotope geochemistry and zircon U–Pb geochronology of the Dabakala area (Côte d'Ivoire): evidence for a 2.3 crustal growth event in the Palaeoproterozoic of West Africa? *Precambrian Research*, **127**, 329–354.

HEIN, K. A. A., MOREL, V., KAGONÉ, O., KIEMDE, F. & MEYERS, K. 2004. Birimian lithological succession and structural evolution in the Goren segment of the Boromo-Goresn Greenstone Belt, Burkina Faso. *Journal of African Earth Sciences*, **39**, 1–23.

HIRDES, W., DAVIS, D. W., LÜDTKE, G. & KONAN, G. 1996. Two generations of Birimian (Paleoproterozoic) volcanic belts in northeastern Côte d'Ivoire (West Africa): consequences for the 'Birimian controversy'. *Precambrian Research*, **80**, 173–191.

HURLEY, P. M., ALMEIDA, F. F. M. *ET AL*. 1967. Test of continental drift by comparison of radiometric ages. *Science*, **157**, 495–500.

HURLEY, P. M., MELCHER, G. C., PINSON, W. H. & FAIRBAIRN, H. W. 1968. Some orogenic episodes in South America by K–Ar and whole-rock Rb–Sr dating. *Canadian Journal of Earth Sciences*, **5**, 633–638.

KLEIN, E. L. & MOURA, C. A. V. 2001. Age constraints on granitoids and metavolcanic rocks of the São Luís craton and Gurupi Belt, northern Brazil: implications for lithostratigraphy and geological evolution. *International Geology Review*, **43**, 237–253.

KLEIN, E. L. & MOURA, C. A. V. 2003. Síntese geológica e geocronológica do Cráton São Luís e do Cinturão Gurupi na região do Rio Gurupi (NE-Pará/NW-Maranhão). *Geologia, Universidade de São Paulo*, **3**, 97–112.

KLEIN, E. L., MOURA, C. A. V. & PINHEIRO, B. L. S. 2005*a*. Paleoproterozoic crustal evolution of the São Luís craton, Brazil: evidence from zircon geochronology and Sm–Nd isotopes. *Gondwana Research*, **8**, 177–186.

KLEIN, E. L., MOURA, C. A. V., KRYMSKY, R. & GRIFFIN, W. L. 2005*b*. The Gurupi belt in northern Brazil: lithostratigraphy, geochronology, and geodynamic evolution. *Precambrian Research*, **141**, 83–105.

LAHONDÈRE, D., THIÉBLEMONT, D., TEGYEY, M., GUERROT, C. & DIABATE, B. 2002. First evidence of early Birimian (2.21 Ga) volcanic activity in Upper Guinea: the volcanics and associated rocks of the Niani suite. *Journal of African Earth Sciences*, **35**, 417–431.

LEDRU, P., JOHAN, V., MILÉSI, J. P. & TEGYEY, M. 1994. Markers of the last stages of the Palaeoproterozoic collision: evidence for a 2 Ga continent involving circum-South Atlantic provinces. *Precambrian Research*, **69**, 169–191.

LESQUER, A., BELTRÃO, J. F. & ABREU, F. A. M. 1984. Proterozoic links between northeastern Brazil and West Africa: a plate tectonic model based on gravity data. *Tectonophysics*, **110**, 9–26.

LOWELL, G. R. & VILLAS, R. N. N. 1983. Petrology of nepheline syenite gneiss from Amazonian Brazil. *Geological Journal*, **18**, 53–75.

MATOS, R. D. 2000. Tectonic evolution of the equatorial South Atlantic. *In*: MOHRIAK, W. & TALWANI, M. (eds) *Atlantic rifts and continental margins*. American Geophysical Union, Geophysical Monographs, **115**, 331–354.

MOURA, C. A. V., ABREU, F. A. M., KLEIN, E. L., PALHETA, E. S. M. & PINHEIRO, B. L. S. 2003. Geochronology of the São Luís craton and the Gurupi Belt, Brazil. *In*: *IV South American Symposium on Isotope Geology, Salvador, Brazil, Short Papers*, 225–228.

NOMADE, S., CHEN, Y. ET AL. 2003. The Guiana and the West African Shield Palaeoproterozoic grouping: new palaeomagnetic data for French Guiana and the Ivory Coast. *Geophysical Journal International*, **154**, 677–694.

NUNES, K. C. 1993. Interpretação integrada da Bacia do Parnaíba com ênfase nos dados aeromagnéticos. *In*: *Congresso Internacional da Sociedade Brasileira de Geofísica, 2, Resumos Expandidos*, **1**, 152–157.

PAIXÃO, M. A. P., NILSON, A. A., DANTAS, E. & GIÓIA, S. 2002. Corpos ofiolíticos da Faixa Araguaia e suas implicações tectônicas. *In*: *Congresso Brasileiro de Geologia 41, João Pessoa, Brazil, Abstracts*. Sociedade Brasileira de Geologia, 326.

PASTANA, J. M. N. 1995. *Programa Levantamentos Geológicos Básicos do Brasil. Programa Grande Carajás. Turiaçu/Pinheiro, folhas SA.23-V-D/SA.23-Y-B. Estados do Pará e Maranhão*. Companhia de Pesquisa de Recursos Minerais, Brasília.

PIMENTEL, M. M., LAUX, J. H., ARMSTRONG, R. A., ARMELE, A. & FERREIRA FILHO, C. F. 2005. Geochronology of mafic-ultramafic complexes in central Brazil: significance for the assembly of Western Gondwana. *In*: PANKHURST, R. J. & VEIGA, G. D. (eds) *Gondwana 12: Geological and Biological Heritage of Gondwana, Abstracts*. Academia Nacional de Ciencias, Córdoba, Argentina, 294.

PINHEIRO, B. L. S., MOURA, C. A. V. & KLEIN, E. L. 2003. Estudo de proveniência em arenitos das formações Igarapé de Areia e Viseu, nordeste do Pará, com base em datação de monocristais de zircão por evaporação de chumbo. *In*: *8 Simpósio de Geologia da Amazônia, Manaus, Brazil, Extended abstracts* (CD-ROM).

PLETSCH, T., ERBACHER, J. ET AL. 2001. Cretaceous separation of Africa and South America: the view from the West African margin (ODP Leg 159). *Journal of South American Earth Sciences*, **14**, 147–174.

POUCLET, A., VIDAL, M., DELOR, C., SIMÉON, Y. & ALRIC, G. 1996. Le volcanisme birimien du nord-est de la Côte d'Ivoire, mise en evidence de deux phases volcano-tectoniques distinctes dans l'évolution géodynamique du Paléoprotérozoïque. *Bulletin de la Societé Géologique de France*, **167**, 529–541.

REZENDE, O. M. & PAMPLONA, H. R. P. 1970. Estudo do desenvolvimento do Arco Ferrer-Urbano Santos. *Boletim Técnico da Petrobras*, **13**, 5–14.

ROGERS, J. J. W. & SANTOSH, M. 2002. Configuration of Columbia, a Mesoproterozoic Supercontinent. *Gondwana Research*, **5**, 5–22.

SANDWELL, D. T. & SMITH, W. H. F. 1995. *Map of marine gravity anomaly from satellite altimetry*. [scale 1:31,000,000]. Scripps Institute of Oceanography, La Jolla, California.

SYLVESTER, P. J. & ATTOH, K. 1992. Lithostratigraphy and composition of 2.1 Ga greenstone belts of the West African craton and their bearing on crustal evolution and the Archaean-Proterozoic boundary. *Journal of Geology*, **100**, 377–393.

THIÉBLEMONT, D., GOUJOU, J. C. ET AL. 2004. Archaean evolution of the Leo Rise and its Eburnian reworking. *Journal of African Earth Sciences*, **39**, 97–104.

TOHVER, E., D'AGRELLA-FILHO, M. S. & TRINDADE, R. I. F. 2006. Paleomagnetic record of Africa and South America for the 1200–500 Ma interval, and evaluation of Rodinia and Gondwana assemblies. *Precambrian Research*, **147**, 193–222.

TORQUATO, J. R. & CORDANI, U. G. 1981. Brazil-Africa geological links. *Earth-Science Reviews*, **17**, 155–176.

TROMPETTE, R. 1997. Neoproterozoic (~600 Ma) aggregation of Western Gondwana: a tentative scenario. *Precambrian Research*, **82**, 101–112.

VIDAL, M. & ALRIC, G. 1994. The Palaeoproterozoic (Birimian) of Haute Comoé in the West African craton in Côte d'Ivoire: a transtentional back-arc basin. *Precambrian Research*, **65**, 207–229.

VIDAL, M., DELOR, C., POUCLET, A., SIMÉON, Y. & ALRIC, G. 1996. Evolution géodynamique de l'Afrique de l'Ouest entre 2.2 GA et 2 Ga: le style archéen des ceintures vertes et des ensembles sédimentaires Birimiens du nord-est de la Côte-d'Ivoire. *Bulletin de la Societé Géologique de France*, **167**, 307–319.

VILLENEUVE, M. & CORNÉE, J. J. 1994. Structure, evolution and palaeogeography of the West African craton and bordering belts during the Neoproterozoic. *Precambrian Research*, **69**, 307–326.

Similarities and differences between the Brazilian and African counterparts of the Neoproterozoic Araçuaí–West Congo orogen

A. C. PEDROSA-SOARES[1], F. F. ALKMIM[2], L. TACK[3], C. M. NOCE[1], M. BABINSKI[4], L. C. SILVA[5] & M. A. MARTINS-NETO[2]

[1]*Universidade Federal de Minas Gerais, IGC-CPMTC, Campus Pampulha, 31270-901 Belo Horizonte, MG, Brazil (e-mail: pedrosa@pesquisador.cnpq.br)*

[2]*Universidade Federal de Ouro Preto, DEGEO, Campus do Cruzeiro, 35400-000 Ouro Preto, MG, Brazil*

[3]*Royal Museum for Central Africa, 3080 Tervuren, Belgium*

[4]*Universidade de São Paulo, IG-USP, 05508-080 São Paulo, Brazil*

[5]*Serviço Geológico do Brasil-CPRM, 70830-030 Brasília, DF, Brazil*

Abstract: The Araçuaí–West Congo orogen encompasses orogenic domains located to the SE of the São Francisco Craton in Brazil, and to the SW of the Congo Craton in Africa. From the opening of the precursor basin to the last orogenic processes, the evolution of the orogen lasted from the very beginning of the Neoproterozoic up to the Cambrian–Ordovician boundary. After the spreading of the South Atlantic Ocean in Cretaceous time, the Araçuaí–West Congo orogen was split into two quite different but complementary counterparts. The Brazilian side (Araçuaí orogen) inherited two thirds of the whole orogenic edifice, including all the Neoproterozoic ophiolite slivers, the entire magmatic arc and syn-collisional to post-collisional magmatism, and the suture zone. The African counterpart (West Congo Belt), a fold–thrust belt free of Neoproterozoic ophiolite and Pan-African orogenic magmatism, inherited the thick pile of bimodal volcanic rocks of the Early Tonian rift stage, implying that the precursor basin was an asymmetrical rift with the thermal–magmatic axis located in the West Congo Belt. Both counterparts of the Araçuaí–West Congo orogen include Neoproterozoic glaciogenic deposits, allowing tentative lithostratigraphic correlations, but identification of the ice ages remains uncertain because the lack of sufficient well-constrained geochronological data.

The Araçuaí–West Congo orogen encompasses the Neoproterozoic orogenic domains located to SE of the São Francisco Craton in Brazil, and to SW of the Congo Craton in Africa (Fig. 1). We use the names Araçuaí orogen and West Congo Belt for the Brazilian and African counterparts of this orogen, respectively.

Prior to the opening of the South Atlantic Ocean in the Cretaceous the São Francisco Craton was connected to the Congo Craton by means of the Bahia–Gabon continental bridge (Fig. 1). Considering that the youngest orogenic event in the cratonic bridge occurred around 2 Ga, the continental link between the São Francisco and Congo cratons must have been formed during the Palaeoproterozoic and remained until the onset of Atlantic opening (Porada 1989; Ledru *et al.* 1994; Trompette 1994; Feybesse *et al.* 1998; Brito Neves *et al.* 1999; Corrêa-Gomes *et al.* 2000; Alkmim *et al.* 2001; Pedrosa-Soares *et al.* 2001*a*; Zhao *et al.* 2002; Silva *et al.* 2002*a*, 2005; Barbosa & Sabaté 2004; D'Agrella *et al.* 2004).

To the south of the Bahia–Gabon cratonic bridge, the Araçuaí–West Congo orogen evolved inside an embayment carved into the São Francisco–Congo palaeocontinent (Brito-Neves *et al.* 1999; Pedrosa-Soares *et al.* 2001*a*; Tack *et al.* 2001; Alkmim *et al.* 2006). The evolution of the orogen lasted from the opening of the precursor basin at the beginning of the Neoproterozoic until late orogenic processes near the Cambrian–Ordovician boundary.

During the Cretaceous opening of the South Atlantic Ocean, the Araçuaí–West Congo orogen was split up in two quite different but complementary counterparts. Despite a long erosion history, which started by the end of Neoproterozoic, the Araçuaí–West Congo orogen still preserves rock assemblages from all development stages from a basin system to an orogenic edifice.

Critical rock assemblages for stratigraphic correlation and tectonic interpretation have been characterized in the Araçuaí orogen as a result of

From: PANKHURST, R. J., TROUW, R. A. J., BRITO NEVES, B. B. & DE WIT, M. J. (eds) *West Gondwana: Pre-Cenozoic Correlations Across the South Atlantic Region.* Geological Society, London, Special Publications, **294**, 153–172. DOI: 10.1144/SP294.9 0305-8719/08/$15.00

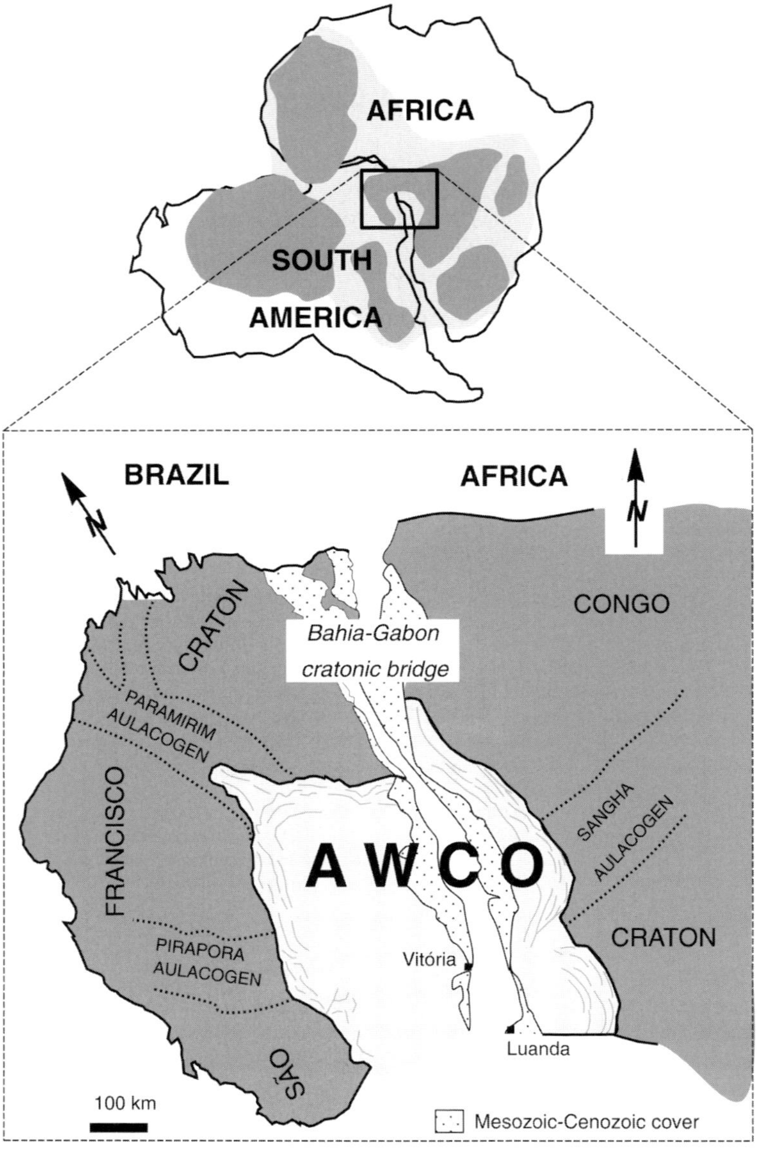

Fig. 1. Location of the Araçuaí–West Congo orogen in relation to the São Francisco and Congo cratons.

field mapping, and geochemical and U–Pb geochronological studies performed in recent years. Development of the West Congo Belt, both in space and time, has been reassessed recently by Tack *et al.* (2001, 2004) and Frimmel *et al.* (2006).

In this paper, we present an updated synthesis with new data for the Araçuaí orogen, including descriptions, previously unavailable in the international literature, of formations of the Macaúbas Group that are keystones for any attempted correlation with the West Congo Belt. We also highlight the differences between the Brazilian and African counterparts of the Araçuaí–West Congo orogen, and suggest correlations between some units of the Macaúbas and West Congolian groups, followed by a discussion concerning the age of glacial events. The following descriptions generally refer to the protoliths rather than their metamorphic equivalents. Geographical coordinates refer to the present-day position of continents.

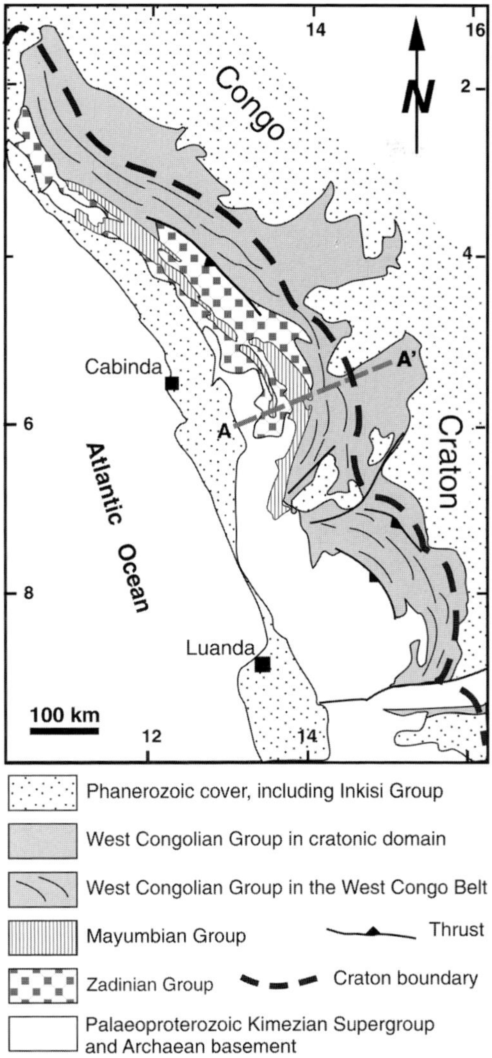

Fig. 2. Geological sketch map of the West Congo Belt (modified from Tack *et al.* 2001). A–A′, section shown in Figure 3.

The West Congo Belt

The 900 km long West Congo Belt, located between latitudes 1°S and 10°S, corresponds to the African side of the Araçuaí–West Congo orogen (Fig. 2). It is a NE-verging fold–thrust belt that involved the Neoproterozoic West Congo Supergroup and partially reworked the Archaean–Palaeoproterozoic basement (Fig. 3), but lacks Neoproterozoic ophiolite and Pan-African orogenic igneous bodies (Cahen *et al.* 1984; Maurin *et al.* 1991; Trompette 1994, 1997; Tack *et al.* 2001, 2004; Frimmel *et al.* 2006). The basement, which is not focused on here, includes Archaean complexes and the Palaeoproterozoic Kimezian Supergroup, which nevertheless have correlatives in the Araçuaí orogen (Vicat & Pouclet 2000; Tack *et al.* 2001; Noce *et al.* 2007). Interestingly, Mesoproterozoic rocks are totally lacking in the West Congo Belt (Tack *et al.* 2004). The Neoproterozoic West Congo Supergroup comprises the Zadinian, Mayumbian and West Congolian groups (Figs 2, 3 and 4).

The Zadinian and Mayumbian groups represent the filling of an Early Tonian continental rift (Tack *et al.* 2001). The Zadinian Group consists of rift-related siliciclastic sediments and peralkaline rhyolites, covered by a thick succession of mafic volcanic rocks (Fig. 4). A U–Pb SHRIMP zircon age of 999 ± 7 Ma for an anorogenic, peralkaline (Noqui-type) granite intruded into the lowest part of the Zadinian Group and one of *c.* 920 Ma for anorogenic granites hosted by the Mayumbian Group constrain the age of the Zadinian Group to the very Early Tonian (Tack *et al.* 2001). The Mayumbian Group, which is up to 4 km thick, consists mostly of felsic volcanic rocks intruded by abundant cogenetic bodies of monzo-syenogranite to minor alkali-feldspar granite, and subordinate sedimentary intercalations (Fig. 4). Mayumbian volcanic rocks have yielded U–Pb SHRIMP zircon ages of 920 ± 8 Ma for the lower part of the pile and 912 ± 7 Ma for the upper section of the group (Tack *et al.* 2001). Together, the Zadinian mafic magmatism and overlying felsic Mayumbian magmatism are a typical example of a bimodal magmatic suite formed during continental rifting.

The West Congolian Group consists of continental rift and passive margin sequences, and carbonate platform deposits (Figs 3 and 4). It includes two diamictite units (the Lower and Upper Mixtite formations) that have been considered to record glacial events in the West Congo Belt (Alvarez & Maurin 1991; Trompette 1994; Alvarez 1995; Tack *et al.* 2001; Frimmel *et al.* 2006).

The lower part of the West Congolian Group, the Sansikwa Subgroup, is a siliciclastic succession starting with conglomerate, followed by argillite, quartz arenite and arkose (Fig. 4). This succession, regarded as a continental rift fill, contains detrital zircons from Archaean to Tonian sources, including the 999 ± 7 Ma (Noqui-type) anorogenic granites and the Mayumbian Group (Tack *et al.* 2001; Frimmel *et al.* 2006). The U–Pb SHRIMP data from detrital zircons constrain the maximum sedimentation age of the Sansikwa Subgroup as 923 ± 43 Ma, in agreement with its stratigraphic relationship to the Mayumbian Group (Frimmel *et al.* 2006).

The oldest diamictite (the Lower Mixtite Formation) covers the rift deposits represented by the

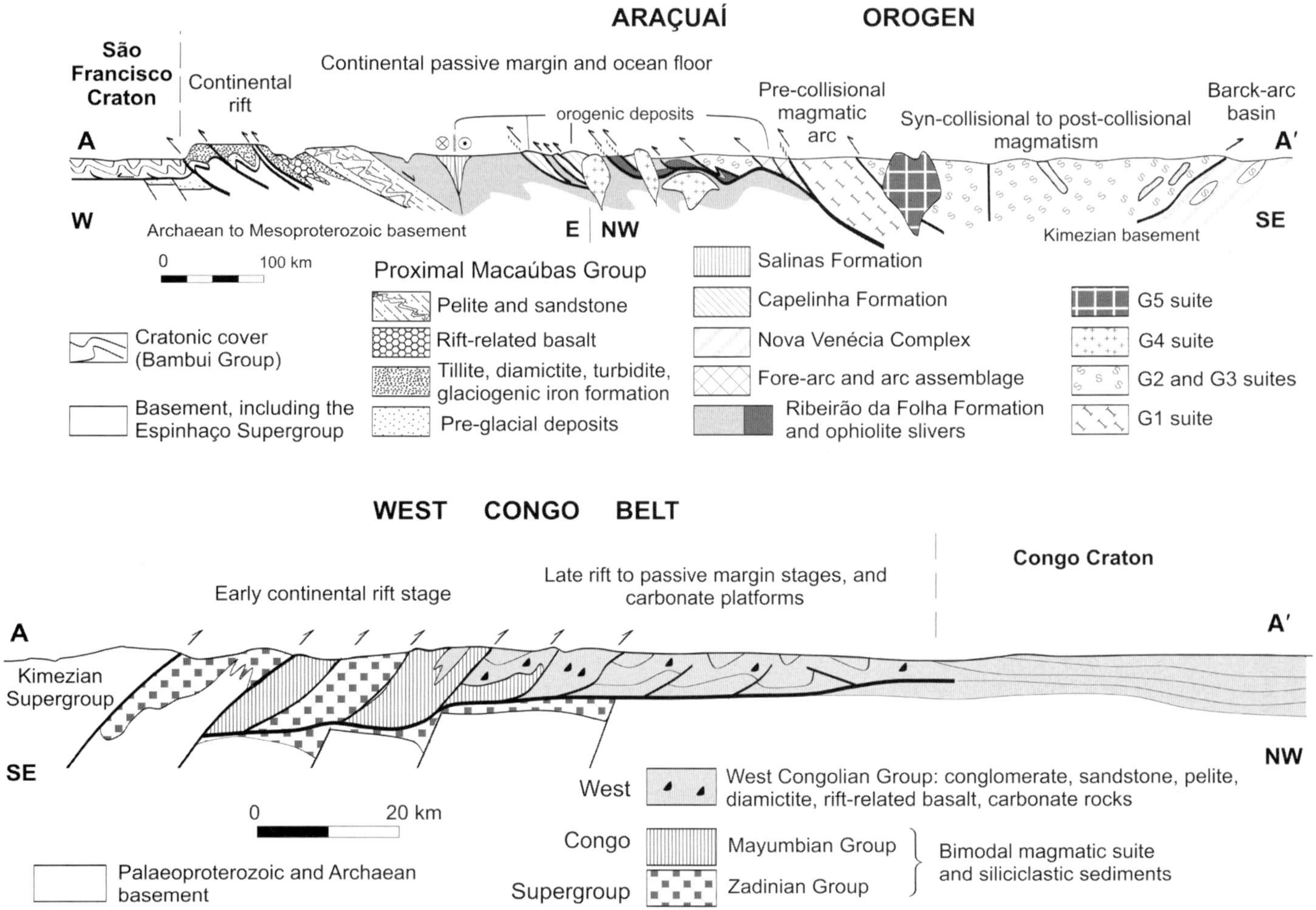

Fig. 3. Geological sketch sections of the Araçuaí orogen and West Congo Belt. A–A′, section locations in Figures 2 and 5.

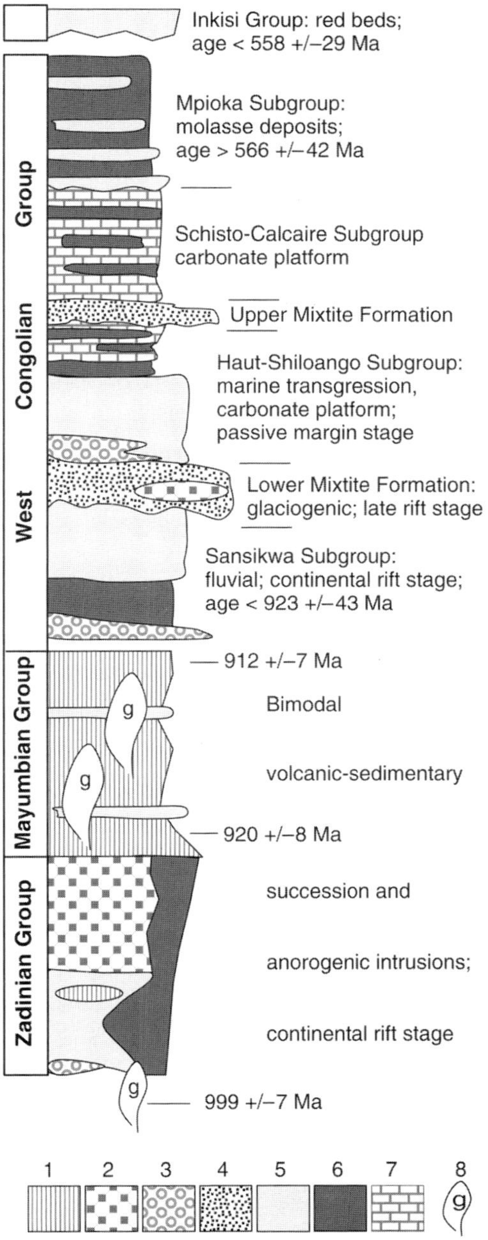

Fig. 4. Stratigraphic sketch column of the West Congo Supergroup and Inkisi Group, in the West Congo Belt (age references cited in text): 1, felsic volcanic rocks; 2, mafic volcanic rocks; 3, conglomerates; 4, diamictites; 5, sandstones; 6, pelites; 7, carbonates; 8, anorogenic granites.

Sansikwa Subgroup (Fig. 4). This formation includes tholeiitic basalts with pillow structures, associated with feeder dolerite dykes and sills indicating an extensional setting for the basin during its deposition (Tack *et al.* 2001; Frimmel *et al.* 2006). The geochemical signature of the mafic rocks suggests volcanism on a thin continental crust, related to a late rift stage (De Paepe *et al.* 1975; Kampunzu *et al.* 1991). The Lower Mixtite Formation has been ascribed to the Sturtian glaciation (*c.* 750 Ma), but no unquestionable age for this glaciogenic sedimentation is so far available (Tack *et al.* 2001; Frimmel *et al.* 2006).

The Haut Shiloango Subgroup overlies the Lower Mixtite Formation and consists of a varied succession of conglomerate, quartz arenite, argillite, calc-arenite, calc-pelite and carbonate rocks, representing the first marine transgression in the West Congo Belt succession (Tack *et al.* 2001; Frimmel *et al.* 2006). The detrital zircon population of an arkose of the upper Haut Shiloango Subgroup has yielded ages from Archaean to late Neoproterozoic (Frimmel *et al.* 2006); these authors suggest that the subgroup represents a passive margin, open marine, post-Sturtian carbonate platform.

The younger diamictite (the Upper Mixtite Formation) locally covers the Haut Shiloango Subgroup and is capped by a thick carbonate–pelite sequence, the Schisto-Calcaire Subgroup (Fig. 4). The sedimentation age of the Upper Mixtite Formation is unknown, but it must be older than the overlying Inkisi Group, which could be as old as 558 ± 29 Ma (Frimmel *et al.* 2006). The latter authors suggest a possible correlation of the upper diamictite with the Marinoan glaciation (*c.* 636 Ma).

The Schisto-Calcaire Subgroup consists of shallow-water, near-shore carbonate rocks, containing stromatolite bioherms and the filamentous cyanobacterium *Obruchevella* in its upper part (Alvarez 1995). The lowermost formation (Kimpese dolomite) of this subgroup has been regarded as a cap carbonate, deposited after the glaciation related to the Upper Mixtite Formation (Frimmel *et al.* 2006). The Schisto-Calcaire Subgroup was eventually covered by siliciclastic molasse deposits (Mpioka Subgroup), which are older than 566 ± 42 Ma (Tack *et al.* 2001; Frimmel *et al.* 2006). The Inkisi red beds, formerly interpreted as the uppermost unit of the West Congolian Group, are considered to form a large intracratonic, post-orogenic, Palaeozoic (though pre-Karoo), foreland blanket of cover deposits (the Inkisi Group; Frimmel *et al.* 2006).

The Araçuaí orogen

We use the name Araçuaí orogen for the Brazilian counterpart of the Araçuaí–West Congo orogen, because it inherited two thirds of the whole orogenic edifice, including their main geotectonic components. The Araçuaí orogen is 1000 km long and up to 500 km wide, encompassing the entire region between the São Francisco Craton and the Atlantic continental margin, to the north of latitude 21°S (Fig. 5). This figure shows an updated version of the geological map of the Araçuaí orogen (Pedrosa-Soares *et al.* 2001*a*, 2005), highlighting the Neoproterozoic and Cambrian units (Figs 5 and 6). It does not show the orogenic region to the north of latitude 1°S (see Alkmim *et al.* 2006) nor that part of the Atlantic coast dominated by Cenozoic cover. The basement of the Araçuaí orogen, not considered here, comprises Archaean to Mesoproterozoic units, including the Statherian to Mesoproterozoic Espinhaço Supergroup (Martins-Neto 2000; Pedrosa-Soares & Wiedemann-Leonardos 2000; Noce *et al.* 2007).

The Araçuaí orogen can be subdivided into two very distinct tectonic domains (Pedrosa-Soares & Wiedemann-Leonardos 2000; Pedrosa-Soares

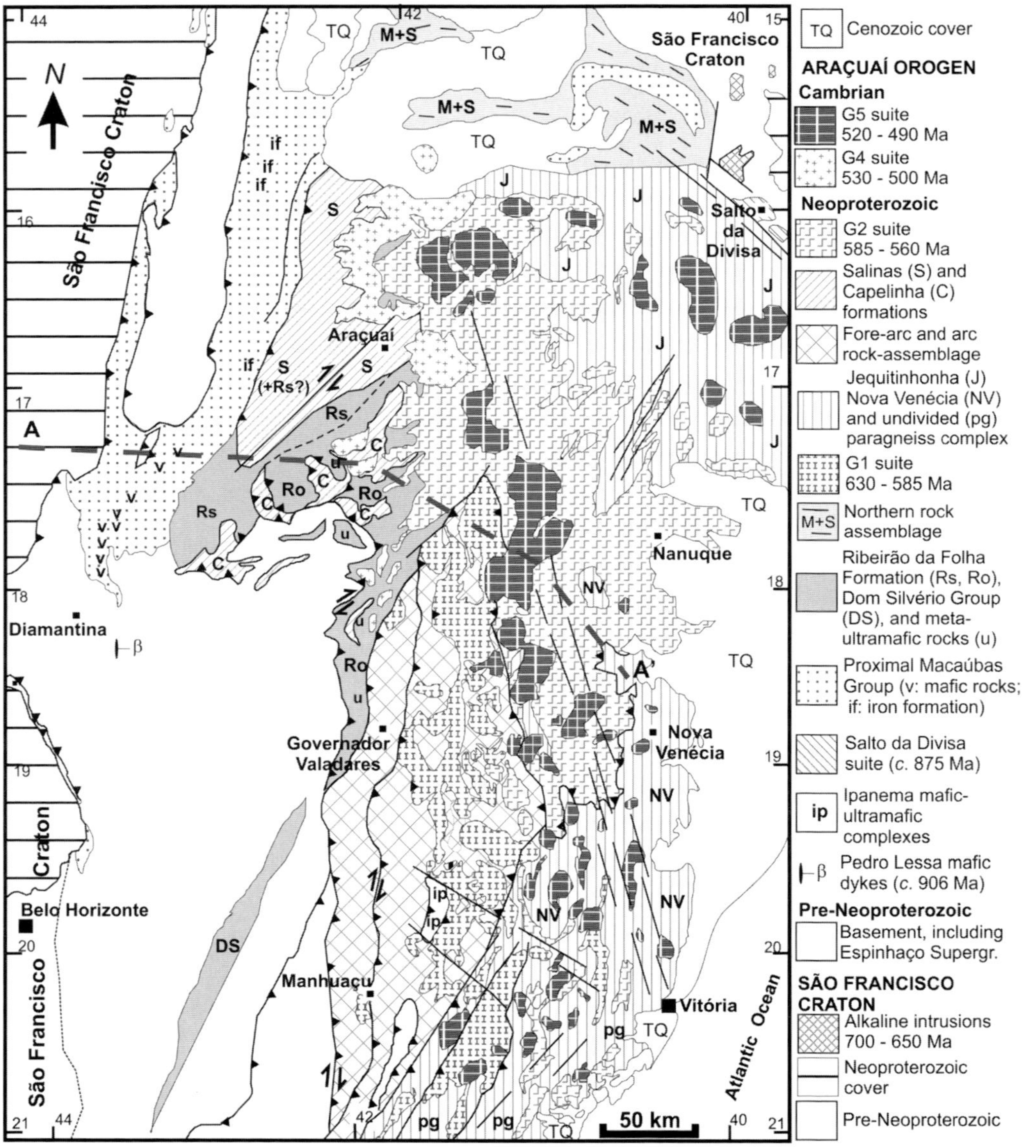

Fig. 5. Geological map of the Araçuaí orogen (modified from Pedrosa-Soares *et al.* 2001*a*, 2005). A–A′, section shown in Figure 3.

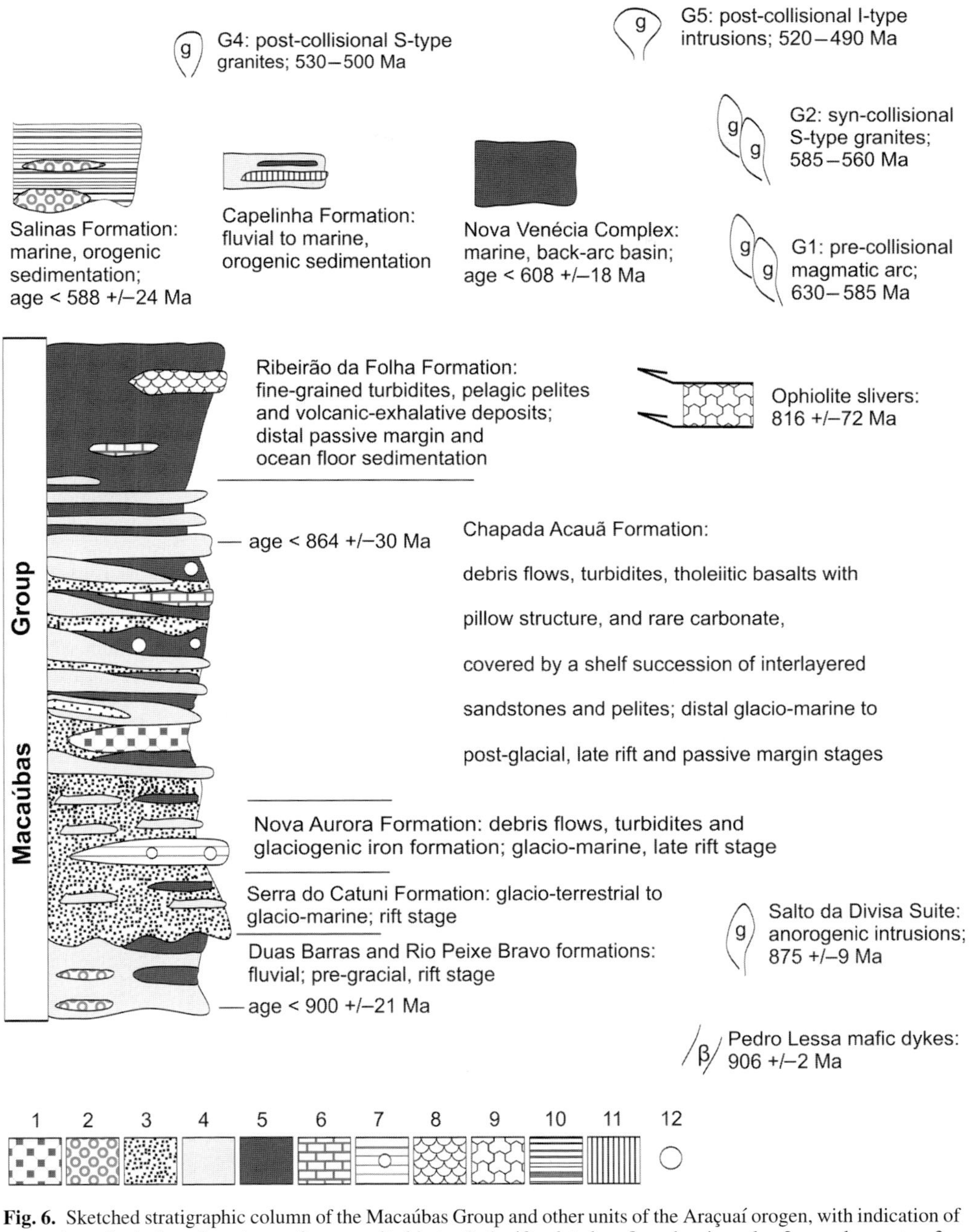

Fig. 6. Sketched stratigraphic column of the Macaúbas Group and other units of the Araçuaí orogen, with indication of some chronological markers (age references cited in text): 1, rift-related mafic volcanic rocks; 2, conglomerates; 3, diamictites; 4, sandstones; 5, pelites; 6, carbonate rocks; 7, glaciogenic iron formation; 8, sulphide-bearing cherts and diopsidites with massive sulphide bodies, and banded iron formations (volcanic-exhalative rocks); 9, mafic and ultramafic rocks of ophiolite slivers; 10, greywacke–pelite association; 11, detrital iron formation; 12, dropstones.

et al. 2001*a*; Alkmim *et al.* 2006). The western external domain, adjacent to the eastern edge of the São Francisco Craton, is a west-verging fold–thrust belt, with increasing intensity of metamorphism from west to east (Figs 3 and 5). The northern segment of the external tectonic domain inflects to the east, outlining the northern curvature of the orogen, where it shows north-verging structures and an increase in metamorphic grade from north to south. The internal tectonic domain is the Neoproterozoic–Cambrian high-grade metamorphic core of the orogen, which includes the pre-collisional magmatic arc, a huge anatectic zone dominated by syn-collisional S-type granites, and post-collisional intrusions. The internal tectonic domain is generally west-verging, but syn-collisional tectonic transport to the east prevails in the northern back-arc zone, between latitudes 17°30′ and 20°S (Pedrosa-Soares *et al.* 2001*a*, 2006; Alkmim *et al.* 2006). Together, this east-verging zone of the Araçuaí orogen and the West Congo Belt exhibit a typical bivergent orogenic architecture (Fig. 3).

The Neoproterozoic Precursor Basin

Proximal units of the Macaúbas Group and Late Tonian anorogenic magmatism record the continental rift stage of the Araçuaí orogen (Pedrosa-Soares *et al.* 1992; Noce *et al.* 1997; Uhlein *et al.* 1998, 1999; Pedrosa-Soares *et al.* 2000; Silva *et al.* 2002*b*, 2007; Angeli *et al.* 2004; Babinski *et al.* 2005; Gradim *et al.* 2005). South of latitude 17°S (Fig. 5), the rift evolved into a passive margin with a narrow ocean basin, represented by the distal Macaúbas Group (Ribeirão da Folha Formation) and associated mafic–ultramafic ophiolite slivers (Pedrosa-Soares *et al.* 1992, 1998, 2001*a*, 2005; Martins-Neto *et al.* 2001; Suita *et al.* 2004; Alkmim *et al.* 2006).

Two magmatic episodes suggest age limits for the rift stage of the Macaúbas basin (Figs 5 and 6). The older one refers to a meta-dolerite dyke from Pedro Lessa dated at 906 ± 2 Ma (zircon and baddeleyite, U–Pb, Machado *et al.* 1989). The layered mafic–ultramafic complexes of Ipanema seem to be correlative with this older rift magmatism (Angeli *et al.* 2004). The A-type, fluorite-bearing granite of Salto da Divisa has yielded a zircon U–Pb SHRIMP age of 875 ± 9 Ma (Silva *et al.* 2002*b*, 2007). The intrusive bodies of the Salto da Divisa region form a bimodal (gabbro–granite) anorogenic suite that was partially deformed and metamorphosed during the Brasiliano orogeny (Paixão & Perrella 2004; Sampaio *et al.* 2004). However, neither the Pedro Lessa mafic dykes nor the A-type Salto da Divisa suite show any direct field relations with the Neoproterozoic Macaúbas Group, occurring as intrusions in the pre-Neoproterozoic units (Fig. 5).

The proximal Macaúbas Group comprises key units for stratigraphic correlation between the Araçuaí orogen and the West Congo Belt. Although the regional metamorphism increases from greenschist facies close to the craton edge, to amphibolite facies in the distal Ribeirão da Folha Formation, the protoliths and many primary features of the Macaúbas rocks can still be recognized. Figure 6 summarizes the stratigraphy of the Neoproterozoic Macaúbas Group (after Karfunkel & Hoppe 1988; Pedrosa-Soares *et al.* 1992, 2001; Grossi-Sad *et al.* 1997; Noce *et al.* 1997; and Uhlein *et al.* 1999). A dolomite, known as the Domingas Formation and formerly interpreted as the oldest unit of the Macaúbas Group (Grossi-Sad *et al.* 1997; Noce *et al.* 1997), is now considered to be Mesoproterozoic in age (Santos *et al.* 2004). The Macaúbas Group ranges in thickness from some tens to many hundreds of metres in the proximal units, and up to 10 km thick in the passive margin sequence.

The oldest deposits of the Macaúbas Group consist of sandstone and arkose with conglomerate lenses of the Duas Barras Formation, and its lateral equivalent, a succession of sandstone and pelite (Rio Peixe Bravo Formation). These formations, which lack evidence of glaciation, represent fluvial sedimentation during the continental rift stage of the Macaúbas Basin (Noce *et al.* 1997; Martins-Neto *et al.* 2001; Martins 2006). U–Pb SHRIMP ages of detrital zircons constrain the maximum sedimentation age of the Duas Barras Formation as 900 ± 21 Ma (Babinski *et al.* unpublished data). The Duas Barras Formation is covered by the most proximal glaciogenic deposits of the Macaúbas Basin (Serra do Catuni Formation), involved in the Araçuaí orogen (Fig. 6).

The Serra do Catuni Formation consists of tillite with minor fluvial sandstone, varvite and esker deposits (i.e., glacio-terrestrial sediments), to the west, and proximal glacio-marine diamictite with lenses of sandstone and pelite, to the east, deposited during the rift stage of the Macaúbas Basin (Karfunkel & Hoppe 1988; Pedrosa-Soares *et al.* 1992; Grossi-Sad *et al.* 1997; Noce *et al.* 1997; Uhlein *et al.* 1998, 1999; Martins-Neto & Alkmim 2001; Martins-Neto *et al.* 2001; Martins-Neto & Hercos 2002; Martins 2006). This glaciogenic formation was thrust over the Neoproterozoic carbonate–pelite cover (Bambui Group) of the São Francisco Craton (Fig. 3). Therefore, the Serra do Catuni Formation might be older than 740 ± 22 Ma, the sedimentation age of the lowermost carbonate unit of the cratonic cover (Sete Lagoas cap carbonate, Pb–Pb whole-rock isochron, Babinski & Kaufman 2003). A thick glacio-marine unit, the Nova Aurora Formation, overlies the pre-glaciogenic Rio Peixe Bravo Formation and is a distal correlative of the Serra do Catuni Formation (Fig. 6). The Nova Aurora Formation comprises diamictite, thick layers of iron formation with few scattered clasts (dropstones), and minor graded sandstone and pelite (Pedrosa-Soares *et al.* 1992; Grossi-Sad *et al.* 1997; Noce *et al.* 1997; Pedrosa-Soares & Oliveira 1997; Uhlein *et al.* 1999).

The uppermost and distal diamictite-bearing unit of the Macaúbas Group overlies the Serra do Catuni and Nova Aurora formations, and is called Chapada Acauã Formation (Fig. 6). It records the late continental rift and passive margin stages of the Macaúbas Basin. The lower part consists of marine debris flows (diamictite and minor clast-supported conglomerate) with lenses of sandstone and pelite, followed by turbidites (graded sandstone and pelite with dropstones) intercalated with lenses of diamictite (Mourão & Pedrosa-Soares 1992; Pedrosa-Soares *et al.* 1992; Grossi-Sad *et al.* 1997; Noce *et al.* 1997; Pedrosa-Soares & Grossi-Sad 1997; Uhlein *et al.* 1999). Mafic volcanic rocks, metamorphosed to greenschists, occur in the lower portion of the Chapada Acauã Formation (Gradim *et al.* 2005). These greenschists show pillow structures and other features of sub-aquatic flows. They have tholeiitic basalt protoliths with a dominant within-plate signature. Inherited zircon grains from older felsic rocks yielded U–Pb SHRIMP ages from Archaean to Late Mesoproterozoic (the youngest at 1156 ± 21 Ma), with a Sm–Nd T_{DM} model age of *c.* 1.5 Ga (Babinski *et al.* 2005). However, samples with oceanic signature and slightly positive $\varepsilon Nd_{(900Ma)}$ of $+0.23$, together with the inherited zircon grains, suggest a transitional mafic magma that migrated through a thinned continental crust. Accordingly, these greenschists provide strong evidence of volcanism in an extensional marine basin floored by thin continental crust, during the very late rift stage of the Macaúbas Basin (Gradim *et al.* 2005). The oldest (and single?) carbonate of the proximal Macaúbas Group, a lens of limestone interbedded with diamictite and pelite, occurs at the top of the diamictite-bearing pile of the Chapada Acauã Formation (Pedrosa-Soares & Grossi-Sad 1997). The uppermost part of this formation is a post-glacial shelf succession of interlayered sandstones and pelites (Pedrosa-Soares *et al.* 1992; Noce *et al.* 1997; Martins-Neto *et al.* 2001). Detrital zircon grains from a sandstone layer mostly yielded U–Pb SHRIMP ages between 1000 Ma and 950 Ma, but the almost concordant youngest zircon grain constrains the maximum sedimentation age of the upper Chapada Acauã Formation at 864 ± 30 Ma (Pedrosa-Soares *et al.* 2000). The most probable sources for the 1000–900 Ma detrital zircons are the Early Tonian felsic magmatic rocks of the West Congo Belt, whereas a probable source for the *c.* 864 Ma zircons would be the anorogenic granites of Salto da Divisa (Fig. 5), implying erosion of the rift shoulders and/or internal horsts. The upper Chapada Acauã Formation represents the proximal passive margin sedimentation of the Macaúbas Basin.

The distal part of the Macaúbas Group is a passive margin sequence free of diamictite, called the Ribeirão da Folha Formation (Figs 5 and 6). This was formerly included in the Salinas Formation (Pedrosa-Soares *et al.* 1992, 2001*a*), but this name was recently redefined to only designate an orogenic metasedimentary succession (Lima *et al.* 2002; Santos 2007). The Ribeirão da Folha Formation is subdivided into two units (Rs and Ro, respectively, in Fig. 5). The proximal unit (Rs) is a succession of fine-grained turbidites with thin calc-silicate (marl) and limestone lenses. It overlies the uppermost diamictites and is a deep-water correlative of the shelf sandstone–pelite unit of the Chapada Acauã Formation (Pedrosa-Soares *et al.* 1992; Mourão & Pedrosa-Soares 1992; Noce *et al.* 1997; Pedrosa-Soares & Grossi-Sad 1997; Uhlein *et al.* 1999).

The distal unit (Ro) of the Ribeirão da Folha Formation is locally associated with thrust slices of oceanic meta-mafic and meta-ultramafic rocks, constituting a tectonically dismembered ophiolite complex (Pedrosa-Soares *et al.* 1992, 1998, 2001*a*; Pedrosa-Soares & Wiedemann-Leonardos 2000; Aracema *et al.* 2000; Suita *et al.* 2004; Queiroga *et al.* 2006). This unit (Ro) is rich in peraluminous micaschists with intercalations of sulphide-bearing cherts and diopsidites associated with massive sulphide bodies, banded iron formations of oxide, silicate and sulphide types, and graphite schists (Figs 5 and 6). From west to east, this unit was metamorphosed in the garnet, staurolite, kyanite and sillimanite zones, recording maximum temperature and pressure around 600 °C and 5.5 kbar (Pedrosa-Soares *et al.* 1996; Queiroga *et al.* 2006). Mineralogical and geochemical attributes suggest volcanic-exhalative protoliths with diverse contributions of argillaceous and carbonate fractions for the sulphide-bearing rocks and the silicate-type iron formation. The peraluminous micaschists and graphite schists represent pelagic pelites (Pedrosa-Soares 1995; Pedrosa-Soares *et al.* 1998; Queiroga 2006). Sm–Nd (T_{DM}) model ages of *c.* 2.2 Ga (Pedrosa-Soares *et al.* unpublished data) suggest a predominant Palaeoproterozoic source for the Ribeirão da Folha pelites.

The meta-mafic rocks associated with the Ribeirão da Folha Formation are coarse-grained massive ortho-amphibolites, representing gabbroic protoliths, and medium- to fine-grained banded ortho-amphibolites that seem to be derived from dolerites and basalts (Pedrosa-Soares *et al.* 1998; Suita *et al.* 2004). Detailed geochemical studies of the Ribeirão da Folha ortho-amphibolites, including Sm–Nd isotopic analysis, were published by Pedrosa-Soares *et al.* (1998). These ortho-amphibolites have a geochemical signature akin to modern ocean-floor mafic rocks and are comparable to similar rocks of other Neoproterozoic ophiolites (Pedrosa-Soares *et al.* 1998; Pedrosa-Soares & Wiedemann-Leonardos 2000; Pedrosa-Soares *et al.* 2001*a*; Suita *et al.* 2004). Together with $\varepsilon Nd_{(800\ Ma)}$ from +3.4 to +4.6, some ortho-amphibolite chemical attributes suggest a depleted mantle source that underwent limited metasomatism because of its prior subcontinental position, but this inheritance ceased on the spreading of the basin, giving place to normal-MORB melts (Pedrosa-Soares 1995; Pedrosa-Soares *et al.* 1998). The recent discovery of plagiogranite veins in some ortho-amphibolite outcrops is a further evidence of oceanic crust (Suita *et al.* 2004; Queiroga ongoing PhD thesis). Considering the associated rock assemblages and geochemical attributes, the Ribeirão da Folha oceanic mafic rocks clearly differ from the Chapada Acauã rift-related basalts (Gradim *et al.* 2005).

Meta-ultramafic rocks constitute tectonic slabs as thick as 1 km that were transported by thrusts over the distal Ribeirão da Folha Formation (Fig. 3). Mylonitic foliation, stretched fragments of ultramafic composition, and hydrothermal alteration characterize the ductile shear zones that outline the contacts between these slabs and the country rocks (Carvalho *et al.* 1992; Pedrosa-Soares *et al.* 1992; Aracema *et al.* 2000). The most common meta-ultramafic rocks are tremolite schists, talc-anthophyllite schists and serpentinites with chromite-bearing meta-peridotite pods. Geochemical data suggest ophiolitic pyroxenite to peridotite protoliths and are comparable to similar rocks in other Neoproterozoic ophiolites (Carvalho *et al.* 1992; Aracema *et al.* 2000). Chemical data from chromite crystals, including their low contents of TiO_2 (<0.3%), also point to an ophiolitic origin (Pedrosa-Soares *et al.* 1998).

Well-constrained ages of the ophiolite slivers are critical to understanding the evolution timing of the Macaúbas Basin. A lack of zircon grains (in almost one ton of crushed samples) and the extremely low uranium contents of the titanite crystals have hindered attempts to obtain U–Pb geochronological data for the magmatic crystallization and metamorphism of the ortho-amphibolites; nevertheless, both factors are indirect evidence of oceanic origin. A Sm–Nd whole-rock isochron age of 816 ± 72 Ma suggests the possible antiquity of the ortho-amphibolite protoliths (Pedrosa-Soares *et al.* 1998). Some meta-ultramafic rock samples plot close to the ortho-amphibolite Sm–Nd isochron, also suggesting an age of around 800 Ma (Pedrosa-Soares *et al.* 1992). Sm–Nd T_{DM} model ages from 16 samples of meta-ultramafic and meta-mafic rocks cluster around 1 Ga, limiting the maximum age of the ophiolite slivers to Neoproterozoic (Pedrosa-Soares *et al.* unpublished data). Indeed, the oceanic lithosphere in the Macaúbas Basin must be younger than 875 ± 9 Ma (the age of Salto da Divisa rift-related granites). The Serra do Catuni Formation (the oldest diamictite unit of the Macaúbas Group) was thrust over the cratonic carbonate platform (Figs 3, 5 and 6), so that the oldest age of the cratonic carbonate platform constrains the minimum age of the ophiolite slivers to around *c.* 740 Ma. The inferred time span from *c.* 875 Ma to *c.* 740 Ma is reasonable for basin evolution from continental rift to ocean floor.

The distal Ribeirão da Folha Formation (Ro) correlates with the Dom Silvério Group, an oceanic rock assemblage made up of metamorphosed pelagic pelite, banded iron formations, cherts, manganese-rich sedimentary rocks, sandstone, and associated mafic and ultramafic rocks (Cunningham *et al.* 1998; Pedrosa-Soares *et al.* 2001*a*). The Dom Silvério Group also records an intermediate-pressure metamorphic regime with maximum *P–T* conditions around 550 °C and 5 kbar (Peres *et al.* 2004 and references therein). Sm–Nd model ages (*c.* 2.2 Ga) of the Dom Silvério pelites are similar to those of the Ribeirão da Folha Formation (Brueckner *et al.* 2000).

The unit here referred to as the northern rock-assemblage consists mainly of metamorphosed pelites and sandstones (Fig. 5). The pelitic schists record the southward increase in metamorphic grade from biotite to sillimanite zones, typical of the northern part of the Araçuaí orogen (Almeida *et al.* 1978). Formerly, this metasedimentary rock-assemblage was correlated with the Palaeoproterozoic–Mesoproterozoic Espinhaço Supergroup (Barbosa & Dominguez 1996 and references therein) or with the Neoproterozoic Macaúbas Group (Almeida *et al.* 1978; Silva *et al.* 1987; Pedrosa-Soares *et al.* 1992, 2001*a*). After 1:100.000 scale geological mapping, Sampaio *et al.* (2004) also correlated the northern metasedimentary rock-assemblage with the Macaúbas Group. Based on regional field sections, we suggest that the northern rock-assemblage could also include orogenic deposits belonging to the Salinas Formation (see Orogenic Stages' below).

The Jequitinhonha Complex comprises a thick paragneiss association, consisting of migmatized biotite gneisses, with variable contents of garnet, cordierite, sillimanite and graphite, thick intercalations of graphite-rich gneiss, and minor quartzite and calc-silicate rocks (Fig. 5). The paragneiss protoliths are interpreted as marine arkosic greywackes to peraluminous pelites, deposited under oxidizing conditions, with horizons of carbonaceous black mud, that were deposited in a restricted marine environment (Pedrosa-Soares & Wiedemann-Leonardos 2000; Pedrosa-Soares *et al.* 2001*a*).

These authors interpreted the Jequitinhonha Complex as a passive margin succession deposited in a northern gulf of the Macaúbas Basin, during the marine transgression coeval with oceanic opening of this basin. Samples from the Jequitinhonha Complex yielded Sm–Nd T_{DM} model ages ranging from 1.6 Ga to 1.8 Ga (Celino *et al.* 2000; Daconti 2004), suggesting mixed material supplied from Palaeoproterozoic and younger sources. The younger sedimentary sources could be magmatic rocks related to the Tonian Macaúbas–West Congo rift, but could also include sediments derived from the *c.* 630–585 magmatic arc of the Araçuaí orogen. If the younger sources were only rocks of this magmatic arc, the Jequitinhonha Complex would be an orogenic succession correlated with the Nova Venécia Complex (see Orogenic Stages' below), but the Jequitinhonha Complex probably includes both passive margin and orogenic deposits (Sampaio *et al.* 2004).

The undivided paragneiss complex located in the SE region of the Araçuaí orogen consists mainly of peraluminous and semipelitic gneisses with thick intercalations of marble and minor quartzite, ortho-amphibolite and probable felsic volcanic rocks (Fig. 5). This complex has been interpreted as a passive margin sequence, but it could include arc-related deposits (Wiedemann *et al.* 1997; Pedrosa-Soares & Wiedemann-Leonardos 2000; Campos *et al.* 2004; Heilbron *et al.* 2004, and references therein).

Orogenic stages

The Brasiliano event in the Araçuaí orogen has been subdivided into four stages: pre-collisional (*c.* 630–585 Ma), syn-collisional (*c.* 585–560 Ma), late collisional (*c.* 560–530 Ma) and post-collisional (*c.* 530–490 Ma) (Pedrosa-Soares & Wiedemann-Leonardos 2000; Pedrosa-Soares *et al.* 2001*a*, 2005; Heilbron *et al.* 2004; Silva *et al.* 2005).

The pre-collisional stage is mainly represented by the I-type G1 plutonic suite (Figs 3 and 5). This suite represents the roots of a magmatic arc formed in a continental active margin setting (Nalini-Junior *et al.* 2000*a*, 2005; Pedrosa-Soares & Wiedemann-Leonardos 2000; Pedrosa-Soares *et al.* 2001*a*). It consists of batholiths and stocks composed of tonalite, granodiorite and minor diorite, with dioritic to mafic enclaves, deformed during the Brasiliano orogeny. Chemical compositions of more than 100 samples from several G1 plutons outline a consistent metaluminous, calc-alkaline, volcanic arc signature (Nalini-Junior *et al.* 2000*a*, 2005; Pedrosa-Soares & Wiedemann 2000; Pedrosa-Soares *et al.* 2001*a*; Campos *et al.* 2004; Martins *et al.* 2004, and references therein). Rb–Sr and Sm–Nd isotope data (εNd values of −5 to −13 and T_{DM} model ages of 1.2 to 2 Ga) suggest significant melting of a deep Palaeoproterozoic continental crust, with minor involvement of mantle components related to subduction of Neoproterozoic oceanic lithosphere (Nalini-Junior *et al.* 2000*a*, 2005; Campos *et al.* 2004; Martins *et al.* 2004). U–Pb zircon and monazite ages constrain the evolution of the G1 suite to between *c.* 630 and *c.* 585 Ma (Söllner *et al.* 1991; Nalini-Junior *et al.* 2000*a*, 2005; Noce *et al.* 2000; Whittington *et al.* 2001; Campos *et al.* 2004; Silva *et al.* 2002*b*, 2005). The location of this magmatic arc relative to the zone with oceanic slivers suggests subduction to the east and a suture zone roughly located along the 42° W meridian, south of latitude 17° S (Figs 3 and 5).

The Nova Venécia Complex represents sedimentation of pelite-rich deposits in the back-arc basin of the Araçuaí orogen (Figs 3, 5 and 6). This complex comprises migmatized, high-grade paragneisses rich in biotite, garnet, cordierite and/or sillimanite, with lenses of calc-silicate rocks (Silva *et al.* 1987; Pedrosa-Soares *et al.* 2006). U–Pb SHRIMP ages of *c.* 630–590 Ma for detrital zircons from a paragneiss sample collected to the east of the city of Nova Venécia constrain the maximum sedimentation age at 608 ± 18 Ma (revised age, data from Noce *et al.* 2004), and the most probable source is the magmatic arc of the Araçuaí orogen (Noce *et al.* 2004). The high-grade metamorphism and an episode of partial melting are syn-kinematic with the regional foliation and took place at around 580–570 Ma, but the complex was also involved in late orogenic episodes of granite genesis (Munhá *et al.* 2005; Pedrosa-Soares *et al.* 2006; Jacobsohn 2007).

The Capelinha and Salinas formations are interpreted as orogenic sedimentary deposits (Pedrosa-Soares *et al.* 2001*a*; Lima *et al.* 2002; Santos 2007). Both formations unconformably overlie the Ribeirão da Folha Formation but also show tectonic contacts with it (Figs 3, 5 and 6). They also record regional metamorphism from greenschist to amphibolite facies (Costa 1989; Pedrosa-Soares *et al.* 1992, 1996; Grossi-Sad *et al.* 1997; Queiroga 2006; Queiroga *et al.* 2006). The Capelinha Formation consists of sandstone, pelite and detrital iron formation, deposited in probably fluvial to marine environments (Grossi-Sad *et al.* 1997; Noce *et al.* 1997; Queiroga 2006). The Salinas Formation formerly included the occurrence area of the Ribeirão da Folha Formation (Pedrosa-Soares *et al.* 2001*a*), but Lima *et al.* (2002) have restricted the name to a metamorphosed greywacke–pelite succession with intercalations of clast-supported conglomerate locally rich in clasts of volcanic rocks, and calc-silicate (meta-marl) lenses (Fig. 5). Well-preserved sedimentary features (such as convolute seismites and intra-formational brecciated beds) suggest marine sedimentation in a tectonically active environment (Lima *et al.* 2002; Santos 2007). U–Pb SHRIMP data for detrital zircons from greywacke samples suggest a maximum sedimentation age of 588 $\pm$ 24 Ma for the Salinas Formation (revised age after data from Lima *et al.* 2002). Zircons from cobbles of felsic volcanic rocks of the Salinas conglomerates yielded U–Pb SHRIMP magmatic ages ranging from *c.* 600 Ma to *c.* 630 Ma (Pedrosa-Soares *et al.* unpublished data). Bearing in mind the regional scenario, the only possible source for these cobbles of volcanic rocks, as well as for the *c.* 588 Ma detrital zircons, is the magmatic arc of the Araçuaí orogen.

The fore-arc and arc rock-assemblage comprises metasedimentary and metavolcanic–sedimentary units, strongly sheared mafic and granitoid bodies, slices of basement units, and intrusions of distinct ages (Figs 3 and 5). It might include Neoproterozoic orogenic units, as well as passive margin deposits and ophiolite slivers. The northern sector of this assemblage depicts a bivergent zone with west-verging low- to medium-angle thrusts in its western part, and east-verging thrusts on the eastern side. The southern sector is dominated by transcurrent zones and west-verging thrusts. The Brasiliano regional metamorphism varies from amphibolite to granulite facies and does not show the regular west-to-east increase in grade expected for this sector of the Araçuaí orogen. The northern-central part of the fore-arc and arc rock-assemblage, dominated by micaschists, meta-greywackes and quartzites of the Rio Doce Group, is at lower metamorphic grade than the western and eastern gneissic units. The tonalitic to granitic mylonite-gneisses of Governador Valadares have yielded zircon U–Pb SHRIMP ages around 560 Ma (Silva *et al.* 2002, 2005) and an ortho-amphibolite located west of Manhuaçu has a Sm–Nd T_{DM} age of *c.* 890 Ma (Fischel *et al.* 1998). The plagioclase-rich sillimanite-garnet paragneisses of the Andrelândia Group (Noce *et al.* 2003, 2006), found in the western and southern zones of the fore-arc and arc rock-assemblage, have Sm–Nd T_{DM} ages of around 1.3 Ga (Fischel *et al.* 1998). Meta-greywackes of the Rio Doce Group, located in the northeastern part of the assemblage, yielded a homogeneous detrital zircon population with U–Pb SHRIMP ages of around 2.1 Ga (Vieira 2005). The latter author also points to the occurrence of dacitic volcanic rocks associated with the Rio Doce Group.

The main deformation and regional metamorphism related to the Brasiliano orogeny, as well as the generation of most S-type granites, took place in the syn-collisional stage (*c.* 585–560 Ma). The syn-collisional G2 suite mainly consists of foliated, S-type, garnet $\pm$ cordierite $\pm$ sillimanite granites and two-mica granites, with common mylonitic features. Xenoliths and roof pendants of metasedimentary rocks in several stages of assimilation and of variable size are very common (Nalini-Junior *et al.* 2000*b*; Celino *et al.* 2000; Pedrosa-Soares & Wiedemann-Leonardos 2000; Pedrosa-Soares *et al.* 2000, 2006; Campos *et al.* 2004; Martins *et al.* 2004). The available U–Pb ages suggest that the main episode of G2 granite generation took place at around 575–560 Ma (Söllner *et al.* 1991; Silva *et al.* 2002*b*, 2005; Campos *et al.* 2004, Pedrosa-Soares *et al.* 2006; Jacobsohn 2007). Large batholiths dominated by mylonitic garnet–biotite granites can show parts with well-preserved magmatic features, both with similar magmatic crystallization ages (Pedrosa-Soares *et al.* 2006; Vauchez *et al.* 2007).

The late collisional stage apparently lasted from *c.* 560 to *c.* 530 Ma, but these age limits are poorly constrained by geochronological data. The S-type G3 Suite is late to post-collisional. It consists of garnet- and/or cordierite-rich leucogranite, which generally occur as veins and small intrusions cutting G2 granites. Zircon and monazite U–Pb data from G3 leucogranite samples have yielded ages from *c.* 550 Ma to 520 Ma (Whittington *et al.* 2001; Campos *et al.* 2004; Silva *et al.* 2005; Pedrosa-Soares *et al.* 2006; Jacobsohn 2007).

The post-collisional stage (*c.* 530–490 Ma) is related to the gravitational collapse of the Araçuaí orogen (Pedrosa-Soares & Wiedemann-Leonardos 2000; Pedrosa-Soares *et al.* 2001*a*). A major shear zone, trains of east-verging folds and antithetic west-dipping normal faults are the main structures related to the extensional collapse (Marshak *et al.* 2006). G4 and G5 are plutonic suites associated with this stage (Fig. 5), both lack the regional foliation. The G4 suite includes relatively shallow (5 to 15 km) S-type granitic intrusions with variable contents of biotite, muscovite and/or garnet. They have circular surface outcrops and sometimes form clusters of amalgamated plutons with locally preserved pegmatoidal roofs (Pedrosa-Soares *et al.* 1987; Grossi-Sad *et al.* 1997; Whittington *et al.* 2001). The G5 suite predominantly consists of I-type granitic intrusions, which may include charnockitic, enderbitic and/or mafic portions, and minor mafic bodies with subordinate granitic and/or charnockitic facies. Magma mingling and mixing features are very common. The granitic rocks are generally porphyritic to sub-porphyritic and have a high-K calc-alkaline signature (Wiedemann *et al.* 2002; Campos *et al.* 2004; Martins *et al.* 2004). The most important gem-rich pegmatites of the Eastern Brazilian Gem Province are related to the G4 and G5 suites (Pedrosa-Soares *et al.* 2001*b*).

Correlations and final remarks

The Brazilian and African counterparts of the Araçuaí–West Congo orogen are different, but were complementary in a palaeogeographical fit at the time just prior to spreading of the South Atlantic Ocean (Fig. 7). After the Cretaceous, the Brazilian counterpart, or Araçuaí orogen, inherited the western fold–thrust belt, all the ophiolite slivers, the suture zone, the entire magmatic arc and all syn-collisional to post-collisional igneous rocks, but few of the rift-related magmatic rocks. The African counterpart, or West Congo Belt, on the other hand, preserved the eastern fold–thrust belt with a thick pile of bimodal volcanic rocks and anorogenic granites that record the Early Tonian rift magmatism, but lacked ophiolite and orogenic magmatism. Both counterparts involve Neoproterozoic glaciogenic rocks.

The Araçuaí–West Congo orogen is an example of a confined orogen; because it developed inside an embayment carved into the São Francisco–Congo palaeocontinent, its precursor basin was not completely ensialic, and the resulting orogen contains ophiolite slivers and a pre-collisional magmatic arc (Pedrosa-Soares *et al.* 2001*a*, 2004; Rogers & Santosh 2004). The Araçuaí–West Congo orogen thus represents the closure of an inland-sea basin partly floored by oceanic crust, like a large gulf or a Red Sea-type basin (Pedrosa-Soares *et al.* 1992, 2001*a*, 2004; Alkmim *et al.* 2006). Therefore, the Brazilian and African counterparts of the precursor basin must have developed in close proximity to each other, allowing stratigraphic correlations based on lithological assemblages and related depositional environments (Figs 7 and 8).

The West Congo Belt encompasses the entire pile of the Early Tonian rift stage, represented by the Zadinian and Mayumbian groups, which have no correlatives on the Brazilian side (Fig. 8). This suggests an asymmetrical rift with the thermal–magmatic axis in the West Congo Belt (Fig. 9). The rift-related Salto da Divisa plutonic suite (875 ± 9 Ma) is younger than the upper Mayumbian rhyolites (912 ± 7 Ma), suggesting a westward migration of the rift thermal axis from *c.* 912 Ma to *c.* 875 Ma.

The fluvial units (Duas Barras and Rio Peixe Bravo formations) of the Macaúbas Group can be correlated with similar deposits of the Sansikwa Subgroup, representing rift-related pre-glacial sedimentation (Fig. 8). In the Macaúbas Basin, glaciogenic sedimentation started with the rift-related diamictite deposits of the Serra do Catuni Formation. Ages of mafic dykes and anorogenic granites suggest that this rift stage took place at around 906–875 Ma. Therefore, the oldest Neoproterozoic glaciogenic sedimentation involved in the Araçuaí orogen might be older than the Sturtian glacial event (*c.* 750 Ma). The Jequitaí Formation, a glacio-terrestrial to glacio-marine cover of the southern São Francisco Craton, has been interpreted as a correlative of the Serra do Catuni Formation (Karfunkel & Hoppe 1988; Buchwaldt *et al.* 1999; Uhlein *et al.* 1999; Karfunkel *et al.* 2002; Martins-Neto & Hercos 2002).

The lower Chapada Acauã Formation of the Araçuaí orogen can be correlated with the Lower Mixtite Formation of the West Congo Belt (Fig. 8). Both glaciogenic piles include submarine pillow basalts with a geochemical signature akin to that of mafic volcanism in an extensional basin floored by thinned continental crust, suggesting a very late rift stage. These units might be correlatives of the Carrancas diamictite, a glaciogenic unit made up of thin (a few metres) and discontinuous lenses covered by the carbonate–pelite sequence of the Bambui Group (Romeiro-Silva *et al.* 1998; Martins-Neto & Alkmim 2001; Martins-Neto 2005; Coelho 2007). This diamictite must be older than 740 ± 22 Ma, the age of the lowermost cap carbonate of the Bambui Group (Babinski & Kaufman 2003). Therefore, the Carrancas diamictite, as well as the glaciogenic lower Chapada Acauã and Lower Mixtite formations, might represent the Sturtian glacial event. The Haut Shiloango Subgroup and the upper Chapada Acauã Formation represent proximal passive margin deposits associated with the post-glacial marine transgression. The distal passive margin and ocean floor deposits (Ribeirão da Folha Formation), as well as the associated ophiolite slivers, are restricted to the Brazilian counterpart of the Araçuaí–West Congo orogen.

The Upper Mixtite Formation, which has been correlated with the Marinoan glaciation (*c.* 636 Ma), seems to lack correlatives in the Araçuaí orogen and southern São Francisco Craton. The Schisto-Calcaire Subgroup has been interpreted as a post-Marinoan carbonate platform. This subgroup might be interpreted as a foreland basin contemporaneous with the Ediacaran (Nova Venécia, Salinas, Capelinha) orogenic sedimentation found in the Araçuaí orogen.

Accurate U–Pb ages indicate that a time interval of *c.* 510 Ma elapsed from the earliest rifting of the precursor basin (*c.* 1 Ga) to the latest post-collisional (490 Ma) processes in the Araçuaí–West Congo orogen. Such a long time period would not be expected for a narrow basin partially floored by oceanic lithosphere to become an orogen. Bearing in mind the confined nature of the Araçuaí–West Congo orogen, this long-lasting history might be explained by the combination of two assumptions: (i) the precursor basin of the Araçuaí–West Congo orogen was a stagnant basin

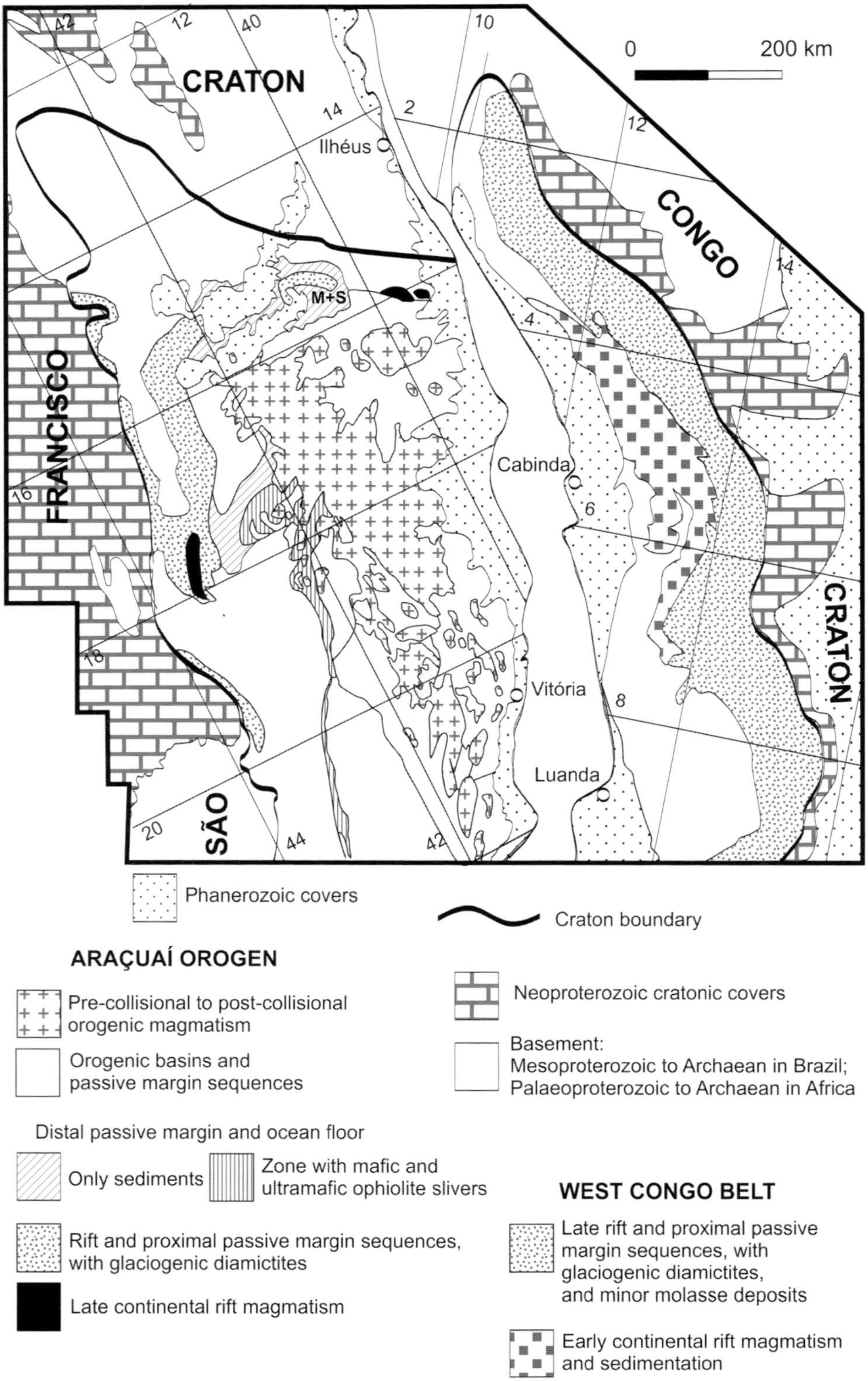

Fig. 7. Correlation sketch map for the Araçuaí–West Congo orogen (M + S, see text).

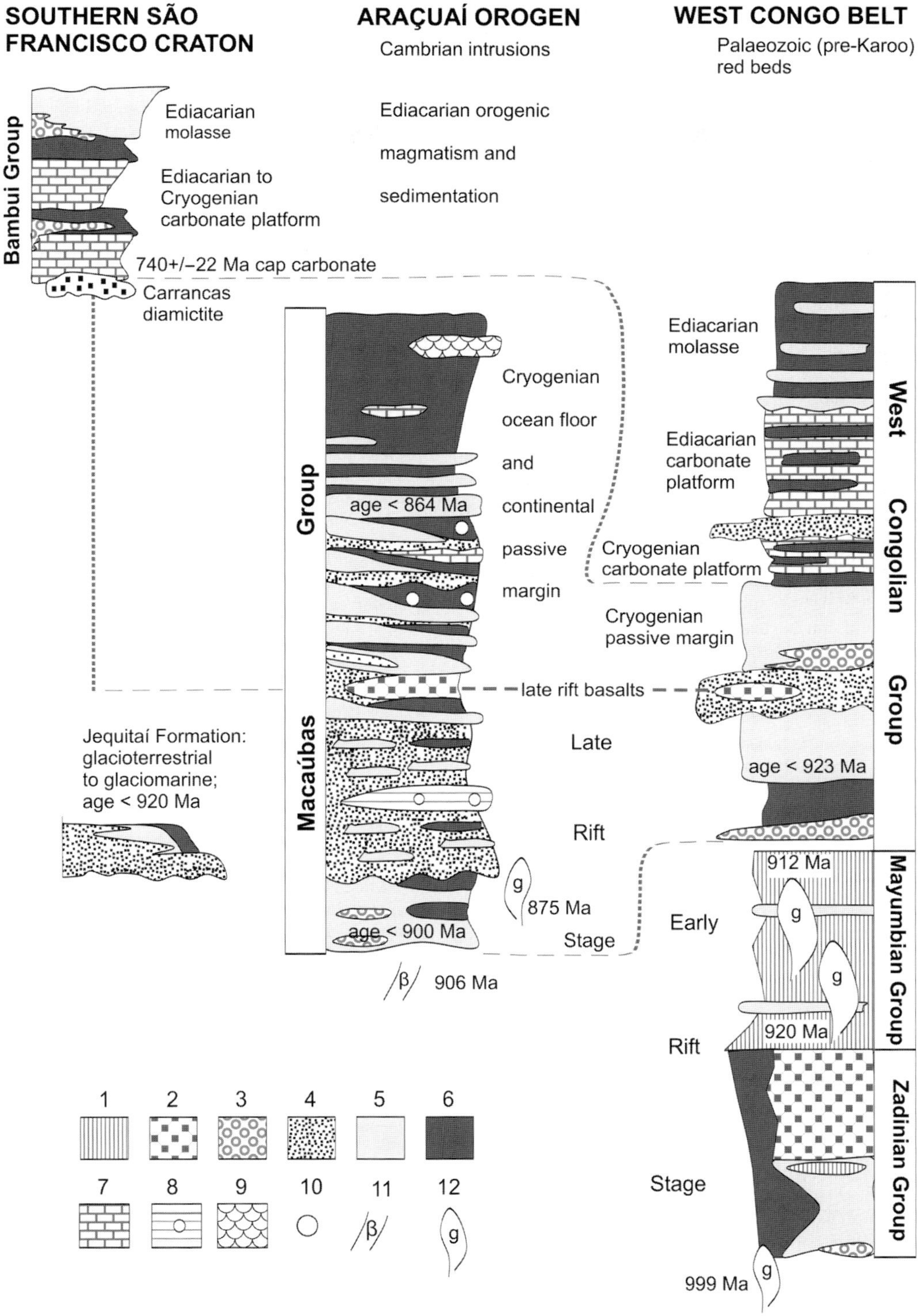

Fig. 8. Correlation stratigraphic columns for the Araçuaí–West Congo orogen (age references cited in text). 1, rift-related felsic volcanic rocks; 2, rift-related mafic volcanic rocks; 3, conglomerates; 4, diamictites; 5, sandstones; 6, pelites; 7, carbonate rocks; 8, glaciogenic iron formation; 9, volcanic-exhalative rocks; 10, dropstones; 11, mafic dykes; 12, rift-related anorogenic granites.

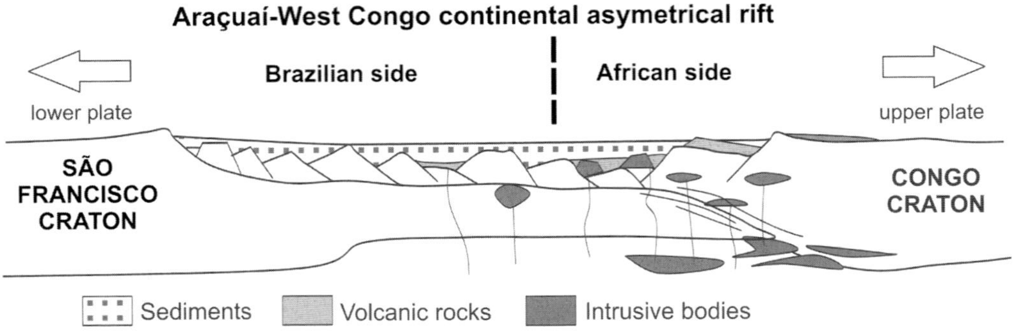

Fig. 9. Cartoon illustrating a section of the Tonian, asymmetric continental rift precursor to the Araçuaí–West Congo orogen, before Cryogenian oceanic opening.

owing to its confinement in a cratonic embayment; (ii) basin closure was triggered by distant collisions that catalyzed subduction of the narrow oceanic lithosphere. The Araçuaí–West Congo orogen also inspired the reasoning for the mechanism of orogenic closure called 'nutcracker tectonics' (Alkmim *et al.* 2006). The concept of a confined orogen can explain orogenic zones succeeding restricted basins such as gulfs and other inland seas, and can be useful for understanding the tectonics, palaeogeography, climate and life evolution in regions not completely separated by oceans and/or not isolated by extensive mountain belts.

The authors acknowledge financial support provided by CNPq (Conselho Nacional de Desenvolvimento Científico e Tecnológico), FINEP (Financiadora de Estudos e Projetos), FAPEMIG (Fundação de Amparo à Pesquisa de Minas Gerais), FAPESP (Fundação de Amparo à Pesquisa do Estado de São Paulo). We also thank two anonymous reviewers for corrections and suggestions.

References

ALKMIM, F. F., MARSHAK, S. & FONSECA, M. A. 2001. Assembling West Gondwana in the Neoproterozoic: Clues from the São Francisco craton region, Brazil. *Geology*, **29**, 319–322.

ALKMIM, F. F., MARSHAK, S., PEDROSA-SOARES, A. C., PERES, G. G., CRUZ, S. C. & WHITTINGTON, A. 2006. Kinematic evolution of the Araçuaí–West Congo orogen in Brazil and Africa: Nutcracker tectonics during the Neoproterozoic assembly of Gondwana. *Precambrian Research*, **149**, 43–63.

ALMEIDA, F. F. M., HASUI, Y., RODRIGUES, E. P. & YAMAMOTO, J. 1978. A Faixa de Dobramentos Araçuaí na região do Rio Pardo. *In*: *Congresso Brasileiro de Geologia 30, Recife, Anais*, **1**. Sociedade Brasiliera de Geologia, 270–283.

ANGELI, N., TEIXEIRA, W., HEAMAN, L., FLEET, M. E., MOORE, M. & SATO, K. 2004. Geochronology of the Ipanema Layered Mafic-Ultramafic Complex, Minas Gerais, Brazil: Evidence of extension at the Meso-Neoproterozoic time boundary. *International Geology Review*, **46**, 730–744.

ARACEMA, L. W., NEVES, A. C., FERREIRA, J. C., PEDROSA-SOARES, A. C., LOBATO, L. M. & NOCE, C. M. 2000. Novas evidências de remanescentes oceânicos na Faixa Araçuaí: As rochas metaultramáficas de São José da Safira. *Geonomos*, **8**, 55–61.

ALVAREZ, P. 1995. Evidence for a Neoproterozoic carbonate ramp on the northern edge of the Central African craton: Relations with Late Proterozoic intracratonic troughs. *Geologische Rundschau*, **84**, 636–648.

ALVAREZ, P. & MAURIN, J. C. 1991. Evolution sédimenataire et tectonique du basin protérozoïque supérieur de Comba (Congo): Stratigraphie séquentielle du Supergroupe Ouest-Congolien et modélle d'amortissement sur déchrochements dans le contexte de tectogénèse panafricaine. *Precambrian Research*, **50**, 137–171.

BABINSKI, M. & KAUFMAN, A. J. 2003. First direct dating of a Neoproterozoic post-glacial cap carbonate. *In*: *IV South American Symposium on Isotope Geology, Salvador, Brazil, Short Papers*, 321–323.

BABINSKI, M., GRADIM, R. J., PEDROSA-SOARES, A. C., ALKMIM, F. F., NOCE, C. M. & LIU, D. 2005. Geocronologia U–Pb (SHRIMP) e Sm–Nd de xistos verdes basálticos do Orógeno Araçuaí: Implicações para a idade do Grupo Macaúbas. *Revista Brasileira de Geociências*, **35** (4-supplement), 77–81.

BARBOSA, J. S. F. & DOMINGUEZ, J. M. L. 1996. *Texto Explicativo do Mapa Geológico da Bahia escala 1:1.000.000.* Companhia Bahiana de Pesquisa Mineral (SGM-CBPM), Salvador, Brazil.

BARBOSA, J. S. F. & SABATÉ, P. 2004. Archean and Paleoproterozoic crust of the São Francisco craton, Bahia, Brazil: Geodynamic features. *Precambrian Research*, **133**, 1–27.

BUCHWALDT, R., TOULKERIDIS, T. *ET AL.* 1999. Age determination and age-related provenance analysis of the Proterozoic glaciation event in central-eastern Brazil. *In*: *II South American Symposium on Isotope Geology, Villa Carlos Paz, Argentina, Actas*, 387–390.

BRITO-NEVES, B. B., CAMPOS-NETO, M. C. & FUCK, R. 1999. From Rodinia to Western Gondwana: An approach to the Brasiliano-Pan African cycle and orogenic collage. *Episodes*, **22**, 155–199.

BRUECKNER, H. K., CUNNINGHAM, D., ALKMIM, F. F. & MARSHAK, S. 2000. Tectonic implications of Precambrian Sm–Nd dates from the southern São Francisco craton and adjacent Araçuaí and Ribeira belts, Brazil. *Precambrian Research*, **99**, 255–269.

CAHEN, L., SNELLING, N. J., DELHAL, J. & VAIL, J. R. 1984. *The Geochronology of Africa*. Clarendon Press, Oxford.

CAMPOS, C. M., MENDES, J. C., LUDKA, I. P., MEDEIROS, S. R., MOURA, J. C. & WALLFASS, C. 2004. A review of the Brasiliano magmatism in southern Espírito Santo, Brazil, with emphasis on postcollisional magmatism. *Journal of the Virtual Explorer*, **17**, http://virtualexplorer.com.au/journal/2004/17/campos.

CARVALHO, C. T., LOBATO, L. M. & PEDROSA-SOARES, A. C. 1992. O corpo ultramáfico do Ribeirão da Folha, MG: Aspectos petrográficos e geoquímica da alteração hidrotermal. *Revista Escola de Minas de Ouro Preto*, **45**, 170–172.

CELINO, J. J., BOTELHO, N. F. & PIMENTEL, M. M. 2000. Genesis of Neoproterozoic Granitoid Magmatism in the Eastern Araçuaí Fold Belt, Eastern Brazil: Field, Geochemical and Sr–Nd Isotopic Evidence. *Revista Brasileira de Geociências*, **30**, 135–139.

COELHO, J. C. C. 2007. *Transecto Geológico na Bacia do São Francisco, Minas Gerais*. MSc thesis, Departamento de Geologia, Universidade Federal de Ouro Preto, Brazil.

CORRÊA-GOMES, L. C. & OLIVEIRA, E. P. 2000. Radiating 1.0 Ga mafic dyke swarms of Eastern Brazil and Western Africa: Evidence for post-assembling extension in the Rodinia supercontinent? *Gondwana Research*, **3**, 325–332.

COSTA, A. G. 1989. Evolução petrológica para uma seqüência de rochas metamórficas regionais do tipo baixa pressão na região de Itinga, NE de Minas Gerais. *Revista Brasileira de Geociências*, **19**, 440–448.

CUNNINGHAM, D., ALKMIM, F. F. & MARSHAK, S. 1998. A structural transect across the coastal mobile belt in the Brazilian Highlands (latitude 20°S): The roots of a Precambrian transpressional orogen. *Precambrian Research*, **92**, 251–275.

D'AGRELLA FILHO, M. S., PACCA, I. I. G., TRINDADE, R. I. F., TEIXEIRA, W., RAPOSO, M. I. B. & ONSTOTT, T. C. 2004. Paleomagnetism and $^{40}Ar^{39}Ar$ ages of mafic dykes from Salvador (Brazil): new constraints on the São Francisco craton APW path between 1080 and 1010 Ma. *Precambrian Research*, **132**, 55–77.

DACONTI, B. C. 2004. *Contexto geológico, controle e correlação regional das mineralizações de grafita da região de Almenara, Província Grafítica do Nordeste de Minas Gerais*. MSc thesis, Instituto de Geociências, Universidade Federal de Minas Gerais, Belo Horizonte.

DE PAEPE, P., HERTOGEN, J. & TACK, L. 1975. Mise en évidence de laves en coussins dans les faciès volcaniques basiques du massif de Kimbungu (Bas-Zaïre) et implications pour le magmatisme ouest-congolien. *Annales Société Géologique Belgique*, **98**, 251–270.

FEYBESSE, J. L., JOHAN, V. *ET AL*. 1998. The West Central African Belt: A model of 2.5–2.0 Ga accretion and two-phase orogenic evolution. *Precambrian Research*, **87**, 161–216.

FISCHEL, D. P., PIMENTEL, M. M., FUCK, R. A., COSTA, A. G. & ROSIÈRE, C. A. 1998. Geology and Sm–Nd isotopic data for the Mantiqueira and Juiz de Fora Complexes (Ribeira Belt) in the Abre Campo-Manhuaçu region, Minas Gerais, Brazil. *In*: *14th International Conference on Basement Tectonics, Ouro Preto, Brazil, Abstracts*, 21–23.

FRIMMEL, H. E., TACK, L., BASEI, M., NUTMAN, A. & BOVEN, A. 2006. Provenance and chemostratigraphy of the Neoproterozoic West Congolian Group in the Democratic Republic of Congo. *Journal of African Earth Sciences*, **46**, 221–239.

GRADIM, R. J., ALKMIM, F. F., PEDROSA-SOARES, A. C., BABINSKI, M. & NOCE, C. M. 2005. Xistos Verdes do Alto Araçuaí, Minas Gerais: Vulcanismo Básico do Rifte Neoproterozóico Macaúbas. *Revista Brasileira de Geociências*, **35** (4-supplement), 59–69.

GROSSI-SAD, J. H., LOBATO, L. M., PEDROSA-SOARES, A. C. & SOARES-FILHO, B. S. 1997. *Projeto Espinhaço em CD-ROM*. CODEMIG, Belo Horizonte, Brazil.

HEILBRON, M., PEDROSA-SOARES, A. C., CAMPOS-NETO, M. C., SILVA, L. C., TROUW, R. & JANASI, V. A. 2004. Brasiliano orogens in Southeast and South Brazil. *Journal of the Virtual Explorer*, **17**, http://virtualexplorer.com.au/journal/2004/17/heilbron

JACOBSOHN, T. 2006. *Comportamento dos sistemas isotópicos em processos tectono-metamórficos de médio e alto grau: Faixa Araçuaí, MG-ES*. PhD thesis, Instituto de Geociências, Universidade de São Paulo.

KARFUNKEL, J. & HOPPE, A. 1988. Late Precambrian glaciation in central-eastern Brazil: Synthesis and model. *Palaeogeography Palaeoclimatology Palaeoecology*, **65**, 1–21.

KARFUNKEL, J., HOPPE, A. & NOCE, C. M. 2002. Serra da Água Fria e vizinhancas, MG: vestígios de glaciação neoproterozóica. *In*: SCHOBBENHAUS, C., CAMPOS, D. A., QUEIROZ, E. T., WINGE, M. & BERBERT-BORN, M. (eds) *Sítios Geológicos e Paleontológicos do Brasil*, **1**. Departamento Nacional da Produção Mineral-DNPM, Brasília, Brazil, 165–173.

KAMPUNZU, A. B., KAPENDA, D. & MANTEKA, B. 1991. Basic magmatism and geotectonic evolution of the Pan African belt in central Africa: Evidence from the Katangan and West Congolian segments. *Tectonophysics*, **190**, 363–371.

LEDRU, P. J., JOHAN, V., MILÉSI, J. P. & TEGUY, M. 1994. Markers of the last stage of the Paleoproterozoic collision: Evidence for a 2 Ga continent involving circum-South Atlantic provinces. *Precambrian Research*, **69**, 169–191.

LIMA, S. A. A., MARTINS-NETO, M. A., PEDROSA-SOARES, A. C., CORDANI, U. G. & NUTMAN, A. 2002. A Formação Salinas na área-tipo, NE de Minas Gerais: Uma proposta de revisão da estratigrafia da Faixa Araçuaí com base em evidências sedimentares, metamórficas e idades U–Pb SHRIMP. *Revista Brasileira de Geociências*, **32**, 491–500.

MACHADO, N., SCHRANK, A., ABREU, F. R., KNAUER, L. G. & ALMEIDA-ABREU, P. A. 1989. Resultados preliminares da geocronologia U–Pb na Serra do Espinhaço Meridional. *Boletim do Núcleo Minas Gerais–Sociedade Brasileira de Geologia*, **10**, 171–174.

MARSHAK, S., ALKMIM, F. F., WHITTINGTON, A. & PEDROSA-SOARES, A. C. 2006. Extensional collapse in the Neoproterozoic Araçuaí orogen, eastern Brazil: A setting for reactivation of asymmetric crenulation cleavage. *Journal of Structural Geology*, **28**, 129–147.

MARTINS, M. S. 2006. *Geologia dos diamantes e carbonados aluvionares da bacia do Rio Macaúbas, MG.* PhD thesis, Instituto de Geociências, Universidade Federal de Minas Gerais, Belo Horizonte.

MARTINS, V. T. S., TEIXEIRA, W., NOCE, C. M. & PEDROSA-SOARES, A. C. 2004. Sr and Nd characteristics of Brasiliano/Pan African granitoid plutons of the Araçuaí orogen, southeastern Brazil: Tectonic implications. *Gondwana Research*, **7**, 75–89.

MARTINS-NETO, M. A. 2000. Tectonics and sedimentation in a Paleo/Mesoproterozoic rift-sag basin (Espinhaço basin, southeastern Brazil). *Precambrian Research*, **103**, 147–173.

MARTINS-NETO, M. A. 2005. A Bacia do São Francisco: Arcabouços estratigráfico e estrutural com base na interpretação de dados de superfície e subsuperfície. 3rd Simpósio sobre o craton do São Francisco, Salvador, Brazil, Sociedade Brasileira de Geologia, Anais, 283–286.

MARTINS-NETO, M. A. & ALKMIM, F. F. 2001. Estratigrafia e evolução tectônica das bacias neoproterozóicas do Paleocontinente São Francisco e suas margens: Registro da quebra de Rodínia e colagem de Gondwana. *In*: PINTO, C. P. & MARTINS-NETO, M. A. (eds) *Bacia do São Francisco: Geologia e Recursos Naturais.* Sociedade Brasileira de Geologia, Belo Horizonte, 31–54.

MARTINS-NETO, M. A. & HERCOS, C. M. 2002. Sedimentation and tectonic setting of Early Neoproterozoic glacial deposits in southeastern Brazil. *In*: ALTERMANN, W. & CORCORAN, P. L. (eds) *Precambrian Sedimentary Environments: A Modern Approach to Ancient Depositional Systems.* International Association of Sedimentologists, Special Publications, **33**, 383–403.

MARTINS-NETO, M. A., PEDROSA-SOARES, A. C. & LIMA, S. A. A. 2001. Tectono-sedimentary evolution of sedimentary basins from Late Paleoproterozoic to Late Neoproterozoic in the São Francisco craton and Araçuaí fold belt, eastern Brazil. *Sedimentary Geology*, **141/142**, 343–370.

MAURIN, J. C., BOUDZOUMOU, F. *ET AL.* 1991. La chaine protérozoique ouest-congolienne et son avant-pays au Congo: nouvelles données géochronologiques et structurales, implications en Afrique centrale. *Comptes Rendus de l' Académie des Sciences Paris*, **312**, 1327–1334.

MOURÃO, M. A. A. & PEDROSA-SOARES, A. C. 1992. Evidências de sedimentação turbidítica no Grupo Macaúbas, Faixa Araçuaí. *Revista Escola de Minas de Ouro Preto*, **45**, 94–96.

MUNHÁ, J. M., CORDANI, U., TASSINARI, C. & PALÁCIOS, T. 2005. Petrologia e termocronologia de gnaisses migmatíticos da Faixa de Dobramentos Araçuaí (Espírito Santo, Brasil). *Revista Brasileira de Geociências*, **35**, 123–134.

NALINI-JUNIOR, H. A., BILAL, E., PAQUETTE, J. L., PIN, C. & MACHADO, R. 2000*a*. Geochronologie U–Pb et géochimie isotopique Sr–Nd des granitoides neoproterozoiques des suites Galileia et Urucum, vallée du Rio Doce, Sud-Est du Brésil. *Comptes Rendus de l' Academie de Sciences Paris*, **331**, 459–466.

NALINI-JUNIOR, H. A., BILAL, E. & CORREIA NEVES, J. M. 2000*b*. Syncollisional peraluminous magmatism in the Rio Doce region: mineralogy, geochemistry and isotopic data of the Urucum suite (eastern Minas Gerais State, Brazil). *Revista Brasileira de Geociências*, **30**, 120–125.

NALINI-JUNIOR, H. A., MACHADO, R. M. & BILAL, E. 2005. Geoquímica e petrogênese da Suíte Galiléia: exemplo de magmatismo tipo-I, metaluminoso, pré-colisional, neoproterozóico da região do Médio Vale do Rio Doce. *Revista Brasileira de Geociências*, **35** (4-supplement), 23–34.

NOCE, C. M., PEDROSA-SOARES, A. C. *ET AL.* 1997. Nova Subdivisão Estratigráfica Regional do Grupo Macaúbas na Faixa Araçuaí: O Registro de uma Bacia Neoproterozóica. *Boletim do Núcleo Minas Gerais–Sociedade Brasileira de Geologia*, **14**, 29–31.

NOCE, C. M., MACAMBIRA, M. B. & PEDROSA-SOARES, A. C. 2000. Chronology of Neoproterozoic-Cambrian granitic magmatism in the Araçuaí Belt, Eastern Brazil, based on single zircon evaporation dating. *Revista Brasileira de Geociências*, **30**, 25–29.

NOCE, C. M., ROMANO, A. W., PINHEIRO, C., MOL, V. & PEDROSA-SOARES, A. C. 2003. Geologia das Folhas Ubá e Muriaé 1:100.000. *In*: PEDROSA-SOARES, A. C., NOCE, C. M., TROUW, R. A. J. & HEILBRON, M. (eds) *Projeto Sul de Minas.* CODEMIG, Belo Horizonte, Brazil, 623–659.

NOCE, C. M., PEDROSA-SOARES, A. C., PIUZANA, D., ARMSTRONG, R., LAUX, J. H., CAMPOS, C. & MEDEIROS, S. R. 2004. Ages of sedimentation of the kinzigitic complex and of a late orogenic thermal episode in the Araçuaí orogen, northern Espírito Santo State, Brazil: Zircon and monazite U–Pb SHRIMP and ID-TIMS data. *Revista Brasileira de Geociências*, **349**, 587–592.

NOCE, C. M., COSTA, A. G., PIUZANA, D., VIEIRA, V. S. & CARVALHO, C. 2006. *Geologia da Folha Manhuaçu 1:100.000.* Programa Geologia do Brasil, CPRM-Serviço Geológico do Brasil, Rio de Janeiro.

NOCE, C. M., PEDROSA-SOARES, A. C., SILVA, L. C., ARMSTRONG, R. & PIUZANA, D. 2007. Evolution of polyciclic basement complexes in the Araçuaí orogen, based on U–Pb SHRIMP data: Implications for Brazil-Africa links in Paleoproterozoic time. *Precambrian Research,* doi: 10.1016/j.precamres.2007.06.001.

PAIXÃO, M. M. & PERRELLA, P. 2004. *Mapeamento geológico da área do contato entre o Maciço Granítico Salto da Divisa e o Complexo Jequitinhonha, Nordeste de Minas Gerais.* Undergraduate Thesis, Instituto de Geociências, Universidade Federal de Minas Gerais, Belo Horizonte, Brasil.

PEDROSA-SOARES, A. C., MONTEIRO, R., CORREIA-NEVES, J. M., LEONARDOS, O. H. & FUZIKAWA, K. 1987. Metasomatic evolution of granites, northeast Minas Gerais, Brazil. *Revista Brasileira de Geociências*, **17**, 512–518.

PEDROSA-SOARES, A. C., NOCE, C. M., VIDAL, P., MONTEIRO, R. & LEONARDOS, O. H. 1992. Toward a new tectonic model for the Late Proterzoic Araçuaí (SE Brazil)—West Congolian (SW Africa) Belt. *Journal of South American Earth Sciences*, **6**, 33–47.

PEDROSA-SOARES, A. C. 1995. *Potencial aurífero do Vale do Araçuaí, Minas Gerais: história da exploração, geologia e controle tectono-metamórfico*. PhD thesis, Universidade de Brasília, Brazil.

PEDROSA-SOARES, A. C. & GROSSI-SAD, J. H. 1997. Geologia da Folha Minas Novas. *In*: GROSSI-SAD, J. H., LOBATO, L. M., PEDROSA-SOARES, A. C. & SOARES-FILHO, B. S. (eds) *Projeto Espinhaço em CD-ROM*. CODEMIG, Belo Horizonte, 952–1052.

PEDROSA-SOARES, A. C. & OLIVEIRA, M. J. R. 1997. Geologia da Folha Salinas. *In*: GROSSI-SAD, J. H., LOBATO, L. M., PEDROSA-SOARES, A. C. & SOARES-FILHO, B. S. (eds) *Projeto Espinhaço em CD-ROM*. CODEMIG, Belo Horizonte, 419–542.

PEDROSA-SOARES, A. C., LEONARDOS, O. H., FERREIRA, J. C. & REIS, L. B. 1996. Duplo regime metamórfico na Faixa Araçuaí: Uma re-interpretação à luz de novos dados. *In*: *Congresso Brasileiro de Geologia 39, Salvador, Anais*, **6**. Sociedade Brasileira de Geologia, 5–8.

PEDROSA-SOARES, A. C. & WIEDEMANN-LEONARDOS, C. M. 2000. Evolution of the Araçuaí Belt and its connection to the Ribeira Belt, Eastern Brazil. *In*: CORDANI, U., MILANI, E., THOMAZ-FILHO, A. & CAMPOS, D. A. (eds) *Tectonic Evolution of South America. 31st International Geological Congress, Rio de Janeiro*, 265–285.

PEDROSA-SOARES, A. C., VIDAL, P., LEONARDOS, O. H. & BRITO-NEVES, B. B. 1998. Neoproterozoic oceanic remnants in eastern Brazil: Further evidence and refutation of an exclusively ensialic evolution for the Araçuaí–West Congo orogen. *Geology*, **26**, 519–522.

PEDROSA-SOARES, A. C., CORDANI, U. & NUTMAN, A. 2000. Constraining the age of Neoproterozoic glaciation in eastern Brazil: First U–Pb (SHRIMP) data from detrital zircons. *Revista Brasileira de Geociências*, **30**, 58–61.

PEDROSA-SOARES, A. C., NOCE, C. M., WIEDEMANN, C. M. & PINTO, C. P. 2001*a*. The Araçuaí–West Congo orogen in Brazil: An overview of a confined orogen formed during Gondwanland assembly. *Precambrian Research*, **110**, 307–323.

PEDROSA-SOARES, A. C., PINTO, C. P. ET AL. 2001*b*. A Província Gemológica Oriental do Brasil. *In*: CASTAÑEDA, C., ADDAD, J. E. & LICCARDO, A. (eds) *Gemas de Minas Gerais*. Sociedade Brasileira de Geologia, Belo Horizonte, 16–33.

PEDROSA-SOARES, A. C., NOCE, C. M. ET AL. 2004 . Confined orogen: The concept, a Precambrian Example and Tectonic Implications. *In*: *32nd International Geological Congress, Firenze, Italy, Abstracts, Session 284, CD-ROM*, 1255–1256.

PEDROSA-SOARES, A. C., NOCE, C. M., SILVA, L. C., CORDANI, U. G., ALKMIM, F. F. & BABINSKI, M. 2005. Orógeno Araçuaí: Estágio atual do conhecimento geotectônico. *In*: *3rd Simpósio sobre o Cráton do São Francisco, Salvador, Brazil, Anais*. Sociedade Brasileira de Geologia, Anais, 243–246.

PEDROSA-SOARES, A. C., CASTAÑEDA, C. ET AL. 2006. Magmatismo e Tectônica do Orógeno Araçuaí no Extremo Leste de Minas Gerais e Norte do Espírito Santo. *Geonomos*, **14**, 99–113.

PERES, G. G., ALKMIM, F. F. & JORDT-EVANGELISTA, H. 2004. The southern Araçuaí belt and the Dom Silvério Group: Geologic architecture and tectonic significance. *Anais Academia Brasileira de Ciências*, **76**, 771–790.

PORADA, H. 1989. Pan-African rifting and orogenesis in southern to equatorial Africa and Eastern Brazil. *Precambrian Research*, **44**, 103–136.

QUEIROGA, G. N. 2006. *A seção sedimentar sulfetada do ofiolito de Ribeirão da Folha, Orógeno Araçuaí, MG*. MSc thesis, Instituto de Geociências, Universidade Federal de Minas Gerais, Belo Horizonte.

QUEIROGA, G. N., PEDROSA-SOARES, A. C., QUÉMÉNEUR, J. & CASTAÑEDA, C. 2006. A unidade metassedimentar do ofiolito de Ribeirão da Folha, Orógeno Araçuaí, Minas Gerais: petrografia, geotermobarometria e calcografia. *Geonomos*, **14**, 9–12.

ROGERS, J. W. & SANTOSH, M. 2004. *Continents and Supercontinents*. Oxford University Press.

ROMEIRO-SILVA, P. C., BRAUN, O., ZALÁN, P. V., PAULA, L. & MARTINS, M. 1998. São Francisco Basin: Exploration Frontier for Gas in Central Brazil. *In*: *AAPG International Conference, Extended Abstracts*, 990–991.

SAMPAIO, A. R., MARTINS, A. M. ET AL. 2004. *Projeto Extremo Sul da Bahia: Geologia e Recursos Minerais. Salvador, Brazil*. Série Arquivos Abertos da Companhia Bahiana de Pesquisa Mineral, **19**.

SANTOS, R. F. 2007. *A Formação Salinas da Faixa Araçuaí, Minas Gerais: acervo estrutural e significado tectônico*. MSc thesis, Departamento de Geologia, Universidade Federal de Ouro Preto, Ouro Preto.

SANTOS, R. V., ALVARENGA, C. ET AL. 2004. Carbon isotopes of Mesoproterozoic-Neoproterozoic sequences from southern São Francisco craton and Araçuaí Belt: Paleogeographic implications. *Journal of South American Earth Sciences*, **18**, 27–39.

SILVA, J. M., LIMA, M., VERONESE, V. F., RIBEIRO-JUNIOR, R. & SIGA-JÚNIOR, O. 1987. *Folha SE.24 Rio Doce, Levantamento de Recursos Naturais, Projeto Radambrasil*. IBGE, Rio de Janeiro.

SILVA, L. C., ARMSTRONG, R. ET AL. 2002*a*. Reavaliação da evolução geológica em terrenos pré-cambrianos brasileiros com base em novos dados U–Pb SHRIMP, Parte I: limite centro-oriental do Cráton do São Francisco na Bahia. *Revista Brasileira de Geociências*, **32**, 161–172.

SILVA, L. C., ARMSTRONG, R. ET AL. 2002*b*. Reavaliação da evolução geológica em terrenos pré-cambrianos brasileiros com base em novos dados U–Pb SHRIMP, parte II: Orógeno Araçuaí, Cinturão Móvel Mineiro e Cráton São Francisco Meridional. *Revista Brasileira de Geociências*, **32**, 513–528.

SILVA, L. C., MCNAUGHTON, N. J., ARMSTRONG, R., HARTMANN, L. & FLETCHER, I. 2005. The Neoproterozoic Mantiqueira Province and its African connections. *Precambrian Research*, **136**, 203–240.

SILVA, L. C., PEDROSA-SOARES, A. C. & TEIXEIRA, L. R. 2007. Tonian rift-related, A-type continental plutonism in the Araçuaí orogen, Eastern Brazil: new evidences for the breakup stage of the São Francisco-Congo Paleocontinent. *Gondwana Research*, doi: 10.1016/j.precamres.2007.06.002.

SÖLLNER, H. S., LAMMERER, B. & WEBER-DIEFENBACH, K. 1991. Die Krustenentwicklung in der Küstenregion nördlich von Rio de Janeiro, Brasilien. *Münchner Geologische Hefte*, **4**, 1–100.

SUITA, M. T. F., PEDROSA-SOARES, A. C., LEITE, C., NILSON, A. A. & PRICHARD, H. 2004 Complexos Ofiolíticos do Brasil e a Metalogenia Comparada das Faixas Araçuaí e Brasília. *In*: PEREIRA, E., CASTROVIEJO, R. & ORTIZ, F. (eds) *Complejos Ofiolíticos en Iberoamérica: guías de prospección para metales preciosos*.Ciencia y Tecnología para el Desarrollo-CYTED, Madrid, 101–132.

TACK, L., WINGATE, M. T. D., LIÉGEOIS, J. P., FERNANDEZ-ALONSO, M. & DEBLOND, A. 2001. Early Neoproterozoic magmatism (1000–910 Ma) of the Zadinian and Mayumbian Groups (Bas-Congo): Onset of Rodinian rifting at the western edge of the Congo craton. *Precambrian Research*, **110**, 277–306.

TACK, L., NKULA, V. K. *ET AL*. 2004. *Guide Book to the Field Trip in the Pan African West Congo belt of the Bas-Congo region (Democratic Republic of the Congo)*. IGCP 470 Meeting, Kinshasa, Democratic Republic of the Congo.

TROMPETTE, R. 1994. *Geology of Western Gondwana (2000–500 Ma). Pan-African-Brasiliano aggregation of South America and Africa*. A. A. Balkema, Rotterdam.

TROMPETTE, R. 1997. Neoproterozoic (~ 600 Ma) aggregation of Western Gondwana: a tentative scenario. *Precambrian Research*, **82**, 101–112.

UHLEIN, A., TROMPETTE, R. & EGYDIO-SILVA, M. 1998. Proterozoic rifting and closure, SE border of the São Francisco craton, Brazil. *Journal of South American Earth Sciences*, **11**, 191–203.

UHLEIN, A., TROMPETTE, R. & ALVARENGA, C. 1999. Neoproterozoic glacial and gravitational sedimentation on a continental rifted margin: The Jequitaí-Macaúbas sequence (Minas Gerais, Brazil). *Journal of South American Earth Sciences*, **12**, 435–451.

VAUCHEZ, A., EGYDIO-SILVA, M., BABINSKI, M., TOMMASI, A., UHLEIN, A. & LIU, D. 2007. Deformation of a pervasively molten middle crust: insights from the Neoproterozoic Ribeira–Acaçuai orogen (SE Brazil). *Terra Nova*, **19**, 278–286.

VIEIRA, V. S. 2005. *Significado do Grupo Rio Doce no Contexto do Orógeno Araçuaí*. PhD qualifying volume, Instituto de Geociências, Universidade Federal de Minas Gerais, Belo Horizonte.

VICAT, J. P. & POUCLET, A. 2000. Palaeo- and Neoproterozoic granitoids and rhyolites from the West Congolian Belt (Gabon, Congo, Cabinda, north Angola): chemical composition and geotectonic implications. *Journal of African Earth Sciences*, **31**, 597–617.

WHITTINGTON, A. G., CONNELLY, J., PEDROSA-SOARES, A. C., MARSHAK, S. & ALKMIM, F. F. 2001. Collapse and melting in a confined orogenic belt: preliminary results from the Neoproterozoic Aracuai belt of eastern Brazil. *In*: *AGU Fall Meeting, Abstract T32B-0895, American Geophysical Union*, **82**, 1181–1182.

WIEDEMANN, C., MENDES, J. C., MOURA, J. C., COSTA-NASCIMENTO, R. C. & LUDKA, I. P. 1997. Granitoids of the Espírito Santo Magmatic Arc. *In*: *2nd International Symposium on Granites and Associated Mineralizations, Salvador, Brazil, Excursion Guide*. Sociedade Brasileira de Geologia, 57–76.

WIEDEMANN, C. M., CAMPOS, C. M., MEDEIROS, S. R., MENDES, J. C., LUDKA, I. P. & MOURA, J. C. 2002. Architecture of Late orogenic Plutons in the Araçuaí-Ribeira Folded Belt, Southeast Brazil. *Gondwana Research*, **19**, 381–399.

ZHAO, G., CAWOOD, P. A., WILDE, S. A. & SUN, M. 2002. Review of global 2.1–1.8 Ga orogens: Implications for a pre-Rodinia supercontinent. *Earth Science Reviews*, **59**, 125–162.

Sedimentary provenance and palaeoenvironment of the Baixo Araguaia Supergroup: constraints on the palaeogeographical evolution of the Araguaia Belt and assembly of West Gondwana

C. A. V. MOURA, B. L. S. PINHEIRO, A. C. R. NOGUEIRA,
P. S. S. GORAYEB & M. A. GALARZA

Universidade Federal do Pará, Centro de Geociências, C.P. 8608, 65075-110, Belém, Pará, Brazil (e-mail: candido@ufpa.br)

Abstract: Provenance studies on metasedimentary rocks of the Baixo Araguaia Supergroup of the Brasiliano Araguaia Belt, central Brazil, yield $^{207}Pb/^{206}Pb$ zircon evaporation ages for detrital zircons from quartzites concentrated around 1000–1200 Ma and 2800–2900 Ma; Sm–Nd T_{DM} model ages of schists and phyllites scatter around 1600–1700 Ma. Facies analysis of low-grade metasedimentary rocks from drill cores suggests a sedimentary environment of basin floor and lower- to upper-slope turbidites. Nearby sources are indicated by the textural and mineralogical immaturity; together with structural geological data indicating tectonic transport of the supracrustal pile towards the NW, this suggests probable provenance from the southeastern portion of the Araguaia Belt and not from the Amazonian Craton as usually believed. The Goiás Massif, Goiás Magmatic Arc, São Francisco Craton and Paranapanema block are considered to be the best candidates. They may have formed a larger continental mass during West Gondwana amalgamation, prior to their collision with the Amazonian Craton to form the Araguaia Belt. Final timing of this collision is constrained by *c.* 550 Ma syntectonic granites. Similar ages for high-grade gneisses in the Rokelide Belt suggest coeval collision and coetaneous metamorphism of the Araguaia and Rokelide belts, but more geological and geophysical data are required for a decisive correlation between these belts.

The relationship between the Brasiliano (Pan-African) belts in South America and Africa has received the attention of those working with palaeogeographical reconstruction of Gondwana (Hoffman 1991; Rogers *et al.* 1995; Unrug 1996; Trompette 1997; Cordani *et al.* 2003). Acquisition of geochronological data for rocks from Brazil contributed to the geological correlation between the geotectonic units in the coastal areas of the South American and the African continents (Renne *et al.* 1990). The Neoproterozoic belts in the interior of South America are also important for Gondwana reconstruction, since some are considered to have extended over the African continent (Brito Neves & Cordani 1991; Trompette 1997). Among these, the Araguaia Belt, which lies over the eastern margin of the Amazonian Craton (Fig. 1) and forms the northern portion of the Paraguay–Araguaia belt (Almeida *et al.* 1981), presumably continues to the north of Brazil and into Africa (as the Rokelide Belt), with a total length exceeding 4000 km (Brito Neves & Cordani 1991). To the east of the Araguaia Belt, rocks were affected by Neoproterozoic events related to the amalgamation of West Gondwana, whereas to the west (Amazonian Craton) they were not involved in this process. Thus, the study of the palaeogeographical evolution of the Araguaia Belt is important for deciphering the palaeogeography of the different blocks that were welded to form West Gondwana and the docking of these landmasses with the Amazonian Craton. Provenance investigation and facies analysis of the metasedimentary rocks of the Baixo Araguaia Supergroup in the Araguaia Belt were carried out to understand its palaeogeographical evolution in the context of West Gondwana assembly. $^{207}Pb/^{206}Pb$ evaporation ages of detrital zircon grains from quartzites and Sm–Nd crustal residence ages of the main lithotypes of this belt were determined. In support of these methods, core-based facies analysis was carried out in the very low metamorphic grade carbonate and siliciclastic rocks.

Geology of the Araguaia Belt

The Araguaia Belt is composed of metamorphosed psammitic and pelitic successions, with minor contributions of carbonate rocks, mafic and ultramafic rocks, and granite intrusions (Alvarenga *et al.* 2000). The Araguaia Belt is 1200 km long and more than 100 km wide, and displays a general north–south orientation. To the east the Palaeozoic sedimentary rocks of the Parnaíba Basin cover the belt, while on the western side the low-grade to anchizone metamorphic rocks of the Araguaia

From: Pankhurst, R. J., Trouw, R. A. J., Brito Neves, B. B. & de Wit, M. J. (eds) *West Gondwana: Pre-Cenozoic Correlations Across the South Atlantic Region.* Geological Society, London, Special Publications, **294**, 173–196. DOI: 10.1144/SP294.10 0305-8719/08/$15.00

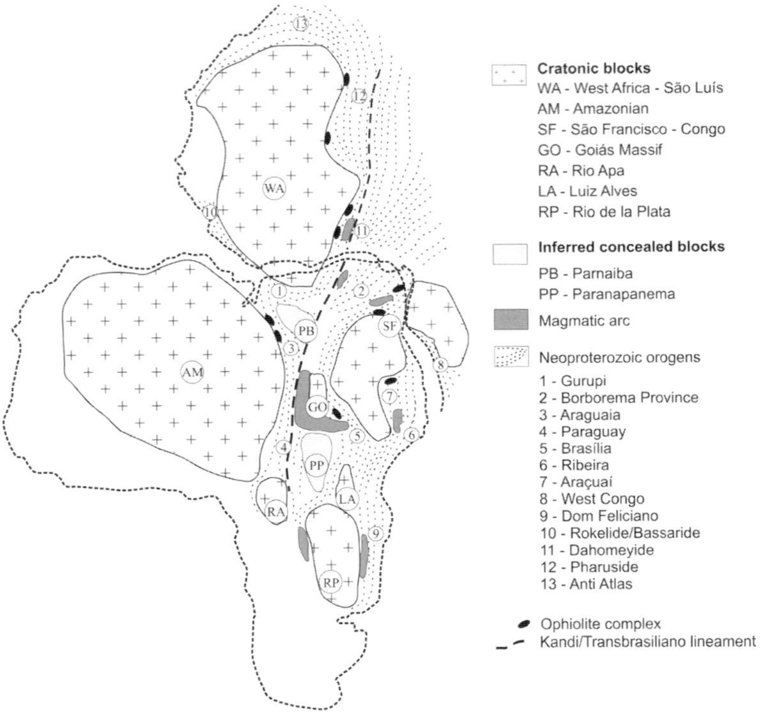

Fig. 1. Palaeogeographical reconstruction showing the Pan-African/Brasiliano belts, and the main Archaean and Palaeoproterozoic cratonic regions and blocks in South America and the West Africa Craton (from Klein & Moura 2008).

Belt unconformably overlie, or are thrust over, the Archaean rocks of the Amazonian Craton. Palaeoproterozoic gneisses and granulites terrane bound the Araguaia Belt to the SE (Fig. 2).

The metasedimentary rocks of the Araguaia Belt are included in the Baixo Araguaia Supergroup (Abreu 1978), divided into the Estrondo and Tocantins groups. The Estrondo Group occurs in the eastern side of the Araguaia Belt (Fig. 2). It is composed of quartzite, meta-conglomerate and mica schist (Morro do Campo Formation). Mica schists with variable amounts of biotite, muscovite, kyanite, staurolite and garnet, calc-schists, marbles, feldspathic schists and amphibolite compose the Xambioá Formation. The Tocantins Group lies on the western side of the Araguaia Belt (Fig. 2) and includes the Pequizeiro and Couto Magalhães formations; the former is composed mainly of chlorite–muscovite–quartz schist with minor intercalations of phyllite and quartzite and the latter is dominated by phyllite and slate, interlayered with minor amounts of quartzite, meta-arkose and metalimestone (Gorayeb 1981; Hasui *et al.* 1984*a*; Dall'Agnol *et al.* 1988).

Alkaline felsic rocks, mafic and ultramafic bodies, and granitic plutons are associated with this supracrustal succession (Hasui *et al.* 1984*a*, *b*; Dall'Agnol *et al.* 1988; Herz *et al.* 1989). The alkaline complexes are composed of syenite and nepheline syenite gneisses; they occur in the southern segment of the belt and in the adjacent basement (Monte Santo and Serra da Estrela bodies). These rocks represent the alkaline magmatism that was concomitant with crustal rifting and formation of the Araguaia basin (Alvarenga *et al.* 2000), where the sediments that formed the rocks of the Baixo Araguaia Supergroup accumulated. A $^{207}Pb/^{206}Pb$ single-zircon evaporation age of 1006 ± 86 Ma for syenitic gneiss sampled from the Serra da Estrela body has been interpreted as the age of this alkaline magmatic event (Arcanjo & Moura 2000) (Table 1).

Mafic and ultramafic bodies are associated with both basement and supracrustal rocks, although the largest ultramafic bodies occur in the western part of the belt (Fig. 2). These ultramafic bodies were tectonically emplaced into supracrustal rocks and are composed of serpentinized peridotites and

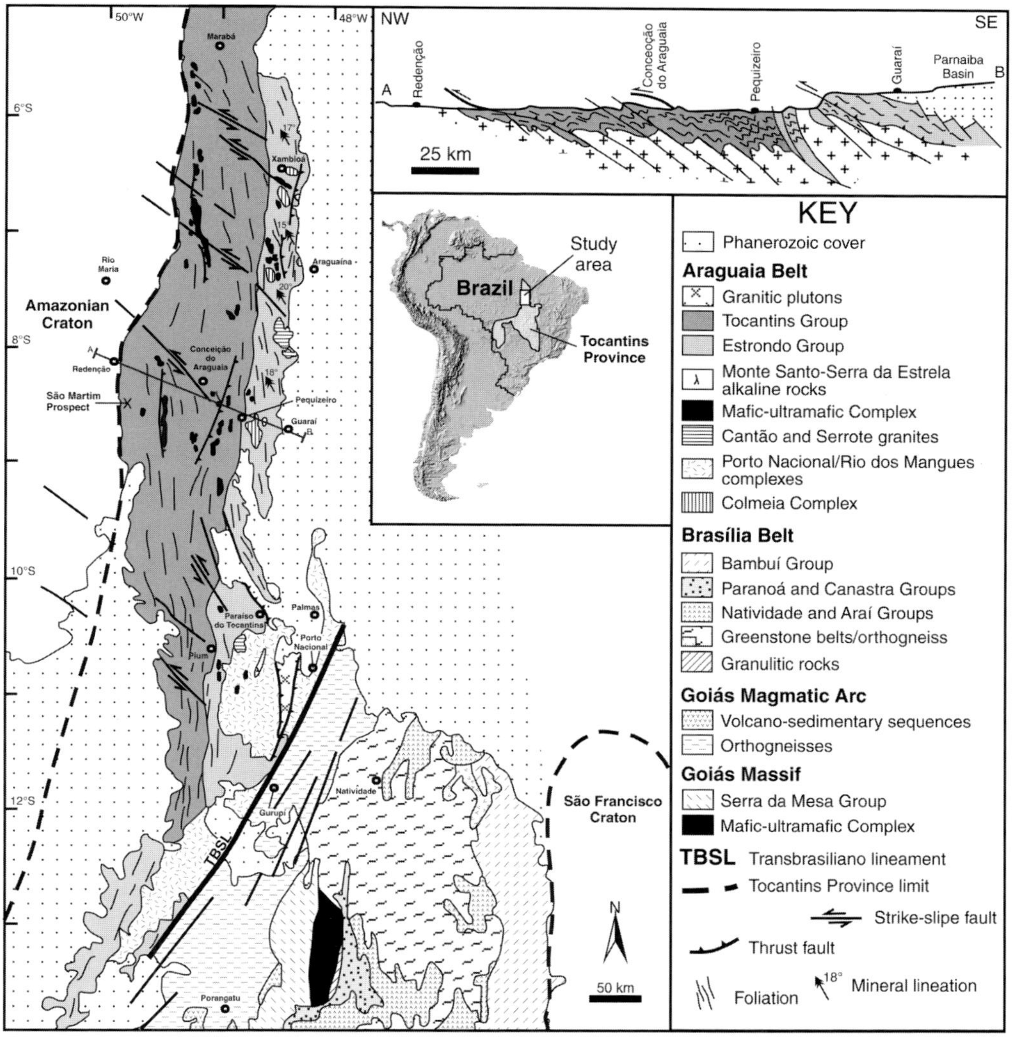

Fig. 2. Simplified geological map of the Araguaia Belt and surrounding areas, after Alvarenga *et al.* (2000) and Pimentel *et al.* (2004).

dunites, minor chromitites and their metamorphic products (steatite, talc schist, tremolite–actinolite schist and chloritites), in addition to chert and jaspilite (Gorayeb 1989). Pillow basalts have also been reported (Kotschoubey *et al.* 1996). This rock association has been interpreted as remnants of ophiolite complexes, suggesting the presence of oceanic crust in the Araguaia Belt (Paixão & Nilson 2002; Kotschoubey *et al.* 2005). A Sm–Nd age of 757 $\pm$ 49 Ma for associated mafic dykes has been interpreted as the age of the oceanic crust formed in the Neoproterozoic (Paixão *et al.* 2002), but zircons from the pillow basalts gave $^{207}Pb/^{206}Pb$ evaporation ages around 2050 Ma (Gorayeb *et al.* 2002). A number of metagabbro bodies with scapolite emplaced into mica schists have been mapped in the Xambioá region and one of these bodies has yielded a $^{207}Pb/^{206}Pb$ zircon age of 817 $\pm$ 5 Ma (Table 1; Gorayeb *et al.* 2004). This value is older than the Sm–Nd age reported for the ophiolitic succession, but reinforces the Neoproterozoic age of the mafic and ultramafic magmatism in the Araguaia Belt. Notwithstanding, the age of this magmatism is still uncertain and needs further investigation.

Granite plutons occur along the eastern part of the Araguaia Belt, associated with the highest metamorphic grade domains (amphibolite facies). This

Table 1. *Geochronological data for the rocks of the Araguaia Belt and surrounding regions*

Rock unit	Rock type	Method	Material dated	Age (Ma)	Comment	Ref.
Araguaia Belt						
Estrondo Group	Schist	K–Ar	Biotite	553 ± 17		1
Estrondo Group	Schist	K–Ar	Muscovite	533 ± 16		1
Estrondo Group	Amphibolite	K–Ar	Hornblende	565 ± 20		1
Estrondo Group	Amphibolite	K–Ar	Hornblende	558 ± 32		1
Colméia Complex	Biotite gneiss	K–Ar	Biotite	535 ± 17		1
Colméia Complex	Biotite gneiss	K–Ar	Muscovite	531 ± 13		1
Colméia Complex	Tonalitic gneiss	Pb-evap.	Zircon	2867 ± 12	$^{207}Pb/^{206}Pb$	2
Colméia Complex	Trondhjemitic gneiss	Pb-evap.	Zircon	2858 ± 20	$^{207}Pb/^{206}Pb$	2
Colméia Complex	Granitic gneiss	Pb-evap.	Zircon	2855 ± 12	$^{207}Pb/^{206}Pb$	2
Cantão Gneiss	Granitic gneiss	Pb-evap.	Zircon	1858 ± 68	$^{207}Pb/^{206}Pb$	2
Rio dos Mangues Complex	Granodioritic gneiss	Pb-evap.	Zircon	2014 ± 18	$^{207}Pb/^{206}Pb$	3
Rio dos Mangues Complex	Calc-silicate gneiss	Pb-evap.	Zircon	2083 ± 14	$^{207}Pb/^{206}Pb$	3
Porto Nacional Complex	Enderbite	Pb-evap.	Zircon	2153 ± 1	$^{207}Pb/^{206}Pb$	4
Porto Nacional Complex	Mafic granulite	Pb-evap.	Zircon	2125 ± 3	$^{207}Pb/^{206}Pb$	4
Serrote Granite	Granite (deformed)	Pb-evap.	Zircon	1851 ± 20	$^{207}Pb/^{206}Pb$	5
Serra da Estrela gneiss	Syenitic gneiss	Pb-evap.	Zircon	1006 ± 86	$^{207}Pb/^{206}Pb$	3
Xambica Intrusive Suite	Metagabbro	Pb-evap.	Zircon	817 ± 5	$^{207}Pb/^{206}Pb$	6
Santa Luzia Granite	Granodiorite	Pb-evap.	Zircon	655 ± 24	$^{207}Pb/^{206}Pb$	7
Ramal do Lontra Granite	Granodiorite	Pb-evap.	Zircon	549 ± 5	$^{207}Pb/^{206}Pb$	8
Cantão Gneiss	Granitic gneiss	Rb–Sr	Whole rock–biotite–K feldspar	479 ± 3	Internal isochron	9
Serrote Granite	Granite (deformed)	Rb–Sr	K feldspar–biotite	536 ± 18	Two points	5
Colméia Complex	Tonalitic gneiss	Sm–Nd	Whole rock	3140 – 3360	Range of T_{DM}	10
Colméia Complex	Trondhjemitic gneiss	Sm–Nd	Whole rock	3100 – 3290	Range of T_{DM}	10
Cantão Gneiss	Granitic gneiss	Sm–Nd	Whole rock	2820 – 2910	Range of T_{DM}	10

Amazonian Craton (southeastern portion)						
Pium Complex	Enderbite	U–Pb	Zircon	3002 ± 12	SHRIMP	11
Lagoa Seca Group	Metarhyiodacite	U–Pb	Zircon	2909 ± 29	TIMS ID	12
Rio Maria Granodiorite	Granodiorite	U–Pb	Zircon	2874 ± 10	TIMS ID	12
Arco Verde Metaonalite	Tonalite	U–Pb	Zircon	2979 ± 25	TIMS ID	12
Plaque Granitic Suite	Granite	Pb-evap.	Zircon	2736 ± 24	$^{207}Pb/^{206}Pb$	13
Grão Pará Group	Felsic volcanic	U–Pb	Zircon	2759 ± 2	TIMS ID	14
Luanga Layered Complex	Anorthositic gabbro	U–Pb	Zircon	2761 ± 3	TIMS ID	14
Grupo Igarapé Pojuca	Amphibolite	U–Pb	Zircon	2732 ± 2	TIMS ID	14
Old Salobo Granite	Granite	U–Pb	Zircon	2573 ± 2	TIMS ID	14
Central Carajás Granite	Granite	U–Pb	Zircon	1880 ± 2	TIMS ID	14
Cigano Granite	Granite	U–Pb	Zircon	1883 ± 2	TIMS ID	14
Musa Granite	Granite	U–Pb	Zircon	1883 ± 5	TIMS ID	14
Goiás Massif						
Metavolcanic rocks	Mafic-ultramafic rocks	Sm–Nd	Whole rock	2895 ± 98	Isochron	15
Uvá Complex	Tonalite	U–Pb	Zircon	2934 ± 5	SHRIMP	16
Posselândia Diorite	Diorito	U–Pb	Zircon	2146 ± 2	TIMS ID	17
Juscelandia Sequence	Subvolcanic felsic rock	U–Pb	Zircon	1263 ± 15	SHRIMP	18
Niquelândia Complex	Gabbro	U–Pb	Zircon	1248 ± 23	SHRIMP	19
Basement rocks of the northern Brasília Belt						
Ribeirão da Areias Complex	Tonalite	U–Pb	Titanite	2455 ± 14	SHRIMP	20
Manto Verde Pluton	Tonalite	U–Pb	Zircon	2206 ± 5	SHRIMP	20
São Martins Pluton	Tonalite	U–Pb	Zircon	2204 ± 4	SHRIMP	20
Goiás Magmatic Arc						
Arenópolis Gneiss	Orthogneiss	U–Pb	Zircon	890 ± 7	TIMS ID	21
Mara Rosa Gneiss	Tonalitic gneiss	U–Pb	Zircon	856 ± 13	TIMS ID	21
Arenopolis Sequence	Metarhyolite	U–Pb	Zircon	929 ± 9	TIMS ID	21

References: 1 Macambira (1983); 2 Moura & Gaudette (1999); 3 Arcanjo & Moura (2000); 4 Gorayeb *et al.* (2000); 5 Sousa & Moura (1995); 6 Gorayeb *et al.* (2004); 7 Moura & Gaudette (1993); 8 Alves (2006); 9 Lafon *et al.* (1990); 10 Moura (1992); 11 Pidgeon *et al.* (1998); 12 Macambira & Lancelot (1996); 13 Avelar *et al.* (1999); 14 Machado *et al.* (1991); 15 Arndt *et al.* (1989); 16 Queiroz (2000); 17 Jost *et al.* (1993); 18 Moraes *et al.* (2006); 19 Suita *et al.* (1994); 20 Cruz (2001); 21 Pimentel *et al.* (1991).

magmatism is considered as a product of partial melting of the supracrustal rocks during the peak of the regional metamorphism (Dall'Agnol *et al.* 1988; Abreu & Gorayeb 1994; Alvarenga *et al.* 2000). The Santa Luzia granite, in the southern portion of the belt, NE of the city of Paraíso do Tocantins (Fig. 2), has given a $^{207}Pb/^{206}Pb$ zircon evaporation age of 655 ± 24 Ma (Moura & Gaudette 1993), which has been used as a first estimate of the age of this granite genesis episode (Alvarenga *et al.* 2000). However, this age may be older than the true age since Teixeira *et al.* (2002) demonstrated the presence of inherited zircon crystals in the Santa Luzia granite, giving $^{207}Pb/^{206}Pb$ evaporation ages that range from 540 to 2500 Ma. A more recent $^{207}Pb/^{206}Pb$ single-zircon evaporation age of 549 ± 5 Ma for the Ramal do Lontra granite (Xambioá region) was obtained by Alves (2006), and may be a more realistic age for this syntectonic granitic magmatism (Table 1).

Basement inliers have been recognized in the core of dome-like structures along the eastern side of the northern segment of the Araguaia Belt (Hasui *et al.* 1984*a*; Herz *et al.* 1989). In the southern segment, Palaeoproterozoic gneissic and granulitic complexes, located to the west of the Transbrasiliano lineament, have been considered as part of the basement of the Araguaia Belt (Hasui *et al.* 1984*b*) (Fig. 2). However, the ages of these basement rocks units are quite distinct. The basement inliers of the northern segment of the Araguaia Belt are composed mainly of Archaean TTG orthogneisses (Colméia complex) with $^{207}Pb/^{206}Pb$ single-zircon evaporation ages around 2860 Ma, and Palaeoproterozoic granitic plutons (Cantão granite) of 1850 Ma (Moura & Gaudette 1999) (Fig. 2, Table 1). An Archaean TTG terrane (2.9–2.87 Ga) intruded by 1880 Ma granitic plutons is well documented in the adjacent southeastern portion of the Amazonian Craton (Macambira & Lafon 1995) (Table 1) which led Moura & Gaudette (1999) to suggest that the basement rocks of the northern segment of the Araguaia Belt could be inliers of the Amazonian Craton. The basement in the southern portion of the belt consists of Palaeoproterozoic tonalitic and calc-silicate gneisses (Rio dos Mangues complex) and granulite rocks (Porto Nacional complex). Both complexes have given $^{207}Pb/^{206}Pb$ single-zircon evaporation ages close to 2100 Ma (Arcanjo & Moura 2000; Gorayeb *et al.* 2000) (Table 1). Another granitic pluton (Serrote granite), with an age of around 1850 Ma, is intruded into the Palaeoproterozoic basement gneisses (Sousa & Moura 1995). Thus the basement rocks cropping out to the southern segment of the belt are part of a Palaeoproterozoic crustal block, whose history and timing of docking to the Amazonian Craton is not well understood yet. In spite of the distinct ages of these two basement segments (Archaean and Palaeoproterozoic), both show the effects of Neoproterozoic tectonic-thermal event related to the framework of Araguaia Belt, indicated by similar structures generally concordant with the supracrustal metamorphic rocks (Gorayeb & Alves 2003).

Major north–south trending structures are imprinted on both supracrustal and basement rocks of the Araguaia Belt. Varying directions may be observed near dome-like structures and near NW–SE and NNE–SSW trending shear zones. The dome-like structures have been correlated with thrust shear-zones involving basement and supracrustal rocks (Herz *et al.* 1989; Alvarenga *et al.* 2000). Thrust shear-zones, and mineral-stretching lineations plunging gently (5–20°) to the SE (110–130°) on low to medium angle east-dipping foliation planes, suggest tectonic transport towards the NW (Abreu & Gorayeb 1994; Alvarenga *et al.* 2000).

The Barrovian-type regional metamorphism affecting the rocks of the Araguaia Belt increases gradually from incipient in the west to middle-high amphibolite facies in the east; north–south isograds and metamorphic zones may be recognized along the belt. The pelitic sequences show the following sequential mineral assemblages towards the east: sericite–chlorite, muscovite–chlorite–epidote, muscovite–biotite ± chlorite, muscovite–biotite–garnet, biotite–garnet–kyanite, biotite–muscovite–garnet–staurolite and, finally, restricted areas of partial melting generating quartz-feldspar veins and small granitic bodies (Abreu & Gorayeb 1994; Alvarenga *et al.* 2000).

K–Ar mineral ages obtained by Macambira (1983) for the supracrustal and basement rocks are still the best estimates for the age of the metamorphic event (Table 1). K–Ar ages between 520 and 560 Ma determined on biotite, muscovite and hornblende from schists and amphibolites of the Estrondo Group may record cooling ages following regional metamorphism. The imprint of the Neoproterozoic metamorphic event on the basement rocks of the Araguaia Belt is evidenced by K–Ar ages around 530 Ma in biotite and muscovite from the Archaean basement gneisses (Macambira 1983), as well as by Rb–Sr mineral ages in the Archaean gneiss and Palaeoproterozoic metagranites (Table 1) (Lafon *et al.* 1990; Moura 1992; Sousa & Moura 1995).

Analytical methods

Sedimentary facies analysis was carried out on drill core SMD-08 of the São Martins prospect, provided by the Western Mining Company, following the systematics of Walker & James (1992). This method emphasizes facies description and the understanding of the sedimentary processes and facies associations, in order to characterize the environment and depositional systems (Miall 1985, 1991). The siliciclastic rocks were classified according to Folk (1974), and the carbonate nomenclature proposed by Dunham (1962) and Embry & Klovan (1971) was adopted.

Geochronological investigations were carried out on detrital zircon grains from two quartzite samples of the Morro do Campo Formation (Estrondo Group), by the single-zircon evaporation $^{207}Pb/^{206}Pb$ method (Kober 1987). The zircon grains were separated from almost 30 kg of sample, and concentrated using well-known heavy mineral separation techniques involving the pulverization, panning and sieving of the sample, and separation of zircon with heavy liquids. Zircon grains were picked, randomly, from the 0.25–0.177 mm fraction, under a binocular microscope. Analytical data were obtained at the Isotope Geology Laboratory of the Federal University of Pará (Pará-Iso), Brazil, using a Finnigan MAT 252 thermo-ionization mass spectrometer. Isotope data were acquired dynamically by ion-counting, with the intensity of the ^{207}Pb beam between 30,000 and 100,000 counts per second. The intensities of the different Pb isotopes were measured in the mass sequence 206–207–208–206–207–204; five mass scans define one block of data with nine $^{207}Pb/^{206}Pb$ ratios. Discrepant isotopic values were eliminated using the Dixon test. Two blocks of data were measured and the average $^{207}Pb/^{206}Pb$ ages of these blocks were considered to represent the age of the zircon for that particular evaporation step. In principal, three evaporation steps are performed, at temperatures of 1450, 1500 and 1550 °C, but in this case Pb contents were too low for evaporation at 1550 °C. Since the ages obtained at 1500 °C were always older than those at 1450 °C, the former were considered to represent the age of mineral crystallization.

For the determination of Sm–Nd T_{DM} model ages, rock samples (3 kg) were fragmented, crushed, pulverized and split in the laboratory. Around 100 mg of pulverized sample were weighed with spike ($^{149}Sm-^{150}Nd$) and dissolved with $HF-HNO_3$ in a Parr bomb, at 150 °C in the oven. After one week, the solution was dried out and the residue attacked again with $HF-HNO_3$ on a hot plate at 100 °C. Two further dissolution steps were conducted with HCl (6.2 N, 2 N) on a hot plate, and finally the residue was dissolved in 2 N HCl. Rare earth elements were separated by cation exchange on Dowex AG1x8 resin using HCl and HNO_3 as elluents. The evaporated residue was dissolved in 7 N HNO_3–methanol for chromatographic separation of Sm and Nd on anion exchange resin Dowex AG1x4. Sm and Nd were mounted on Ta filaments and analysed in the Pará-Iso Finnigan MAT 252, using a Ta–Re double filament arrangement. Analyses of La Jolla performed during the course of this study gave $^{143}Nd/^{144}Nd$ of 0.511854 ± 0.000010 (2σ on 3 analyses). Nd and Sm blanks were <170 pg. The $^{143}Nd/^{144}Nd$ ratios were normalized to $^{146}Nd/^{144}Nd = 0.7219$. The Sm–Nd crustal residence ages (T_{DM}) were calculated using the depleted mantle model of DePaolo (1988).

Results

Core-based facies analysis

The very low-grade metamorphosed carbonate and siliciclastic rocks of the Couto Magalhães Formation (Tocantins Group), which overlies the Archaean rocks of the eastern part of the Amazonian Craton, were used for facies analysis. In the Araguaia Belt these rocks are poorly exposed and, usually, intensely weathered or covered by laterites, which has hindered outcrop-based facies analysis, so drill core samples from the São Martim prospect (Fig. 2) were used. Drill core SMD-08 was selected because it could be continuously sampled to a depth of 570 m and preserves several sedimentary structures indicative of deep-sea deposits These deposits are probably related to a slope–apron setting and consists of two associations: (i) basin floor and (ii) lower- to upper-slope turbidites, constituted by fine-grained sandstone, mudstone with even parallel lamination and abundant deformational structures, and massive diamictite (floatstones) associated with subordinate limestone.

The basin-floor association is the lower part of the sedimentary succession, discordantly overlying Archaean banded iron formation. It is 30 m thick and is predominantly characterized by pelagic nodular lime mudstone interbedded with minor shale with massive bedding and even parallel laminations (Fig. 3a). The lower- to upper-slope turbidites consist of fine-grained sandstone, mudstone with even, parallel lamination and abundant deformational structures, and massive diamictite associated with subordinate limestone (Fig. 3b). Both mudstone and fine-grained sandstone display convolute lamination and microfaults, while load-cast structures, breccias and ejection dykes are observed in the sandstone. Coarsening-upward siltstone and sandstone overlie the lower succession. The siltstone has massive bedding and parallel lamination, which changes upwards to cross-lamination and wavy beds. The sandstone shows cross-lamination truncated by wavy and parallel lamination; load casts, fine-grained intraclasts,

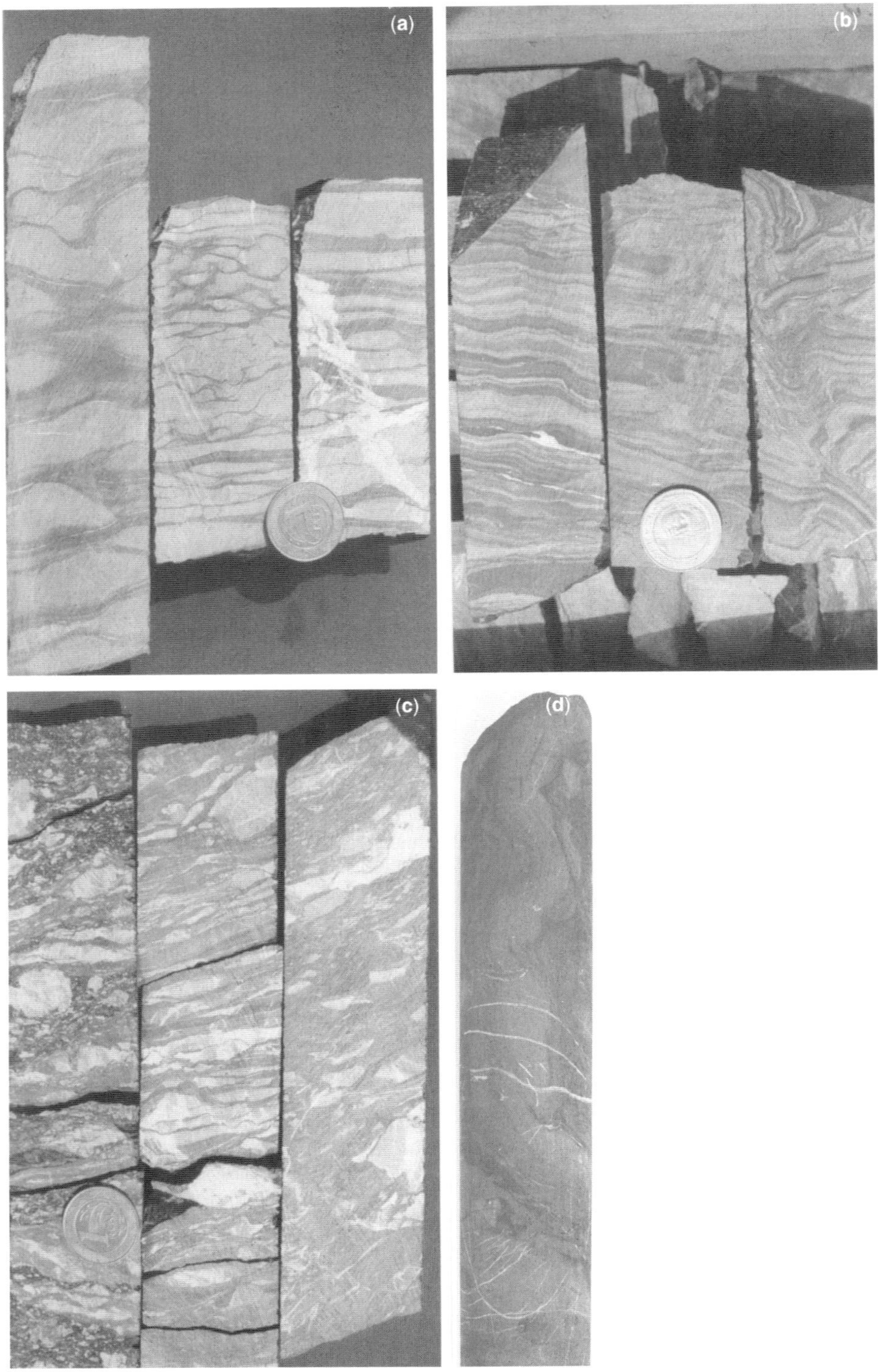

Fig. 3. Sedimentary features of very low-grade metasedimentary rocks of the Tocantins Group in the São Martim prospect drill core: (**a**) nodular lime mudstone; (**b**) siltstone and claystone rhythmite; (**c**) conglomerate and breccias with flow foliation, rotated clasts, microfaults and boudinage; (**d**) fine-grained argillaceous sandstone with contorted beds and slump folds.

injection dykes and convolute lamination are also present. These structures, along with fractures and syn-sedimentary faulting, indicate that liquefaction and fluidization processes affected these rocks. The fining-upward cycles of this succession suggests that turbidity currents were responsible for its formation, particularly the lower-slope turbidites. The upper part of the succession is composed of breccias and mudstones. The breccias exhibit flow foliation, contorted beds, and rotated intraclasts (Fig. 3c). The black matrix can be either clastic or carbonaceous with rounded and angulose clasts. Slump, load-cast and flame structures are common (Fig. 3d). The mudstone shows pseudo-bedding due to frequent stylolith surfaces. Gravitational flux of detritus seems to have been the major process forming the upper succession of the SMD-08 drill core. Plastic flow and pressure solution are probably related to diagenesis and can be partly attributed to the deformation associated with the formation of the Araguaia Belt. The upper part of the succession is interpreted as upper slope turbidite. Figure 4 shows a composite stratigraphic column of the SMD-8 drill core.

The facies associations of the Couto Magalhães Formation indicate two depositional environments. One is characterized by abundant occurrence of slumps and fluid-escape structures suggesting a slope environment; the other is of fine-grained distal sediments characteristic of lower slope and proximal basin floor environments, in which the presence of convolutions, slumps and load-cast structures is less evident. There is also a good correspondence between the coarsening-upward cycles and Bouma intervals, with or without the Ta and Te intervals. These characteristics may be attributed to turbidity currents with little capacity to erode the substratum, probably generated in the transition from the lower slope to the basin floor environment, where these sediments would be modified as the flow waned allowing the Ta interval to be largely preserved. The low mineralogical and textural maturity of these rocks led Figueiredo *et al.* (2006) to suggest a source area near to the basin. They also considered that the slope/fan deposits filled a deep foreland basin whose source area was to the east. The average thickness of the sedimentary cycles is around 17 m and may reach 40 m. This suggests a considerable accommodation space of the lower slope towards the basin floor, compatible with a deep foreland basin.

Single-zircon Pb-evaporation ages

Two quartzite samples of the Morro do Campo Formation (Estrondo Group) were collected for single-grain evaporation $^{207}Pb/^{206}Pb$ ages of detrital zircons. Sample BP-08 was collected from the northern part of the Araguaia Belt, near the city of Xambioá. From this sample, 69 zircon grains were randomly selected for mass spectrometry, and 50 gave Pb signals suitable for isotope analysis and age calculation. Most of the analysed grains were rounded to sub-rounded, and a few preserved prismatic faces. The zircon grains are mostly colourless to yellowish, and some show metamict portions and inclusions. The $^{207}Pb/^{206}Pb$ ages were interpreted as minima for the detrital zircon grains, and vary from 1623 $\pm$ 16 Ma to 3087 $\pm$ 17 Ma (Table 2). A frequency distribution diagram for these zircon ages indicates a positively skewed distribution with a mode in the 2800–2900 Ma interval (Fig. 5a); they suggest a major contribution from source rocks of Meso- and Neoarchaean ages with minor input from Palaeoproterozoic sources.

Sample BP-33 was collected near the city of Paraíso do Tocantins, in the southern region of the Araguaia Belt. Zircon grains are rounded to sub-rounded with few grains showing prismatic features. Fifty-four grains were analysed and 48 gave Pb signals suitable for mass spectrometric analysis and age calculations. The $^{207}Pb/^{206}Pb$ ages of these grains vary between 697 $\pm$ 28 Ma and 2796 $\pm$ 08 Ma (Table 3). A frequency histogram shows a bimodal distribution with the main mode in the 1000–1100 Ma interval and a secondary mode at 1800–1900 Ma (Fig. 5b). These ages suggest major inputs from Mesoproterozoic sources with some contribution from Palaeoproterozoic and Neoproterozoic sources.

Some limitations on the analytical method used to date the zircon grains have to be considered before further interpretations are advanced. Firstly, the number of dated zircon grains is not sufficient to perform a quantitative analysis. Secondly, the analytical procedure does not allow the analysis of the smaller grains ($<$0.125 mm) and this size population could not be represented. Finally, $^{207}Pb/^{206}Pb$ ages in minerals are apparent ages and ought to be interpreted as minimum ages, regardless of the analytical method used (Pb-evaporation, laser ablation, isotope dilution or even SHRIMP). Numerous papers comparing the $^{207}Pb/^{206}Pb$ single-zircon evaporation ages with U–Pb ages obtained by isotope dilution and SHRIMP have shown that these ages are comparable within the limits of the errors (Kober 1986; Ansdell & Kyser 1991; Kröner *et al.* 1994; Gaudette *et al.* 1998). However, the significance of the ages depends on the nature of the analysed grains. Metamict zircon grains or crystals, for instance, are not recommended for any of the above techniques. Zircon with mineral inclusions and inherited portions should be avoided in isotope dilution and Pb-evaporation analyses. The random selection of the zircon grains used for provenance studies implies the potential existence of unwanted grains in the selected minerals. This means that some of the zircon ages obtained in this work are minimum ages that may be close to the true ages, while others may not. Most of the $^{207}Pb/^{206}Pb$ data obtained for the Neo- and Mesoproterozoic zircon grains are minimum ages, but probably very near the true age, since the ages obtained in

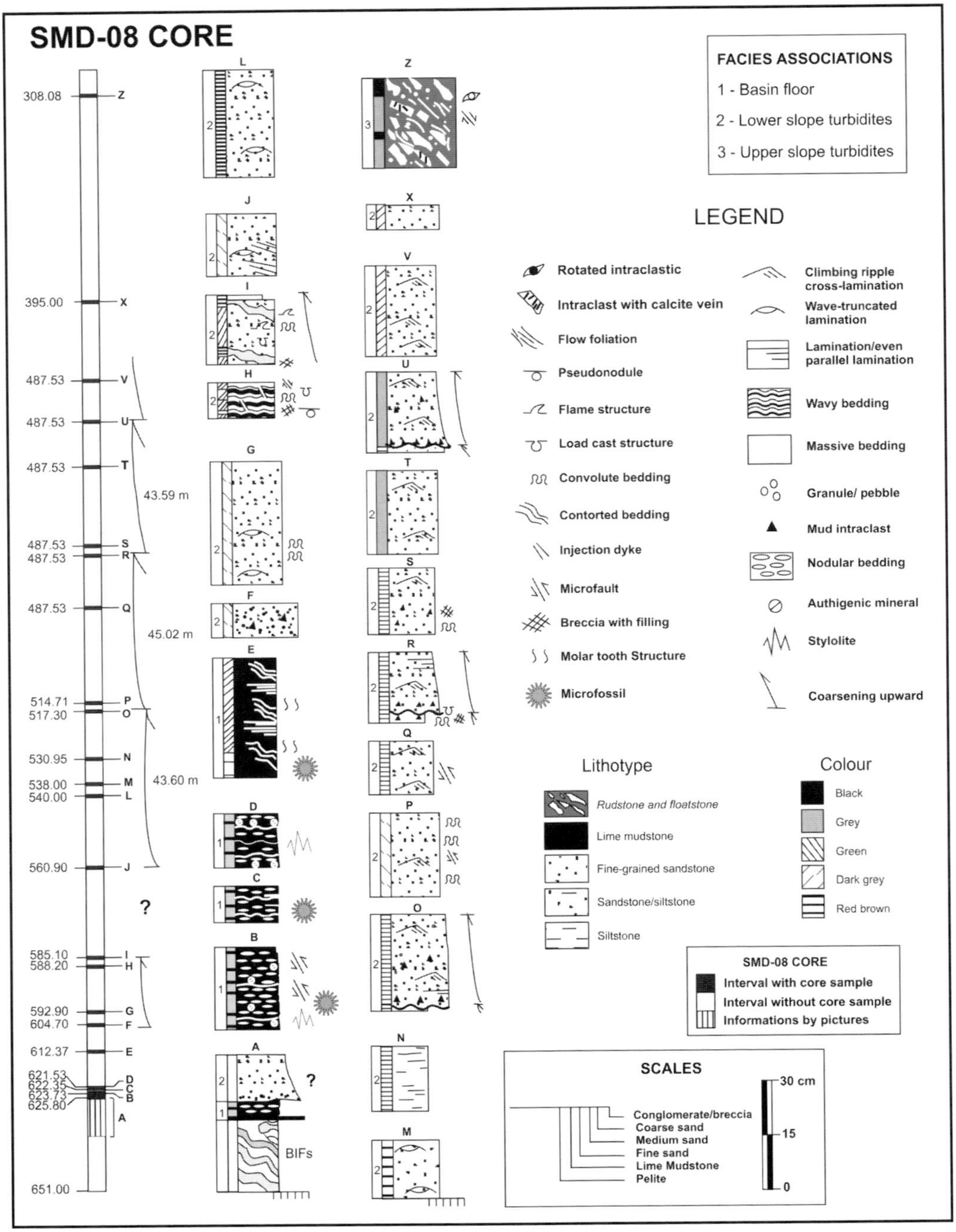

Fig. 4. Schematic stratigraphic column of the metasedimentary rocks of the Tocantins Group in the São Martim prospect.

Table 2. *$^{207}Pb/^{206}Pb$ ages for detrital zircon grains from Xambioá region quartzite*

Sample	$^{204}Pb/^{206}Pb$	$^{207}Pb/^{206}Pb$ ($\pm 2\sigma$)	Age Ma ($\pm 2\sigma$)
BP08/01	0.000159	0.18687 (037)	2715 (03)
BP08/02	0.000078	0.19881 (004)	2817 (03)
BP08/04	0.000186	0.18224 (049)	2674 (04)
BP08/05	0.000014	0.20440 (053)	2862 (04)
BP08/07	0.000065	0.18951 (056)	2738 (05)
BP08/08	0.000201	0.09995 (085)	1623 (16)
BP08/10	0.000024	0.20022 (044)	2828 (04)
BP08/11	0.000018	0.12875 (031)	2081 (04)
BP08/12	0.000026	0.12984 (071)	2096 (10)
BP08/14	0.000094	0.20790 (042)	2890 (03)
BP08/15	0.000189	0.18108 (202)	2663 (18)
BP08/16	0.000050	0.21159 (226)	2918 (17)
BP08/17	0.000210	0.16934 (047)	2551 (05)
BP08/18	0.000021	0.20269 (092)	2848 (07)
BP08/19	0.000360	0.19226 (062)	2762 (05)
BP08/22	0.000354	0.17925 (069)	2646 (06)
BP08/23	0.000055	0.23511 (254)	3087 (17)
BP08/25	0.000360	0.21138 (174)	2917 (13)
BP08/28	0.000254	0.15071 (102)	2354 (12)
BP08/29	0.000017	0.20506 (052)	2867 (04)
BP08/32	0.000080	0.11369 (091)	1859 (14)
BP08/33	0.000362	0.20124 (211)	2837 (17)
BP08/34	0.000198	0.18985 (089)	2741 (08)
BP08/35	0.000192	0.21854 (069)	2970 (05)
BP08/36	0.000036	0.21308 (038)	2929 (03)
BP08/37	0.000157	0.14737 (037)	2316 (04)
BP08/38	0.000323	0.21250 (066)	2925 (05)
BP08/39	0.000051	0.21654 (278)	2955 (21)
BP08/40	0.000795	0.20638 (021)	2878 (17)
BP08/41	0.000364	0.20262 (204)	2848 (16)
BP08/42	0.000074	0.12352 (165)	2008 (24)
BP08/43	0.000335	0.19458 (006)	2782 (05)
BP08/44	0.000278	0.19175 (094)	2758 (08)
BP08/45	0.000290	0.18955 (216)	2739 (19)
BP08/46	0.000072	0.20402 (106)	2859 (08)
BP08/47	0.000085	0.20153 (073)	2839 (06)
BP08/48	0.000122	0.18900 (061)	2734 (05)
BP08/50	0.000013	0.20370 (051)	2856 (04)
BP08/53	0.000104	0.16293 (096)	2487 (10)
BP08/56	0.000804	0.19388 (553)	2776 (47)
BP08/57	0.000134	0.20042 (164)	2830 (13)
BP08/60	0.000118	0.15327 (081)	2383 (09)
BP08/62	0.000515	0.17981 (778)	2651 (72)
BP08/63	0.000219	0.18037 (114)	2657 (10)
BP08/64	0.000026	0.20121 (096)	2836 (08)
BP08/65	0.000048	0.18737 (045)	2720 (04)
BP08/66	0.000080	0.19911 (235)	2819 (19)
BP08/67	0.000088	0.20447 (151)	2863 (12)
BP08/68	0.000069	0.20283 (013)	2849 (10)
BP08/69	0.000135	0.17832 (113)	2638 (11)

$^{207}Pb/^{206}Pb$ ratios corrected for common Pb using the two-stage model of Stacey & Kramers (1975).

the 1450 °C step of evaporation overlap, within the limits of errors, with the ages yielded by the highest temperature step (1500 °C). On the other hand, some of the Palaeoproterozoic and Archaean zircon grains probably display minimum ages that may not be very close to the true age because, generally, the ages obtained in the step of evaporation of 1500 °C are considerably older than the ages

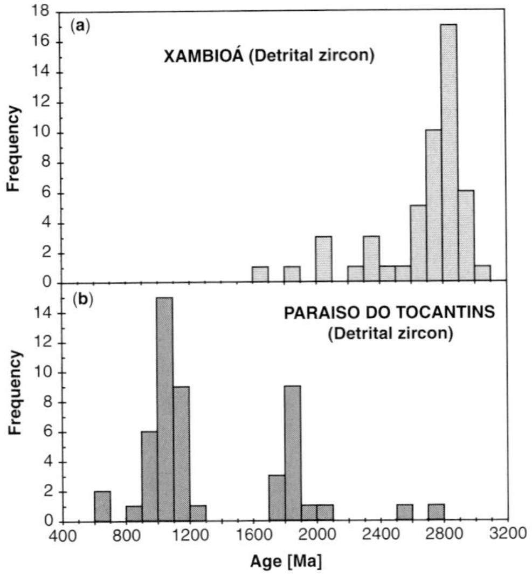

Fig. 5. Frequency histogram of the $^{207}Pb/^{206}Pb$ evaporation ages of detrital zircon grains of quartzites: (**a**) Xambioá region (sample BP-08); (**b**) Paraíso do Tocantins region (sample BP-33).

given by the lower temperature step (1450 °C). Despite these limitations, the single-zircon $^{207}Pb/^{206}Pb$ evaporation ages presented here permit the identification of contributions from different sources for the studied quartzites, and represent a valuable tool for provenance investigation (Fig. 5a, b).

Sm–Nd model ages

Sm–Nd crustal residence ages (T_{DM}) were determined in 44 samples of metasedimentary rocks collected from the Tocantins (20 samples) and Estrondo (24 samples) groups. Sampling was conducted in outcrops along the belt in order to cover the different rock units, but three samples of the Couto Magalhães Formation (Tocantins Group) were collected from drill cores (SMD-03, SMD-08) of the São Martim propect (Fig. 2). The Tocantins Group samples were mica schist, phyllite, slate and meta-siltstone, whereas the Estrondo Group samples were mica schist, graphite schist, and mica schists with garnet, staurolite and kyanite. In some cases, weathered samples were collected because fresh rocks could not be found. Sampling of weathered rocks is not critical for T_{DM} age calculations since, in general, the Sm–Nd system is not fractionated by supergene processes, and the sediments maintain the T_{DM} ages of their sources (Goldstein & Jacobsen 1988; Goldstein *et al.* 1997).

The Sm–Nd isotope data are shown in Table 4, along with the rock types and the determined T_{DM} model ages. The values of the fractionation factor $f(Sm/Nd)$ are also given as a parameter to evaluate the geological meaning of the calculated T_{DM} ages. For crustal rocks that have evolved from juvenile compositions in one stage, the value of this parameter is expected to be between -0.60 and -0.35 (Sato & Siga Jr. 2000). In general, the $f(Sm/Nd)$ values obtained in this study were in this interval, but some of the weathered rock samples are slightly above this range. Two samples were not considered for T_{DM} ages calculations because their $f(Sm/Nd)$ values were far from the expected interval.

The Sm–Nd T_{DM} ages of the metasedimentary rocks of the Baixo Araguaia Supergroup along the Araguaia Belt are shown in Figure 6. The frequency histogram of the Sm–Nd T_{DM} ages of these metasedimentary rocks has a bimodal distribution, with ages scattering around 1600–1700 Ma interval and, less frequently, 2400–2700 Ma (Fig. 7a). The crustal residence ages of the rocks from the northern and central-southern portions of the Araguaia Belt are mainly concentrated in the younger age interval (Fig. 7b, c). The older age interval (>2.4 Ga) is more frequent in the rocks of the northern segment. The younger T_{DM} ages from samples of the Estrondo Group are more abundant and define a main mode in the 1600–1700 Ma interval, with ages ranging from 1500 to 2200 Ma (Fig. 7d). Six samples display ages older than 2500 Ma. Similarly, the younger T_{DM} ages are also more common in samples of the Tocantins Group and are distributed in the range 1300–1800 Ma, without defining a main mode (Fig. 7e); four samples fall in the 2000–2500 Ma interval and only two samples show ages older than 2500 Ma.

The εNd values of these samples were calculated at 900 Ma (εNd_{900}) as the probable maximum age of deposition of the metasedimentary rocks of the Araguaia Belt, based on the 1000–1100 Ma mode of the zircon ages of the quartzites in the southern segment of the belt and the occurrence of scapolite meta-gabbros with a $^{207}Pb/^{206}Pb$ zircon age of 817 ± 5 Ma (Gorayeb *et al.* 2004), associated with metasedimentary rocks in the northern portion of the belt. The values of εNd_{900} are all negative, varying from -3.15 to -23.8 (Table 4). The T_{DM} ages (1.45–3.15 Ga) and εNd_{900} values suggest that the metasedimentary rocks of the Araguaia Belt may store contributions from sources with both long and short crustal residence ages, and probably represent a mixture of these in different proportions.

Table 3. *$^{207}Pb/^{206}Pb$ ages for detrital zircon grains from Paraíso do Tocantins region quartzite*

Sample	$^{204}Pb/^{206}Pb$	$^{207}Pb/^{206}Pb$ ($\pm 2\sigma$)	Age Ma ($\pm 2\sigma$)
BP33/01	0.000217	0.07474 (004)	1062 (11)
BP33/03	0.000475	0.11111 (057)	1818 (09)
BP33/04	0.000082	0.07318 (124)	1019 (34)
BP33/05	0.000180	0.07847 (233)	1159 (59)
BP33/06	0.000124	0.10893 (003)	1782 (05)
BP33/07	0.000000	0.07614 (192)	1099 (51)
BP33/08	0.000113	0.07570 (085)	1087 (23)
BP33/09	0.000090	0.11185 (168)	1830 (27)
BP33/10	0.000063	0.07433 (072)	1051 (19)
BP33/12	0.000242	0.06808 (186)	871 (56)
BP33/13	0.000004	0.07585 (018)	1091 (05)
BP33/15	0.000091	0.07650 (065)	1108 (17)
BP33/16	0.000192	0.07564 (135)	1086 (36)
BP33/17	0.000346	0.07182 (039)	981 (11)
BP33/18	0.000026	0.07607 (008)	1097 (21)
BP33/19	0.000000	0.07812 (044)	1150 (32)
BP33/20	0.000344	0.07680 (083)	1116 (11)
BP33/22	0.000159	0.06266 (044)	697 (28)
BP33/23	0.000247	0.07615 (045)	1099 (12)
BP33/24	0.000015	0.07721 (095)	1127 (12)
BP33/25	0.000191	0.19632 (049)	2796 (08)
BP33/26	0.000136	0.07514 (044)	1073 (13)
BP33/27	0.000050	0.07281 (052)	1009 (12)
BP33/28	0.000049	0.12622 (075)	2046 (07)
BP33/30	0.000536	0.07142 (075)	970 (21)
BP33/34	0.000139	0.07179 (016)	980 (05)
BP33/35	0.000198	0.12247 (146)	1993 (21)
BP33/36	0.000095	0.07246 (055)	999 (15)
BP33/37	0.000059	0.07740 (129)	1132 (33)
BP33/38	0.000140	0.07491 (034)	1066 (09)
BP33/39	0.000065	0.07302 (017)	1015 (05)
BP33/40	0.000016	0.16897 (086)	2548 (08)
BP33/41	0.000249	0.10919 (036)	1786 (06)
BP33/43	0.000044	0.11231 (102)	1837 (16)
BP33/44	0.000128	0.11450 (038)	1872 (06)
BP33/45	0.000000	0.08029 (174)	1204 (43)
BP33/46	0.000058	0.11376 (046)	1861 (07)
BP33/47	0.000062	0.07061 (016)	946 (05)
BP33/49	0.000975	0.06123 (123)	647 (43)
BP33/50	0.000190	0.07476 (106)	1062 (29)
BP33/51	0.000232	0.07496 (015)	1068 (40)
BP33/53	0.000062	0.07225 (033)	993 (09)
BP33/54	0.000021	0.11171 (229)	1828 (37)
BP33/55	0.000044	0.11425 (103)	1869 (16)
BP33/56	0.000080	0.11289 (061)	1847 (10)
BP33/57	0.000071	0.10582 (099)	1729 (17)
BP33/59	0.000079	0.07688 (027)	1118 (07)
BP33/60	0.000400	0.07685 (067)	1118 (17)

$^{207}Pb/^{206}Pb$ ratios corrected for common Pb using the two-stage model of Stacey & Kramers (1975).

Discussion

Provenance and Palaeogeography

The $^{207}Pb/^{206}Pb$ evaporation ages in single zircon grains indicate an Archaean source for the quartzites of the northern portion of the Araguaia Belt (Xambioá region), with minor contribution of Palaeoproterozoic detritus (Fig. 5a). On the other hand, the $^{207}Pb/^{206}Pb$ evaporation ages from single zircon grains of the quartzite from the Paraíso do Tocantins region are mainly

Table 4. *Sm–Nd data for metasedimentary rocks of the Baixo Araguaia Supergroup, Araguaia Belt*

Sample	Lithotype	Sm	Nd	$f_{(Sm/Nd)}$	$^{147}Sm/^{144}Nd$	$^{143}Nd/^{144}Nd$	$\varepsilon Nd_{(900)}$	T_{DM} (Ga)
BP/01	Garnet–biotite schist	5.03	26.63	−0.42	0.11428	0.511761	−7.63	1.97
BP/02	Garnet–biotite schist	6.43	32.64	−0.39	0.11913	0.511788	−7.67	2.02
BP/03	Biotite–muscovite schist	4.99	26.79	−0.43	0.11262	0.511782	−7.03	1.90
BP/04	Biotite schist	5.82	29.33	−0.39	0.12003	0.511928	−5.03	1.81
BP/05	Biotite schist	6.49	32.82	−0.39	0.11956	0.512013	−3.32	1.67
BP/06	Staurolite–garnet mica schist	6.68	36.75	−0.44	0.10985	0.511066	−20.71	2.93
BP/07	Biotite schist	3.82	19.47	−0.4	0.11858	0.511538	−12.49	2.42
BP/09	Biotite–muscovite schist	4.93	24.47	−0.38	0.12184	0.511992	−3.99	1.74
BP/10	Kyanite–biotite schist	6.45	31.73	−0.38	0.12289	0.512024	−3.49	1.71
BP/11	Biotite–muscovite schist	6.3	31.22	−0.38	0.12204	0.512021	−3.45	1.70
BP/12	Biotite–muscovite schist	3.07	15.64	−0.4	0.11861	0.511429	−14.62	2.61
BP/13	Graphite schist with garnet	5.79	28.02	−0.37	0.12483	0.511484	−14.27	2.7
BP/14	Mica schist (weathered)	7.27	43.01	−0.48	0.10222	0.511816	−5.17	1.68
BP/15	Mica schist (weathered)	1.98	8.99	−0.32	0.1332	0.511642	−12.14	2.68
BP/16	Phyllite (weathered)	3.57	14.83	−0.26	0.14562	0.511958	−7.4	2.44
BP/17	Phyllite (weathered)	6.22	35.26	−0.46	0.10672	0.512027	−1.46	1.45
BP/18	Phyllite (weathered)	2.93	13.34	−0.32	0.13259	0.511941	−6.27	2.08
BP/19	Magnetite phyllite (weathered)	3.63	19.89	−0.44	0.11034	0.51091	−23.82	3.19
BP/20	Phyllite (weathered)	2.46	8.42	−0.1	0.17704	0.511929	−11.59	–
BP/21	Phyllite (weathered)	5.97	32.75	−0.44	0.11026	0.512002	−2.46	1.53
BP/22	Phyllite (weathered)	12.9	71.77	−0.45	0.10863	0.512021	−1.9	1.48
BP/23	Mica schist (weathered)	4.54	23.7	−0.41	0.11581	0.511981	−3.51	1.65

BP/24	Mica schist (weathered)	4.61	23.16	−0.39	0.12039	0.512007	−3.53	1.69
BP/25	Garnet schist (weathered)	2.88	14.06	−0.37	0.12401	0.511878	−6.47	1.98
BP/26	Mica schist (weathered)	7.81	38.71	−0.38	0.12196	0.511559	−12.47	1.98
BP/27	Mica schist (weathered)	6.01	31.85	−0.42	0.11403	0.511945	−4.01	2.48
BP/28	Garnet schist (weathered)	9.64	35.62	−0.17	0.16354	0.511361	−21.14	–
BP/29	Graphite schist	11.08	48.01	−0.29	0.13948	0.511824	−9.31	2.52
BP/30	Mica schist (weathered)	13.13	81.45	−0.5	0.09748	0.511891	−3.15	1.51
BP/31	Mica schist (weathered)	2.83	14.74	−0.41	0.11602	0.511418	−14.54	2.55
BP/32	Staurolite–garnet schist (weathered)	4.6	22.71	−0.38	0.12237	0.511756	−8.67	2.15
BP/34	Mica schist (weathered)	6.97	35.33	−0.39	0.11925	0.511912	−5.26	1.82
BP/35	Mica schist (weathered)	2.02	7.94	−0.22	0.15409	0.511884	−9.82	3.02
BP/36	Magnetite phyllite (weathered)	9.69	59.76	−0.5	0.09799	0.511999	−1.1	1.38
BP/37	Chlorite–biotite schist	6	29.77	−0.38	0.12183	0.511942	−4.97	1.83
BP/38	Phyllite (weathered)	5.9	32.88	−0.45	0.10848	0.511942	−3.43	1.59
BP/39	Phyllite (weathered)	9.43	43.93	−0.34	0.12973	0.511746	−9.71	2.37
BP/40	Slate	13.56	61.23	−0.32	0.13387	0.511859	−7.98	2.27
BP/41	Mica schist (weathered)	4.9	24.96	−0.4	0.11862	0.511988	−3.7	1.69
BP/42	Biotite schist	4.6	23.98	−0.41	0.1159	0.511998	−3.19	1.63
BP/43	Mica schist (weathered)	19.91	98.09	−0.38	0.12274	0.511963	−4.66	1.81
SMD-03A	Metasiltite	5.5	29.74	−0.43	0.11173	0.511929	−4.05	1.66
SMD-03B	Metasiltite	6.86	35.2	−0.4	0.11782	0.511966	−4.03	1.71
SMD − 08	Metasiltite	7.06	36.68	−0.41	0.1164	0.511944	−4.3	1.72

concentrated between 1000 Ma and 1200 Ma, suggesting a dominant contribution from Mesoproterozoic sources (Fig. 5b). Palaeoproterozoic (1800–1900 Ma) and Neoproterozoic ages are also recorded. The $^{207}Pb/^{206}Pb$ evaporation ages in single zircon grains from the quartzites of the northern and southern segments of the Araguaia Belt are quite distinct, and suggest that they had different provenance. The most appropriate and natural candidate for a source area is the adjacent Amazonian Craton, constituted mainly by Archaean and Palaeoproterozoic terranes (Tassinari & Macambira 1999). The ages of the rock units in the southeastern region of the Amazonian Craton,

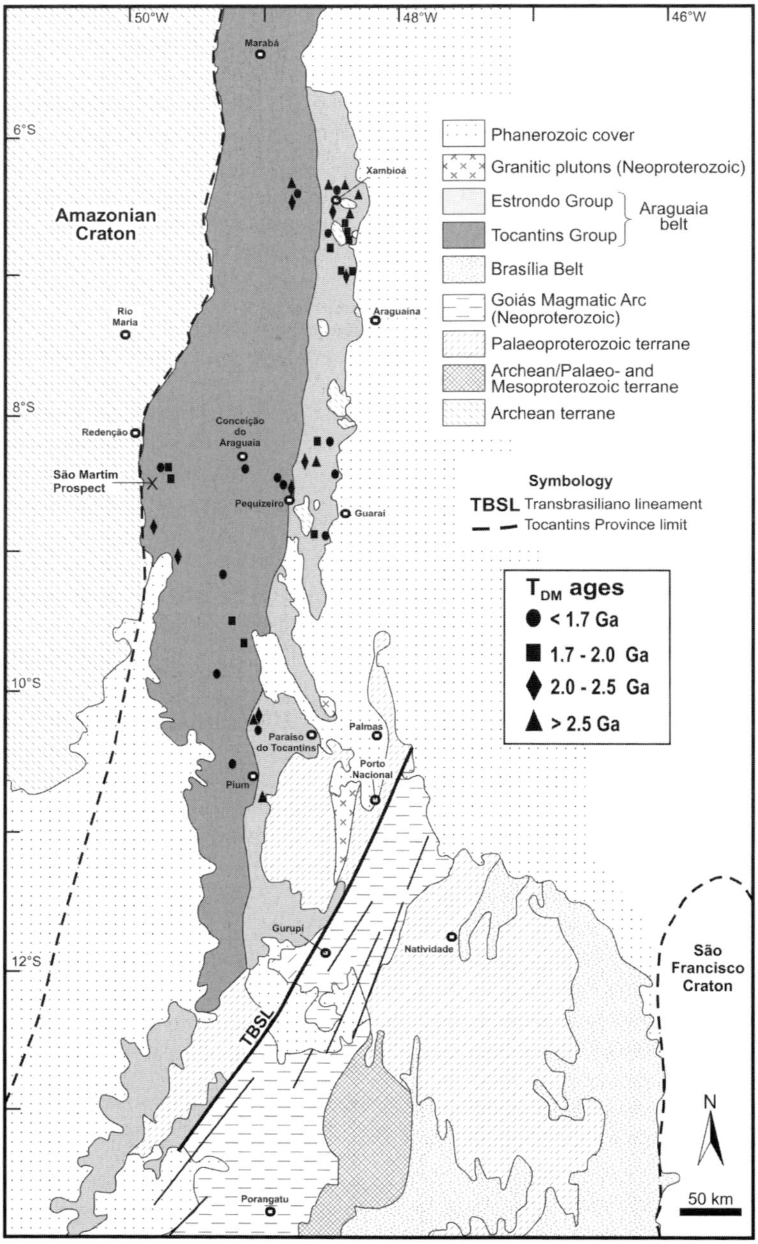

Fig. 6. Representation of the Sm–Nd T_{DM} ages on a map of the Baixo Araguaia Supergroup along the Araguaia Belt. The Archaean and Proterozoic terranes of the surrounding areas of the belt are also indicated.

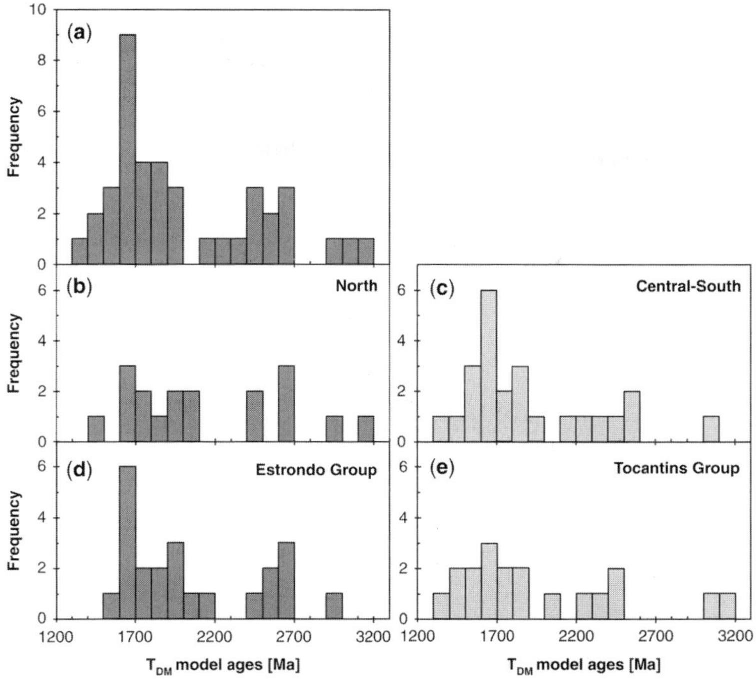

Fig. 7. Frequency histograms of Sm–Nd T_{DM} model ages of metasedimentary rocks of the Baixo Araguaia Supergroup, Araguaia Belt: (**a**) all samples together; (**b**) northern segment (Xambioá region); (**c**) central-southern segment (Conceição do Araguaia regions); (**d**) Estrondo Group; (**e**) Tocantins Group.

for instance, range from 2500 to 3000 Ma (Table 1). In spite of this, it may be an oversimplification to consider the Amazonian Craton the only source for the quartzites of the Estrondo Group, since this cratonic region could hardly supply the 1000–1200 Ma detrital zircons found in the quartzites of the southern segment of the Araguaia Belt. Tassinari & Macambira (1999) recognized the Rondonian–San Ignácio (1500–1300 Ma) and the Sunsas (1250–1000 Ma) geochronological provinces in the western part of the Amazonian Craton, but these Mesoproterozoic provinces are very far from the Araguaia Belt and a closer source has to be found to account for the higher proportion of 1000–1200 Ma detrital zircon grains present in these quartzites.

The input of sediments from sources located to the east of the Araguaia Belt needs to be considered to explain the detrital zircon ages found in the quartzites of the Araguaia Belt. The number of recognized Archaean blocks or microplates in the South American Platform has increased with the improvement of geochronological data. The extent of the Palaeoproterozoic orogenic collage of these Archaean blocks and microplates suggests a large continental mass, now dispersed throughout South America and Africa (Brito Neves 1999). Detrital zircon ages of around 1100 Ma were obtained by Valeriano *et al.* (2004) for the metasedimentary rocks of the Brasilia Belt; they suggested that Mesoproterozoic terranes underlying the Bambui Group in much of the southern portion of the São Francisco Craton, could be the main source. Other possible sources for these detrital zircons are the rock units incorporated into the Goiás Massif, where mafic–ultramafic complexes and volcano-sedimentary sequences with zircon ages around 1250 Ma have been described (Suita *et al.* 1994; Moraes *et al.* 2006).

The Sm–Nd crustal residence ages help to constrain the provenance of the sediments of the Araguaia Belt. As noted above, their T_{DM} ages have a main mode of 1600–1700 Ma, more evident for the rocks of the Estrondo Group than for those of the Tocantins Group, but for both units most are lower than 2.0 Ga (Fig. 7a, d, e). These T_{DM} ages are probably a result of mixing of older and younger crustal detritus. The older sources may be of Archaean and/or Palaeoproterozoic ages, and the younger sources must be Neoproterozoic (900–600 Ma) and/or Mesoproterozoic (1200 Ma or less) juvenile crust. The possible Archaean sources are the Amazonian and São Francisco cratons (Figs 2 & 6) and the Goiás

Massif, where Archaean ages have been documented (Table 1). The probable Palaeoproterozoic sources are in the Amazonian and São Francisco cratons, and the Palaeoproterozoic gneisses and granulites complexes to the east of the Araguaia Belt and in the northern segment of the Brasilia Belt (Figs 2 & 6). Palaeoproterozoic rock units have also been identified in the Goiás Massif (Table 1). Unfortunately, Palaeozoic sedimentary rocks hide the relationship between the Araguaia Belt and other Precambrian terranes located to the east, where other Proterozoic terranes, covered by sediments since Phanerozoic times, could also be the source of the sediments of the Baixo Araguaia Supergroup. This may include the concealed Parnaíba block (Nunes 1993) and other terranes in the Borborema Province (Fig. 1).

As mentioned above, in order to explain the T_{DM} ages around 1600–1700 Ma, a contribution from a younger juvenile source is necessary. Available nearby sources are mainly the terranes of the Neoproterozoic Goiás Magmatic Arc (Table 1, Figs 2 & 6). This contribution is demonstrated by the Nd-isotope evolution diagram of terranes presently located SE of the Araguaia Belt (Fig. 8a). The eventual contribution from other possible Proterozoic landmasses

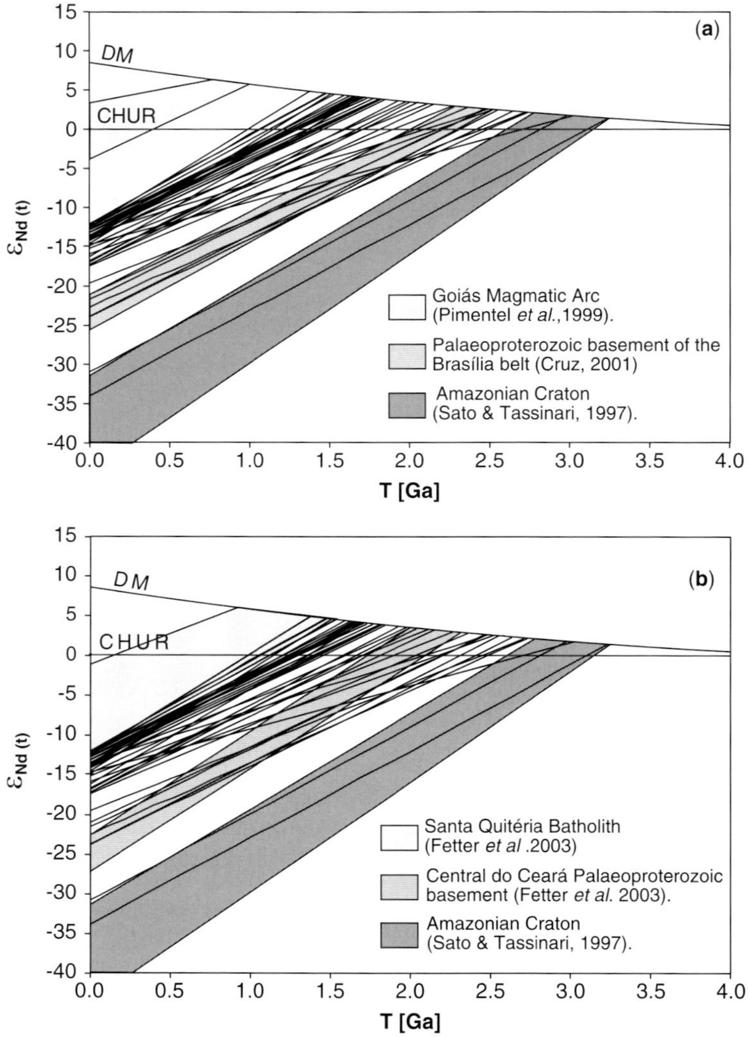

Fig. 8. Nd-isotope evolution diagrams for the metasedimentary rocks of the Baixo Araguaia Supergroup and possible source areas: (**a**) Amazonian Craton, Proterozoic basement of the Brasília Belt and Goiás Magmatic Arc; (**b**) Amazonian Craton and terranes of the Borborema province, NE Brazil.

located to the east, is illustrated by data of the Borborema Province (Fig. 8b). In both diagrams, data from the Amazonian Craton are used as reference for the evolution of the Archaean rocks. These diagrams illustrate the interpretation that the T_{DM} ages of the metasedimentary rocks of the Baixo Araguaia Supergroup represent mixing between Neoproterozoic sources and older crust (Archaean and/or Palaeoproterozoic).

Further constraints on the possible sources of the metasedimentary rocks of the Baixo Araguaia Supergroup come from the available structural data. Low-angle mineral and stretching lineation (10–20°/110) recorded in both metasedimentary rocks and basement orthogneisses indicate tectonic transport from SE to NW. Thus areas located to the SE of the Araguaia Belt are the most probable sources. The Goiás Massif, the Goiás Magmatic Arc, the São Francisco Craton, and even the concealed Paranapanema block seem to be the most likely candidates (Figs 1 & 2).

Facies analysis of the rocks of the Couto Magalhães Formation carried out on the SMD-08 drill core indicates deep-sea deposits probably related to a slope–apron setting. The low mineralogical and textural maturity of these rocks led Figueiredo *et al.* (2006) to suggest a proximal source area to the east. This interpretation is supported by the T_{DM} model ages for the rocks of this core, which range between 1660 and 1720 Ma (Table 4). These T_{DM} ages are consistent with the interpretation of mixing with material from a younger (Meso?–Neoproterozoic) juvenile crust, which was available to the east of the Araguaia Belt. The Archaean and Palaeoproterozoic rocks of the Amazonian Craton could not be the main source of these metasedimentary rocks, since the São Martin prospect is very close to this cratonic region.

According to Valeriano *et al.* (2004) the geotectonic structure of the Paraguay–Araguaia Belt may have been formed 50 to 100 million years after that of the Brasília Belt, which resulted from the collision of the Paranapanema block, the Goiás Massif and the terranes of the Goiás Magmatic Arc with the São Francisco Craton. Thus, the tectonic evolution of the Araguaia Belt may have been preceded by the construction of a palaeocontinental collage of different Archaean and Palaeoproterozoic crustal blocks, in addition to some Neoproterozoic juvenile terranes (magmatic arcs). This palaeocontinent formation probably occurred at 600–700 Ma according to Valeriano *et al.* (2004). Tectonic events in this age interval are recorded in the Neoproterozoic belts surrounding the eastern margin of the West African Craton (Villeneuve & Cornée 1994; Trompette 1997) and, in Brazil, in the northwestern part of the Borborema Province (Fetter *et al.* 2003) and in the Goiás Magmatic Arc (Pimentel *et al.* 2000). Deep-water sediments accumulated in the associated marginal ocean of this palaeocontinent, with a mixed T_{DM} signature as a result of different contributions of Neoproterozoic, Mesoproterozoic and Palaeoproterozoic juvenile crustal rocks, along with recycled Archaean sediments or crustal rocks. The subsequent oblique collision of this palaeocontinent with the Amazonian Craton caused the northwestward tectonic transport of the deep-water deposits along with slabs of Neoproterozoic oceanic crust. These successions were thrust over the eastern margin of the Amazonian Craton and metamorphosed, resulting in the formation of the Araguaia Belt and as this landmass docked against the Amazonian Craton. The minimum age of this collision, and formation of the Araguaia Belt, may be constrained by the 549 ± 5 Ma syntectonic Ramal do Lontra granite (Alves 2006) associated with the rocks of the Estrondo Group.

This scenario is in agreement with a palaeogeographical reconstruction of South American cratonic fragments during West Gondwana amalgamation that includes the existence of the Brasiliano Ocean between the São Francisco and the Amazonian cratons (Cordani *et al.* 2003). This large ocean also separated smaller crustal fragments (e.g., Goiás Massif) and magmatic arcs (e.g., Goiás Magmatic Arc). The closure of this ocean culminated with the amalgamation of these landmasses to the Amazonian Craton. Deep seismic refraction data has identified the subsurface structure of these different terranes and suggested possible underthrusting of the mafic lower crust of the Araguaia Belt beneath the Goiás Magmatic Arc (Soares *et al.* 2006).

Correlation of the Araguaia and Rokelide belts

A number of authors have suggested correlation of Araguaia and Rokelide belts (Brito Neves & Cordani 1991; Trompette 1994; Almeida *et al.* 2000; Brito Neves *at al.* 2001). However, extensive Phanerozoic cover and limited geological knowledge on the West African side still keep this suggestion in the speculative field. At present, tentative examination of this correlation relies on studies conducted mainly in the 1980s, since few geological data have been generated in these belts in recent years that could help to address this subject.

The Rokelide Belt (Fig. 1) is the southern branch of a larger orogen that bordered the western edge of the West African Craton, including the Bassaride and Mauritanide belts. These orogenic belts record a polyphase thermo-tectonic evolution showing two Pan-African and Hercynian orogenic events

(Villeneuve & Dallmeyer 1987). The Bassaride Belt records the older event (*c.* 660 Ma), named Pan-African I, whereas the younger Pan-African II event (*c.* 550 Ma), is best identified in the Rokelide Belt (Villeneuve & Dallmeyer 1987). The Mauritanide Belt formed at the same time as the Bassaride Belt, but was largely reworked by the Hercynian orogen (Villeneuve & Cornée 1994).

The main geological unit of the Rokelide Belt is the Rockel River Group, which is divided into six formations (Williams & Culver 1982). The basal Tabe Formation lies unconformably on Archaean gneiss and is composed of tillites, turbidites and lacustrine rhythmites with dropstones. It is succeeded by the Makani (orthoquartzite and subarkose) and Teye (shale and turbidite) formations. Overlying these units are the Taia (shale and mudstone) and Mabole (shale, siltstone, arkose and orthoquartzite) formations. Within these uppermost formations there are lenses of calc-alkaline lavas and pyroclastic rocks (Kasewe Hills Formation).

The Rokel River Group outcrop is an inclined synclinorium, with a partially thrust-faulted western margin. The major fold axis direction is NNW with axial planes dipping SW. Low-angle westerly dipping thrusts parallel the planes of major folds. Greenschist-facies metamorphism is restricted to the western part of the synclinorium and to zones of cataclastic deformation in the contact with the high-grade gneisses and granulites of the Kasila Group. This unit, along with the lower-grade supracrustal rocks of the Marampa Group, occurs to the west of the Rockel River Group. They are considered to be reworked Archaean rocks, thermally affected and deformed during the Rokelide event (Williams & Culver 1982). However, U–Pb zircon geochronology conducted by Delor *et al.* (2005) in the Kasila Group in Guinea demonstrated that some of the Rokelide high-grade gneisses are products of the Neoproterozoic tectono-thermal event and not simply reactivation of Archaean high-grade terranes. These authors reported an age of 558 ± 7 Ma for a Kasila Group mafic granulite that was interpreted as the age of granulite metamorphism. Additionally, they reported ages of 573 ± 8 Ma and 571 ± 7 Ma for tonalite and pyroxene-bearing granites, respectively, and interpreted them as the ages of magma emplacement.

This brief description of the geology of the Rokelide Belt highlights some similarities and differences to that of the Araguaia Belt. The most striking similarity is the age of formation of these belts (*c.* 550 Ma). They also share the NNW trend of the structures, although they have opposite vergence: those in the Araguaia Belt display a westward vergence (toward the Amazonian Craton), while in the Rokelide Belt the vergence is eastward (toward the West African Craton). The metamorphism in these belts has also opposite polarity. The metamorphic grade decreases westwards in the Araguaia Belt, while it increases in this direction in the Rokelide Belt. Glacial deposits have not been reported so far from the Araguaia Belt, although they have been described in the Paraguay Belt (Trompette 1994; Nogueira *et al.* 2003). The sources of the metasedimentary rocks of the Araguaia Belt were probably those terranes presently located to the east/southeast of the belt. Provenance information is not available for the Rokelide Belt, but Williams & Culver (1982) considered that the uppermost deltaic deposits were derived from the east. Finally, the tectonic setting of these belts may not have been the same. The Rokelide Belt is considered to have formed in an intracontinental trough (Villeneuve & Cornée 1994). The same type of environment was also proposed for the Araguaia Belt (Villeneuve & Cornée 1994; Alvarenga *et al.* 2000), but the provenance study conducted here demonstrates that a more complex evolution has to be considered for the Araguaia Belt, since the source areas of the sediments are an amalgamation of terranes of Archaean to Neoproterozoic ages. A larger ocean (Brasiliano Ocean) separating the São Francisco Craton from the Amazonian Craton is suggested in the palaeogeographical reconstruction of Gondwana presented by Cordani *et al.* (2003).

The geochronological data available for the Araguaia and Rokelide belts indicate that the Pan-African II event (*c.* 550 Ma) was widespread along this portion of Gondwana, and they share the same metamorphic timing. Considering the geographical distribution of these belts, this is a strong argument to suggest the continuation of the Araguaia Belt into the African continent where the Rokelide Belt would be its counterpart. The geological differences (e.g., vergence, metamorphic polarity, tectonic setting) need to be further discussed as more geological and geophysical information becomes available.

Conclusions

Detrital zircon crystals from quartzite from the northern segment of the Araguaia Belt (Xambioá region) display Archaean ages with a mode in the 2800–2900 Ma interval. In contrast, detrital zircon from quartzite of the southern segment (Paraíso do Tocantins region) indicates Mesoproterozoic rocks (1000–1200 Ma) as the main sources, with some Palaeoproterozoic and Neoproterozoic contributions. Sm–Nd crustal residence ages of the metasedimentary rocks of the Baixo Araguaia Supergroup have a bimodal distribution, with abundant ages around 1600–1700 Ma and, less frequently, 2400–2700 Ma. These ages are interpreted as mixed ages due to contributions from Archaean/Palaeoproterozoic and younger (Meso-Neoproterozoic) terranes. The Neoproterozoic contribution is also recorded by detrital zircon ages in the southern segment of the Araguaia Belt. Palaeoenvironmental interpretations indicate deep-sea deposits probably related to a slope–apron setting that consists of two facies associations: (i) basin floor, characterized by predominantly pelagic nodular lime mudstone and shales and (ii) lower- to upper-slope turbidites, constituted by fine-grained sandstone, mudstone with even, parallel lamination and abundant deformational structures, and massive diamictite (floatstones) associated with subordinate limestone.

Based on these data, it is suggested that the Archaean rocks of the Amazonian Craton were not the main source of the metasedimentary rocks of the Baixo Araguaia Supergroup. The major sources were those areas currently at the eastern/southeastern side of the Araguaia Belt, such as the São Francisco Craton, Paranapanema block, Goiás Massif and Goiás Magmatic Arc. The amalgamation of these blocks represented an earlier stage of West Gondwana assembly that formed the Brasilia Belt. This collage resulted in a palaeocontinent and an associated oceanic basin that accumulated deep-water sediments. The subsequent oblique collision of this palaeocontinent with the Amazonian Craton caused northwestward tectonic transport of these deep-water deposits along with oceanic crust slabs. These successions were metamorphosed and thrust over the eastern margin of the Amazonian Craton, resulting in the formation of the Araguaia Belt and docking of this landmass against the Amazonian Craton; 550 Ma syntectonic granite magmatism in the Araguaia Belt constrains the age of this collision. A U–Pb zircon age of 558 ± 7 Ma for high-grade gneisses of the Rokelide Belt suggests that this tectono-thermal event was widespread along this portion of West Gondwana, and that the Araguaia Belt shares the metamorphic timing of the Rokelide Belt, although different tectonic settings may be argued for their evolution. Geochronological data suggest the continuation of the Araguaia Belt into the African continent having the Rokelide Belt as a counterpart. Nonetheless, more geological and geophysical data are required to better understanding this correlation.

This work was partially funded by CNPq project (Procs. 478173/2003-2 and 478395/2004-3), and PRONEX/CNPq/UFPA project (66.2103/1998). The Western Mining Company (WMC) made available the drill core of the São Martim Prospect.

References

ABREU, F. A. M. 1978. O Supergrupo Baixo Araguaia. *In*: *Congresso Brasileiro de Geologia 30, Recife, Anais*. Sociedade Brasileira de Geologia, **2**, 539–545.

ABREU, F. A. M. & GORAYEB, P. S. S. 1994. Tectônica e inversão metamórfica no Cinturão Araguaia. *In*: *Simpósio de Geologia da Amazônia 4, Belém, Anais*. Sociedade Brasileira de Geologia–Núcleo Norte, 1–4.

ALMEIDA, F. F. M., HASUI, Y., BRITO NEVES, B. B. & FUCK, R. A. 1981. Brazilian structural provinces; an introduction. *Earth-Science Reviews*, **17**, 1–29.

ALMEIDA, F. F. M., BRITO NEVES, B. B. & DAL RÉ CARNEIRO, C. 2000. The origin and evolution of the South American Platform. *Earth-Science Reviews*, **50**, 77–111.

ALVARENGA, C. J. S., MOURA, C. A. V., GORAYEB, P. S. S. & ABREU, F. A. M. 2000. Paraguay and Araguaia belts. *In*: CORDANI, U. G., MILANI, E. J., THOMAZ FILHO, A. & CAMPOS, D. A. (eds) *Tectonic Evolution of South America. 31st International Geological Congress, Rio de Janeiro*, 183–193.

ALVES, C. L. 2006. *Petrologia, geoquímica e geocronologia do Granito Ramal do Lontra e sua relação com a tectônica e metamorfismo do Cinturão Araguaia, Xambioá (TO)*. Masters Dissertation, Universidade Federal do Pará.

ANSDELL, K. M. & KYSER, T. K. 1991. Plutonism, deformation and metamorphism in the Proterozoic Flin Flon greenstone belt, Canada: limits on timing provided by the single zircon Pb-evaporation technique. *Geology*, **19**, 518–521.

ARCANJO, S. H. S. & MOURA, C. A. V. 2000. Geocronologia das rochas do embasamento do setor meridional do Cinturão Araguaia na região de Paraíso do Tocantins (TO). *Revista Brasileira de Geociências*, **29**, 453–460.

ARNDT, N. T., TEIXEIRA, W. & WHITE, W. M. 1989. Bizarre geochemistry of komatiites from Crixás Greenstone belt, Brazil. *Contributions to Mineralogy and Petrology*, **74**, 549–564.

AVELAR, V. G., LAFON, J. M., CORREIA, F. C., JR. & MACAMBIRA, E. M. B. 1999. O magmatismo arqueano da região de Tucumã-Província Mineral de Carajás: novos resultados geocronológicos. *Revista Brasileira de Geociências*, **30**, 665–670.

BRITO NEVES, B. B. 1999. América do Sul: quatro fusões, quarto fissões e o processo acrescionário Andino. *Revista Brasileira de Geociências*, **29**, 379–392

BRITO NEVES, B. B. & CORDANI, U. G. 1991. Tectonic evolution of South America during the Late Proterozoic. *Precambrian Research*, **53**, 23–40.

BRITO NEVES, B. B., VAN SCHMUS, W. R. & FETTER, A. H. 2001. Noroeste da África-Nordeste do Brasil (Província Borborema) ensaio comparativo e problemas de correlação. *Revista do Instituto de Geociências–USP*, **1**, 59–78.

CORDANI, U. G., D'AGRELLA-FILHO, M. S., BRITO NEVES, B. B. & TRINDADE, R. I. F. 2003. Tearing up Rodinia: the Neoproterozoic paleogeography of South America cratonic fragments. *Terra Nova*, **15**, 350–359.

CRUZ, E. L. C. C. 2001. *A gênese e o contexto tectônico da mina Córrego Paiol, terreno Almas–Conceição: um depósito de ouro hospedado em anfibolito do embasamento da Faixa de Dobramento Brasília*. PhD Thesis. Universidade de Brasília.

DALL'AGNOL, R., TEIXEIRA, N. P., MACAMBIRA, J. B., KOTSCHOUBEY, B., GORAYEB, P. S. S. & SANTOS, M. D. 1988. Petrologia dos gnaisses e micaxistos da porção norte da Faixa de Dobramentos Araguaia, Goiás-Brasil. *In*: *Latino-American Geological Congress 7, Belém, Anais*, **1**. Sociedade Brasileira de Geologia, 1–19.

DELOR, C., LAFON, J. M., MILÉSI, J. P. & FANNING, M. 2005. First evidence of 560–575 Ma granulites and syn-tectonic magmatism in the Rokelides belt: geology, geochronology and geodynamic implications. *In*: *19th Colloquium of African Geology, El Jadida, Marrocos, Abstracts*, 58.

DEPAOLO, D. J. 1981. Neodymium isotopes in the Colorado Front Range and crust–mantle evolution in the Proterozoic. *Nature*, **291**, 193–197.

DUNHAM, R. J. 1962. Classification of carbonate rocks according to depositional texture. *In*: HAM, W. E. (ed.) *Classification of carbonate rocks*. American Association of Petroleum Geologists, Memoirs, **1**, 108–121.

EMBRY, A. F. & KLOVAN, J. E. 1971. A Late Devonian reef tract on Northeastern Banks Island, NWT. *Canadian Petroleum Geology Bulletin*, **19**, 730–781.

FETTER, A. H., SANTOS, T. J. S., VAN SCHMUS, W. R., HACKSPACHER, P. C., BRITO NEVES, B. B., ARTHAUD, M. H., NOGUEIRA NETO, J. A. & WERNICK, E. 2003. Evidence of Neoproterozoic continental arc magmatism in the Santa Quitéria batholith of Ceará state, NW Borborema Província, NE Brasil: implication for the assembly of West Gondwana. *Gondwana Research*, **6**, 265–273.

FIGUEIREDO, F. T., PETRI, S., HIDALGO, R. L., SILVA, J. B. C., JR. & NOGUEIRA, A. C. R. 2006. A Formação Couto Magalhães, Neoproterozóico da Faixa Araguaia: fácies e estratigrafia do furo SMD-08, Redenção, Pará. *In*: *Simpósio de Geologia da Amazônia 9, Belém, Anais*. Sociedade Brasileira de Geologia–Núcleo Norte, CD-ROM.

FOLK, R. L. 1974. *Petrology of sedimentary rocks*. Hemphill, Austin, USA.

GAUDETTE, H. E., LAFON, J. M., MOURA, C. A. V., MACAMBIRA, M. J. B. & SCHELLER, T. 1998. Comparison of single filament Pb evaporation/ionization zircon ages with conventional U-Pb results: examples from the Precambrian of Brazil. *Journal of South American Earth Sciences*, **11**, 351–363.

GOLDSTEIN, S. L. & JACOBSEN, S. B. 1988. Nd and Sr isotopic systematics of river water suspended material: implication for crustal evolution. *Earth and Planetary Sciences Letters*, **87**, 249–265.

GOLDSTEIN, S. L., ARNDT, N. T. & STALLARD, R. F. 1997. The history of a continent from U–Pb ages of zircon from Orinoco River sand and Sm-Nd isotopes in Orinoco basin sediments. *Chemical Geology*, **139**, 271–286.

GORAYEB, P. S. S. 1981. *Evolução geológica da região de Araguacema-Pequizeiro (GO), Brasil*. Masters Dissertation, Universidade Federal do Pará.

GORAYEB, P. S. S. 1989. Corpos serpentiníticos da Faixa Araguaia na Região de Araguacema-Pequizeiro-Conceição do Araguaia (Goiás-Pará). *Revista Brasileira de Geociências*, **19**, 51–62.

GORAYEB, P. S. S. & ALVES, C. L. 2003. Superposição tectono-metamórfica do Cinturão Araguaia no embasamento arqueano na região de Xambioá e Araguanã-TO. *In*: *9 Simpósio Nacional de Estudos Tectônicos, Búzios, Brasil, Anais*, 84–86.

GORAYEB, P. S. S., MOURA, C. A. V. & BARROS, G. R. 2000. Pb–Pb zircon ages of the Porto Nacional high-grade metamorphic terrain, northern portion of the Goiás Massif, Central Brazil. *Revistá Brasileira de Geociências*, **30**, 190–194.

GORAYEB, P. S. S., MOURA, C. A. V. & CALADO, W. M. 2002. Idades paleoproterozóicas em zircão dos pillows basaltos da Serra do Tapa (Cinturão Araguaia)–Município de Sapucaia (PA). *In*: *Simpósio de Geologia da Amazônia 4, Belém, Anais*. Sociedade Brasileira de Geologia–Núcleo Norte, CD-ROM.

GORAYEB, P. S. S., MOURA, C. A. V. & CALADO, W. M. 2004. Suíte Intrusiva Xambica: um magmatismo toleítico Neoproterozóico pré-tectônico no Cinturão Araguaia. *In*: *Congresso Brasileiro de Geologia 42, Araxá, Anais*. Sociedade Brasileira de Geologia, CD-ROM.

HASUI, Y., COSTA, J. B. S. & ABREU, F. A. M. 1984*a*. Província Tocantins-Setor setentrional. *In*: ALMEIDA, F. F. A. & HASUI, Y. (eds) *O Precambriano no Brasil*. Edgard Blücher, São Paulo, 137–204.

HASUI, Y., COSTA, J. B. S., GORAYEB, P. S. S., LEMOS, R. L. L., GAMA, T., JR. & BEMERGUY, R. L. 1984*b*. Geologia da região de Paraíso do Norte de Goiás, GO. *In*: *Congresso Brasileiro de Geologia 33, Rio de Janeiro, Anais*, **5**, Sociedade Brasileira de Geologia. 2220–2233.

HERZ, N., HASUI, Y., COSTA, J. B. S. & MATTO, M. A. S. 1989. The Araguaia Fold Belt, Brazil: a reactivated Brasiliano-Pan-Africano cycle (550 Ma) geosuture. *Precambrian Research*, **42**, 371–386.

HOFFMAN, P. F. 1991. Did the breakout of Laurentia turn Gondwanaland inside out? *Science*, **252**, 1409–1412.

JOST, H., PIMENTEL, M. M., FUCK, R. A., DANNI, J. C. M. & HEAMAN, L. 1993. Idade U–Pb do Diorito Posselândia, Hidrolina, Goiás. *Revista Brasileira de Geociências*, **23**, 352–355.

KLEIN, E. L. & MOURA, C. A. V. 2008. São Luís Craton and Gurupi Belt (Brazil): possible links with the West African Craton and surrounding Pan-African belts. *In*: PANKHURST, R. J., TROUW, R. A. J., BRITO NEVES, B. B. & DE WIT, M. J. (eds) *West Gondwana: Pre-Cenozoic Correlations across the South Atlantic Region*. Geological Society, London, Special Publications, **294**, 137–151.

KOBER, B. 1986. Whole grain evaporation $^{207}Pb/^{206}Pb$ age investigations on a single zircon using a double filament source. *Contributions to Mineralogy and Petrology*, **93**, 482–490.

KOBER, B. 1987. Single grain evaporation combined with Pb^{+} emitter bedding for $^{207}Pb/^{206}Pb$ investigation using thermal ionization mass spectrometry, and implications for zirconology. *Contributions to Mineralogy and Petrology*, **96**, 63–71.

KOTSCHOUBEY, B., HIERONIMUS, B., RODRIGUES, O. B. & AMARAL, R. T. 1996. Basaltos e serpentinitos da área da Serra do Tapa (PA). Prováveis testemunhos de um complexo ofiolítico pouco evoluído e desmembrado. *In*: *Congresso Brasileiro de Geologia 39, Salvador, Anais*, **6**. Sociedade Brasileira de Geologia, 25–28.

KOTSCHOUBEY, B., HIERONYMOUS, B. & ALBUQUERQUE, C. A. R. 2005. Disrupted peridotites and basalts from Neoproterozoic Araguaia belt (northern Brazil): remnants of a poorly evolved oceanic crust? *Journal of South American Earth Science*, **20**, 211–230.

KRÖNER, A., JAECKEL, P. & WILLIAMS, I. S. 1994. Pb-loss patterns in zircons from a high-grade metamorphic terranes as revealed by different methods: U–Pb and Pb–Pb ages for igneous and metamorphic zircons from northern Sri Lanka. *Precambrian Research*, **66**, 151–181.

LAFON, J. M., MACAMBIRA, J. B., MACAMBIRA, M. J. B., MOURA, C. A. V., GAUDETTE, H. E. & SOUZA, A. C. C. 1990. A Faixa de Dobramentos Araguaia (TO): novos dados geocronológicos. *In*: *Congresso Brasileiro de Geologia 36, Natal, Anais*, **5**. Sociedade Brasileira de Geologia, 2550–2563.

MACAMBIRA, J. B. 1983. *Geologia e ocorrências minerais da braquianticlinal do Lontra (GO)*. Masters Dissertation, Universidade Federal do Pará.

MACAMBIRA, M. J. B. & LAFON, J. M. 1995. Geocronologia da Província Mineral de Carajás; síntese dos dados e novos desafios. *Boletim do Museu Paraense Emílio Goeldi, série Ciências da Terra, Belém*, **7**, 263–288.

MACAMBIRA, M. J. B. & LANCELOT, J. R. 1996. Time constraints for the formation of the Archean Rio Maria crust southeastern Amazonian craton, Brasil. *International Geology Reviews*, **38**, 1134–1142.

MACHADO, N., LINDENMAYER, Z. G., KROGH, T. E. & LINDENMAYER, D. 1991. U–Pb geochronology of Archean magmatism and basement reactivation in the Carajás área, Amazon shield, Brazil. *Precambrian Research*, **49**, 329–254.

MIALL, A. D. 1985. Architectural-element analysis: a new method of facies analysis applied to fluvial deposits. *Earth-Science Reviews*, **22**, 105–118.

MIALL, A. D. 1991. Hierarchies of architectural units in terrigenous clastic rocks and their relationship to sedimentation rate. *In*: MIALL, A. D. & TYLER, N. (eds) *The three-dimensional facies architecture of terrigenous clastic sediments and its implications for hydrocarbon discovery and recovery*. Society for Sedimentary Geology, Tulsa, Concepts in Sedimentology and Paleontology, **3**, 6–12.

MORAES, R., FUCK, R. A., PIMENTEL, M. M., GIOIA, S. M. C. L., HOLLANDA, M. H. B. & ARMSTRONG, R. 2006. The bimodal rift-related Juscelândia volcanosedimentary sequence in central Brazil: Mesoproterozoic extension and Neoproterozoic metamorphism. *Journal of South American Earth Sciences*, **20**, 287–301.

MOURA, C. A. V. 1992. *Geochronology and geochemistry of the basement orthogneisses of the Araguaia belt, Brazil*. PhD Thesis. University of New Hampshire, USA.

MOURA, C. A. V. & GAUDETTE, H. E. 1993. Evidence of Brasiliano/Pan-African deformation in the Araguaia belt: Implication for Gondwana evolution. *Revista Brasileira de Geociências*, **23**, 117–123.

MOURA, C. A. V. & GAUDETTE, H. E. 1999. Zircon ages of the basement orthogneisses from the northern segment of the Araguaia Belt, Brazil. *In*: SINHA, A. K. (ed.) *Basement Tectonics 13*. Kluwer Academic Publishers, Dordrecht, 155–178.

NOGUEIRA, A. C. R., RICCOMINI, C., SIAL, A. N., MOURA, C. A. V. & FAIRCHILD, T. R. 2003. Soft-sediment deformation at the base of the Neoproterozoic Puga cap carbonate (southwester Amazon craton, Brazil): confirmation of rapid icehouse to greenhouse transition in snowball Earth. *Geology*, **31**, 613–616.

NUNES, K. C. 1993. Interpretação integrada da Bacia do Parnaíba com ênfase nos dados aeromagnéticos. *In*: *3rd International Congress of the Brazilian Geophysical Society, Rio de Janeiro, Anais*, **1**, 152–157.

PAIXÃO, M. A. P. & NILSON, A. A. 2002. Fragmentos ofiolíticos da Faixa Araguaia: caracterização geológica e implicações tectônicas. *In*: KLEIN, E. L., VASQUEZ, M. L. & ROSA-COSTA, L. T. (eds) *Contribuições à Geologia da Amazônia*, **3**. Sociedade Brasileira de Geologia–Núcleo Norte, Belém, 85–103.

PAIXÃO, M. A. P., NILSON, A. A., DANTAS, E. & GIÓIA, S. 2002. Corpos ofiolíticos da Faixa Araguaia e suas implicações tectônicas. *In*: *Congresso Brasileiro de Geologia 41, João Pessoa, Anais*. Sociedade Brasileira de Geologia, 326.

PIDGEON, R., MACAMBIRA, M. J. B. & LAFON, J. M. 1998. The Th–U–Pb isotopic systems and internal structures of complex zircons from an enderbite from the Pium Complex, Carajás Province, Brazil: evidence for the ages of granulite facies metamorphism and the protolith of the enderbite. *Chemical Geology*, **166**, 159–171.

PIMENTEL, M. M., HEAMAN, L. & FUCK, R. A. 1991. U–Pb zircon and sphene geochronology of late Proterozoic volcanic arc rock units from southwestern Goiás, central Brasil. *Journal of South American Earth Sciences*, **4**, 329–339.

PIMENTEL, M. M., FUCK, R. A., JOST, H., FERREIRA-FILHO, C. F. & ARAUJO, S. M. 2000. The basement of the Brasilia fold belt and the Goiás Magmatic Arc. *In*: CORDANI, U. G., MILANI, E. J., THOMAZ FILHO, A. & CAMPOS, D. A. (eds) *Tectonic Evolution of South America. 31st International Geological Congress, Rio de Janeiro*, 195–229.

PIMENTEL, M. M., FUCK, R. A. & JOST, H. 2004. O embasamento da Faixa Brasília e o Arco Magmático de Goiás. *In*: MANTESSO-NETO, V., BARTORELLI, A., DAL RE CARNEIRO, C. & BRITO NEVES, B. B. (eds) *Geologia do continente Sul-Americano: Evolução da Obra de Fernando Flávio Marques de Almeida*. Beca Produções Culturais Ltda., São Paulo, 355–368.

QUEIROZ, C. L. 2000. *Evolução tectono-estrutural dos terrenos granito-greenstone belt de Crixás, Brasil Central. Brasília*. PhD Thesis, Universidade de Brasília.

RENNE, P. R., ONTSTOTT, T. C., D'AGRELA-FILHO, M. S., PACCA, I. G. & TEIXEIRA, W. 1990. $^{40}Ar/^{39}Ar$ dating of the 1.0–1.1 Ga magnetizations from the São Francisco and Kalahari Cratons: tectonic implications for Pan-African and Brasiliano mobile belts. *Earth and Planetary Science Letters*, **101**, 349–366.

ROGERS, J. J. W., UNRUG, R. & SULTAN, M. 1995. Tectonic Assembly of Gondwana. *Journal of Geodynamics*, **19**, 1–34.

SATO, K. & SIGA, O., JR. 2000. Evidence of the superproduction of the continental crust during Paleoproterozoic in South American Platform. Implications regarding the interpretataive value of the Sm–Nd model ages. *Revista Brasileira de Geociências*, **30**, 126–129.

SATO, K. & TASSINARI, C. C. G. 1997. Principais eventos de acresção continental no Craton Amazônico baseados em idades modelo Sm-Nd, calculadas em evoluções de estágio único e estágio duplo. *In*: COSTA, M. L. & ANGELICA, R. S. (eds) *Contribuições à Geologia da Amazônia*. Sociedade Brasileira de Geologia–Núcleo Norte, Belém, **1**, 91–142.

SOARES, J. E., BERROCAL, J., FUCK, R. A., MOONEY, W. D. & VENTURA, D. B. R. 2006. Seismic characteristics of central Brazil crust and upper mantle: a deep seismic refraction study. *Journal of Geophysical Research*, **111**, 1–31.

SOUSA, D. J. L. & MOURA, C. A. V. 1995. Estudo geocronológico do Granito Serrote, Paraíso do Tocantins (TO). *In*: *Congresso Brasileiro de Geoquímica 5, Niterói, Anais*. Sociedade Brasileira de Geoquímica, CD-ROM.

STACEY, J. S. & KRAMERS, J. D. 1975. Approximation of terrestrial lead isotope evolution by two stage model. *Earth and Planetary Sciences Letters*, **26**, 207–221.

SUITA, M. T. F., KAMO, S. L., KROGH, T. E., FYFE, W. S. & HARTMANN, L. A. 1994. U–Pb ages from the high grade Barro Alto mafic–ultramafic Complex (Goiás, central Brazil): Middle Proterozoic continental mafic magmatism and the Upper Proterozoic continental collision. *In*: *Eighth International Conference on Geochronology, Cosmochronology and Isotope Geology, Berkeley, 1994, US Geological Survey Circular*, **1107**, 309.

TASSINARI, C. C. G. & MACAMBIRA, M. J. B. 1999. Geochronological Provinces of the Amazonian Craton. *Episodes*, **22**, 174–182.

TEIXEIRA, K. S. L., MOURA, C. A. V., COSTI, H. T. & RAMOS, J. F. R. 2002. Utilização de MEV-EDS no apoio à seleção de monocristais de zircão para datação por evaporação de chumbo. *In*: *Congresso Brasileiro de Geologia 41, João Pessoa, Anais*. Sociedade Brasileira de Geologia, 128.

TROMPETTE, R. 1994. Geology of the Western Gondwana (2000–500 Ma): Pan-African-Brasiliano aggregation of South America and Africa. A.A. Balkema, Rotterdam.

TROMPETTE, R. 1997. Neoproterozoic ($\approx$ 600 Ma) aggregation of Western Gondwana: a tentative scenario. *Precambrian Research*, **82**, 101–112.

UNRUG, R. 1996. The assembly of Gondwanaland. *Episodes*, **19**, 11–20.

VALERIANO, C. M., MACHADO, N., SIMONETTI, A., VALLADARES, C. S., SEER, H. J. & SIMÕES, L. S. A. 2004. U–Pb geochronology of the southern Brasília belt (SE-Brazil): sedimentary provenance, Neoproterozoic orogeny and assembly of West Gondwana. *Precambrian Research*, **130**, 27–55.

VILLENEUVE, M. & CORNÉE, J. J. 1994. Structure, evolution and paleogeography of the West African craton and bordering belt during the Neoproterozoic. *Precambrian Research*, **69**, 307–326.

VILLENEUVE, M. & DALMEYER, R. D. 1987. Geodynamic evolution of the Mauritanides, Bassarides and Rokelides orogens (West Africa). *Precambrian Research*, **37**, 19–28.

WALKER, R. G. & JAMES, N. P. 1992. *Facies Models: Response to Sea Level Change. Geo Text*, **1**. Geological Association of Canada, Newfoundland.

WILLIAMS, H. R. & CULVER, S. J. 1982. The Rokelides of West Africa–Pan African aulacogen or back-arc basin? *Precambrian Research*, **18**, 261–273.

Tectonic evolution of the Brasília Belt, Central Brazil, and early assembly of Gondwana

C. M. VALERIANO[1,4], M. M. PIMENTEL[2,4], M. HEILBRON[1,4], J. C. H. ALMEIDA[1,4] & R. A. J. TROUW[3,4]

[1]*Universidade do Estado do Rio de Janeiro, Faculdade de Geologia, Rua São Francisco Xavier 524/4006-A, Maracanã 20559-900, Rio de Janeiro, RJ, Brazil (e-mail: claudio.valeriano@pq.cnpq.br)*

[2]*Universidade de Brasília, Instituto de Geociências, Instituto de Geociências—Universidade de Brasília Campus Universitário Darcy Ribeiro, 70910-900, Brasília, DF, Brazil*

[3]*Universidade Federal do Rio de Janeiro, Instituto de Geociências, Centro de Ciências Matemáticas e da Natureza, Bloco F, Ilha do Fundão 21941-590, Rio de Janeiro, RJ, Brazil*

[4]*Conselho Nacional de Desenvolvimento Científico e Tecnológico—CNPq*

Abstract: The Brasília Belt comprises terranes and thrust-sheets that were tectonically transported towards the western passive margin of the São Francisco–Congo palaeocontinent during an orogenic episode resulting from collision of the Paranapanema and Goiás blocks and the Goiás magmatic arc against São Francisco–Congo at 0.64–0.61 Ga. The tectonic zones of the belt are, from east to west: a foreland zone with Archaean–Palaeoproterozoic granite–greenstone basement covered by Neoproterozoic anchimetamorphic sedimentary rocks (Bambuí Group); a low metamorphic grade thrust-fold belt of proximal shelf successions, mostly siliciclastic, containing rare basement slivers; metamorphic nappes in upper greenschist to granulite facies of distal shelf and slope metasediments and subordinate tholeiitic metabasalts; the Goiás massif, possibly a microcontinent; and the Goiás magmatic arc. The accretion of these terranes against the western margin of the São Francisco–Congo palaeocontinent took place during an early phase of Gondwana supercontinent amalgamation, when terranes accreted around São Francisco–Congo to create a proto-West Gondwana landmass, around which subsequent collisional and accretionary events followed, such as those in the Borborema–Trans-Saharan province (*c.* 0.62–0.60 Ga); in the Ribeira–Araçuaí belt (*c.* 0.58 Ga); along the Araguaia and Paraguay belts (collision of Amazonia, *c.* 0.54–0.52 Ga); and the accretion of Cabo Frio terrane in the Ribeira Belt (*c.* 0.53–0.50 Ga).

The assembly of Gondwana (Fig. 1) took place through a long succession of Neoproterozoic collisional events, referred to as the Brasiliano or Pan-African event (Brito Neves *et al.* 1999; Meert 2003). Post-orogenic regional uplift and cooling took place during the Cambrian–Ordovician. A transition to relatively stable platform conditions in South America did not occur until the Silurian–Devonian with the development of the large Paraná, Parnaíba and Amazon intracratonic sag-type sedimentary basins (Milani & Thomaz Filho 2000; Brito Neves 2002).

The Brasiliano orogenic belts accommodated most of the lithospheric convergence either through subduction of oceanic lithosphere in pre-collisional phases, or through folding and thrusting along the former continental margins during collisional processes. Within the orogenic belts, reworked domains of older rock units (Archaean to Mesoproterozoic) are common and referred to as massifs, either interpreted as reworked basement of the palaeocontinental margin, or as exotic terranes.

This work presents a brief description of the tectonic elements of the Brasília Belt, central Brazil, followed by a synthesis of the tectonic evolution in the context of Gondwana assembly. Available geological information shows that continental rifting of the São Francisco–Congo palaeocontinent, and passive margin construction around it, started at *c.* 1.0 Ga, and that the oceanic basin closed after continental collision at *c.* 0.64 Ga. Available geochronological data show that this orogenic event took place relatively early in the history of Gondwana, with the collision of the Paranapanema block, the Goiás massif and Goiás magmatic arcs against the western margin of São Francisco–Congo. This and other early collisional events formed a continental nucleus around which other masses aggregated to form the final outline of Gondwana at *c.* 0.50 Ga.

From: PANKHURST, R. J., TROUW, R. A. J., BRITO NEVES, B. B. & DE WIT, M. J. (eds) *West Gondwana: Pre-Cenozoic Correlations Across the South Atlantic Region.* Geological Society, London, Special Publications, **294**, 197–210. DOI: 10.1144/SP294.11 0305-8719/08/$15.00

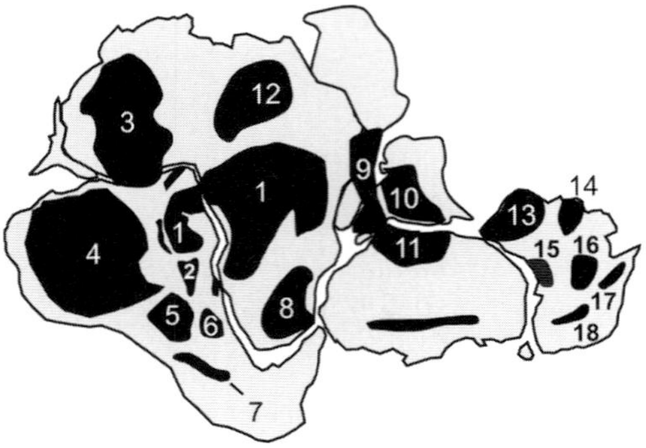

Fig. 1. The Gondwana supercontinent, a mosaic of continental fragments that collided mostly during the Neoproterozoic (modified from Unrug 1996): 1, São Francisco–Congo; 2, Paranapanema; 3, West Africa; 4, Amazonia; 5, Rio Apa; 6, Rio de la Plata; 7, Pampia; 8, Kalahari; 9, Dharwar; 10, Bundelkhand; 11, East Antarctica; 12, Uweinat–Nile; 13, West Australian; 14, Kimberleys; 15, Gawler; 16, Arunta; 17, NE Australia; 18, Willyama.

The Tocantins Province

The Tocantins Province (Fig. 2) is a branched orogenic system developed between the Amazonia, São Francisco–Congo and Paranapanema palaeocontinental blocks. The Paranapanema block is completely covered by Phanerozoic rocks of the Paraná Basin. Although its limits have been only recently inferred from gravimetric data (Mantovani & Brito Neves 2005), the presence of a collider located beneath the Paraná Basin was long suspected in view of the extensive nappe system cropping out between the Paraná Basin and the southern São Francisco Craton (Almeida *et al.* 1981; Brito Neves *et al.* 2000).

Three orogenic belts compose the Tocantins Province: the Araguaia Belt runs north–south and the Paraguay Belt follows a sinuous southwesterly trend. Both have tectonic vergence, respectively towards the eastern and SE margin of Amazonia. The Brasília Belt also runs north–south but has an opposite vergence towards the western margin of the São Francisco–Congo craton.

Following the original proposal of Rodinia (Hoffman 1991), suggesting that a supercontinent formed at 1.0 Ga and rifted soon after, most paleogeographic reconstructions include Amazonia as part of Rodinia, connected to Laurentia and Baltica along orogenic belts of Grenvillian age (*c.* 1.0 Ga). In contrast, the position of São Francisco–Congo with respect to Rodinia is not clear and lacks reliable palaeomagnetic constraints (Meert & Torsvik 2003). However, most palaeogeographic reconstructions constrained by palaeomagnetic data seem to favour the former existence of a vast oceanic basin facing the (present) western margin of the São Francisco–Congo palaeocontinent (e.g., Weil *et al.* 1998; Pesonen *et al.* 2003; Pisarevsky *et al.* 2003). Subduction of oceanic lithosphere to the west of São Francisco–Congo is geologically supported the by the presence of the large Goiás magmatic arc (Pimentel & Fuck 1992), which probably continues northeaswards through the Borborema province in NE Brazil and into the Trans-Saharan orogenic province in West Africa. Additionally, ophiolitic rock associations such as the Quatipuru and Serra do Tapa complexes are found in the Araguaia Belt (Fonseca *et al.* 2004). In the Brasília Belt, metamorphosed ophiolitic mélanges associated with the Araxá Group have been described (Strieder & Nilson 1992; Seer 1999).

The Brasília Belt

The Brasília Belt comprises terranes and crustal thrust-sheets that converged towards the east against the western São Francisco–Congo platform (Fig. 3). The belt has a curved shape with two differently oriented branches: the northern Brasília Belt trends northeasward and the southern Brasília Belt trends southeaswards. The pronounced concavity of the belt facing east is the result of the accommodation of the accreted terranes around the protuberant western margin of the São Francisco Craton, which acted as an indentor. The original outline of the continental margin of the São Francisco–Congo craton is inferred from gravimetric data that show a conspicuous gravimetric discontinuity below the allochthons, suggesting a much wider continent than usually represented (Lesquer *et al.* 1981; Haralyi & Hasui 1982). An outline with a sharp corner pointing to the west at 17°S is supported by more recent gravimetric coverage presented by Ussami (1993) and Alkmim *et al.* (1993). Therefore, large portions of the western part of the palaeocontinent are presently covered by thin-skinned thrust-sheets and nappes, especially along the southern Brasília Belt.

The tectonic zonation of the Brasília Belt

The tectonic zonation of the Brasília Belt presented below is marked by a foreland zone of the São Francisco palaeocontinent in the east, passing westward through the external and internal allochthonous zones, both former integrants of the Neoproterozoic passive margin of São Francisco–Congo, to the exotic terranes of the Central Goiás microcontinent and the Goiás magmatic arc.

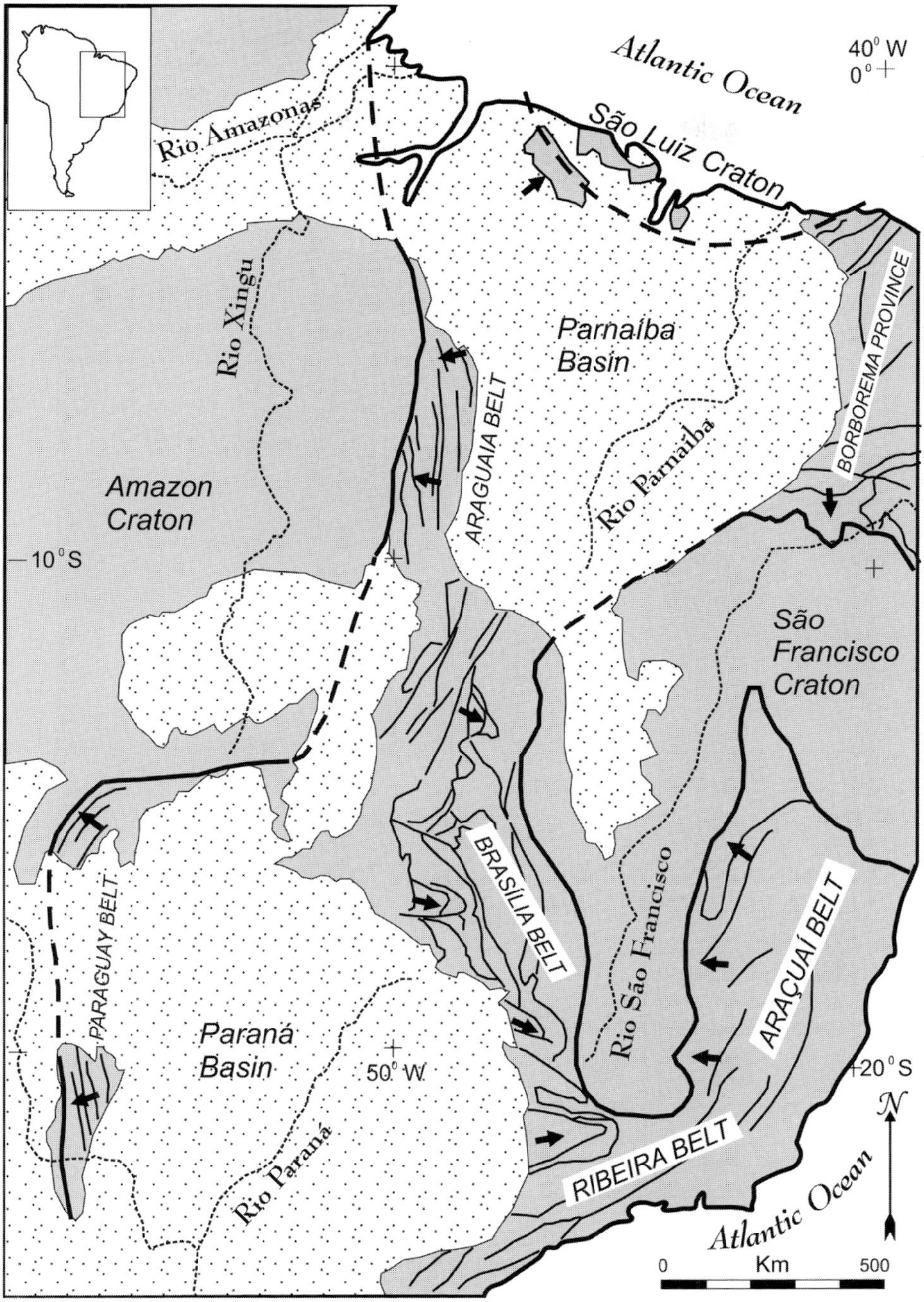

Fig. 2. The Tocantins Province and the tectonic elements of central Brazil, sketched from Delgado & Pedreira (1995). Heavy lines depict the outlines of cratonic blocks. Arrows indicate main direction of tectonic transport along Neoproterozoic orogenic belts. Legend: grey, Precambrian crystalline basement; stippled, unmetamorphosed Phanerozoic cover.

The foreland zone

The foreland zone is a domain of shallow sub-horizontal thrust-sheets of Neoproterozoic anchimetamorphic sediments of the Bambuí Group covering a crystalline basement of Archaean–Palaeoproterozoic rocks (Dardenne 2000). Both in the belt and in the cratonic zone

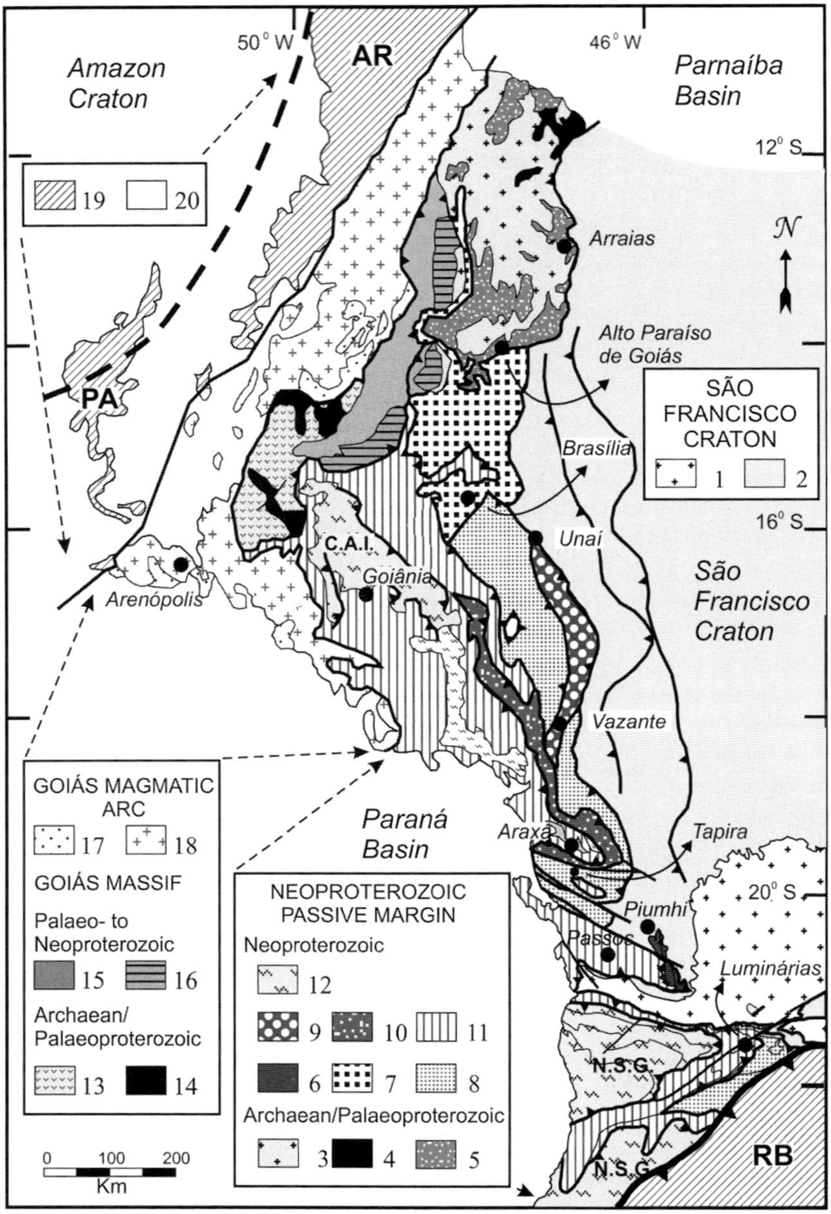

Fig. 3. Tectonic units of the Brasília Belt (compiled from Dardenne 2000; Pimentel *et al.* 2000; Valeriano *et al.* 2000; Seer 1999; Silva 2003; Valeriano *et al.* 2004*b*). **Foreland zone (São Francisco Craton)**: 1, Archaean/Palaeoproterozoic granite–greenstone and gneiss–migmatite terrain; 2, autochthonous/para-autochthonous metasedimentary cover (S. João del Rei, Carandaí, Andrelândia, Bambuí groups). Brasíla Belt (3–18): **External zone**: 3, Archaean/Palaeoproterozoic granite–greenstone and gneiss–migmatite association; 4, Archaean/Palaeoproterozoic greenstone belts; 5, Palaeo- to Mesoproterozoic rift successions (Araí Group); 6, Ilicínea–Piumhi thrust system. Neoproterozoic passive margin successions: 7, Paranoá Group; 8, quartzite-phyllite units (Canastra, Andrelândia groups) and occasional basement slivers; 9, Vazante Group 10- Ibiá Group; **Internal zone**: 11, Araxá and Andrelândia groups and associated tholeiitic mafic rocks, ophiolitic melange complexes, basement slivers and syn-collisional leucogranites; 12, granulitic nappes (C.A.I, Anápolis–Itauçu Complex; N.S.G., Socorro–Guaxupé nappe; **Goiás massif**: 13, Archaean/Palaeoproterozoic granite-gneiss–migmatite complexes; 14, Archaean/Palaeoproterozoic greenstone belts; 15, Mesoproterozoic volcano-sedimentary successions (Juscelândia, Palmeirópolis, Serra da Mesa); 16, Meso- to Neoproterozoic layered basic-ultrabasic complexes; **Goiás magmatic arc**: 17, Neoproterozoic supracrustal (metavolcanic/sedimentary) rocks; 18, Neoproterozoic orthogneisses and granitoid rocks; 19, Paraguay (PA), Araguaia (AR) and Ribeira (RB) belts; 20, Phanerozoic cover.

the basement associations are exposed only in the south. In the autochthonous domains, the São Francisco cratonic basement is discordantly covered by autochthonous glacial rudites (Jequitaí Formation) and cap carbonate-pelitic rocks (Sete Lagoas Fm) of the Bambuí Group.

Two contrasting basement lithologic associations occur in the area. One consists of Archaean greenstone belts, such as the Rio das Velhas Supergroup (*c.* 3.2 to 2.7 Ga), and roughly contemporaneous domed complexes of Neoarchaean tonalite–trondhjemite–granodiorite (TTG) orthogneisses, e.g. the Bonfim, Belo Horizonte and Campo Belo complexes (Machado & Carneiro 1992; Machado *et al.* 1992; Teixeira *et al.* 1996; Carneiro *et al.* 1998). The other type consists of Palaeoproterozoic orogenic domains defined as the Mineiro Belt, where Archaean basement rocks and Palaeoproterozoic supracrustal sequences (e.g. Minas Supergroup) were metamorphosed at *c.* 2.1 Ga, with associated granitoid rocks (Teixeira & Figueiredo 1991; Noce *et al.* 2000; Alkmim & Marschak 2001). Younger intrusive complexes are more common in the south, with plutons of gabbroic to granitic compositions dated at *c.* 1.9 Ga.

Basement rocks occur also in the southern portions of the Brasília Belt in anticlinal zones between synformal nappe systems that advanced to the east. One example is the Campos Gerais Complex cropping out along a NW-trending anticlinal corridor between the Neoproterozoic Passos nappe to the north and the Varginha–Guaxupé nappe to the south (Schmidt 1983). This severely sheared granite–greenstone basement complex shows lateral continuity to the east with the autochthonous exposed basement of the southern São Francisco Craton. Orthogneissic complexes of tonalitic to monzogranitic composition were mylonitized to various degrees and a few greenstone belt remnants were recognized, such as the Alpinópolis and Morro do Ferro greenstone belts. The mylonite fabric dips steeply to the SSW and displays left-lateral kinematic indicators. Another example of an anticlinal basement zone occurs in the southernmost Brasília Belt, represented by Archaean TTG orthogneisses (the Amparo Complex and the Serra Negra tonalites) that crop out between the Varginha-Guaxupé nappe to the north, and the Socorro nappe in the south (Campos Neto 2000).

The external zone

The northern and southern portions of the external zone of the belt represent two contrasting styles of proximal continental passive margins. In the northern part of the belt, the external zone between Alto Paraíso de Goiás and Natividade (Fig. 3) consists of a large crustal block of exposed Palaeoproterozoic basement, partially covered by gently folded rift sequence of the Araí Group metasediments (1.77 Ga and younger). This block is bounded to the east by thrusting onto the footwall rocks of the Bambuí Group. To the south of Alto Paraíso de Goiás (Fig. 3), the Araí Group is covered discordantly by the Neoproterozoic Paranoá Group, which is also thrusted over the Bambuí Group to the east. South of Unaí, the allochthonous nature of the supracrustal sequences is progressively more evident in the external zone, with the development of a thin-skinned thrust-fold system directed towards SE over the pelitic and carbonatic rocks of the Bambuí Group.

These two zones are briefly described below.

Basement domain of the northern external zone. The basement rocks cropping out north of Alto Paraíso de Goiás (Fig. 3) are predominantly composed of granites and orthogneisses, with associated volcano-sedimentary sequences of Palaeoproterozoic age (Riachão do Ouro Group and Ticunzal Formation) and mafic–ultramafic bodies (Pimentel *et al.* 2004).

This basement association is partially covered by gently folded strata of the Araí Group. This includes quartzites and conglomerates, predominating in the basal levels, followed by predominant calc-pelites towards the top (Dardenne 2000), with a maximum thickness of 1500 m. Felsic metavolcanic rocks are intercalated in the basal clastic units and are broadly contemporaneous with *c.* 1.7 Ga anorogenic tin-bearing plutonic suites intruded into the basement of the Araí Group (Pimentel *et al.* 1991). The Araí Group and the associated magmatism are products of a major Statherian rifting event which affected the São Francisco–Congo palaeocontinent, producing the correlative sedimentary sequences of the Espinhaço and Mayombe supergroups, respectively in the São Francisco and Congo cratons (Trompette 1994; Martins-Neto 2000).

Thrust-fold belts of the southern external zone. The external domain in the southern part of the Brasília Belt is an imbricated stack of thrust sheets of Neoproterozoic passive margin successions, mostly in greenschist facies but up to the garnet zone. The metasedimentary units of the external allochthons were deposited in a proximal shelf environment, composed either of siliciclastic or carbonatic successions. Carbonatic rocks are best represented by the Vazante Group, a pelite–dolomite shallow water sequence that hosts important zinc and lead deposits. The siliciclastic successions are represented by the Paranoá, Canastra and basal Andrelândia groups (Dardenne 2000; Paciullo *et al.* 2000). In these units, quartzites are associated to metapelitic phyllites and schists, with minor carbonatic rocks. A study of detrital zircons from quartzites in the Canastra and Andrelândia groups has revealed Archaean–Palaeoproterozoic provenance from the cratonic basement and a variety of Mesoproterozoic detrital zircons defining a conspicuous mode at *c.* 1.3 Ga (Valeriano *et al.* 2004*a*). Nd model ages of metapelites from the Canastra and Paranoá groups concentrate in the 1.9–2.3 Ga interval (Pimentel *et al.* 2001), indicating provenance from the São Francisco–Congo basement. The Ibiá Group is a distinct unit, of more distal environment, composed of basal polymict diamictites, interpreted as of glacial origin, and covered by monotonous calcareous chloritic phyllites and schists with intercalations of fine-grained metasandstone (Seer *et al.* 2001).

Provenance studies based on Sm–Nd and U–Pb zircon data on metapelitic rocks of the Araxá and Ibiá groups, on the other hand, suggest that at least some of the rock units included in these two groups most probably represent Neoproterozoic back–arc deposits (Pimentel *et al.* 2001; Piuzana *et al.* 2003*a*).

The deformational history observed in the external zone typically starts with syn-metamorphic development of a gently dipping foliation associated with sub-horizontal shear, followed by upright folding associated with the final tectonic thrust-slice imbrications. Several WNW-directed sinistral strike-slip fault zones trend obliquely to the belt and acted as transfer zones between the east-verging synformal thrust-systems.

Basement thrust-slices in the external thrust–fold systems are restricted to the southernmost sector of the Brasília Belt. They probably represent basement high blocks formed during the continental rifting stage (*c.* 0.90 Ga), bounded by normal faults that spawned (or were reactivated as) thrust faults during orogenic inversion at *c.* 0.64 Ga. South of the town of Piumhi (Fig. 3), the Ilicinia–Piumhi thrust system is an imbricated set of thrust sheets that intercalated Neoproterozoic quartzite–phyllite shelf units and an Archaean granite–greenstone belt association (Valeriano *et al.* 2000, 2004*b*). Further to the south, in a frontal position with respect to the Varginha–Guaxupé–Luminárias nappe system (Campos Neto & Caby 1999, 2000), thrust sheets of basement granite–greenstone rocks also occur imbricated with Neoproterozoic quartzites and schists of the proximal Andrelândia Group (Ribeiro *et al.* 1995; Trouw *et al.* 2000; Heilbron *et al.* 2004).

The metamorphic nappes of the internal zone

The internal zone of the southern Brasília Belt comprises systems of nappes overriding the thrust–fold belts of the external zone. This outcrop pattern of the upper nappe units is segmented into three nappe systems by subvertical lateral ramp fault zones (Fig. 3).

The northernmost and largest structure extends from Goiânia towards the SE to the town of Araxá, where the dominant structure is a NW-trending synformal structure and where the Araxá Group is thrust over the low metamorphic grade rocks of the Ibiá and Canastra groups. The southern limb of this synformal structure is bounded by an ESE-trending subvertical fault zone (Seer 1999; Seer & Dardenne 2000). To the south, the Passos nappe also has a synformal structure with a WNW-trending axial plane. The northern limb of the Passos nappe is bounded by a subvertical WNW-trending lateral ramp shear zone, whereas the southern limb is gently to moderately dipping. The southernmost synformal structure of the internal Brasília Belt has an east–west trending axial plane passing through the town of Luminárias, where the upper granulitic Guaxupé nappe overlies lower nappes of highly deformed Neoproterozoic metasedimentary rocks. Further to the south, in the Guaxupé-Luminárias nappe system, the structures that might have been developed in the context of the thrust-stacking of the Brasília Belt were intensely and progressively overprinted by the deformational and metamorphic processes related to the Ribeira Belt, a younger belt that is orthogonal to the Brasília Belt (Trouw *et al.* 2000; Hackspacher *et al.* 2004).

The metasedimentary rocks of the metamorphic nappes of the internal zone belong to the Araxá (Simões 1995; Seer 1999; Seer *et al.* 2001) and Andrelândia groups (Ribeiro *et al.* 1995), the latter in the southern portion of the belt. These Neoproterozoic units comprise metapelitic schists with quartzite and paragneiss intercalations, with subordinated calc-silicate and tholeiitic metabasic rocks. The high proportion of pelitic schists, the characteristic thin-bedded and fine-grained quartzites and the frequent intercalations of calc-silicate rocks suggest deposition in distal shelf and/or slope environments. The geochemical compositions of the tholeiitic metabasic rocks intercalated in the Araxá and Andrelândia groups are similar to those of modern within-plate basalts and to oceanic ridge basalts. This range of compositions is also compatible with a continental rift environment that evolved to passive margin setting with thin lithosphere and emplacement of asthenospheric MORB-like magmatism (Valeriano & Simões 1997). From Araxá to the NW, there seems to be a progressively larger proportion of rock units formed in an oceanic environment. In the Araxá nappe, amphibolites of tholeiitic composition make up large proportions of the lower beds of the Araxá Group

(Seer 1999; Seer *et al.* 2001). Further examples of oceanic environments to the north are associations of mafic to ultramafic tectonites, carbonaceous schists and banded iron formation intercalated within the mica-schists of the Araxá Group of Serra das Caldas and Abadiânia regions, described by Drake (1980) and Strieder & Nilson (1992) and interpreted as ophiolitic melanges.

Within the three synformal nappe systems an inverted metamorphic gradient is observed. The lowermost rocks display greenschist facies and the uppermost ones reached either amphibolite or granulite facies. The granulitic rocks of the uppermost structural units are referred to as the Anápolis–Itauçu granulite complex (Pimentel *et al.* 1999*a*, *b*; Piuzana *et al.* 2003*a*, *b*). This is an association of Neoproterozoic rocks of varied nature, including differentiated mafic–ultramafic bodies, orthogranulites of granitic to tonalitic compositions and aluminous metasedimentary granulites. Locally, ultra-high temperatures were estimated to have reached 1030–1050 °C, at pressures of *c.* 10 kbar (Moraes *et al.* 2002). In the Passos nappe (Simões 1995) and further south in the Socorro–Guaxupé nappe system (Campos Neto & Caby 1999, 2000; Garcia & Campos Neto 2003), metamorphic conditions reached *c.* 800–900 °C, at pressures up to 15 kbar. The metamorphic conditions of intermediate to high pressure regime indicate that the sedimentary pile of the distal passive western margin of São Francisco–Congo underwent subduction to the SW, beneath the Paranapanema block, followed by rapid tectonic exhumation, in the form of subhorizontal nappes, during the collisional phase of the orogeny (*c.* 0.64 Ga).

The Goiás massif

Stretching for 600 km in the SW–NE direction in central Brazil, the Goiás massif is a probable microcontinent, mostly composed of relatively old rocks (3.0 to 0.8 Ga). It is bounded to the east by the distal passive margin of São Francisco–Congo and to the west by the Goiás magmatic arc (Fig. 4).

The southwestern portion of the Goiás massif is a granite–greenstone terrane of Neoarchaean age composed of dome-shaped TTG gneiss complexes (Queiroz *et al.* 2000), separated by Archaean keel-shaped complexes of greenstone-belt nature. These NW-trending greenstone belts (Serra de Santa Rita, Crixás, Guarinos and Pilar de Goiás) contain metamorphosed komatiitic to basaltic lavas and associated metasedimentary rocks (carbonaceous metapelites and meta-rhythmites associated with banded iron formation) and commonly host gold deposits, mostly located in the vicinities of the towns of Crixás and Goiás. This Archaean nucleus was strongly reworked

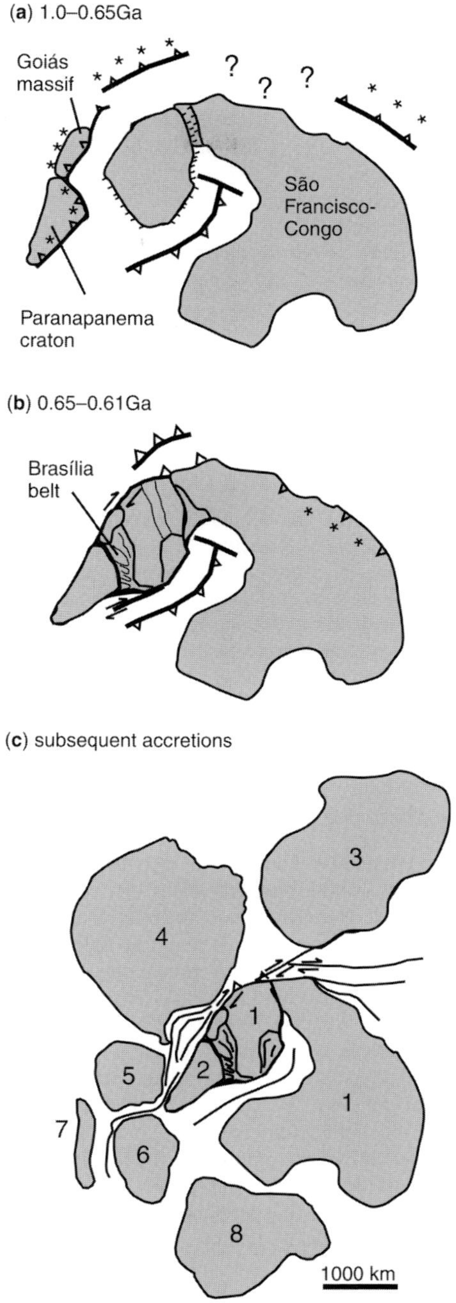

Fig. 4. Schematic geodynamic reconstruction of early Gondwana amalgamation. (**a**) subduction zones around São Francisco–Congo and the formation of magmatic arcs (stars); (**b**) collision of Paranapanema, the Goiás massif and Goiás magmatic arc; (**c**) subsequent accretion of continental blocks (numbering as in Fig. 1): 1, São Francisco–Congo; 2, Paranapanema; 3, West Africa; 4, Amazonia; 5, Rio Apa; 6, Rio de la Plata; 7, Pampia; 8, Kalahari.

and suffered anatexis in the Palaeoproterozoic (Transamazonian-age, 2.2–2.0 Ga, Jost *et al.* 2005), and in the Neoproterozoic (Brasiliano, 0.63 Ga, Pimentel *et al.* 2003).

Northeast of this Archaean granite–greenstone terrane, three large mafic–ultramafic differentiated complexes stretch for 300 km in the NNE direction, in the form of tectonic lenses, and are referred to as the Niquelândia, Barro Alto and Cana Brava complexes. They were thrust eastwards over underlying basement gneisses and reached the granulite facies of metamorphism (Moraes *et al.* 2006). Recent TIMS U–Pb zircon dating carried out by Laux *et al.* (2006) on samples from the Cana Brava Complex indicates crystallisation at 0.78 Ga, rapidly followed by metamorphism. It is interesting to note that major plume-related tholeiitic igneous activity with a strikingly similar age (Harlan *et al.* 2003; Li *et al.* 2003, 2004) has been recorded in North America and China, interpreted as related to the break-up of Rodinia.

The mafic–ultramafic complexes are tectonically overlain by volcano-sedimentary sequences (the Juscelândia, Palmeirópolis and Indaiatuba complexes) that were metamorphosed to the amphibolite facies (Moraes *et al.* 2006). An extensional tectonic setting is envisaged for these units based on the bimodal character of the associated magmatic rocks (Suita *et al.* 1996; Strieder & Suita 1999).

The Goiás magmatic arc

Since Pimentel & Fuck (1992) recognized that large extents of granitic-gneiss terrains of Central Brazil are in fact Neoproterozoic magmatic arc assemblages, and not basement massifs as previously believed, a large amount of geochemical and isotopic data on these rocks has been published (Pimentel *et al.* 1999*a*, *b*, 2000, 2004; Laux *et al.* 2005).

The Goiás magmatic arc extends NE–SW and its ends are covered by the Phanerozoic rocks of the Parnaíba and Paraná basins. The arc is composed of both plutonic and supracrustal rocks, from gabbroic to granitic composition, with a large proportion of metamorphosed tonalites, diorites and granodiorites. Juvenile Nd and Sr isotope signatures are well documented in several areas such as the Arenópolis and Mara Rosa regions, pointing to an early phase (*c.* 0.89–0.80 Ga) of intra-oceanic arc magmatism, although contamination with older continental crust is also observed locally, especially in younger (*c.* 0.65–0.60 Ga) meta-igneous rock units. The sequence of events has been recently summarised by Pimentel *et al.* (2004), as follows. Between *c.* 0.89 Ga and 0.80 Ga, immature intra-oceanic arc systems developed, with predominance of dioritic to tonalitic plutonism and associated calc-alkaline volcanism. At *c.* 0.80 Ga, the intrusion of extensive layered mafic–ultramafic bodies (the Cana Brava, Niquelândia and Barro Alto complexes) took place. These mafic–ultramafic complexes record granulite facies metamorphism at 0.78–0.76 Ga. Arc magmatism then continued until the main collisional event at *c.* 0.64 Ma.

Tectonic evolution of the Brasília Belt

The tectonic evolution of the Brasília Belt in the context of the assembly of Gondwana is presented below, divided into two stages: from continental rifting (1.1–0.9 Ga) until the onset of the main collisional stage (0.65 Ga), and the collisional stage itself until regional cooling at 0.60 Ga.

Continental break-up and drift of the São Francisco–Congo palaeocontinent

Regardless of whether São Francisco–Congo belonged to Rodinia or not, successful continental rifting is recorded between 1.0 and 0.8 Ga, both within the cratonic or preserved continental blocks, and within the Neoproterozoic orogenic belts. This extensional event may be associated with continental break-up that evolved to the isolation and subsequent drift of the São Francisco–Congo palaeocontinent. The construction of Neoproterozoic passive margins was contemporaneous with sedimentation in intracratonic settings within São Francisco–Congo, represented by the São Francisco and West Congo supergroups (Trompette 1994; Martins-Neto & Hercos 2002). After orogenic inversion during Gondwana assembly these passive margin successions were deformed and metamorphosed, thus representing the external and internal zones of the orogenic belts surrounding the western margin of São Francisco–Congo. In the Brasília Belt, the stratigraphy of the Neoproterozoic passive margins units is partially obliterated by the thrust tectonics imposed by the orogeny. Ages of sedimentation are constrained by geochronological data.

It has been suggested that a mantle plume triggered the Tonian rifting event, leading to continental break-up, as indicated by associated tholeiitic continental mafic magmatism (Corrêa-Gomes & Oliveira 2000; Mazzucchelli *et al.* 2001). Tholeiitic magmatism occurs both within the São Francisco–Congo cratonic domain and in the adjacent Neoproterozoic orogenic belts. In the São Francisco cratonic area mafic dyke swarms (Teixeira 1989) are observed to intrude only granite–greenstone Archaean–Palaeoproterozoic basement and the Espinhaço Supergroup (*c.* 1.75 Ga and younger) and are discordantly covered by the Neoproterozoic strata of the São Francisco Supergroup (Macaúbas and Bambuí groups). In the Brasília Belt lenses of greenschists and amphibolites of tholeiitic affinity are described within the Neoproterozoic Araxá Group (Valeriano & Simões 1997) and Andrelândia Group (Ribeiro *et al.* 1995), further south. In the Araçuaí Belt, a disrupted and metamorphosed ophiolite succession has been recognised within a distal metapelitic unit. Amphibolites with MORB affinity were dated at *c.* 0.82 Ga by a Sm–Nd whole-rock isochron (Pedrosa-Soares *et al.* 1998).

In the West Congo Belt (Bas Congo region, SW Africa), the counterpart of the Araçuaí Belt in pre-Atlantic Ocean times, broadly coeval tholeiitic magmatism is represented by the Gangila continental flood basaltic sequence with U–Pb ages between *c.* 1.0 Ga and 0.92 Ga (Tack *et al.* 2001).

Neoproterozoic magmatic arcs and subduction of Goianide–Pharusian oceanic lithosphere

While the continental drift of São Francisco–Congo and passive margin construction was in progress, subduction of oceanic lithosphere was taking place to the west and north, in the Goianide–Pharusian ocean. The former suture of this ocean is indicated by the alignment of the Goiás magmatic arc in the Tocantins province, with a continuation to the NE in the Borborema Province (the Ceará domain), and into the Trans-Saharan Province in West Africa (Trompette 1994).

Pre-collisional arc magmatism is documented in northern Chad (Pouclet *et al.* 2006), where orogenic domains north of the Congo Craton represent the continuation of the Borborema orogenic province of NE Brazil. Low metamorphic grade metavolcanic and metasedimentary successions from the Rei Bouba Group were interpreted as representing a back-arc basin setting coeval with the calc-alkaline magmatic arc plutonic rocks of the Poli Complex and Sinassi massif. An indication of the age of arc magmatism is given by a U–Pb age of *c.* 0.83 Ga, obtained from a rhyolite sample of the Rei Bouba Group (Toteu *et al.* 1987). In NE Brazil, the Ceará Domain crops out opposite the point where the northeastern Goiás magmatic arc is covered by the Parnaíba Basin, following the same NE–SW structural trend. Arc-related rocks as old as *c.* 0.78 Ga were dated by Fetter *et al.* (2003), who envisaged subduction of the southeastern plate beneath the northwestern one.

In a recent summary of the available isotopic database of the large Goiás magmatic arc, Laux *et al.* (2005) point out that the oldest intra-oceanic subduction-related rocks were generated between 0.89 and 0.80 Ga. In the three arc complexes cited above, subsequent magmatism lasted until the collisional episode of 0.65–0.61 Ga, represented by more contaminated magmas of continental arc setting.

On the opposite side of the San Franciscan peninsula, the Rio Negro magmatic arc (Tupinambá *et al.* 2000) in the Ribeira Belt, and similar assemblages in the Araçuaí Belt (Pedrosa Soares *et al.* 2001), testify partly contemporary subduction of oceanic lithosphere. Pre-collisional lithological associations include calc-alkaline plutonic orthogneisses of tonalite to granite compositions, and subordinate gabbros, with U–Pb ages of crystallization between 0.79 and 0.62 Ga (Heilbron & Machado 2003).

The collisional episode in the Brasília Belt

Following subduction of the southwestern São Francisco–Congo oceanic lithosphere to the west, collisional tectonics started with partial subduction of the distal sedimentary pile of the passive margin beneath the Paranapanema block and other accreting terranes that converged from the west, such as the Goiás massif and Goiás magmatic arc. During this early collisional stage, the supracrustal rocks of the Araxá and Andrelândia groups underwent intense low-angle shearing associated with isoclinal recumbent folding and development of pervasive foliation under a metamorphic regime of intermediate to high pressure-gradient (Simões 1995; Campos Neto & Caby 1999; Garcia & Campos Neto 2003).

The age interval between 0.65 and 0.63 Ga for the metamorphic peak, associated with partial melting and emplacement of syn-tectonic granitic bodies, is well constrained by U–Pb geochronology in most units of the Brasília Belt. In the Anápolis–Itauçu granulite complex, high-grade metamorphism was dated between 0.65 and 0.64 Ga, based on U–Pb zircon ages of metamorphic zircons in para-granulites and of syn-metamorphic ortho-granulites (Piuzana *et al.* 2003*b*). In the Araxá nappe, the Serra Velha granite (Seer *et al.* 2005) is a syn-collisional peraluminous granitic body intruded in the Araxá Group rocks that has yielded a U–Pb zircon age of 639 ± 1 Ma (Valeriano *et al.* 2004*a*). In the Passos nappe, a U–Pb age of 631 ± 4 Ma was obtained by Valeriano *et al.* (2004*a*) for anatexis of the Araxá Group schists. In the southernmost internal domain of the Brasília Belt, U–Pb zircon ages of syn-tectonic granitoid orthogneisses of the Socorro–Guaxupé nappe concentrate in the interval of 0.65–0.61 Ga (Ebert *et al.* 1996; Töpfner 1996; Campos Neto & Caby 1999) and are broadly coincident with monazite microprobe ages between 0.63 and 0.61 Ma from rocks of the same tectonic unit (Vlach & Gualda 2000).

This collisional event also affected the Goiás massif and the Goiás magmatic arc. In the former terrane, a U–Pb zircon age of 626 ± 7 Ma of an anatectic leucogranite was obtained by Pimentel *et al.* (2003) in the Archaean Caiçara tonalitic complex (2.93 Ga). In the Goiás arc concordant U–Pb ages were obtained on titanite and on zircons from numerous granitoid bodies (Pimentel *et al.* 2004).

The late collisional stage is marked by tectonic exhumation and emplacement of the metamorphic nappes over the external units of the belt (Campos Neto & Caby 2000). Cooling of metamorphic rocks of the Passos nappe is indicated by a second generation of monazite U–Pb ages around *c.* 0.61 Ga and by K–Ar ages of *c.* 0.60 Ga in biotite and muscovite (Valeriano *et al.* 2000, 2004*b*).

Indications of syn-orogenic sedimentation are found in the foreland and in the internal zones. In the first case, thrust-stacking in the southern Brasília Belt triggered formation of a foreland basin that onlaps the rocks of the Bambuí Group. Infill of the foredeep is represented by the immature clastic rocks of the Samburá and Três Marias formations, the latter being distal with respect to the orogen. Deposition of the Samburá Formation took place along sub-aqueous fan deposits of polymict meta-conglomerates within metapelites (Castro & Dardenne 2000). The sedimentation of the Samburá Formation probably occurred within the interval between 0.60 Ga and 0.58 Ga, in which K–Ar ages on biotite and white mica of the external allochthons are concentrated (Valeriano 1992; Valeriano *et al.* 2000), reflecting cooling by tectonic exhumation.

Young modes of U–Pb ages of detrital zircons and of Nd model ages indicate syn-orogenic contribution to the sedimentation .of some of the metasedimentary units of the internal zone of the belt. For example, micaschists considered to belong to the Araxá Group in the Goiânia region form the upper metamorphic thrust-sheets to the east of the Goiás magmatic arc. A young mode of detrital zircon ages in the range *c.* 0.67–0.68 Ga (Piuzana *et al.* 2003*a*) points to a mixed provenance for this unit, from a Palaeoproterozoic source and from the Neoproterozoic magmatic arc. Nd model ages (T_{DM}) of around 1.3 Ga have been documented in metapelites of the Ibiá and Araxá groups (Pimentel *et al.* 1999*b*, 2001) and were interpreted as resulting from sedimentary mixing of two components, one of cratonic origin (1.9–2.1 Ga) and the other from Neoproterozoic magmatic arc sources. These data indicate that part of what has been considered to belong to the Araxá Group may have been deposited in a fore-arc setting just before and/or during accretion of the magmatic arc, and that it was tectonically emplaced as metamorphic thrust sheets.

Concluding remarks

The amalgamation of Gondwana resulted in development of accretionary orogenic belts all around São Francisco–Congo (Fig. 1). However, among these surrounding belts, collisional events older than 0.60 Ga, such as that of the Brasília Belt, are restricted to the northern and northeastern margin. In particular, they are developed along the Borborema–Trans-Saharan orogenic system and in the East African orogen. Orogeny along the Sergipano and Oubanguides orogenic belts, facing respectively the northern margin of São Francisco–Congo in NE Brazil and Cameroon (Oliveira *et al.* 2006), took place at *c.* 0.63–0.60 Ga (Toteu *et al.* 2006) with south-verging nappe emplacement and regional metamorphism in response to accretion of the Pernambuco–Alagoas massif and African correlatives. A similar collage of tectonic terranes developed in the northeastern margin of São Francisco–Congo in the East African orogen, between 0.75 and 0.62 Ga (Meert 2003).

From *c.* 0.60 to 0.50 Ga, subsequent terrane accretion and continental collision around São Francisco–Congo followed a centripetal path of continental growth. For example, collision of the Rio Negro magmatic arc and related igneous complexes against the southern and eastern margins of São Francisco peninsula took place at *c.* 0.59–0.57 Ga along the Ribeira and Araçuaí Belts. This collisional event developed a pervasive SE-dipping low angle foliation and contemporaneous granitoid intrusions of batholithic dimensions of metaluminous and peraluminous affinities, respectively derived from the basement and the metasedimentary pile (Machado *et al.* 1996; Pedrosa Soares *et al.* 2001; Heilbron & Machado 2003).

Still later orogenic events took place along the Araguaia and Paraguay belts (at *c.* 0.54 Ga) as the result of the collision of Amazonia against the previously accreted terranes of the western margin of São Francisco–Congo palaeocontinent. In the Ribeira Belt, a younger orogenic event between 0.53 and 0.50 Ga (Schmitt *et al.* 2004) is related to the docking of the Cabo Frio terrane and with the development of superimposed subvertical folding and steep dextral-oblique shear zones that controlled the emplacement of late collisional granites.

Careful reading and comments made by Reinhardt Fuck and Alan Vaughan on the submitted version of the manuscript brought substantial improvement to this article. The authors are also thankful to Brazilian funding agencies

CNPq and FINEP for making possible the acquisition of much of the data compiled here.

References

ALKMIM, F. F. & MARSHAK, S. 2001. The Transamazonian orogeny in the Quadrilátero Ferrífero, Minas Gerais, Brazil: Paleoproterozoic collision and collapse in the southern São Francisco Craton region. *Precambrian Research*, **90**, 29–58.

ALKMIM, F. F., BRITO NEVES, B. B. & ALVES, J. A. C. 1993. Arcabouço tectônico do Cráton do São Francisco — uma revisão. *In*: DOMINGUEZ, M. L. & MISI, A. (eds) *O Cráton do São Francisco: trabalhos apresentados na reunião preparatória do II Simposio sobre o Cráton do São Francisco*. SBG-SGM-CNPq, Salvador, Brazil, 45–62.

ALMEIDA, F. F. M., HASUI, Y., BRITO NEVES, B. B. & FUCK, R. A. 1981. Brazilian structural provinces: an introduction. *Earth-Science Reviews*, **17**, 1–29.

BRITO NEVES, B. B. 2002. Main stages of the development of the sedimentary basins of South America and their relationship with the tectonics of supercontinents. *Gondwana Research*, **5**, 175–196.

BRITO NEVES, B. B., CAMPOS NETO, M. C. & FUCK, R. A. 1999. From Rodinia to Western Gondwana: an approach to the Brasiliano-Pan African Cycle and orogenic collage. *Episodes*, **22**, 155–166.

BRITO NEVES, B. B., ALMEIDA, F. F. M. & CARNEIRO, C. D. R. 2000. The origin and evolution of the South American Platform. *Earth-Science Reviews*, **50**, 77–111.

CAMPOS NETO, M. C. 2000. Orogenic systems from southwestern Gondwana. An approach to Brasiliano-Pan African Cycle and orogenic collage in souteastern Brazil. *In*: CORDANI, U. G., MILANI, E. J., THOMAZ FILHO, A. & CAMPOS, D. A. (eds) *Tectonic Evolution of South America. 31st International Geological Congress, Rio de Janeiro*, 335–365.

CAMPOS NETO, M. C. & CABY, R. 1999. Neoproterozoic high-pressure metamorphism and tectonic constraint from nappe system south of the São Francisco craton, southeast Brazil. *Precambrian Research*, **97**, 3–26.

CAMPOS NETO, M. C. & CABY, R. 2000. Lower crust extrusion and terrane accretion in the Neoproterozoic nappes of southeast Brazil. *Tectonics*, **19**, 669–687.

CARNEIRO, M. A., TEIXEIRA, W., CARVALHO, I. M., JR. & FERNANDES, R. A. 1998. Ensialic tectonic setting of the Archean Rio das Velhas Greenstone Belt: Nd and Pb isotopic evidence from the Bonfim Metamorphic Complex, Quadrilátero Ferrífero, Brazil. *Revista Brasileira de Geociências*, **28**, 189–200.

CASTRO, P. T. A. & DARDENNE, M. A. 2000. The sedimentology, stratigraphy and tectonic context of the São Francisco Supergroup at the southeast boundary of the São Francisco Craton. *Revista Brasileira de Geociências*, **30**, 439–441.

CORRÊA-GOMES, L. C. C. & OLIVEIRA, E. P. 2000. Radiating 1.0 Ga mafic dyke swarms of Eastern Brazil and Western Africa: evidence of post-assembly extension in the Rodinia supercontinent? *Gondwana Research*, **3**, 325–332.

DARDENNE, M. A. 2000. The Brasília fold belt. *In*: CORDANI, U. G., MILANI, E. J., THOMAZ FILHO, A. & CAMPOS, D. A. (eds) *Tectonic Evolution of South America*. 31st International Geological Congress, Rio de Janeiro, 231–263.

DELGADO, I. M. & PEDREIRA, A. J. 1995. *Tectono-Geological Map of Brazil, 1:7,000,000 scale*. Geological Survey of Brazil—Companhia de Pesquisa de Recursos Minerais-CPRM, Rio de Janeiro.

DRAKE, A. A., JR. 1980. The Serra de Caldas window, Goiás. *In*: *Tectonic studies in the Brazilian shield. United States Geological Survey Professional Paper*, **1191-A-B**, 1–11.

EBERT, H. D., CHEMALE, F., JR., BABINSKI, M., ARTUR, A. C. & VAN SCHMUS, W. R. 1996. Tectonic setting and U/Pb zircon dating of the plutonic Socorro Complex in the Transpressive Rio Paraíba do Sul Shear Belt, SE Brazil. *Tectonics*, **15**, 688–699.

FETTER, A. H., SANTOS, T. J. S. *ET AL*. 2003. Evidence for Neoproterozoic continental arc magmatism in the Santa Quitéria Batholith of Ceará State, NW Borborema Province, NE Brazil: Implications for the assembly of West Gondwana. *Gondwana Research*, **6**, 265–273.

FONSECA, M., DE OLIVEIRA, C. & EVANGELISTA, H. 2004. The Araguaia Belt, Brazil: part of a Neoproterozoic continental-scale strike-slip fault system. *In*: WEINBERG, R., TROUW, R., FUCK, R. & HACKSPACHER, P. (eds) *The 750–550 Ma Brasiliano Event of South America. Journal of the Virtual Explorer, Electronic Edition*, **17**, Paper 6.

GARCIA, M. G. M. & CAMPOS NETO, M. 2003. Contrasting metamorphic conditions in the Neoproterozoic collision-related nappes, south of São Francisco Craton, SE Brazil. *Journal of South American Earth Sciences*, **15**, 853–870.

HACKSPACHER, P., FETTER, A., TEIXEIRA, W., DANTAS, E., EBERT, H., TROUW, R. & VASCONCELOS, P. 2004. Final stages of the Brasiliano Orogenesis in SE Brazil: U–Pb and 40Ar/39Ar evidence for overprinting of the Brasília Belt by the Ribeira Belt Tectonics. *In*: WEINBERG, R., TROUW, R., FUCK, R. & HACKSPACHER, P. (eds) *The 750–550 Ma Brasiliano Event of South America. Journal of the Virtual Explorer, Electronic Edition*, **17**, Paper 2.

HARALYI, N. L. E. & HASUI, Y. 1982. The gravimetric information and the Archean-Proterozoic structural framework of eastern Brazil. *Revista Brasileira de Geociências*, **12**, 160–166.

HARLAN, S. S., HEAMAN, L., LECHEMINANT, A. N. & PREMO, W. R. 2003. Gunbarrel mafic magmatic event: a key 780 Ma time marker for Rodinia plate reconstructions. *Geology*, **31**, 1053–1056.

HEILBRON, M. & MACHADO, N. 2003. Timing of terrane accretion in the Neoproterozoic-Eopaleozoic Ribeira orogen (SE Brazil). *Precambrian Research*, **125**, 87–112.

HEILBRON, M., PEDROSA-SOARES, A. C., CAMPOS NETO, M. C., SILVA, L. C., TROUW, R. A. J. & JANASI, V. A. 2004. Provínica Mantiqueira. *In*: MANTESSO NETO, V., BARTORELLI, A., CARNEIRO, C. D. R. & BRITO-NEVES, B. B. (eds) *Geologia do Continente Sul-Americano: evolução e obra de Fernando Flávio Marques de Almeida*, Beca, São Paulo, 204–234.

HOFFMAN, P. F. 1991. Did the breakout of Laurentia turn Gondwanaland inside out? *Science*, **252**, 1409–1412.

JOST, H., FUCK, R. A. *ET AL.* 2005. Geologia e geocronologia do Complexo Uvá, Arqueano de Goiás. *Revista Brasileira de Geociências*, São Paulo, **35**, 563–576.

LAUX, J. H., PIMENTEL, M. M., DANTAS, E. L., ARMSTRONG, R. & JUNGES, S. L. 2005. Two Neoproterozoic crustal accretion events in the Brasília belt, central Brazil. *Journal of South American Earth Sciences*, **18**, 183–198.

LAUX, J. H., PIMENTEL, M. M., DANTAS, E. L. & LIMA, F. T. M. 2006. Idade Neoproterozóica para o Complexo máfico-ultramáfico de Cana Brava, Goiás Brasil. *In*: *Congresso Brasileiro de Geologia 43, Aracajú, Anais*. Sociedade Brasileira de Geologia, 128.

LESQUER, A., ALMEIDA, F. F. M., DAVINO, A., LACHAUD, J. C. & MAILLARD, P. 1981. Signification structurale des anomalies gravimétriques de la partie sud du craton de São Francisco (Brésil). *Tectonophysics*, **76**, 273–293.

LI, Z. X., LI, X. H., KINNY, P. D., WANG, J., ZHANG, S. & ZHOU, H. 2003. Geochronology of Neoproterozoic syn-rift magmatism in the Yangtze Craton, South China and correlations with other continents: evidence for a mantle superplume that broke up Rodinia. *Precambrian Research*, **122**, 85–10.

LI, X. H., EVANS, A. D. & ZHANG, S. 2004. A 90° spin on Rodinia: possible causal links between the Neoproterozoic supercontinent, superplume, true polar wander and low-latitude glaciation. *Earth and Planetary Science Letters*, **220**, 409–421.

MACHADO, N. & CARNEIRO, M. A. 1992. A major Archean tectonothermal event in the São Francisco shield, Brazil: U/Pb evidence from the Quaedrilátero Ferrífero. *Canadian Journal of Earth Sciences*, **29**, 2341–2346.

MACHADO, N., NOCE, C. M., LADEIRA, E. A. & OLIVEIRA, O. B. 1992. U–Pb geochronology of Archean magmatism and Proterozoic metamorphism in the Quadrilátero Ferrífero, Southern São Francisco Craton, Brazil. *Geological Society of America Bulletin*, **104**, 1221–1227.

MACHADO, N., VALLADARES, C., HEILBRON, M. & VALERIANO, C. M. 1996. U–Pb Geochronology of central Ribeira belt: implications for the evolution of Brasiliano orogeny. *Precambrian Research*, **79**, 347–361.

MANTOVANI, M. S. M. & BRITO NEVES, B. B. 2005. The Paranapanema lithospheric block: its importance for Proterozoic (Rodinia, Gondwana) supercontinent theories. *Gondwana Research*, **8**, 303–315.

MARTINS-NETO, M. A. 2000. Tectonics and sedimentation in a Proterozoic Rift-Sag Basin (Espinhaço Basin, Southeastern Brazil). *Precambrian Research*, **103**, 147–173.

MARTINS-NETO, M. A. & HERCOS, C. M. 2002. Sedimentation and tectonic setting of Early Neoproterozoic glacial deposits in southeastern Brazil. *In*: ALTERMANN, W. & CORCORAN, P. (eds) *Precambrian sedimentary environments: a modern approach to ancient depositional systems*. Special Publications of the International Association of Sedimentologists, **33**. Blackwell, Oxford, 383–404.

MAZZUCCHELLI, M., RIVALENTI, G., LEAL, A. B. M., GIRARDI, V. A. V., BRITO NEVES, B. B. & TEIXEIRA, W. 2001. Petrology of metabasite dikes in the Diamantina region, Minas Gerais, Brazil. *Periodico di Mineralogia*, Rome, **70**, 231–254.

MEERT, J. G. 2003. A synopsis of events related to the assembly of eastern Gondwana. *Tectonophysics*, **362**, 1–40.

MEERT, J. G. & TORSVIK, T. H. 2003. The making and unmaking of a supercontinent: Rodinia revisited. *Tectonophysics*, **375**, 261–288.

MILANI, E. J. & THOMAZ FILHO, A. 2000. Sedimentary basins of South America. *In*: CORDANI, U. G., MILANI, E. J., THOMAZ FILHO, A. & CAMPOS, D. A. (eds) *Tectonic Evolution of South America. 31st International Geological Congress, Rio de Janeiro*, 389–452.

MORAES, R., BROWN, M., FUCK, R. A., CAMARGO, M. A. & LIMA, T. M. 2002. Characterization and P–T evolution of melt-bearing ultrahigh-temperature granulites: an example from the Anápolis–Itauçu Complex of the Brasília Fold Belt, Brazil. *Journal of Petrology*, **43**, 1673–1705.

MORAES, R., FUCK, R. A., PIMENTEL, M. M., GIOIA, S. M. C. L., HOLLANDA, M. H. B. M. & ARMSTRONG, R. 2006. The bimodal rift-related Juscelândia volcanosedimentary sequence in central Brazil: Mesoproterozoic extension and Neoproterozoic metamorphism. *Journal of South American Earth Sciences*, **20**, 287–301.

NOCE, C. M., TEIXEIRA, W., QUEMENEUR, J. J. G., MARTINS, V. S. T. & BOLZACHINI, E. 2000. Isotopic signatures of Paleoproterozoic granitoids from the southern São Francisco Craton, NE Brazil, and implications for the evolution of the Transamazonian Orogeny. *Journal of South American Earth Sciences*, **13**, 225–239.

OLIVEIRA, E. P., TOTEU, S. F. *ET AL.* 2006. Geologic correlation between the Neoproterozoic Sergipano belt (NE Brazil) and the Yaoundé belt (Cameroon, Africa). *Journal of African Earth Sciences*, **44**, 470–478.

PACIULLO, F. V. P., RIBEIRO, A. & ANDREIS, R. R. 2000. The Andrelândia basin, a Neoproterozoic intraplate continental margin, southern Brasília belt. *Revista Brasileira de Geociências*, **30**, 200–202.

PEDROSA-SOARES, A. C., VIDAL, P., LEONARDOS, O. H. & BRITO-NEVES, B. B. 1998. Neoproterozoic oceanic remnants in eastern Brazil: further evidence and refutation of an exclusively ensialic evolution for the Araçuaí-West Congo belt. *Geology*, **26**, 519–522.

PEDROSA-SOARES, C. M., NOCE, C. M., WIEDEMANN, C. M. & PINTO, C. P. 2001. The Araçuaí–West-Congo Orogen in Brazil: an overview of a confined orogen formed during Gondwanaland assembly. *Precambrian Research*, **110**, 307–323.

PESONEN, L. J., ELMING, S.-A. *ET AL.* 2003. Paleomagnetic configuration of continents during the Proterozoic. *Tectonophysics*, **375**, 289–324.

PIMENTEL, M. M. & FUCK, R. A. 1992. Neoproterozoic crustal accretion in central Brazil. *Geology*, **20**, 375–379.

PIMENTEL, M. M., HEAMAN, L., FUCK, R. A. & MARINI, O. J. 1991.U–Pb Zircon Geochronology of Precambrian tin-bearing continental-type acid magmatism in central Brazil: *Precambrian Research*, **52**, 321–335.

PIMENTEL, M. M., FUCK, R. A. & BOTELHO, N. F. 1999*a*. Granites and the geodynamic history of the Neoproterozoic Brasília Belt, Central Brazil: a review. *Lithos*, **46**, 1–21.

PIMENTEL, M. M., FUCK, R. A. & FISCHEL, D. P. 1999*b*. Estudo isotópico Sm–Nd regional da Porção Central da Faixa Brasília, Goiás: implicações para idade e origem dos granulitos do Complexo Anápolis-Itauçu e rochas metassedimentares do Grupo Araxá. *Revista Brasileira de Geociências*, **29**, 271–276.

PIMENTEL, M. M., FUCK, R. A., JOST, H., FERREIRA FILHO, C. F. & ARAÚJO, S. M. 2000. The basement of the Brasília Fold Belt and the Goiás Magmatic Arc. *In*: CORDANI, U. G., MILANI, E. J., THOMAZ FILHO, A. & CAMPOS, D. A. (eds) *Tectonic Evolution of South America. 31st International Geological Congress, Rio de Janeiro*, 195–229.

PIMENTEL, M. M., DARDENNE, M. A. ET AL. 2001. Nd isotopes and the provenance of the detrital sediments of the Neoproterozoic Brasília Belt, central Brazil. *Journal of South American Earth Sciences*, **14**, 571–585.

PIMENTEL, M. M., JOST, H., FUCK, R. A., ARMSTRONG, R. A., DANTAS, E. L. & POTREL, A. 2003. Neoproterozoic anatexis of 2.9 Ga old granitoids in the Goiás-Crixás Archean Block, Central Brazil: evidence from new SHRIMP U–Pb data and Sm–Nd isotopes. *Geologia USP, Série Científica*, **3**, 1–12.

PIMENTEL, M. M., JOST, H. & FUCK, R. A. 2004. O embasamento da Faixa Brasília e o Arco Magmático de Goiás. *In*: MANTESSO NETO, V., BARTORELLI, A., CARNEIRO, C. D. R. & BRITO-NEVES, B. B. (eds) *Geologia do Continente Sul-Americano: evolução e obra de Fernando Flávio Marques de Almeida*. Beca, São Paulo, 355–368.

PISAREVSKY, S. A., WINGATE, M. T. D., POWELL, C., MCA., JOHNSON, S. & EVANS, D. A. 2003. Models of Rodinia assembly and fragmentation. *In*: YOSHIDA, M., WINDLEY, B. F. & DASGUPTA, S. (eds) *Proterozoic East Gondwana: Supercontinent Assembly and Breakup*. Geological Society, London, Special Publications, **206**, 35–55.

PIUZANA, D., PIMENTEL, M. M., FUCK, R. A. & ARMSTRONG, R. 2003*a*. SHRIMP U–Pb and Sm–Nd data from the Araxá Group and associated magmatic rocks: constraints for the age of sedimentation and geodynamic context of the southern Brasília Belt, central Brazil. *Precambrian Research*, **125**, 139–160.

PIUZANA, D., PIMENTEL, M. M., FUCK, R. A. & ARMSTRONG, R. 2003*b*. Neoproterozoic granulite facies metamorphism and coeval granitic magmatism in the Brasília Belt, Central Brazil: regional implications of new SHRIMP U–Pb and Sm–Nd data. *Precambrian Research*, **125**, 245–273.

POUCLET, A., VIDAL, M., DOUMNANG, J.-C., VICAT, J.-P. & TCHAMENI, R. 2006. Neoproterozoic crustal evolution in Southern Chad: Pan-African ocean basin closing, arc accretion and late- to post-orogenic granitic intrusion. *Journal of African Earth Sciences*, **44**, 543–560.

QUEIROZ, C. L., MCNAUGTHTON, N. J., FLETCHER, I. R., JOST, H. & BARLEY, M. E. 2000. Polymetamorphic history of the Crixás-Açu Gneisses, Central Brasil: SHRIMP U–Pb evidence from titanite and zircon. *Revista Brasileira de Geociências*, **30**, 40–44.

RIBEIRO, A., TROUW, R. A. J., ANDREIS, R. R., PACIULLO, F. V. P. & VALENÇA, J. G. 1995. Evolução das bacias proterozóicas e o termo-tectonismo Brasiliano na margem sul do Cráton do São Francisco. *Revista Brasileira de Geociências*. **25**, 235–248.

ROGERS, J. J. W. & SANTOSH, M. 2002. Configuration of Columbia, a Mesoproterozoic supercontinent. *Gondwana Research*, **5**, 5–22.

SCHMIDT, W. 1983. *Die Geologie der Araxá Gruppe in Sudwest Minas Gerais, Brasilien, unter besonderer Berucksichtigung des Grunstein-gurtels von Fortaleza de Minas*. PhD thesis, Freiburg University, Germany.

SCHMITT, R. S., TROUW, R. A. J., VAN SCHMUS, W. R. & PIMENTEL, M. M. 2004. Late amalgamation in the central part of Western Gondwana: new geochronological data and the characterization of a Cambrian collisional orogeny in the Ribeira belt (SE Brazil). *Precambrian Research*, **133**, 29–61.

SEER, H. J. 1999. *Evolução tectônica dos grupos Araxá, Ibiá e Canastra na Sinforma de Araxá, Araxá, Minas Gerais*. PhD Thesis, University of Brasília.

SEER, H. J. & DARDENNE, M. A. 2000. Tectonostratigraphic terrane analysis on Neoproterozoic times: the case study of the Araxá Synform, Minas Gerais, Brazil—implications to the final collage of the Gondwanaland. *Revista Brasileira de Geociências*, **30**, 78–81.

SEER, H. J., BROD, J. A., FUCK, R. A., PIMENTEL, M. M., BOAVENTURA, G. R. & DARDENNE, M. A. 2001. Grupo Araxá em sua área tipo: um fragmento de crosta oceânica Neoproterozóica na Faixa de Dobramentos Brasília. *Revista Brasileira de Geociências*, **31**, 389–400.

SEER, H. J., BROD, J. A., VALERIANO, C. M. & FUCK, R. A. 2005. Leucogranitos intrusivos no Grupo Araxá: registro de um evento magmático durante colisão neoproterozóica na porção meridional da Faixa Brasília. *Revista Brasileira de Geociências*, **35**, 33–42.

SILVA, C. H. 2003. *Evolução geológica da Faixa Brasília na região de Tapira, Sudoeste de Minas Gerais*. PhD Thesis, Universidade Estadual Paulista-Rio Claro, Brazil.

SIMÕES, L. S. A. 1995. *Evolução tectonometamórfica da Nappe de Passos, Sudoeste de Minas Gerais*. PhD Thesis, Universidade Estadual Paulista-Rio Claro, Brazil.

STRIEDER, A. J. & NILSON, A. A. 1992. Melange ofiolítica nos metassedimentos Araxá de Abadiânia (GO) e implicações tectônicas regionais. *Revista Brasileira de Geociências*, **22**, 204–215.

STRIEDER, A. J. & SUITA, M. T. F. 1999. Neoproterozoic tectonic evolution of the Tocantins Structural Province, Central Brazil. *Journal of Geodynamics*, **28**, 67–289.

SUITA, M. T. F. 1996. *Geoquímica e metalogenia de elementos do grupo da platina (EGP + Au) em complexos máfico-ultramáficos do Brasil: critérios e guias com ênfase no complexo máfico-ultramáfico acamadado de alto grau de Barro Alto (CBA, Goiás)*. PhD thesis, Instituto de Geociências, Universidade Federal do Rio Grande do Sul, Porto Alegre, Brazil.

TACK, L., WINGATE, M. T. D., LIÉGEOIS, J.-P., FERNANDEZ-ALONSO, M. & DEBLOND, A. 2001. Early Neoproterozoic magmatism (1000–910 Ma) of the Zadinian and Mayumbian Groups (Bas-Congo): onset of Rodinia rifting at the western edge of the Congo craton. *Precambrian Research*, **110**, 277–306.

TEIXEIRA, W. 1989. Dykes in the southern part of the São Francisco craton: a tectonic review based on K–Ar geochronology. *Boletim IG-USP, Série Científica*, **20**, 25–30.

TEIXEIRA, W. & FIGUEIREDO, M. C. H. 1991. An outline of Early Proterozoic crustal evolution in the São Francisco craton, Brazil: a review. *Precambrian Research*, **53**, 1–22.

TEIXEIRA, W., CARNEIRO, M., NOCE, C. M., SATO, K. & TAYLOR, P. N. 1996. Pb, Sr and Nd isotope constraints on the Archean evolution of gneissic-granitoid complexes in southern São Francisco Craton, Brazil. *Precambrian Research*, **78**, 151–164.

TÖPFNER, C. 1996. *Brasiliano-granitoide in den Bundestaten São Paulo und Minas Gerais, Brasilien- eine vergleischende studie*. Münchner Geologische Hefte, **A17**.

TOTEU, S. F, MICHARD, A., BERTRAND, J. M. & ROCCI, G. 1987. U/Pb dating of Precambrian rocks from northern Cameroon, orogenic evolution and chronology of the Pan-African belt of Central Africa. *Precambrian Research*, **37**, 71–87.

TOTEU, S. F., FOUATEU, R. Y *ET AL*. 2006. U–Pb dating of plutonic rocks involved in the nappe tectonic in southern Cameroon: consequence for the Pan-African orogenic evolution of the central African fold belt. *Journal of African Earth Sciences*, **44**, 479–493.

TROMPETTE, R. 1994. *Geology of Western Gondwana*. A. A. Balkema, Rotterdam.

TROUW, R. A. J., HEILBRON, M. *ET AL*. 2000. The central segment of the Ribeira belt. *In*: CORDANI, U. G., MILANI, E. J., THOMAZ FILHO, A. & CAMPOS, D. A. (eds) *Tectonic Evolution of South America. 31st International Geological Congress, Rio de Janeiro*, 287–310.

TUPINAMBÁ, M., TEIXEIRA, W. & HEILBRON, M. 2000. Neoproterozoic Western Gondwana assembly and subduction related plutonism: the role of the Rio Negro Complex in the Ribeira belt, SE Brazil. *Revista Brasileira de Geociências*, **30**, 7–11.

USSAMI, N. 1993. Estudos geofísicos no Cráton do São Francisco: estágio atual e perspectivas. *In*: DOMINGUEZ, J. M. L. & MISI, A. (eds) *O Cráton do São Francisco*. SBG/SGM/CNPq, Salvador, Brazil, 35–43.

VALERIANO, C. M. 1992. *Evolução tectônica da extremidade meridional da Faixa Brasília, região da Represa de Furnas, Sudoeste de Minas Gerais*. PhD thesis, Universidade de São Paulo, Instituto de Geociências, São Paulo.

VALERIANO, C. M. & SIMÕES, L. S. A. 1997. Geochemistry of proterozoic mafic rocks from the Passos Nappe (Minas Gerais, Brazil): tectonic implications to the evolution of the southern Brasília Belt. *Revista Brasileira de Geociências*, **27**, 99–110.

VALERIANO, C. M., SIMÕES, L. S. A., TEIXEIRA, W. & HEILBRON, M. 2000. Southern Brasília belt (SE Brazil): tectonic discontinuities, K–Ar data and evolution during the Neoproterozoic Brasíliano orogeny. *Revista Brasileira de Geociências*, **30**, 195–199.

VALERIANO, C. M., MACHADO, N., SIMONETTI, A., VALLADARES, C. S., SEER, H. J. & SIMÕES, L. S. A. 2004*a*. U–Pb geochronology of the southern Brasília belt (SE-Brazil): sedimentary provenance, Neoproterozoic orogeny and assembly of West-Gondwana. *Precambrian Research*, **130**, 27–55.

VALERIANO, C. M., DARDENNE, M. A., FONSECA, M. A., SIMÕES, L. S. A. & SEER, H. J. 2004*b*. A evolução tectônica da Faixa Brasília. *In*: MANTESSO NETO, V., BARTORELLI, A., CARNEIRO, C. D. R. & BRITO-NEVES, B. B. (eds) *Geologia do Continente Sul-Americano: evolução e obra de Fernando Flávio Marques de Almeida*. Beca, São Paulo, 575–593.

VLACH, S. R. F. & GUALDA, G. A. R. 2000. Microprobe monazite dating and the ages of some granitic and metamorphic rocks from southeastern Brazil. *Revista Brasileira de Geociências*, **30**, 214–218.

WEIL, A. B., VAN DER VOO, R., MCNIOCAILL, C. & MEERT, J. G. 1998. The Proterozoic supercontinent Rodinia: paleomagnetically derived reconstructions for 1100 to 800 Ma. *Earth and Planetary Science Letters*, **154**, 13–24.

Correlation of Neoproterozoic terranes between the Ribeira Belt, SE Brazil and its African counterpart: comparative tectonic evolution and open questions

M. HEILBRON[1,4], C. M. VALERIANO[1,4], C. C. G. TASSINARI[2,4], J. ALMEIDA[1,4], M. TUPINAMBÁ[1], O. SIGA JR[2,4] & R. TROUW[3,4]

[1]*TEKTOS Research Group, Faculdade de Geologia, Universidade do Estado do Rio de Janeiro (UERJ), Rua São Francisco Xavier 524/4006-A, Maracanã, 20559-900, Rio de Janeiro, Brazil (e-mail: heilbron@uerj.br)*

[2]*CPGeo/IG-USP, Centro de Pesquisas Geocronológicas, Instituto de Geociências, Universidade de São Paulo (USP), Rua do Lago, Cidade Universitária, São Paulo, SP, Brazil.*

[3]*Departamento de Geologia, Universidade Federal do Rio de Janeiro (UFRJ), Brazil*

[4]*Pesquisador do Conselho Nacional de Desenvolvimento Científico e Tecnológico – CNPQ, Brazil*

Abstract: Four main classes of tectonic entities may be considered for the Ribeira Belt and southwest African counterparts: (1) cratonic fragments older than 1.8 Ga and their passive margin successions, (2) reworked basement terranes with Mesoproterozoic and/or Neoproterozoic deformed cover, (3) magmatic arc associations, (4) terranes with Palaeoproterozoic basement and deformed Neoproterozoic back-arc successions. Based on comparative investigation, a tectonic model of polyphase amalgamation is proposed with *c.* 790 and 630–610 Ma major episodes of intra-oceanic and cordilleran arc magmatism along both sides of the Adamastor Ocean. Subsequent diachronous collision of the arc terranes and small plates followed at *c.* 630, 600, 580 and 530 Ma. The tectonic complexity reflects an accretionary evolution from Cryogenian to Cambrian times. The São Francisco–Congo and Angola palaeo-continents did probably not behave as one consolidated block, but rather may have accommodated considerable convergence during the Brasiliano/Pan-African episodes. The final docking of Cabo Frio and Kalahari in the Cambrian was coeval with the arrival of Amazonia on the opposite side, resulting in lateral reactivation and displacement between the previously amalgamated pieces. The transition between the Cambrian and the Ordovician is marked by the extensional collapse of the metamorphic core zones of the orogens.

The Ribeira Belt is one of the best studied Neoproterozoic belts of West Gondwana (Figs 1, 2 & 3), extending for 1400 km along the Atlantic coast of SE Brazil (Hasui *et al.* 1975; Almeida 1977; Almeida *et al.* 1981; Campos Neto 2000; Trouw *et al.* 2000; Heilbron *et al.* 2004*a*, *b*). The African counterpart, in Angola and Namibia, is represented from north to south by the West Congo Belt, the Angola Craton and the Kaoko Belt.

Tectonostratigraphic terranes of SE Brazil and SW African counterparts do not provide a clear match when detailed palaeogeographical reconstructions of West Gondwana are considered. Comparison of both sides of the Atlantic yields reasonable continuity between terranes of the northern segment of the Araçuaí Belt and of the West Congo Belt. South of this area, however, there is mismatch between the Ribeira Belt (SE Brazil), including the Rio Negro magmatic arc (790–620 Ma) and the Cabo Frio terrane, and the southern segment of the intracontinental West Congo Belt and the basement of the Angola Craton, south of Luanda.

The main purpose of this contribution is to summarize and compare the geological evolution of the Ribeira Belt and SW African pre-Atlantic counterparts. Several important open points arising from this comparison are addressed: (a) how do the pieces of the Neoproterozoic collage fit (b) how to accommodate juvenile arcs and a multiple collision history with the frequently postulated narrow ocean between the cratonic blocks of South America and Africa during the Neoproterozoic (c) did the São Francisco–Congo and Angola cratons compose a monolithic block since the Palaeoproterozoic (d) when was the assembly of West Gondwana finally completed.

From: PANKHURST, R. J., TROUW, R. A. J., BRITO NEVES, B. B. & DE WIT, M. J. (eds) *West Gondwana: Pre-Cenozoic Correlations Across the South Atlantic Region.* Geological Society, London, Special Publications, **294**, 211–237. DOI: 10.1144/SP294.12 0305-8719/08/$15.00

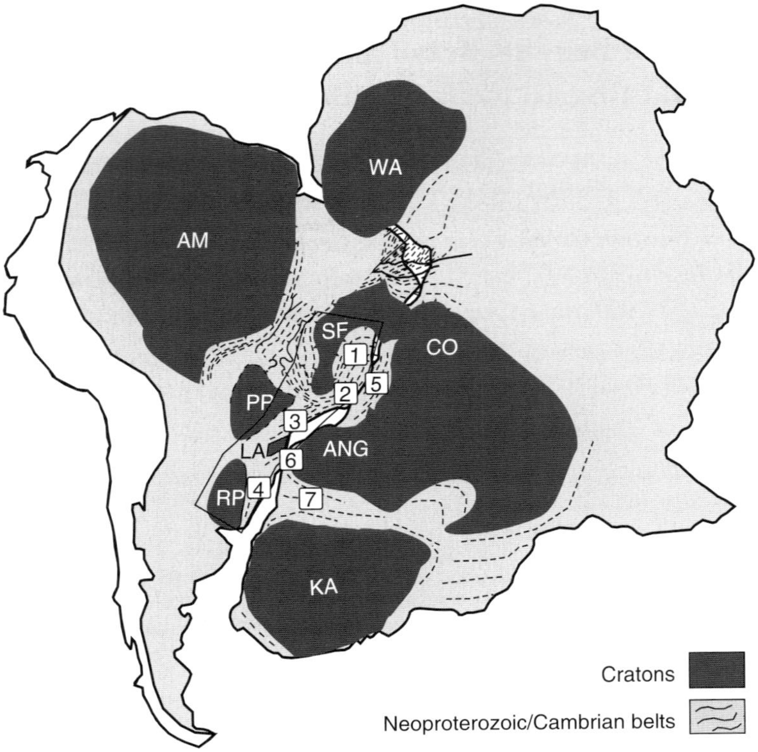

Fig. 1. Location of Neoproterozoic mobile belts and cratons in South America and Africa (Western Gondwana), modified from Trompette (1998). Neoproterozoic **belts**: 1, Araçuaí Belt; 2, Central Ribeira Belt; 3, Southern Ribeira Belt; 4, Dom Feliciano Belt; 5, West Congo Belt; 6, Kaoko Belt; 7, Damara Bel. Major cratons: AM, Amazonia; SF, São Francisco; LA, Luis Alves; RP, Rio de la Plata; WA, West Africa; CO, Congo; ANG, Angola; KA, Kalahari. The polygon indicates the region detailed in Fig. 2.

Tectonic framework of the Ribeira Belt

The Ribeira Belt is defined as a NE-trending orogenic system that resulted from the collision interaction between the São Francisco–Congo palaeo-continent with the southwestern part of the Angola Craton (Fig. 1). One or more microplates in between were also involved (Campos Neto 2000; Trouw *et al.* 2000; Heilbron *et al.* 2000). The convergence is expressed by the Pan-African/ Brasiliano orogenic episodes, with its main development during the Neoproterozoic–Cambrian and with latest tectonic stages reaching Early Ordovician times.

The major tectonic framework of the Ribeira Belt (Figs 2, 3 & 4) comprises several tectono-stratigraphic terranes (Howell 1989): Occidental, Paraíba do Sul–Embú, Oriental and Cabo Frio terranes in Rio the Janeiro and southern Espírito Santo states (northern and central Ribeira Belt, Heilbron *et al.* 2004*a*, *b*, and Socorro, Apiaí, Embú, Curitiba terranes and Luis Alves Craton in São Paulo and Paraná states (southern Ribeira Belt, Campos Neto 2000). The terranes are limited by thrust faults or by dextral transpressive shear zones. The connection between the central and southern segments of the Ribeira Belt still has important open questions. The almost orthogonal junction between the Brasília and the Ribeira belts (Figs 2 & 3) is marked by an interference zone (Fig. 3) where deformational phases and metamorphic episodes are superposed (Trouw *et al.* 2000).

The Occidental terrane and the lower nappes of the southern Brasília Belt (Fig. 2) are regarded as the reworked southern border of the São Francisco palaeo-continent. In consequence, the original extension of the São Francisco–Congo palaeo-continent in Brazil encompassed these terranes and occupied a much larger area than that of the present São Francisco Craton.

The accretionary history of the belt is characterized by complex diachronous docking of Neoproterozoic magmatic arcs and older cratonic

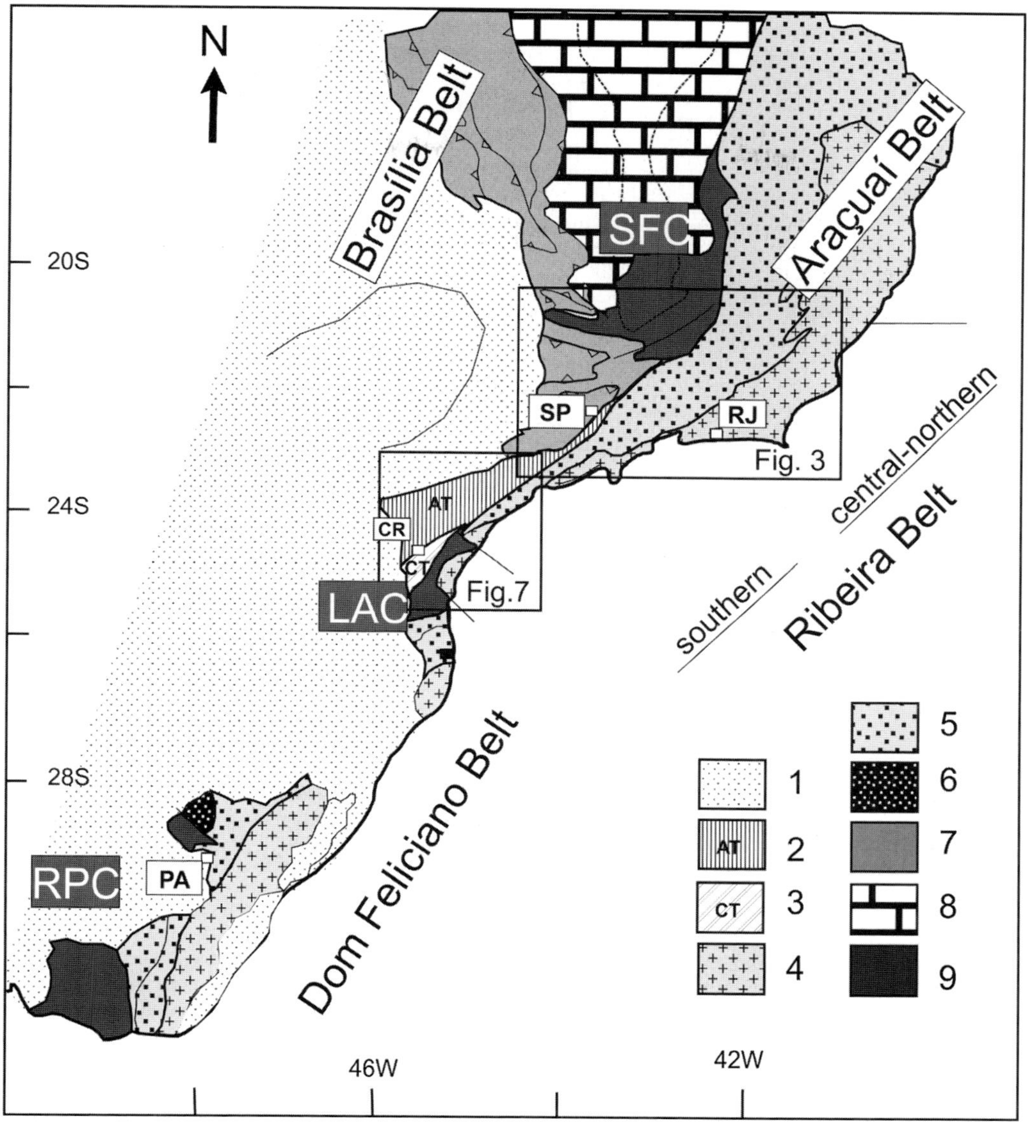

Fig. 2. Tectonic units of southern Brazil and Uruguay, modified from Heilbron *et al.* (2004*b*). Cratonic fragments: SFC, São Francisco; LAC, Luis Alves; RPC, Rio de la Plata. 1, post-Cambrian sedimentary basins, 2–4, terranes of the Ribeira Belt; 2, Apiaí terrane; 3, Curitiba terrane; 4, Oriental terrane; 5, Occidental, Paraíba do Sul and Embú terranes; 6, São Gabriel Belt; 7, Brasília Belt; 8, cratonic cover; 9, basement of São Francisco and other cratons. The rectangles indicate the central-northern and southern segments of the Ribeira Belt presented in Figs 3 & 7. Major cities: RJ, Rio de Janeiro; SP, São Paulo; CR, Curitiba; PA, Porto Alegre.

fragments throughout the southern and southeastern sectors of the São Francisco palaeo-continent. Terranes and microplates (Figs 3 & 4) were progressively accreted during four major tectonic episodes.

(a) The oldest collisional event is recorded in the interference zone with the Brasília Belt (Figs 2 & 3) and resulted from the accretion of the Socorro Nappe with NE–E vergence onto the southern part of the São Francisco palaeo-continent (Trouw *et al.* 2000; Campos Neto *et al.* 2000). High pressure granulites developed during this episode that, in fact, represents the main collision in the southern Brasília Belt at *c.* 640–610 Ma and resulted in the development of the southern Brasília Belt (Valeriano *et al.* 2008). The tectonic scenario is possibly related to the interaction of the Luis Alves

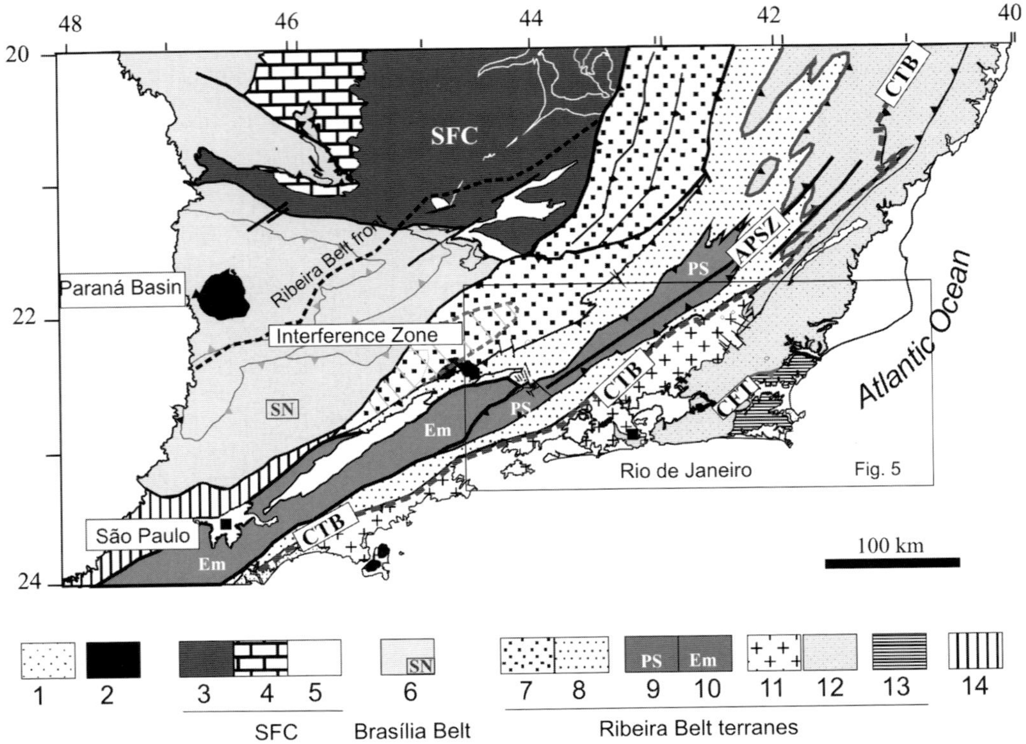

Fig. 3. Tectonic map of central Ribeira Belt and interference zone with the Brasília Belt, compiled and modified from Trouw *et al.* (2000), Campos Neto *et al.* (2000); Heilbron *et al.* (2000, 2004*a*, *b*) and Pedrosa Soares *et al.* (2001) 1, Phanerozoic cover; 2, Upper Cretaceous alkaline plutons; 3–5, São Francisco Craton; 3, Palaeoproterozoic to Archaean basement; 4, cratonic cover; 5, Mesoproterozoic to Neoproterozoic metasediments of the autochthonous domain; 6, *c.* 640–610 Ma east-verging Brasília Belt, including Soccoro nappe (SN); 7–13, terranes of Ribeira Belt; 7, Andrelândia and 8, Juiz de Fora domains of Occidental terrane of the Ribeira Belt; 9, Paraíba do Sul terrane; 10, Embú terrane; 11, Neoproterozoic magmatic arc and 12, Neoproterozoic metasedimentary successions of the Oriental terrane; 13, Cabo Frio terrane; 14, Apiaí terrane. CTB, Central Tectonic Boundary; CFT, Cabo Frio Thrust; APSZ, Além Paraíba Shear Zone. Rectangle indicates the detailed geological map presented in Figure 5. Dotted line indicates the thermal and deformational front of the Ribeira Belt at the interference zone with Brasília Belt. Light-grey shaded area approximately indicates relicts of high pressure rocks related to the evolution of the Brasília Belt.

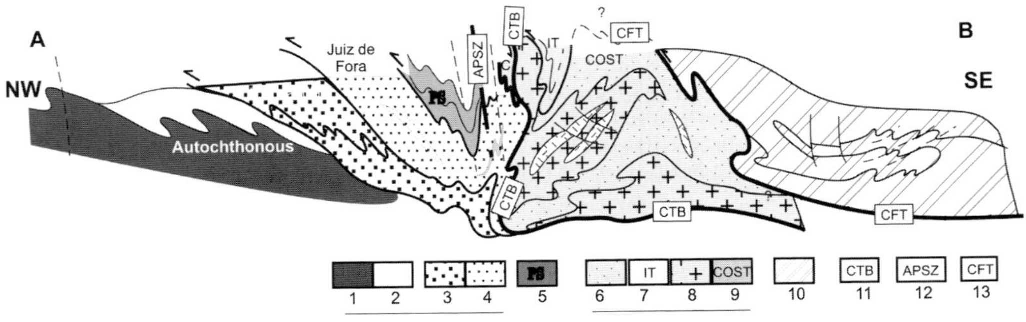

Fig. 4. Transverse composed tectonic section of central Ribeira Belt, modified from Heilbron *et al.* (2004*b*) 1, basement and 2, cover of the Autochthonous domain; 3, Andrelândia and 4, Juiz de Fora domains of the Occidental terrane; 5, Paraíba do Sul terrane; 6–9; Oriental terrane; 6, Cambuci domain; 7, Italva domain; 8, Rio Negro Arc and 9, metasedimentary successions of the Costeiro domain; 10, Cabo Frio terrane; 11, Central Tectonic Boundary; 12, Além Paraíba Shear Zone; 13, Cabo Frio Thrust.

microplate with the Paranapanema block, presently hidden under the Paraná Basin (Mantovani & Brito Neves 2005; Mantovani *et al.* 2005), and the São Francisco Craton.

(b) The second collisional episode took place between *c.* 605 and 580 Ma. The thermal and deformational effects of this episode overprint the southern segment of the Brasília Belt. The main metamorphism recorded in the Paraíba do Sul and Embú terranes (Figs 3 & 4) is dated within this episode, while the interaction of the Curitiba terrane and the Luis Alves Craton (Fig. 2) is estimated to have developed at *c.* 600 Ma. The interaction of the Apiaí terrane with its surrounding terranes is marked by younger vertical shear zones that obliterate the geological relationship with the other terranes of the Ribeira Belt.

(c) The third collisional event is related to the docking of the Oriental terrane, which contains the Neoproterozoic Rio Negro arc, against the São Francisco palaeo-continent at *c.* 580–550 Ma.

(d) Finally, collision of the Cabo Frio terrane took place during the Cambrian, at *c.* 530–510 Ma (Schmitt *et al.* 2004). In response to the compression, the previously amalgamated terranes were overprinted by important NE–SW dextral shear zones that displace the contacts between the terranes and locally reactivate older thrust zones.

In the following section the terranes of the Ribeira Belt are briefly described. They are separated in two groups related respectively to the northern-central and the southern segments of the belt. A description of the rock units and their local names are summarized in Tables 1 and 2.

Terranes of northern and central Ribeira Belt

The Occidental terrane: reworked passive margin of the São Francisco palaeocontinent

The Occidental terrane (Figs 3 & 4) comprises reworked Palaeoproterozoic/Archaean basement units and a Neoproterozoic passive margin succession (Table 1, Ribeiro *et al.* 1995; Trouw *et al.* 2000, Heilbron *et al.* 2004*a*).

The basement associations comprise two distinct units (Figs 5 & 6, Table 1). The first one is made up of Palaeoproterozoic orthogneisses with amphibolites, migmatites and minor granulites. The second one (Figs 5 & 6, Table 1) crops out eastwards of the previous unit and is composed of Palaeoproterozoic ortho-granulites with a wide compositional variation. Geochemical and geochronological data suggest a complex evolution with *c.* 2.4 Ga MORB-like basic granulites, *c.* 2.1–2.0 intermediate to acidic rocks of a juvenile magmatic arc setting, and *c.* 1.7 Ga alkaline basic granulites.

The Neoproterozoic passive margin (Andrelândia Megasequence, Paciullo *et al.* 2000) is mainly a metamorphosed siliciclastic succession, subdivided into two sequences separated by a major unconformity (Figs 5 & 6, Table 1). The basal sequence comprises quartz–feldspar paragneisses, quartzites and mica schists containing mafic and ultramafic lenses and pods. A palaeo-environmental interpretation suggests that shelf deposits of the basal sequence grade to the top into a more distal deep marine turbidites with transitions to ocean floor assemblages. The presence of dropstones in the upper sequence and regional correlations suggest the influence of glaciation (Paciullo *et al.* 2000). The U–Pb age of the youngest detrital zircon (*c.* 900 Ma, Valeriano *et al.* 2004) is used considered the best estimate for the maximum depositional age of the Andrelândia Megasequence and thus of the passive margin (Valeriano *et al.* 2004; Valladares *et al.* 2004).

The Occidental terrane records the metamorphic and deformational effects of all collisional episodes related to West Gondwana amalgamation. In the region south of the São Francisco Craton, at the so-called interference zone (Fig. 3), a superposition of the main metamorphic episode of the Brasília (*c.* 640–610 Ma) and the *c.* 605–580 Ma episode related with the development of the Ribeira is recorded. Southeast of the São Francisco Craton (Figs 2 & 3), metamorphic episodes of *c.* 605–600, 580–550 and 530–510 Ma were detected (Machado *et al.* 1996*a*; Trouw *et al.* 2000; Heilbron *et al.* 2000, 2004*a*; Campos Neto & Caby 2004; Valeriano *et al.* 2004, 2008).

Paraíba do Sul and Embú terranes (docked at c. 605–580 Ma)

Two major lithological associations were mapped within the Paraíba do Sul terrane (Figs 5 & 6, Table 1): (a) the first one is represented by Palaeoproterozoic hornblende-bearing orthogneisses; (b) the second one is a metasedimentary siliciclastic sequence with dolomitic marble and calc-silicate lenses.

The east–west trending Embú terrane is limited in the north and south by kilometre-wide thick dextral shear zones (Figs 3 & 7). Along strike this terrane wedges out against the Occidental terrane to the east and against the Apiaí terrane to the west, both with still poorly studied tectonic contacts (Fig. 7). The contact with the Paraíba do Sul terrane is marked by an expressive dextral shear zone (Campos Neto 2000; Heilbron *et al.* 2004*b*). The Embú terrane also comprises two associations:

Table 1. *Basement and cover units of the Northern and central Ribeira Belt*

Terranes and docking period	Structural domains	Basement	Deformed cover
Cabo Frio *c.* 535–510 Ma		**Região dos Lagos Complex** (*9): *c.* 1.9 Ga tonalitic to granitic orthogneisses, with diorite enclaves and abundant amphibolite lenses. Sr and Nd data suggest that both are reworked Archaean crust with a juvenile contribution.	**Búzios–Palmital Successions** (*10):(kyanite)–sillimanite–garnet–biotite gneisses with abundant intercalations of calc-silicate layers and amphibolites Minor intercalations of garnet-quartz gneisses and feldspar-rich quartzites.
Oriental *c.* 590–550 Ma	Italva		**Italva Group** (*8): banded biotite gneisses, platform calcite marbles, amphibolites (*c.* 840 Ma) and amphibole schists. The succession is interpreted as originating in a shelf environment with basaltic volcanism, metamorphosed to amphibolite facies.
	Costeiro		**Costeiro Complex** (*8): high-grade pelitic and psammitic paragneisses with minor quartzites and calc-silicate rocks.
	Cambuci		**Cambuci sequence** (*8): highly deformed and migmatitic garnet–biotite gneisses with lenses of dolomite olivine marble and calc-silicate rocks with lenses of mafic rocks, metamorphosed to garnet–diopside granulites. U–Pb data from detrital zircons indicate sources from the basement and from the Neoproterozoic Rio Negro magmatic arc that crops out in the Costeiro Domain.
Embu *c.* 605–570 Ma and 790 Ma?		Approximately 2. 1–2.0 Ga orthogneisses with granitic to tonalitic compositions (*6).	**Embú Complex** (*7): still undated unit with fine-grained biotite gneisses, mica schists with intercalations of immature quartzites, calc-silicate rocks and amphibolites.
Paraíba do Sul *c.* 605–570 Ma		**Quirino Complex** (*4): *c.* 2.19–2.17 Ga hornblende-bearing orthogneisses of tonalite to granodiorite compositions containing enclaves of ultramafic, mafic and calc-silicate rocks.	**Paraíba do Sul Complex or Group** (*5): still undated metasedimentary sequence essentially composed of biotite gneisses and pelitic sillimanite–biotite gneisses, rich in garnet and tourmaline with dolomitic marble, calc-silicate rock, sillimanite quartz schist and garnet-rich meta-chert.

Occidental reworked São Francisco palaeocontinent	Juiz de Fora	**Juiz de Fora Complex** (*2): ortho-granulites that display a wide compositional variation. Geochemical and geochronological data suggest complex evolution with *c.* 2.4 Ga MORB-like basic granulites, *c.* 2.1–2.0 Ma intermediate to acidic rocks of juvenile magmatic arc affinity, and *c.* 1.7 Ga alkaline basic granulites.	**Andrelândia Megasequence** (*3): metamorphosed siliciclastic passive margin unit subdivided into two sequences separated by a major unconformity. Maximum sedimentation age of *c.* 900 Ma. *Serra do Turvo sequence*: garnet-bearing psammo-pelitic schists and gneisses. K-feldspar, kyanite, sillimanite and orthopyroxene may occur, depending on metamorphic facies. *Carrancas sequence*: quartz–feldspar paragneisses with quartzites and mica schists containing mafic and ultramafic lenses and pods. *Amphibolites* interlayered with the passive margin sequences indicate a progressive transition from continental to oceanic environment.
	Andrelândia	**Mantiqueira Complex** (*1): *c.* 2.2–2.1 Ga orthogneisses with amphibolites, and minor granulites. Minor Archaean inliers of *c.* 2.7–2.6. Important *c.* 2.06–2.05 metamorphic event (Transamazonian orogeny) (cordilleran-type convergent tectonic setting) Basic rocks of intraplate and extensional settings still undated.	

*Selected references: 1 Machado *et al.* (1992, 1996*a*), Fischel *et al.* (1998), Heilbron *et al.* (2000, 2003), Duarte *et al.* (2004); Silva *et al.* (2005); 2 Heilbron *et al.* (1988), Machado *et al.* (1996*a*, *b*), Fischel *et al.* (1998); 3 Trouw *et al.* (2000), Paciullo *et al.* (2000), Valladares *et al.* (2004), Valeriano *et al.* (2004); 4 Valladares *et al.* (2003); 5 Eirado Silva *et al.* (2007); 6 Babinski *et al.* (2001); 7 Tassinari & Campos Neto (1998), Eirado Silva *et al.* (2007); 8 Grossi Sad & Dutra (1988), Tupinambá *et al.* (2000), Heilbron & Machado (2003); 9 Heilbron *et al.* (1982), Heilbron *et al.* (2000), Schmitt *et al.* (2004).

Table 2. *Meso- to Neoproterozoic units of the Apiaí terrane*

Age Ma	Units in Figure 7	No. of domain in Fig. 7	Selected references
c. 600	**Lageado Group**: meta-conglomerates and metapelites/meta-psammites, with subordinate meta-limestones and basic volcanic rocks.	3	Campanha *et al.* 2004; Hackspacher *et al.* 2000
628–605	**São Roque Group**: upper unit: meta-arenite (ortho-quartzites and meta-arkose) with intercalations of metabasic rocks, pillow lava and vesicular volcanic flows. Basal unit: stromatolitic carbonate rocks with tholeiitic metabasic rocks (pillow structures), meta-arkose and quartzite, locally with meta-conglomerate and metapelite lenses.	3	Bergmann 1988
630–620	**Cunhaporanga** and **Três Córregos** calcalkaline batholiths.	1 to 4	Guimarães 2000; Prazeres Filho *et al.* 2003; Prazeres Filho 2005
645–628	**Abapã Sequence**: meta-arkoses, metavolcanic rocks, meta-conglomerates, metapelites.	1	
1030–908	**Itaiacoca Sequence**: carbonate rocks with columnar stromatolites.	1	Sallun Filho & Fairchild, 2004; Siga *et al.* 2003
1395	**Serra do Itaberaba Group**: intermediate to acidic rocks and graphite schist and iron formation, unconformably overlain by metapelite beds with metavolcanic rocks.	3	Juliani *et al.* 2000
1500–1450	**Betara, Perau, Votuverava formations**: clastic metasediments (metapelites, meta-rhythmites, meta-sandstones, meta-conglomerates), manganese and graphitic metasediments, banded iron formations, meta-chert, and metavolcanic rocks.	4	Basei *et al.* 2003
1600–1500	**Água Clara Fm**: carbonate-rich succession interlayered with pelitic schists and metabasic rocks.	2	Weber *et al.* 2004

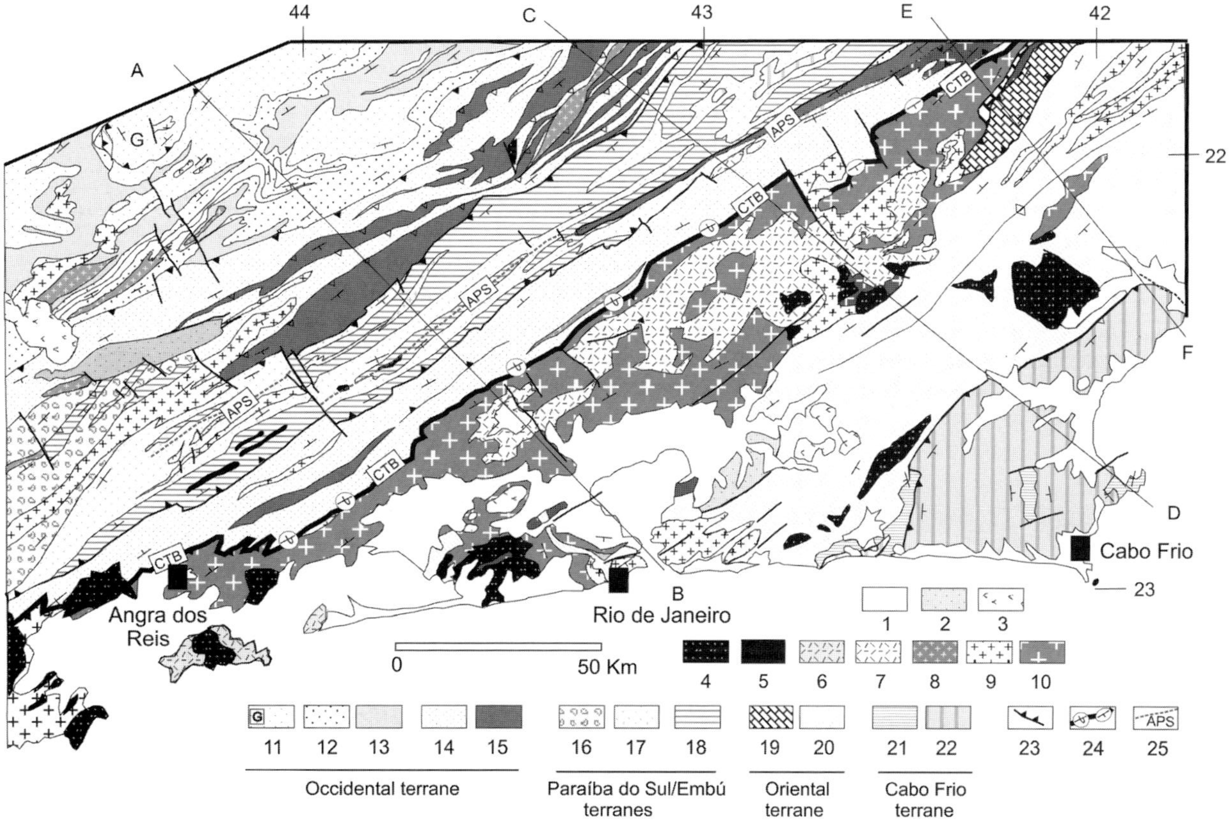

Fig. 5. Geological map of the central Ribeira Belt modified from Heilbron *et al.* (2004*b*). 1, Quaternary cover; 2, Tertiary rift basins; 3, KT alkaline rocks; (4–9) Neoproterozoic granites and orthogneisses; 4, post-collisional biotite granites (510–480 Ma (G6); 5, syn-535–520 Ma leucogranites (G5); 6, charnockites (G4); 7, late collisional 560 Ma granites (G3); 8, porphyritic I-type syn-590–560 Ma granites (G2I); 9, syn-590–560 Ma leucogranites and S-type to hybrid granites and charnockites (G2S); 10, Rio Negro magmatic arc and related suites (790–620 Ma, G1 pre-collisional); Occidental terrane (11–15); Andrelândia Megasequence (11–12); 11, upper sequence (Rio do Turvo) locally reaching high-*P* granulite facies (G); 12, lower sequence (Carrancas); 13, Archaean to Palaeoproterozoic orthogneisses and amphibolites (Mantiqueira complex); 14, distal facies of the Andrelândia Megasequence in the Juiz de Fora domain; 15, Palaeoproterozoic ortho-granulites (Juiz de Fora complex); (1618) Embú and Paraíba do Sul terranes: 16, Embú complex; 17, Paraíba do Sul complex; 18, Palaeoproterozoic orthogneisses (Quirino complex); Oriental terrane (19–20); 19, shelf marbles and amphibolites (Italva succession); 20, high-grade paragneisses (Costeiro succession); Cabo Frio terrane (21–22): 22, (Ky)–grt–sil paragneisses (Búzios and Palmital successions); 22, *c*. 1.9 Ga orthogneisses and amphibolites (Região dos Lagos complex); 23, major thrust zones; 24, Central Tectonic Boundary (CTB) in reverse and normal position; 25, dextral subvertical shear zones as Além Paraíba shear zone (APS). AB, CD and EF are structural sections shown in Figure 6.

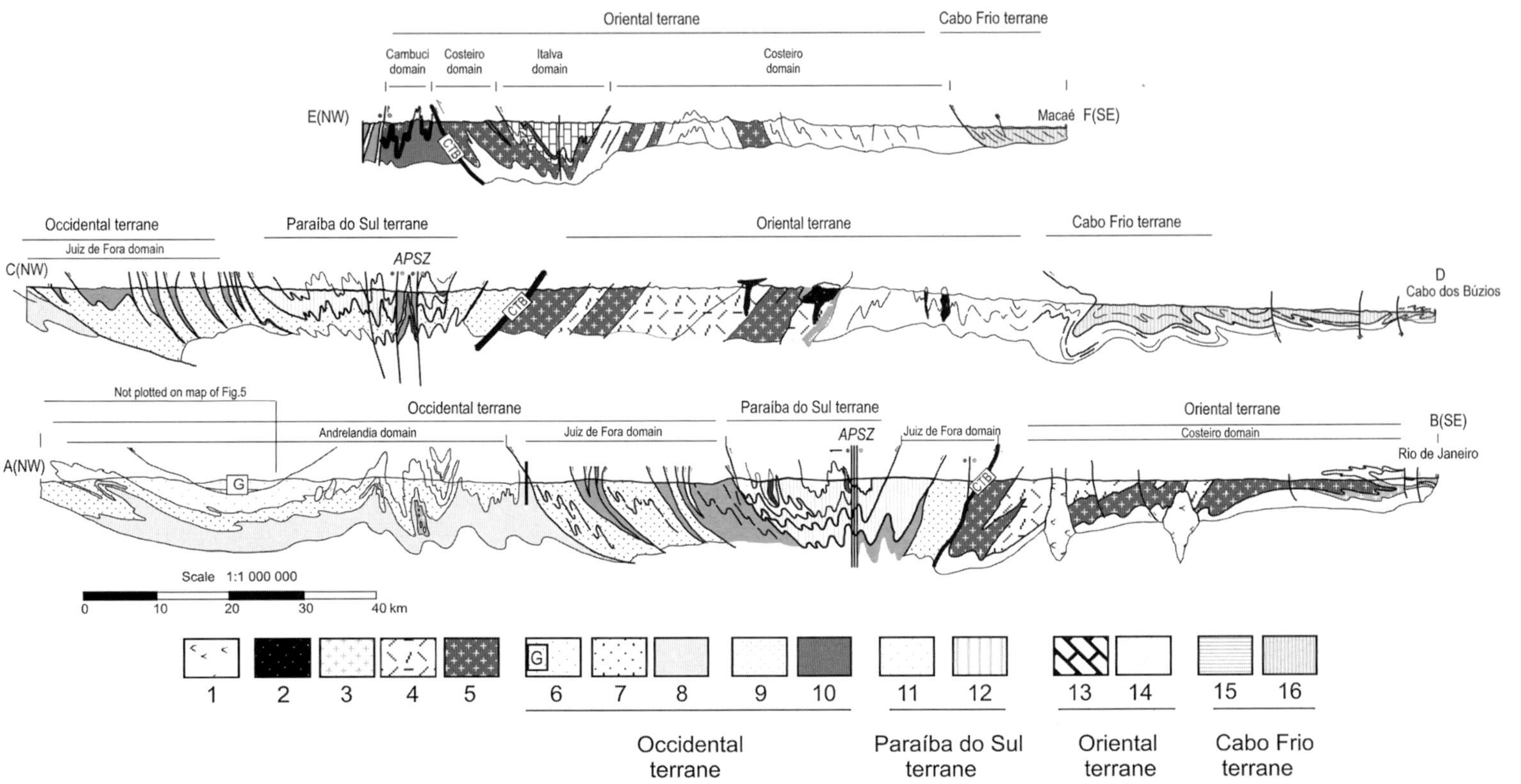

Fig. 6. Detailed structural sections of the Central Ribeira Belt. 1, KT alkaline rocks; (2–5) Neoproterozoic granites and orthogneisses; 2, post-collision biotite granites (510–480 Ma, G6); 3, biotite granites (*c.* 560 Ma, G3), 4, porphyritic I-type granite (*c.* 580–560 Ma, G3), leucogranites and S-type to hybrid granites and charnockites (*c.* 580–560 Ma, G2S); 5, Rio Negro magmatic arc and related suites (790–620 Ma, G1, pre-collisional); (6–10) Occidental terrane; 6 and 7, upper and lower units of the Andrelândia Megasequence at Andrelândia domain; 8, Archaean to Palaeoproterozoic orthogneisses and amphibolites (Mantiqueira complex); 9, distal facies of the Andrelândia Megasequence in the Juiz de Fora domain; 10, Palaeoproterozoic ortho-granulites (Juiz de Fora complex); (11–12) Paraíba do Sul terrane; 11, Paraíba do Sul complex; 12, Palaeoproterozoic orthogneisses (Quirino complex); Oriental terrane (13–14): 13, shelf marbles and amphibolites (Italva succession); 14, high-grade paragneisses (Costeiro succession); Cabo Frio terrane (15–16); 15, (ky)–grt–sil paragneisses (Búzios and Palmital successions); 16, orthogneisses and amphibolites (Região dos Lagos complex, *c.* 1.9 Ga. CTB, Central Tectonic Boundary; APSZ, Além Paraíba Shear Zone.

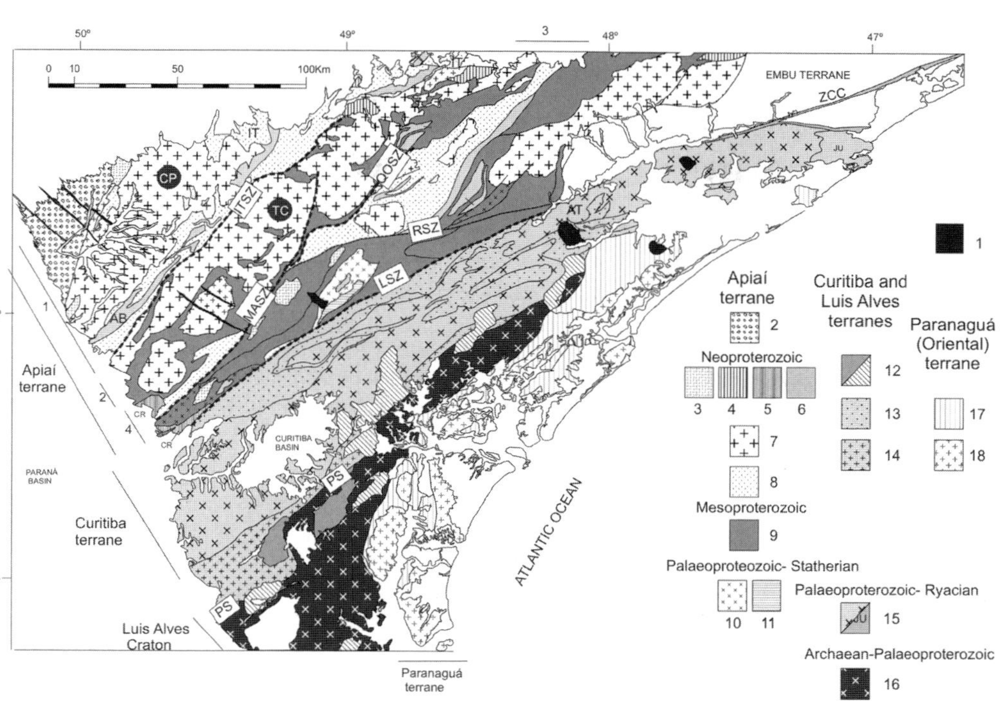

Fig. 7. Geological map of southern Ribeira Belt. Data from Campos Neto *et al.* (2000), Campanha (1991), Campanha & Sadowsky (1999), and modified from Heilbron *et al.* (2004*a*). 1, Phanerozoic cover; (2–11) Apiaí terrane; 2, molasse sequences of Castro Group and Camarinha Formation; 3, A-type granites; 4, Itú Suite; 5, Curitiba and Camarinha basins; 6, youngest Neoproterozoic successions; 7, Neoproterozoic calc-alkaline granites; 8, oldest Neoproterozoic successions; 9, Mesoproterozoic successions; 10, Statherian formations; 11, Statherian peralkaline orthogneisses (12–16). Curitiba terrane and Luis Alves Craton; 12, extensional basins and peralkaline granites; 13, Neoproterozoic successions; 14, Neoproterozoic calc-alkaline granites (CP, Cunhaporanga; TC, Três Córregos); 15, Rhyacian orthogneisses (AT, Atuba complex; JU,/Juréia granulite); 16, orthogneisses and granulites of the Luis Alves terrane; (17–18) Paranaguá terrane; 17, Neoproterozoic calc-alkaline granites; 18, Neoproterozoic successions. Major shear zones: IT, Itapirapuã; MASZ, Morro Agudo; QOSZ, Quarenta Oitavas; RSZ, Ribeira; LSZ, Lancinha; ZCC, Cubatão; PS, Piên Suture. Numbers 1 to 4 of the Apiaí terrane are referred in Table 2.

(a) Palaeoproterozoic orthogneisses with granitic to tonalitic compositions and (b) an undated metasedimentary association (Table 1). The terrane records an old metamorphic episode of *c.* 790 Ma (Vlach 2001; Cordani *et al.* 2002). The geodynamic context of this event is not recorded in adjacent terranes and is not well understood.

Tectonic kinematic indicators suggest that the Paraíba do Sul and Embú terranes docked laterally against the Apiaí and Occidental terranes (Campos Neto 2000; Heilbron *et al.* 2004*b*). The majority of the available geochronological data indicate the period of major tectonic activity and collision magmatism within the 605–570 Ma interval (Janasi & Ulbrich 1991; Machado *et al.* 1996*b*; Janasi *et al.* 2003; Mendes *et al.* 2006).

Oriental terrane (docked at c. 580–550 Ma)

The Oriental terrane is subdivided into three thrust slices (Heilbron & Machado 2003), listed below from bottom to top along a NW–SE section (Figs 5, 6, Table 1).

(a) the Cambuci domain is composed of highly deformed migmatitic garnet–biotite gneisses with lenses of olivine-bearing marble and calc-silicate rocks. The sedimentary protoliths of these metamorphic rocks are interpreted as having been deposited in a fore-arc setting.

(b) the Costeiro domain represents a magmatic arc setting. Orthogneisses of the Rio Negro complex and high-grade pelitic paragneisses comprise the two rock associations of this domain (Figs 5 & 6, Table 1). The orthogneisses comprise calc-alkaline tonalite to granite and gabbros. Geochemical and isotopic data suggest at least two stages of development for this arc: *c.* 790 Ma and *c.* 635–620 Ma (Heilbron & Machado 2003; Tupinambá *et al.* 2000). Isotopic data also indicates at least two different groups. The most primitive one is related to the oldest medium-K suite (group 1), with T_{DM} model ages between 1.23 and

0.95 Ga and positive εNd_t values (from +4 to 0). The youngest high-K suite (group 2) yield T_{DM} model ages between 1.23 and 1.75 Ga, and εNd_t values ranging from 0 to −12. The data suggest a progressive evolution from a intra-oceanic to a cordilleran style. Strontium isotope ratios are also consistent with values of 0.7042 in the more primitive rocks of group 1, to 0.7333 within the most evolved samples of group 2 (Heilbron *et al.* 2005).

(c) The Italva domain is the upper thrust slice and is represented by a low-grade metasedimentary succession rich in shelf carbonates, interpreted as formed in passive margin or back-arc setting. The depositional age of this succession is constrained by a U–Pb zircon age of *c.* 840 Ma from an intercalated amphibolite (Heilbron & Machado 2003).

The Central Tectonic Boundary (CTB, Almeida *et al.* 1998; Almeida 2000) is a major tectonic discontinuity that separates the Oriental terrane from the Occidental terrane (Figs 3, 4 & 5). The CTB is a folded shear zone (Figs 5 & 6) which shows a complex long-term structural evolution developed under high-temperature conditions. Mylonitic fabrics and intrusive granite bodies of multiple generations are observed and used as relative chronological markers for deformational and magmatic events.

The main deformation of the Oriental terrane is characterized by two progressive tectonic episodes. The first one is represented by a pervasive low angle schistosity (S_1) sub-parallel to bedding and was overprinted by an axial-surface foliation (S_2), related to isoclinal folds (F_2). All available U–Pb data for syn-collisional granites and metamorphic minerals fall within the 580–550 Ma interval (Heilbron & Machado 2003; Silva *et al.* 2005; Tupinambá *et al.* 2000).

Superimposed deformational and metamorphic features upon the Central Tectonic Boundary were caused by later docking of the Cabo Frio terrane, as described below. These include D_3 deformational phase oblique and open folding and the generation of subvertical shear zones throughout the terrane.

The Cabo Frio terrane, collided at c. 530–510 Ma

Two major stratigraphic units were mapped in the Cabo Frio terrane (Heilbron *et al.* 1982): the older one comprises Palaeoproterozoic (*c.* 1.9 Ga) orthogneisses with amphibolite intrusions (Schmitt *et al.* 2004); the younger one is a high-grade metasedimentary succession composed of pelitic to psammitic paragneisses with lenses of amphibolite and calc-silicate rock. U–Pb (SHRIMP) dating of detrital zircons (Schmitt *et al.* 2003) reveal Archaean *c.* 2.5 Ga, Palaeoproterozoic (*c.* 2.0 Ga), and Neoproterozoic (*c.* 1.0 Ga and *c.* 800–600 Ma) sources. Heilbron & Machado (2003) suggested, based on the age, geographic location and composition (pelites, carbonates and basalts) of these successions, that sedimentation of this unit took place in a Neoproterozoic back-arc basin related to the Rio Negro magmatic arc.

The Cabo Frio terrane collided with the belt at *c.* 530–510 Ma. This Cambrian episode has been referred to as the Búzios orogeny (Schmitt *et al.* 2004) and generated important low angle structures in the Cabo Frio terrane (Heilbron *et al.* 1982). The kyanite/K-feldspar mineral assemblage observed in meta-pelitic rocks, with superimposed growth of sillimanite, characterizes *P–T* metamorphic conditions as at least 9 kbar and 780 °C.

This late tectonic episode resulted in the superposition of folding and dextral lateral shear zones that overprinted all previously amalgamated terranes (Oriental, Paraíba do Sul–Embú and Occidental terranes). One outstanding example of these dextral shear zones is the Além Paraíba shear zone (Campanha 1981), that runs from São Paulo to Espírito Santo states (Figs 5 and 6) with a kilometre wide mylonite zone. This deformation is associated with a metamorphic episode of *c.* 535–520 Ma (M_2 of Machado *et al.* 1996*a*) and with the emplacement of leucogranites along the shear zones. It is important to stress that metamorphic ages in the previously amalgamated terranes are 5 Ma older than the collisional event reported for the Cabo Frio terrane.

Terranes of the southern Ribeira Belt

The Apiaí terrane

The Apiaí terrane is located between the southern termination of the Brasília Belt and the western side of the Curitiba and Paraíba do Sul–Embú terranes (Fig. 3). The Apiaí terrane is subdivided into four tectonic domains (numbered in Fig. 7), each one bounded by expressive shear zones of thrust and strike-slip motion (Fiori 1992; Hackspacher *et al.* 2000; Campanha & Sadowsky 1999).

The Palaeoproterozoic rock sequences of the Apiaí terrane are represented by anorogenic syenogranites and metabasic rocks of *c.* 1.75 Ga (domain 2 of Fig. 7; Kaulfuss 2001; Cury *et al.* 2002; Siga *et al.* 2005).

A remarkable feature of the Apiaí terrane is the widespread occurrence of Mesoproterozoic successions. They fall in several time intervals (*c.* 1600–1500, 1500–1400, 1030–908 Ma) and crop out in all the structural domains presented on Figure 7. The successions are rich in carbonatic rocks, with siliciclastic and meta-volcanic units with both basic to acid compositions. A more detailed description of the units and local names is presented in Table 2.

Several younger Neoproterozoic successions also occur in the Apiaí terrane. They are represented by both siliciclastic–metavolcanic and carbonatic–metavolcanic units (see Table 2 and Fig. 7). Available geochronological data suggests deposition around *c.* 645–628, 620–605 and 600 Ma, partially coeval with the generation of arc-related calc-alkaline batholiths of *c.* 620–600 Ma (Fig. 7), suggesting back-arc or intra-arc tectonic setting for these sequences (Hackspacher *et al.* 2000).

Main deformation, magmatism and metamorphism took place between 620 and 600 Ma. The timing of displacement along the dextral shear zones that separate the tectonic domains of the Apiaí terrane is constrained by a recently described late-tectonic sequence of *c.* 590–575 Ma (Campanha *et al.* 2005).

Curitiba terrane and Luis Alves Craton

The Curitiba terrane (Figs 2 & 7) is made up of Palaeoproterozoic basement rocks overlain by a passive margin metasedimentary succession. Both were affected by the Neoproterozoic metamorphic episode of *c.* 600 Ma. The Palaeoproterozoic succession is composed of amphibole-rich orthogneisses with some nuclei of granulite facies rocks of *c.* 2.1 Ga (Siga *et al.* 1995; Basei *et al.* 2000, 2003). The metasedimentary successions are a low-grade sequence of dolomite meta-limestones with stromatolitic bioherms, meta-pelites, meta-rhythmites and meta-sandstones, and a more metamorphic sequence of biotite schists, calc-schists, quartzites, schists, marbles, and amphibolites (Faleiros & Campanha 2005).

The Curitiba and Luis Alves terranes are separated by the NE-trending Pien suture zone (Fig. 7), characterized by *c.* 610 Ma arc-related rocks, regarded as an evidence of northwestward subduction (Basei *et al.* 2000). The Luis Alves terrane is a predominantly granulite-facies terrane that records at least two metamorphic episodes of *c.* 2.35 and 2.31 Ga (Hartmann *et al.* 2000). Late-tectonic leucogranites of *c.* 2.01 Ga complete the rock associations.

Late to post-collision basins of Late Ediacaran to Cambrian age are related to the latest phases of the transpressive shear zones (Basei *et al.* 1998; Cordani *et al.* 1999; Teixeira 2000; Teixeira *et al.* 2004).

Tectonic units of western Africa in Angola and northern Namibia

This section presents a brief review of the tectonic organization of SW Africa, mainly based on the literature.

A major tectonic boundary is an east–west shear zone that transects Angola along the 9°S parallel, south of Luanda (Fig. 8). For this reason, the geology north and south of this zone is described separately. Information on the major tectonic units of these two crustal blocks is given in Table 3.

North of Luanda: the West Congo Belt

The present review is based on Carvalho *et al.* (2000); Tack *et al.* (2001); Milesi *et al.* (2006) and Toteu *et al.* (1994, 2001).

The major tectonic units north of Luanda are: the basement of the Congo Craton, the West Congo Belt and the coastal area, relatively poorly studied (Fig. 8).

The West Congo Belt is envisaged as a Neoproterozoic intra-continental rift inverted at *c.* 560 Ma (Tack *et al.* 2001). The rift was initially filled with incipient sedimentation and rhyolites of *c.* 997 Ma, which is the age of associated peralkaline granites. This was followed by thick continental flood basalts and subsequent felsic lavas of *c.* 924–912 Ma. This volcanic succession was then covered by the metasediments of the West Congolian Group.

To the west, the rift sequence is thrust by reworked basement gneisses and migmatites of 2.15 to 2.05 Ga intruded by 2.05–1.97 Ga granites. The timing of the Neoproterozoic thermal overprint upon the basement association is bracketed by whole rock Rb–Sr isochrons of 684 Ma (Carvalho *et al.* 2000) and by the age of metamorphism of *c.* 560 Ma (Tack *et al.* 2001).

South of Luanda: the Angola Craton

The present review is based on the compilation of the geological map of Angola and on Cahen *et al.* (1984), Seth *et al.* (1998), Carvalho *et al.* (2000), Passchier *et al.* (2002), Goscombe *et al.* (2003, 2005), Kröner *et al.* (2004) and Milesi *et al.* (2006).

South of Luanda, the major tectonic units are the Angola Craton, which together with the Congo Craton is one of the main cratonic blocks of Africa, and the Neoproterozoic Kaoko and Damara belts, respectively located to the west and to the south of this cratonic block.

According to the detailed description by Carvalho *et al.* (2000) the Angola Craton can be subdivided into many tectonic zones. Simplification by

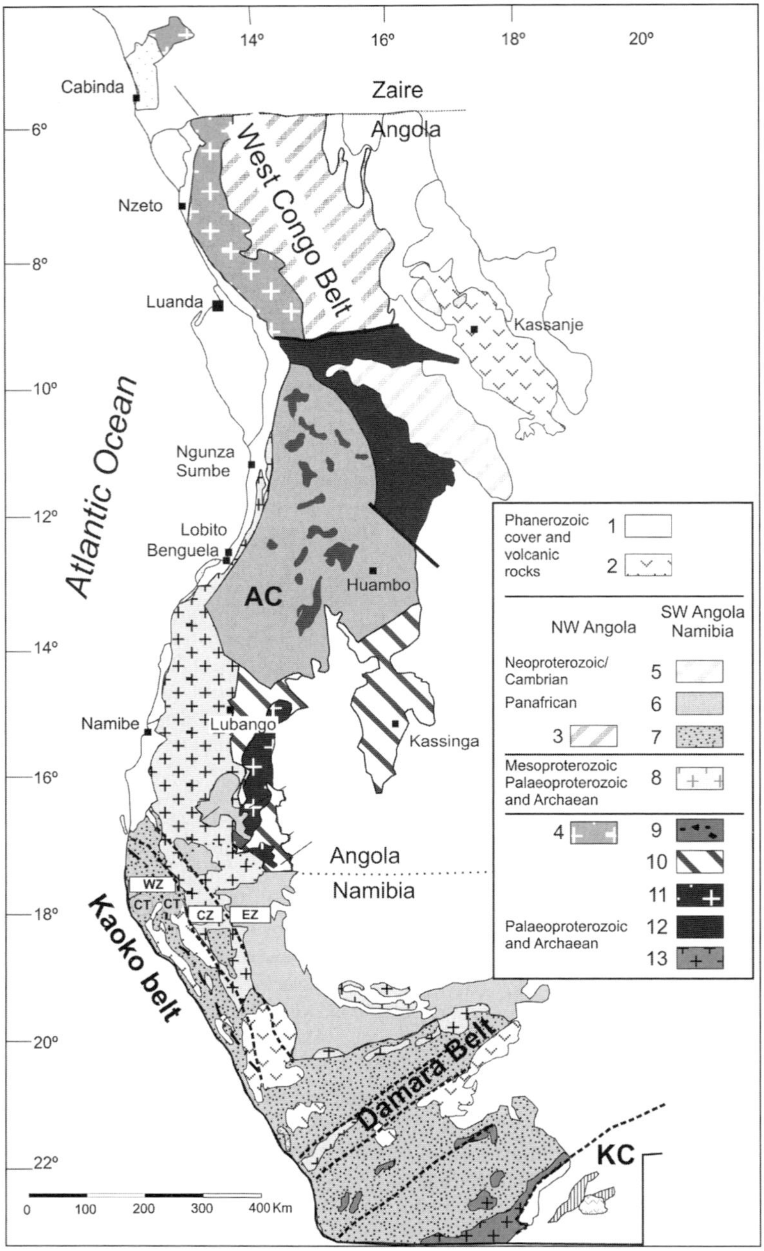

Fig. 8. Simplified tectonic map of western Angola and northwestern Namibia, based on Carvalho *et al.* (2000), Tack *et al.* (2001) and Goscombe *et al.* (2005). 1, Phanerozoic cover; 2, Karroo and Etendeka sequences; (3–4) Units of northern Angola; 3, rift and sag Neoproterozoic deformed successions of the West Congo Belt; 4, Palaeoproterozoic granitoid rocks reworked by the Pan-African event; (5–13) tectonic units of SW Angola and northern Namibia; 5, Neoproterozoic cratonic successions; 6–7, passive margin and foreland Neoproterozoic successions of Kaoko Belt; 8, zone with Mesoproterozoic, Palaeoproterozoic and Archaean granitoid rocks (Polyorogenic unit of Angola); 9, Eburnian Central Zone; 10, Eburnian Cassinga and Lubango zone with Archaean inliers; 11, gabbro–anorthosite complex of southern Angola; 12, Archaean Central Shield of Angola; 13, basement of Kalahari Craton. Tectonic zoning of the Kaoko Belt: WZ, Western Zone subdivided in the Coastal terrane (CT) and the (OC) Orogenic Core; CZ, Central Zone; EZ, Eastern Zone; AC, Angola Craton; KC, Kalahari Craton.

Table 3. *Major tectonic episodes in Angola and northern Namibia*

Northern Angola segment		Southern Angola and Northern Namibia segment		
Tectonic unit	Major episodes	Tectonic unit	Internal zones	Major episodes
Congo craton	2.05 Ga 2.90 Ga		Poly-orogenic coastal region	1.21.1 Ga basic dykes 1.41.3 and 1.521.45 Ga bimodal granites 1.7 Ga 2.2 To 1.8 Ga. Older inliers
West Congo Belt	924–917 Ma bimodal magmatism 997 Ma rhyolites and peralkaline granites	AngolaKasai craton	Gabbro–anorthosite Lubango–Cassinga Zone Central Eburnian Zone	>2.16 2.2 to 1.8 Ga. Older inliers
Coastal region	1.9 Ga granites 2.1 Ga orthogneisses Older supracustal lenses	Central shield	2.68 to 2.52 Ga 2.92.8 Ga	
		Kaoko Belt	Eastern Zone (foreland) Central Zone (escape zone) Western Zone (orogen core zone)	535–505 Ma 580–550 Ma 535–505 Ma 580–550 Ma
			Coastal terrane (part of Western zone)	535–505 Ma 580–550 Ma 655–645 Ma

Based on Carvalho *et al.* (2000), Seth *et al.* (1998, 2002), Tack *et al.* (2001), Passchier *et al.* (2002), Goscombe *et al.* (2003, 2005), Kröner *et al.* (2004) and Milesi *et al.* (2006).

merging similar terranes (Fig. 8) results in the following major tectonic units summarized here: (a) Central Shield, (b) Central Eburnian zone, (c) Lubango–Cassinga zone, (d) Gabbro–Anorthosite complexes of SW Angola and (e) Coastal Polyorogenic zone.

The oldest tectonic zone is the Central Shield and minor outcrops located in the east part of the Angola, which are separated by the Phanerozoic Congo Basin (Fig 8). They comprise a gabbro–norite–charnockite complex of *c.* 2.9–2.8 Ga and intrusive granites and migmatites of *c.* 2.68 to 2.52 Ga. In west Angola the Archaean rock associations are characterized by a gneiss–migmatite–granite complex unconformably overlain by meta-volcano-sedimentary rocks of the Jamba Group. The Archaean rocks comprise a gabbro–norite and charnockitic complex, dated by Delhal *et al.* (1976) with a Rb–Sr isochron at 2822 $\pm$ 66 Ma. These rocks are intruded by granitoid rocks, granite–tonalite gneiss and migmatites dated by Carvalho *et al.* (2000) at 2560 $\pm$ 50 Ma.

The Eburnian (corresponding to a Palaeoproterozoic event in South America) units comprise the Chivanda, Bale and Oendolongo groups, which are composed of Palaeoproterozoic metasedimentary sequences and of granitoid rocks. In the Central Eburnian Zone (Fig. 8, central Angola), the latter are predominantly represented by granitoid suites and by granite-gneisses with Rb–Sr ages around 2.2 Ga (Torquato *et al.* 1979). The supracrustal rocks are characterized by widespread metavolcano-sedimentary sequences with metamorphic ages around 2.1 Ga (Cahen *et al.* 1984). Some of these units are covered by mafic and felsic volcanic rocks.

According to Carvalho & Alves (1990), the gabbro–anorthosite complex of SW Angola (Fig. 8) is older than 2160 Ma, either belonging to the initial stage of the Eburnian orogeny or of pre-Eburnian age. Porphyroblastic granitoid rocks, referred to as the Quibala-type granites, with an age of 1850 Ma (Carvalho *et al.* 1979), and syenites with ages of 1960 Ma (Cahen *et al.* 1984) complete the Eburnian geological association.

The Cassinga–Lubango zone (Fig. 8) is characterized by Archaean to early Palaeoproterozoic gneissic and volcano sedimentary kilometre size inliers within late Palaeoproterozoic granites.

During the end of the Palaeoproterozoic and the Mesoproterozoic widespread anorogenic magmatic activity took place from NW Namibia to the central eastern region of Angola. The magmatic rocks are represented by leucocratic granite, porphyritic biotite granite (Furnas Granite), porphyritic coarse-grained granite (Caraculo–Bibala Granite) and by alkaline granite (Chicala–Cacula Granite). These magmatic occurrences are dated by several Rb–Sr isochrons between 1760 Ma and 1550 Ma (Carvalho *et al.* 1979; Carvalho & Tassinari 1992; Torquato & Carvalho 1992).

The Mesoproterozoic rocks (Kibaran) in the region are represented by the Chela Supergroup, a volcano-sedimentary sequence that includes volcaniclastic rocks, conglomerates, quartzites and dolomitic limestones, locally stromatolitic. Sills and dykes of norite, dolerite, gabbros and basalts have ages around 1.2 Ga. Alkaline and calc-alkaline granitoid rocks have ages of 1.4–1.2 Ga, and around 1.0 Ga (Carvalho *et al.* 2000). Several occurrences of the so-called Red Granites and associated rocks are known in southwestern Angola, many of them intruding the Gabbro–Anorthosite complex. These rocks comprise granite–porphyry–rhyolite associations and their ages range from 1.4 Ga to 1.3 Ga (Carvalho *et al.* 2000). In northern Namibia, the anorogenic magmatism can be subdivided in three major episodes of *c.* 1.7, 1.52–1.45 and 1.4–1.3 Ga, based on U–Pb geochronological data (Seth *et al.* 1998; Kröner *et al.* 2004). In the Lubango zone magmatic rocks predominate and intrude older rocks of the Coastal Poly-orogenic zone.

Northern Kaoko Belt

The following review of the belt is based on Miller & Grote (1998), Seth *et al.* (1998, 2002), Stanistreet & Charlesworth (2001), Passchier *et al.* (2002), Goscombe *et al.* (2003, 2005) and Kröner *et al.* (2004).

The Kaoko Belt runs approximately north–south, parallel to the coast of southern Angola and northern Namibia, where it connects with the NE–SW Damara Belt. The tectonic framework of the belt comprises, from east to west, the following units: (a) the Eastern Zone, regarded as the foreland region, (b) the Central Zone, also known as the Escape Zone, (c) the Western Zone, subdivided into the orogen core and the Coastal terrane (Fig. 8, Goscombe *et al.* 2005). The first three domains are regarded as the reworked passive margin of the Angola Craton, while the Coastal terrane is considered an exotic terrane accreted onto this margin at *c.* 580 to 550 Ma. Deformation was transpressive with a sinistral lateral component. The Neoproterozoic passive margin sedimentary successions (Damara Sequences) are represented by shelf, slope and deep-water carbonate rocks with abundant turbidites. In the foreland zone a molasse sequence overlies the passive margin succession. Detrital zircons and palaeo-environmental analysis indicates that the southern Angola Craton was the major source area. Geochronology by Seth *et al.* (1998) on basement inliers and anticlines reveal the following episodes: crystallization of the orthogneisses at *c.* 2.65 to 2.59 Ga, metamorphism at *c.* 2.29 Ga; generation of igneous rocks at 1.98 to 1.96 Ga, and Mesoproterozoic granitoid rocks between 1.70 and 1.50 Ga.

The Coastal terrane, regarded as exotic (Goscombe *et al.* 2005), is composed of (cordierite)–sillimanite–garnet–biotite gneisses with arc affinity recording an older metamorphic event around 655–645 Ma.

A major tectonic event is related to the accretion of the Coastal terrane to the Angola margin at *c.* 580–550 Ma, resulting in sinistral transpression along the Kaoko Belt (Passchier *et al.* 2002; Goscombe *et al.* 2003, 2005).

A younger tectonic episode around 535–505 Ma is recorded in the southern segment of the Kaoko Belt. All tectonic interpretations point to a frontal convergence between the Kalahari and Angola cratons, resulting in the Damara Belt (Fig. 8, Passchier *et al.* 2002; Goscombe *et al.* 2005).

Neoproterozoic–Cambrian evolution of West Gondwana

A comparative synthesis of the Ribeira and Araçuaí belts and their SW African counterpart is presented (Figs 9, 10 & 11), along with a polyphase evolutionary model for the assembly of the southern sector of West Gondwana from 630 Ma to 500 Ma, based on the available geological and geochronological information.

Pre-Gondwana palaeo-continents: Archaean to Mesoproterozoic history

The major Archaean to Mesoproterozoic palaeocontinental blocks that collided in the Neoproterozoic to form West Gondwana are the São Francisco/Congo, Angola, Rio de la Plata, Paranapanema (presently covered by the Phanerozoic Paraná Basin), and Kalahari cratons. Smaller fragments, possibly micro-continents, were involved: the Curitiba/ Luis Alves, Paraíba do–Embú, Apiaí and Cabo Frio terranes (Figs 9 & 10).

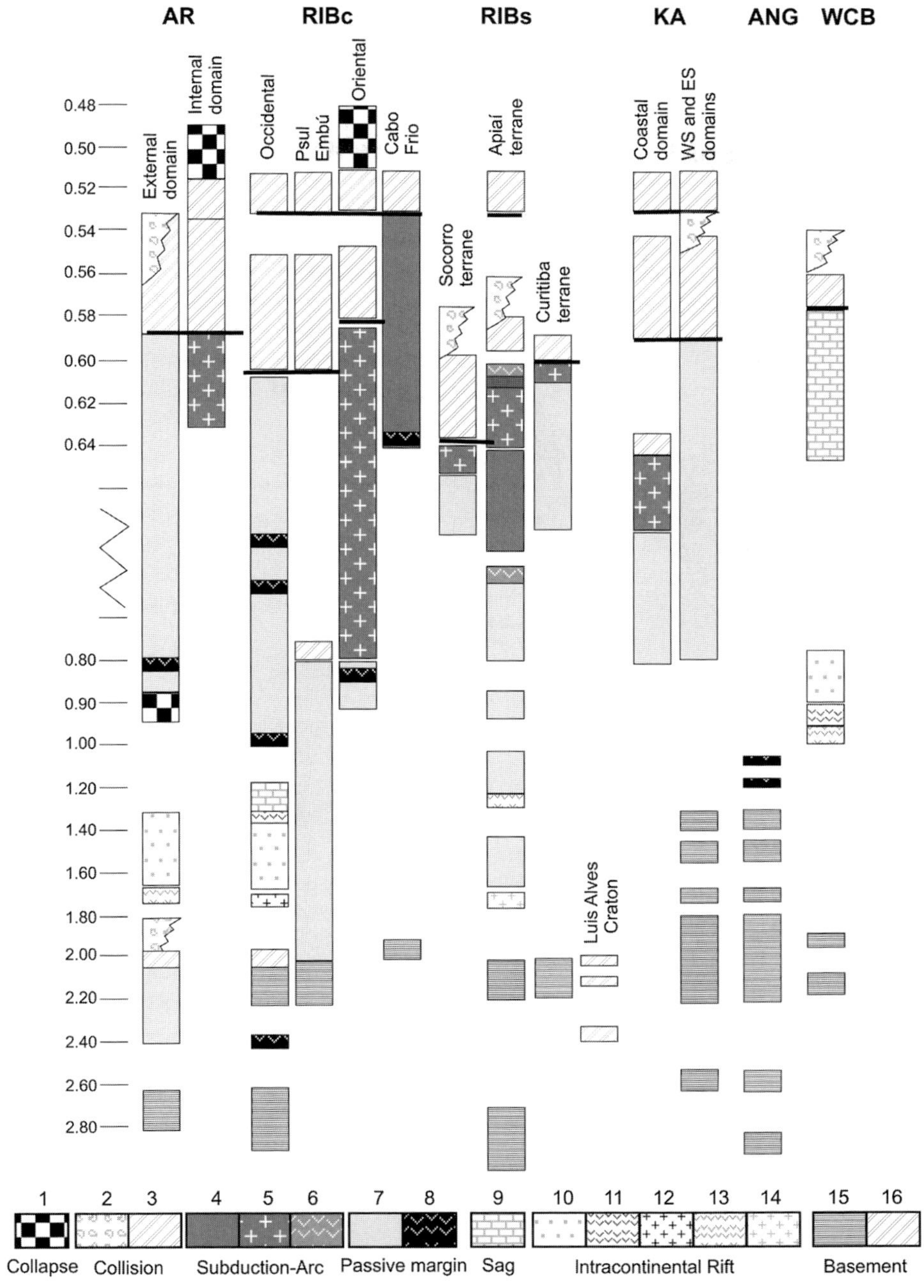

Fig. 9. Comparative tectonic evolution of the Neoproterozoic–Cambrian belts and cratonic blocks of southeastern Brazil and southwestern Africa. AR, Araçuaí Belt; RIBc, central Ribeira Belt; RIBs, southern Ribeira Belt; KA, Kaoko Belt; ANG, Angola Craton; WCB, West Congo Belt. 1, Bimodal post-collisional magmatism (collapse); 2, late collision and successor basins; 3, major collision episodes; 4, back-arc and fore-arc basins; 5, plutonic arc-related rocks; 6, volcanic arc-related rocks; 7, passive margin successions; 8, tholeiitic magmatism and ocean floor associations; 9, sag successions; 10, rift successions; 11, basic to alkaline dykes and plutons; 12, basic to alkaline volcanic rocks; 13, felsic volcanic rocks; 14, felsic plutons; 15, orthogneisses; 16, ortho-granulites. Geochronological data from Machado *et al.* (1996), Tupinambá *et al.* (2000), Pedrosa-Soares *et al.* (2000), Tack *et al.* (2001), Carvalho *et al.* (2000), Heilbron & Machado (2003), Heilbron *et al.* (2000, 2004*a*, *b*), Goscombe *et al.* (2005), Silva *et al.* (2005).

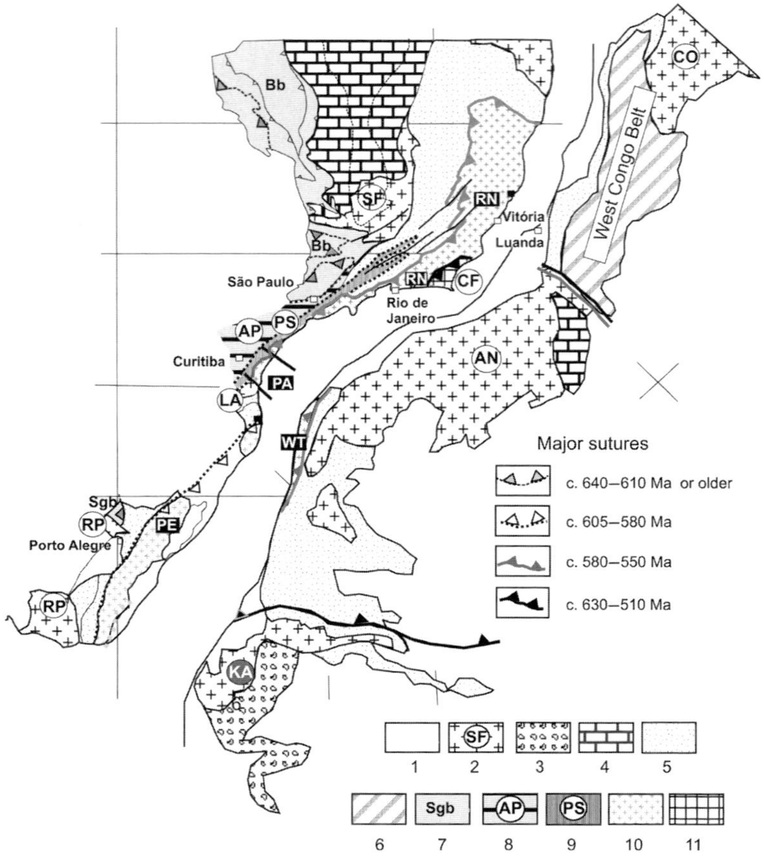

Fig. 10. Integrated tectonic map of SE Brazil and SW Africa, based on the compilations presented in Figs 1–8. 1, Post-Cambrian sedimentary basins; 2, cratons (SF, São Francisco; Co, Congo; LA, Luis Alves; RP, Rio de la Plata; AN, Angola; KA, Kalahari); 3, Mesoproterozoic units; 4, cratonic cover; 5, reworked cratonic margins, including basement and Neoproterozoic passive margins; 6, intra-continental West Congo Belt; 7, Brasília (Bb) and São Gabriel (Sgb) belts; 8, Apiaí terrane; 9, Paraíba do Sul, Embú and Curitiba terranes; 10, magmatic arcs (RN, Rio Negro; PA, Paranaguá; PE, Pelotas; WT, Western terrane); 11, Cabo Frio terrane.

The basement of the São Francisco palaeo-continent exposed in the reworked southern passive margins is typically composed of Palaeoproterozoic plutonic or supracrustal metamorphic rocks (2.2 to 1.9 Ga), but containing well characterized Archaean nuclei (2.8–2.6 Ga). Platform conditions in the São Francisco–Congo palaeo-continent were only achieved after a very important and widespread orogeny around *c.* 2.05 Ga ('Transamazonian' event, Alkmim & Marshak 1998; Teixeira *et al.* 2000). A very similar situation is reported for the basement of the Congo area (Carvalho *et al.* 2000; Toteu *et al.* 2001; Milesi *et al.* 2006), where rocks formed predominantly between *c.* 2.9 Ga and 2.1 Ga, followed by metamorphism at 2.05 Ga. Relicts of Palaeoproterozoic supracrustal successions with sedimentary context suggesting a passive margin setting is also present in the classic area of the 'Quadrilátero Ferrífero' (Iron Quadrangle), Minas Gerais State, Brazil (Alkmin & Marshak 1998).

An additional characteristic of the tectonic history of the São Francisco–Congo palaeo-continent is a very important Statherian (*c.* 1.7 Ga) episode of continental rifting recorded by the diamond-bearing clastic successions and subordinated felsic volcanic rocks of the Espinhaço Supergroup (Martins Neto 2000; Martins Neto *et al.* 2001).

The basement associations in the Paraíba do Sul–Embu and Curitiba terranes are also characterized by *c.* 2.2 to 2.1 Ga orthogneisses and granulites with Nd isotopic signatures indicating derivation from Archaean crust (Figs 9 & 10). In the Cabo Frio terrane the basement association (Região dos Lagos complex) is comparatively younger, with the majority of isotopic ages of crystallization between 2.0 and 1.9 Ga.

The basement of the Apiaí terrane contrasts with the other blocks of eastern Brazil with the occurrence of expressive Mesoproterozoic volcano-sedimentary sequences, besides Palaeoproterozoic orthogneisses (*c.* 2.2–1.7 Ga) and granitic rocks. These Mesoproterozoic sequences are thought to represent both extensional settings, rift to ocean floor, and compressional settings, ocean floor to volcanic arc (Campo Neto 2000).

The Angola and Kalahari cratons in SW Africa display contrasting tectonic associations with predominant Archaean rocks affected by several episodes of magmatism during the late Palaeoproterozoic and Mesoproterozoic (Seth *et al.* 1998; Carvalho *et al.* 2000; Kröner *et al.* 2004).

Neoproterozoic passive margins

The following geological features mark the onset of continental rifting in the São Francisco–Congo and Angola Kasai palaeo-continents during the Mesoproterozoic–Neoproterozoic transition: (a) the intrusion of *c.* 900 Ma mafic dyke swarm in the São Francisco Craton; (b) Tonian alkaline plutons in the Araçuaí Belt (Pedrosa Soares *et al.* 2001); (c) *c.* 999 and 912 Ma expressive bimodal volcanism in the West Congo Belt (Tack *et al.* 2001); and (d) dolerites and gabbros of 1.2–1.1 Ga in southern Angola (Carvalho *et al.* 2000).

This important rifting event was successful enough to evolve into spreading of oceanic lithosphere and consequently to the isolation of several palaeo-continents, such as São Francisco, and smaller fragments that later amalgamated to form Gondwana (Brito Neves *et al.* 1999; Alkmim *et al.* 2001; Cordani *et al.* 2003; Valeriano *et al.* 2004; Trindade *et al.* 2006).

Although no complete ophiolitic relicts have been recognized in the central sector of the Ribeira Belt, MORB-type metabasic rocks dated at 848 ± 11 Ma (U–Pb TIMS, Heilbron & Machado 2003) are interpreted as representing the oldest magmatic event compatible with the generation of oceanic lithosphere between the São Francisco palaeo-continent and the Oriental terrane. This matches the similar 816 ± 72 Ma Sm–Nd isochron age interpreted to date the formation of ocean floor in the Araçuaí Belt (Pedrosa-Soares *et al.* 1998; Silva *et al.* 2007), the along-strike continuation of the Ribeira Belt to the north.

The passive margin successions around the São Francisco palaeo-continent are represented by the siliciclastic Andrelândia Megasequence in the Occidental terrane (Paciullo *et al.* 2000; Trouw *et al.* 2000).

In Africa the Neoproterozoic successions were deposited in the West Congo rift to sag basin. The reworked Palaeoproterozoic basement rocks crop out along the northern coast of Angola (Fig. 8). The Neoproterozoic passive margin of the western Angola–Kasai palaeo-continent is represented by the sedimentary successions (Namibian Series) of the central and external domains of the Kaoko Belt (Figs 9 & 10). The Angola Craton as the source area is clearly indicated by U–Pb ages of detrital zircons (Goscombe *et al.* 2005) and by palaeocurrent distribution in turbidites of the southern part of the Kaoko Belt (Passchier *et al.* 2002).

The maximum age of sedimentation of the West Congolian Group is constrained by U–Pb data to around 650 Ma (Frimmel *et al.* 2006). Orogenic inversion during the Neoproterozoic (*c.* 566 Ma) developed through thrusting of the rift succession over the sag-basin deposits of the West Congolian Group. The Inkisi Group is a post-orogenic succession that completes the depositional history of the region.

Cryogenian to Ediacaran subduction

Eastward subduction of the oceanic lithosphere belonging to the São Francisco palaeo-plate caused the development of a long-lived magmatic arc that developed from *c.* 790 to 610 Ma, presently incorporated in the Ribeira Belt as the Rio Negro complex (Tupinambá *et al.* 2000; Heilbron & Machado 2003). The Rio Negro arc was emplaced into passive margin metasedimentary successions of the Oriental terrane, with tonalites, granodiorites, granites and gabbros with juvenile isotopic signatures. Similar granitoid rocks of *c.* 630–610 Ma in the Apiaí terrane (Figs 9 & 10) were also described as formed in magmatic arc environment by Hackspacher *et al.* (2000), Janasi *et al.* (2001) and Prazeres Filho (2005).

Generation of granite took place in the Coastal terrane of the Kaoko Belt simultaneously to the generation of the Rio Negro arc, at *c.* 655–645 Ma. This was followed by the generation of *c.* 635 Ma granites (Seth *et al.* 1998) and by an episode of high-grade metamorphism (Figs 9 and 10). The country rocks of these arc intrusive complexes in the Coastal terrane in the Kaoko Belt are cordierite–garnet–biotite gneisses, very similar to those of the Oriental terrane in the central Ribeira Belt.

The summary above indicates long-lived subduction of oceanic lithosphere, with generation of intra-oceanic to cordilleran magmatic arcs within and around the Adamastor ocean from *c.* 790 Ma to 600 Ma. This scenario is not compatible with a palaeogeography of a narrow ocean between the São Francisco–Congo and Angola palaeo-continents during the Neoproterozoic.

Subduction was associated with the development of fore-, intra- and back-arc basins that were inverted during several subsequent tectonic episodes. This is exemplified by the volcano-sedimentary record of the Neoproterozoic basins of the Apiaí and Cabo Frio terranes (Hackspacher *et al.* 2000; Campos Neto 2000; Heilbron *et al.* 2000, 2004, 2004*b*; Campos Neto *et al.* 2004).

Late Ediacaran (c. 605–550 Ma) major continental amalgamation

Between 605 and 550 Ma most pieces of West Gondwana were already attached together. Using the São Francisco–Congo craton as reference, this time interval is marked by diachronous collisions of Palaeoproterozoic microplates and Neoproterozoic magmatic arcs.

The available geochronological database indicates that docking of the Socorro terrane (Fig. 9) took place between *c.* 640 and 610 Ma related to the evolution of the southernmost Brasília Belt (Ebert *et al.* 1996; Campos Neto 2000; Valeriano *et al.* 2008). The docking of the Apiaí and Curitiba–Paraíba do Sul–Embú terranes followed at *c.* 605–580 Ma (Basei *et al.* 2000; Heilbron *et al.* 2004*a*). Finally, docking of large magmatic arcs of the Oriental and Coastal terranes took place at *c.* 580–550 Ma, both in the Ribeira and Kaoko belts (Heilbron *et al.* 2003; Goscombe *et al.* 2005).

These tectonic episodes were dominated by the oblique convergence of colliding terranes and resulted in pervasive deformation observed in the Ribeira and Kaoko belts. Deformation began with frontal thrusting with isoclinal to tight folds and a low dipping foliation defined by metamorphic minerals related to the main metamorphic event. The deformation evolved to a late phase with a major right-lateral component in the Ribeira Belt, and left-lateral in the Kaoko Belt (Heilbron *et al.* 2000, 2004*a*, *b*; Passchier *et al.* 2002; Goscombe *et al.* 2005). Coeval metamorphic parageneses in the collisional zones indicate intermediate- to high-pressure regimes in the lower plates (reworked passive margins) and of lower pressure regime in the upper plates (magmatic arc terranes).

Around 560 Ma, closure of the rift and sag successions of the West Congo Belt was in course (Tack *et al.* 2001; Frimmel *et al.* 2006).

Latest stage of West Gondwana amalgamation in the Cambrian (c. 530–510 Ma)

As described below and considering data from other Neoproterozoic belts of West Gondwana (Fig. 1; Brito Neves *et al.* 2000; Pimentel *et al.* 2000; Kröner & Cordani 2003; Valeriano *et al.* 2004; Trindade *et al.* 2006), by around 550 Ma the majority of blocks of the central portion of West Gondwana were attached together. However, an important Cambrian tectonic-metamorphic event is recorded around *c.* 530–510 Ma along the border of proto-West Gondwana. It is recorded in the Ribeira Belt (Búzios orogeny, Schmitt *et al.* 2004) and in the Kaoko and Damara belts (Goscombe *et al.* 2005), but is also detected in the Dom Feliciano (Bossi & Gaucher 2004), Paraguai–Araguaia (Trompette 1998; Alvarenga *et al.* 2000) and Pampean belts (Rapela 2000).

The tectonic setting that emerges from this tectonic correlation is that of another large ocean closed around proto-West Gondwana with the arrival of large blocks such as the Amazonian and Kalahari palaeo-continents, and of minor blocks such as the Cabo Frio terrane in the Ribeira Belt, and the Pampean terrane in Argentina.

Cambrian–Ordovician (c. 510–480 Ma) magmatism: a record of the collapse of Gondwana orogens?

Post-collisional deformation recorded in the Oriental and Cabo Frio terranes in central Ribeira Belt marks the transition to extensional tectonic regimes (Fig. 9), represented by two groups of structures: (a) north–south to NE–SW brittle-ductile subhorizontal shear zones striking parallel to the orogen with down-dip movement and associated down-dip verging folds; (b) dextral transtensional subvertical NW–SE shear zones that are transversal to the orogen (Fig. 5).

In the Oriental terrane, this deformational episode is associated with widespread post-collision calc-alkaline granitic intrusions in the form of circular stocks, sills or dykes, with U–Pb ages between 510 and 480 Ma (Machado *et al.* 1996*a*; Heilbron & Machado 2003). The shear zones behaved as channels for the ascending magmas as shown by frequently observed magmatic flow structures (Wiedemann *et al.* 2002). An important characteristic of these intrusive bodies is the common association with mafic enclaves and magma mingling along the border zones. In the Cabo Frio terrane this tectonic episode is represented by pegmatites striking parallel to NW–SE subvertical shear zones.

The extensional collapse following thrust-stacking was possibly triggered by slab detachment, causing exposition of the high temperature core of the orogen and associated bimodal magmatism (Heilbron *et al.* 2000; Heilbron & Machado 2003).

An integrated tectonic model

The comparative tectonic evolution presented above suggests that the history of amalgamation of the central portion of West Gondwana was complex, with a major subduction period with arc-related rocks both in intra-oceanic and cordilleran settings, followed by the diachronous collision of these arcs and minor continental fragments. The tectonic cartoon of Figure 11 illustrates the major periods of tectonic activity during the Neoproterozoic amalgamation and presents a tectonic model.

The first amalgamation of a major continental mass (proto-West Gondwana) was achieved around 640–610 Ma by the collision of the São Francisco–Congo and Paranapanema palaeocontinents and intervening arcs during evolution of the southern Brasília Belt.

At the same time, east-verging subduction was in course along the eastern and southern margin of this proto-continent (Ribeira and Araçuaí belts) and along the western margin (Coastal terrane) of the Angola-Kasai palaeo-plate, presently in the Kaoko Belt.

The oldest (*c.* 790–630 Ma) juvenile rocks indicate that subduction began in intra-oceanic or peri-oceanic settings in the central Ribeira Belt. Probably, important transform faults allowed lateral displacement between the São Francisco–Congo and Angola–Kasai plates at this time. The successions of active margin and arc related rocks of this episode are well preserved in the Apiaí and Oriental terranes in Brazil, and by the Coastal terrane in Africa (Fig. 11).

The record of the Pelotas arc at the eastern margin of the Luis Alves–Rio de la Plata palaeoplates is also coeval with this major subduction period around the Adamastor Ocean. The geological data suggest also eastwards cordilleran subduction along the Kalahari border or alternatively along some continental fragment of South Africa (Fig. 11a), because available geochronological data indicates that the Kalahari palaeo-continent only collided with proto-Gondwana at around 530 Ma. This complex geometric arrangement suggests a large open Adamastor Ocean, which at that time widened to the south, with small fragments (arcs and microcontinents) moving towards each other, and not a narrow ocean as suggested elsewhere.

A second and third subsequent collisional episodes took place between *c.* 600 and 550 Ma, when the arc terranes (Rio Negro, Pelotas, Paranaguá, Coastal terrane) and basement blocks (Paraíba do Sul–Curitiba, Luis Alves, Rio de la Plata and Angola) were assembled to proto-West Gondwana (Fig. 11b). The dextral and sinistral transpression respectively along the Brazilian and African margins could be controlled by the previous geometry of São Francisco and Angola palaeocontinents, respectively. At that time, differential movement along transform faults resulted in diachronous collision episodes recorded at *c.* 600 Ma (Dom Feliciano and Curitiba–Paraíba do Sul and Apiaí), *c.* 580 Ma (Ribeira and Kaoko belts) and 560 Ma (Araçuaí and West Congo belts).

Finally, a fourth and last collisional episode at *c.* 530–510 Ma resulted in closing of back arc basins and in the accretion of the Cabo Frio and Pampean terranes, and the Kalahari Craton to West Gondwana (Fig. 11c). Together with the arrival of the Amazonian plate, these final collisions reactivated important older sutures and transform zones as major lateral shear zones to accommodate the continental convergence.

Final remarks

The final amalgamation of West Gondwana along the Ribeira Belt was achieved only at *c.* 510 Ma by collision of the Cabo Frio terrane (and the Angola Craton?). A still open question is the relative position of São Francisco–Congo palaeocontinent with respect to Rodinia. In spite of scarce detailed palaeomagnetic data from the southern sector of West Gondwana in Brazil, recent reconstructions suggest that São Francisco–Congo was very distant in longitude from the other blocks that belonged to Rodinia in the Early Neoproterozoic (Cordani *et al.* 2003; Pisarevsky *et al.* 2003; Trindade *et al.* 2006).

Regarding correlation and matching of the Neoproterozoic pieces of both sides of the Atlantic, the best fit is found in northeastern Brazil, around the cratonic 'bridge' between Congo and São Francisco, where the modern passive margin gap is relatively narrow. At the latitude of the Ribeira Belt, the hidden stretch of continental crust is up to *c.* 400 Km wide and matching is not as straightforward. Figures 10 and 11 show that Cryogenian to Ediacaran arc-related accretionary terranes in Brazil faced a large cratonic segment of Angola. To the south, the Neoproterozoic Western terrane of the Kaoko Belt in Africa is partly facing the Luis Alves Craton in Brazil.

One important prediction that emerges from the integrated tectonic model presented above is that the Adamastor Ocean became progressively wider southwards from the cratonic 'bridge' between Congo and São Francisco. This provides the necessary space to accommodate long-lived subduction of oceanic lithosphere and generation of magmatic arcs, for at least 190 Ma, followed by docking of the arcs and other terranes during three collisional events in the Ribeira Belt (*c.* 600, 580 and 520 Ma). In this context, important transform zones (Fig. 11) must have been involved. One candidate that should be investigated in more detail is the Luanda shear zone providing displacement of

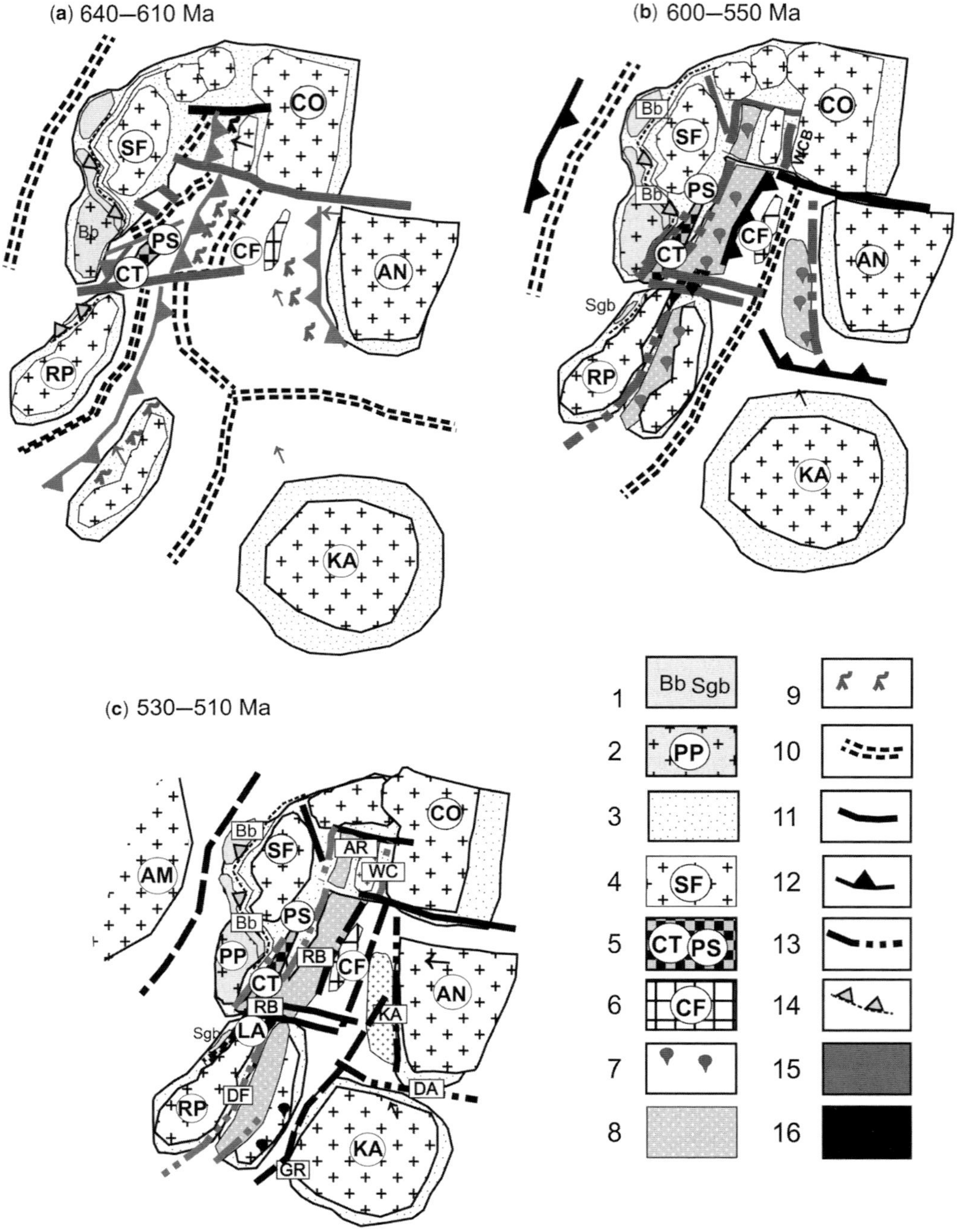

Fig. 11. Combined tectonic evolution of southeastern–southern Brazil and western Africa. (**a**) *c*. 630 Ma after Brasília Belt orogeny, (**b**) after 600 and 580 Ma orogenies, (**c**) after 530–510 Ma orogeny. 1, Older than *c*. 630 Ma belts (Brasília and São Gabriel); 2, Paranapanema Craton; 3, passive margin and reworked cratonic basement; 3, cratons (SF, São Francisco; CO, Congo; AN, Angola; LA, Luis Alves; RP, Rio de la Plata; Ka, Kalahari); 5, Paraíba do Sul–Curitiba terranes; 6, Cabo Frio terrane; 7, syn-to-late collisional granitoid rocks; 8, magmatic arc terranes; 9, pre-collisional granitoid rocks (subduction-related); 10, mid ocean ridge, 11, transform fault zone; 12, subduction zone; 13, collision zone (suture zones); 14, *c*. 640–610 sutures; 15, *c*. 605–550 Ma and *c*. 580–550 Ma collisional episodes; 16, *c*. 530–510 Ma collisions.

the southern Angola cratonic block with respect to the São Francisco–Congo Craton.

The authors thank the FAPERJ, FAPESP, CNPq and FINEP Brazilian agencies for the financial support for field work and geochronology studies. We also thank our colleagues from TEKTOS and CPGeo research groups of Rio de Janeiro State University (UERJ) and University of São Paulo (USP), and Benjamin Bley de Brito Neves, Karel Schulman and Robert Pankhurst for thoughtful reviews that improved the manuscript substantially.

References

ALKMIM, F. F. & MARSHAK, S. 1998. The Transamazonian orogeny in the Quadrilátero Ferrífero, Minas Gerais, Brazil: Paleoproterozoic Collision and Collapse in the Southern São Francisco craton region. *Precambrian Research*, **90**, 29–58.

ALKMIM, F. F., MARSHAK, S. & FONSECA, M. A. 2001. Assembly Western Gondwana in the Neoproterozoic: Clues from the São Francisco craton region, Brazil. *Geology*, **29**, 319–322.

ALMEIDA, F. F. M. 1977. O Cráton do São Francisco. *Revista Brasileira de Geociências*, **7**, 349–364.

ALMEIDA, J. C. H. 2000. *Zonas de cisalhamento dúctil de alto grau do Médio Vale do Rio Paraíba do Sul*. PhD Thesis, Instituto de Geociências, Universidade Estadual Paulista, Rio Claro.

ALMEIDA, F. F. M., HASUI, Y., BRITO NEVES, B. B. & FUCK, R. A. 1981. Brazilian structural provinces: an introduction. *Earth-Science Reviews*, **17**, 1–29.

ALMEIDA, J. C. H., TUPINAMBÁ, M., HEILBRON, M. & TROUW, R. 1998. Geometric and kinematic analysis at the Central Tectonic Boundary of the Ribeira Belt, Southeastern Brazil. *In*: *Congresso Brasileiro de Geologia 40, Belo Horizonte, Anais*. Sociedade Brasileira de Geologia, 32.

ALVARENGA, C. J. S., DE MOURA, C. A. V., GORAYEB, P. S. S. & ABREU, F. A. M. 2000. Paraguay and Araguaia belts. *In*: CORDANI, U. G., MILANI, E. J., THOMAZ FILHO, A. & CAMPOS, D. A. (eds) *Tectonic Evolution of South America, 31st International Geological Congress, Rio de Janeiro*, 183–193.

BABINSKI, M., TASSINARI, C. C. G., NUTMAN, A. P., SATO, K., MARTINS, P. O. R. & IYER, S. S. 2001. U/Pb SHRIMP zircon ages of migmatites from the basement of the Embú Complex, Ribeira Fold Belt, Brazil: indications for *c*. 1.4–1.3 Ga Pb-Pb and Rb–Sr isochron ages of no geological meaning. *In*: III South American Symposium on Isotope Geology, Pucón, Chile, Actas, 91–93.

BASEI, M. A. S., CITRONI, S. B. & SIGA, O., JR. 1998. Stratigraphy and Age of Fini-Proterozoic Basins of Paraná and Santa Catarina States, Southern Brazil. *Boletim do Instituto de Geociêcias da Universidade de São Paulo*, **29**, 195–216.

BASEI, M. A. S., SIGA, O., JR., MASQUELIN, O. M., REIS NETO, J. M. & PRECIOZZI, P. 2000. The Dom Feliciano Belt of Brazil and Uruguay and its foreland domain, the Rio de la Plata craton. *In*: CORDANI, U. G., MILANI, E. J., THOMAZ FILHO, A. & CAMPOS, D. A. (eds) *Tectonic Evolution of South America, 31st International Geological Congress, Rio de Janeiro*, 311–334.

BASEI, M. A. S., SIGA, O., JR., *ET AL*. 2003. Geochronology and Isotope Geology of Voturerava and Perau Mesoproterozoic Basins, Southern Ribeira Belt, Brazil. *In*: *IV South American Symposium on Isotope Geology, Salvador, Brazil, Extended Abstracts*, **2**, 501–504.

BERGMANN, M. 1988. *Caracterização Estratigráfica e estrutural da seqüência vulcano-sedimentar do Grupo São Roque na região de Pirapora do Bom Jesus, Estado de São Paulo*. Masters Dissertation, Instituto de Geociências, Universidade de São Paulo.

BOSSI, J. & GAUCHER, C. 2004. The Cuchilla Dionisio terrane, Uruguay: an allochthonous block accreted in the Cambrian to SW-Gondwana. *Gondwana Research*, **7**, 661–674.

BRITO NEVES, B. B., CAMPOS NETO, M. C. & FUCK, R. A. 1999. From Rodinia to Western Gondwana: an approach to the Brasiliano-Pan African cycle and orogenic collage. *Episodes*, **22**, 155–166.

BRITO NEVES, B. B., ALMEIDA, F. F. M. & CARNEIRO, C. D. R. 2000. The origin and evolution of the South American Platform. *Earth-Science Reviews*, **50**, 77–111.

CAHEN, L., SNELLING, N. J., DELHAL, J. & VAIL, J. R. 1984. *The geochronology and evolution of Africa*. Clarendon Press, Oxford.

CAMPANHA, G. A. C. 1981. O lineamento Além Paraíba na área de Três Rios (RJ), *Revista Brasileira de Geociências*, **11**, 159–171.

CAMPANHA, G. A. C. & SADOWSKY, G. R. 1999. Tectonics of the southern portion of the Ribeira Belt (Apiaí Domain). *Precambrian Research*, **98**, 31–51.

CAMPANHA, G. A. C., BASEI, M. A. S., TASSINARI, C. C., NUTMAN, A. P. & FALEIROS, F. M. 2004. Determinações Radiométricas U–Pb SHRIMP na Faixa Ribeira para a Formação Iporanga (SP). *In*: *Simpósio 40 anos de Geocronologia no Brasil, Boletim de Resumos*, **1**, 60.

CAMPOS NETO, M. C. 2000. Orogenic systems from Southwestern Gondwana, an approach to Brasiliano-Pan African cycle and orogenic collage in Southeastern Brazil. *In*: CORDANI, U. G., MILANI, E. J., THOMAZ FILHO, A. & CAMPOS, D. A. (eds) *Tectonic Evolution of South America, 31st International Geological Congress, Rio de Janeiro*, 335–365.

CAMPOS NETO, M. C. & CABY, R. 2004. Lower crust extrusion and terrane accretion in the Neoproterozoic nappes of southeast Brazil. *Tectonics*, **19**, 669–687.

CAMPOS NETO, M. C., BASEI, M. A. S., VLACH, S. R. F., CABY, R., SZABO, G. A. J. & VASCONCELOS, P. 2004. Migração de Orógenos e Superposição de Orogêneses: Um esboço da Colagem Brasiliana no Sul do craton do São Francisco, SE -Brasil. *Boletim do Instituto de Geociências/USP*, **4**, 13–40.

CARVALHO, H. & ALVES, P. 1990. *Gabbro–anorthosite Complex of SW Angola/NW Namíbia. Notes about the general geology. An essay of genetic interpretation*. Comunicações, Instituto Investigação Científica Tropical, Lisbon, Portugal, **2**.

CARVALHO, H. & TASSINARI, C. C. G. 1992. Idades do magmatismo da região do Caraculo-Bibaia (SW deAngola) e sua implicações na correlação geológica com o

Cinturão Ribeira no sudeste do Brasil. *Revista Brasileira de Geociências*, **22**, 73–81.

CARVALHO, H., FERNADEZ, A. & VIALLETE, Y. 1979. Chronologie absolue du Précambrian du Sud-Oest de l'Angola. *Comptes Rendus de l'Académie des Sciences, poris*, **288**, 1647–1650.

CARVALHO, H., TASSINARI, C., ALVES, F., GUIMARÃES, D. & SIMÕES, M. C. 2000. Geochronological Review of the Precambrian in western Angola: links with Brazil. *Journal of African Earth Sciences*, **31**, 383–402.

CORDANI, U. G., BASEI, M. A., SIGA, O., JR. & NUTMAN, A. P. 1999. Geochronology U–Pb (SHRIMP) of Campo Alegre, Itajaí and Castro basins: Neoproterozoic extensional regimes in the Brazilian south region. *In*: *Geological Society of America Annual Meeting 282, Denver, Colorado, 1999*, 282.

CORDANI, U. G., COUTINHO, J. M. V. & NUTMAN, A. P. 2002. Geochronological constraints on the evolution of the Embú Complex, São Paulo, Brazil. *Journal of South American Earth Sciences*, **14**, 903–910.

CORDANI, U. G., DÁGRELLA-FILHO, M. S., BRITO NEVES, B. B. & TRINDADE, R. I. F. 2003. Tearing up Rodinia: the Neoproterozoic paleogeography of South American cratonic fragments. *Terra Nova*, **15**, 350–359.

CURY, L. F., KAULFUSS, G. A., SIGA, O., JR., BASEI, M. A. S., HARARA, O. M. & SATO, K. 2002. Idades U–Pb (zircões) de 1.75 Ga em granitóides alcalinos deformados dos Núcleos Betara e Tigre: Evidências de regimes extensionais do Estateriano na Faixa Apiaí. *Geologia USP*, **2**, 95–108.

DELHAL, J., LEDENT, D. & TORQUATO, J. R. 1976. Nouvelles donnés géochrologique relativs au complexe gabbro-noritique et charnockitique du bouclier du Kasai et à son prolongement em Angola. *Annales Société Géologique Belgique*, **99**, 211–226.

DUARTE, B. P., VALENTE, S., HEILBRON, M. & CAMPOS NETO, M. C. 2004. Petrogenesis of orthogneisses of the Mantiqueira Complex, Central Ribeira Belt, SE Brazil: an Archean to Paleoproterozoic basement unit reworked during the Pan-African Orogeny. *Gondwana Research*, **7**, 437–450.

EBERT, H. D., CHEMALE, F., BABINSKI, M., ARTUR, A. C. & VAN SCHMUS, W. R. 1996. Tectonic setting and U/Pb zircon dating of the plutonic Socorro Complex in the transpressive Rio Paraíba do Sul Shear Belt, SE Brazil. *Tectonics*, **15**, 688–699.

EIRADO SILVA, L. G., HEILBRON, M. & ALMEIDA, J. C. H. 2006. Os terrenos tectônicos da Faixa Ribeira central na baía da Ilha Grande e Serra da Bocaina, SE Brazil. *Revista Brasileira de Geociências*, **36**, 426–436.

FALEIROS, M. F. & CAMPANHA, G. A. C. 2005. Geometria e Cinemática do Domínio Curitiba (Complexos Turvo-Cajati e Gnáissico-Migmatítico) na Região do Alto Jacupiranguin, Cajati–SP. *In*: *10° Simpósio Nacional de Estudos Tectônicos, Curitiba, Boletim de Resumos*, 428–431.

FIORI, A. P. 1992. O sistema de dobramento Apiaí, Estado do Paraná. *Revista Brasileira de Geociências*, **23**, 5–17.

FISCHEL, D. P., PIMENTEL, M. M., FUCK, R. A., COSTA, A. G. & ROSIERE, C. A. 1998. Geology and Sm-Nd Isotopic Data for the Mantiqueira and Juiz de Fora Complexes (Ribeira Belt) in the Abrecampo-Manhuaçú Region, Minas Gerais, Brazil. *In*: *Abstracts of the 14th International Conference on Basement Tectonics, Ouro Preto, Brazil*, 21–23.

FRIMMEL, H. E., TACK, L., BASEI, M. S., NUTMAN, A. P. & BOVEN, A. 2006. Provenance and chemostratigraphy of the Neoproterozoic West Congolian Group in the Democratic Republic of Congo. *Journal of African Earth Sciences*, **46**, 221–239.

GOSCOMBE, B., HAND, M., GRAY, D. & MAWBY, J. 2003. The metamorphic architecture of a transpressional orogen: The Kaoko Belt, Namibia. *Journal of Petrology*, **44**, 679–711.

GOSCOMBE, B., GRAY, D., *ET AL*. 2005. Event geochronology of the Pan-African Kaoko Belt, Namibia. *Precambrian Research*, **140**, 1–41.

GROSSI SAD, J. H. & DUTRA, C. V. 1998. Chemical composition of supracrustal rocks from Paraíba do Sul group, Rio de Janeiro State, Brazil. *Geochimica Brasiliensis*, **2**, 143–166.

GUIMARÃES, G. B. 2000. *As Rochas Granitóides do Complexo Granítico Cunhaporanga, Paraná: Aspectos Geológicos, Geofísicos,* Geoquímicos e Mineralógicos. PhD Thesis, Instituto de Geociências, Universidade de São Paulo, São Paulo.

HACKSPACHER, P. C., DANTAS, E. L., SPOLADORE, A., FETTER, A. H. & OLIVEIRA, M. A. F. 2000. Evidence of Neoproterozoic back arc basin development in the Central Ribeira Belt, South-eastern Brazil: New geochronological and geochemical constraints from the São Roque-Açungui groups. *Revista Brasileira de Geociências*, **30**, 110–114.

HARTMANN, L. A., LEITE, J. A. D., *ET AL*. 2000. Advances in SHRIMP geochronology and their impact on understanding the tectonic and metallogenic evolution of southern Brazil. *Australian Journal of Earth Sciences*, **47**, 829–844.

HASUI, Y., CARNEIRO, C., DAL, R. & COIMBRA, A. M. 1975. The Ribeira Fold Belt. *Revista Brasileira de Geociências*, **5**, 257–267.

HEILBRON, M., SIMÕES, L. S. A., ALVES, R. P. & CHRISPIM, S. J. 1982. Geologia do Cabo dos Búzios. *Anais da Academia Brasileira de Ciências, Rio de Janeiro*, **54**, 553–562.

HEILBRON, M., DUARTE, B. P. & NOGUEIRA, J. R. 1998. The Juiz de Fora complex of the Central Ribeira Belt, SE Brazil: a segment of Palaeoproterozoic granulitic crust thusted during the Pan-African Orogen. *Gondwana Research*, **1**, 373–382.

HEILBRON, M. & MACHADO, N. 2003. Timing of terrane accretion in the Neoproterozoic-Eopaleozoic Ribeira orogen (SE Brazil). *Precambrian Research*, **125**, 87–112.

HEILBRON, M., MOHRIAK, W., VALERIANO, C. M., MILANI, E., ALMEIDA, J. C. H. & TUPINAMBÁ, M. 2000. From collision to extension: the roots of the south-eastern continental margin of Brazil. *In*: TALWANI, M. & MOHRIAK, W. (eds) *Atlantic Rifts and Continental Margins*. American Geophysical Union, Geophysical Monographs, **115**, 1–34.

HEILBRON, M., MACHADO, N., SIMONETTI, T. & DUARTE, B. 2003. A Palaeoproterozoic orogen reworked within the Neoproterozoic Ribeira Belt, SE Brazil. *In*: *IV South American Symposium on Isotope Geology, Salvador, Brazil, Short Papers*, **1**, 186–189.

HEILBRON, M., PEDROSA-SOARES, A. C., CAMPOS NETO, M., SILVA, L. C., TROUW, R. A. J. & JANASI, V. C. 2004*a*. A Província Mantiqueira: *In*: MANTESSO-NETO, V., BARTORELLI, A., CARNEIRO, C. D. R. & BRITO NEVES, B. B. (eds) *O Desvendar de um Continente: A Moderna Geologia da América do Sul e o Legado da Obra de Fernando Flávio Marques de Almeida*. Beca, São Paulo, 203–234.

HEILBRON, M., PEDROSA-SOARES, A. C., CAMPOS NETO, M., SILVA, L. C., TROUW, R. A. J. & JANASI, V. C. 2004*b*. Brasiliano Belts in SE Brazil. *Journal of Virtual Explorer*, **17**, paper 4.

HEILBRON, M., TUPINAMBÁ, M. & PORTO, R., JR. 2005. Neoproterozoic juvenile accretion at the Ribeira Belt, SE Brazil: implications for Western Gondwana amalgamation. *In*: PANKHURST, R. J. & VEIGA, G. D. (eds) *Gondwana 12—Geological and Biological Heritage of Gondwana, Abstracts*. Academia Nacional de Ciencias, Córdoba, Argentina, 194.

HOWELL, D. G. 1989. *Tectonic of suspect terranes: mountain building and continental growth*. Chapman & Hall, London.

JANASI, V. A. & ULBRICH, H. H. G. J. 1991. Late Proterozoic granitoid magmatism in the State of São Paulo, south-eastern Brazil. *Precambrian Research*, **51**, 351–374.

JANASI, V. A., ALVES, A., VLACH, S. R. F. & LEITE, R. J. 2003. Granitos peraluminosos da porção central da Faixa Ribeira, Estado de São Paulo: sucessivos eventos de reciclagem da crosta continental no Neoproterozóico. *Geologia USP, Série Científica*, **3**, 13–24.

JANASI, V. A., LEITE, R. J. & VAN SCHMUS, W. R. 2001. U–Pb chronostratigraphy of the granitic magmatism in the Agudos Grandes Batholith (west of São Paulo, Brazil)—implications for the evolution of the Ribeira Belt. *Journal of South American Earth Sciences*, **14**, 363–376.

JULIANI, C., HACKSPACHER, P., DANTAS, E. & FETTER, A. H. 2000. The Mesoproterozoic volcano-sedimentary Serra do Itaberaba Group of the central Ribeira Belt, São Paulo State, Brazil: implications for the age of the overlying São Roque Group. *Revista Brasileira de Geociências*, **30**, 82–86.

KAULFUSS, G. A. 2001. *Geocronologia dos núcleos de embasamento Setuva, Betara e Tigre, norte de Curitiba, Paraná*. MSc Thesis, Instituto de Geociências, Universidade de São Paulo.

KRÖNER, A. & CORDANI, U. 2003. African, southern India and South American cratons were not part of Rodinia supercontinent: evidence from filed relationships and geochronology. *Tectonophysics*, **375**, 325–352.

KRÖNER, S., KONOPÁSEK, J., ET AL. 2004. U–Pb and Pb-Pb zircon ages for metamorphic rocks in the Kaoko Belt of Northwestern Namibia: A Palaeo- to Mesoproterozoic basement reworked during the Pan-African Orogeny. *South African Journal of Earth Sciences*, **107**, 455–476

MACHADO, N. & GAUTHIER, G. 1996*b*. Determination of $^{207}Pb/^{206}Pb$ ages on zircon and monazite by laser-ablation ICPMS and application to a study of sedimentary provenance and metamorphism in southeastern Brazil. *Geochimica et Cosmochimica Acta*, **60**, 5063–5073.

MACHADO, N., NOCE, C. M., LADEIRA, E. A. & BELO DE OLIVEIRA, O. 1992. U–Pb geochronology of Archean magmatism and Proterozoic metamorphism in the Quadrilátero Ferrífero, southern São Francisco craton, Brazil. *Geological Society of America Bulletin*, **104**, 1221–1227.

MACHADO, N., SCHRANK, A., NOCE, C. M. & GAUTHIER, G. 1996*a*. Ages of detrital zircon from Archean–Paleoproterozoic sequences: implications for greenstone belt setting and evolution of a Transamazonian foreland basin in Quadrilátero Ferrífero, southeast Brazil. *Earth and Planetary Science Letters*, **141**, 259–276.

MACHADO, N., VALLADARES, C., HEILBRON, M. & VALERIANO, C. 1996*b*. U–Pb geochronology of the central Ribeira Belt (Brazil) and implications for the evolution of the Brazilian Orogeny. *Precambrian Research*, **79**, 347–361.

MANTOVANI, M. S. M. & BRITO NEVES, B. B. 2005. Geophysical boundaries of Paranapanema Proterozoic block: its importance for the Rodinia to Gondwana evolutionary theories. *Gondwana Research*, **8**, 18–22

MANTOVANI, M., QUINTAS, M. C. L., SHUKOSKY, W. & BRITO NEVES, B. B. 2005. Delimation of the Paranapanema Proterozoic Block: A Geophysical Contribution. *Episodes*, **28**, 18–25.

MARTINS-NETO, M. A. 2000. Tectonics and sedimentation in a Proterozoic rift-sag basin (Espinhaço Basin, South-eastern Brazil). *Precambrian Research*, **103**, 147–173.

MARTINS-NETO, M. A., PEDROSA-SOARES, A. C. & LIMA, S. A. A. 2001. Tectonic–sedimentary evolution of sedimentary basins from Late Paleoproterozoic to Late Neoproterozoic in the São Francisco craton and Araçuaí fold belt, eastern Brazil. *Sedimentary Geology*, **141**, 343–370.

MENDES, J. C., ÁVILA, C., MELLO, R., HEILBRON, M. & MOURA, C. 2006. $^{207}Pb/^{206}Pb$-ages of zircons from syn-collisional I-type porphyritic granites of the central Ribeira Belt, SE Brazil. *Gondwana Research*, **9**, 326–335.

MILESI, J. P., TOTEU, S. F., ET AL. 2006. An overview of the geology and major ore deposits of Central Africa: Explanatory note for the 1:4,000,000 map "Geology and major ore deposits of Central Africa". *Journal of African Earth Sciences*, **44**, 571–595.

MILLER, R. MCG. & GROTE, W. 1998. *Geological Map of the Damaran orogen of South West Africa/Namibia 1:5,000,000*. Geological Survey of Namibia, Windhoek.

PACIULLO, F. V. P., RIBEIRO, A., ANDREIS, R. R. & TROUW, R. A. J. 2000. The Andrelândia Basin, a Neoproterozoic intra-plate continental margin, southern Brasília Belt. *Revista Brasileira de Geociências*, **30**, 200–202.

PASSCHIER, C. W., TROUW, R. A. J., RIBEIRO, A. & PACIULLO, F. V. P. 2002. Tectonic evolution of the Southern Kaoko Belt, Namibia. *Journal of African Earth Sciences*, **35**, 61–75.

PEDROSA-SOARES, A. C., VIDAL, P., LEONARDOS, O. H. & BRITO-NEVES, B. B. 1998. Neoproterozoic Oceanic Remnants in Eastern Brazil: Further Evidence and Refutation of an Exclusively Ensialic Evolution for the Araçuaí-West Congo Belt. *Geology*, **26**, 519–522.

PEDROSA-SOARES, A. C., NOCE, C. M., WIEDEMANN, C. & PINTO, C. P. 2001. The Araçuaí–West-Congo Orogen in Brazil: An overview of a confined orogen formed during Gondwanaland assembly. *Precambrian Research*, **110**, 307–323.

PIMENTEL, M. M., FUCK, R. A., JOST, H., FERREIRA FILHO, C. F. & ARAÚJO, S. M. 2000. The basement of the Brasília Fold Belt and the Goiás Magmatic Arc. *In*: CORDANI, U. G., MILANI, E. J., THOMAZ FILHO, A. & CAMPOS, D. A. (eds) *Tectonic Evolution of South America, 31st International Geological Congress, Rio de Janeiro*, 195–229.

PISAREVSKY, S. A., WINGATE, M. T. D., POWELL, C. MCA., JOHNSON, S. & EVANS, D. A. D. 2003. Models of Rodinia assembly and fragmentation. *In*: YOSHIDA, M., WINDLEY, B. F. & DASGUPTA, S. (eds) *Proterozoic East Gondwana: Supercontinent Assembly and Break-up*. Geological Society, London, Special Publications, **206**, 35–55.

PRAZERES FILHO, H. J. 2005. *Caracterização Geológica e Petrogenética do Batolito Granítico Três Córregos (PR-SP): Geoquímica Isotópica (Nd–Sr–Pb), Idades (ID-TIMS/SHRIMP) e O18 em Zircão*. PhD thesis, Instituto de Geociências, Universidade de São Paulo, São Paulo.

PRAZERES FILHO, H. J., BASEI, M. A. S., PASSARELLI, C. R., HARARA, O. M. & SIGA, O., JR. 2003. U–Pb zircon ages of the post-orogenic granitic magmatism in Apiaí folded belt (Paraná State, southern Brazil): Petrological and geotectonic significance. *In*: *IV South American Symposium on Isotope Geology, Salvador, Brazil, Abstracts*, **2**, 656–659.

RAPELA, C. W. 2000. The Sierras Pampeanas of Argentina: paleozoic building of the southern proto-Andes. *In*: CORDANI, U. G., MILANI, E. J., THOMAZ FILHO, A. & CAMPOS, D. A. (eds) *Tectonic Evolution of South America, 31st International Geological Congress, Rio de Janeiro*, 381–387.

RIBEIRO, A., TROUW, R. A. J., ANDREIS, R. R., PACIULLO, F. V. P. & VALENÇA, J. G. 1995. Evolução das bacias Proterozóicas e o termo-tectonismo Brasiliano na margem sul do Cráton do São Francisco. *Revista Brasiliera de Geociências*, **25**, 235–248.

SALLUN FILHO, W. & FAIRCHILD, T. R. 2004. Esreomatólitos do Grupo Itaiacoca ao Sul de Itapeva, Estado de São Paulo, Brasil. *Revista Brasileira de Paleontologia*, **7**, 359–370.

SCHMITT, R. S., PIMENTEL, M., VAN SCHMUS, W. R., TROUW, R. A. J. & ARMSTRONG, R. A. 2003. Marine sedimentation related to the latest stages of Gondwana assembly in the Ribeira Belt: new U/Pb data. *In*: *IV South American Symposium on Isotope Geology, Salvador, Brazil, Short Papers*, 294–297.

SCHMITT, R. S., TROUW, R. A. J., VAN SCHMUS, W. R. & PIMENTEL, M. M. 2004. Late amalgamation in the central part of Western Gondwana: new geochronological data and the characterization of a Cambrian collision orogeny in the Ribeira Belt (SE Brazil). *Precambrian Research*, **133**, 29–61.

SETH, B., KRÖNER, A., MEZGER, K., NEMCHIN, A. A., PIDGEON, R. T. & OKRUSCH, M. 1998. Archaean to Neoproterozoic magmatic events in the Kaoko Belt of NW Namibia and their geodynamic significance. *Precambrian Research*, **92**, 341–363.

SETH, B., JUNG, S. & HOERNES, S. 2002. Isotope constraints on the origin of Pan-African granitoids rocks in the Kaoko Belt, NW Namibia. *South African Journal of Earth Sciences*, **105**, 179–192.

SIGA, O., JR., BASEI, M. A. S., REIS NETO, J. M., MACHIAVELLI, A. & HARARA, O. M. 1995. O Complexo Atuba: um cinturão Paleoproterozóico intensamente retrabalhado no Neoproterozóico. *Boletim IG-USP, Série Científica*, **26**, 69–98.

SIGA, O., JR., BASEI, M. A. S., *ET AL.* 2003. U–Pb (zircon) ages of metavolcanic rocks from the Itaiacocoa Group: tectonic implications. *Geologia, Universidade de São Paulo*, **3**, 39–50.

SIGA, O., JR., CURY, L. F., KAULFUSS, G. A., HARARA, O. M., SATO, K., RIBEIRO, L. M. & BASEI, M. A. S. 2005. Evidências de Regimes Extensionais do Estateriano no Leste Paranaense, com base em Estudos Geocronológicos. *In*: *X Simpósio Nacional de Estudos Tectônicos, Curitiba Short Papers*, **1**, 353–356.

SILVA, L. C., DA MCNAUGHTON, N. J., ARMSTRONG, R., HARTMANN, L. A. & FLETCHER, I. 2005. The Neoproterozoic Mantiqueira Province and its African connections: a zircon-based U–Pb geochronological subdivision for the Brasiliano/Pan-African systems of orogens. *Precambrian Research*, **166**, 203–240.

SILVA, L. C., PEDROSA-SOARES, A. C. & TEIXEIRA, L. 2007. Tonian rift-related, A-type continental plutonism in the Araçuaí Orogen, eastern Brazil: new evidences for the breakup stage of the São Francisco–Congo Paleocontinent. *Gondwana Research,* doi: 10.1016/j.gr.2007.06.002.

STANISTREET, I. G. & CHARLESWORTH, G. 2001. Damaran basement-cored fold nappes incorporating pre-collision basins of Kaoko Belt and controls on Mesozoic supercontinental break-up. *South African Journal of Geology*, **104**, 1–12.

TACK, L., WINGATE, M. P. D., LIÉGOIS, J. P., FERNADEZ-ALONSO, M. & DEBLOND, A. 2001. Early Neoproterozoic magmatism (1000–910 Ma) of the Zadinian and Mayumbian Groups (Bas Congo): onset of the Rodinia rifting at the western edge of the Congo craton. *Precambrian Research*, **110**, 277–306.

TASSINARI, C. C. G. & CAMPOS NETO, M. C. 1988. Precambrian continental crust evolution of South-eastern São Paulo State, Brazil, based on isotopic evidence. *Geochimica Brasiliensis*, **2**, 175–183.

TEIXEIRA, A. L. 2000. *Análise das bacias da transição Proterozóico-Fanerozóico do Estado de São Paulo e adjacências*. PhD Thesis, Instituto de Geociências, Universidade de São Paulo.

TEIXEIRA, A. L., GAUCHER, C., PAIM, P. S. G., FONSECA, M. M., PARENTE, C. V., SILVA FILHO, W. F. & ALMEIDA, A. R. 2004. Bacias do Estágio de transição da Plataforma Sul-Americana. *In*: MANTESSO-NETO, V., BARTORELLI, A., CARNEIRO, C. D. R. & BRITO NEVES, B. B. (eds) *Geologia do Continente Sul-Americano*. Beca, São Paulo, 487–536.

TEIXEIRA, W., SABATÉ, P., BARBOSA, J., NOCE, C. M. & CARNEIRO, M. A. 2000. Archean and Paleoproterozoic tectonic evolution of the São Francisco craton, Brazil. *In*: CORDANI, U. G., MILANI, E. J., THOMAZ FILHO, A. & CAMPOS, D. A. (eds) *Tectonic Evolution of South America, 31st International Geological Congress, Rio de Janeiro*, 101–137.

TORQUATO, J. R. & CARVALHO, J. A. R. 1992. Idade Rb–Sr do granito do Caraculo, uma nova evidência para a existência do Evento Namib no Sudoeste de Angola. *Revista Geologia*, **50**, 157–167.

TORQUATO, J. R., SILVA, A. T. S. F., CORDANI, U. G. & KAWASHITA, K. 1979. Aevolução geológica do Cinturão Móvel do Quipungo no Ocidente de Angola. *Anais da Academia Brasileira de Ciências*, **51**, 133–143.

TOTEU, S. F., VAN SCHMUS, R. W., PENAYE, J. & NYOBE, J. B. 1994. U–Pb and Sm–Nd evidence for Eburnian and Pan-African high-grade metamorphism in cratonic rocks of southern Cameron. *Precambrian Research*, **67**, 321–347.

TOTEU, S. F., VAN SCHMUS, R. W., PENAYE, J. & MICHARD, A. 2001. New U–Pb and Sm–Nd data from north-central Cameroon and its bearing on the pre-Pan-African history of central Africa. *Precambrian Research*, **108**, 45–73.

TRINDADE, R. I. F., D'AGRELLA FILHO, M. S., EPOF, I. & BRITO NEVES, B. B. 2006. Paleomagnetism of Early Cambrian Itabaiana mafic dikes (NE Brazil) and the final assembly of Gondwana. *Earth and Planetary Science Letters*, **244**, 361–377.

TROMPETTE, R. 1998. *Geology of Western Gondwana* (2000–500 Ma). A.A. Balkema, Rotterdam.

TROUW, R. A. J., HEILBRON, M., *ET AL*. 2000. The central segment of the Ribeira Belt. *In*: CORDANI, U. G., MILANI, E. J., THOMAZ FILHO, A. & CAMPOS, D. A. (eds) *Tectonic Evolution of South America, 31st International Geological Congress, Rio de Janeiro*, 287–310.

TUPINAMBÁ, M., TEIXEIRA, W. & HEILBRON, M. 2000. Neoproterozoic Western Gondwana assembly and subduction-related plutonism: the role of the Rio Negro Complex in the Ribeira Belt, South-eastern Brazil. *Revista Brasileira de Geociências*, **30**, 7–11.

VALERIANO, C. M., MACHADO, N., SIMONETTI, A., VALLADARES, C. S., SEER, H. J. & SIMÕES, L. S. 2004. U–Pb Geochronology of the Southern Brasília Belt (SE Brazil): sedimentary provenance, Neoproterozoic orogeny and assembly of Western Gondwana. *Precambrian Research*, **130**, 27–55

VALERIANO, C. M., PIMENTEL, M. M., HEILBRON, M., ALMEIDA, J. C. H. & TROUW, R. A. J. 2008. Tectonic evolution of the Brasília Belt, Central Brazil, and early assembly of Gondwana. *In*: PANKHURST, R. J., TROUW, R. A. J., BRITO NEVES, B. B. & DE WIT, M. J. (eds) *West Gondwana: Pre-Cenozoic Correlations Across the South Atlantic Region*. Geological Society, London, Special Publications, **294**, 197–210.

VALLADARES, C. S., SOUZA, S. F. M. & RAGATKY, D. 2003. The Quirino Complex: a Transamazonian magmatic arc of the central segment of the Brasiliano/Pan-African Ribeira Belt, SE Brazil. *Revista Universidade Rural, Série Ciências Exatas e da Terra*, **22**, 49–61.

VALLADARES, C. S., MACHADO, N., HEILBRON, M. & GAUTHIER, G. 2004. Ages of detrital zircon from siliciclastic successions of the Brasília Belt, southern border of the São Francisco craton, Brazil: implications for the evolution of Proterozoic basins. *Gondwana Research*, **7**, 913–921.

VLACH, S. R. F. 2001. Microprobe monazite constraints for an early (ca. 790 Ma) Brasiliano orogeny: the Embú terrane, South-eastern Brazil. *In*: *III South American Symposium on Isotope Geology, Pucon, Chile, Extended Abstracts*, 265–268.

WEBER, W., SIGA, O., JR., SATO, K., REIS NETO, J. M., BASEI, M. A. S. & NUTMAN, A. 2004. A Formação Água Clara na região de Araçaiba —SP: Registro U–Pb de uma bacia Mesoproterozóica. *Revista do Instituto de Geociências da Universidade de São Paulo, Série Científica*, **4**, 101–110.

WIEDEMANN, C. M., MEDEIROS, S. R., DE LUDKA, I. P., MENDES, J. C. & MOURA, J. C. 2002. Architecture of late orogenic plutons in the Araçuaí-Ribeira fold belt, southeast Brazil. *Gondwana Research*, **5**, 381–400.

West Gondwana amalgamation based on detrital zircon ages from Neoproterozoic Ribeira and Dom Feliciano belts of South America and comparison with coeval sequences from SW Africa

M. A. S. BASEI[1], H. E. FRIMMEL[2], A. P. NUTMAN[3,4] & F. PRECIOZZI[5]

[1]*Instituto de Geociências, Universidade de São Paulo, São Paulo, Rua do Lago 562, SP, Brazil (e-mail: baseimas@usp.br)*

[2]*Department of Geological Sciences, University of Cape Town, Rondebosch 7701, South Africa, Present address: Institute of Mineralogy, University of Würzburg, Am Hubland, D-97074 Würzburg, Germany*

[3]*Research School of Earth Sciences, Australian National University, Canberra, ACT, Australia*

[4]*Beijing SHRIMP Centre, Chinese Academy of Geological Sciences, 26, Baiwanzhuang Road, Beijing, 100037 China*

[5]*Departamento de Geología, Instituto de Geología y Paleontología, Universidad de la República, Iguá 4225, Malvin Norte, CP 11400, Montevideo, Uruguay*

Abstract: Neoproterozoic–Cambrian amalgamation of West Gondwana involved the collision of several terranes of older crust that are now in eastern South America and western Africa. U–Pb (SHRIMP) detrital zircon ages from representative metasedimentary units of the Ribeira and Dom Feliciano belts (South America) and Gariep and Damara belts (Africa) provide constraints on the possible sediment source areas across probable suture zones. Ribeira detrital zircons are Palaeoproterozoic and Archaean. For the Dom Feliciano Belt, a contribution of Meso- and Neoproterozoic zircons is present, which definitely indicate Neoproterozoic sedimentation. It is proposed that the inflow of material to the Ribeira basin was essentially derived from the Paranapanema and Rio de la Plata cratons, whereas for the Damara and Gariep–Rocha belts source areas were from the Namaqua Belt. The Dom Feliciano Belt received sediments from the South American side and to a lesser degree from African sources. These results highlight the differences in the detrital zircon signatures across a proposed West Gondwanan suture, with those in the west being derived from distinctive South American basement sources and those in the east from distinctive African sources.

Crustal evolution in southern Brazil involved the Neoproterozoic Brasiliano orogenic cycle with tectonic events that led to the amalgamation of different terranes during collisional orogeny, culminating in the formation of West Gondwana by the start of the Cambrian (Brito Neves & Cordani 1991; Campos Neto & Figueiredo 1995; Brito Neves *et al.* 1999; Campos Neto 2000). Ophiolitic remnants and magmatic arc roots signal the existence of fossil subduction and collision zones (Brito Neves *et al.* 1999; Campos Neto 2000; Basei *et al.* 2000). In this study, new SHRIMP U–Pb ages for detrital zircon grains from 11 samples from the major metasedimentary units of the Ribeira, Dom Feliciano and Damara belts of the southern portions of South America and South Africa are presented and compared to available analyses for the Kaoko (Goscombe *et al.* 2005) and Gariep (Basei *et al.* 2005) belts. A revised tectonic model is then presented for the formation of West Gondwana, based on the combination of these U–Pb ages with Sm–Nd bulk rock isotope data. These detrital zircon studies support the location of a suture based on other evidence.

Geological setting

The Precambrian–Cambrian geology of southeastern Brazil and Uruguay is marked by the presence of the Paranapanema and Rio de la Plata cratons in the western portion, which are now mostly

From: PANKHURST, R. J., TROUW, R. A. J., BRITO NEVES, B. B. & DE WIT, M. J. (eds) *West Gondwana: Pre-Cenozoic Correlations Across the South Atlantic Region.* Geological Society, London, Special Publications, **294**, 239–256. DOI: 10.1144/SP294.13 0305-8719/08/$15.00

covered by the Palaeozoic sediments of the Paraná Basin. These cratonic domains were the foreland during the Neoproterozoic evolution of the southern portion of the Ribeira Belt and the Dom Feliciano Belt (Fig. 1).

The granitic–migmatitic–granulitic rocks of the allochthonous Luís Alves and Curitiba terranes are continental fragments that separate the Ribeira and Dom Feliciano belts (Basei *et al.* 1999). They contain Archaean rocks with superimposed

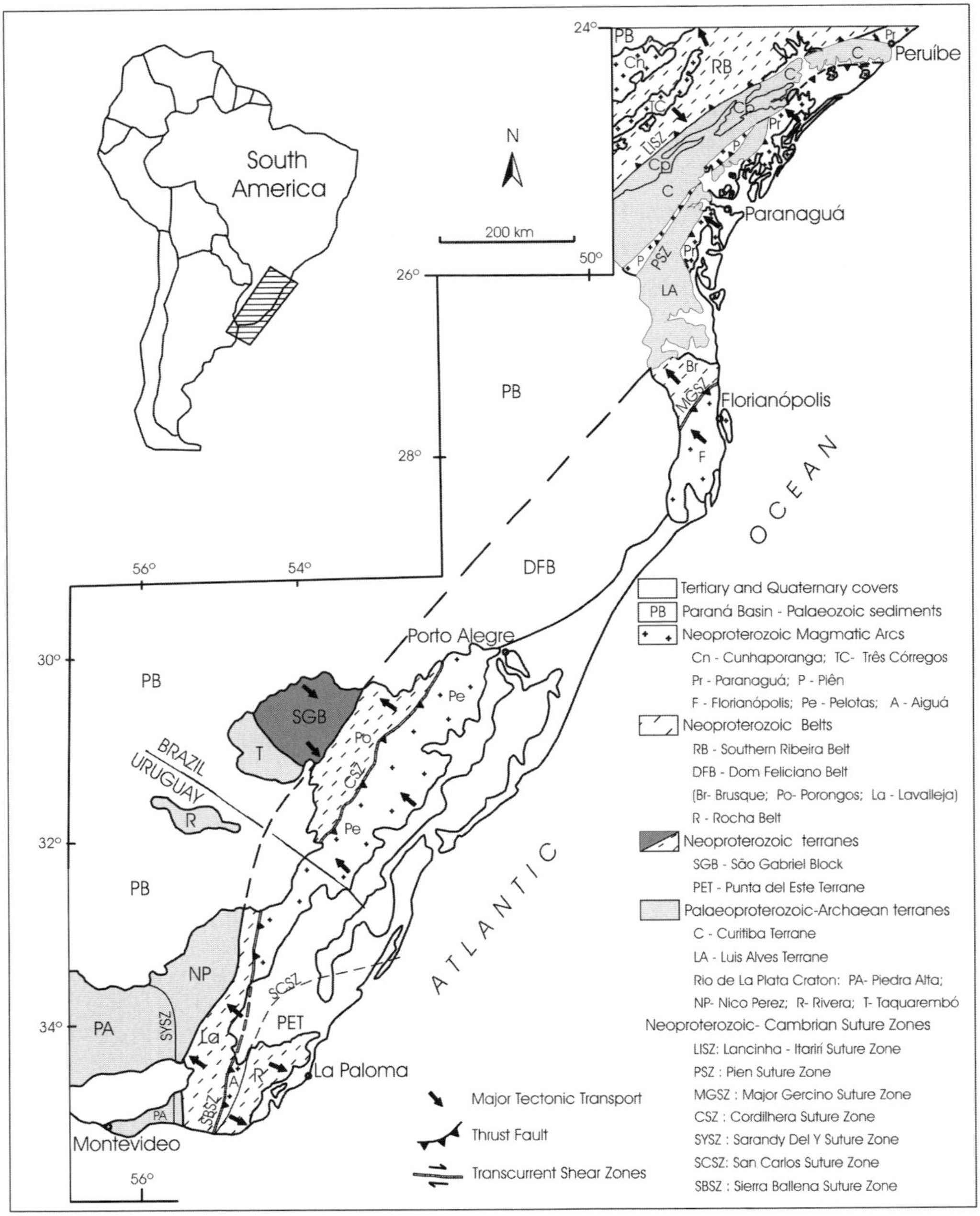

Fig. 1. Distribution of the main tectonic units of the southern Brazil and Uruguay (modified from Basei *et al.* 1999, 2000; Soares *et al.* 2000; Campos Neto 2000; Sanchez-Bettucci *et al.* 2004).

Palaeoproterozoic high-grade metamorphism (Basei *et al.* 1998). Despite the similarities between these continental blocks, the Luís Alves microplate escaped Neoproterozoic tectono-thermal overprint (Palaeoproterozoic K–Ar ages), whereas the Curitiba domain was intensely affected by Neoproterozoic migmatization and crustal melting during the Brasiliano orogeny (Basei *et al.* 1998).

The juxtaposition of the Luís Alves and Curitiba terranes involved the destruction of Neoproterozoic oceanic crust whose remnants are locally preserved in the Pien–São Bento do Sul region (Fig. 1; Harara 2001). Destruction of the oceanic domains at a convergent plate boundary included production of an extensive calc-alkaline granitoid batholith with magmatic arc affinities, which is now found associated with remaining dismembered ophiolitic, mafic–ultramafic complexes (Fig. 1).

Unmetamorphosed volcano-sedimentary basins are well represented by the Campo Alegre and Guaratubinha basins. They are spatially associated with the coeval A-type alkaline-peralkaline plutons (the Serra do Mar Suite). These reflect extensional events that took place in the Luís Alves and Curitiba terranes by the end of the Neoproterozoic (Siga *et al.* 1997, 1999).

The southern branch of the Ribeira Belt

The southernmost part of the Ribeira Belt is characterized by a series of NE–SW trending domains with Meso- to Neoproterozoic supracrustal rocks at low metamorphic grade, intruded by Neoproterozoic granitic batholiths and stocks (Campanha *et al.* 1985; Campanha & Sadowski 1999; Campos Neto 2000; Prazeres 2005). Most of these domains are separated by shear zones that have marked dextral horizontal displacements (Fiori 1992). This represents a polycyclic evolution, during which the Meso- and Neoproterozoic supracrustal units became juxtaposed. Neoproterozoic tectonothermal evolution is marked by northwestwards vergence towards the Paranapanema Craton (Mantovani & Brito Neves 2005). In contrast, Neoproterozoic transport in the southeastern portion of the belt was towards the Curitiba domain (Fiori 1992). From northwest to southeast, five major supracrustal domains are recognized: the Itaiacoca, Água Clara, Lageado, Votuverava and Capiru domains (Fig. 2).

The Itaiacoca Domain. This is an approximately 10 km wide, NE–SW elongated belt (Reis Neto 1994; Prazeres 2005) that occurs between the Cunhaporanga and the Três Córregos batholiths. It is marked by the Itaiacoca Group of shallow-water platform deposits of mostly meta-arkose with subordinate felsic metavolcanic rocks (Siga *et al.* 2003, 2006), plus a lower unit of stromatolitic dolomitic marble with some mafic rocks of volcanic origin (Reis Neto 1994). The NW border of the Itaiacoca Group is intruded by the Cunhaporanga batholith composed of granodiorite, monzogranite and quartz monzonite.

The Água Clara Domain. Lithologically, this domain consists of metacalcarenite, micritic metalimestone, meta-calcisiltite, calc-schist and subordinately mica schist, iron formation plus schists of probable volcanogenic origin (Fassbinder 1996; Weber *et al.* 2004). The regional metamorphism is of low amphibolite facies. The Três Córregos batholith intruded the Agua Clara sediments when they had already been deformed and metamorphosed (Janasi *et al.* 2001; Prazeres *et al.* 2005).

The Lageado Domain. This includes metaconglomerate, metasandstone, meta-rhythmite and calcitic marble of the Lageado Subgroup (Campanha *et al.* 1985) interpreted as shallow-water platform deposits, with alternating thick layers of carbonate and psammo-pelitic units. The metamorphic grade is greenschist facies. There are several post-tectonic granitic stocks, such as the Itaoca granite. The Iporanga and Antinha sedimentary units (Campanha *et al.* 1987; Campanha & Sadowski 1999) are here considered as the uppermost portion of the Lageado Subgroup.

The Votuverava–Perau Domain. The Votuverava Formation is composed of quartzite, rhythmic laminated phyllite, meta-siltite, meta-conglomerate, dolomitic marble, graphitic quartzite beds and metabasite lenses, all affected by greenschist facies metamorphism. These metasedimentary rocks interfinger with metabasic and volcaniclastic rocks and iron-manganese formations, from which a deep-water palaeo-environment is inferred (Fiori 1992; CPRM 1998). Again, post-tectonic Neoproterozoic granitic stocks occur, such as the Morro Grande, Cerne and Varginha intrusions. 2.2 Ga Palaeoproterozoic basement is represented by the Betara and Tigre gneissic–migmatitic nuclei (Cury *et al* 2002), exposed north of the Lancinha–Itariri suture zone (Basei *et al.* 1999).

The Capiru Domain. This comprises metasedimentary rocks covering the northern part of the Atuba Complex (Curitiba microplate), south of the Lancinha–Itariri suture zone (Figs 1 and 2). The Capiru Formation is composed of dolomitic marble (with stromatolite structures locally preserved), quartzite, and subordinated phyllite in the upper portions. Low-grade metamorphism (greenschist facies, chlorite zone) affected the Capiru Formation. Palaeoproterozoic migmatitic gneisses,

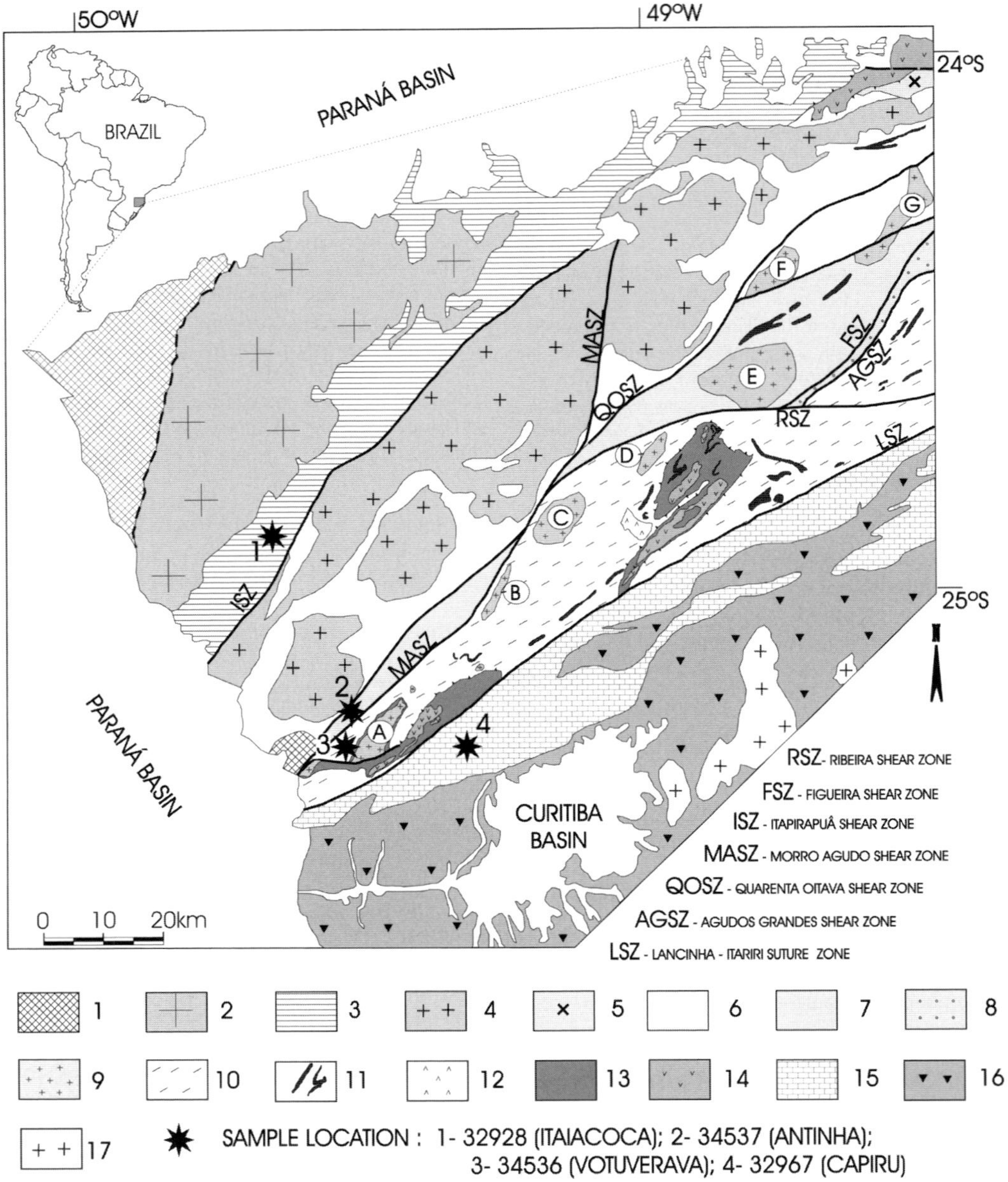

Fig. 2. Simplified geological map of the southern Ribeira Belt (adapted from Campanha *et al.* 1985; Basei *et al.* 1999; Campos Neto 2000; Siga *et al.* 2003; Prazeres 2005): 1, Castro (NW) and Camarinha (SE) post-tectonic basins; 2, Cunhaporanga Batholith (610–590 Ma); 3, Itaiacoca Group (630 Ma); 4, Três Córregos Batholith and associated granitic bodies (610–600 Ma); 5, A-type deformed granitoids (1750 Ma); 6, Água Clara Formation (*c.* 1450 Ma); 7, Lajeado Subgroup and Antinha Formation (*c.* 600 Ma); 8, Iporanga Formation (*c.* 590 Ma); 9, Syn- to post-collisional granitoids (590–570 Ma) (A, Cerne; B, Piedade; C, Morro Grande; D, Varginha; E, Itaóca; F, Apiaí; G, Espírito Santo); 10, Votuverava Formation (1450 Ma); 11, undifferentiated metabasic rocks (1450 Ma); 12, Tunas Syenite (85 Ma); 13, Perau and Betara formations (1450 Ma); 14, Basement inliers—deformed calc-alkaline granitoids (2200 Ma); 15, Capirú Formation (Neoproterozoic?); 16, Atuba Gneissic–Migmatitic Complex (2100 Ma); 17, alkaline–peralkaline granitoids of Serra do Mar Suite (595–570 Ma).

granites, amphibolites and mylonites of the Setuva Anticline are basement rocks correlated with the gneissic–migmatitic rocks of the Curitiba domain (Silva *et al.* 1998; Passarelli *et al.* 2003).

The Dom Feliciano Belt

The 1200 km long Dom Feliciano Belt has three domains (Fig. 3), here described from ESE to NNW. (i) The Eastern Granitoid Belt comprises the Florianópolis (Santa Catarina State), Pelotas (Rio Grande do Sul State) and Aiguá (Uruguay) granitic batholiths. They are all calc-alkaline in composition and represent the roots of a Neoproterozoic magmatic arc. (ii) The Supracrustal Schist Belt (central portion of the Dom Feliciano Belt) is predominantly composed of low-grade metavolcano-sedimentary units. It has been multiply folded, with northwestward tectonic transport, and was intruded by several generations of late- to post-tectonic granitoids that developed contact metamorphic aureoles. (iii) Foreland basins, less affected by deformation and metamorphism than the adjacent schist belt, are represented by the Itajaí Group (Santa Catarina State, Brazil), the Camaquã Basin (Rio Grande do Sul State, Brazil) and several fragments of similar basins that form the El Soldado Group (Uruguay).

The Supracrustal Schist Belt comprises all the metamorphic rocks located between the Eastern Granitic Belt and the foreland basins of the Dom Feliciano Belt (Fig. 3). It occurs discontinuously as a narrow belt with average width of around 40 km. From north to south, there are three metamorphic complexes within the supracrustal belt: Brusque (Santa Catarina State), Porongos (Rio Grande do Sul State) and Lavalleja (Uruguay).

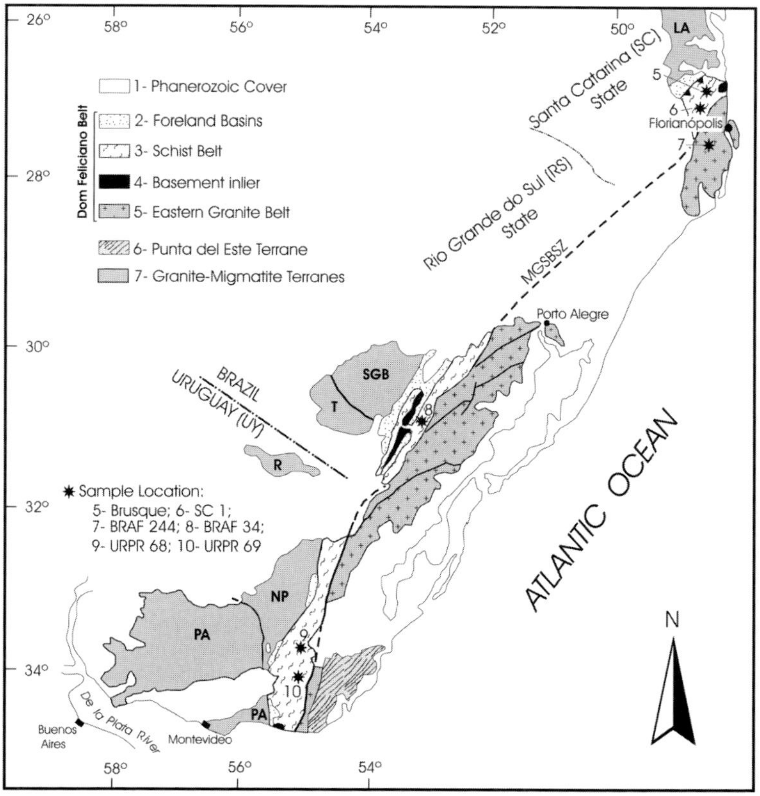

Fig. 3. Geological sketch of the Dom Feliciano Belt (modified from Basei *et al.* 2001): 1, Undifferentiated Phanerozoic cover; 2, Foreland basins (SC, Itajaí; RS, Camaquã; UY, El Soldado – Piriapolis); 3, schist belts and intrusive granitoids (SC, Brusque Metamorphic Complex; RS, Porongos Metamorphic Complex; UY, Lavalleja Metamorphic Complex); 4, reworked Palaeoproterozoic basement inliers (SC, Morro do Boi; RS, Encantadas; UY, Punta Rasa; 5, Eastern Granitoid Belt) (SC, Florianópolis; RS, Pelotas; UY, Aiguá batholiths); 6, Punta del Este Terrane; 7, Granite Migmatite Terranes (LA, Luis Alves; SGB, São Gabriel; R, Rivera; T, Taquarembó; NP, Nico Perez; PA, Piedra Alta.)

These complexes comprise poly-deformed sequences with at least three phases of folding associated with a northwestward tectonic transport that evolved to a predominantly lateral movement (Basei 1985; Fernandes *et al.* 1995; Basei *et al.* 2000). The regional metamorphic grade is greenschist to locally low amphibolite facies.

The Brusque Metamorphic Complex

This complex is composed of two metavolcano-sedimentary domains separated by the Valsungana batholith (Basei *et al.* 2000). In the northern segment, the sedimentary sequence starts with a pelitic–psammitic unit (now garnet-rich mica schist and quartzite) that grade to a psammo-pelitic unit (meta-rhythmites and homogeneous sericite schist), overlain by a metavolcano-sedimentary unit (meta-marl, calcareous schist, metabasic rocks and subordinate grey sericite schists (Basei 1985; Caldasso *et al.* 1995*a*, *b*). The mafic rocks represent syn-sedimentary basic magmatism of tholeiitic to alkaline affinity; structures produced by liquid immiscibility are frequently observed, characterizing them as variolitic basalt (Silva *et al.* 1985; Basei 1985; Sander 1992).

In the southern segment, the basal sequence is composed of a metavolcano-sedimentary unit that possibly represents the rift phase of the Brusque palaeo-basin (Basei *et al.* 2000). In this unit volcano-exhalative deposits are characterized by tourmalinites, associated with metabasalt, banded iron formation, quartzite (chert?) and calc-silicate rocks (Silva *et al.* 1985; Basei *et al.* 1994). Tectonically overlying is a thick psammo-pelitic sequence of micaceous quartzites, quartz–sericite schists and pelitic sericite schists where acid metavolcanic rocks occur locally. Granite magmatism is characterized by homogeneous to slightly deformed bodies of metaluminous to peraluminous composition whose genesis involved marked crustal contributions. They can be classified in three main suites, all post-dating much of the deformation and metamorphism in their host rocks (Castro *et al.* 1999).

The Porongos Metamorphic Complex

In Rio Grande do Sul State, the best exposures of the Porongos Metamorphic Complex are observed in the vicinity of Santana da Boa Vista Dome. They surround the Encantadas Gneisses in a Palaeoproterozoic basement inlier that forms the centre of an antiformal structure. The contact between the complex and the gneisses is tectonic (Jost 1981, 1982). The Porongos Metamorphic Complex contains a lower metasedimentary unit in which meta-arkose and impure quartzite predominate, intercalated with metapelite and rare amphibolite bodies. Quartzitic meta-rhythmites predominate in the intermediate portion, and mica schists form the top, with marble and orthoquartzite intercalations. The Porongos Metamorphic Complex also contains a metavolcanic sequence with meta-andesite, meta-dacite and several types of pyroclastic rocks. Subordinate, metachert, marble, metapelite, graphite schist and rare quartzite also occur (Jost 1981). The tectonic vergence is towards west, well observed in the eastern flank of the Santana Dome. These units were affected by medium pressure metamorphism (2.0–4.8 kbar) varying from chlorite zone (greenschist facies) to staurolite zone (amphibolite facies) (Jost 1982).

In the northern part of the Porongos Metamorphic Complex there is important magmatism dated at 780 Ma (Porcher *et al.* 1999). Its geochemical characteristics suggest evolution from a rift stage (alkaline gneisses) to a subduction stage (calc-alkaline acid volcanic rocks) associated with consumption of oceanic crust (Marques *et al.* 1998*a*, *b*). The Porongos complex supracrustal rocks are also intruded by the Campinas Granitic Suite. These granites occur as small stocks in the eastern part of the schist belt. They are mostly equigranular, isotropic, two-mica leucogranites showing a peraluminous trend, which suggests the involvement of the upper crust in their generation (Frantz & Jost 1983).

The Lavalleja Metamorphic Complex

In Uruguay, the Lavalleja Metamorphic Complex represents the southern extension of the Dom Feliciano Belt supracrustal rocks. It is subdivided from east to west into the Zanja del Tigre, Fuente del Puma and Minas formations (Sanchez-Bettucci 1998). The metamorphic grade decreases from east to NW, from lower amphibolite to sub-greenschist facies.

The Zanja del Tigre Formation is a metavolcano-sedimentary sequence in which gabbro and amphibolite occur within micaschists, garnet-rich schists and marbles. The overlying Fuente del Puma Formation consists of meta-conglomerate, calc-arenite, calcitic dolomite and mica schists, but with lesser amounts of volcanic and hypabyssal components (Sanchez-Bettucci 1998). The Minas Formation only contains metasedimentary rocks, with metapelite, quartzite, meta-arkose and stromatolite-bearing limestones. Some rocks attributed to the Lavalleja Group by Sanchez-Bettucci (1998) are now considered on the basis of fossil content to be younger, and have been correlated with the Vendian El Soldado Group (Gaucher 2000; Gaucher *et al.* 1996, 2004).

As observed in the Brusque Belt in Santa Catarina State, the Lavalleja supracrustal rocks were also intruded by abundant and chemically diverse granites (Mallmann *et al.* 2003; Sanchez-Bettucci *et al.* 2004). The largest bodies are the Maldonado intrusion into the Fuente del Puma Formation, and the Penitente intrusion in the Zanja del Tigre Formation.

Detrital zircon provenance

This study was undertaken to provide a 'fingerprint' of zircon provenance ages in metasedimentary rocks across the whole Neoproterozoic orogen, and thereby identify or confirm fundamental sutures. Due to budgetary constraints, it was only possible to analyse about 20 zircons from each of the eleven samples investigated, rather than the usual target of about 60 grains. With 20 grains analysed, there is a 95% probability of detecting age components present at the 15% level in the detrital zircon population. We consider this adequate for our orogen-wide reconnaissance study.

Sample collection was concentrated in typical units for each belt/unit. As the study aimed to reveal the general picture of the provenance of zircons, we tried to preserve all the different zircon morphologies during crystal concentration and handpicking, including crystal fragments. Zircon separation by standard gravimetric and isodynamic techniques and the mounting of selected zircons into epoxy resin discs were carried out at the Institute of Geological Sciences, University of São Paulo. Cathodoluminescence (CL) imaging and age determinations by SHRIMP took place in the Research School of Earth Sciences, The Australian National University, according to standard procedures (Compston *et al.* 1984; Williams 1998; Stern 1998; Sircombe 2000).

Choice of SHRIMP analytical sites was guided by CL imaging (McClaren *et al* 1994). All SHRIMP zircon analyses are available online at http://www.geolsoc.org.uk/SUP18290. A hard copy can be obtained from the Society Library. Most of the data yielded close to concordant ages. The data are portrayed graphically as histograms, with cumulative frequency curves shown in the background. These figures were generated in the program Isoplot/Ex (Ludwig 2001). The data were filtered prior to plotting, to remove analyses with the most disturbed radiogenic Pb-systematics. This removed analyses that were $<90\%$ concordant (if >1.0 Ga) and with $>2.5\%$ ^{206}Pb of common origin (calculated from measured ^{204}Pb using Cumming & Richards' (1975) Pb evolution curves for common Pb compositions). For grains with ages >1.0 Ga the $^{207}Pb/^{206}Pb$ age was plotted, whereas for grains <1.0 Ga the $^{206}Pb/^{238}U$ age was chosen.

The Ribeira Belt

Four of the five main units that compose the southern Ribeira Belt were studied. These are distributed along a NW–SE trend, including the Itaiacoca, Lageado (Iporanga–Antinha), Votuverava and Capiru units. Sample locations are indicated in Figure 2. Figure 4 shows representative CL images, and ages are displayed in Figure 5.

An arkosic quartzite belonging to the Abapã unit represents the Itaiacoca Group (sample 32928, UTM 619250 W; 7241755 S). Two Archaean grains (dated at 3.2 and 3.5 Ga) were detected, whereas the remainder are Palaeoproterozoic (1.9–2.2 Ga). No Neoproterozoic grains were encountered. On the other hand, the volcanic rocks of the same unit have mostly *c.* 630 Ma zircons that are interpreted as the age of zircon crystallization and consequently the age of deposition of the sedimentary units (Siga *et al.* 2005).

The Iporanga (SP) and Antinha (PR) formations of the Lageado Subgroup were studied. A metamarl of the Antinha unit (sample 34537, UTM 634677 W; 7201947 S) yielded Palaeoproterozoic zircons (1.8–2.2 Ga) with an age similar to that obtained for the detrital zircons of the Itaiacoca Group sample. Four grains of 0.59–0.61 Ga were dated, which are probably of volcanogenic origin (detail in Fig. 4). These data place the sedimentation of the unit in the Neoproterozoic. This age is similar to the age of the Iporanga Formation (Campanha *et al* 2006.) in São Paulo State, which is stratigraphically correlated with the Antinha Formation. The Votuvera Group is represented by a typical rhythmic psammo-pelitic phyllite (sample 32970, UTM 644976 W; 7203750 S). Two Archaean ages were obtained (3.2 and 2.8 Ga), two *c.* 2.4 Ga ages, a main group between 2.2 and 1.9 Ga and two ages close to 1.75 Ga. A Capiru Formation quartzite (sample 32967, UTM 671500; 7206625) gave an essentially unimodal zircon population at *c.* 2.2 Ga.

The Dom Feliciano Belt

Five samples representing the three metasedimentary units of the Dom Feliciano Belt were analysed: two from Brusque (Santa Catarina State), one from Porongos (Rio Grande do Sul) and two from Lavalleja (Uruguay). Sample locations are indicated in Figure 3, ages are summarized in Figure 6, and Figure 7 shows representative CL images.

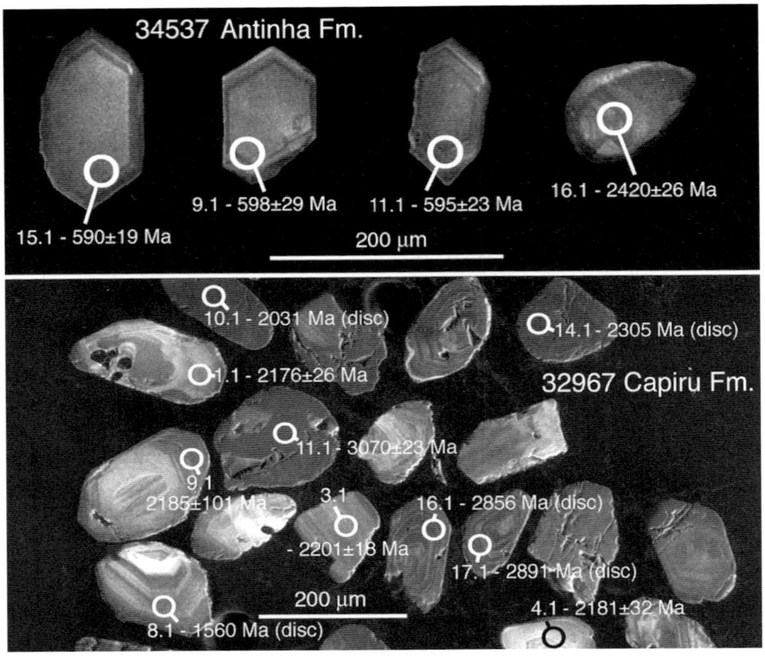

Fig. 4. Cathodoluminescence images of Ribeira Belt zircon grains. The detail shows the characteristics of the Antinha Formation volcanogenic zircon grains.

For the Brusque Metamorphic Complex, 22 zircon dates were obtained for a mica schist with a volcanogenic contribution (sample SC1, UTM 702267 W; 6977450 S) and a garnet–biotite schist (sample 32929, UTM 694676 W; 6992222 S), shown collectively as 'Brusque' on Figure 7. These yielded eight grains between 2.25 and 1.7 Ga, six between 1.5 and 1.3 Ga, four between 1.3 and 1.1 Ga and two between 0.54 and 0.57 Ga. Note the presence of Mesoproterozoic ages and the lack of Archaean ones.

For the Porongos Metamorphic Complex in Rio Grande do Sul State, 23 zircons were dated from a sericitic phyllite, BRAF 34. The sample was collected along the Pelotas–Caçapava highway, close to the bridge over the Rio Camaquã (UTM 298238 W; 6579203 S). Four age groups are present, with eight grains between 2.2 and 1.7 Ga, three between 1.5 and 1.4 Ga, six between 1.3 and 0.9 Ga and five Neoproterozoic grains in the 0.62–0.99 Ga interval (three of which are between 0.62 and 0.70 Ga).

From the Lavalleja Metamorphic Complex two samples were analysed. Quartz–sericite schist of the Fuente del Puma (sample URPR 68, Lat 34°29′25″ S; Long. 55°13′01″ W) was collected from a well-exposed outcrop on Road 60. A broad range of zircon ages was obtained: seven Archaean (2.6–3.4 Ga), six between 1.78 and 2.4 Ga, and five Neoproterozoic ages between 0.60 and 1.06 Ga, with three values close to the youngest age.

For the Zanja del Tigre Formation, considered by Sanchez-Bettucci (1998) as belonging to the Lavalleja Metamorphic Complex, a sample from a rhythmic meta-psammitic rock was collected on Road 12 (sample URPR 69, Lat 34°33′13″ S; 55°05′26″ W). For this sample, eight Archaean grains were found (four >3.0 Ga) and the others were all 1.8–2.3 Ga. A sole 1.4 Ga age was not taken into account, due to the uncertainty associated with its high degree of discordance. Thus, no definite Meso- or Neoproterozoic zircons were found in this sample.

Overview of the detrital zircon ages

Approximately 2.2 Ga detrital zircon is an important component in all of the Ribeira Belt samples. For two samples, ages older than 3.0 Ga were detected from the NW (Itaiacoca) and the SE (Capiru) borders of the belt. Only the Antinha Formation sample yielded *c.* 0.60 Ga zircon ages (interpreted as a volcanogenic contribution). The bimodal pattern of *c.* 2.2 Ga and some older zircon ages is characteristic of the Precambrian domains constituting the Paraná Basin basement in southeast Brazil (Bossi *et al.* 1992, 2004;

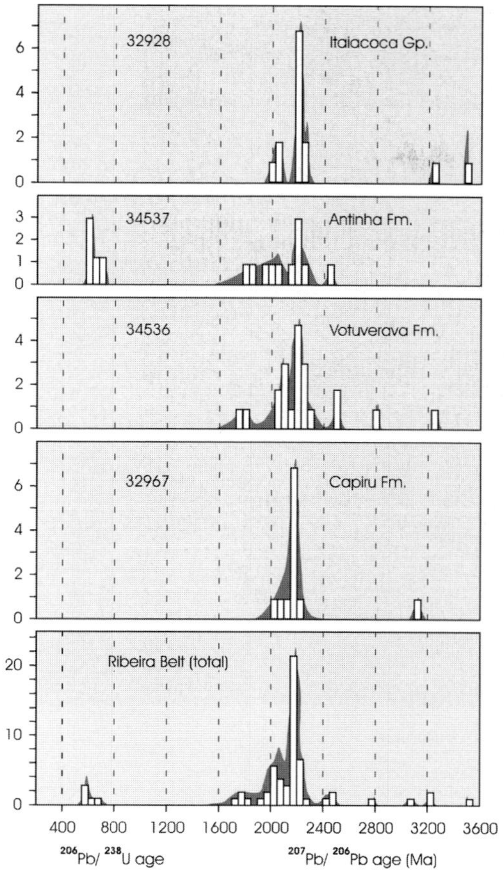

Fig. 5. Histogram of SHRIMP U–Pb and Pb–Pb ages for the Ribeira Belt detrital zircons. The lack of detrital zircons of Mesoproterozoic age is striking. The Neoproterozoic values are due to crystals of volcanic origin.

Hartmann *et al.* 1999, 2000*a*, *b*, *c*, 2001, 2003, 2004; Silva *et al.* 1999, 2000, Mallmann *et al.* 2003; Sanchez-Bettucci *et al.* 2004; Basei *et al.* 2000, 2005). As discussed below, the lack of 1.0–1.3 Ga Mesoproterozoic grains in the Ribeira Belt units is a striking feature.

The results for the Dom Feliciano Belt are similar to those for the Ribeira Belt and show that the Archaean–Palaeoproterozoic terranes were important sources. On the other hand, the *c.* 1.2 Ga Mesoproterozoic detrital zircons of the Dom Feliciano Belt stand out, because source areas of this age are unknown in southeastern Brazil. Moreover, for the sediments of the Lavalleja unit, approximately half of the grains have ages of 2.7–3.5 Ga, and as yet there has been no positive identification of source terranes of this age in southeastern Brazil and Uruguay.

In the three segments of the Dom Feliciano Belt, Neoproterozoic ages were observed, showing that deposition of a great part of the sequences took place during the Neoproterozoic Brasiliano cycle. This resolves a long-lasting controversy over the timing of their deposition.

Comparison with southwestern Africa

The geology of southwestern Africa is marked by the Kaoko, Damara, Gariep and Saldania Neoproterozoic belts (Fig. 8). Except for the internal Damara branch, these belts are parallel to the coast, defining the western margins of the Congo and Kalahari cratons (Fig. 8). In the Gariep Belt, *c.* 0.75 Ga and *c.* 0.58 Ga glacial events have been recognized (Frimmel *et al.* 2002), and the remains of a Neoproterozoic ocean floor have been identified in the Marmora terrane (Frimmel *et al.* 1996). The kinematic history of the Gariep Belt is characterized by nappe and thrust systems with east-southeastward tectonic transport, towards the Kalahari Craton. The maximum age of sedimentation for the basal continental rift phase is 771 ± 6 Ma, which is the youngest U–Pb single zircon age yet obtained from the underlying basement (Frimmel *et al.* 2001). Ar–Ar ages around 575 and 525 Ma, obtained on hornblende and micas, record an early, probably accretion-related metamorphic pulse and the peak of continental collision, respectively (Frimmel 1995; Frimmel & Frank 1998). Post-tectonic alkaline magmatism affected the central part of the belt. The best constraint available on the timing of this magmatism stems from a U–Pb single zircon age of 505 ± 6 Ma, obtained from the largest of the intrusive bodies, the Kuboos granite pluton (Frimmel *et al* 1996).

The ages of Gariep Belt detrital zircons were presented by Basei *et al.* (2005). That study was carried out on samples from the basal and outermost quartzite unit (Stinkfontein Subgroup) and from siliciclastic phyllites of the Oranjemund Formation, which represents syn-orogenic foredeep deposits that were laid down on top of, and in front of, the advancing oceanic crust now preserved as fragments in the Marmora terrane (Frimmel & Foelling 2004). These data are incorporated in the histogram of Figure 9, which also contains unpublished ages from rocks of the Damara Belt. The age pattern for the Gariep detrital zircons indicates that 1.0–1.2 Ga terranes were the main source areas. A few 0.6–0.8 Ga grains were detected in the samples both from the external and internal parts of the belt. This suggests that despite the lithological contrasts, the same detritus sources remained available for these two sectors. Detrital zircons with *c.* 0.6 Ga

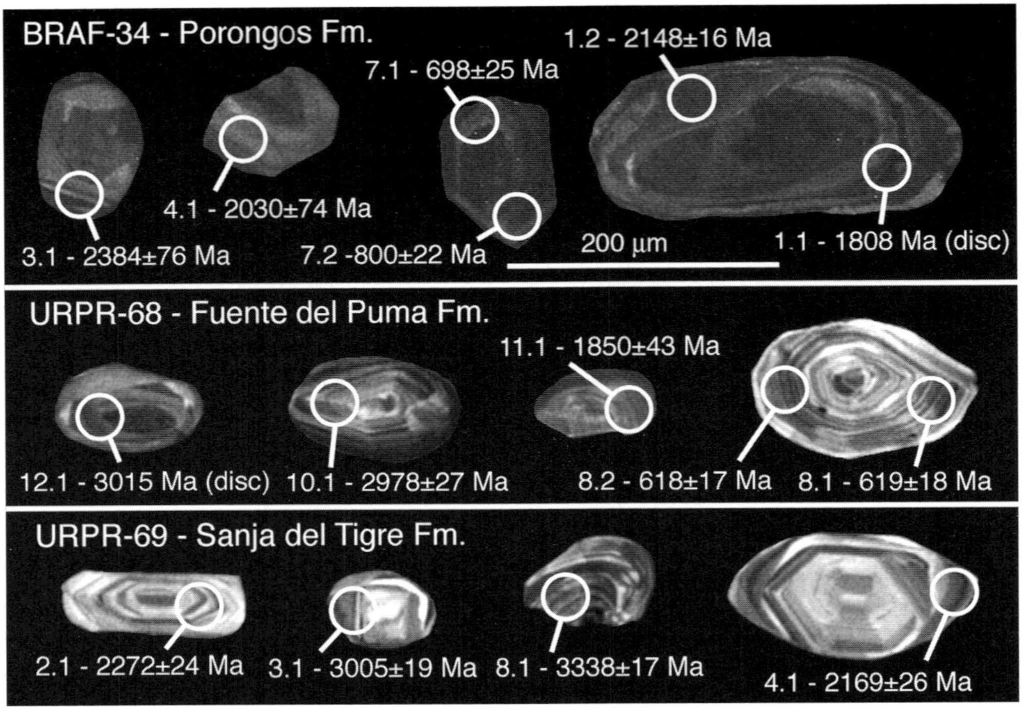

Fig. 6. Cathodo-luminescence images of Dom Feliciano Belt zircon grains.

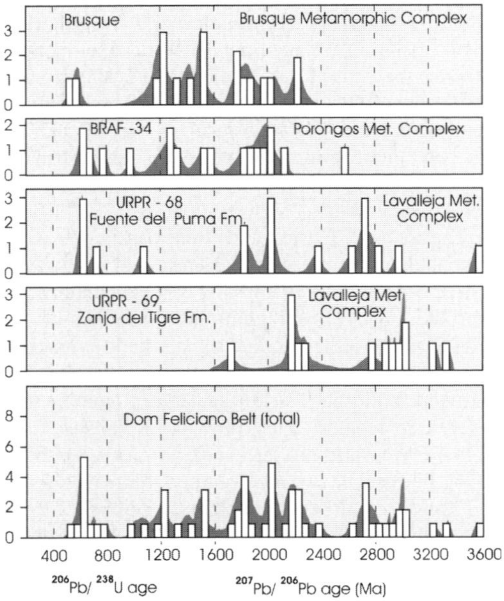

Fig. 7. Histogram of SHRIMP U–Pb and Pb–Pb ages for the Dom Feliciano Belt detrital zircons. Note some detrital zircons with ages around 1.2 Ga. Most of Archaean ages were obtained from samples of the Lavalleja units in Uruguay.

ages are concentrated in the western and youngest units of the Gariep Basin, notably in the Marmora terrane, and could have been derived from the 600 $\pm$ 10 Ma Florianópolis–Pelotas batholith (Phillip 1998; Sial *et al.* 1999; Silva *et al.* 1999, 2005; Basei *et al.* 2000).

For the Nosib Group of the Damara Belt, two grains have ages of *c.* 1.0 Ga, there is a main group at *c.* 1.2 Ga, and only two Palaeoproterozoic (1.8–1.9 Ga) grains were detected. Even though this small number of analyses allows only preliminary interpretations, it is noted that no 0.80 Ga zircon ages were found; detrital zircons of this age represent the Gariep Belt, where they correspond to the Richterveld Suite further south. Ages of the Damara Belt units are summarized in Figure 9 and representative CL images for the Nosib Group are displayed in Figure 10.

Age values similar to those presented in this paper have also been noted in other localities in western and southwestern Africa. Mesoproterozoic ages were obtained for detrital zircons from sandstones of the Damara Belt (sample NK 91, Mulden Group) in the northwestern region of Namibia, with detrital zircon populations indicating two main source areas of *c.* 2.0–1.75 Ga and 1.4–1.0 Ga. Metamorphic overgrowth dated at 572 $\pm$ 4 Ma was identified on

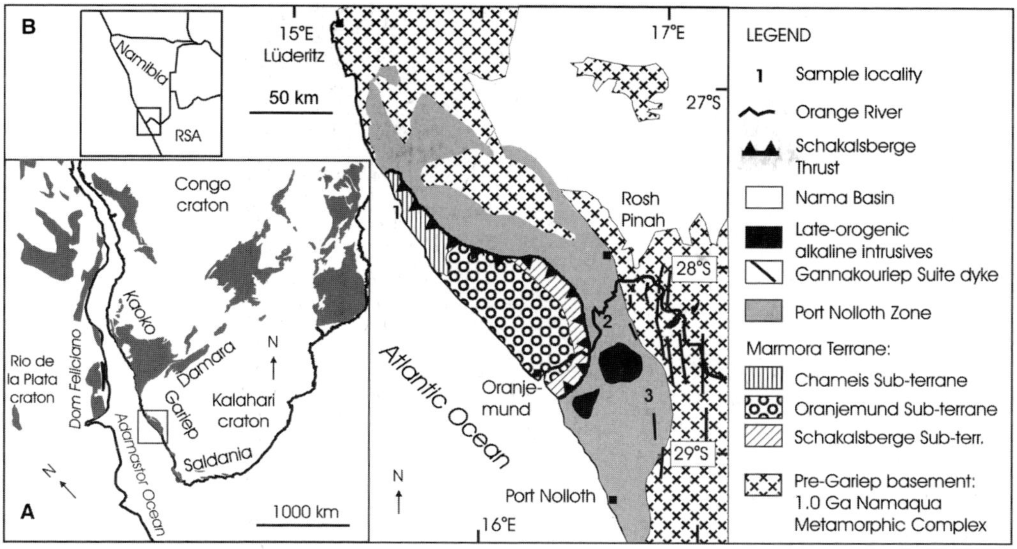

Fig. 8. Geological sketch of the Gariep Belt (after Frimmel *et al.* 2002).

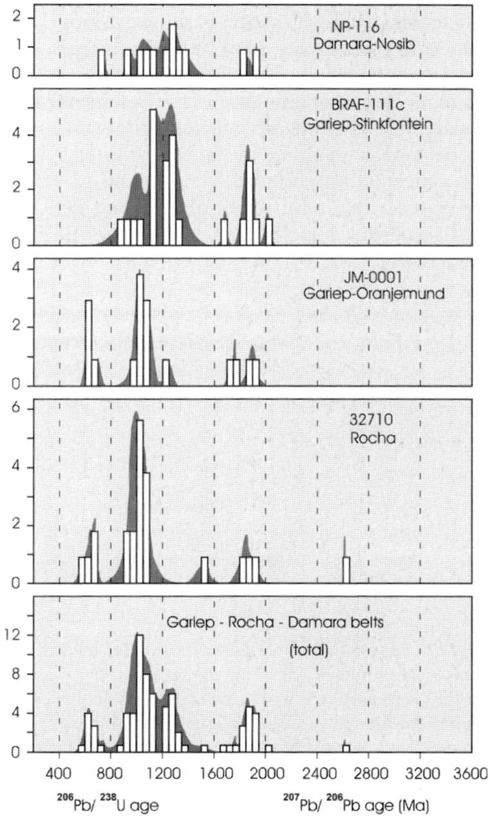

Fig. 9. Integrated histogram of the radiometric ages for detrital zircons from the Gariep and Damara belts.

zircon from the same sample (Goscombe *et al.* 2005). Further north, in the West Congo Belt source areas between 0.8 and 1.2 Ga have been identified for detrital zircons of the Sansikwa, Shiloango and Inkisi subgroups (Frimmel *et al.* 2006). The Inkisi (uppermost), in contrast to the lower groups, shows a predominantly 0.7–0.6 Ga Neoproterozoic detrital zircon population.

No Archaean zircon grains were found in any of the African samples studied, and most grains are <2.0 Ga. This strongly suggests that the basement of the major cratons to the east, such as Kalahari and Congo, were not source areas for the sediments that filled the Damara and Gariep basins. The source for these basins was their own basement, best represented by 1.0 and 1.2 Ga high-grade metamorphic rocks and granites of the Namaqua Metamorphic Complex. Interpretations of the Gariep Basin architecture suggest that the Namaqua complex constituted elevated terrains (Frimmel *et al.* 2002), and thus could be the main source area for the Gariep Basin sediments. At the same time the 'Namaqua Mountains' acted as a barrier for zircons originating in the eastern cratons.

Detrital zircons of 0.9–1.2 Ga are lacking in the southern Ribeira Belt and are rare in the Dom Feliciano Belt but dominate in the African belts. This suggests an African source for such zircons (Fig. 11). On the other hand, in the Central Ribeira Belt segment (Trouw *et al.* 2000; Heilbron *et al.* 2003), located *c.* 700 km NE of the study region in Rio de Janeiro and Minas Gerais states, detrital zircons with ages in the 2.0–2.2 Ga and

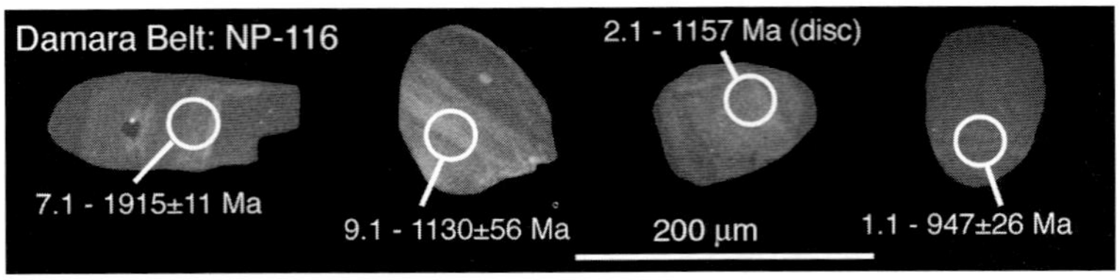

Fig. 10. Cathodo-luminescence images of the Damara Belt zircon grains.

1.0–1.4 Ga intervals are common (Machado & Gauthier 1996; Söllner & Trouw 1997; Valeriano *et al.* 2004; Valladares *et al.* 2004, 2005). Although some of these authors suggest that the possible Mesoproterozoic zircon sources are in the São Francisco Craton, we suggest here the possibility of an African source. Consequently, it is also questioned whether there is a direct correlation between the metasedimentary units that compose the Central Ribeira Belt domain with the occurrences in the south discussed in this paper. In the Búzios region (in the northeast of Rio de Janeiro State), detrital zircons with the same 2.0–2.2 Ga and 1.0–1.4 Ga ages are found in the high-grade metamorphic rocks of the Palmital metasedimentary cover of Palaeoproterozoic gneissic basement (Schmitt *et al.* 2003, 2004). An African source might also be explored for these zircons.

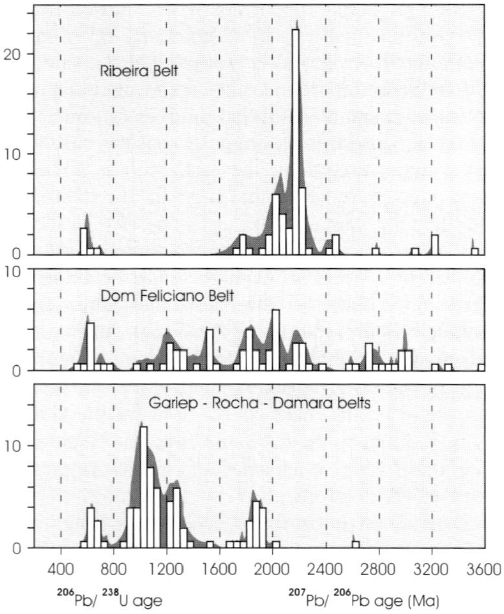

Fig. 11. Integrated histogram comparing the patterns of detrital zircon ages for the Ribeira, Dom Feliciano and Gariep/Damara belts. An eastward increase of age values in the 1.4–1.0 Ga interval is due to African source areas.

Nd model ages Nd (T_{DM}) and zircon provenance patterns

Nd isotopes are considered good indicators of average crustal ages. Regionally uniform Nd isotope signatures can reflect a common geological history and on a broad scale usually allow the identification of distinct crustal domains (Cordani *et al.* 2000).

In the histogram of Figure 12, available Nd (T_{DM}) ages are shown for a roughly west–east transect across the units observed on both sides of the Atlantic Ocean. The histogram shows that there is a conspicuous decrease in model ages eastwards; with the Damara Belt displaying the youngest values. This difference is more striking when some discrepant ages, resulting from problematic analyses (anomalous Sm–Nd ratios, etc.) are eliminated, or when the influence of the basement on the lower units of the metasedimentary pile is taken into account (as exemplified by the Damara Belt). There is a concentration of model ages around 2.0 Ga for the Supracrustal Schist Belt of the Dom Feliciano Belt, whereas for the Eastern Granite Belt the

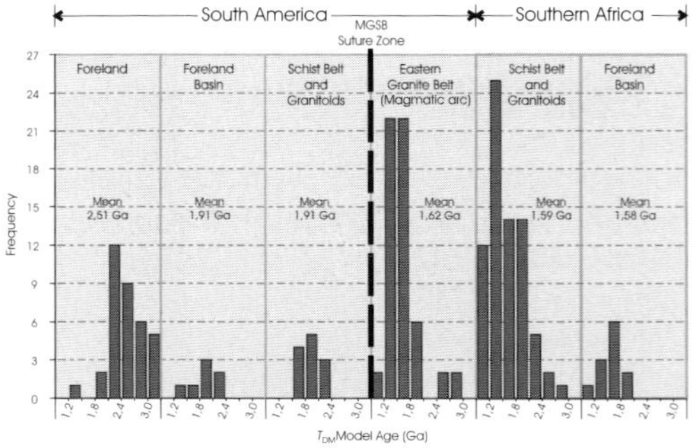

Fig. 12. Pattern of Nd model ages (T_{DM}) for the main geological units that constitute the terrains of southeastern South America and southwestern Africa. The Major Gercino–Sierra Ballena Suture Zone separates both groups of ages. Data sources: Guj (1970), Basei (1985), McDermott (1986), Mantovani *et al.* (1987), May (1990), McDermott & Hawkesworth (1990), Basei *et al.* (1998, 1999, 2000, 2001), Jung *et al.* (1998), Möller *et al.* (1998), Phillip (1998), Cordani & Sato (1999), Silva *et al.* (1999, 2000) and Harara (2001).

average is between 1.3 and 1.6 Ga. For the Damara Belt (mainly its granitoids), the youngest values are 1.1 Ga, but the average is also in the 1.3–1.6 Ga interval. Mesoproterozoic model ages predominate for rocks of the western portion of the Damara Belt (the region between Walvis Bay, Karibib, and the Huab river) with Palmental-type calc-alkaline granitoids, intraplate syenites and granites showing model ages of 1.1–1.5 Ga (McDermott 1986). A very similar pattern was found for the metaluminous A-type granitoids of the Damara central region (Jung *et al.* 1998). Metasedimentary rocks in parts of the Damara (notably Rössing and Kuiseb formations) and Nama (mainly Kuibis and Schwarzrand formations) have Nd model ages similar to those obtained for the Eastern Granite Belt, suggesting that this was an important source area. This corroborates an evolutionary model (Basei *et al.* 2005) that places the Gariep units in a back-arc basin environment and, consequently, the Nama Group on the hinterland side.

This similarity, which may represent an affinity of the source areas for the Eastern Granite Belt and the African portion, can be explained by the participation of similar sources in the generation of these materials. Therefore, the isotopic differences between the Eastern Granite Belt and those further west, even for those of similar crystallization ages, strengthen suggestions that the Major Gercino–Sierra Ballena lineament (Fig. 3) should be viewed as a Neoproterozoic suture (Basei *et al.* 2000, 2005). An important gravimetric anomaly along the lineament also supports this interpretation (Hallinan & Mantovani 1993).

Tectonic model

Several models for tectonic correlation across the southern Atlantic Ocean were proposed by Porada (1979, 1989), Torquato & Cordani (1981), Soares *et al.* (2000), Campos Neto (2000), Basei *et al.* (2000, 2005), Cordani *et al.* 2003, Frimmel & Folling (2004), Schmitt *et al.* (2003, 2004), Valeriano *et al.* (2004) and Silva *et al.* (2005), with the later ones proposed within the framework of West Gondwana assembly. Figure 13 shows hypothetical NW–SE sections from the Ribeira to the Gariep Belt. Figure 13a is a representation at *c.* 0.61 Ga, when subduction zones were active with magmatic arcs being generated. These arcs are now represented by the Cunhaporanga and Três Córregos batholiths in the Ribeira Belt, the Piên batholith between the Curitiba and Luís Alves terranes, and the Florianópolis–Pelotas–Aiguá batholith between the Dom Feliciano and Gariep/Damara belts on the African side. Figure 13b represents the tectonic situation at *c.* 0.53 Ga, just after the collisions during Gondwana assembly, and displays the location of the suture zones.

The westward collision between the Florianópolis–Pelotas–Aiguá batholiths and the Dom Feliciano supracrustal belt occurred at *c.* 0.60 Ga. However, on the African side, only at *c.* 545 Ma did eastward-directed nappes and regional metamorphism affect the supracrustal units (Frimmel & Frank 1998). This event was expressed on the South American side by reactivation of 0.60 Ga structures, and deformation in the foreland basins (Itajaí, Camaquã and Arroio del Soldado). The deformation of African foreland basins (e.g.

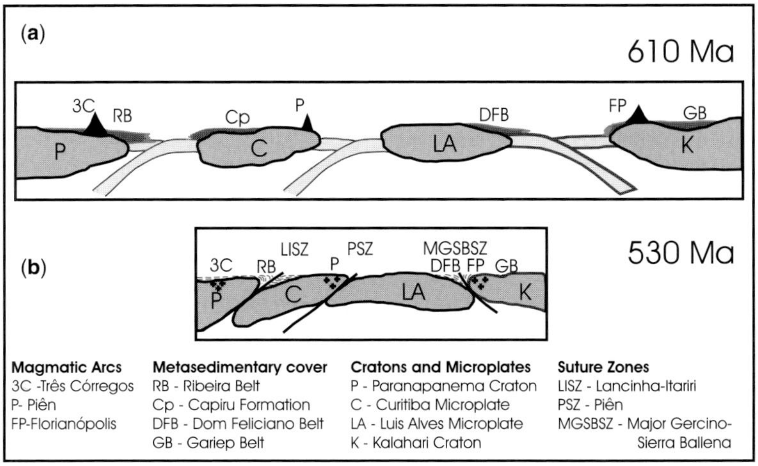

Fig. 13. Simplified tectonic model emphasizing the main geological units that were juxtaposed during collisions associated with the formation of Gondwana. Remnants of the oceanic crust are known only in the Piên region: (**a**) situation before collisions; (**b**) after collisions.

Nama) also started at around 0.54 Ga (older units) but continued through the Cambrian.

The terranes east of the Major Gercino–Sierra Ballena lineament (proposed suture) are interpreted as remnants of African terranes that were juxtaposed during the formation of West Gondwana. After opening of the South Atlantic, only small parts of these belts were preserved in South America.

Testing the tectonic model

As proposed by Basei *et al.* (2000, 2005) and discussed in this work, the Gariep–Rocha–Damara and Dom Feliciano belts represent units deposited at the opposite margins of the ocean that separated the African and South American pre-Gondwana palaeocontinents. In this tectonic context, the Dom Feliciano Belt represents a passive margin associated with the Luis Alves microplate and the Rio de la Plata Craton. However, deposition of the Gariep Belt metavolcano-sedimentary units took place in a back-arc basin related to the Florianópolis–Pelotas–Aiguá magmatic arc that was generated from the eastward subduction of oceanic crust under the Kalahari Craton (Fig. 13a).

If the proposed model is correct, all the units east of the Major Gercino–Sierra Ballena lineament would have African sources. This model, initially based on the pattern of Nd age distribution, is now supported by detrital zircon age data. This was further tested by dating zircons from a Queçaba Formation metasediment. The Quecaba Formation represents one of the few metasedimentary units that overlie the Florianópolis Batholith, east of the Major Gercino–Sierra Ballena lineament. It is in tectonic contact with the batholith rocks, and is composed of phyllites, quartzites and greywackes (Zanini *et al.* 1997). They are foliated and carry greenschist-facies assemblages. According to the model proposed above (Fig. 13), the Queçaba Formation should have an 'African' signature like the Gariep Belt, and should be different from the Dom Feliciano Belt.

As shown in Figure 14, an 'African' signature is evident for the Queçaba Formation, with Mesoproterozoic zircons of mostly *c.* 1.2 Ga. A second age peak falls in the 1.7–1.9 Ga interval, attributed to the Richtersveld terrane that represents pre-Namaqua basement granitoids. It is interesting to note that despite the proximity of the Florianópolis Batholith granitoids, the source area of Queçaba Formation sediments is not related to the erosion of these Neoproterozoic rocks, because no zircons in this age interval have been detected. Additionally, Nd model ages for these metasediments cluster around 1.6 Ga, differing from

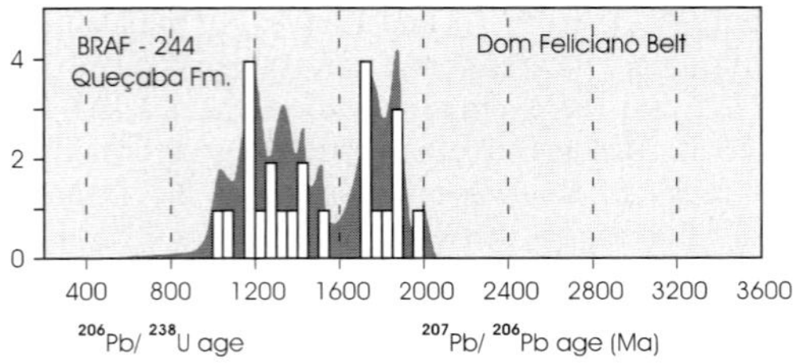

Fig. 14. Histogram of SHRIMP U–Pb ages for the Queçaba Formation detrital zircons. Note the main peak around 1.2 Ga and the lack of Neoproterozoic ages.

Palaeoproterozoic values around 2.0 Ga typical of the Brusque Metamorphic Complex to the west of the Major Gercino–Sierra Ballena lineament. The age pattern for the Queçaba Formation metasediment detrital zircons supports the model of Basei *et al.* (2000, 2005).

Conclusions

(1) The pattern of zircon U–Pb ages for the basement terrains throughout southeastern Brazil and Uruguay is marked by Archaean and Palaeoproterozoic values with noted lack of late Mesoproterozoic grains. Detrital ages of 0.9–1.2 Ga are considered an 'African' signature, typified by the Gariep and Damara belts and inherited from their basement of Namaqua Metamorphic Complex. This signature was only found in the Rocha Group in extreme southeastern Uruguay.

(2) The 'African' signature of the detrital zircon grains of the Rocha Group makes sense if this group is understood as a continuation of the Gariep Belt, now isolated in South America by the opening of the modern Atlantic.

(3) The Dom Feliciano Belt is regarded as a Neoproterozoic passive margin basin developed on the 'Brazilian side' of an ocean that originally separated it from the Gariep Belt (Fig. 13).

(4) The southern units of the Ribeira Belt are derived exclusively from South American sources, such as the Paranapanema Craton.

(5) For the Dom Feliciano Belt, detrital zircon grains from its three segments indicate Neoproterozoic ages, assigning deposition of these supracrustal units to the Brasiliano cycle.

(6) It is possible that the Zanja del Tigre Formation, rather than belonging to the Lavalleja Belt, constitutes part of the basement of this belt, with probable sedimentation in the Mesoproterozoic. This interpretation is based on the detrital zircon ages obtained in this study, on its higher metamorphic grade and its association with granitoids with ages of *c.* 1750 Ma.

The authors wish to thank H. J. Prazeres Filho and C. R. Passarelli for comments on earlier versions of the manuscript. The reviewers D. Gray and M. Heilbron are thanked for many helpful comments and suggestions. The authors thank the research funding agency of São Paulo State – Fundação de Amparo a Pesquisa do Estado de São Paulo (FAPESP process 2005/58688-1) for the financial support for the SHRIMP analyses and field work on both continents. A.P.N.'s participation in the zircon analyses for this project was for salary recovery whilst at RSES, ANU. This is a contribution to IGCP 478.

References

Basei, M. A. S. 1985. *O Cinturão Dom Feliciano em Santa Catarina*. PhD thesis, University of São Paulo, Brazil.

Basei, M. A. S., Campos Neto, M. C. & Siga, O., Jr. 1994. Geologia do Grupo Brusque na Região de Canelinhas, Sc. *In*: *Congresso Brasileiro de Geologia 38, Camboriú-SC, Extended Abstracts*, **1**. Sociedade Brasileira de Geologia, 243–244.

Basei, M. A. S., McReath, I. & Siga, O., Jr. 1998. The Santa Catarina granulite complex of southern Brazil, a review. *Gondwana Research*, **1**, 383–391.

Basei, M. A. S., Siga, O., Jr. *et al*. 1999. Paleoproterozoic granulitic belts of the Brazilian southern region (PR-SC) *In*: *II South American Symposium on Isotope Geology, Villa Carlos Paz, Argentina, Short Papers*, 291–294.

Basei, M. A. S., Siga, O., Jr., Masquelin, H., Harara, O. M., Reis Neto, J. M. & Preciozzi, F. 2000. The Dom Feliciano Belt (Brazil-Uruguay) and its foreland domain, the Rio de la Plata craton: framework, tectonic evolution and correlation with similar provinces of Southwestern Africa. *In*: Cordani, U. G., Milani, E. J., Thomaz Filho, A. & Campos, D. A. (eds) *Tectonic Evolution of South America*. 31st International Geological Congress. Rio de Janeiro, Brazil, 311–334.

Basei, M. A. S., Siga, O., Jr., Harara, O. M., Preciozzi, F., Sato, K. & Kaulfuss, G. 2001. Precambrian terranes of African affinities in the southeastern part of Brazil and Uruguay. *In*: *III South American Symposium on Isotope Geology, Pucón, Chile, Short Papers*, 98–101.

Basei, M. A. S., Frimmel, H. E., Nutman, A. P., Preciozzi, F. & Jacob, J. 2005. The connection between the Neoproterozoic Dom Feliciano (Brazil/Uruguay) and Gariep (Namibia/South Africa) orogenic belts. *Precambrian Research*, **139**, 139–221.

Bossi, J. & Campal, N. 1992. Magmatismo y tectônica transcurrente durante el Paleozoico Inferior en Uruguay. *In*: Gutierrez-Marco, J. G., Saavedra, J. & Rabano, I. (eds) *Paleozoico Inferior de Iberoamérica*. Universidad de Extremadura, Spain, 343–356.

Bossi, J. & Gaucher, C. 2004. The Cuchilla Dionisio terrane, Uruguay: an allochthonous block accreted in the Cambrian to SW-Gondwana. *Gondwana Research*, **7**, 661–674.

Brito Neves, B. B. & Cordani, U. C. 1991. Tectonic evolution of South America during Late Proterozoic. *Precambrian Research*, **33**, 23–40.

Brito Neves, B. B., Campos Neto, M. C. & Fuck, R. 1999. From Rodinia to Eastern Gondwana: an approach to the Brasiliano-Pan African cycle and orogenic collage. *Episodes*, **22**, 155–166.

Caldasso, A. L., da S., Krebs, A. S. J., Silva, M. A. S., da Camozzato, E. & Ramgrab, G. E. 1995*a*. *Programa de Levantamentos Geológicos Básicos, 1: 100.000, Folha Brusque (SG-22-Z-D-II-1), SC*. Companhia de Pesquisa de Recursos Minerais, Brasília.

Caldasso, A. L., da, S., Krebs, A. S. J., Silva, M. A. S., da, Camozzato, E. & Ramgrab, G. E. 1995*b*. *Programa de Levantamentos Geológicos Básicos, 1: 100.000, Folha Botuverá (SG-22-Z-D-I-2), SC*. Companhia de Pesquisa de Recursos Minerais, Brasília.

Campanha, G. A. C., Basei, M. A. S., Tassinari, C. C. G., Nutman, A. P. & Faleiros, S. N. 2006 A 590–575 Ma SHRIMP U-Pb zircon reconnaissance data for the base of Iporanga Formation, Ribeira Belt, SE Brazil: implications for regional Neoproterozoic crustal evolution, possible timing of a glaciation. *In*: *V South American Symposium on Isotope Geology, Punta del Este, Uruguay, Short Papers*, **1**, 65–67.

Campanha, G. A. C. & Sadowski, G. R. 1999. Tectonics of the southern portion of the Ribeira Belt (Apiaí Domain). *Precambrian Research*, **98**, 31–51.

Campanha, G. A. C., Gimenez Filho, A. *et al*. 1985. *Geologia das Folhas Iporanga (SG.22-X-B-V-2) e Gruta do Diabo (SG.22-X-B-VI-1), Estado de São Paulo*. Contrato IPT/Pró-Minério, (Rel. IPT 22352), São Paulo.

Campanha, G. A. C., Bistrichi, C. A. & Almeida, M. A. 1987. Considerações sobre a organização litoestratigráfica e evolução tectônica da Faixa de Dobramentos Apiaí. *In*: *III Simpósio Sul-Brasileiro de Geologia. Curitiba*, **2**. Sociedade Brasileira de Geologia, 725–742.

Campos Neto, M. C. 2000. Orogenic Systems from Southwestern-Gondwana: an approach to Brasiliano-Pan African cycle and orogenic collage in southeastern-Brazil. *In*: Cordani, U. G., Milani, E. J., Thomaz Filho, A. & Campos, D. A. (eds) *Tectonic Evolution of South America. 31st International Geological Congress, Rio de Janeiro, Brazil*, 335–365.

Campos Neto, M. C. & Figueiredo, M. C. H. 1995. The Rio Doce orogeny, southeastern Brazil. *Journal of South American Earth Sciences*, **8**, 143–162.

Castro, N. A., de, Basei, M. A. S. & Crósta, A. P. 1999. The W (Sn–Mo) specialized Catinga and other intrusive granitoids in the Brusque Group, Neoproterozoic of the state of Santa Catarina, southern Brazil. *Revista Brasileira de Geociências*, **29**, 17–26.

Compston, W., Williams, I. S. & Meyer, C. 1984. U–Pb geochronology of zircons from lunar breccia 73217 using a sensitive high mass-resolution ion microprobe. *Journal of Geophysical Research*, **89**, Supplement, 525–534.

Cordani, U. G. & Sato, K. 1999. Crustal Evolution of the South American platform, based on Nd isotopic systematics on granitoid rocks. *Episodes*, **22**, 167–173.

CORDANI, U. G., SATO, K., TEIXEIRA, W., TASSINARI, C. G. C. & BASEI, M. A. S. 2000. Crustal evolution of the South American platform. *In*: CORDANI, U. G., MILANI, E. J., THOMAZ FILHO, A. & CAMPOS, D. A. (eds) *Tectonic Evolution of South America. 31st International Geological Congress, Rio de Janeiro, Brazil*, 9–40.

CORDANI, U. G., BRITO NEVES, B. B. & D'ÁGRELLA-FILHO, M. S. 2003. From Rodinia to Gondwana: a review of the available evidence from South America. *Gondwana Research*, **6**, 275–283.

CPRM 1998. *Projeto Folha Curitiba (SG.22-X-D I) Final Report*. Companhia de Pesquisa de Recursos Minerais, São Paulo.

CUMMING, G. L. & RICHARDS, J. R. 1975. Ore lead isotope ratios in a continuously changing Earth. *Earth and Planetary Science Letters*, **28**, 155–171.

CURY, L. F., KAULFUSS, G. A., SIGA, O., JR., BASEI, M. A. S., HARARA, O. M. M. & SATO, K. 2002 Idades U-Pb (zircões) de 1.75 Ga em granitóides alcalinos deformados dos Núcleos Betara e Tigre: evidências de registros do Estateriano na Faixa Apiaí. *Geologia USP, Série Científica*, **2**, 95–108.

FASSBINDER, E. 1996. A unidade Água Clara no contexto do Grupo Açungui: um modelo transpressivo de colisão oblíqua no Neoproterozóico paranaense. PhD thesis, University of São Paulo.

FERNANDES, L. A. D., MENEGAT, R. *ET AL*. 1995. Evolução Tectônica do cinturão Dom Feliciano no escudo sul-riograndense: Parte I—uma contribuição a partir do registro geológico. *Revista Brasileira de Geociências*, **25**, 351–374.

FIORI, A. P. 1992. Tectônica e estratigrafia do Grupo Açungui, PR. *Boletim IG-USP, Série Científica*, **23**, 55–74.

FRANTZ, J. C. & JOST, H. 1983. Petrologia dos granitos estaníferos do Rio Grande do Sul. *In*: *I Simpósio Sul-Brasileiro de Geologia, Porto Alegre, Brazil*, 49–67.

FRIMMEL, H. E. 1995. Metamorphic evolution of the Gariep Belt. Southern Africa. *Journal of Geology*, **98**, 176–190.

FRIMMEL, H. E. & FÖLLING, P. G. 2004. Late Vendian closure of the Adamastor Ocean: timing of tectonic inversion and syn-orogenic sedimentation in the Gariep Basin. *Gondwana Research*, **7**, 685–699.

FRIMMEL, H. E. & FRANK, W. 1998. Neoproterozoic tectono-thermal evolution of the Gariep Belt and its basement, Namibia/South Africa. *Precambrian Research*, **90**, 1–28.

FRIMMEL, H. E., HARTNADY, C. J. H. & KOLLER, F. 1996. Geochemistry and tectonic setting of magmatic units in the Pan-African Gariep belt, Namibia. *Chemical Geology*, **130**, 101–121.

FRIMMEL, H. E., ZARTMAN, R. E. & SPÄTH, A. 2001. Dating Neoproterozoic continental break-up in the Richtersveld Igneous Complex, South Africa. *Journal of Geology*, **109**, 493–508.

FRIMMEL, H. E., FÖLLING, P. G. & ERIKSSON, P. G. 2002. Neoproterozoic tectonic and climatic evolution recorded in the Gariep Belt, Namibia and South Africa. *Basin Research*, **14**, 55–67.

FRIMMEL, H. E., TACK, L., BASEI, M. A. S., NUTMAN, P. & BOVEN, A. 2006. Provenance and chemostratigraphy of the Neoproterozoic West Congolian Group in the Democratic Republic of Congo. *Journal of African Earth Sciences*, **46**, 221–239.

GAUCHER, C. 2000. Sedimentology, palaeontology and stratigraphy of the Arroyo del Soldado Group (Vendian to Cambrian, Uruguay). *Beringeria*, **26**, 1–120.

GAUCHER, C., SPRECHMANN, P. & SCHIPILOV, A. 1996. Upper and Middle Proterozoic fossiliferous sedimentary sequences of the Nico Perez terrane of Uruguay: Lithostratigraphic units, paleontology, depositional environments and correlations. *Neues Jahrbuch für Geologie und Paläontologie Abhandlung*, **199**, 339–367.

GAUCHER, C., CHIGLINO, L. & PECOITS, E. 2004. Southernmost exposures of the Arroyo del Soldado Group (Vendian to Cambrian, Uruguay): Paleogeographic implications for the amalgamation of W-Gondwana. *Gondwana Research*, **7**, 701–714.

GOSCOMBE, B., GRAY, D., ARMSTRONG, R., FOSTER, D. A. & VOGL, J. 2005. Event geochronology of the Pan-African Kaoko Belt. *Precambrian Research*, **140**, 103–131.

GUJ, P. 1970. *The Damara mobile belt in the southwestern Kaokoveld, South West Africa*. Bulletin of the Precambrian Research Unit, University Cape Town, **10**.

HALINANN, S. E. & MANTOVANI, S. M. M. 1993. Structural Framework of the Southern Brazilian Shield: the perspective from the gravity models. *In*: *III International Geophysical Congress, Anais*, **2**. Sociedade Brasileira de Geofísica, Rio de Janeiro, 1078, 1083.

HARARA, O. M. M. 2001. *Mapeamento e investigação petrológica e geocronológica dos litotipos da região do Alto Rio Negro (Pr-SC): Um exemplo de sucessivas e distintas atividades magmáticas durante o Neoproterozóico*. PhD thesis, Institute of Geosciences, University of São Paulo.

HARTMANN, L. A., LEITE, J. A. D., MCNAUGHTON, N. J. & SANTOS, J. O. S. 1999. Deepest exposed crust of Brazil-SHRIMP establishes three events. *Geology*, **27**, 947–950.

HARTMANN, L. A., PIÑEYRO, D., BOSSI, J. & LEITE, J. A. D. 2000*a*. Zircon U–Pb SHRIMP dating of Palaeoproterozoic Isla Mala granitic magmatism in the Rio de la Plata Craton, Uruguay. *Journal of South American Earth Sciences*, **13**, 105–113.

HARTMANN, L. A., LEITE, J. A. D., SILVA, L. C., REMUS, M. V. D. & MCNAUGHTON, N. J. 2000*b*. Advances in SHRIMP geochronology and their impact on understanding the tectonic and metallogenic evolution of southern Brazil. *Australian Journal of Earth Sciences*, **47**, 829–844.

HARTMANN, L. A., SANTOS, J. O. S., MCNAUGHTON, N. J., VASCONCELLOS, M. A. Z. & SILVA, L. C. 2000*c*. SHRIMP dates recurrent granulite facies metamorphism in the Santa Catarina granulites, southern Brazil. *Anais Academia Brasileira de Ciências*, **72**, 559–572.

HARTMANN, L. A., CAMPAL, N., SANTOS, J. O., MACNAUGHTON, N. J. & SCHIPILOV, A. 2001. Archean crust in the Rio de la Plata Craton, Uruguay: SHRIMP U–Pb reconnaissance geochronology. *Journal of South American Earth Sciences*, **14**, 557–570.

HARTMANN, L. A., BITENCOURT, M. F., SANTOS, J. O. S., MCNAUGHTON, N. J., RIVERA, C. B. & BETIOLLO, L. 2003 Prolonged Paleoproterozoic magmatic participation in the Neoproterozoic Dom Feliciano belt, Santa Catarina, Brazil, based on zircon U–Pb SHRIMP geochronology. *Journal of South American Earth Sciences*, **16**, 477–492.

HARTMANN, L. A., PHILIPP, R. P., LIU, D., WAN, Y., WANG, Y., SANTOS, J. O. S. & VASCONCELLOS, M. A. Z. 2004. Paleoproterozoic provenance of detrital zircon, Porongos Complex quartzites, southern Brazilian Shield. *International Geology Review*, **46**, 127–157.

HEILBRON, M. & MACHADO, N. 2003. Timing of terrane accretion in the Neoproterozoic-Eopaleozoic Ribeira belt SE Brazil. *Precambrian Research*, **125**, 87–112.

JANASI, V. A., LEITE, R. J. & VAN SCHMUS, W. R. 2001. U–Pb chronostratigraphy of the granitic magmatism in the Agudos Grandes Batholith (west of São Paulo, Brazil)—implications for the evolution of the Ribeira Belt. *Journal of South American Earth Sciences*, **14**, 363–376.

JOST, H. 1981. *Geology and Metallogeny of the Santana da Boa Vista region, South Brazil.* PhD thesis, University of Athens, Georgia, USA.

JOST, H. 1982. Condições de metamorfismo de uma fração da faixa de Dobramentos Tijucas no Rio Grande do Sul. *Acta Geológica Leopoldênsia*, **4**, 27–60.

JUNG, S., MEZGER, K. & HOERNES, S. 1998. Petrology and geochemistry of syn- to post-collisional metaluminous A-type granites—a major and trace element and Nd–Sr–Pb–O–isotope study from the Proterozoic Damara Belt, Namibia. *Lithos*, **45**, 147–175.

LUDWIG, K. R. 2001. *Using Isoplot/Ex. A geochronological toolkit for Microsoft Excel.* Berkeley Geochronology Center, Special Publications No. 1, Berkeley, USA.

MACHADO, N. & GAUTHIER, G. 1996. Detrmination of $^{207}Pb^{206}Pb$ ages on zircon and monazite by laser-ablation ICPMS and application to a study of metasedimentary provenance and metamorphism in south-eastern Brazil. *Geochemica et Cosmochimica Acta*, **60**, 5063–5073.

MALLMANN, G., CHEMALE, Jr. F., ARMSTRONG, R. & KAWASHITA, K. 2003. Sm–Nd and U–Pb SHRIMP zircon studies of the Nico Perez Terrane, reworked Rio de la Plata Craton, Uruguay. *In*: *IV South American Symposium on Isotope Geology, Salvador, Short Papers*, **1**, 207–209.

MANTOVANI, M. S. M. & BRITO NEVES, B. B. 2005. The Paranapanema Listhospheric Block: Its importance for Proterozoic (Rodinia, Gondwana) supercontinent theories. *Gondwana Research*, **8**, 303–315.

MANTOVANI, M. S. M., HAWKESWORTH, C. J. & BASEI, M. A. S. 1987. Nd and Pb isotope studies bearing on the crustal evolution of southeastern Brazil. *Revista Brasileira de Geociências*, **17**, 263–269.

MAY, S. E., 1990. *Pan-African Magmatism and Regional Tectonics of South Brazil.* PhD thesis, The Open University, UK.

MCDERMOTT, P. F. 1986. *Granite petrogenesis and crustal evolution studies in the Damara Pan-African orogenic belt, Namibia.* PhD thesis, The Open University, UK.

MCDERMOTT, P. F. & HAWKESWORTH, C. J. 1990. Intracrustal recycling and upper-crustal evolution: a case study from the Pan–African Damara mobile belt, central Namibia. *Chemical Geology*, **83**, 263–280.

MCCLAREN, A. C., FITZGERALDS, J. D. & WILLIAMS, I. S. 1994. The microstructure of zircon and its influence on the age determination from Pb/U isotopic ratios measured by ion microprobe. *Geochimica et Cosmochimica Acta*, **58**, 993–1005.

MARQUES, J. C., JOST, H., ROISENBERG, A. & FRANTZ, J. C. 1998*a*. Eventos ígneos da Suíte Metamórfica Porongos na área da Antiforme Capané, Cachoeira do Sul, southern Brazil, *Revista Brasileira de Geociências*, **28**, 419–430.

MARQUES, J. C., JOST, H., ROISENBERG, A. & FRANTZ, J. C. 1998*b*. Rochas metasedimentares, geologia estrutural e metamorfismo da suíte metamórfica Porongos na área da Antiforme Capané, Cachoeira do Sul–RS. *Revista Brasileira de Geociências*, **28**, 467–47.

MÖLLER, A., MEZGER, K. & SCHENK, V. 1998. Crustal age domains and the evolution of the continental crust in the Mozambique belt of Tanzania: combined Sm–Nd, Rb–Sr and Pb-Pb isotopic evidence. *Journal of Petrology*, **39**, 749–783.

PASSARELLI, C. R., PRAZERES FILHO, H. J., SIGA, O., JR., BASEI, M. A. S. & CAMPOS NETO, M. C. 2003. Geochronology and isotope geology of the Precambrian terranes of Southern São Paulo State, Brazil. *In*: *IV South American Symposium on Isotope Geology, Salvador, Short Papers*, **2**, 635–638.

PHILLIP, R. P. 1998. A Evolução geológica e Tectônica do Batolito Pelotas no Rio Grande do Sul. PhD thesis, University of São Paulo.

PORADA, H. 1979. The Damara-Ribeira orogen of the PanAfrica-Brasiliano cycle in Namibia (Southwest Africa) and Brazil as interpreted in terms of continental collision. *Tectonophysics*, **57**, 237–265.

PORADA, H. 1989. Pan–African rifting and orogenesis in southern to equatorial Africa and eastern Brazil. *Precambrian Research*, **44**, 103–136.

PORCHER, C. C., MCNAUGHTON, N. J., LEITE, J. A. D., HARTMANN, L. A. & FERNANDES, L. A. D. 1999. Idade Shrimp em zircão: vulcanismo ácido do Complexo Metamórfico Porongos. *In*: *I Simpósio sobre Vulcanismo e Ambientes Associados, Abstracts, Gramado*, 110.

PRAZERES FILHO, H. J. 2005. *Caracterização Geológica e Petrogenética do Batolito Granítico Três Córregos (Pr-Sc): geoquímica isotópica (Nd-Sr-Pb), idades (ID-TIMS/SHRIMP) e ^{18}O em Zircão.* PhD thesis, Institute of Geosciences, University of São Paulo.

REIS NETO, J. M. 1994. *Faixa Itaiacoca: registro de uma colisão entre dois blocos continentais no Neoproterozóico.* PhD thesis, University of São Paulo.

SANCHEZ-BETTUCCI, L. 1998. *Evolución Tectónica del cinturón Dom Feliciano en la región Minas-Piriápolis, Republica Oriental del Uruguay.* PhD thesis, University of Buenos Aires, Argentina.

SANCHEZ-BETTUCCI, L., OYHANTÇABAL, P., LOUREIRO, J., RAMOS, V. A., PRECIOZZI, F. & BASEI, M. A. S. 2004. Mineralizations of the Lavalleja Group (Uruguay), a probable Neoproterozoic volcano-sedimentary sequence. *Gondwana Research*, **7**, 745–751.

SANDER, A. 1992. *Petrologia e litogeoquímica de uma parcela da sequência vulcano-sedimentar do complexo metamórfico Brusque na região do Ribeirão do Ouro, SC.* PhD thesis, Federal University of Rio Grande do Sul, Brazil.

SCHMITT, R. S., PIMENTEL, M. M., VAN SCHMUS, W. R., TROUW, R. A. J. & ARMSTRONG, R. A. 2003. Marine sedimentation related to the latest stages of Gondwana assembly in the Ribeira Belt: new U–Pb data. *In*: *IV South American Symposium on Isotope Geology, Salvador, Short Papers*, **1**, 294–297.

SCHMITT, R. S., TROUW, R. A. J., VAN SCHMUS, W. R. & PIMENTEL, M. M. 2004. Late amalgamation in the central part of West Gondwana: new geochronological data and the characterization of a Cambrian orogeny in the Ribeira Belt – SE Brazil. *Precambrian Research*, **133**, 29–61.

SIAL, A. N., DALL'AGNOL, R., FERREIRA, V. P., NARDI, L. V. S., PIMENTEL, M. M. & WIEDEMANN, C. M. 1999. Precambrian granitic magmatism in Brazil. *Episodes*, **22**, 191–198.

SIGA, O., JR., BASEI, M. A. S. *ET AL.* 1997. Ages and tectonic setting of alkaline — peralkaline granitoids of Paraná and Santa Catarina states, southern Brazil. *In*: *South American Symposium on Isotope Geology, Campos do Jordão, Brazil, Short Papers*, 301–303.

SIGA, O., JR., BASEI, M. A. S. *ET AL.* 1999. Post-orogenic magmatism and sedimentation in Neoproterozoic extensional regimes in the Brazilian southern region. *In*: *II South American Symposium on Isotope Geology, Villa Carlos Paz, Argentina, Short Papers*, 367–370.

SIGA, O., JR., BASEI, M. A. S. *ET AL.* 2003. U–Pb (zircon) Ages of metavolcanic rocks from the Itaiacoca Group: tectonic implications. *Revista Geologia — Institute of Geosciences-USP, Serie Cientifica*, **3**, 39–50.

SIGA, O., JR., BASEI, M. A. S. *ET AL.* 2006. Geochronology of the Itaiacoca Belt (Parana – Brazil): Tectonic Implications. *In*: *V South America Symposium on Isotope Geology, Punta del Este, Uruguay, Short Papers*, **1**, 186–189.

SILVA, L. C., OLIVEIRA, J. M. P., AUMOND, J. J., LOPES, R. M. M., EIPPER, J. & FERRO, G. 1985. Caracterização petrográfica da Sequência (Meta) Vulcano-sedimentar Rio do Oliveira (Cinturão do Itajaí Mirim, SC) *In*: *II Simpósio Sul-Brasileiro de Geologia, Florianópolis*. Sociedade Brasileira de Geologia, 11–23.

SILVA, P. C. S., VASCONCELLOS, C. V. S., YAMATO, A. A. & PEDREIRA, A. J. 1998. O Grupo Açungui na Folha Curitiba (SG22-X-D-I). *In*: *Congresso Brasileiro de Geologia 40, Belo Horizonte, Anais*, **1**. Sociedade Brasileira de Geologia, 40.

SILVA, L. C., HARTMANN, L. A., MCNAUGHTON, N. J. & FLETCHER, I. R. 1999. Shrimp U/-Pb Zircon dating of Neoproterozoic granitic magmatism and collision in the Pelotas Batholith, southernmost Brazil. *International Geology Review*, **41**, 531–551.

SILVA, L. C., HARTMANN, L. A., MCNAUGHTON, N. J. & FLETCHER, I. 2000. Zircon U–Pb SHRIMP dating of a Neoproterozoic overprint in Paleoproterozoic granitic-gneissic terranes, southern Brazil. *American Mineralogist*, **85**, 649–667.

SILVA, L. C., MCNAUGHTON, N. J., ARMSTRONG, R., HARTMANN, L. A. & FLETCHER, I. R. 2005. The Neoproterozoic Mantiqueira Province and its African connections: a zircon-based U–Pb geochronologic subdivision for the Brasiliano/Pan-African system of orogens. *Precambrian Research*, **136**, 203–240.

SIRCOMBE, K. N. 2000. Quantitative comparison of large data sets of geochronological data using multivariate analysis: a provenance study example from Australia. *Geochimica et Cosmochimica Acta*, **64**, 1593–1616.

SÖLLNER, F. & TROUW, R. A. J. 1997. The Andrelandia depositional cycle (Minas Gerais, Brazil), a post-Transamazonic sequence south of the São Francisco Craton: evidence from U–Pb dating on zircons of a metasediment. *Journal of South American Earth Sciences*, **10**, 21–28.

SOARES, P. C., FIORI, A. P. & ROSTIROLLA, S. P. 2000. A geotectonic view of the Ribeira and Dom Feliciano belts. *Revista Brasileira de Geociências*, **30**, 130–134.

STERN, R. A. 1998. High resolution SIMS determination of radiogenic trace-isotope ratios in minerals. *In*: CABRI, L. J. & VAUGHAN, D. J. (eds) *Modern Approches to Ore and Environmental mineralogy*. Short Course Series, Mineralogical Association of Canada, 241–268.

TORQUATO, J. R. & CORDANI, U. C. 1981. Brazil-Africa geological links. *Earth Science Reviews*, **17**, 155–176.

TROUW, R. A. J., HEILBRON, M. *ET AL.* 2000. The central segment of the Ribeira belt. *In*: CORDANI, U. G., MILANI, E. J., THOMAZ FILHO, A. & CAMPOS, D. A. (eds) *Tectonic Evolution of South America, 31st International Geological Congress, Rio de Janeiro, Brazil*, 287–310.

VALERIANO, C. M., MACHADO, N., SIMONETTI, A., VALLADARES, C. S., SEER, H. J. & SIMÕES, L. S. 2004. U–Pb Geochronology of the Southern Brasília belt (SE Brasil): sedimentary provenance, Neoproterozoic orogeny and assembly of Western Gondwana. *Precambrian Research*, **130**, 27–55.

VALLADARES, C. S., MACHADO, N., HEILBRON, M. & GAUTHIER, G. 2004. Ages of detrital zircon from siliciclastic successions of the Brasília belt, southern border of the São Francisco craton, Brazil: implications for the evolution of Proterozoic basins. *Gondwana Research*, **7**, 913–921.

VALLADARES, C. S., MACHADO, N., HEILBRON, M. & GAUTHIER, G. 2005. A Mesoproterozoic source at southern São Francisco Craton. *In*: *III Simposio sobre o Craton do Sao Francisco, Short Papers*, 263–266.

WEBER, W., SIGA, O., JR., SATO, K., REIS NETO, J. M., BASEI, M. A. S. & NUTMAN, A. P. 2004. A Formação Água Clara na região de Araçaíba-SP: registro de uma bacia Mesoproterozóica. *Boletim do Instituto de Geociências-USP*, **4**, 101–110.

WILLIAMS, I. S. 1998. U–Th–Pb geochronology by ion microprobe. *In*: MCKIBBEN, M. A., SHANKS, W. C. III. & RIDLEY, W. I. (eds) *Applications of microanalytical techniques to understanding mineralising processes.* Society of Economic Geologists, Reviews in Economic Geology, **7**, 1–35.

ZANINI, L. F. P., BRANCO, P. M., CAMOZZATO, E. & RAMGRAB, G. E. 1997. *Programa de Levantamentos Geológicos Básicos 1: 100.000, Folhas Florianópolis e Lagoa*. CPRM, Brasília.

A Damara orogen perspective on the assembly of southwestern Gondwana

D. R. GRAY[1], D. A. FOSTER[2], J. G. MEERT[2], B. D. GOSCOMBE[3], R. ARMSTRONG[4], R. A. J. TROUW[5] & C. W. PASSCHIER[6]

[1]*School of Earth Sciences, University of Melbourne, Melbourne, 3010, Vic. Australia (Now at Geostructures Pty Ltd, Melbourne) (e-mail: drgray@unimelb.edu.au)*

[2]*Department of Geological Sciences, University of Florida, Gainesville, FL, 32611-2120, USA*

[3]*School of Earth and Environmental Sciences, The University of Adelaide, Adelaide 5005, SA. Australia*

[4]*School of Earth Sciences, The Australian National University, Canberra, Australia*

[5]*Instituto de Geociências, Universidade Federal do Río de Janeiro, 21949-900 Río de Janeiro, Brazil*

[6]*Institut für Geoswissenschaften, Johannes Gutenberg University, 55099 Mainz, Germany*

Abstract: The Pan-African Damara orogenic system records Gondwana amalgamation involving serial suturing of the Congo–São Francisco and Río de la Plata cratons (North Gondwana) from 580 to 550 Ma, before amalgamation with the Kalahari–Antarctic cratons (South Gondwana) as part of the 530 Ma Kuunga–Damara orogeny. Closure of the Adamastor Ocean was diachronous from the Araçuaí Belt southwards, with peak sinistral transpressional deformation followed by craton overthrusting and foreland basin development at 580–550 Ma in the Kaoko Belt and at 545–530 Ma in the Gariep Belt. Peak deformation/metamorphism in the Damara Belt was at 530–500 Ma, with thrusting onto the Kalahari Craton from 495 Ma through to 480 Ma. Coupling of the Congo and Río de la Plata cratons occurred before final closure of the Mozambique and Khomas (Damara Belt) oceans with the consequence that the Kuunga suture extends into Africa as the Damara Belt, and the Lufilian Arc and Zambezi Belt of Zambia. Palaeomagnetic data indicate that the Gondwana cratonic components were in close proximity by *c.* 550 Ma, so the last stages of the Damara–Kuunga orogeny were intracratonic, and led to eventual outstepping of deformation/metamorphism to the Ross–Delamerian orogen (*c.* 520–500 Ma) along the leading edge of the Gondwana supercontinental margin.

Understanding supercontinent reconstruction requires detailed knowledge of the orogens that bind the former continental fragments together. Apart from knowledge of palaeomagnetic poles for the constituent cratonic masses, this includes the component lithofacies, the gross crustal architecture, the geometry of the major fault and shear zones as well as the thermal and temporal aspects of deformation, metamorphism and magmatism. For West Gondwana supercontinent construction (Fig. 1) this requires understanding of Brasiliano/Pan-African orogenesis, as West Gondwana is made of a mosaic of cratons linked by a complex set of Pan-African/Brasiliano fold belts (Fig. 2).

The Pan-African Damara orogen of Namibia (Fig. 3) reflects part of the West Gondwana suture. It provides connection between the Brasiliano orogens of South America through the Ribeira and Dom Feliciano belts of southern Brazil (Fig. 2) and is related to convergence between the Río de la Plata and the Congo and Kalahari cratons of South America and Southern Africa (e.g., Prave 1996). The Damara orogen consists of three component arms that define a three-pronged orogenic system or collisional triple junction (Coward 1981, 1983; Hoffman *et al.* 1994). These component fold belts are the NNW-trending northern coastal arm or Kaoko Belt, the S-trending southern coastal arm or Gariep Belt and the ENE-trending Inland Branch or Damara Belt (e.g., Kröner 1977; Martin & Porada 1977*a*, *b*; Miller 1983*a*). The Damara Belt extends under cover into Botswana and ultimately links with the Lufilian Arc and the Zambezi, Mozambique and Lurio belts (see Goscombe *et al.* 2000; Hanson 2003).

From: PANKHURST, R. J., TROUW, R. A. J., BRITO NEVES, B. B. & DE WIT, M. J. (eds) *West Gondwana: Pre-Cenozoic Correlations Across the South Atlantic Region.* Geological Society, London, Special Publications, **294**, 257–278. DOI: 10.1144/SP294.14 0305-8719/08/$15.00

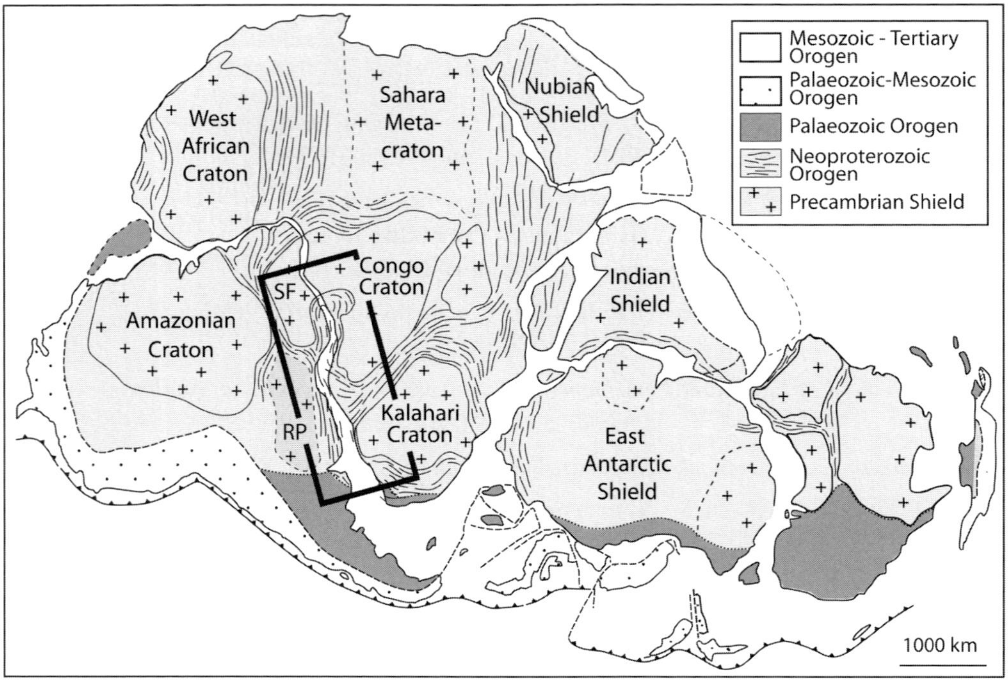

Fig. 1. Map of Gondwana showing the positions of the cratonic nuclei and the orogenic belts that weld the supercontinent together. The younger orogens occur along the supercontinent margins. The map region shown in Figure 2 is outlined by the heavy-lined box. SF, São Francisco Craton; RP, Río de la Plata Craton.

Questions remain regarding the timing and circumstances of accretion of the cratonic continental fragments, the relative positions of the cratonic fragments over time, and the presence and widths of ocean basins between the fragments. Tectonic scenarios range from ensimatic models with ocean basins that developed with oceanic lithosphere (e.g., Barnes & Sawyer 1980; Kasch 1983*a*; John *et al.* 2003) through to ensialic models of failed Cambrian intracratonic rifting (e.g., Martin & Porada 1977*a*, *b*; Trompette 1997). Despite similar questions and discussions in the detailed works on the Damara orogen published in the early 1980s (e.g., Martin & Eder 1980; Miller 1983*b*) the nature, size and substrate to the respective ocean basins, their tectonic settings of ocean closure, and the presence, or lack of subduction systems, as well as the directions of subduction, are still uncertain.

This paper revisits these issues in the light of the most recent geological, geochronological and thermochronological data for the Damara orogen. As part of this analysis the paper investigates the geological components of the Damara orogen and summarises the most recent data on (1) the structural style and crustal architecture, (2) the metamorphism, (3) geochronological and thermochronological constraints and (4) deformation kinematics. It is a review paper that attempts to link these data with the time-equivalent belts of South America. It also updates and revises the tectonic evolution of the various belts that make up the Damara orogen, particularly in the context of Gondwana amalgamation.

Background

Connections between the orogenic components of Africa and South America were first recognised by du Toit (1937), as part of his Samfrau orogenic Zone of Permo-Triassic age. Porada (1979, 1989) investigated more fully the genetic links between the different parts of the Pan-African Damara orogen and the Brasiliano Ribeira orogen with a detailed review of Damara orogen geological relationships, including regional stratigraphy, structure and metamorphism.

Porada (1989) argued that the Damara orogenic system originated as a three-pronged continental rift system at *c.* 1000 Ma, where the Damara Belt was considered as a failed rift or aulocogen. This scenario includes two episodes; the Katangan at 900–750 Ma and the Damaran at 750–500 Ma. More recently, Trompette (1997) argued for West

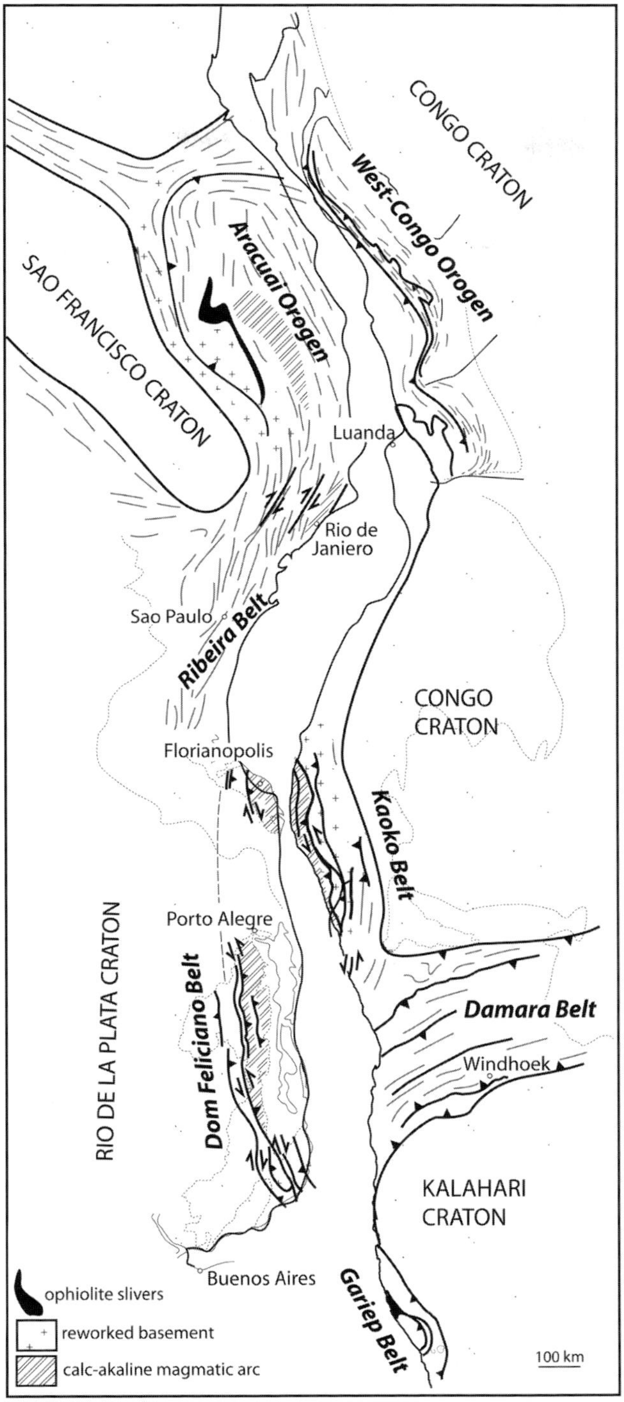

Fig. 2. Map of the Brasiliano and Pan-African orogens defining the amalgamation sutures of West Gondwana between the South American and African cratonic nuclei. The composite orogenic system is made up different component belts and orogens, including from north to south the Araçauí–West Congo orogen, the Ribeira Belt, the Dom Feliciano–Kaoko Belts and the Gariep Belt. Geological relationships including fault traces from the Araçauí orogen are from Pedrosa-Soares *et al.* (2001), the Ribeira Belt from Heilbron & Machado (2003), the Dom Feliciano Belt from Basei *et al.* (2000) and Frantz & Botelho (2000), and the Kaoko Belt from Goscombe *et al.* (2003*a*, *b*; 2005*a*, *b*).

Gondwana supercontinent aggregation from 900 to 600 Ma, involving a two-stage evolution with intracratonic rifting (ensialic) at *c*. 600 Ma followed by basin closure at 520 Ma.

The timing of ocean basin closure has been disputed. Stannistreet *et al.* (1991) proposed that the Khomas Ocean between the Congo and Kalahari cratons (i.e., the Damara Belt) closed before the southern part of the Adamastor Ocean between the Río de la Plata and Kalahari cratons (Gariep Belt), in contrast to Prave (1996) who used sedimentological evidence to argue the opposite. Ocean closure, particularly for the Adamastor Ocean is generally accepted as being diachronous, closing initially in the north (Kaoko Belt) and migrating southwards with a 'zip closure' action (e.g., Germs & Gresse 1991; Gresse & Germs 1993; Stannistreet *et al.* 1991; Frimmel & Frank 1998; Maloof 2000). Most recent geochronology/thermochronology of the Damara orogen (e.g., Goscombe *et al.* 2005*b*; Gray *et al.* 2006) linked with existing data (e.g., Frimmel & Frank (1998) for the Gariep; Kukla (1993) and Jung & Mezger (2003) for the Damara Belt) supports closure of the northern Adamastor Ocean resulting in the Kaoko Belt, then the southern Adamastor Ocean producing the Gariep Belt and finally the Khomas Ocean, suturing along the Damara Belt.

Recent provenance studies utilising U–Pb analyses of detrital zircon populations have established linkages between the various lithostratigraphic units on both sides of the Atlantic Ocean and have helped to establish or confirm tectonic evolutionary scenarios. For example, Frimmel *et al.* (1996) argued for west-directed subduction beneath the Río de la Plata Craton, which has been supported by the provenance data of Basei *et al.* (2005). Similar detrital zircon populations in the Rocha (Dom Feliciano Belt), Oranjemund and Stinkfontein groups (Gariep Belt) establish basin/sedimentation linkages that require subduction in the southern Adamastor Ocean beneath the Río de la Plata Craton.

Recent palaeomagnetic studies and/or reviews of Gondwana palaeomagnetism (Rapalini 2006; Tohver 2006) suggest that West Gondwana was a coherent block by 550 Ma, as there is a single Apparent Polar Wander Path for its components from this time onwards, requiring continent–continent collisions for the Damara and Gariep belts at this time. However, the detailed geochronology/thermochronology presented in Goscombe *et al.* (2005*b*) and Gray *et al.* (2006) reviewed in this paper greatly refines this and allows a new revised look at the tectonics of West Gondwana amalgamation.

Damara Orogen crustal architecture: overview

Lithostratigraphy

The major geological components of the Damara orogen (Fig. 3) are the Archaean–Proterozoic basement inliers, the Damara Sequence passive margin carbonates that rimmed the ocean basins between the cratons (Otavi facies), the deeper water turbidites within the ocean basins (Swakop facies) and the foreland basin deposits (molasse) of the Mulden and upper Nama (Fish River Sub-Group) groups of northern and southern Namibia, respectively. The basement comprises continental-scale ovoid cratonic nuclei, partly contained within Namibia (Fig. 3a) and now preserved either as large inliers, the Kunene and Kamanjab inliers of the Congo Craton in northern Namibia and basement of the Kalahari Craton in the Southern Margin Zone of the Damara Belt and bordering the eastern margin of the Gariep Belt in southern Namibia (Figs 3 & 4). Basement is also exposed in the cores of smaller, elongated domes within the Central Zone of the Damara Belt and in antiformal nappes and thrust slivers in the Kaoko Belt (Fig. 4*c*).

Deposition of the Damara Sequence spanned the Neoproterozoic between at least 770 and 600 Ma (Miller 1983*a*; Prave 1996; Hoffman *et al.* 1994). The basal Damara Sequence is represented by rift-related siliciclastic rocks of the Nosib Group, comprised of quartzites, conglomerates and arenites. Quartz-syenite, alkaline ignimbrite and alkali-rhyolite units in the upper Nosib Group have U–Pb and Pb–Pb zircon ages ranging from 757 ± 1 to 746 ± 2 (Hoffmann *et al.* 1994; Hoffman *et al.* 1998: de Kock *et al.* 2000), constraining the minimum age of the Nosib Group to be approximately 750 Ma (Prave 1996; Hoffman *et al.* 1998). The overlying Otavi Group is dominated by turbiditic greywacke with pelitic schists and psammites and rare mafic schists. Within this succession are two turbiditic carbonate formations, parts of which are correlated with regional diamictite horizons that are elsewhere interpreted as 750–735 Ma and 700 Ma in age (Hoffman *et al.* 1994; Frimmel 1995; Hoffman *et al.* 1998; Folling *et al.* 1998). The uppermost Otavi Group is the widespread Kuiseb Formation, which is comprised of turbiditic greywacke and pelite schists with thin calc-silicate bands (Fig. 3).

Structure

The belts that make up the Damara orogen (Fig. 3), or the arms of the collisional triple junction, have distinct structural trends and style (Figs 4 & 5).

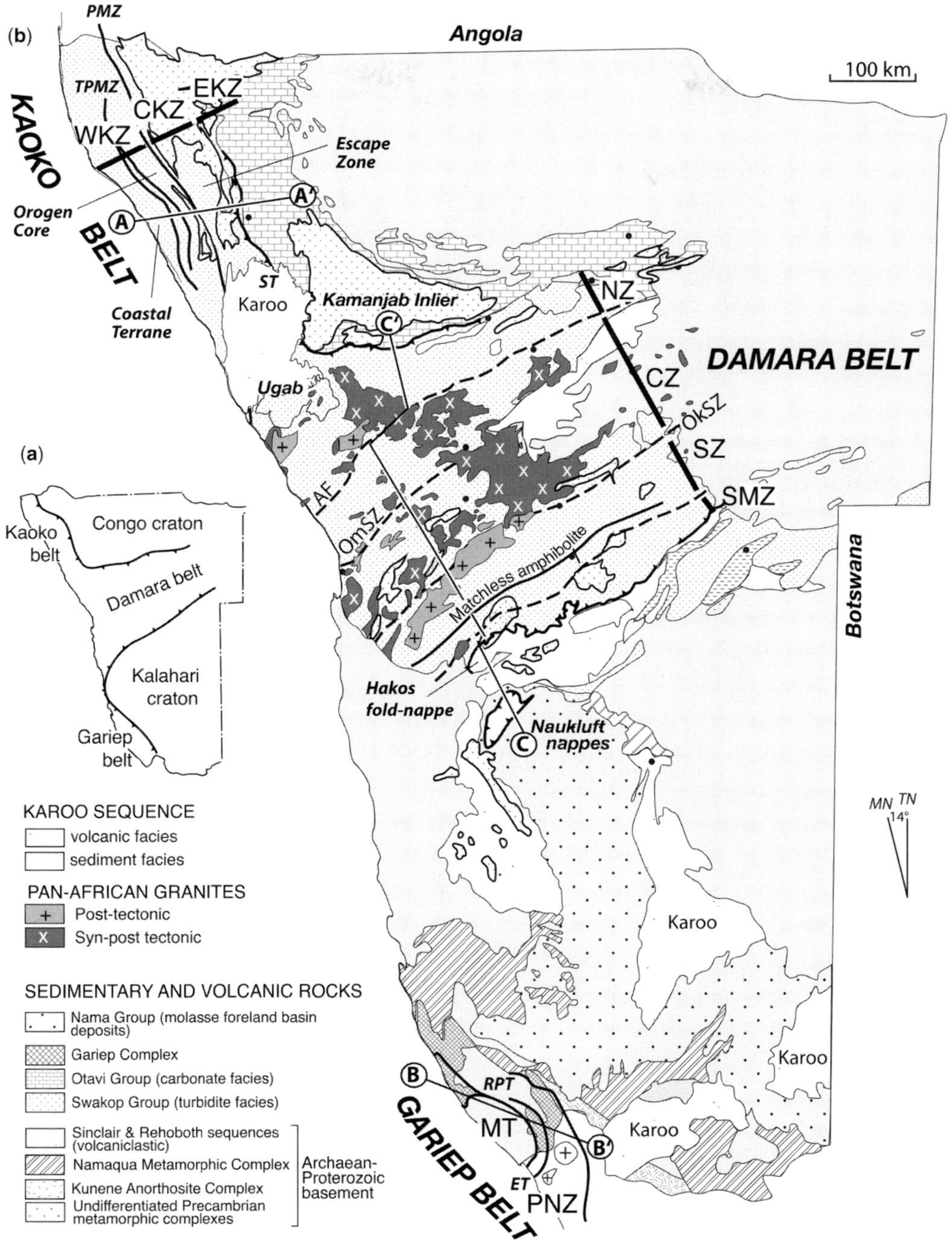

Fig. 3. Geological map of the Damara orogen showing the main geological units, the major faults, and the distribution of plutonic rocks and Swakop Group turbidites (map modified from geological map of Namibia). Inset (**a**) shows the relative positions of the component fold belts and the Congo and Kalahari cratons. The locations of profiles A–A′, B–B′ and C–C′ from Figure 4 are shown. WKZ, Western Kaoko Zone; CKZ, Central Kaoko Zone; EKZ, Eastern Kaoko Zone; TPMZ, Three Palms Mylonite Zone; PMZ, Purros Mylonite Zone; ST, Sesfontein Thrust (Kaoko Belt). AF, Autseib Fault; OmSZ, Omaruru Shear Zone; OkSZ, Okahandja Shear Zone; NZ, Northern Zone; CZ, Central Zone; SZ, Southern Zone; SMZ, Southern Margin Zone (Damara Belt). MT, Marmora Terrane; PNZ, Port Nolloth Zone; RPT, Rosh Pinah Thrust; ET, Eksteenfontein Thrust (Gariep Belt).

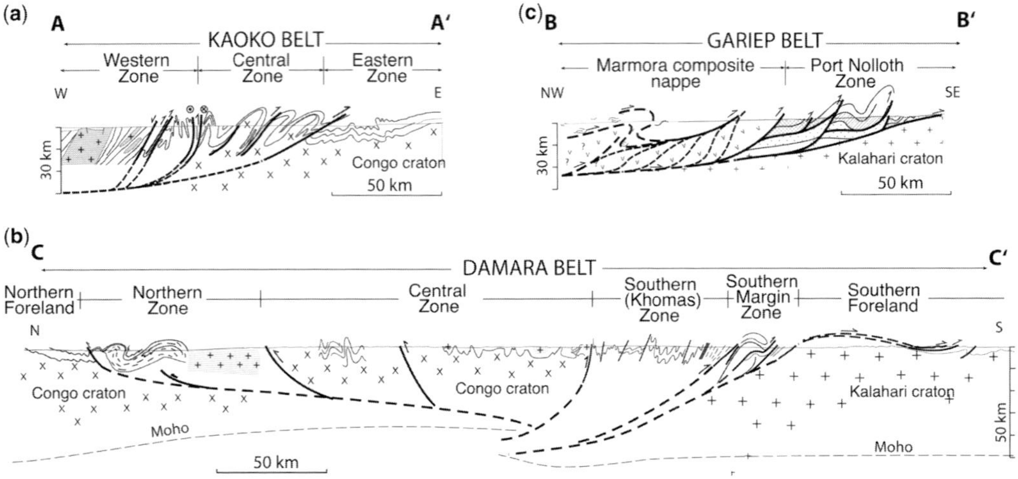

Fig. 4. Simplified structural profiles across the Kaoko, Gariep and Damara belts of the Damara orogen. (**a**) Crustal architecture of the Kaoko Belt (modified from Goscombe *et al.* 2003*a*). (**b**) Crustal architecture of the Gariep Belt (modified from Von Veh 1983 *in* Frimmel & Frank 1998). (**c**) Crustal architecture of the Damara Belt of the Damara orogen (modified from Miller & Grote 1988: profiles on Damara orogen 1:500,000 Map sheets). For location of the profiles see Figure 1. Note (a) and (c) are transpressional belts underlain by inferred west-dipping décollements and (b) shows an asymmetric orogen profile with an inferred former subduction interface, now thrust/shear zone system, penetrating to Moho depth beneath the Southern Zone.

Structural grain is NNW-trending in the Kaoko and Gariep Belts, but is ENE-trending in the Damara Belt (Fig. 5). Both coastal arms are sinistral transpressional belts (Kaoko Belt: Dürr & Dingeldey 1996; Maloof 2000; Passchier *et al.* 2002; Goscombe *et al.* 2003*a*, *b* and Gariep Belt: Davies & Coward 1982; Frimmel 1995; Hälbich & Alchin 1995), whereas the Damara Belt is a divergent orogen that formed during high-angle convergence between the Congo and Kalahari cratons (Coward 1981; Miller 1983*a*; Porada *et al.* 1983). The junction between the southern Kaoko Belt and the Damara Belt (Fig. 3) is the distinctive Ugab Zone with complex fold interference (Coward 1983; Porada *et al.* 1983; Maloof 2000; Passchier *et al.* 2002; Goscombe *et al.* 2004).

The Kaoko Belt is dominated by two NNW-trending crustal-scale shear zones and interlinking shear zones that define orogen-scale shear lenses and similar-trending arcuate shear zones define the major boundaries in the Gariep Belt (Fig. 5). The Damara Belt consists of fault- and shear-bounded zones of varying structural style from north to south: a fold-thrust belt displaying complex fold interference, a granite-dominated inner-zone with elongate, WNW-trending basement cored domes and Damara Sequence basins, and in the south a transposed schist belt and another marginal fold-thrust zone with basement-cored fold nappes (Fig. 5).

Each belt of the Damara orogen is dominated by craton-vergent, imbricate thrust–shear zone systems (Fig. 4). Both the Kaoko and Gariep belts have crustal architectures with inferred west-dipping décollements (Fig. 4a, c). In the Kaoko Belt the steeply west-dipping mylonite zones and inclined E-vergent basement-cored fold-nappes are considered to be rooted in this décollement (Goscombe *et al.* 2005*a*). The Gariep Belt geometry (Fig. 4c) has a composite, obducted ophiolite thrust-nappe, overlying imbricate faults in the passive margin sequence (Frimmel 1995). In contrast, the Damara Belt is an asymmetric, doubly vergent orogen (Fig. 4b). The southern margin is defined by a wide zone of intense, north-dipping, shear-dominated transposed fabrics (Southern Zone) and basement-cored fold-nappes bordering the Kalahari Craton (Southern Margin Zone). The Northern Zone is a craton-vergent, fold-thrust belt without a strongly sheared transposed zone. The Northern and Southern Margin zones must have décollements dipping away from the respective cratons (Fig. 4b).

Metamorphism

The Damara orogen shows contrasting styles of metamorphism. The Kaoko Belt consists of high-grade amphibolite to granulite facies metamorphic rocks (Orogen Core of the Western Kaoko Zone) juxtaposed against intermediate-pressure amphibolite facies rocks (Escape Zone of the Central

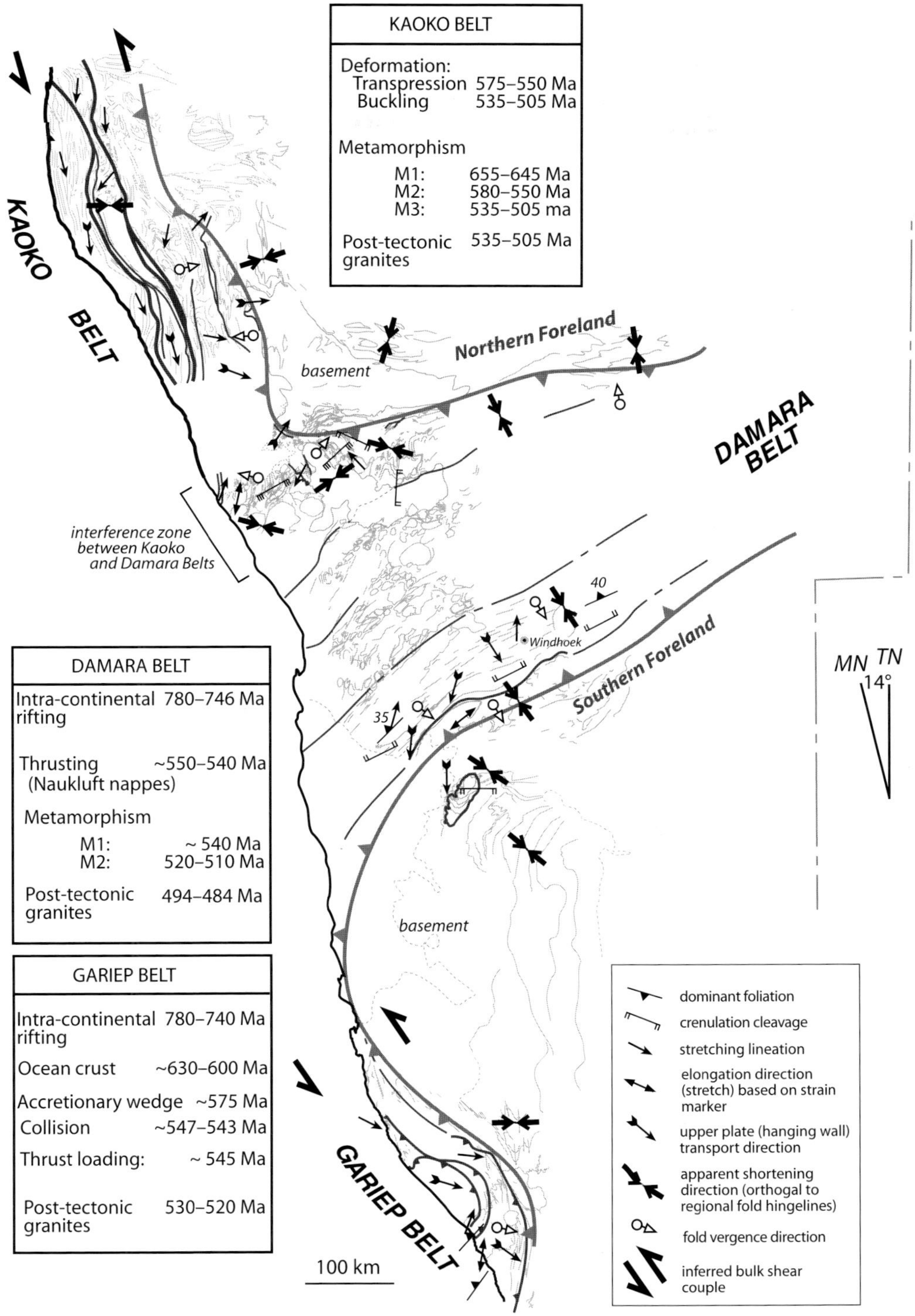

Fig. 5. Summary map of deformation kinematic data for the Damara orogen with insets providing a summary of the timing of key geological processes for each of the component fold belts. Kinematic data is based on the author's unpublished Namibian dataset. Note the fold vergence direction is drawn orthogonal to regional fold hinge lines.

Kaoko Zone) in the footwall of the Purros Mylonite Zone (PMZ, Fig. 3) and low-grade greenschist-facies rocks of the foreland or Eastern Kaoko Zone below the Sesfontein Thrust (ST, Fig. 3) (Goscombe *et al.* 2003*a*, 2005*a*; Will *et al.* 2004). The belt shows marked thermal partitioning into a heterogeneous though largely high-grade and high average thermal gradient Orogen Core bounded by major shear zones, and an inverted Barrovian-series margin of intermediate pressure with basement-cored fold-nappes thrust onto the Congo Craton. Peak metamorphic conditions for the high-grade parts of the Orogen Core were 800–840 °C and 6–8 kbar, and between 500 and 690 °C and 8–9 kbar in the Escape Zone. The Coastal Terrane of the western Kaoko Zone (CT, Fig. 3) experienced two metamorphic events: an early high-grade migmatitic event of *c.* 725 °C and 7 kbar and, during transpressional reworking, conditions of 550 °C and 4.5 kbar (Goscombe *et al.* 2005*a*).

The Gariep Belt is mostly of low metamorphic grade (Frimmel 2000), with greenschist-to-transitional amphibolite facies conditions in the imbricated passive margin sequences of the Port Nolloth Zone (PNZ, Fig. 3) (Fig. 4c). Temperatures ranged from 400 °C to 500 °C and pressures from 2.5 to *c.* 3 kbar (Frimmel 2000). The Chameis Complex mélange of the Marmora Terrane (MT, Fig. 3) records sub-blueschist, subduction-related metamorphism and peak temperatures of 500 °C to 550 °C and pressures of *c.* 6 kbar (Frimmel 2000).

The Damara Belt consists of a central high-T/low-P, granite-dominated belt, flanked by the Northern, Southern, and Southern Margin zones (NZ, SZ and SMZ, Fig. 3), which have intermediate-T/intermediate-P metamorphism (Kasch 1983*a*). The granite-dominated Central Zone (CZ, Fig. 3) underwent peak temperatures of *c.* 750 °C and pressures of *c.* 5.0–6.0 kbar (Kasch 1983*a*; Jung *et al.* 2000). Post-kinematic granites are largely confined to the Central and Northern Zones of the Damara Belt (Fig. 3). These granitoids are typically composite bodies, some concentrically zoned, with at least three intrusive phases ranging from syenite to biotite granite and late-stage aplite dykes. The Southern Zone underwent peak temperatures of *c.* 600 °C and pressures of *c.* 10 kbar (Kasch 1983*a*). The Northern Zone of the Damara Belt shows along-strike variation in metamorphism during north-south convergence, with low-P contact metamorphism with anticlockwise P–T paths dominating in the west (Ugab, Fig. 3) and higher-P (Barrovian series) metamorphism with clockwise P–T paths in the east (Goscombe *et al.* 2005*a*). The eastern part of the Northern Zone has peak metamorphic conditions of 635 °C and 8.7 kbar and experienced deep burial, high-P/moderate-T Barrovian metamorphism (Goscombe *et al.* 2005*a*).

Orogen kinematics

Structurally the Gariep Belt shows bulk SE-directed transport (Fig. 5) partitioned into (a) strike-slip faulting (Davies & Coward 1982) and longitudinal or NW–SE stretching in the northern part and in major shear zones of the Marmora sheet (Davies & Coward 1982; Gresse 1994) and (b) very strong axial elongation or NE–SW stretching in the southern arc (Gresse 1994). In the outer Gariep Belt, particularly near the contact between the Holgat and Stinkfontein groups (Port Nolloth Zone), the development of sheath folds during transposition (Gresse 1994) reflects very high shear strains.

Folds in the Gariep Belt change character and vergence around the Gariep Arc (see Fig. 18 of Gresse 1994). In the NE outer arc, defined by the NE-trending Rosh Pinah thrust (RPT, Fig. 3), the folds are more open, east-vergent and associated with east-directed thrusting. Southwards, around the arc where the Eksteenfontein thrust (ET, Fig. 3) is north-trending, these folds become tighter and isoclinal, and have SE-vergence. Here, the early folds are overprinted by NE-vergent F_2 folds (Gresse 1994), associated with a north- to NW-trending crenulation cleavage suggestive of a late component of margin-orthogonal compression (cf. fig. 4 of Frimmel 2000).

In the Kaoko Belt a zone of craton-vergent, basement-cored, isoclinal fold-nappes in the Central Kaoko Zone (Fig. 3) or Escape Zone appear to extrude from the dominant, medial Purros Mylonite Zone (PMZ, Fig. 3) (Goscombe *et al.* 2005*a*, *b*). These fold-nappes coincide with a swing in the lineation pattern to higher angles (up to 70–80°) to the grain of the orogen, reflecting a component of high-angle escape towards the orogen margin (Dürr & Dingeldey 1996; Goscombe *et al.* 2003*a*, 2005*a*). The Orogen Core, or eastern part of the Western Kaoko Zone inboard of the Three Palms Mylonite Zone (TPMZ, Fig. 3), contains shear-zone bounded domains of sheared migmatites with steep foliations and sub-horizontal lineations, a single domain of lower-grade chevron-folded turbidites and reworked basement gneiss slivers (Goscombe *et al.* 2003*a*, *b*). Coastal Terrane migmatitic gneisses and orthogneisses were down-graded and heterogeneously reworked by steep mid-amphibolite facies foliations and discrete shear zones (Goscombe *et al.* 2005*a*).

The Damara Belt shows high-angle convergence (Fig. 5) and lacks evidence of oblique or transcurrent movements, despite arguments for sinistral movements and top-to-the-SW tectonic transport by Downing & Coward (1981) and Coward (1981, 1983). Shear bands, developed in Kuiseb Formation schist and units of the Southern Margin Zone indicate north-over-south movement in a north–south transport direction (Fig. 5). Variably noth-dipping, asymmetric crenulations and mesoscopic folds reflect a bulk south-directed shear strain (Fig. 5). High-strain at the basement/cover contact is shown by deformed conglomerates in the cover (Chuos Formation), down-dip stretching lineations and mylonitic basement. The frontal lobes of the Hakos fold-nappe (Fig. 3) display prolate strains with the stretch direction at high angles to the transport direction as shown by shear bands.

The Central Zone of the Damara Belt displays contrasting kinematic behaviour with orogen-parallel stretch and shortening at high angles to the orogen at different levels (Oliver 1994; Kisters *et al.* 2004). During high-grade metamorphism and migmatization, the deeper levels of the Central Zone underwent pure shear deformation, with lateral orogen-parallel stretch (Kisters *et al.* 2004). This is in marked contrast to interpretations of SW-directed orogen-parallel extrusion (e.g., Downing & Coward 1981; Oliver 1994), where the domes were interpreted as large, SW-facing sheath folds rooted in the northeast Central Zone and requiring top-to-the-SW transport in a crustal scale shear zone (Downing & Coward 1981; Coward 1981, 1983). At shallower crustal levels the Central Zone has undergone crustal thickening, orogen-normal shortening by folding and NE-directed thrusting (Kisters *et al.* 2004).

Within the Southern and Southern Margin zones major south-directed bulk shear strain deformation was responsible for crustal-scale underthrusting of the Kalahari Craton northwards (Fig. 4c), as well as continued thrusting and crustal thickening along the margins of the orogen. Crustal thickening and burial along this margin led to the Barrovian metamorphism. Significant magmatic underplating related to extension in the lower part of the overriding plate, led to marked magmatism and younger, high-T/low-P metamorphism in the Central Zone.

Temporal aspects of deformation, metamorphism and magmatism of the Damara Orogen: review

Geochronological studies in the Kaoko Belt (Goscombe *et al.* 2005*b*), in the Damara Belt (Jung & Mezger 2003) and in the Gariep Belt (Frimmel & Frank 1998), as well as a geochronological/thermochronological study across the Damara orogen (Gray *et al.* 2006) provide a more comprehensive picture of the tectonothermal evolution of the orogen (Fig. 6).

The Kaoko Belt preserves evidence for three distinct metamorphic episodes: M_1 (655–645 Ma, restricted to the westernmost Coastal Terrane), M_2 (580–570 Ma) and M_3 (530–505 Ma) (see Goscombe *et al.* 2005*a*). Collision and docking of the outboard Coastal Terrane with the rest of the Kaoko Belt occurred after 645 Ma, but prior to 580 Ma at the onset of transpressional orogenesis and M_2 metamorphism (Goscombe *et al.* 2005*b*). During transpression the Coastal Terrane rock sequences were reworked at lower strains and lower metamorphic grade compared to the rest of the Kaoko Belt (Goscombe *et al.* 2005*b*). Transpressional orogenesis in the Kaoko Belt and Ugab Zone had ceased by *c.* 535 Ma, with cratonization marked by intrusion of post-kinematic granite and pegmatite between 535 Ma and 505 Ma (Goscombe *et al.* 2005*b*).

The Damara and Gariep belts both show younger deformation and metamorphism than the Kaoko Belt (Fig. 6; Gray *et al.* 2006). Continued high-angle convergence through 530 Ma in the Damara Belt coincides with large-scale open east–west trending folds in the Kaoko Belt (Goscombe *et al.* 2003*a*, *b*).

The Gariep Belt underwent thrust-nappe emplacement onto the Kalahari Craton at *c.* 550–540 Ma (Frimmel & Frank 1998). Oceanic sequences in the Marmora Terrane preserve (1) an earlier seafloor metamorphism suggesting that Adamastor Ocean seafloor spreading was occurring at *c.* 630 Ma and (2) subduction-related metamorphism at *c.* 580–570 Ma, suggesting that the Adamastor Ocean was closing at this time (Frimmel & Frank 1998). The Gariep Belt was cratonized by 520 Ma, with erosion into the Nama foreland basin commencing at *c.* 540 Ma (Gresse & Germs 1993; Gresse 1994; Frimmel 2000). It was intruded by post-kinematic granites at *c.* 507 Ma (Frimmel 2000), although east-directed thrusting continued inboard within the Nama foreland basin through 496 Ma (Gresse *et al.* 1988).

The Damara Belt shows a more complex high-T metamorphic history from 540 to 510 Ma with metamorphism coincident with pulses of magmatism (Jung & Mezger 2003). Intrusion of post-kinematic A-type granites from 495 Ma to 486 Ma (McDermott *et al.* 2000) was followed by cooling and exhumation of the Damara Belt through 470 Ma (Gray *et al.* 2006).

Damara Orogen tectonic evolution: problems and issues

Problems pertaining to Damara orogen evolution relevant to Gondwana amalgamation relate to: (1) the positions of the respective cratons through time, (2) the sizes of the ocean basins between them and (3) the positions and directions of

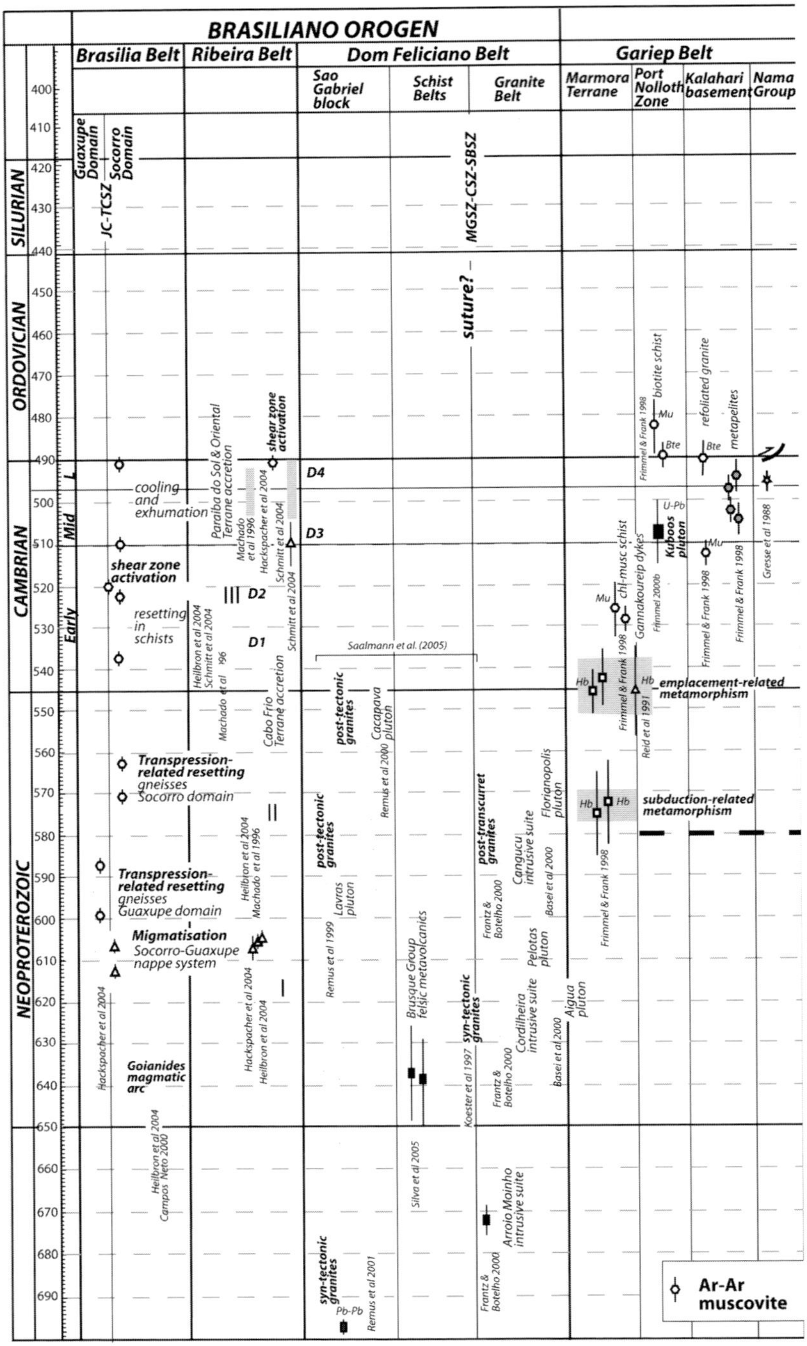

Fig. 6. Damara orogen and part of the Brasiliano time-space plot of recently published geochronological data including the $^{40}Ar/^{39}Ar$ data from Gray *et al.* (2006), and U–Pb data on zircon, monazite and titanite. Sources are listed in the figure. Unsourced data is from Gray *et al.* (2006). The diagram highlights the major periods of magmatism, metamorphism/deformation, thrusting and cooling and exhumation. Fault and shear zone abbreviations are as listed. WKZ, Western Kaoko Zone; CKZ, Central Kaoko Zone; EKZ, Eastern Kaoko Zone; TPMZ, Three Palms Mylonite Zone; VMZ, Village Mylonite Zone; PMZ, Purros Mylonite Zone; AMZ, Ahub Mylonite Zone; ST, Sesfontein Thrust; GMZ, Guantegab Mylonite Zone; OmL, Omaruru Lineament (Shear Zone); OkL, Okahandja Lineament (Shear Zone); JC-TCSZ, Jacutinga-Três Corações Shear Zones; MGSZ-CSZ-SBSZ, Major Gercino Shear Zone–Cordilhera Shear Zone–Sierra Ballena Shear Zone.

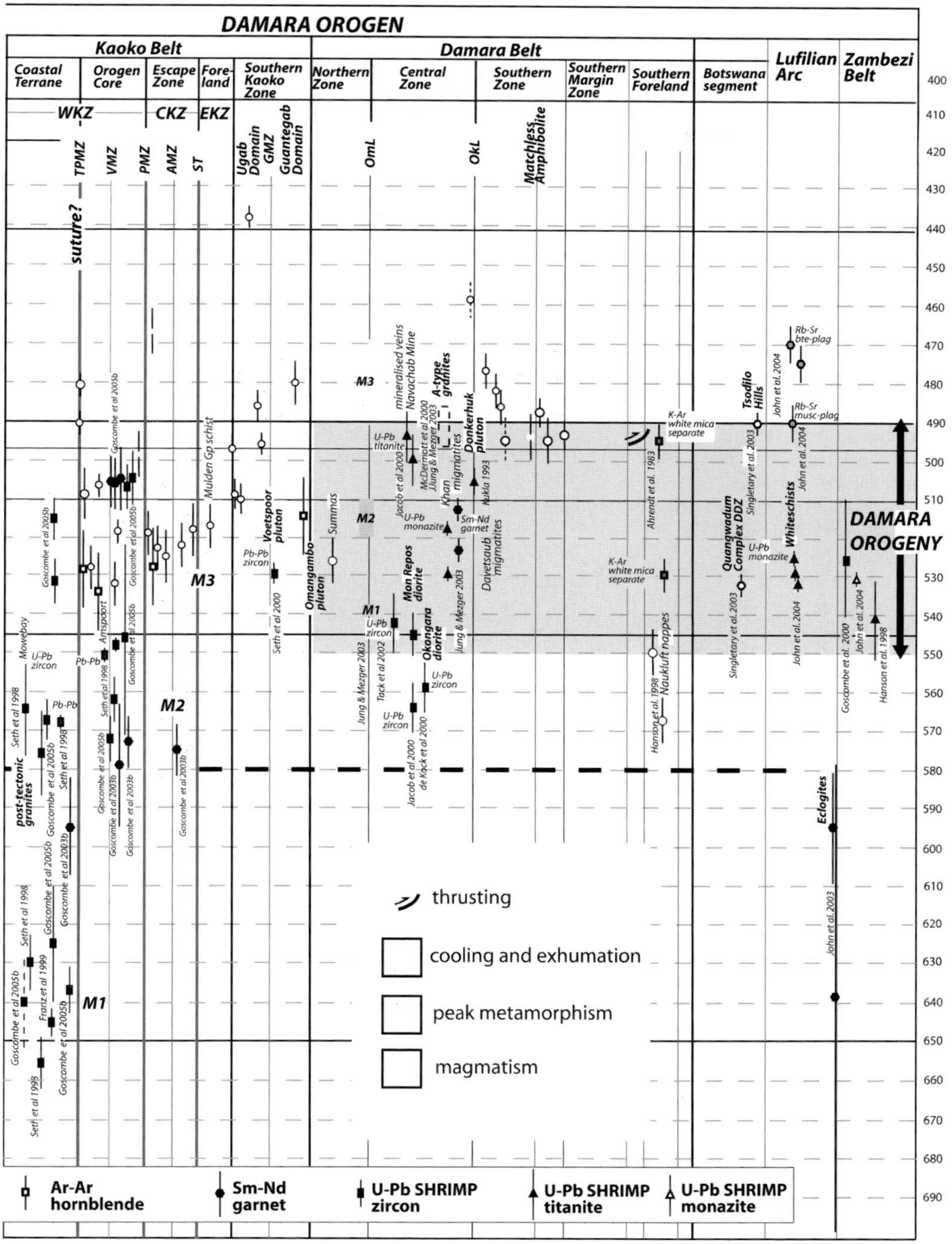

Fig. 6. (*Continued.*)

subduction zones that closed the ocean basins. Answers to problems (1) and (2) will require better definition of palaeomagnetic poles in the future, and particularly for that of the Kalahari Craton in the period 750–550 Ma. The presence and/or lack of subduction zones to close the intervening ocean basins will now be addressed.

Intracratonic orogeny with exclusive ensialic evolution has been applied to the Damara orogen, particularly for the Damara Belt (Kröner 1977; Martin & Porada 1977*a*, *b*; Porada 1979). In this model the strongly deformed and metamorphosed Matchless Amphibolite is questioned as an ophiolite remnant, despite MORB-type geochemistry (Barnes & Sawyer 1980) and a chert-Cu/Zn mineralization association (Killick 2000) typical of oceanic lithosphere. The lack of subduction-related metamorphism is also cited as evidence against ocean closure due to subduction, although the presence of eclogites in the Zambezi Belt (John *et al.* 2003) and white schists in the Lufilian Arc (John *et al.* 2004), part of the continuation of the Damara Belt into Zambia, provide alternative evidence.

In the Zambezi Belt, in contrast to Hanson *et al.* (1994) and Hanson (2003), John *et al.* (2003) argued for the presence of a large (>1000 km wide) ocean basin with MORB-type eclogites and meta-gabbros subducted to a depth of *c.* 90 km during basin closure. The timing of the eclogite facies metamorphism is 595 ± 10 Ma, suggesting that subduction was occurring at this time, some 60 Ma earlier than the *c.* 530 Ma peak of metamorphism in the central Damara Belt.

The long, apparently continuous, linear trace of the Matchless Amphibolite within intensely deformed Kuiseb Formation schist of the Southern Zone in the Damara Belt is unusual but may have similarities to the fault-bounded Dun Mountain ophiolite belt and Haast Schist of New Zealand (Gray *et al.* 2007). The transposed layering and pronounced schistosity in the Kuiseb Formation schist is almost identical to that of the central Otago part of the Haast Schist suggesting deformation under similar conditions in a scenario where the turbidite is on the down-going plate of an oceanic subduction system (see Coombs *et al.* 1976). In the Otago Schist an intermediate-*T*/intermediate-*P* (Barrovian-style) metamorphism linked to wedge thickening (Mortimer 2000) has almost totally eradicated the earlier subduction-related, intermediate- to high-*P* metamorphism (see Yardley 1982). The older metamorphism is only preserved as crossite relics in the cores of younger amphibole and albite porphyroblasts (see Fig. 2*c* of Yardley 1982). Widespread metamorphic overprinting at higher temperatures appears typical of Barrovian-style thickened and metamorphosed accretionary wedges, and is therefore likely to have obliterated any older intermediate-*P* to high-*P* metamorphism in the Kuiseb Formation schists of the Southern Zone.

The kinematics of the Southern Zone schists, by comparison with the Otago Schist Belt of New Zealand, combined with geochemistry of the more primitive diorites and syenites that are part of the Central Zone early magmatic history supports northward subduction of the Khomas Ocean lithosphere beneath the attenuated leading edge of the Congo Craton; as originally suggested by Barnes & Sawyer (1980) and Kasch (1983*b*).

For the Kaoko Belt, the lack of ophiolite sequences or high-*P* metamorphism has led to intracratonic fold belt interpretations (Dürr & Dingeldy 1996; Konopasek *et al.* 2005). More recently, the recognition of arc affinity for the Coastal Terrane has led to proposal of a subduction-related tectonic evolution, with subduction inferred to be both west-directed (Machado *et al.* 1996; Masberg *et al.* 2005) and further outboard east-directed subduction (Basei *et al.* 2000; Goscombe & Gray 2007).

If an ocean basin closed between the African and South American components of the Brasiliano/Pan-African orogenic system within the Kaoko Zone, the suture would have to be at the proto-Three Palms Mylonite Zone. Evidence for major crustal displacements with juxtaposition of distinctly different aged basement either side of the Purros Mylonite Zone (Goscombe *et al.* 2003*a*, *b*), combined with the lack of ophiolite slivers and high-*P* metamorphism, suggests that both shear zones are part of a broad, complex 'suturing' zone behind the former arc (i.e. in a back-arc position), between the arc and the African continental margin. This 'suturing' involved high-*T* metamorphism of turbidites deposited on the attenuated leading edge of the Congo Craton and included deformation and reworking of the cratonic basement (Goscombe & Gray 2007). This magmatic arc could well represent the continuation of the magmatic arc recognised in the Oriental Terrane of the Ribeira Belt (Heilbron *et al.* 2004) with similar ages and/or it could be part of the granite belt of the Dom Feliciano Belt.

Another issue for the Kaoko Belt is the inferred 750–600 Ma timing of foreland basin evolution for the Congo Craton (Prave 1996). This is problematical, in that the age of the Mulden Group is inconsistent with the 580–550 Ma and 530 Ma periods of deformation that established the tectonothermal character of the Kaoko Belt (Goscombe *et al.* 2005*a*, *b*). The Mulden Group was folded and metamorphosed prior to the late-stage thrusting event (Sesfontein Thrust), and a 750–600 Ma depositional age clearly predates the timing of peak (M_2) metamorphism. If the published Mulden Group age range is correct then the sedimentary facies and erosional hiatus of Prave (1996) must reflect instability associated with the initial collision of the Coastal Terrane, and therefore the Mulden Group sediments should contain a significant component of the 650 Ma detrital zircons.

An ensimatic, subduction-related origin has been accepted for the Gariep Belt, largely due to the Chameis Complex mélange of the Marmora Terrane with its mafic and ultramafic blocks, some of which contain Na-rich amphibole (Frimmel & Hartnady 1992). Although not strictly blueschist metamorphism, intermediate pressure (*c.* 6 kbar) and low temperature metamorphic conditions combined with the facies association of mélange (Chameis Complex), turbidites (Oranjemund Formation) and metavolcanic rocks (Grootderm Formation) support this contention (see descriptions and discussions in Frimmel 2000).

The direction of subduction has been discussed (see Frimmel *et al.* 1996), and recent provenance work on detrital zircon populations (Basei *et al.* 2005, this volume) supports west-directed subduction beneath the Río de la Plata Craton. This establishes continuity of a linked west- or north-directed subduction system that closed the former Adamastor Ocean and subsequently the Khomas Ocean to form the Gariep Belt and then the Damara Belt.

Damara Orogen tectonic evolution in a global context

West Gondwana amalgamation is shown in a series of global reconstructions for different time periods (modified from Collins & Pisarevsky 2005) incorporating temporal and tectonothermal constraints from the Damara orogen. In these tectonic reconstructions fragments of ophiolite and calc- alkaline volcanic rocks have been used as indicators of ocean closure, the ages of metamorphism and deformation indicate periods of accretion and crustal thickening, and the age of post-kinematic magmatism indicates the timing of cratonization.

780 Ma to 740 Ma (Fig. 7)

In our 750 Ma reconstruction the cratonic nuclei that eventually come together to form West Gondwana are separated by some 30° latitude with an ocean of unknown dimensions inferred between the Congo and Kalahari cratonic fragments. Such a reconstruction either contradicts or conflicts with previous interpretations of intracratonic rifting between these cratons, as represented by the Nosib rhyolites of the Congo margin and the Rosh Pinah volcanic rocks of the Kalahari margin (see fig. 7 of Frimmel & Frank 1998).

This is, however, a difficult time for which to fully constrain the palaeogeography. There are no reliable late Neoproterozoic poles from the Kalahari Craton. Collins & Pisarevsky (2005) argued for a Kalahari–West Australia connection based partly on the presence of overlapping Grenvillian-age

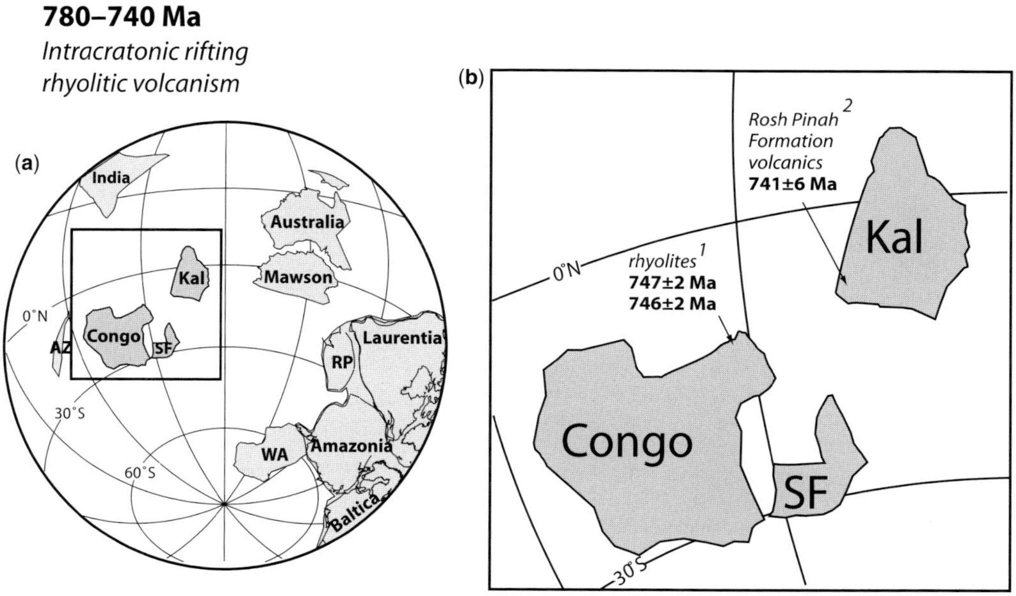

Fig. 7. Global reconstruction of continents at 780–740 Ma with enlargement (**b**) showing the key geological constraints during this time period from the Congo, Kalahari and São Francisco cratons prior to the development of the Pan-African/Brasiliano orogenic system. Continental fragments are AZ, Azania; SF, São Francisco; RP, Río de la Plata; WA, West Africa; Kal, Kalahari. Data references: 1, Hoffman *et al.* (1996); 2, Frimmel *et al.* (1996).

events in the Northampton Block (Australia) and the Namaqua–Natal belts (Kalahari Craton). Both the Kalahari Craton and Australia have reliable palaeomagnetic poles of Grenvillian age (1050–1100 Ma; see Meert & Torsvik 2003; Pesonen *et al.* 2003). These Grenvillian poles show a latitudinal offset between the Northampton Block and Kalahari of more than 30 degrees. Thus, in our reconstruction, we show the Kalahari Craton in proximity to the Congo–São Francisco craton, but detached from it. The position of the Congo–São Francisco cratons is based on the 755 Ma Mbozi Complex pole (Meert *et al.* 1995).

655 Ma to 600 Ma (Fig. 8)

Subduction-related closure begins in the northern Adamastor Ocean as evidenced by calc-alkaline magmatism in the Araçuaí and Ribeira belts between 625 Ma and 585 Ma, in the Dom Feliciano Belt from 620 Ma to *c.* 580 Ma (Basei *et al.* 2000), and from 655 Ma to 625 Ma in the Coastal Terrane of the Kaoko Belt (Masberg *et al.* 2005; Goscombe *et al.* 2005*b*). Collisional orogenesis was taking place in the Brasilano orogen at *c.* 640 Ma with nappe emplacements over the São Francisco Craton between 640 Ma and 630 Ma (Valeriano *et al.* 2004, 2008), due to collision with the Paranapanema block, now hidden under the Paraná Basin.

At *c.* 630 Ma seafloor spreading was underway in the southern Adamastor Ocean as recorded by seafloor metamorphism in Marmora Terrane of the Gariep Belt (Frimmel & Frank 1998).

580 Ma to 550 Ma (Fig. 9)

Arc–continent collision occurred in the Ribeira Belt (Río Negro arc: 595–560 Ma; Heilbron *et al.* 2004) and in the Kaoko Belt (Coastal Terrane: pre-580 Ma; Goscombe *et al.* 2005*b*). Peak metamorphism in the Kaoko Belt occurred at *c.* 580–570 Ma with transpressional reworking from 570–550 Ma (Goscombe *et al.* 2005*b*).

At this time (580–570 Ma) subduction-related metamorphism was taking place in the southern Adamastor Ocean (Marmora Terrane, Gariep Belt; Frimmel & Frank 1998) with subduction-related ocean closure in the Khomas Ocean (560–550 Ma) suggested by mafic magmatism (diorites).

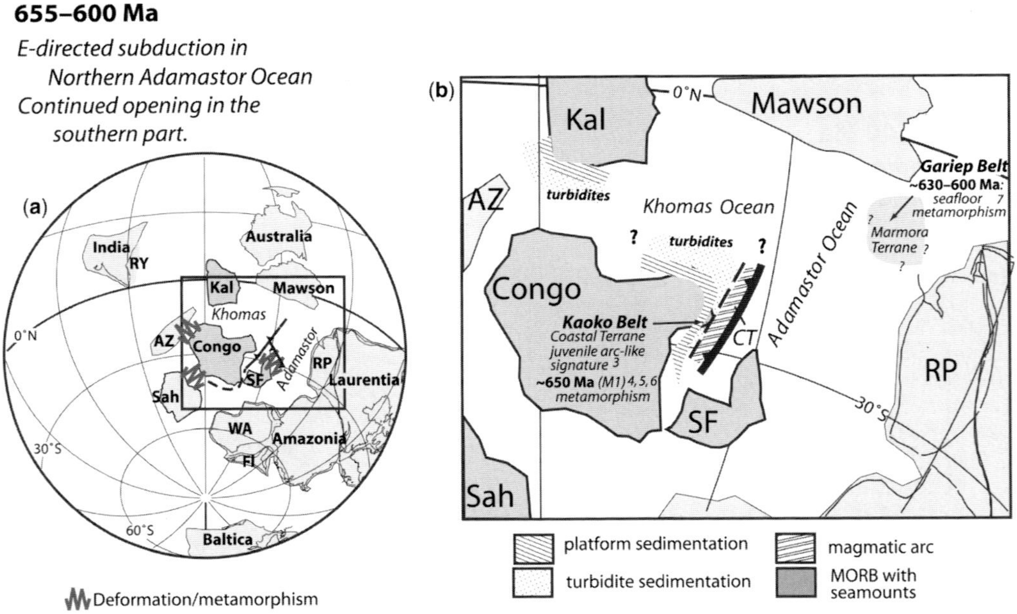

Fig. 8. Global reconstruction of continents at 655–600 Ma with enlargement (**b**) showing palaeogeographic lithofacies distributions and the key geological constraints during this time period from the Congo, Kalahari and São Francisco cratons prior to the development of the Pan-African/Brasiliano orogenic system. Traces of subduction zones are shown by heavy lines with barbs, where the barbs are drawn on the upper plate side and designate the subduction zone dip. CT, Coastal Terrane of the Kaoko Belt (shown as magmatic arc); Sah, Saharan Craton; MT, Marmora Terrane of the Gariep Belt (shown as oceanic lithosphere). Other abbreviations as in Figure 7. Data references: 3, Masberg *et al.* (2005); 4, Franz *et al.* (1999); 5, Seth *et al.* (1998); 6, Goscombe *et al.* (2005*b*); 7, Frimmel & Frank (1998).

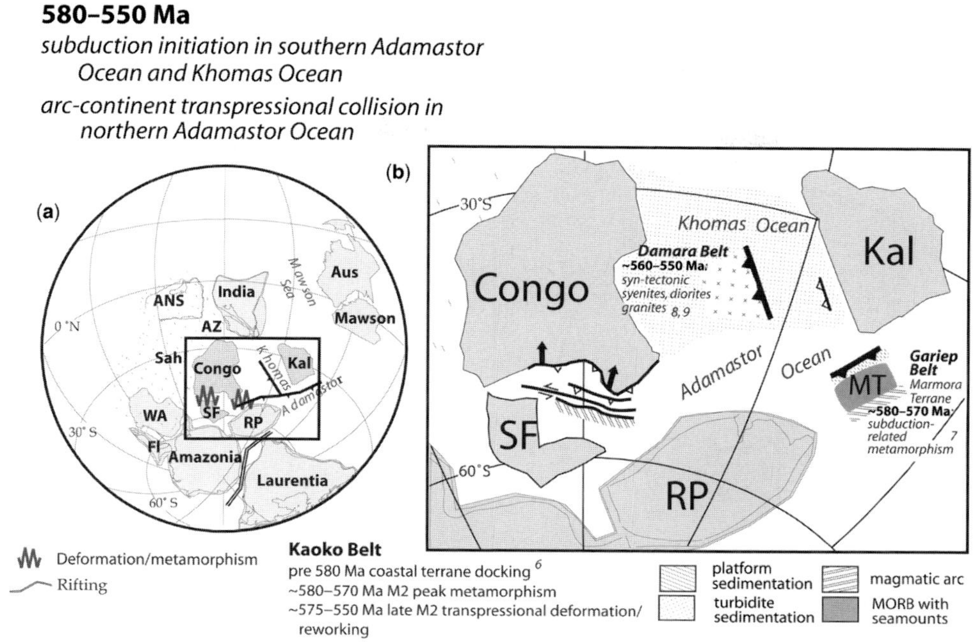

Fig. 9. Global reconstruction of continents at 580–550 Ma with enlargement (**b**) showing palaeogeographic lithofacies distributions and the key geological constraints during this time period from the Congo, Kalahari and São Francisco cratons prior to the development of the Pan-African/Brasiliano orogenic system. Traces of subduction zones are shown by heavy lines with barbs, where the barbs are drawn on the upper plate side and designate the subduction zone dip. An, Antarctica. Other abbreviations as in Figures 7 and 8. Data references: 6, Goscombe *et al.* (2005*b*); 7, Frimmel & Frank (1998); 8, Jacob *et al.* (2000); 9, de Kock *et al.* (2000).

550 Ma to 500 Ma (Fig. 10)

Closure of the southern Adamastor Ocean occurred from *c.* 550 to 540 Ma (Frimmel & Frank 1998) with oblique transpressional obduction of the Marmora Terrane oceanic suite over the imbricated passive margin sequence (Port Nolloth Zone, Gariep Belt) and initiation of Nama sequence foreland basin sedimentation (Gresse & Germs 1994).

Peak deformation/metamorphism took place in the Damara Belt through the Lufilian Arc into the Zambezi Belt at *c.* 530–520 Ma (Goscombe *et al.* 2000; Jung & Mezger 2003; Singletary *et al.* 2003; John *et al.* 2003, 2004). The Damara Belt shows marked magmatism and high-*T*/low-*P* metamorphism at this time (Kasch 1983*a*). At the margins of the orogen, over-thrusting and related crustal thickening caused intermediate-*T*/intermediate-*P* (Barrovian style) metamorphism (Northern Zone: Goscombe *et al.* 2004; Southern Zone: Kasch 1983*a*, *b*; Kukla 1993) with thrusting of the passive margin sequences back over the cratonic nuclei (Naukluft Nappes: *c.* 500 Ma; Ahrendt *et al.* 1977). Effects of the Damara Belt collisional deformation are seen as broad warpings and a younger thermal and magmatic event (M_3: 530–505 Ma) in the Kaoko Belt (Goscombe *et al.* 2005*b*).

In the Cabo Frio domain of the Ribeira Belt relatively high pressure and high temperature metamorphism at 530–510 Ma is interpreted as related to collision (Schmitt *et al.* 2004).

505 Ma to 480 Ma (Fig. 11)

Inboard transmission of stress from the outboard, Gondwana margin (Ross–Delamerian) subduction system caused continued thrusting (Naukluft Nappes: 500–495 Ma; Ahrendt *et al.* 1983) and syn-tectonic sedimentation in the Nama foreland basin (Ahrendt *et al.* 1983; Gresse *et al.* 1988; Gresse & Germs 1993) and in the Camaquã and Itajaí basins of Brazil (Gresse *et al.* 1996). It also led to shear zone reactivation in the Kaoko Belt (490–467 Ma; Gray *et al.* 2006) and Gariep Belt (506–495 Ma; Frimmel & Frank 1998).

Emplacement of post-tectonic A-type granites occurred in the Central Zone (McDermott *et al.*

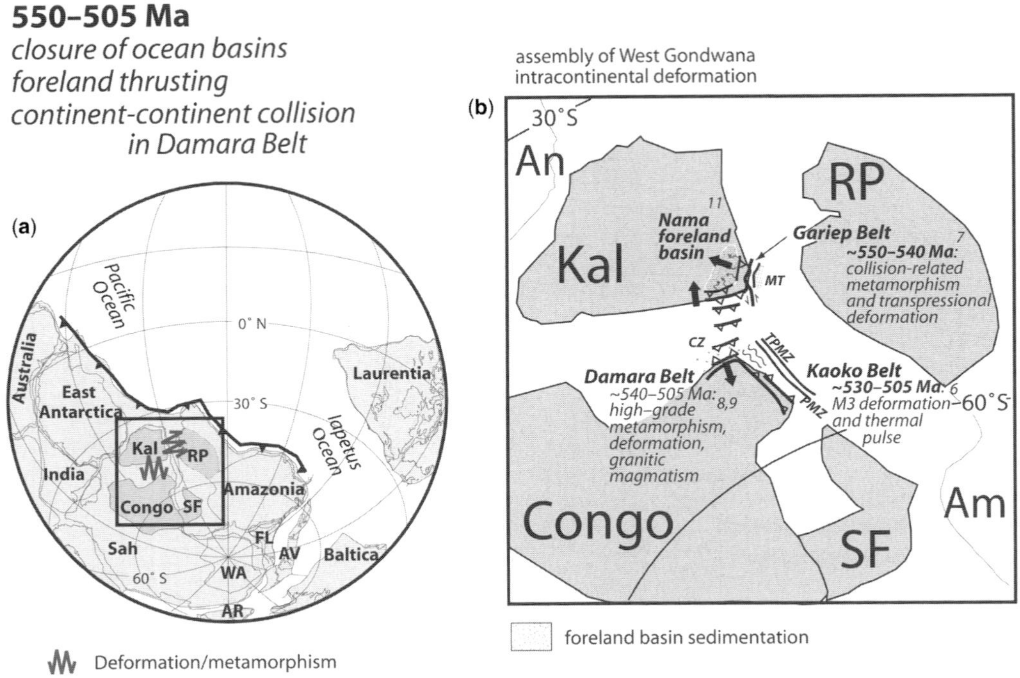

Fig. 10. Global reconstruction of continents at 550–505 Ma with enlargement (**b**) showing palaeogeographical lithofacies distributions and the key geological constraints during this time period from the Congo, Kalahari and São Francisco cratons during the development of the Pan-African/Brasiliano orogenic system. Traces of subduction zones are shown by heavy lines with barbs, where the barbs are drawn on the upper plate side and designate the subduction zone dip. The thinner heavy lines in (b) are fault traces. PMZ, Purros Mylonite Zone; TPMZ, Three Palms Mylonite Zone of the Kaoko Belt; AR, Armorica; AV, Avalonia; Fl, Florida. Other abbreviations as in Figures 7–9. Data references: 6, Goscombe *et al.* (2005); 7, Frimmel & Frank (1998); 8, Jacob *et al.* (2000); 9, de Kock *et al.* (2000); 10, Jung & Mezger (2003); 11, Gresse & Germs (1993).

2000) with continued cooling and exhumation in the Damara Belt through 480 Ma (Gray *et al.* 2006).

Significance for Gondwana assembly

From a western African perspective, the assembly of Gondwana shows complex suturing that does not reflect a simple final amalgamation of East and West Gondwana (Fig. 12). It is perhaps better described as an amalgamation of North (São Francisco–Congo–India) and South (Kalahari–Antarctica) Gondwana during the Kuunga orogeny (550–530 Ma), as proposed by Meert (2003) and Boger & Miller (2004) for the assembly of eastern Gondwana. Geochronological data from South America and southwestern Africa (Fig. 6) suggest closure of a Khomas–Mozambique ocean from 530 to 500 Ma, as part of a combined Damara–Kuunga orogeny. The composite Kuunga–Damara orogen incorporates the Damara orogen of Namibia, the Lufilian Arc and Zambezi Belt of Zambia, and joins the Lurio Belt of Mozambique and a belt made up of the Napier Complex of Antarctica, and the Eastern Ghats of India. It has dimensions comparable to the younger Ross–Delamerian orogen (Fig. 12).

Global reconstructions based on palaeomagnetic data suggest larger separations, and therefore significant ocean basins between the Río de la Plata, Congo and Kalahari cratonic nuclei that eventually define West Gondwana. This has a requirement of ensimatic subduction-related ocean closures, rather than ensialic, intracratonic evolutions that were originally proposed to explain many of the Brasiliano/Pan-African orogens. The position of the Kalahari Craton however, remains controversial. In the Collins & Pisarevsky (2005) reconstructions the Kalahari Craton abuts against the West Australian side of the Australian craton (Fig. 6), whereas according to Meert (2003, Fig. 1) it is situated outboard of a conjoined East Antarctica–Laurentia surrounded by Congo–São Francisco and Río de la Plata. From an African perspective

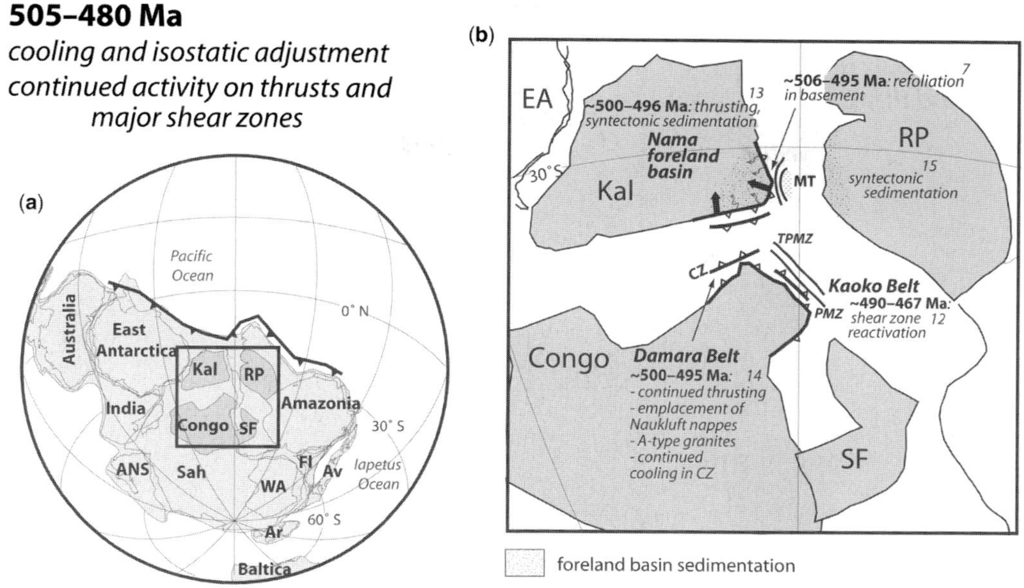

Fig. 11. Global reconstruction of continents at 505–480 Ma with enlargement (**b**) showing palaeogeographical lithofacies distributions and the key geological constraints during this time period from the Congo, Kalahari and São Francisco cratons during the development of the Pan-African/Brasiliano orogenic system. Reconstruction (**a**) is based on Figure 6 of Grunow *et al.* (1996). Traces of subduction zones are shown by heavy lines with barbs, where the barbs are drawn on the upper plate side and designate the subduction zone dip. The thinner heavy lines in (**b**) are fault traces. ANS, Arabian/Nubian Shield; Av, Avalonia; Ar, Armorica; EA, East Antarctica; Fl, Florida; CZ, Central Zone (Damara Belt). Other abbreviations as in Figs 7–10. Data references: 6, Goscombe *et al.* (2005*b*); 11, Gresse & Germs (1993); 12, Gray *et al.* (2006); 13, Gresse *et al.* (1988); 14, Ahrendt *et al.* (1977); 15, Gresse *et al.* (1996).

this provides a better fit for Gondwana assembly, as shown in Figures 7–11.

The West Gondwana suture between Africa and South America reflects the closure of the Adamastor Ocean, and provides the most detailed evolution sequences for SW Gondwana assembly (Fig. 12b). The Brasiliano orogens of South America show more complicated tectonic evolution with multiple tectonothermal events (see also Fig. 6), although the Dom Feliciano and Ribeira belts flanking the Río de la Plata Craton experienced collisional orogenesis with a transpressional component at the same time as the main phase deformation in the Kaoko Belt (Frantz & Botelho 2000; Heilbron & Machado 2003; Heilbron *et al.* 2004; Goscombe *et al.* 2005*b*). The collisional stage in the Ribeira orogen was at 590–560 Ma and is characterised by terranes juxtaposed by relatively steeply dipping, dextral transcurrent shear zones (Heilbron *et al.* 2004). In the Kaoko Belt collision immediately pre-dates main phase orogenesis in the period from 580–550 Ma (Goscombe *et al.* 2005*b*).

The linkage between the Brasiliano and Damara orogens is a *c.* 680–580 Ma magmatic arc component along the 2800 km long composite orogenic system (Fig. 2). In the former Adamastor Ocean, records of arc magmatism suggest a more complex tectonic evolution than perhaps a simple southwards migration of ocean closure, although this appears to be the case in the Ribeira–Dom Feliciano/Kaoko–Gariep part of the orogenic system.

Arc magmatism varies from 680–670 Ma in the Brasiliano belts along the west side of the São Francisco Craton (Heilbron *et al.* 2004), to 650–640 Ma in the Coastal Terrane of the Namibian Kaoko Belt (Seth *et al.* 1998; Franz *et al.* 1999), and from 620–580 Ma in the granite belt of the Dom Feliciano Belt (Basei *et al.* 2000). Southward migration of arc magmatism is further suggested by southward younging of the granite batholiths within the 'granite belt' of the Dom Feliciano Belt; the northernmost Florianopolis Batholith has an age of *c.* 620 Ma, the centrally located Pelotas Batholith an age of *c.* 610 Ma, and the southernmost Aguia Batholith an age of*c.* 580 Ma (Basei *et al.* 2000).

A 630–585 calc-alkaline magmatic arc in the Araçuaí Belt suggests that the main arc system may have followed the Brasiliano trend around the São Francisco Craton, rather than through the Araçuaí orogen, which shows a younger subduction-related closure of a Red Sea-type rift

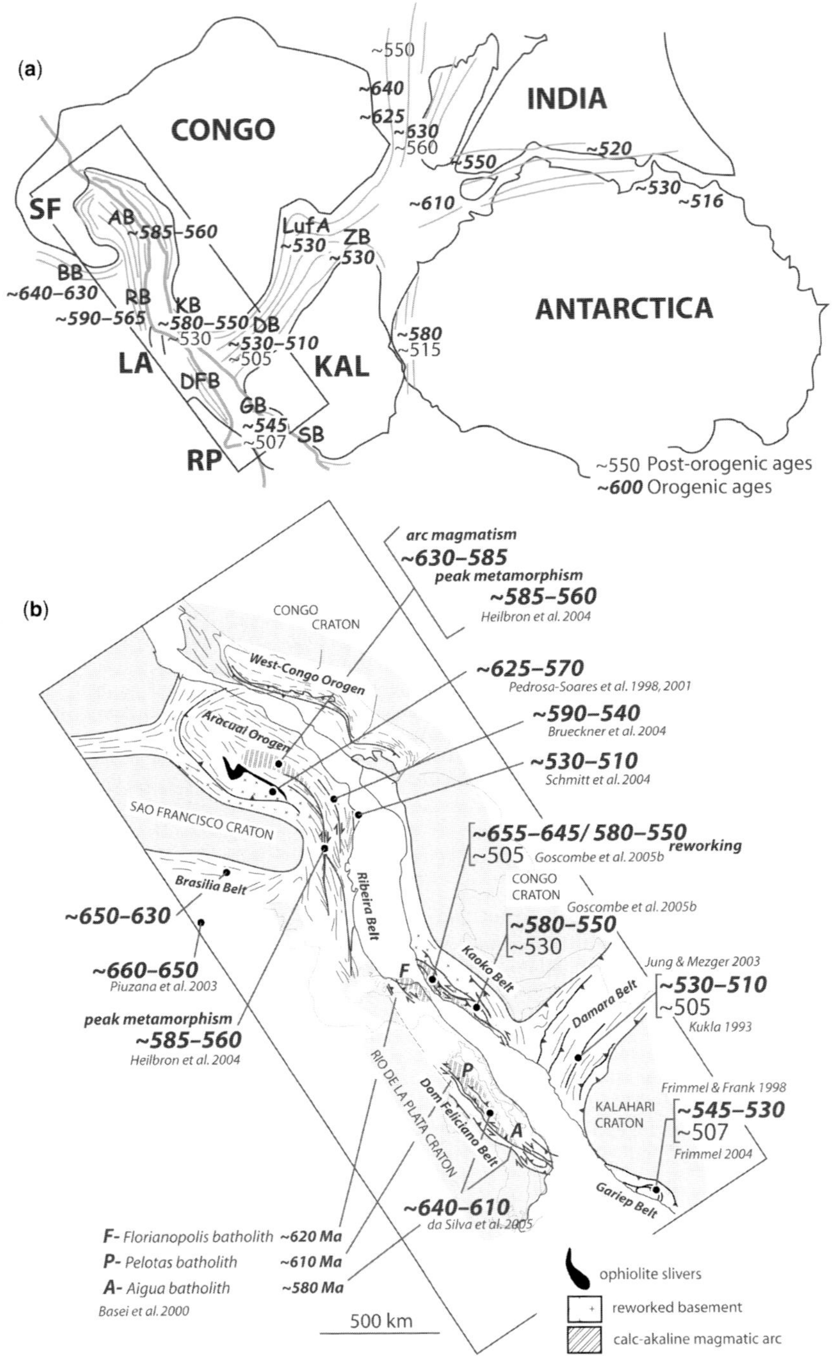

Fig. 12. Ages of orogenic suturing across Gondwana (modified from fig. 10 of Meert, 2003). The Gondwana reconstruction shows the various component orogens, the orogenic ages (bold italic) reflecting the timing of peak metamorphism/deformation, and post-orogenic ages (normal font) reflecting post-tectonic magmatism and therefore the timing of cratonisation, for each of the component belts. The inset (**b**) is an enlargement of the West Gondwana suture resulting from the closure of the Adamastor Ocean. The cratons include SF (São Francisco), LA (Luis Alves), RP (Río de la Plata), Kal (Kalahari), Congo, India and Antarctica. Component belts are BB, Brasília Belt; AB, Araçuaí Belt; RB, Ribeira Belt; KB, Kaoko Belt; DFB, Dom Feliciano Belt; GB, Gariep Belt; SB, Saldania Belt; DB, Damara Belt; LufA, Lufilian Arc; ZB, Zambezi Belt. Data sources are shown on the figure.

arm of the Adamastor Ocean (cf. Alkmin *et al.* 2006).

In summary, the closing of the Adamastor and Khomas oceans between three continental or cratonic blocks, the Río de la Plata, Congo and Kalahari cratons, resulted in a three-fold orogenic system or collisional triple junction during the welding of the Gondwana supercontinent. The differences in timing between deformation, metamorphism, and magmatism of the component belts of the Damara orogen provide a history of Gondwana suturing that is more refined than the palaeomagnetic data indicating that these cratonic nuclei were together by 550 Ma.

Support for the research was from an Australian Research Council (ARC) Large Grant A00103456 awarded to D.R.G and from NSF Grant 0440188 awarded to D.A.F. The manuscript was completed by D.R.G. as part of an Australian Professorial Research Fellowship supported on ARC Discovery Grant DP0210178 (awarded to D.R.G.). B.G. was supported as a Post-Doctoral Fellow from ARC Large Grant A00103456, and he acknowledges support from the University of Melbourne, the University of Adelaide and Monash University. We thank Gabbi Schneider (Director), Charlie Hoffmann (Deputy Director), Mimi Dunaiski and Thomas Bekker of the Namibian Geological Survey for support, and acknowledge discussions on Namibian geology with Charlie Hoffmann, Roy Miller, Paul Hoffman and Thomas Bekker. Nuno Machado and Chris Hawkesworth provided useful reviews that helped improve the manuscript.

References

AHRENDT, H., HUNZIKER, J. C. & WEBER, K. 1977. Age and degree of metamorphism and time of nappe emplacement along the southern margin of the Damara orogen/Namibia (SW- Africa). *Geologisches Rundschau*, **67**, 719–742.

ALKMIN, F. F., MARSHAL, S., PEDROSA-SOARES, A. C., PERES, G. G., CRUZ, S. C. P. & WHITTINGTON, A. 2006. Kinematic evolution of the Aracuai-West Congo orogen in Brazil and Africa: nutcracker tectonics during the Neoproterozoic assembly of Gondwana. *Precambrian Research*, **149**, 43–64.

BARNES, S. & SAWYER, E. 1980. An alternative model for the Damara mobile belt: ocean crust subduction and continental convergence. *Precambrian Research*, **13**, 297–336.

BASEI, M. A. S., SIGA, O., MASQUELIN, H., HARARA, O. M., REIS NETO, J. M. & PRECIOZZI, F. 2000. The Dom Feliciano Belt of Brazil and Uruguay and its foreland domain. *In*: CORDANI, U. G., MILANI, E. J., THOMAS FILHO, A. & CAMPOS, D. A. (eds) *Tectonic evolution of South America, 31st International Geological Congress, Río de Janiero, Brazil*, 311–334.

BASEI, M. A. S., FRIMMEL, H. E., NUTMAN, A. P., PRECIOZZI, F. & JACOB, J. 2005. A connection between the Neoproterozoic Dom Feliciano (Brazil/Uruguay) and Gariep (Namibia/South Africa) orogenic belts-evidence from a reconnaissance provenance study. *Precambrian Research*, **139**, 195–221.

BOGER, S. D. & MILLER, J. MCL. 2004. Terminal suturing of Gondwana and the onset of the Ross-Delamerian orogeny: the cause and effect of an Early Cambrian reconfiguration of plate motions. *Earth and Planetary Science Letters*, **219**, 35–48.

COLLINS, A. S. & PISAREVSKY, S. A. 2005. Amalgamating eastern Gondwana: the evolution of the circum-Indian orogens. *Earth Science Reviews*, **71**, 229–270.

COOMBS, D. S., LANDIS, C. A., NORRIS, R. J., SINTON, J. M., BORNS, D. J. & CRAW, D. 1976. The Dun Mountain Ophiolite belt, New Zealand, its tectonic setting, constitution, and origin, with special reference to the southern portion. *American Journal of Science*, **276**, 561–603.

COWARD, M. P. 1981. The junction between Pan African Mobile Belts in Namibia: its structural history. *Tectonophysics*, **76**, 59–73.

COWARD, M. P. 1983. The tectonic history of the Damaran Belt. *In*: MILLER, R. MCG. (ed.) *Evolution of the Damara orogen of South West Africa/Namibia*. Geological Society South Africa Special Publications, **11**, 409–421.

DAVIES, C. J. & COWARD, M. P. 1982. The structural evolution of the Gariep arc in southern Namibia (South-west Africa). *Precambrian Research*, **17**, 173–198.

DOWNING, K. N. & COWARD, M. P. 1981. The Okahandja Lineament and its significance for Damaran Tectonics in Namibia. *Neues Jahrbuch für Geologie und Mineralogie*, **70**, 972–1000.

DE KOCK, G. S., EGLINGTON, B., ARMSTRONG, R. A., HERMER, R. E. & WALRAVEN, F. 2000. U–Pb and Pb–Pb ages of the Naauwpoort Rhyolite, Kawakeup leptite and Okongava Diorite: implications for the onset of rifting and of orogenesis in the Damara Belt, Namibia. *In*: MILLER, R. MCG. (ed.) *Henno Martin Commemorative Volume*. Communications of the Geological Survey of Namibia, **12**, 81–88.

DÜRR, S. B. & DINGELDEY, D. P. 1996. The Kaoko belt (Namibia): Part of a late Neoproterozoic continental-scale strike-slip system. *Geology*, **24**, 503–506.

DU TOIT, A. L. 1937. *Our Wandering Continents*. Oliver & Boyd.

FOLLING, P. G., FRIMMEL, H. E. & ZARTMAN, R. E. 1998. Chemostratigraphical correlation and Pb–Pb dating of metasedimentary sequences in the Gariep Belt; evidence for two different diamictites. *Journal of African Earth Sciences*, **27**, 76–77.

FRANTZ, J. C. & BOTELHO, N. F. 2000, Neoproterozoic granitic magmatism and evolution of the eastern Dom Feliciano Belt in southeastern Brazil: a tectonic model. *Gondwana Research*, **3**, 7–19.

FRANZ, L., ROMER, R. L. & DINGELDEY, D. P. 1999. Diachronous Pan-African granulite-facies metamorphism (650 Ma and 550 Ma) in the Kaoko belt, NW Namibia. *European Journal of Mineralogy*, **11**, 167–180.

FRIMMEL, H. E. 1995. Metamorphic evolution of the Gariep Belt. *South African Journal of Geology*, **98**, 176–190.

FRIMMEL, H. E. 2000. The Pan-African Gariep Belt in southwestern Namibia and western South Africa.

In: MILLER, R. MCG. (ed.) *Henno Martin Commemorative Volume*. Communications of the Geological Survey of Namibia, **12**, 197–209.

FRIMMEL, H. E. & FRANK, W. 1998. Neoproterozoic tectono-thermal evolution of the Gariep Belt and its basement, Namibia and South Africa. *Precambrian Research*, **90**, 1–28.

FRIMMEL, H. E. & HARTNADY, C. J. H. 1992. Blue amphiboles and their significance for the metamorphic history of the Pan-African Gariep belt, Namibia. *Journal of Metamorphic Geology*, **10**, 651–669.

FRIMMEL, H. E., HARTNADY, C. J. H. & KOLLER, F. 1996. Geochemistry and tectonic setting of magmatic units in the Pan-African Gariep Belt, Namibia. *Chemical Geology*, **130**, 101–121.

GERMS, G. J. B. & GRESSE, P. G. 1991. The foreland basin of the Damara and Gariep orogens in Namaqualand and southern Namibia: stratigraphic correlations and basin dynamics. *South African Journal of Geology*, **94**, 159–169.

GOSCOMBE, B. D. & GRAY, D. R. 2007. The Coastal Terrane of the Kaoko Belt, Namibia: outboard arc-terrane and tectonic significance. *Precambrian Research*, **155**, 139–158.

GOSCOMBE, B. D., ARMSTRONG, R. & BARTON, J. M. 2000. Geology of the Chewore Inliers: Constraining the Mesoproterozoic to Palaeozoic evolution of the Zambezi Belt. *Journal of African Earth Sciences*, **30**, 589–627.

GOSCOMBE, B., HAND, M. & GRAY, D. R. 2003*a*. Structure of the Kaoko Belt, Namibia: progressive evolution of a classic transpressional orogen. *Journal of Structural Geology* **25**, 1049–1081.

GOSCOMBE, B., HAND, M., GRAY, D. R. & MAWBY, J. 2003*b*. The metamorphic architecture of a transpressional orogen: the Kaoko Belt, Namibia. *Journal of Petrology*, **44**, 679–711.

GOSCOMBE, B., GRAY, D. R. & HAND, M. 2004. Variation in metamorphic style along the northern margin of the Damara orogen, Namibia. *Journal of Petrology*, **45**, 1261–1295.

GOSCOMBE, B., GRAY, D. R. & HAND, M. 2005*a*. Extrusional tectonics in the core of a transpressional orogen: the Kaoko Belt, Namibia. *Journal of Petrology*, **46**, 1203–1241.

GOSCOMBE, B., GRAY, D. R., ARMSTRONG, R., FOSTER, D. A. & VOGL, J. 2005*b*. Event geochronology of the Pan-African Kaoko Belt, Namibia. *Precambrian Research*, **140**, 1–41.

GRAY, D. R., FOSTER, D. A., GOSCOMBE, B. D., PASSCHIER, C. W. & TROUW, R. A. J. 2006. $^{40}Ar/^{39}Ar$ thermochronology of the Pan-African Damara orogen, Namibia, with implications for tectonothermal and geodynamic evolution. *Precambrian Research*, **150**, 49–72.

GRAY, D. R., FOSTER, D. A., MAAS, R., SPAGGIARI, C. V., GREGORY, R. T., GOSCOMBE, B. D. & HOFFMANN, K. H. 2007. Continental growth and recycling by accretion of deformed turbidite fans and remnant ocean basins: examples from Neoproterozoic and Phanerozoic orogens. *In*: HATCHER, R. D., JR., CARLSON, M. P., MCBRIDE, J. H. & MARTÍNEZ CATALÁN, J. R. (eds) *The 4D Framework of Continental Crust*. Geological Society of America, Memoirs, doi: 10.1130/2007.1200(05).

GRESSE, P. G. 1994. Strain partitioning in the southern Gariep arc as reflected by sheath folds and stretching directions. *South African Journal of Geology*, **97**, 52–61.

GRESSE, P. G. & GERMS, G. J. B. 1993. The Nama foreland basin: sedimentation, major unconformity and bounded sequences and multisided active margin advance: *Precambrian Research*, **63**, 247–272.

GRESSE, P. G., FITCH, F. J. & MILLER, J. A. 1988. $^{40}Ar/^{39}Ar$ dating of the Cambro-Ordovician Vanrhynsdorp tectonite in southern Namaqualand. *South African Journal of Geology*, **91**, 257–263.

GRESSE, P. G., CHEMALE, F., DA SILVA, L. C., WALRAVEN, F. & HARTMANN, L. A. 1996. Late- to post-orogenic basins of the Pan-African—Brasiliano collision orogen in southern Africa and southern Brazil. *Basin Research*, **8**, 157–171.

GRUNOW, A., HANSON, R. & WILSON, T. R. 1996. Were aspects of Pan African deformation linked to Iapetus opening? *Geology*, **24**, 1063–1066.

HÄLBICH, I. W. & ALCHIN, D. J. 1995. The Gariep Belt; stratigraphic-structural evidence for obliquely transformed grabens and back-folded thrust-stacks in a combined thick-skin thin-skin structural setting. *Journal of African Earth Sciences*, **21**, 9–33.

HANSON, R. E. 2003. Proterozoic geochronology and tectonic evolution of southern Africa. *In*: YOSHIDA, M., WINDLEY, B. & DASGUPTA, S. (eds) *Proterozoic East Gondwana: Supercontinent Assembly and Breakup*. Geological Society, London, Special Publications, **206**, 427–463.

HANSON, R. E., WILSON, T. J. & MUNYANYIWA, H. 1994. Geologic evolution of the Neoproterozoic Zambezi orogenic Belt in Zambia. *Journal of African Earth Sciences*, **18**, 135–150.

HEILBRON, M. & MACHADO, N. 2003. Timing of terrane accretion in the Neoproterozoic-Eopaleozoic Ribeira orogen (SE Brasil). *Precambrian Research*, **125**, 87–112.

HEILBRON, M., PEDROSA-SOARES, A., NETO, M., DA SILVA, L., TROUW, R. & JANASI, V. 2004. Brasiliano orogens in Southeast and South Brazil. *In*: WEINBERG, R., TROUW, R., FUCK, R. & HACKSPACHER, P. (eds) *The 750–550 Ma Brasiliano Event of South America. Journal of the Virtual Explorer, Electronic Edition*, Volume 17, Paper 4.

HOFFMAN, P. F., SWART, R., ECKHARDT, E. F. & GUOWEI, H. 1994. *Damara orogen of northwest Namibia*. Geological Excursion Guide Geological Survey of Namibia.

HOFFMAN, P. F., HAWKINS, D. P., ISACHSEN, C. E. & BOWRING, S. A. 1996. Precise U–Pb zircon ages for early Damaran magmatism in the Summas Mountains and Welwitschia Inlier, northern Damara Belt, Namibia. *Communications of the Geological Survey of Namibia*, **11**, 47–52.

HOFFMAN, P. F., KAUFMAN, A. J., HALVERSON, G. P. & SCHRAG, D. P. 1998. A Neoproterozoic snowball Earth. *Science*, **281**, 1342–1346.

JACOB, R. E., MOORE, J. M. & ARMSTRONG, R. A. 2000. Zircon and titanite age determinations from igneous rocks in the Karibib District, Namibia: implications for Navachab vein-style gold mineralization. *In*: MILLER,

R. McG. (ed.) *Henno Martin Commemorative Volume*. Communication of the Geological Survey of Namibia, **12**, 157–166.

JOHN, T., SCHENK, V., HASSE, K., SCHERER, E. & TEMBO, F. 2003. Evidence for a Neoproterozoic ocean in south-central Africa from mid-oceanic-ridge-type geochemical signatures and pressure-temperature estimates of Zambian eclogites. *Geology*, **31**, 243–246.

JOHN, T., SCHENK, V, MEZGER, K & TEMBO, F. 2004. Timing and PT evolution of whiteschist metamorphismin the Lufilian Arc–Zambezi Belt orogen (Zambia): implications for the assembly of Gondwana. *Journal of Geology*, **112**, 71–90.

JUNG, S. & MEZGER, K. 2003. U–Pb garnet chronometry in high-grade rocks; case studies from the central Damara orogen (Namibia) and implications for the interpretation of Sm–Nd garnet ages and the role of high U-Th inclusions. *Contributions to Mineralogy and Petrology*, **146**, 382–396.

JUNG, S., HOERNES, S. & MEZGER, K. 2000. Geochronology and petrogenesis of Pan-African syn-tectonic S-type and post-tectonic A-type granite (Namibia)—products of melting of crustal sources, fractional crystallization and wall rock entrainment. *Lithos*, **50**, 259–287.

KASCH, K. W. 1983*a*. Regional P-T variations in the Damara orogen with particular reference to early high-pressure metamorphism along the southern margin. *In*: MILLER, R. McG. (ed.) *Evolution of the Damara orogen of South West Africa/Namibia*. Geological Society of South Africa Special Publications, **11**, 243–253.

KASCH, K. W. 1983*b*. Continental collision, suture progradation and thermal relaxation: a plate tectonic model for the Damara orogen in central Namibia. *In*: MILLER, R. McG. (ed.) *Evolution of the Damara orogen of South West Africa/Namibia*. Geological Society of South Africa Special Publications, **11**, 73–80.

KILLICK, A. M. 2000. The Matchless Belt and associated sulphide mineral deposits, Damar orogen, Namibia. *In*: MILLER, R. McG. (ed.) *Henno Martin Commemorative Volume*. Communications of the Geological Survey of Namibia, **12**, 197–209.

KISTERS, A. F. M., JORDAAN, L. S. & NEUMAIER, K. 2004. Thrust-related dome structures in the Karibib district and the origin of orthogonal fabric domains in the south Central Zone of the Pan-African Damara belt, Namibia. *Precambrian Research*, **133**, 283–303.

KONOPASEK, J., KRÖNER, S., KITT, S. L., PASSCHIER, C. W. & KRÖNER, A. 2005. Oblique collision and evolution of large-scale transcurrent shear zones in the Kaoko Belt, NW Namibia. *Precambrian Research*, **136**, 139–157.

KRÖNER, A. 1977. Precambrian mobile belts in southern and eastern Africa—ancient sutures or sites of ensialic mobility. A case for crustal evolution towards plate tectonics. *Tectonophysics*, **40**, 101–135.

KUKLA, C. 1993. Strontium isotope heterogeneities in amphibolite facies, banded metasediments—a case study from the Late Proterozoic Kuiseb Formation of the southern Damara orogen, Central Namibia. *Communications of the Geological Survey of Namibia Memoir*, **15**, 139 pp.

MCDERMOTT, F., HARRIS, N. B. W. & HAWKESWORTH, C. J. 2000. Geochemical constraints on the petrogenesis of Pan-African A-type granites in the Damara Belt, Namibia. *In*: MILLER, R. McG. (ed.) *Henno Martin Commemorative Volume*. Communications of the Geological Survey of Namibia, **12**, 139–148.

MACHADO, N., VALLADRES, C., HEILBRON, M. & VALERIANO, C. 1996. U–Pb geochronology of the central Ribeira belt (Brazil) and implications for the Brazilian orogeny. *Precambrian Research*, **79**, 347–361.

MALOOF, A. C. 2000. Superposed folding at the junction of the inland and coastal belts, Damara orogen, Namibia. *In*: MILLER, R. McG. (ed.) *Henno Martin Commemorative Volume*. Communications of the Geological Survey of Namibia, **12**, 89–98.

MARTIN, H. & EDER, F. W. 1980. (eds) *Intracontinental Fold Belts*. Berlin, Springer-Verlag.

MARTIN, H. & PORADA, H. 1977*a*. The intracratonic branch of the Damara orogen in southwest Africa I. Discussion of geodynamic models. *Precambrian Research*, **5**, 311–338.

MARTIN, H. & PORADA, H. 1977*b*. The intracratonic branch of the Damara orogen in southwest Africa II. Discussion of relationships within the Pan-African mobile belt system. *Precambrian Research*, **5**, 339–357.

MASBERG, P., MIHM, D. & JUNG, S. 2005. Major and trace element and isotopic (Sr, Nd, O) constraints for Pan-African crustally contaminated grey granite gneisses from the southern Kaoko belt, Namibia. *Lithos*, **84**, 25–50.

MEERT, J. G. 2003. A synopsis of events related to the assembly of eastern Gondwana. *Tectonophysics*, **362**, 1–40.

MEERT, J. G. & TORSVIK, T. H. 2003. A synopsis of events related to the assembly of eastern Gondwana. *Tectonophysics*, **375**, 261–288.

MEERT, J. G., VAN DER VOO, R. & AYUB, S. 1995. Palaeomagnetic investigation of the Neoproterozoic Gagwe lavas and Mbozi Complex, Tanzania and the assembly of Gondwana. *Precambrian Research*, **75**, 225–244.

MILLER, R. McG. 1983*a*. The Pan-African Damara orogen of Namibia. *In*: MILLER, R. McG. (ed.) *Evolution of the Damara orogen of south west Africa/Namibia*. Geological Society of South Africa Special Publications, **11**, 431–515.

MILLER, R. McG. (ed.) 1983*b*. *Evolution of the Damara orogen of south west Africa/Namibia*. Geological Society of South Africa Special Publications, **11**.

MILLER, R. McG. & GROTE, W. 1988. Geological Map of the Damara orogen, Namibia (scale 1:500,000). *Geological Survey of Namibia, Windhoek*.

MORTIMER, N. 2000. Metamorphic discontinuities in orogenic belts: example of garnet–biotite–albite zone in the Otago Schist, New Zealand. *International Journal of Earth Science*, **89**, 295–306.

OLIVER, G. J. H. 1994. Mid-crustal detachment and domes in the central zone of the Damaran orogen, Namibia. *Journal of African Earth Sciences*, **19**, 331–344.

PASSCHIER, C. W., TROUW, R. A. J., RIBEIRO, A. & PACIULLO, F. V. P. 2002. Tectonic evolution of the

southern Kaoko Belt, Namibia. *Journal of African Earth Sciences*, **35**, 61–75.
PEDROSA-SOARES, A. C., NOCE, A. C., WIDEMANN, C. M. & PINTO, C. P. 2001. The Araçuaí–West Congo orogen in Brazil: an overview of a confined orogen formed during Gondwanaland assembly. *Precambrian Research*, **110**, 307–323.
PESONEN, L. J., ELMING, S.-A. *ET AL.* 2003. Palaeomagnetic configuration of continents during the Proterozoic. *Tectonophysics*, **375**, 289–324.
PORADA, H. 1979. The Damara-Ribeira orogen of the Pan-African-Brasiliano cycle in Namibia (Southwest Africa) and Brazil as interpreted in terms of continental collision. *Tectonophysics*, **57**, 237–265.
PORADA, H. 1989. Pan-African rifting and orogenesis in southern to Equatorial Africa and eastern Brazil. *Precambrian Research*, **44**, 103–136.
PORADA, H., AHRENDT, H., BEHR, J. & WEBER, K. 1983. The join of the coastal and intracontinental branches of the Damara orogen, Namibia, South West Africa. *In*: MARTIN, H. & EDER, F. W. (eds) *Intracontinental Fold Belts*. Berlin, Springer-Verlag, 901–912.
PRAVE, A. R. 1996. Tale of three cratons: Tectonostratigraphic anatomy of the Damara orogen in northwestern Namibia and the assembly of Gondwana. *Geology*, **24**, 1115–1118.
RAPALINI, A. E. 2006. New late Proterozoic palaeomagnetic pole for the Río de la Plata Craton: implications for Gondwana. *Precambrian Research*, **147**, 223–233.
SCHMITT, R. S., TROUW, R. A. J., VAN SCHMUS, W. R. & PIMENTEL, M. M. 2004. Late amalgamation in the central part of West Gondwana: new geochronological data and the characterisation of a Cambrian collisional orogeny in the Ribeira Belt (SE Brazil). *Precambrian Research*, **133**, 29–61.
SETH, B., KRÖNER, A., MEZGER, K., NEMCHIN, A. A., PIDGEON, R. T. & OKRUSCH, M. 1998. Archaean to Neoproterozoic magmatic events in the Kaoko belt of NW Namibia and their geodynamic significance. *Precambrian Research*, **92**, 341–363.
SINGLETARY, S. J., HANSON, R. E. *ET AL.* 2003. Geochronology of basement rocks in the Kalahari Desert, Botswana, and implications for regional; Proterozoic tectonics. *Precambrian Research*, **121**, 47–71.
STANNISTREET, I. G., KUKLA, P. A. & HENRY, G. 1991. Sedimentary basinal sequences to a Late Precambrian Wilson cycle: the Damara orogen and Nama foreland, Namibia. *Journal of African Earth Sciences*, **13**, 141–156.
TOHVER, E., D'AGRELLA-FILHO, M. S. D. & TRINDADE, R. I. F. 2006. Palaeomagnetic record of Africa and South America for the 1200–500 Ma interval, and evaluation of Rodinia and Gondwana assemblies. *Precambrian Research*, **147**, 193–222.
TROMPETTE, R. 1997. Neoproterozoic (*c.* 600 Ma) aggregation of western Gondwana: a tentative scenario. *Precambrian Research*, **82**, 101–112.
VALERIANO, C. M., MACHADO, N., SIMONETTI, A., VALLADRES, C. S., SEER, H. J. & SIMOES, L. A. 2004. U–Pb geochronology of the southern Brasilia belt (SE-Brazil): sedimentary provenance, Neoproterozoic orogeny and assembly of West Gondwana. *Precambrian Research*, **130**, 27–55.
VALERIANO, C. M., PIMENTEL, M. M., HEILBORN, M., ALMEIDA, J. C. H. & TROUW, R. A. J. 2008. Tectonic evolution of the Brasília Belt, Central Brazil, and early assembly of Gondwana. *In*: PANKHURST, R. J., TROUW, R. A. J., BRITO NEVES, B. B. & DE WIT, M. J. (eds) *West Gondwana: Pre-Cenozoic Correlations Across the South Atlantic Region.* Geological Society, London, Special Publications, **294**, 197–210.
WILL, T. M., OKRUSCH, M. & GRUNER, B. B. 2004. Barrovian and Buchan type metamorphism in the Pan-African Kaoko belt, Namibia: implications for its geotectonic position within the framework of Western Gondwana. *South African Journal of Geology*, **107**, 431–454.
YARDLEY, B. W. D. 1982. The early metamorphic history of the Haast Schists and related rocks of New Zealand. *Contributions to Mineralogy and Petrology*, **81**, 317–327.

Cambrian orogeny in the Ribeira Belt (SE Brazil) and correlations within West Gondwana: ties that bind underwater

R. S. SCHMITT[1], R. A. J. TROUW[2], W. R. VAN SCHMUS[3] & C. W. PASSCHIER[4]

[1]*Faculdade de Geologia, Universidade do Estado do Rio de Janeiro, Maracanã, Rio de Janeiro, CEP: 20550-013, Brazil (e-mail: renataschmitt@uol.com.br)*

[2]*Depto. de Geologia, Universidade Federal do Rio de Janeiro, Ilha do Fundão, Rio de Janeiro, CEP: 21900-900, Brazil*

[3]*Department of Geology, University of Kansas, 120 Lindley Hall, Lawrence, KS 66045, USA*

[4]*Institut für Geowissenschaften, Johannes Gutenberg Universität, Becherweg 21, 55099, Mainz, Germany*

Abstract: A 530–490 Ma tectono-metamorphic event, the Búzios orogeny, is recognized within the Ribeira Belt, along the coast of SE Brazil. Tectonic evolution started with a Late Neoproterozoic marine basin and volcanic activity at *c.* 610 Ma. The rocks in this basin were affected by high-grade metamorphism at *c.* 530 Ma, coeval with deformational phases D_1–D_2, which generated compressive low-angle tectonic structures with top-to-NW tectonic transport. Large recumbent folds with NW–SE axes parallel to the main stretching lineation formed during D_3 as the Cabo Frio tectonic domain, the focus of this study, collided with the Oriental terrane to the NW. D_4 sub-vertical shear zones are limited in extent. A new U–Pb age of 501 ± 6 Ma is reported for zircon from an amphibolite-facies shear zone related to either D_3 or D_4. Post-tectonic 440 Ma pegmatites mark the final stage of tectono-magmatic activity. The Cabo Frio tectonic domain has African affinities and is exotic to the Ribeira Belt. Middle Cambrian deformational and metamorphic ages are also reported from the 'Angolan' Pan-African belt, the southern Kaoko and Damara belts in Namibia, and the Cuchilla Dionisio–Punta Del Este terrane in Uruguay. The occurrence of Cambrian metamorphic rocks along the present African and South American coastlines shows that Mesozoic rifting closely follows Palaeozoic sutures of West Gondwana.

Although most collisional events that led to the amalgamation of Gondwana occurred at the end of the Neoproterozoic, growing evidence points to important tectonism extending into the Cambrian (Schmitt *et al.* 2004; Collins & Pisarevsky 2005; Trindade *et al.* 2006). Particularly in the coastal areas of SE South America and SW Africa, in the Dom Feliciano, Ribeira and Kaoko belts (Fig. 1), tectono-metamorphic events in the age range of 530–480 Ma have been recognized (Kukla 1993; Campos Neto & Figueiredo 1995; Jung *et al.* 2000; Jung & Mezger 2003; Bossi & Gaucher 2004; Schmitt *et al.* 2004; Goscombe *et al.* 2005; Delor *et al.* 2006). During this time interval, West Gondwana was consolidating and, among other tectonic events, Late Neoproterozoic–Cambrian sedimentary foreland successions were deformed and metamorphosed at low grade (Germs 1995; Gaucher *et al.* 2003). The recognition of medium- to high-pressure metamorphism at *c.* 525 Ma in these belts suggests that the Cambrian history of Gondwana probably includes collisional events in addition to the well known extensional magmatism exposed in low-pressure and high-temperature terranes (e.g., de Wit *et al.* 2001; Pedrosa-Soares *et al.* 2001; Heilbron & Machado 2003; Rapela *et al.* 2003). The occurrence of Cambrian metamorphic rocks along the present coastlines of Africa (Angolan, Kaoko and Damara belts) and South America (Ribeira and Dom Feliciano belts) shows that Mesozoic South Atlantic Ocean rifting closely follows Palaeozoic sutures of West Gondwana (Fig. 1) (Kukla 1993; Bossi & Gaucher 2004; Schmitt *et al.* 2004).

We present a discussion of data from the coastal Ribeira Belt, SE Brazil, and compare both sides of the Atlantic Ocean, being aware that the submerged continental crust in the shelves may hide information: 'the ties bind underwater'. We use geochronological data (U–Pb in single zircons) together with the results of recent petrographical and structural mapping of an onshore tectonic domain including islands along the coast of Rio de Janeiro State in Brazil and correlation with offshore geophysical information. Emphasis is placed on the sequence of deformation of the Cambrian tectonic events recorded in the Ribeira Belt, together with a review of related published data.

From: PANKHURST, R. J., TROUW, R. A. J., BRITO NEVES, B. B. & DE WIT, M. J. (eds) *West Gondwana: Pre-Cenozoic Correlations Across the South Atlantic Region.* Geological Society, London, Special Publications, **294**, 279–296. DOI: 10.1144/SP294.15 0305-8719/08/$15.00

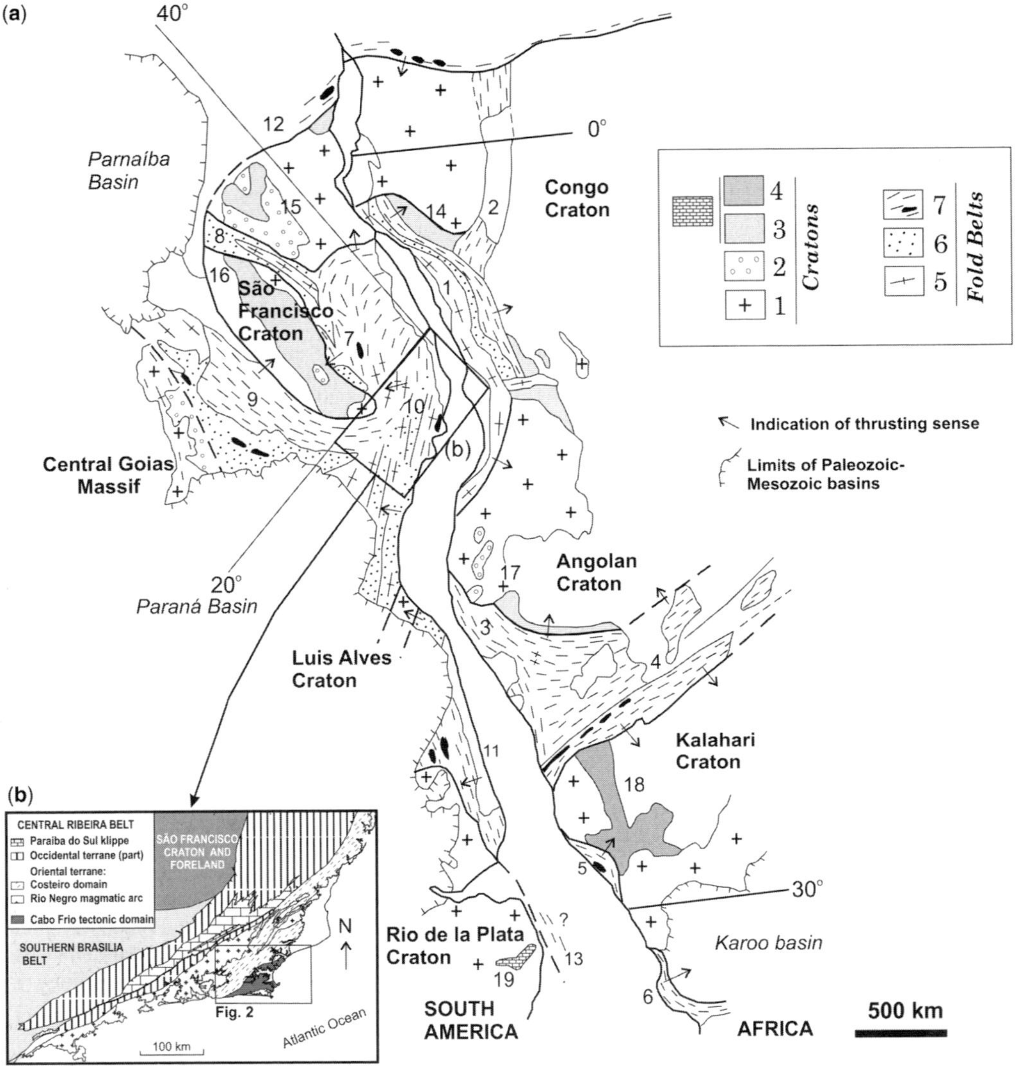

Fig. 1. (**a**) Schematic map of Pan-African/Brasiliano belts fringing the South Atlantic Ocean (modified after Trompette 1994). Main fold belts: (1) West Congo; (2) Central Africa; (3) Kaoko; (4) Damara; (5) Gariep; (6) Saldania; (7) Araçuaí; (8) Paramimim aulacogen; (9) Brasília; (10) Ribeira; (11) Dom Feliciano; (12) NE Brazilian province; (13) assumed Brasiliano metasedimentary rocks found in a drill-hole east of Mar del Plata. Main Pan-African/Brasiliano cratonic sequences adjacent to the peri-South Atlantic orogens: (14) West Congo Basin; (15 and 16) Bambuí sequence and correlates; (17) Chela Supergroup; (18) Nama Supergroup; (19) Sierra de Tandíl. Lithostratigraphic legend: Cratons: (1) basement; (2) Early–Middle Proterozoic cover rocks; (3) Neoproterozoic cover rocks; (4) latest Proterozoic and early Cambrian cover rocks. Fold Belts: (5) reworked basement; (6) Early–Middle Proterozoic metasedimentary rocks; (7) Neoproterozoic metasedimentary rocks with mafic and ultramafic intercalation. Location of insert (b) is indicated by the rectangle. (**b**) Simplified tectonic map of the Central Ribeira Belt, the southern Brasília Belt and the southern part of the Sao Francisco craton (modified after Trouw *et al.* 2000 and Heilbron & Machado 2003). Note the location of Figure 2.

Geological setting

The Cabo Frio tectonic domain (CFTD) crops out in the southeastern part of the Neoproterozoic–Palaeozoic NE–SW-trending Ribeira Belt, along the Atlantic margin of SE Brazil (Figs 1 & 2; Brito Neves & Cordani 1999; Trouw *et al.* 2000). The Ribeira Belt extends along the southern and southeastern border of the São Francisco Craton, continuing to the north as the Araçuaí Belt and to the SW towards the Curitiba microplate, north of the Luis Alves block (Fig. 1). To the west, the central part of the Ribeira Belt develops an interference zone with the slightly older Brasília Belt (Fig. 1).

In a geological section from NW to SE, four tectonic domains have been described in the Ribeira Belt (Fig.1; Heilbron *et al.* 2000; Heilbron & Machado 2003): (a) the Occidental terrane, composed of slices of reworked São Francisco Craton basement, interleaved with Meso–Neoproterozoic deformed passive margin sediments; (b) the Paraíba do Sul klippe, a tectonic slice of granulite facies rocks overlying the Occidental terrane; (c) the Oriental terrane, composed of paragneisses (Costeiro domain) intruded by the Rio Negro arc with crystallization ages between 780 and 620 Ma (Tupinambá *et al.* 2000), and several post-620 Ma plutonic rocks; (d) the Cabo Frio tectonic domain. The origin of the Oriental terrane and the Cabo Frio tectonic domain is not well constrained. They may be related to the São Francisco Craton margin, to the Congo Craton margin in W–SW Africa or even to a separate microcontinent (Valladares *et al.* 2008).

Heilbron & Machado (2003) suggested that the Oriental terrane collided with the Occidental terrane at *c.* 580 Ma as a result of the southeastward subduction of the São Francisco plate. The Cabo Frio tectonic domain registers younger tectono-metamorphic ages, *c.* 525 Ma, probably due to late docking of the Congo plate during the final stages of Gondwana amalgamation, defined as the Búzios orogeny by Schmitt *et al.* (2004). This compressive tectonic event is detailed below with special attention to the chronology of deformation and the relation with other SW Gondwana tectonic domains.

The Cabo Frio tectonic domain is separated from the Neoproterozoic Oriental terrane by a major NE–SW orientated thrust zone, dipping 35° to the SE, which is well defined in the northern portion of the domain (Fig. 2). Following the thrust along its strike to the SW, it is strongly affected by a late NE–SW shear zone, and the limit between the CFTD and the Oriental terrane is only inferred (Fig. 2). To the east the CFTD is hidden under the Atlantic Ocean (Fig. 2). It comprises a Palaeoproterozoic allochthonous basement tectonically interleaved with a mafic gneiss and paragneiss association of late Neoproterozoic age (Schmitt *et al.* 2004; Figs 2 and 3).

The basement is predominantly composed of felsic orthogneisses (with tonalitic to syenogranitic composition), called the Região dos Lagos unit. These metagranitoids have U–Pb zircon crystallization ages ranging from 2.0 to 1.95 Ga (upper discordia intercept, Schmitt 2001; Schmitt *et al.* 2004). There is more than one generation of mafic rocks (Forte de São Mateus unit; Figs 2 and 3), occurring as xenoliths, tectonic lenses and dykes within these felsic orthogneisses. A single U–Pb zircon age of 1.96 Ga was obtained for the crystallization of the mafic protolith (upper discordia intercept, Schmitt *et al.* 2004).

The mafic gneiss and paragneiss association (representing metavolcano-sedimentary protoliths) is at least 500 m thick and contains four compositional groups: aluminous rocks with kyanite/sillimanite, calc-silicate rocks, quartz–feldspathic rocks, and meta-basites. The compositional layering of the association is interpreted to reflect sedimentary–volcanic bedding. Two main sedimentary–volcanic successions were recognized (Fig. 3), based on the lithotypes. The Búzios succession is a thick, aluminous metasediment package (sillimanite–kyanite–garnet–biotite gneiss) with numerous calc-silicate and mafic intercalations. The Palmital succession is constituted mainly by quartzo-feldspathic paragneisses (>300 m), with some sillimanite-bearing aluminous intercalations, calc-silicate rocks, feldspathic quartzite layers (up to 3 m thick) and minor quartzites. The massive layers of meta-basite from the Búzios succession have chemical signatures similar to E-MORB type magmatic rocks (Schmitt *et al.* 2008). These successions are interpreted as representing a volcano-sedimentary basin developing in an ocean-floor environment, named the Buzios–Palmital basin. U–Pb SHRIMP analyses on detrital zircons from the Palmital succession (Schmitt *et al.* 2003, 2004) revealed crystallization ages of 2.6, 2.0, 1.2, 1.0, 0.8 and 0.6 Ga for the source rocks. The 1.2 and 1.0 Ga source rocks are recognized on the African side of the Atlantic Ocean, but not in the adjacent Brazilian terranes (see discussion). The youngest detrital zircon is 620 Ma old, which is interpreted as the maximum depositional age of the sediment. Sm–Nd data for whole-rock samples of the metavolcano-sedimentary unit (Fonseca 1993; Schmitt *et al.* 2004) indicate T_{DM} model ages ranging from 1.75 to 1.0 Ga, the last one representative of the mafic gneisses. A Sm–Nd whole-rock isochron of the ortho-amphibolite layers yielded an age of 608 Ma for crystallization, which roughly coincides with the youngest detrital zircon age (Schmitt *et al.* 2008).

Structural analysis

The deformational Structures in the Cabo Frio tectonic domain are interpreted as the record of at least four deformational phases related to the Búzios orogeny. During these phases, the

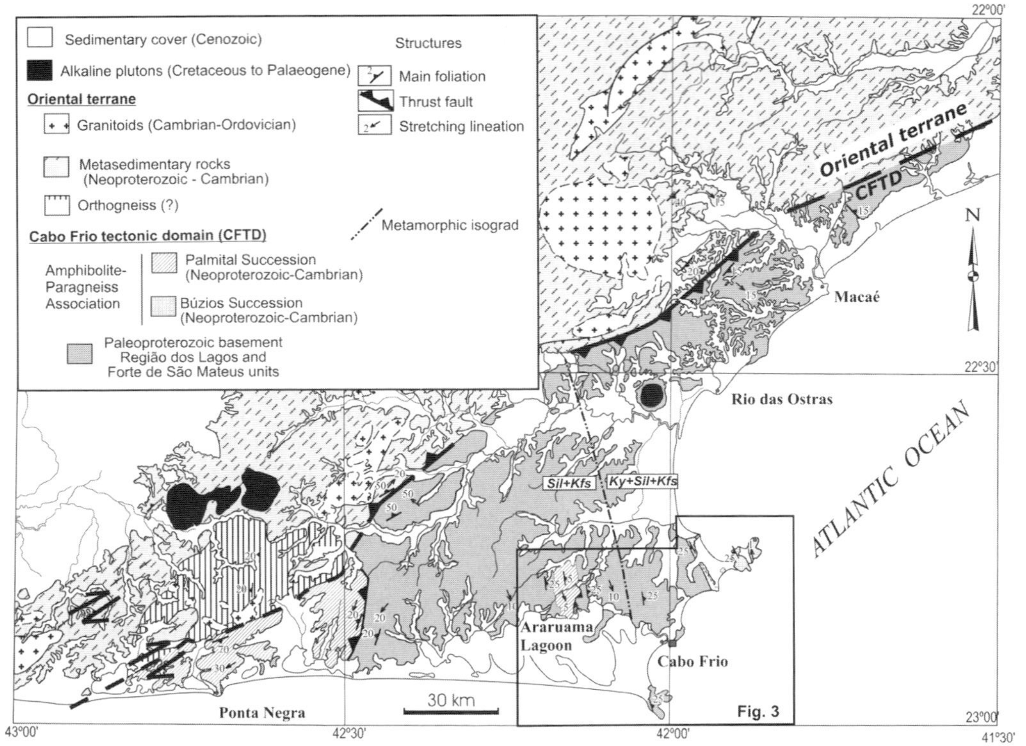

Fig. 2. Tectonic map of the Cabo Frio tectonic domain and adjacent areas of the Oriental terrane. Modified from Reis & Mansur (1995), Fonseca (1998) and Schmitt (2001). Note location of Figure 3.

lithostratigraphic units showed ductile and ductile–brittle behaviour associated with high-grade regional metamorphism. Both basement and the metavolcano-sedimentary association present geometrically similar structures. The first three deformational phases (D_1, D_2 and D_3) generated a penetrative low- to medium-angle foliation and mineral lineation, associated with NW–SE subhorizontal thrusting. Deformational phase D_4 is related to the same regional shortening event; producing high-angle transpressional shear zones associated with a variably orientated stretching lineation.

In the basement the deformation is strongly partitioned into low- and high-strain domains (Fig. 3). The low-strain domains preserve primary structures such as cross-cutting contacts between distinct magmatic facies, angular xenoliths and east–west alignment of K-feldspar phenocrysts. A pre-intrusive deformation is indicated by angular amphibolite xenoliths that exhibit metamorphic foliation and amphibole lineation. The 2.0 Ga igneous rocks that contain the xenoliths do not have this foliation, which therefore represents a Palaeoproterozoic or older tectono-metamorphic event.

D_1/D_2

D_1 and D_2 structures are grouped together because of their similar geometry and deformation style. The local presence of intrafolial folds within the main metamorphic foliation (S_2 or $S_{1/2}$) justifies the recognition of two separate phases. Frequently S_1 and S_2 are parallel; therefore D_2 is interpreted as a progressive phase in the same ductile regime as D_1. $S_{1/2}$ in the basement is parallel to compositional banding between the different lithotypes, including the ortho-amphibolites and leucosome veins (Fig. 4a). In the metavolcano-sedimentary sequence $S_{1/2}$ is subparallel to the original compositional layering (S_0) and also reveals a stromatitic structure. Biotite and quartz–feldspar aggregates largely define the foliation (Fig. 4b). The $S_{1/2}$ foliation generally dips in various directions at a very low angle and, in some regions, is subhorizontal (Fig. 5a, b). A subhorizontal NW–SE to north–south trending stretching lineation was identified in all lithotypes (Fig. 5d, e). This penetrative lineation, usually defined by elongated quartz and feldspar aggregates, shows a regular pattern with minor variation in orientation. The amphibolites

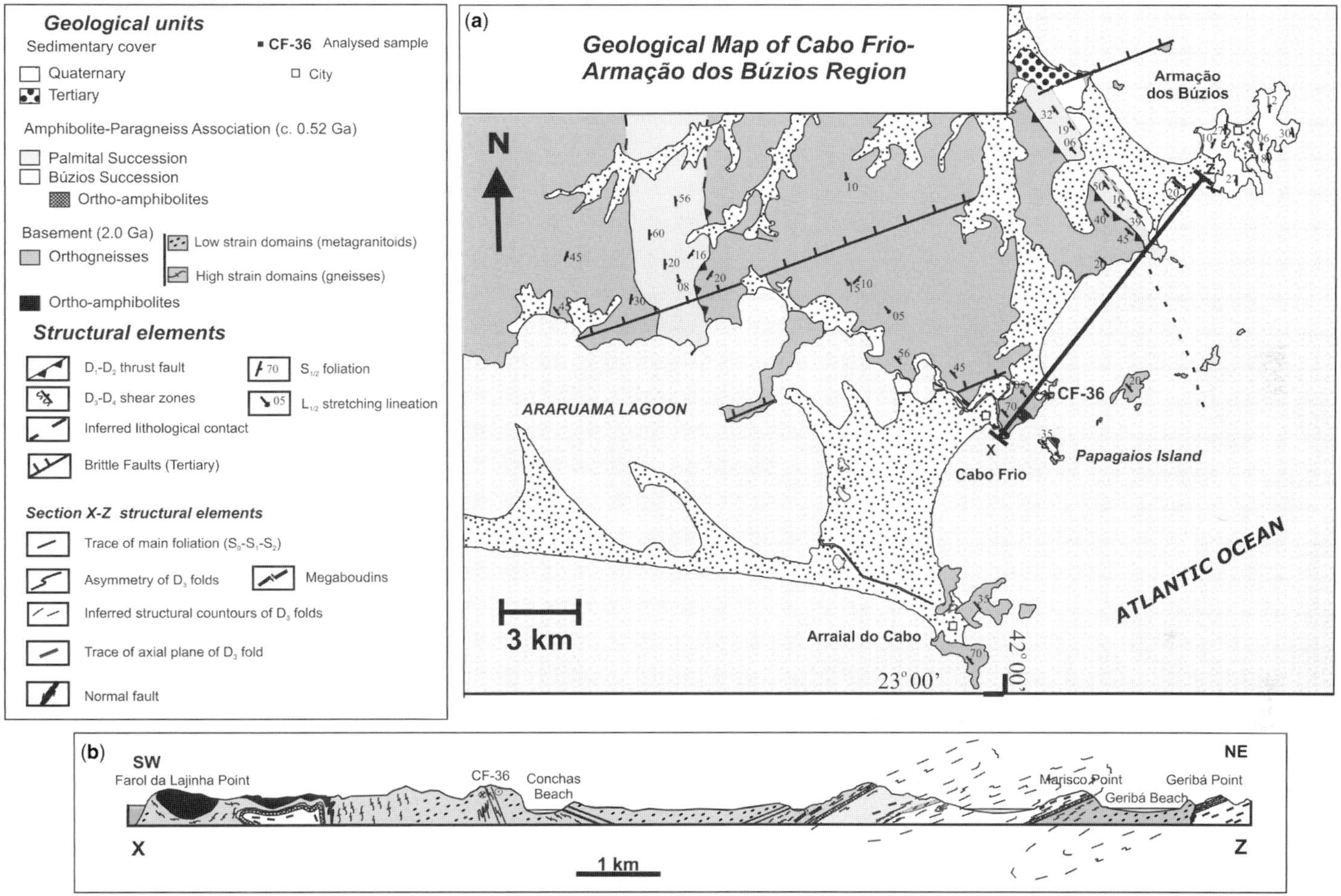

Fig. 3. (**a**) Geological map of Cabo Frio and Buzios area and location of section (b). (**b**) A SW–NE cross-section.

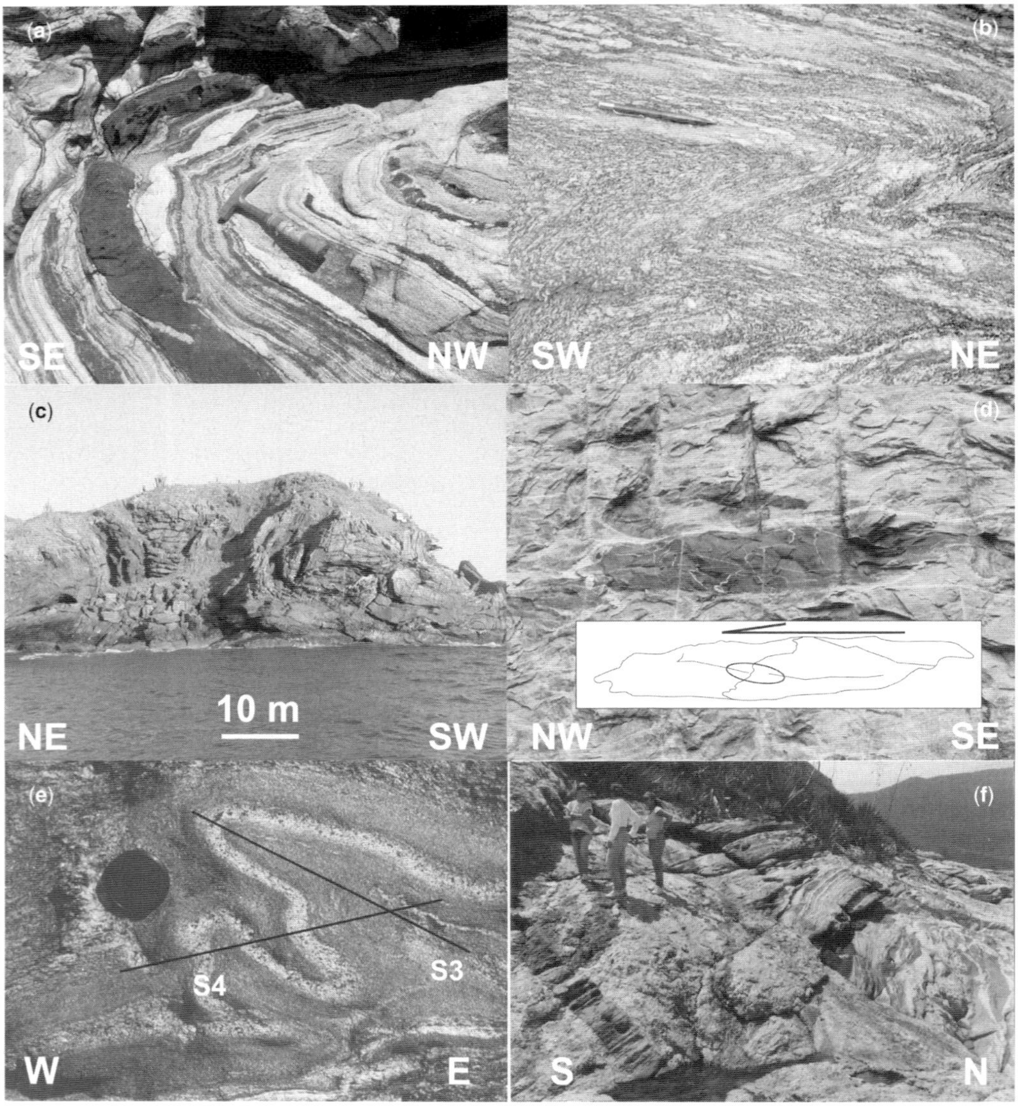

Fig. 4. (**a**) Basement orthogneiss. $S_{1/2}$ composed of meta-tonalite, leucosome veins and ortho-amphibolite layers. Note the mafic boudins within the foliation folded in D_3 or D_4 folds. (**b**) $S_{1/2}$ foliation in paragneiss from the Palmital succession. Note the sillimanite lineation folded in D_3 folds. (**c**) $S_{1/2}$ main foliation folded in recumbent D_3 folds within the amphibolite–paragneiss association, Papagaios Island (see location in Fig. 3). (**d**) Ortho-amphibolite boudin within the basement orthogneiss. Note the leucocratic veins in the boudin indicating movement top-to-NW. Orientation of the strain ellipse indicated. (**e**) Refolding pattern in a sillimanite–biotite paragneiss from Palmital succession. Note D_3 fold asymmetry with west-vergence and the interference pattern related to D_4. Pattern identified only in the western part of the Cabo Frio tectonic domain. (**f**) Post-tectonic pegmatite with east–west strike dated at 440 Ma by Schmitt *et al.* (2004). Note a strip of xenolith from the basement in the middle of the dyke. The main foliation seen is $S_{1/2}$ in the basement.

contain an amphibole mineral lineation parallel to the mineral lineation in the orthogneisses. Within the paragneisses, kyanite and sillimanite crystals are aligned and sometimes boudinaged parallel to this NW–SE to north–south trend (Fig. 5f). All stretching and mineral lineations plunge mainly either to NW or to SE at low angles (0° to 10°). An L-tectonite fabric is common in the basement sheets. Some dioritic enclaves are also stretched parallel to the lineation. In thin section, a

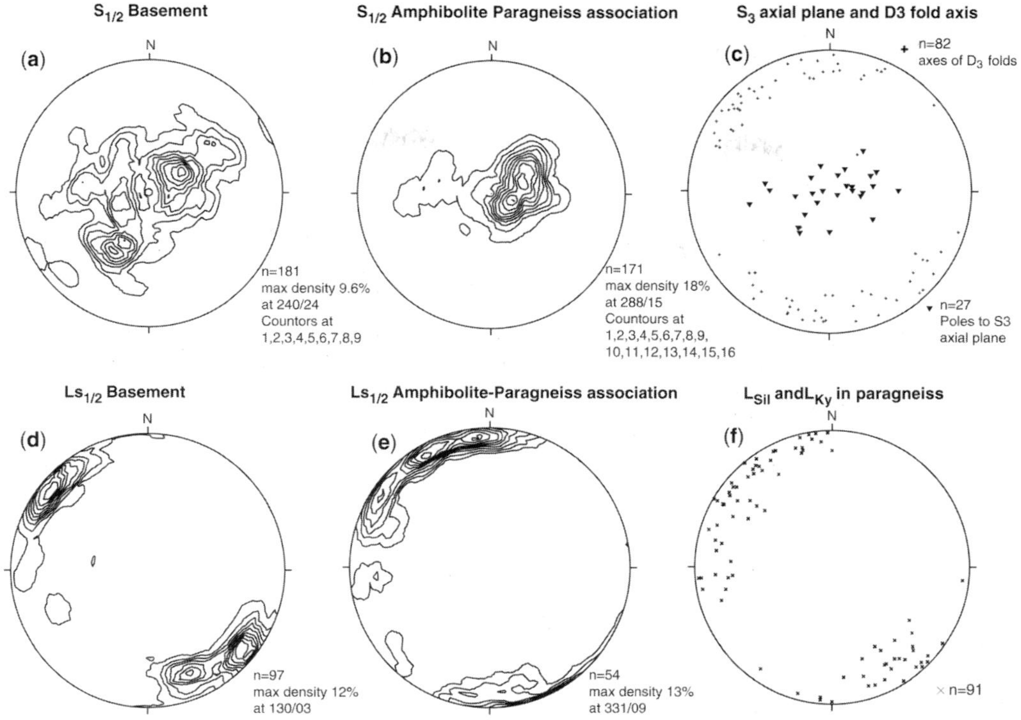

Fig. 5. Stereograms of the main structural elements of the Cabo Frio tectonic domain, related to D_1, D_2 and D_3. (**a**) Poles to S_1/S_2 foliation in the basement. (**b**) Poles to S_1/S_2 foliation in the amphibolite–paragneiss association. (**c**) Poles to S_3 foliation and axial plane of D_3 folds. Crosses represent F_3 fold axes. (**d**) Stretching lineation related to D_1/D_2 in the basement. (**e**) Stretching lineation related to D_1/D_2 in the amphibolite–paragneiss association. (**f**) Sillimanite and kyanite mineral lineations in the paragneisses.

granoblastic texture usually obliterates any previous mylonitic fabric, due to static recrystallization. Quartz ribbons, biotite alignment and K-feldspar aggregates orientated preferentially parallel to the NW–SE trend can locally be recognized.

Boudinage related to D_1/D_2 can be observed in the basement ortho-amphibolites and in the calc-silicate and meta-mafic rocks from the mafic gneisses and paragneiss association (Fig. 4a, e). The boudins are mostly extended in north–south or NW–SE, and east–west or NE–SW directions, developing a typical chocolate tablet pattern. The boudin necks are filled with leucosome (K-feldspar, quartz and garnet in the paragneisses; plagioclase and amphibole in the ortho-amphibolites). Due to subsequent D_3 deformation, it is difficult to estimate the extension, but a rough estimate is at least 10%.

There are very few kinematic indicators related to the $L_{1/2}$ stretching lineation. The sense of movement is interpreted as top-to-NW based on garnet porphyroblasts with sigma-type asymmetric strain shadows, C′ shear planes dipping 30° to NW and syntectonic veins in different directions. An example of these veins is an amphibolite boudin in the basement that shows two sets of quartz–feldspathic veins: one is folded by shortening and the other, perpendicular, is stretched (Fig. 4d). This geometric arrangement reveals the oblique position of the strain ellipsoid, indicating shear with top-to-NW. According to the literature this is a common kinematic indicator in high-grade rocks, where such indicators, in general, are rare (Passchier 1990; Passchier & Coelho 2006).

Most of the contacts between the basement and the mafic gneiss and paragneiss association are inverted. The basement rocks are strongly deformed in 50 to 150 m thick zones along these contacts (Fig. 3b), which are interpreted as D_1–D_2 ductile thrust zones. These detachment surfaces and shear zones are located preferentially along the compositional contacts (e.g., basement/metapelites) and pre-existing structures such as bedding. Within the allochthonous basement sheets, isolated mylonitic shear zones with some L-tectonites are preserved.

The stromatitic structure in both units associated with $S_{1/2}$ suggests that the metamorphic peak was contemporaneous with the D_1 and D_2 deformational phases. The P–T conditions were of at least 9 kbar and 780 °C, transitional from amphibolite to granulite facies (Schmitt *et al.* 2004). These conditions were reached in the eastern part of the Cabo Frio tectonic domain, defined by the coexistence of kyanite and K-feldspar (Fig. 2). To the west, where only sillimanite occurs, the metamorphic peak probably did not pass the upper amphibolite facies field.

D_3

Spectacular meso- to macroscopic, tight to isoclinal, NW–SE to NE–SW trending folds with subhorizontal axes and axial planes were generated during D_3 (Figs 4b, c and 5c). The main foliation $S_{1/2}$, the thrust contacts, and the boudinage structures were folded in these recumbent folds, with axes parallel to the main mineral lineation ($L_{1/2}$). An axial plane cleavage (S_3) is defined by biotite aggregates (Fig. 5c). A few leucosome veins are orientated parallel to the F_3 axial planes, cross-cutting the leucosome veins generated during D_1–D_2. The folds are cylindrical and some garnet-gneiss layers are continuous for several kilometres. Extension along the fold axes is evidenced by boudinaged layers and a quartz–feldspar stretching lineation (L_3).

The vergence of these D_3 folds varies from ENE in the eastern portion of the Cabo Frio tectonic domain to WSW in the western portion (Fig. 3b). This opposite vergence probably means that the main transport was top-to-NW, parallel to the D_3 fold axes, possibly related to the geometry of a major basement sheet in a sheath- or curtain-fold model (Passchier & Trouw 2005). Locally there is also an east–west mineral lineation (L_3) formed by fibrolitic sillimanite (Fig. 5f), plunging 5–25°, almost orthogonal to $L_{1/2}$, and restricted to the F_3 axial plane foliation. Kinematic indicators such as S/C structures, rotated porphyroblasts and mica fish indicate top-to-west and top-to-east transport in alternating fold limbs. Locally, some fold limbs are disrupted subparallel to the axial plane foliation along small scale shear zones (1 cm to 2 metres wide) without lineation and usually filled with quartz–feldspar veins.

During D_3, the metamorphic conditions were at lower P and/or T than during D_1–D_2, following a clockwise return path in the P–T diagram (Schmitt *et al.* 2004), mainly related to decompression. This clockwise path is corroborated by the growth of sillimanite parallel to L_3 and the folding of the leucosome associated with D_1–D_2. However, the presence of quartz–feldspar veins with composition similar to the D_1–D_2 leucosome and parallel to the F_3 axial planes suggests that there was still melt present during this phase.

The regional thrust contact between the Cabo Frio tectonic domain and the Oriental terrane to the NW is interpreted as related to D_3 (Fig. 2). In this shear zone, kinematic indicators including S/C structures, disrupted fold limbs and C′ shear bands show top-to-NW transport (Fig. 2). The presence of sillimanite oriented down-dip in the plane of the shear zone indicates that the metamorphic grade was still in the amphibolite facies. Shear sense indicators are frequent and well preserved in this D_3 thrust, whereas the D_1/D_2 shear zones show very few kinematic indicators, probably because the higher metamorphic conditions during D_1/D_2 favoured complete recrystallisation and crystal growth.

D_3 is largely responsible for the present arrangement of the lithostratigraphic units. It is considered progressive with respect to D_1–D_2 for the following reasons: (1) the L_3 stretching lineation and fold axes are parallel to similar D_1/D_2 structures; (2) the structural style of folding is very similar; (3) the metamorphic P–T path is consistent in the sense that the pressure dropped, as indicated by pseudomorphs of kyanite substituted by sillimanite.

D_4

The lower contact of the Cabo Frio tectonic domain, a D_3 thrust, is rotated to an opposite dip direction, towards the SW, from the Silva Jardim to Ponta Negra regions (Fig. 2). This rotation by folding is attributed to a transpressional dextral sub-vertical NE–SW shear zone of at least 5 to 10 km wide and more than 50 km long, that also affects the 550 Ma granitoids of the Oriental terrane (Mendes *et al.* 2006) (Fig. 2). Mineral and stretching lineations plunging 10–30° to SW developed parallel to the axes of tight to open folds, with axial planes dipping steeply to NW and SE (Fig. 6c). The main $S_{1/2}$ foliation of the CFTD lithotypes is folded and re-orientated so as to dip 50–90° to SE and NW (Ferrari *et al.* 1982) (Figs 4e and 6b). An axial plane cleavage is locally developed in the fold hinges, defined by biotite and often exhibiting parallel quartz-feldspar veins, and named S_4. The metapelites contain an S_4 crenulation cleavage. In high strain zones there is a transposition foliation parallel to the axial plane of the F_4 folds. Sillimanite and amphibole were realigned and quartz and feldspar were stretched (L_4) in D_4 shear zones (Fig. 6c). In some folds the L_3 sillimanite lineation can still be recognized in addition to new growth of sillimanite along L_4 (Rocha 2002). The dextral shear sense is evident from asymmetric boudins and garnet porphyroblasts with asymmetric strain shadows, drag of $L_{1/2}$ into the shear zone, and development of local top-to-east thrust faults. There is an interference pattern between F_3 folds with vergence to

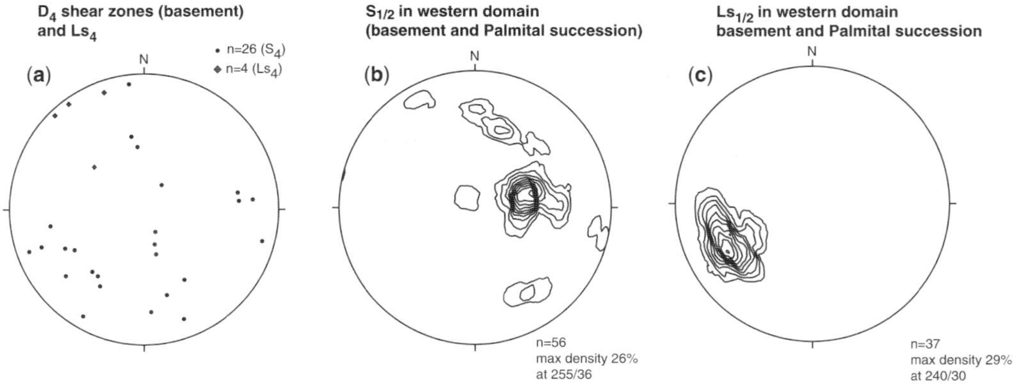

Fig. 6. Stereograms of the structural elements of the Cabo Frio tectonic domain, related to D_4 and D_5. (**a**) Poles to S_4 and stretching lineations (Ls_4); (**b**) $S_{1/2}$ foliation poles from the basement and the Palmital succession in the western domain; (**c**) $Ls_{1/2}$ stretching lineation from the basement and the Palmital succession in the western domain.

the west and F_4 folds, usually with vergence to the east (Fig. 4e). The growth of sillimanite, the stretching lineation of quartz–feldspar and the leucosome veins, all indicate that metamorphic conditions were still in the amphibolite facies during the D_4 transpressional dextral shear.

In the Cabo Frio region, these refolding structures and NE–SW sub-vertical shear zones are not present. On the other hand, $S_{1/2}$ and S_3 are cut by sub-vertical NW–SE oriented shear zones, which could be attributed to a D_4 deformational phase (Figs 2 & 7b). These mylonitic shear zones were only observed in the basement sheets where they are 50 cm to 10 m thick. Some of them are associated with tight to isoclinal upright folds with NW–SE inclined axes. They dip normally 40–70° to NE and show a weak oblique mineral lineation (Figs 6a & 7b). Drag structures and rotated porphyroblasts suggest a dextral sense of shear. Although it is obvious in the field that these zones are concentrated strain zones with an apparent mylonitic foliation and stretched layers, enclaves and increased leucosome proportion, in thin sections the minerals show a granoblastic fabric due to static recrystallisation. This S_4 foliation is only present in the shear zones. Gentle to open folds on a regional scale, related to D_4, might be responsible for some dispersion in the concentration pattern in the $S_{1/2}$ stereograms (Fig. 5a, b).

Time constraints on the deformation

Schmitt *et al.* (2004) presented concordant U–Pb zircon ages from leucosome veins of the basement and of the mafic gneiss and paragneiss association. The veins are syn-D_1/D_2 and yielded U–Pb zircon ages of 525 ± 9 Ma (paragneisses) and 517 ± 5 Ma (basement). We infer from these data that the metamorphic peak (syn-D_1/D_2) occurred at *c.* 520 Ma (Middle Cambrian). These data match the concordant U–Pb ages obtained from metamorphic zircons of a granulite facies paragneiss and of an ortho-amphibolite (Schmitt *et al.* 2004).

The youngest age determined for the evolution of the Cabo Frio tectonic domain was obtained from a post-tectonic sub-vertical east–west orientated pegmatite, which yielded a U–Pb zircon age of 440 ± 10 Ma (Schmitt *et al.* 2004), confirmed by Ar–Ar dating (Fonseca 2004) (Fig. 4f). Therefore the ages of the D_3 and D_4 phases must fit in the range 520–440 Ma.

Monazites and titanites have yielded concordant U–Pb ages between 515 and 510 Ma (Schmitt *et al.* 2004). These data indicate that during this period the rocks were still at high temperatures (*c.* 700 °C), but the ages cannot be linked to a specific deformational phase since the relation between these minerals and structures of the successive deformational phases is not clear. However, D_3, interpreted as a progressive phase after D_1–D_2, probably occurred after 520 Ma and before 505 Ma. Geochronological data from the Ponta Negra area in the western part of the Cabo Frio tectonic domain indicate younger ages (Schmitt *et al.* 2004): an interval of 505–490 Ma, recorded in zircons and monazites, is interpreted as corresponding to the D_4 deformational activity that affected this area.

In order to define the age of D_3 and D_4, we present here new geochronological data from a mylonitic shear zone within a basement thrust sheet (Figs 3 and 7b). In general, new zircon may form during high-grade metamorphism either by growth as a result of metamorphic reactions from the breakdown of other minerals, or by crystallization of zircon from partial melt. Even without new growth, pre-existing zircon from the protolith may

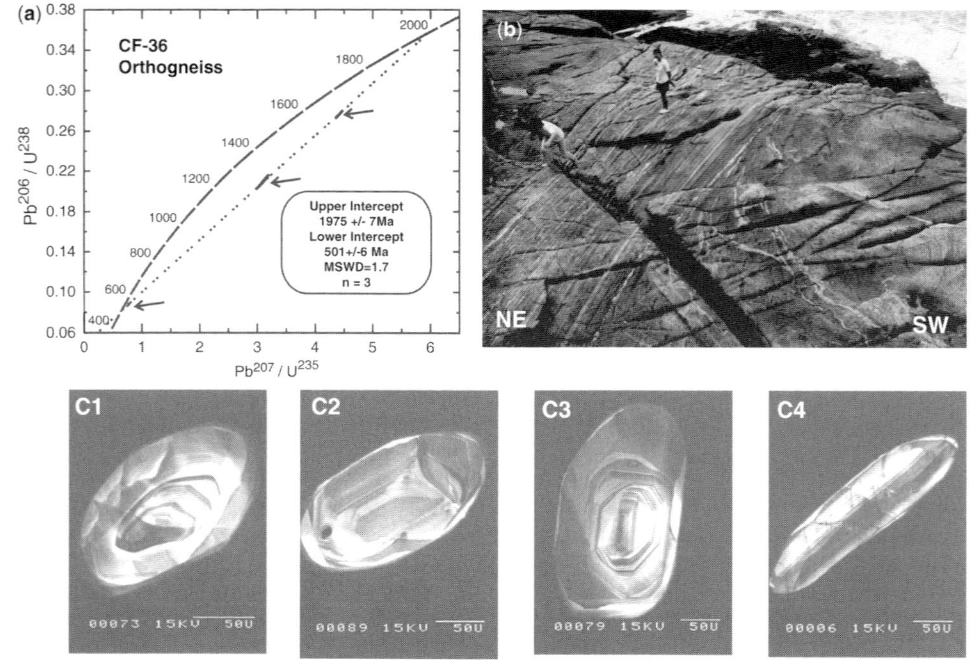

Fig. 7. (**a**) Plot of U–Pb data from a mylonitic orthogneiss of Região dos Lagos Unit using three single crystals (zircons) from a shear zone. Uncertainties at 2-sigma (n = number of fractions analysed). (**b**) Sampled NW–SE shear zone, a transpressional dextral shear zone within the Palaeoproterozoic basement. Note the drag structure marked by the veins and foliation outside of the shear zone. C1–4 – cathodo-luminescence images of the dated zircons: C1, zircon core with igneous zoning is cut by the metamorphic overgrowth; C2, cat's-eye zircon with 60% of metamorphic overgrowth; C3, igneous zoning in the core and large overgrowth rims (40%); C4, metamorphic acicular zircon crystal.

undergo partial resetting resulting from one or more Pb-loss processes (Mezger & Krogstadt 1997). The metamorphic crystals described here may result from the increase in temperature, but could also have been generated by crystallization from the partial melt, considering that it was not possible to separate the thin leucosome veins from the host rock.

A mylonitic orthogneiss sample (CF-36) was obtained from a 7 m thick high-grade NW–SE dextral sub-vertical shear zone (Figs 3a, b and 7b). The strain concentration in the shear zone is clear at the outcrop, with a sharp contact between meta-granodiorite and the mylonitic orthogneiss (Fig. 7b). The meta-granodiorite contains dioritic enclaves with slightly elliptical shape in the low-strain domains. Within the shear zone, this rock is transformed into a thinly banded mylonitic orthogneiss, with a poorly developed oblique stretching lineation; the enclaves are stretched to an average ratio 20:1. Static recrystallization is widespread, as can be observed in thin section. The percentage of leucosome is larger in the shear zone, suggesting that the deformation favoured partial melting or vice versa. This relation is recognized throughout the Cabo Frio tectonic domain.

Two populations of zircons were identified under the microscope and dated by the ID-TIMS method in single crystals at the Isotope Geochemistry Laboratory at Kansas University, USA (Fig. 7a, Table 1). The procedures and laboratory conditions are similar to those described by Schmitt *et al.* (2004). A population of bi-pyramidal to 'cat's-eye' shaped crystals was identified as light pink, transparent and inclusion-free zircon. Another population is composed of acicular transparent and pale pink crystals, also inclusion-free. A much more variable pattern of structures was recognized using cathodo-luminescence images. Igneous and metamorphic features alternate through the crystals, as is expected for high-grade rocks (Poller 2003). The cat's-eye population shows complex internal patterns, with igneous zoning recognized in the cores and metamorphic growth observed in the rims (Fig. 7c, 1–3). The zircons from the acicular population show almost no igneous features (Fig. 7c, 3), similar in this respect to those with the metamorphic morphology. The bi-pyramidal

Table 1. *U–Pb Data for Sample CF-36, Mylonitic metagranodiorite (NW-SE shear zone) (zircon)*

Sample fraction†	Size (mg)	Observed			Radiogenic ratios§			Calculated ages (Ma)‖		
		U (ppm)‡	Pb (ppm)‡	^{206}Pb/^{204}Pb	^{207}Pb*/^{235}U	^{206}Pb*/^{238}U	^{207}Pb*/^{206}Pb*	^{206}Pb*/^{238}U	^{207}Pb*/^{235}U	^{207}Pb*/^{206}Pb*
nm(0)[1]	0.009	413.53	74.848	65.1967	0.770132	0.087652	0.063724	541.64	579.84	732 ± 15
M(0)[1]	0.014	66.988	19.895	711.185	4.41537	0.27654	0.1158	1573.9	1715.2	1892 ± 2
M(1)[1]	0.002	255.89	72.842	203.463	3.09725	0.209237	0.107359	1224.8	1432	1755 ± 6

*Radiogenic Pb. nm, non-magnetic, M, magnetic; numbers in parentheses indicate side tilt used on Franz separator at 1.5A; [1] = numbers of grains analyzed.
‡Total U and Pb concentrations, corrected for analytical blank.
§Pb corrected for blank and non-radiogenic Pb.
‖Ages using decay constants recommended by Steiger & Jäger (1977); uncertainties at 2-sigma.
Upper Intercept = 1975 ± 6.5 Ma, Lower Intercept = 501 ± 6 Ma (2-sigma); MSWD = 1.73; P = 0.19 (Model 1)
Locality: Coastal outcrop Conchas beach (Fig. 3); Lat: 22°52′ 22″S Long: 41°59′ 4″W

cat's-eye shaped crystals produced discordant ages, close to a Palaeoproterozoic upper intercept at 2.0 Ga, interpreted as approximating the crystallization age of the granodiorite (Fig. 7a), and consistent with the previous results given by Schmitt *et al.* (2004). Data for the population of tiny acicular crystals are also discordant, but are distributed closer to the lower intercept of 501 ± 6 Ma, probably indicating formation due to high-grade metamorphism. The upper and lower intercepts are quite sharply defined, with MWSD = 1.7, indicating that the rocks underwent only one metamorphic event with major Pb-loss, as also suggested by all the other discordia obtained in the basement (Zimbres *et al.* 1990; Schmitt *et al.* 2004). However, the present discordia is defined only on the basis of three points and the best way of confirming this result might be through *in situ* analysis of metamorphic zircon using the U–Pb SHRIMP method.

From the foregoing it can be concluded that the shear zone was formed at about 501 ± 6 Ma. The field evidence indicates that this shear zone is related to either D_3 or D_4. Although the geometry and kinematics fit with the D_4 structures, this outcrop does not present the typical F_3 recumbent folds, but only isoclinal folds with axial planes parallel to the steep shear zone (Fig. 3b). Therefore it is also possible that the shear zone belongs to D_3, if this area developed sub-vertical ductile structures due to a local deviation of the regional D_3 strain pattern (Fig. 3b).

In either case (D_3 or D_4), the shear zone would be 15 to 20 Ma younger than the metamorphic peak during the D_1–D_2 progressive phases, dated at *c.* 520 Ma in the Búzios area, and would be coeval with D_4 structures dated in the Ponta Negra region.

Discussion

The data presented here corroborate the detailed characterization of the Búzios orogeny (Schmitt *et al.* 2004), according to which the lithostratigraphic units of the Cabo Frio tectonic domain register a main crustal thickening event between 525 and 520 Ma, during the D_1/D_2 deformational phases. According to the cooling curve determined with U–Pb, K–Ar and Sm–Nd data, after 510 Ma the cooling rate dropped drastically from 20 °C to 10 °C/Ma (Schmitt *et al.* 2004). 510 Ma is also the youngest mineral crystallization age obtained from monazites (in metapelites) and titanites (in amphibolites) in the central area of the Cabo Frio tectonic domain. This change in cooling rate could be related to D_3, when the Cabo Frio tectonic domain was juxtaposed to the Oriental terrane.

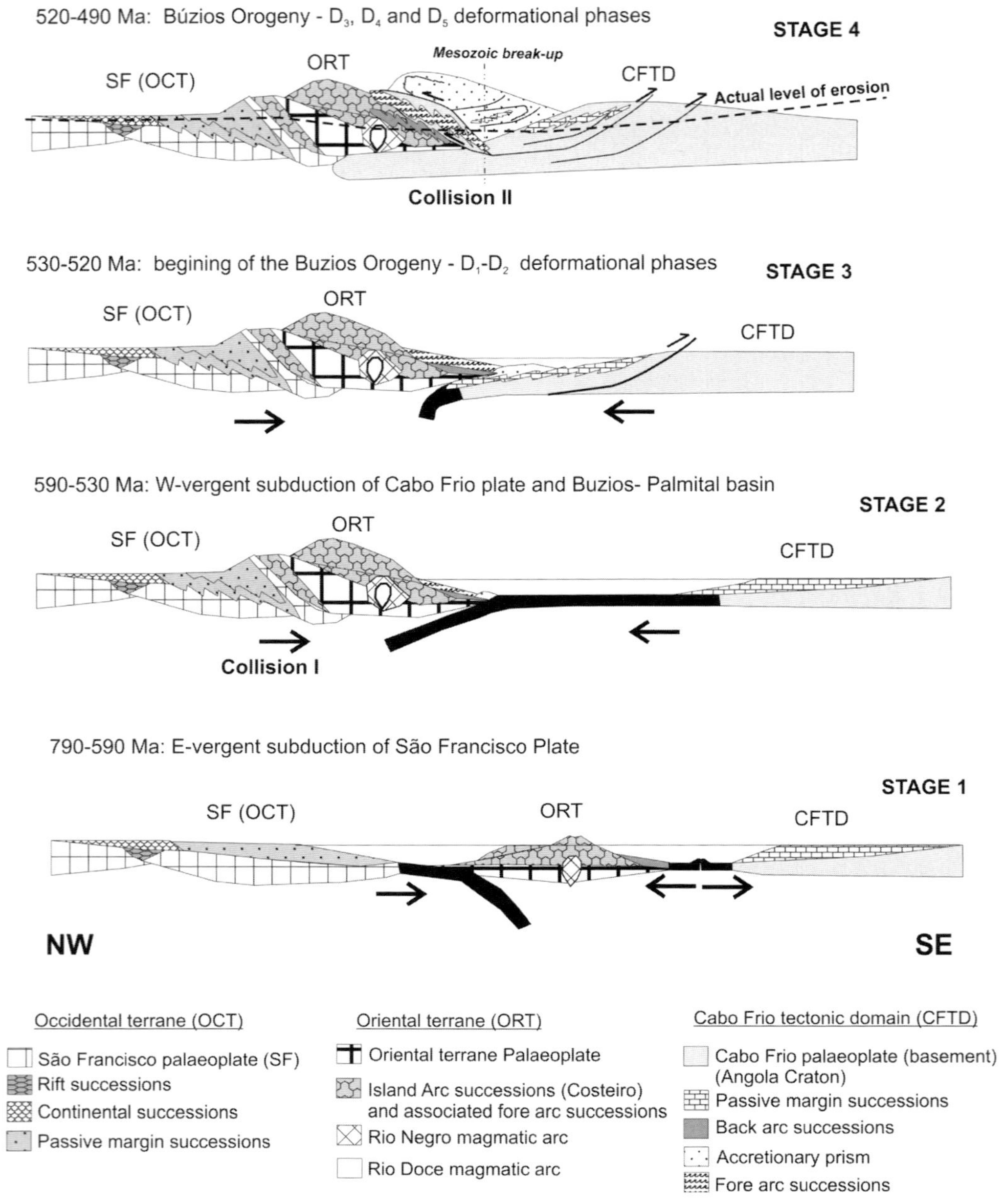

Fig. 8. Tectonic cartoons showing the evolution of the Cabo Frio tectonic domain. Stage 1 is taken from Heilbron & Machado (2003), as well as the positions of the Occidental terrane (OCT) and the Oriental terrane (ORT) at stages 2, 3 and 4.

If the new age of 501 ± 6 Ma presented here is related to a D_4 NW–SE orientated shear zone in the coastal outcrops, it would be coeval with the D_4 transpressional NE–SW orientated shear zone in the western area (Fig. 2). At this stage, the central and eastern parts of the Cabo Frio tectonic domain were already cooling at a rate of about 10 °C/Ma. After 480 Ma, the cooling rate diminished to 5 °C/Ma (Schmitt *et al.* 2004).

During the Búzios orogeny, the other tectonic domains of the Ribeira Belt were submitted to deformational phases D_{3Rb} (Rb–Sr age 540–520 Ma) and D_{4Rb} (Rb–Sr age 520–480 Ma) (Heilbron *et al.* 2000). These are attributed respectively to a post-collisional stage (with respect to the collision between the Oriental and Occidental terranes at *c.* 590–560 Ma), and a transitional stage related to thermal relaxation and orogenic collapse. The development of strike-slip sub-vertical ductile D_{3Rb} shear zones and associated folding with steep axes is associated with the intrusion of cal-calkaline plutons in a metamorphic regime of high temperature and low pressure. The D_{4Rb} thermal relaxation and transtensional zones are associated with calcalkaline and tholeiitic rocks, orientated NE–SW, and weakly foliated in steep ductile–brittle shear zones with the formation of open to gentle folds (Heilbron *et al.* 2000). Most D_{3Rb} and $D_{4\ Rb}$ shear zones are orientated NW–SE, N–S or NE–SW.

In relation to the Búzios orogeny, D_{3Rb} and D_{4Rb} are pre- to syn-collisional and syn- to late-collisional respectively. It has to be expected that the tectonic effects of this younger tectono-metamorphic event should be identified in both D_{3Rb} and D_{4Rb} structures. Between 525 and 510 Ma, D_1 to D_3 deformational phases of the Cabo Frio tectonic domain developed in response to a NW–SE orientated frontal collision. By this time, the possible effects in the remaining belt could be: formation of NE–SW tight to open folds, generation of NW–SE mineral lineations, development of strike-slip and thrust shear zones and others structures. Most of these features are recognized in D_{3Rb} structures. After 510 Ma, when the Cabo Frio tectonic domain was already attached to the adjacent Oriental terrane, the main compression shifted to east–west, as indicated by the NE–SW dextral D_4 shear zone. This movement could reactivate and develop strike-slip dextral NE–SW shear zones, strike-slip sinistral NW–SE shear zones, east–west faults with north–south extension, and also folds with north–south orientated axes. All these structures were identified in the southwestern contact zone of the Cabo Frio tectonic domain and the Oriental terrane (Fig. 2).

It is expected that the Búzios orogeny, representing a collisional event, had a compressional effect in the neighbouring Oriental terrane, corresponding to D_{3Rb}, D_{4Rb} or both. This effect is in fact recognized, but the direct association of D_{3Rb} and D_{4Rb} with the Búzios orogeny needs confirmation by new geochronological constraints on the timing of deformation in the Oriental terrane.

Within the Ribeira Belt, the Cabo Frio tectonic domain appears to be exotic on account of its younger metamorphic ages, its Palaeoproterozoic basement (absent from the adjacent Oriental terrane), and the age of the source rocks dated through detrital zircons (Schmitt *et al.* 2004). The metasedimentary rocks of the Oriental terrane (Fig. 2) are interpreted as a distal passive margin succession intruded by the *c.* 790 and 630 Ma Rio Negro cordilleran arc (Heilbron & Machado 2003). This could well be related to the same marine basin recognized in the Cabo Frio tectonic domain as the Buzios–Palmital successions. Valladares *et al.* (2005) proposed that both the Oriental terrane and the Cabo Frio tectonic domain would be parts of the same microcontinent. Considering our geochronological data, the 630 Ma Rio Negro arc would be contemporaneous with sedimentation in this basin, which is constrained by the youngest detrital zircon of *c.* 630 Ma (Schmitt *et al.* 2004), and also with the age of ocean floor remnants (Sm–Nd isochron age of 608 Ma, Schmitt *et al.* 2008). Heilbron & Machado (2003) interpreted the Buzios–Palmital sequence as formed in a basin developed in a back arc environment related to the Rio Negro arc. This model is only valid up to 590 Ma, at which time the Oriental terrane and its arc collided with the margin of the São Francisco palaeo-continent. Between 590 and 530 Ma, another subduction zone must have developed in order to enable the collision that resulted in the Búzios orogeny (Fig. 8) and demonstrated by the high P–T conditions in the Buzios–Palmital successions and basement rocks, but the direction of this subduction is not constrained at present. The D_1–D_2 tectonic structures indicate tectonic transport with top-to-NW, though this does not necessarily imply subduction towards the SE. New data from the Angolan part of this belt (Delor *et al.* 2006) are even less conclusive about the direction of this subduction. The north–south Angolan Pan-African belt, only 20 km wide, is composed of a 1.95 Ga strongly deformed basement with Cambrian (522 Ma) leucosome, similar to the Cabo Frio tectonic domain. This belt presents east-verging deformational structures, towards the Congo Craton.

There are no reports of Neoproterozoic–Cambrian granites in the coastal belt of Angola. On the other hand, the Oriental terrane in the Ribeira Belt is constituted by large 580 to 550 Ma plutons. If subduction was directed towards the NW, beneath the Oriental terrane, this extensive set of granitoids would represent the magmatic arc related to this subduction. In this case the Oriental terrane would contain an older magmatic arc (Rio Negro arc; 630 Ma, Tupinambá *et al.* 2000) and a younger superposed one (Fig. 8). This young magmatic arc has been described as the Rio Doce arc (Campos Neto & Figueiredo 1995; Campos Neto 2000). The Oriental terrane (Heilbron *et al.* 2000; Heilbron & Machado 2003) is referred to as Serra do Mar terrane by Campos Neto & Figueiredo (1995) and Campos Neto (2000). This hypothesis of subduction towards the NW would explain the differences in $P-T$ paths between the Oriental terrane and the Cabo Frio tectonic domain (Schmitt *et al.* 2004): the former contains typically high-temperature- low pressure metamorphism (with a peak at 550 Ma), whereas the second consists of high-temperature and medium- to high-pressure rocks (peak at 525 Ma). If the Buzios–Palmital basin corresponds to a restricted Late Neoproterozoic ocean, the angle of subduction would probably have been low, since the subducted plate would have been relatively hot. Low-angle subduction tends to cause compressive structures in the upper plate as a high-stress subduction-zone setting (Condie 1997). Such structures were in fact recognized in several 580–530 Ma plutons in the Oriental terrane (Mendes *et al.* 2006). The main collision of the Búzios orogeny at 525 Ma is responsible for thrusting and inversion of stratigraphy, including the penetration of Palaeoproterozoic basement slices into the sedimentary sequences, which does not occur in the overlying Oriental terrane (Fig. 2). In the Oriental terrane, a widespread magmatic event occurred between 530 and 480 Ma. The plutons related to this event may be considered syn-collisional with regard to the Búzios orogeny. Further north, in Espirito Santo State, similar plutons are labelled as G4 and G5 magmatic suites (Pedrosa-Soares & Wiedemann-Leonardos 2000), characterized by a bimodal character with S- and I-type granitoids associated with diorites and gabbros.

Correlation with adjacent areas in SW Gondwana: the ties bind underwater

Middle Cambrian deformational and metamorphic ages have also been reported from other coastal terranes that were close in Gondwana reconstructions (Fig. 1): the recently dated 'Angolan' Pan-African belt (Delor *et al.* 2006), the southern Kaoko Belt and the Damara Belt (inland branch) in Namibia (Kukla 1993; Jung *et al.* 2000; Jung & Mezger 2003; Goscombe *et al.* 2005) and the Cuchilla Dionisio terrane (Uruguay) (Bossi & Gaucher 2004).

Rough estimates indicate that, on average, some 250 km of continental crust is hidden in the Brazilian and African coastal shelves (Chang *et al.* 1992; Austin & Uchupi 1982), corresponding to almost the total width of the onshore Ribeira Belt (Fig. 2). Therefore, the Cabo Frio tectonic domain could be much wider than it appears and should contain a lot more evidence relating to the unknown Palaeozoic evolution of Gondwana, just as do other submerged terranes.

The abrupt bend of the Brazilian coastline from NE–SW to east–west in Rio de Janeiro State has a direct relation to the Cabo Frio tectonic domain. The Cabo Frio tectonic domain coincides partially with an important tectonic structure named the Cabo Frio Structural High, which separated the Campos and Santos sedimentary basins from Cretaceous time onward (Mohriak *et al.* 1995). Geophysical data from the offshore area (magnetometry; Zalan & Oliveira 2005) integrated with onshore data indicate that the southern limit of the Cabo Frio tectonic domain continues offshore into the Santos Basin, as shown by a strong magnetic anomaly. It is interrupted by the platform hinge of São Paulo State and probably continues into the slope. Recent work in seven coastal islands indicates that the same units as mapped onshore and the same Cambrian deformational structures are present in the islands (Fig. 4c) (Schmitt *et al.* 2005). Furthermore, the mafic gneisses that probably represent relics of a Late Neoproterozoic oceanic crust are even more abundant offshore. High-resolution geophysical data within the Cabo Frio tectonic domain reveal that alignments of magnetic anomalies could be interpreted as related either to Mesozoic dolerites or to the mafic gneisses of the Búzios succession (Schmitt *et al.* 2005).

The tectonic boundary between the Cabo Frio tectonic domain and the Oriental terrane to the NE continues into the West Congo Belt of Africa (Fig. 1). Towards the SW, the Cabo Frio tectonic domain could be matched with the Curitiba microplate (Basei *et al.* 2000) in southern Brazil, or with the Cuchilla Dionisio terrane/Punta Del Este terrane (Bossi & Gaucher 2004; Basei *et al.* 2005) in eastern Uruguay. Assuming that the boundary is a continuous linear fault, the Cabo Frio tectonic domain would be continuous with the Curitiba microplate, which occurs along the northern border of the Palaeoproterozoic Luis Alves block (Fig. 1) and is composed of reworked Palaeoproterozoic basement interleaved with supracrustal rocks (Basei *et al.* 2000), in a similar way to the Cabo Frio tectonic domain.

The African counterpart is located along the southern Angolan coast, which is geologically poorly known, with most geochronological data from the late seventies to early eighties (Torquato & Cordani 1981) and limited recent data (Delor *et al.* 2006). It is referred to as the Angolan shield, with ages of 2.0 Ga obtained by the Rb–Sr method. It is important to recall that the Rb–Sr whole rock isochrons of the basement of the Cabo Frio tectonic domain also produced Palaeoproterozoic ages (Fonseca 1993), although the domain was partially melted and deformed during the Cambrian (Schmitt *et al.* 2004). In satellite images it is possible to discern a north–south orientated belt in southern Angola, with its continuation in Namibia known as the Kaoko Belt. Delor *et al.*

(2006) presented new data that support the characterization of an 'Angolan Pan-African belt'; their rock descriptions and ages strongly reinforce the hypothesis that the Cabo Frio tectonic domain is continuous with this Angolan Pan-African belt.

In the much better known Kaoko Belt of northwest Namibia, three tectono-metamorphic cycles have been recognized (Goscombe *et al.* 2005; Gray *et al.* 2008): M_1 (655–645 Ma), within a coastal exotic terrane; M_2 (580–550 Ma), attributed to the collision of this exotic terrane with the passive margin of the Congo and/or Angola Craton; and M_3 (535–508 Ma), related to late-kinematic magmatism and low-grade buckling attributed to collision between the Kalahari and Congo/Angola cratons. Although these events match surprisingly well with the Ribeira Belt evolution, there is no direct evidence yet for matching any of these Kaoko Belt domains with the Cabo Frio tectonic domain. The detrital zircons from the passive margin sedimentary sequences in the Kaoko Belt show the same source ages as the detrital zircons from the Buzios–Palmital basin, with the exception of the Neoproterozoic sources, which are not present in Kaoko Belt data (Goscombe *et al.* 2005). The M_3 period coincides with the Búzios orogeny and is well developed in the southern Kaoko Belt at the triple junction with the inland Damara Belt (Miller *et al.* 1983; Passchier *et al.* 2002; Passchier *et al.* 2007). This area, the Lower Ugab domain, presents syn-D_2 syenitic plutons with Cambrian ages (Seth *et al.* 2000; Passchier *et al.* 2008). This deformation is probably related to the evolution of the Damara Belt, which presents Cambrian syn-tectonic plutons and high-grade metamorphism (e.g., Miller 1983; Kukla 1993; Jung *et al.* 2000; Jung & Mezger 2003). Besides this, 1.95 Ga U–Pb ages on zircons are reported for the basement inliers of the Kaoko Belt, similar to the crystallization ages of the CFTD basement (Seth *et al.* 1998).

Basei *et al.* (2005, 2008) correlate the Gariep Belt, further south in Namibia, with the Dom Feliciano Belt in Brazil (Fig. 1). These authors show that provenance ages from the Gariep Belt match with detrital zircons from the Punta Del Este terrane, an exotic terrane along the Uruguayan coast. Both presented 'Grenvillian' (*c.* 1000 Ma) ages for the sources, which are not recognized in southeastern South America. They interpret the Punta Del Este terrane as an African domain. Following this line of thought, the Cabo Frio tectonic domain could be also considered as part of the 'African-side geology', taking into account the detrital zircons U–Pb SHRIMP ages (Schmitt *et al.* 2004) and the similar basement ages (Seth *et al.* 1998; Delor *et al.* 2006). Provenance information on zircons from the neighbouring Oriental terrane of the Ribeira Belt is mainly based on Pb–Pb ages (Valladares *et al.* 2005), which are difficult to correlate with SHRIMP data, especially when dealing with high-grade rocks.

The 530 Ma compressive deformational phases from these coastal domains may be related to the palaeogeographical location of pre-Gondwana terranes. It is proposed in the literature that Congo–Rio de la Plata suturing pre-dated Congo–Kalahari suturing during the assembly of Gondwana (Prave 1996). Collins & Pisarevsky (2005) also proposed that the Kalahari palaeocontinent collided with the Congo/Tanzania/Bangweulu block along the Zambezi and Damara belts and with the Rio de la Plata Craton along the Gariep Belt in latest Neoproterozoic/Cambrian times. This late amalgamation of the Kalahari palaeocontinent with the rest of proto-Gondwana is also suggested to have caused the latest Neoproterozoic to Cambrian Rio Doce and Buzios orogeny in the Ribeira Belt.

The best evidence for correlation beyond the Atlantic margin is the chronology of deformational phases that could clarify the kinematic frame of the last pulses of Gondwana agglomeration. This tectonic activity is also coeval with orogenies along the margins of Gondwana, e.g. the Pampean orogeny (530 Ma) in Argentina (Rapela *et al.* 1998, 2008), east of the Andes and the Ross orogeny in Antarctica (Cawood 2005; Cawood & Buchan 2007). During this period, Gondwana was not only surrounded by marginal accretionary orogenies but also suffered final adjustments and collisions in its interior.

Cathodo-luminescence images were obtained through collaboration with Ulrike Poller and Wolfgang Todt at the Max Planck Institüt für Chemie, in Mainz, Germany. Helpful discussions with them are also gratefully acknowledged. This is a contribution to International Geological Correlation Programme (IGCP) Project 478 'Neoproterozoic–Early Paleozoic Events in SW-Gondwana'. We thank Peter Cawood and André Steenken for their thoughtful reviews.

References

Austin, J. A. jr. & Uchupi, E. 1982. Continental-oceanic transition off southwest Africa. *American Association of Petroleum Geologists, Bulletin*, **66**, 1328–1347.

Basei, M. A. S., Siga, O. jr., Masquelin, H., Harara, O. M., Reis Neto, J. M. & Preciozzi, F. 2000. The Dom Feliciano belt of Brazil and Uruguay and its foreland domain, the Rio de la Plata Craton–framework, tectonic evolution and correlation with similar provinces of southwestern Africa. *In*: Cordani, U. G., Milani, E. J., Thomaz-Filho, A. & Campos, D. A. (eds) *Tectonic Evolution of South America. 31st International Geological Congress, Rio de Janeiro, Brazil*, 311–334.

BASEI, M. A. S., FRIMMEL, H., NUTMAN, A. P., PRECIOZZI, F. & JACOB, J. 2005. A connection between Neoproterozoic Dom Feliciano Belt (Brazil/Uruguay) and Gariep (Namíbia/South Africa) orogenic belts – evidence from a reconnaissance provenance study. *Precambrian Research*, **139**, 195–221.

BASEI, M. A. S., FRIMMEL, H. E., NUTMAN, A. P. & PRECCIOZI, F. 2008. West Gondwana amalgamation based on detrital zircon ages from Neoproterozoic Ribeira and Dom Feliciano belts of South America and a comparison with coeval sequences from southwestern Africa. *In*: PANKHURST, R. J., TROUW, R. A. J., BRITO NEVES, B. B. & DE WIT, M. J. (eds) *West Gondwana: Pre-Cenozoic Correlations across the South Atlantic Region*. Geological Society, London, Special Publications, **294**, 239–256.

BOSSI, J. & GAUCHER, C. 2004. The Cuchilla Dionisio Terrane, Uruguay: An allochthonous block accreted in the Cambrian to SW-Gondwana. *Gondwana Research*, **7**, 661–674.

BRITO NEVES, B. B. & CORDANI, U. C. 1991. Tectonic evolution of South America during Late Proterozoic. *Precambrian Research*, **33**, 23–40.

CAMPOS NETO, M. C. 2000. Orogenic Systems from Southwestern-Gondwana: an approach to Brasiliano- Pan-African cycle and orogenic collage in southeastern-Brazil. *In*: CORDANI, U. G., MILANI, E. J., THOMAZ-FILHO, A. & CAMPOS, D. A. (eds) *Tectonic Evolution of South America. 31st International Geological Congress, Rio de Janeiro, Brazil*, 335–365.

CAMPOS NETO, M. C. & FIGUEIREDO, M. C. H. 1995. The Rio Doce Orogeny, southeastern Brazil. *Journal of South American Earth Sciences*, **8**, 143–162.

CAWOOD, P. A. & BUCHAN, C. 2007. Linking accretionary orogenesis with supercontinent assembly. *Earth Science Reviews*, **82**, 217–256.

CAWOOD, P. A. 2005. Terra Australis orogen: Rodinia breakup and development of the Pacific and Iapetus margins of Gondwana during the Neoproterozoic and Paleozoic. *Earth-Science Reviews*, **69**, 249–279.

CHANG, H. K., KOWSMANN, R. O., FIGUEIREDO, A. M. F. & BENDER, A. A. 1992. Tectonics and stratigraphy of the east Brazil rift system: an overview. *Tectonophysics*, **213**, 97–138.

CONDIE, K. C. 1997. *Plate tectonics and crustal evolution* (4th ed.). Pergamon Press.

COLLINS, A. S. & PISAREVSKY, S. A. 2005. Amalgamating eastern Gondwana: The evolution of the Circum-Indian orogens. *Earth Science Reviews*, **71**, 229–270.

DE WIT, M. J., BOWRING, S. A., ASHWAL, L. D., RANDRIANASOLO, L. G., MOREL, V. P. I. & RAMBELOSON, R. A. 2001. Age and tectonic evolution of Neoproterozoic ductile shear zones in southwestern Madagascar, with implications for Gondwana studies. *Tectonics*, **20**, 1–45.

DELOR, C., LAFON, J. M. *ET AL*. 2006. Unravelling Precambrian crustal growth of Central West Angola: Neoarchean to Suderian inheritance main Osirian accretion and discovery of the "Angolan" Panafrican Belt. *In*: *21st Colloquium of African Geology, Mozambique, Abstracts*, 40–41.

FERRARI, A. L., BRENNER, T. L., DALCOMO, M. T. & NUNES, H. R. C. 1982. O Pré-cambriano das folhas de Itaboraí, Maricá, Saquarema e Baía da Guanabara. *In*: *Congresso Brasileiro Geologia 32, Salvador, Anais*, **1**. Sociedade Brasileira de Geologia, 103–114.

FONSECA, A. C. 1993. *Esboço geocronológico da região de Cabo Frio, Estado do Rio de Janeiro*. PhD thesis, University of São Paulo, Brazil.

FONSECA, M. J. 1998. *Mapa Geológico do estado do Rio de Janeiro*. 1:400.000 Departamento Nacional de Produção Mineral, Rio de Janeiro.

FONSECA, M. J. 2004. *Pegmatitos da parte centro-leste do estado do Rio de Janeiro e seu contexto tectônico*. PhD Thesis, Federal University of Rio de Janeiro, Brazil.

GAUCHER, C., BOGGIANI, P. C., SPRECHMANN, P., SIAL, A. N. & FAIRCHILD, T. 2003. Integrated correlation of the Vendian to Cambrian Arroyo del Soldado and Corumbá Groups (Uruguay and Brazil): palaeogeographic, palaeoclimatic and palaeobiologic implications. *Precambrian Research*, **120**, 241–278.

GERMS, G. 1995. The Neoproterozoic of southwest Africa, with emphasis on platform stratigraphy and paleontology. *Precambrian Research*, **73**, 137–151.

GOSCOMBE, B., GRAY, D., ARMSTRONG, R., FOSTER, D. A. & VOGL, J. 2005. Event geochronology of the Pan-African Kaoko Belt, Namibia. *Precambrian Research*, **140**, 1–41.

GRAY, D. R., FOSTER, D. A. *ET AL*. 2008. A Damara Orogen perspective on the assembly of southwestern Gondwana. *In*: PANKHURST, R. J., TROUW, R. A. J., BRITO NEVES, B. B. & DE WIT, M. J. (eds) *West Gondwana: Pre-Cenozoic Correlations across the South Atlantic Region*. Geological Society, London, Special Publications, **294**, 257–278.

HEILBRON, M. & MACHADO, N. 2003. Timing of terrane accretion in the Neoproterozoic-Eopaleozoic Ribeira belt SE Brazil. *Precambrian Research*, **125**, 87–112.

HEILBRON, M., MOHRIAK, W. U., VALERIANO, C. M., MILANI, E. J., ALMEIDA, J. & TUPINAMBÁ, M. 2000. From collision to extension: The roots of the southeastern continental margin of Brazil. *In*: MOHRIAK, W. U. & TALWANI, M. (eds) *Atlantic Rifts and continental margins*. American Geophysical Union, Geophysical Monographs, **115**, 1–32.

JUNG, S. & MEZGER, K. 2003. Petrology of basement-dominated terranes: I. Regional metamorphic T-t path from U–Pb monazite and Sm–Nd garnet geochronology (Central Damara Orogen, Namibia). *Chemical Geology*, **198**, 223–247.

JUNG, S., HOERNES, S. & MEZGER, K. 2000. Geochronology and petrogenesis of Pan-African, syn-tectonic, S-type and post-tectonic A-type granite (Namibia): products of melting of crustal sources, fractional crystallization and wall rock entrainment. *Lithos*, **50**, 259–287.

KUKLA, C. 1993. *Strontium isotope heterogeneities in amphibolite facies, banded metasediments – a case study from the Late Proterozoic Kuiseb Formation of the Southern Damara orogen, Central Namibia*. Geological Survey of Namibia, Memoirs, **15**.

MENDES, J. C., SCHMITT, R. S., PENHA, H., LUDKA, I. P. & DANTAS, E. L. 2006. Calc-Alkaline magmatism in a Low Pressure High Temperature Neoproterozoic

Terrane, SE Brazil. *In*: *V South American Symposium on Isotope Geology, Punta del Este, Uruguay, Expanded Abstracts*, **1**, 123–125.

MEZGER, K. & KROGSTADT, E. J. 1997. Interpretation of discordant U–Pb zircon ages: An evaluation. *Journal of Metamorphic Geology*, **15**, 127–140.

MILLER, R. McG. 1983. The Pan-African Damara Orogen of South West Africa/Namibia. *In*: MILLER, R. MCG. (ed.) *Evolution of the Damara Orogen*. Geological Society of South Africa, Special Publications, **11**, 431–515.

MILLER, R. MCG., FREYER, E. E. & HÄLBICH, I. W. 1983. A turbidite succession equivalent to the entire Swakop Group. *In*: MILLER, R. MCG. (ed.) *Evolution of the Damara Orogen*. Geological Society of South Africa, Special Publications, **11**, 65–71.

MOHRIAK, W. U., MACEDO, J. M. *ET AL*. 1995. Salt tectonics and structural styles in the deep-water province of the Cabo Frio region, Rio de Janeiro, Brazil. *In*: JACKSON, M. P. A., ROBERTS, D. G. & SNELSON, S. (eds) *Salt tectonics: a global perspective*. American Association of Petroleum Geologists, Memoirs, **65**. 273–304.

PASSCHIER, C. W. 1990. Reconstruction of deformation and flow parameters from deformed vein sets. *Tectonophysics*, **180**, 185–199.

PASSCHIER, C. W. & COELHO, S. 2006. An outline of shear analysis in high-grade rocks. *Gondwana Research*, **10**, 66–76.

PASSCHIER, C. W. & TROUW, R. A. J. 2005. *Microtectonics* 2nd Edition. Springer, New York.

PASSCHIER, C. W., TROUW, R. A. J., RIBEIRO, A. & PACIULLO, F. V. P. 2002. Tectonic evolution of southern Kaoko Belt, Namibia. *Journal of African Earth Sciences*, **35**, 61–75.

PASSCHIER, C. W., TROUW, R. A. J., GOSCOMBE, B., GRAY, D. & KRÖNER, A. 2007. Intrusion mechanisms in a turbidite sequence: the Voetspoor and Doros plutons in NW Namibia. *Journal of Structural Geology*, **29**, 481–496.

PEDROSA-SOARES, A. C. & WIEDEMANN-LEONARDOS, C. M. 2000. Evolution of the Araçuaí Belt and its connection to the Ribeira belt, eastern Brazil. *In*: CORDANI, U. G., MILANI, E. J., THOMAZ-FILHO, A. & CAMPOS, D. A. (eds) *Tectonic Evolution of South America, 31st International Geological Congress, Rio de Janeiro*, Brazil, 265–285.

PEDROSA-SOARES, A. C., NOCE, C. M., WIEDEMANN, C. M. & PINTO, C. P. 2001. The Aracuai-West-Congo orogen in Brazil: an overview of a confined orogen formed during Gondwanaland assembly. *Precambrian Research*, **110**, 307–323.

POLLER, U. 2003. A combination of single zircon dating by TIMS and cathodoluminescence investigations on the same grain: the CLC method U–Pb geochronology for metamorphic rocks. *In*: PAGEL, M., BARBIN, V., BLANC, P. & OHNENSTETTER, D. (eds) *Cathodoluminescence in Geosciences*. Berlin, Springer, 401–414.

PRAVE, A. R. 1996. Tale of three cratons: Tectonostratigraphic anatomy of the Damara orogen in northwestern Namibia and the assembly of Gondwana. *Geology*, **24**, 1115–1118.

RAPELA, C. W., PANKHURST, R. J., CASQUET, C., BALDO, E., SAAVEDRA, J. & GALINDO, C. 1998. Early evolution of the proto-Andean margin of South America. *Geology*, **26**, 707–710.

RAPELA, C. W., PANKHURST, R. J., FANNING, C. M. & GRECCO, L. E. 2003. Basement evolution of the Sierra de la Ventana Fold Belt: new evidence for Cambrian continental rifting along the southern margin of Gondwana. *Journal of the Geological Society, London*, **160**, 613–628.

RAPELA, C. W., CASQUET, C. *ET AL*. 2008. The Río de la Plata craton and the assembly of SW Gondwana. *Earth-Science Reviews*, **83**, 49–82.

REIS, A. P. & MANSUR, K. L. 1995. Sinopse geológica do Estado do Rio de Janeiro—Mapa Geológico 1:400.000. 1995. 90 f. Geological Survey of Rio de Janeiro State, Brazil. Departamento de Recursos Minerais do Estado do Rio de Janeiro (DRM-RJ), Niterói (RJ).

ROCHA, F. P. 2002. *Mapeamento geológico da região de Marica, Estado do Rio de Janeiro. Masters Dissertation*, Federal University of Rio de Janeiro, Brazil.

SCHMITT, R. S. 2001. *A orogenia Búzios- caracterização de um evento tectono-metamórfico cambro-ordoviciano no Domínio Tectônico de Cabo Frio, sudeste da Faixa Ribeira*. PhD thesis, Federal University of Rio de Janeiro.

SCHMITT, R. S., PIMENTEL, M. M., VAN SCHMUS, W. R., TROUW, R. A. J. & ARMSTRONG, R. A. 2003. Marine sedimentation related to the latest stages of Gondwana assembly in the Ribeira Belt: new U–Pb data. *In*: *IV South American Symposium on Isotope Geology, Salvador, Short Papers Volume*, **1**, 294–297.

SCHMITT, R. S., TROUW, R. A. J., VAN SCHMUS, W. R. & PIMENTEL, M. M. 2004. Late amalgamation in the central part of West Gondwana: new geochronological data and the characterization of a Cambrian orogeny in the Ribeira Belt—SE Brazil. *Precambrian Research*, **133**, 29–61.

SCHMITT, R. S., TROUW, R. A. J., MEDEIROS, S. R. & DANTAS, E. L. 2008. Age and geotectonic setting of Late-Neoproterozoic juvenile mafic gneisses and associated paragneisses association from the Ribeira belt (SE brazil) based on geochemistry and Sm–Nd data-implications on Gondwana assembly. *Gondwana Research*, in press.

SETH, B., KRONER, A., MEZGER, K., NEMCHIN, A. A., PIDGEON, R. T. & OKRUSCH, M. 1998. Archean to neoproterozoic magmatic events in the Kaoko belt of NW Namibia and their geodynamic significance. *Precambrian Research*, **92**, 341–363.

SETH, B., OKRUSH, M., WILDE, M. & HOFFMANN, K. H. 2000. The Voetspoor intrusion, Southern Kaoko Zone, Namibia: mineralogical, geochemical and isotopic constraints for the origin of a syenitic magma. *In*: MILLER, R. MCG. (ed.) *Henno Martin Memorial Volume*. Communications of the Geological Survey of Namibia, **12**, 125–137.

STEIGER, R. H. & JÄGER, E. 1977. Subcommission on geochronology: convention on the use of decay constants in geocosmochronology. *Earth & Planetary Science Letters*, **36**, 359–362.

TORQUATO, J. R. & CORDANI, U. C. 1981. Brazil-Africa geological links. *Earth Science Reviews*, **17**, 155–176.

TRINDADE, R. I. F., DÁGRELLA FILHO, M. S., EPOF, I. & BRITO NEVES, B. B. 2006. Paleomagnetism of Early Cambrian Itabaiana mafic dikes (NE Brazil) and the final assembly of Gondwana. *Earth and Planetary Science Letters*, **244**, 361–377.

TROMPETTE, R. 1994. *Geology of Western Gondwana (2000–500 Ma).* A.A. Balkema Rotterdam.

TROUW, R. A. J., HEILBRON, M. *ET AL.* 2000. The central segment of the Ribeira belt. *In*: CORDANI, U. G., MILANI, E. J., THOMAZ-FILHO, A. & CAMPOS, D. A. (eds) *Tectonic Evolution of South America, 31st International Geological Congress, Rio de Janeiro*, Brazil, 287–310.

TUPINAMBÁ, M., TEIXEIRA, W. & HEILBRON, M. 2000. Neoproterozoic western Gondwana assembly and subduction-related plutonism: the role of the Rio Negro Complex in the Ribeira Belt. *Revista Brasileira de Geociências*, **30**, 7–11.

VALLADARES, C. S., MACHADO, N., HEILBRON, M., DUARTE, B. P. & GAUTHIER, G. 2005. Sedimentary Provenance in the Central Ribeira belt based on LaserAblation ICPMS $^{207}Pb/^{206}Pb$ Zircon Ages. *In*: PANKHURST, R. J. & VEIGA, G. (eds) *Gondwana 12: Geological and Biological Heritage of Gondwana.* Academia Nacional de Ciéncias, Córdoba, Argentina, 360–360.

VALLADARES, C. S., MACHADO, N., HEILBRON, M., DUARTE, B. P. & GAUTHIER, G. 2008. Sedimentary Provenance in the Central Ribeira belt based on Laser Ablation ICPMS $^{207}Pb/^{206}Pb$ Zircon Ages. *Gondwana Research*, in press.

ZALAN, P. V. & OLIVEIRA, J. A. B. 2005. Origem e evolução estrutural do sistema de Riftes Cenozóicos do sudeste do Brasil. *Boletim de Geociências da Petrobras*, **13**, 269–300.

ZIMBRES, E., KAWASHITA, K. & VAN SCHMUS, W. R. 1990. Evidências de um núcleo Transamazônico na Região de Cabo Frio, RJ e sua correlação com o cráton de Angola, Africa. *In*: *Congresso Brasileiro de Geologia 36, Natal, Short Papers*, **6**. Sociedade Brasileira de Geologia, 2735–2743.

The Neoproterozoic Quatipuru ophiolite and the Araguaia fold belt, central-northern Brazil, compared with correlatives in NW Africa

M. A. P. PAIXÃO[1], A. A. NILSON[1] & E. L. DANTAS[2]

[1]*Department of Mineralogy and Petrology, Institute of Geosciences, University of Brasília, Brasília, DF 70910-900, Brazil (e-mail: pires_paixao@ig.com.br)*

[2]*Geochronology Laboratory, Institute of Geosciences, University of Brasília, Brasília, DF 70910-900, Brazil*

Abstract: The Araguaia Belt, the northern branch of Neoproterozoic Tocantins tectonic province, developed during West Gondwana amalgamation as a result of collision between the Amazon and West African and/or São Francisco/Congo cratons. The external zone of the belt consists of ophiolitic slices and fragments, sedimentary rocks derived from magmatic arc sources, volcanic rocks, and part of a passive continental margin with low-grade metamorphic rocks, while the internal zone corresponds to a pile of low- to medium-grade metasedimentary rocks. The largest and best preserved ophiolitic bodies occur in the southern part of the belt, where the Quatipuru and Morro do Agostinho ophiolites are composed predominantly of mantle peridotites (mainly residual harzburgite) representing the base of the Moho transition zone. They contain chromitite pods, dunitic lensoid bodies and a suite of mafic–ultramafic dykes and/or sills resulting from partial melting, magma impregnation and diapiric up-rise. A Sm–Nd isochron age of 757 ± 49 Ma indicates oceanic crust formation during the Early Neoproterozoic. NW African correlatives of the Araguaia Belt, the Mauritanide–Bassaride–Rokelide belt, show similarities with respect to lithostratigraphic units, the ages of basement and supracrustal rocks, the presence of Neoproterozoic ophiolitic slices and fragments, suture zones characterized by high gravity anomalies and centrifugal tectonic vergence. We conclude that these belts were probably formed around the same Neoproterozoic ocean or several small coeval oceans.

The evolutionary history of the western Gondwana supercontinent involved continental collision between South American and African crustal blocks, starting at around 850–750 Ma (Porada 1989; Castaing *et al.* 1994; Trompette 1994; Brito Neves *et al.* 1999; Cordani *et al.* 2003) and extending until 550–500 Ma. However, the location of crustal block boundaries, generally represented by nappe systems and ophiolitic sequences marking suture zones, remains one of the most intriguing problems for reconstruction models of correlation between them. Ophiolites represent ancient oceanic crust, formed in a mid-ocean ridge or back-arc environment, but they are difficult to recognize when they have undergone intense tectonic fragmentation and dismemberment.

The presence of Neoproterozoic suture zones has been recorded in the Mauritanide–Bassaride–Rokelide orogenic belt bordering the western portion of the West African Craton (Villeneuve & Dallmeyer 1987; Léchorché *et al.* 1989). This orogenic belt displays a sinuous form in a general north–south direction, extending through more than 2000 km with a variable width of up to 120 km. The belt appears to have developed as a result of successive collisions of magmatic arcs, accretionary mélanges and/or previously amalgamated terranes (Hefferan *et al.* 2000). A rifting event has been proposed at around 700 Ma on the western border of the West African Craton (Villeneuve & Dallmeyer 1987) as well as subsequent convergence with arc magmatism and two collisional events (660–640 Ma and 550 Ma), of which the younger one is related to west-dipping subduction and docking between the Guyana Craton and the southwest part of the West African Craton (Villeneuve & Cournée 1994). However, palaeogeographical reconstruction models suggest that West Gondwana was not formed until after 630 Ma. One of the principal requirements for such models is to know where the suture zones in the South American counterpart are located and what do they represent (Tohver *et al.* 2002; Kröner & Cordani 2003).

Geological reconstructions of Neoproterozoic fold belts in West Gondwana (Porada 1989; Trompette 1994) show a possible link between the Araguaia fold belt in northern Brazil and the Mauritanide–Rokelide belt in northwestern Africa (Fig. 1). The identification of dismembered ophiolites in the Neoproterozoic Araguaia fold belt (Paixão & Nilson 2002; Kotschoubey *et al.* 2005) is an important recent contribution to a better understanding of its geological evolution. This evolution

From: PANKHURST, R. J., TROUW, R. A. J., BRITO NEVES, B. B. & DE WIT, M. J. (eds) *West Gondwana: Pre-Cenozoic Correlations Across the South Atlantic Region.* Geological Society, London, Special Publications, **294**, 297–318. DOI: 10.1144/SP294.16 0305-8719/08/$15.00

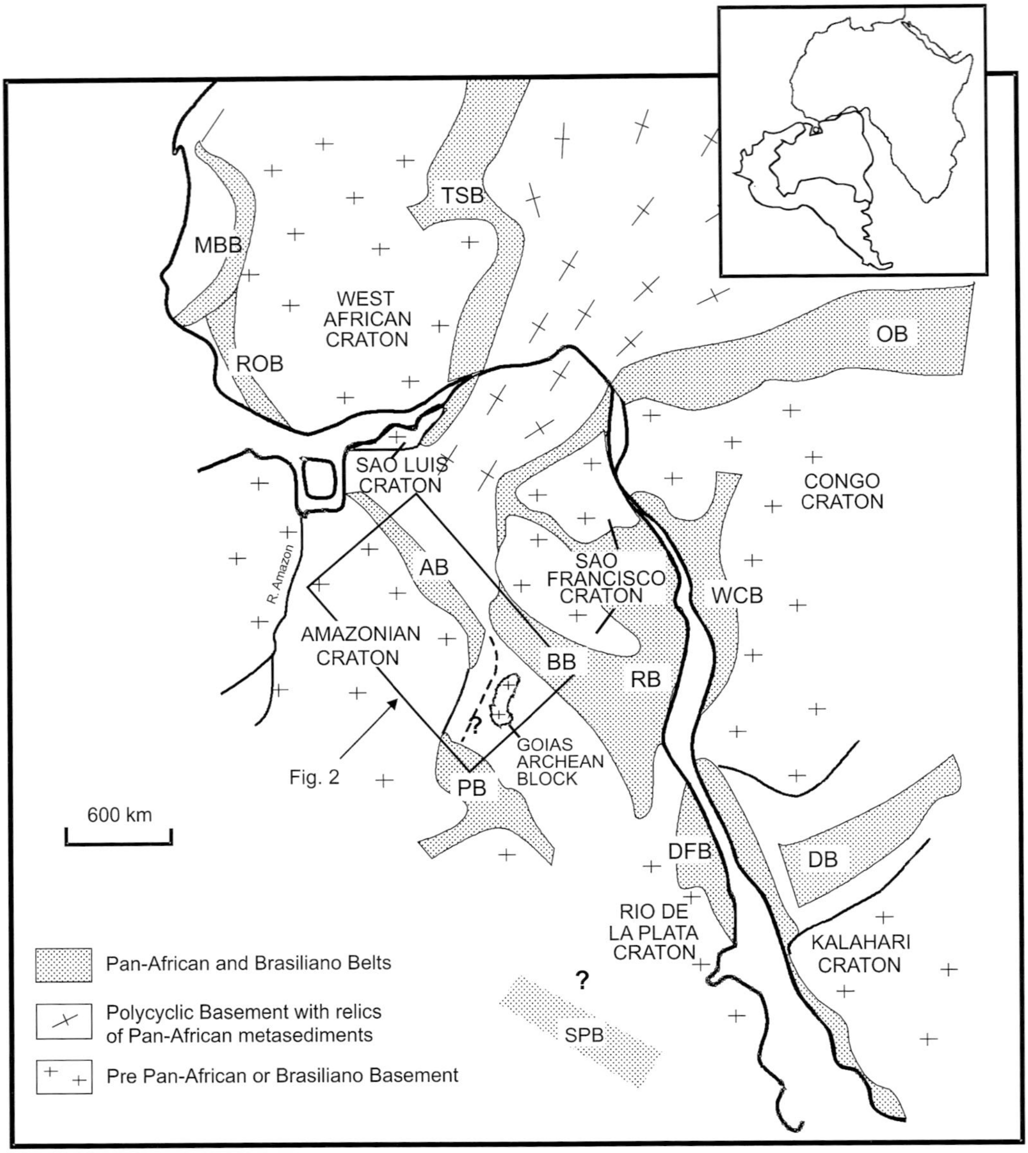

Fig. 1. Structural relationships between the Pan-African fold belts of western Africa and the Brasiliano fold belts. MBB, Mauritanide–Bassaride belt; ROB, Rokelide Belt; TSB, Trans-Saharan Belt; AB, Araguaia Belt; PB, Paraguay Belt; BB, Brasília Belt; RB, Ribeira Belt; DFB, Dom Feliciano Belt; SPB, Sierras Pampeanas Belt; OB, Obanguides Belt; WCB, West Congo Belt; DB, Damara Belt. Modified from Trompette (1997).

would have started with crustal rifting leading to the formation of the Araguaia ocean basin and later to ocean closure during collision of an eastern crustal block with the Amazon Craton. However, the exact age, size, and main geochemical characteristics of the Araguaia basin oceanic lithosphere are still an open question. Correlations between these and other Brasiliano and Pan-African fold belts have been attempted, but the lack of key geological information and the scarcity of geochronological data have hampered reliable reconstructions.

This paper describes structural, textural and petrological features of the different rock units, and isotopic data for the Quatipuru and Morro do Agostinho ophiolites and associated rocks. A review and comparison of geological data pertaining to the

Braziliano (Araguaia–Paraguay) and Pan-African (Mauritanide–Rokelide, and perhaps Dahomeyide) belts reveals clear and partial similarities in lithostratigraphy, type of basement and supracrustal rocks, calc-alkaline intrusive and extrusive rocks, ophiolite type and age, suture zones, glacial deposits and other features.

Geological setting

The Araguaia Belt is a Neoproterozoic geotectonic unit consisting of metasedimentary and meta-igneous rocks that extends north–south for more than 1200 km in length and is 100 km wide, bordering the eastern edge of the Amazon Craton (Almeida *et al.* 1986), while its northern and western limits are covered by Palaeozoic sediments of the Parnaíba Basin (Fig. 2). The southeastern contact is not well defined.

The Araguaia Belt is divided into different structural domains (Costa *et al.* 1998; Fonseca *et al.* 1999). The eastern domain corresponds to basement terrains composed of granulites and gneisses exposed in antiformal structures (e.g., the Xambioá and Colméia domes), in addition to granitoid bodies and the supracrustal rocks of the Estrondo Group. Low-grade metamorphic rocks represent the western domain and distinguish it from rocks of the Tocantins Group.

The basement in the western part of the Araguaia Belt is characterized by two orthogneissic suites: (a) Colméia Complex TTG-type Archaean gneiss domes (e.g., the Colméia and Xambioá domes, *c.* 2.85 Ga old), and (b) Palaeoproterozoic Cantão gneiss (1.85 Ga), geochemically similar to anorogenic granites of the southeastern Amazon Craton, thus possibly representing reworking of a part of this cratonic region (Dall'Agnol *et al.* 1987; Gorayeb *et al.* 2000). To the southwestern border of the belt, a small Archaean fragment belonging to the Serra Azul shear belt is recognized by Pimentel *et al.* (2000) and is characterized by orthogneiss with a Sm–Nd isochron age of 3058 ± 120 Ma. Such a crustal fragment could be a part of the Goiás Archaean block or of the southeastern margin of the Amazon Craton. It is interpreted as part of the basement of the Araguaia Belt.

The Araguaia Belt is represented by the Baixo Araguaia Supergroup (Abreu 1978), consisting of the Estrondo and Tocantins groups. The Estrondo Group consists of the following lithotypes: quartzite and muscovite quartzite with associated kyanite quartzite, magnetite quartzite and oligomictic meta-conglomerate (Morro do Campo Formation), followed upwards by muscovite–biotite schists and calc-schists, minor marble, staurolite, kyanite or fibrolite schists (Xambioá Formation) and feldspathic schists with quartzite, biotite schist and calc-schist intercalations (Canto da Vazante Formation, Abreu 1978). The Tocantins Group consists of slate, meta-siltstone, meta-arkose, meta-greywacke and quartzite with associated ophiolitic mafic–ultramafic bodies (Couto Magalhães Formation), chlorite schists, quartz–chlorite schists, sericite–chlorite schists and metabasite bodies (Pequizeiro Formation), and a top unit consisting of meta-greywackes and basalt flows (Tucuruí Formation, Trouw *et al.* 1976).

Costa & Hasui (1997) suggested that the evolution of the Araguaia Belt began with the formation of a semi-graben type basin, where deposition of the Baixo Araguaia sedimentary pile took place in a passive-margin transgressive sequence. Moura *et al.* (2000), propose that the formation of the Araguaia basin started at approximately 1.0 Ga through a crustal rifting event, as evidenced by the occurrence of felsic alkaline plutons. Several authors (e.g., Kotschoubey *et al.* 1996; Osborne 2001) agree that the ocean-basin forming event reached the oceanization stage as represented by ophiolitic mafic–ultramafic bodies such as the Serra do Tapa and Quatipuru ophiolites.

The Brasiliano event which was responsible for the inversion of the Araguaia basin was also marked by the intrusion of granitic bodies, such as the 660 Ma old Santa Luzia Granite (Moura & Gaudette 1993; Moura *et al.* 2000). Associated granitic dykes yield a 513 ± 17 Ma age (Cambro-Ordovician). Granitic plutons such as the Lajeado, Matança and Palmas plutons in the Porto Nacional region in southeastern Tocantins State yield an age close to 550 Ma. Some granites show mylonite borders related to the Porto Nacional Shear Zone (Gorayeb *et al.* 2000).

Detrital zircons from Tocantins Group rocks, (meta-rhythmites and meta-greywackes), suggest sedimentary provenance involving reworking of Brasiliano and Palaeoproterozoic sources, the latter possibly indicating that the intra-oceanic arc was close to a continental border (Osborne 2001). Detrital zircons from the rocks that represent the passive continental margin portion of the Tocantins Group (carbonate rocks and banded iron formation) show that the basin closed after 544 Ma (Osborne 2001).

The Estrondo Group consists dominantly of psammo-pelitic rocks that characteristically show contributions from the basement inliers. Zircon ages obtained in Morro do Campo Formation formerly indicated contribution from both Archaean (2909 ± 5 and 2668 ± 2Ma) and Palaeoproterozoic sources (1748 ± 5 and 1747 ± 6 Ma), but none of Brasiliano provenance. Nevertheless, other units (e.g., the Xambioá Formation) have Sm–Nd model ages which point to the contribution of Brasiliano rocks, indicating that Tocantins and

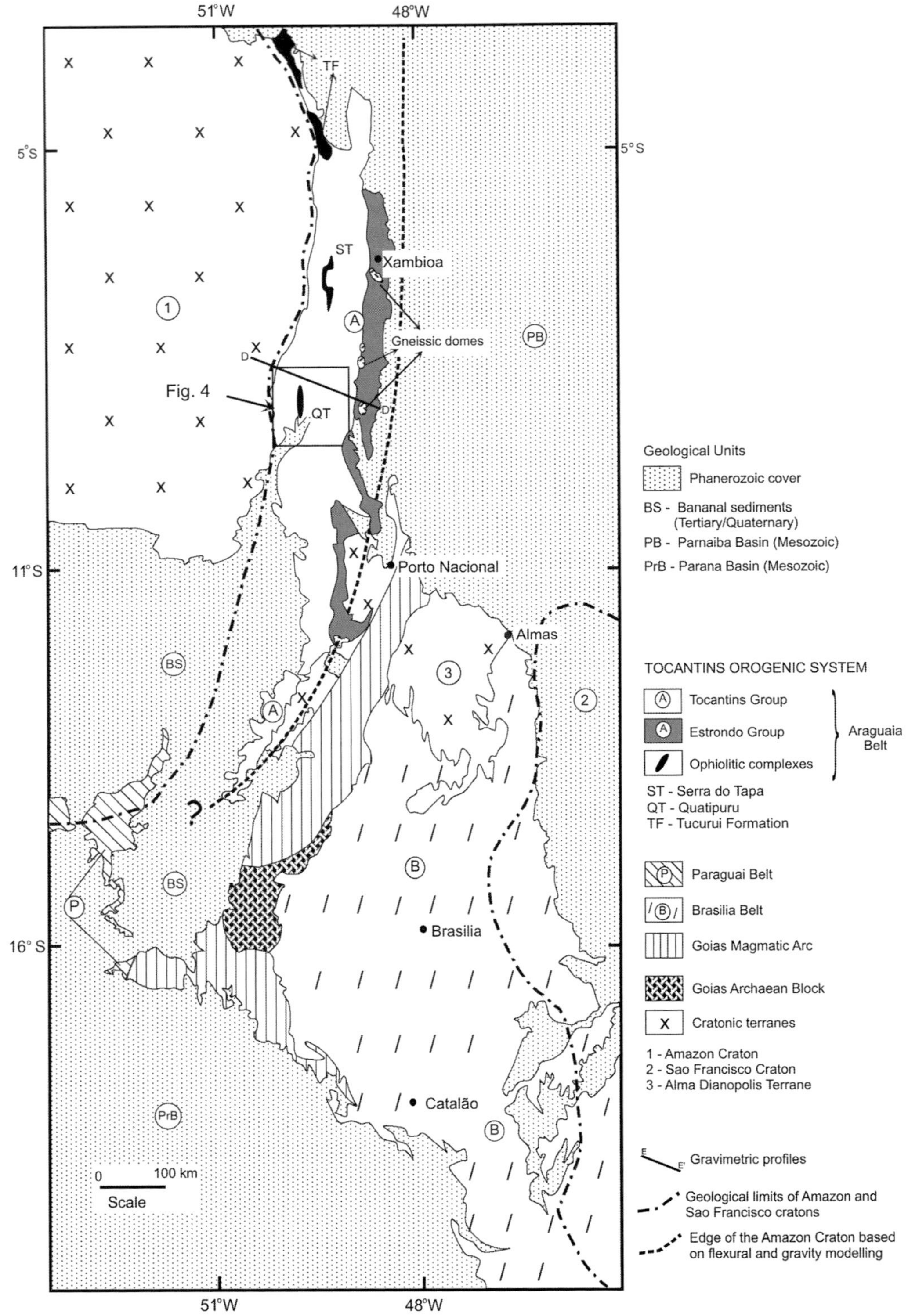

Fig. 2. Geological setting of the Brasiliano fold belts in the Tocantins tectonic province. The major ophiolitic fragments are the Serra do Tapa (ST) and Quatipuru (QT) complexes, in the north and south, respectively. The inset delineates Figure 4. Adapted from Alvarenga *et al.* (2000), Hasui *et al.* (1984), Almeida *et al.* (1986) and Ussami & Molina (1999).

Estrondo groups were either part of the same basin or different basins that were temporally correlated (Osborne 2001).

Tectonic models imply that early structures associated with the inversion of the Araguaia basin consist of a first phase of thrusts with a sinistral oblique component verging towards the WNW and a second phase of overthrusts with associated lateral ramps (Hasui & Costa 1990; Abreu *et al.* 1994; Costa & Hasui 1997; Fonseca *et al.* 2004). A second generation of structures is represented by Brasiliano age transcurrent, ductile–brittle shear zones.

Evidence of continental collision in the Araguaia Belt is characterized by slip-line features, developed close to the border with the Amazon Craton. Such features are defined by linear structures identified in satellite images (e.g., elongated hills) and by curvilinear magnetic anomalies present in magnetometric maps (Fig. 3). These indicate that crustal block (Amazon Craton) behaved as a rigid indenter and had a straight limit in relation to the Araguaia Belt (Fig. 3). This is in agreement with the Amazon plate border defined by Ussami & Molina (1999) using gravimetric methods. The Quatipuru and Morro do Agostinho ophiolites, as well as other ophiolitic bodies, occur along the boundaries of gravimetrically-defined crustal blocks and correspond to magnetic anomalies that

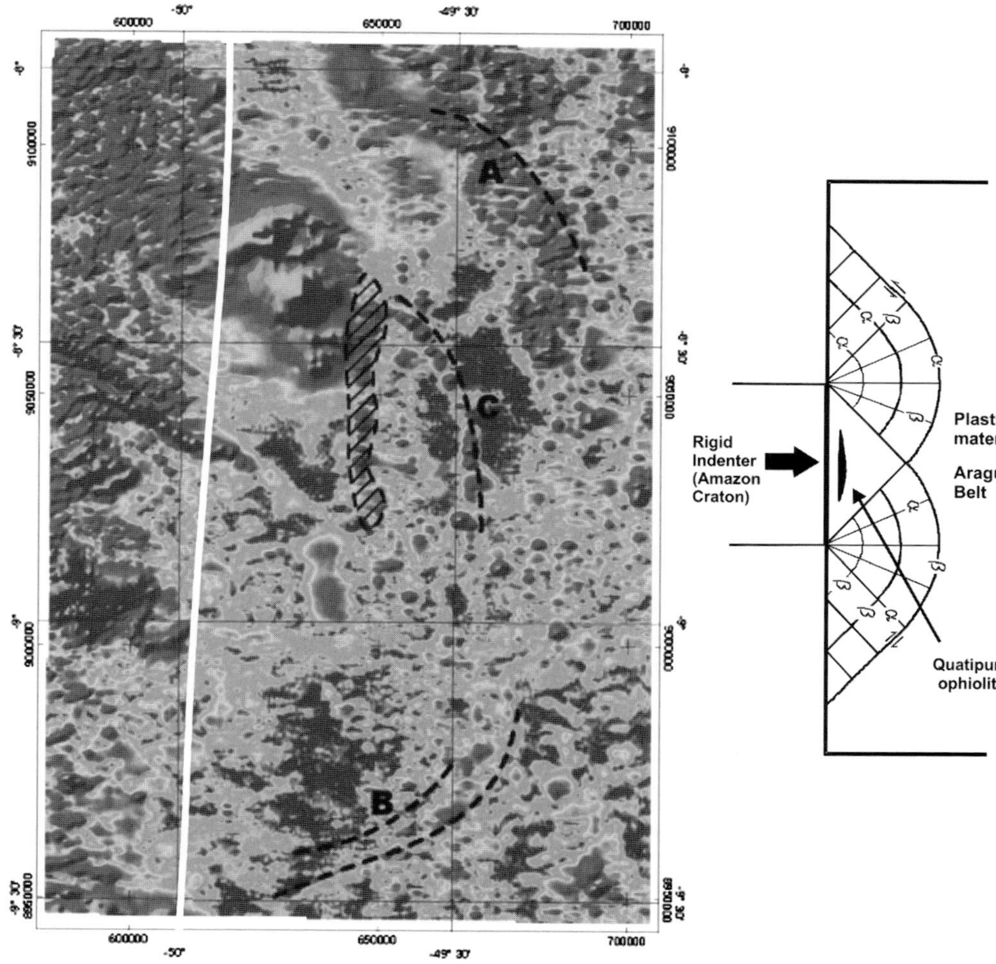

Fig. 3. Magnetometric (analytical sign) map of the Quatipuru region. Letter A corresponds to São José Hill (metasedimentary rocks from Tocantins group), while the hachured area represents the Quatipuru ophiolite; both have topographic relief and a high magnetic signature in this map (dark grey). The other structures indicated by letters B and C, are not identified on the ground or in satellite images. All structures define slip-line features indicating a straight limit to the Amazon Craton in this region. Modified from Paixão & Nilson (2001*b*). The schematic model on right corresponds to a slip-line model of a rigid indenter with straight limit (Molnar & Tapponier 1975).

extend for approximately 1000 km (Almeida *et al.* 1986).

Ophiolitic bodies in the Araguaia Belt

The main ophiolitic slices of the Araguaia Belt are located in the southern portion of this belt, between parallels 6°30′S and 9°00′S. These complexes consist of serpentinized and/or metamorphosed (magnesian schists) peridotite, basaltic lavas and ferruginous silexite[1] forming tectonic slices of decametric to kilometric scale. The silexite unit may occur as isolated bodies or in association with serpentinized peridotite and/or magnesian schist.

In a general way the ophiolite complexes are preferentially aligned in a north–south direction, with inflections generating an anastomosing pattern. Such structural features are generally concordant with equivalent structures in the country rocks.

The mafic–ultramafic bodies in the Araguaia belt have been variously interpreted as alpine bodies (Cordeiro & McCandless 1976; Nilson 1984), as magmatic intrusions emplaced along deep-seated faults, related to the Tocantins–Araguaia suture (Almeida 1974; Gorayeb 1989), and as thrust slices of a probable volcano-sedimentary terrane (Souza *et al.* 1995). However, most authors now agree that they are slices of ophiolite (Trouw *et al.* 1976; Hasui *et al.* 1977; Kotschoubey *et al.* 1996; Osborne 2001; Paixão & Nilson 2001*a*, *b*; Kotschoubey *et al.* 2005). Paixão & Nilson (2001*a*) characterize some of these ophiolites as remnants of the Moho transition zone in Neoproterozoic oceanic lithosphere.

The age of the ophiolites is regarded as Neoproterozoic, based on U–Pb ages (*c.* 630 Ma) for magmatic zircons from rhyolitic tuffs which, according to Osborne (2001), were related to ocean-basin volcanism in the proximity of the Quatipuru Complex. Below we present a Sm–Nd isochron age of 757 ± 49 Ma for narrow gabbroic dykes cross-cutting harzburgite of the Quatipuru ophiolite.

The largest ophiolites of the Araguaia Belt are represented by the Quatipuru and Serra do Tapa complexes (Fig. 2). The principal concentration of ophiolitic slices is located in the southern part of this belt, where the largest and/or best preserved are represented by the Quatipuru and Morro do Agostinho complexes (Fig. 4). They appear to be part of a single complex that was tectonically dismembered into two portions. Morro do Agostinho is characterized by a distinct association of serpentinized harzburgite and pillow basalts; in contrast the Quatipuru Complex consists of serpentinized harzburgite and a suite of scattered mafic–ultramafic dykes and/or sills and chromitite pods.

The Quatipuru ophiolite

The Quatipuru Complex strikes north–south; it is 40 km long and only about 1.5 km wide, showing some NE and NW inflections (Fig. 5), and dips approximately 45°E in structural parallelism with the country rocks. It shows boudinage features in the central-north and extreme southern portions.

The country rocks pertaining to the Tocantins Group are represented by a metasedimentary pile consisting of a metric to decametric alternation of incipiently metamorphosed meta-siltstone, meta-sandstone, slate, meta-greywacke and meta-rhythmite (turbidite), as well as rare meta-limestone lenses. The country rocks show sub-greenschist to greenschist-facies metamorphism (chlorite and muscovite in meta-greywacke), and structural elements (foliation and folds) with tectonic vergence towards the Amazon Craton.

The Quatipuru Complex consists of serpentinized peridotites in the central part involved with a ferruginous silexite envelope. The envelope is wider in the eastern part of the complex and consists of massive, brecciated and veined parts, with local mylonite bands. These rocks are grey-coloured when fresh (hematite) and reddish brown when weathered (goethite) due the strong iron oxide/hyroxide impregnation. The presence of tectonic features (foliation and fractures) and fresh pyrite grains distinguish it from birbirite (derived by weathering of ultramafic rocks).

The serpentinized peridotites of the complex correspond to a lenticular litho-structural arrangement of predominant harzburgite with small sparse (decametric) lensoid dunite intercalations. In addition, the peridotites host chromitite pods and a suite of mafic and ultramafic dykes/sills.

The harzburgites exhibit a diffuse foliation and mantle structures, such as proto-granular texture (Mercier & Nicolas 1975) (Fig. 6) and local websteritic banding. The original modal composition, estimated from the inferred primary mineralogy, is olivine 70–71%, orthopyroxene 28–29%, and chromite 1–2%.

The dunite lenses are small (up to 5 m thick and 30 m long) and orientated parallel or sub-parallel to the harzburgite foliation. Their modal composition is olivine 97–98%, chromite 2–3% and orthopyroxene <1%. Locally such lenses contain chromitite micro-pods and magma impregnation

[1]Silexite is a light grey, usually massive, coarse-grained quartz rock (>95% quartz) with irregular interlocking grains, sometimes showing mylonitic foliation; it may form an envelope around the peridotite bodies.

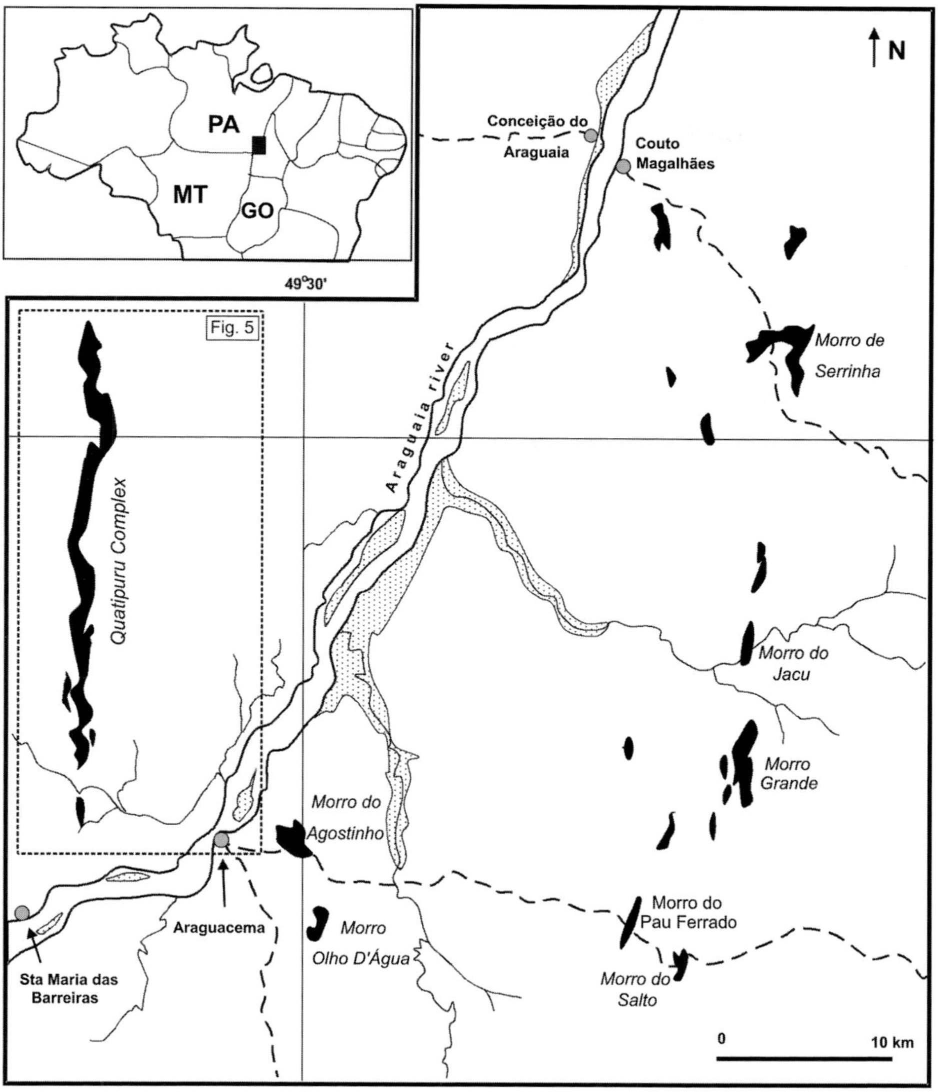

Fig. 4. Location map of the major ophiolitic bodies in the southern portion of the Araguaia Belt (adapted from Gorayeb 1989).

structures, features commonly described in ophiolitic complexes and in the oceanic lithosphere (Boudier & Nicolas 1995).

The harzburgite–dunite contact consists of a narrow transition, interpreted as the result of partial melting, with dissolution of orthopyroxene grains generating a dunitic residue, as proposed by Kelemen *et al.* (1995). These authors interpret such dunite residue as having formed through reactive porous flow between ascending melts and harzburgite host beneath a mid-ocean ridge axis in asthenospheric mantle.

Harzburgites and dunites from the Quatipuru ophiolite show a characteristic chemical signature of residual mantle, as demonstrated by the comparison with peridotite samples from modern oceanic lithosphere and peridotite from the Maqsad diapir (Semail ophiolite, Oman) (Fig. 7).

The chromitites appear as pods and lenses, varying from 1 to 10 m in length (parallel to the main foliation), 0.3 to 4.5 m thick and 5 m wide along the foliation dip. Commonly, these chromitite pods show a dunite envelope and three distinct textural types: disseminated, massive and nodular (Fig. 8). Both the

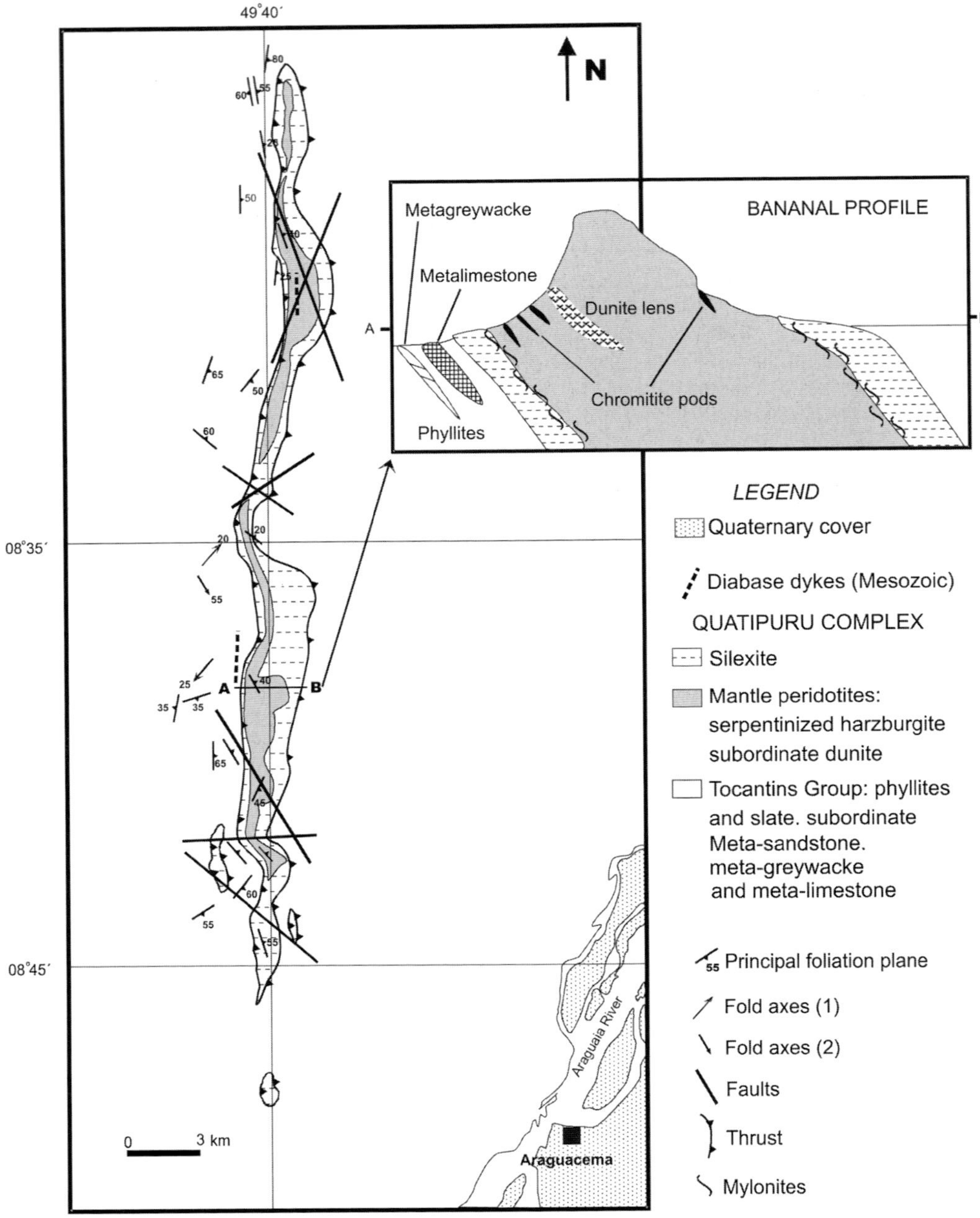

Fig. 5. Geological map of the Quatipuru Complex.

nodular texture and the dunitic envelope are characteristic features of ophiolite complexes (Nicolas 1989). In the discriminant diagram for podiform and stratiform chromites (Fig. 9), Quatipuru chromites plot in the podiform composition field typical of ophiolite complexes such as Troodos (Cyprus) and Semail (Oman) ophiolites.

Dyke suite

One of the most interesting features of the Quatipuru ophiolite is the occurrence of a suite of scattered narrow ultramafic and mafic dykes and sills intruding harzburgites and dunites. It is divided in two groups: (1) ultramafic dykes

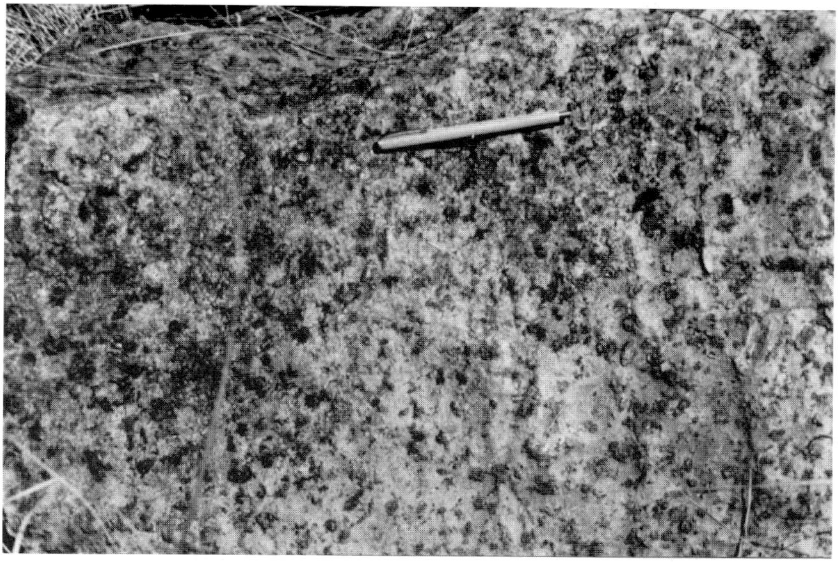

Fig. 6. Harzburgite with proto-granular texture. Dark grey grains correspond to bastite (orthopyroxene) immersed in a serpentine matrix (olivine). Pen is 12.5 cm long.

(and sills) and (2) gabbroic dykes (and sills). The ultramafic group consists of pegmatoid orthopyroxenite, clinopyroxenite and wehrlite. The second group is represented by olivine gabbro and diabase, which are the most abundant and occur throughout the complex. Temporal relations between them are, from oldest to youngest: pegmatoid orthopyroxenite, pegmatoid clinopyroxenite, wehrlite, olivine gabbro and diabase.

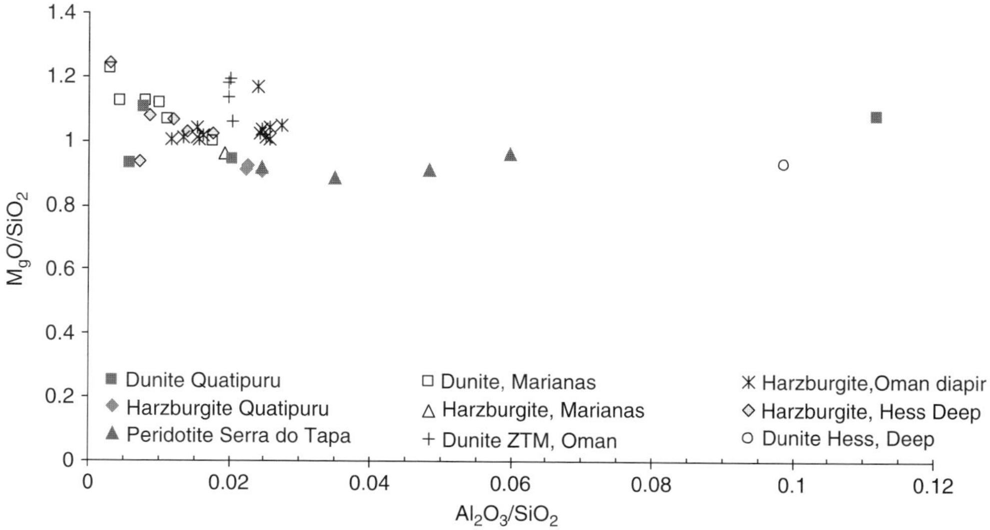

Fig. 7. Geochemical plot of peridotite samples from Quatipuru Complex compared to peridotites from modern oceanic lithosphere, Maqsad diapir (Semail ophiolite) and Serra do Tapa. Note the similarity with the compositional trend of refractory abyssal peridotites, characterized by low Al_2O_3/SiO_2 (0.01–0.03) and high MgO/SiO_2.

Fig. 8. Nodular chromite with discoidal nodules. Pen is 14 cm long.

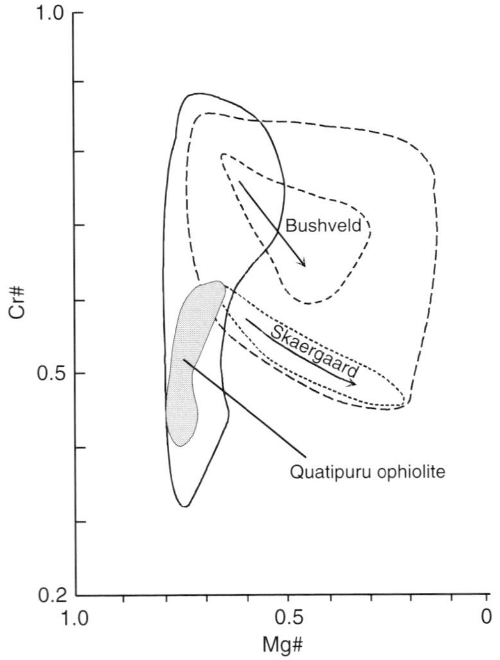

Fig. 9. Cr# (Cr/(Cr + Al)) and Mg# (Mg/(Mg + Fe) plot for podiform and stratiform chromites (fields with solid and dashed line, respectively). Small fields with traced line correspond to chromites of the Bushveld and Skaergaard layered complexes and the arrows indicate differentiation trend (data from Wall 1975 and Jackson 1969). Podiform chromite field from Thayer (1970).

Pegmatoid orthopyroxenite sills average 0.5 m thick and occur only locally. They are usually associated with orthopyroxenite pockets and centimetric websteritic banding. Both these sills and banding exhibit tight folds formed in response to asthenospheric–lithospheric mantle flow (Fig. 10), as proposed by Suhr (1992) for similar features in the Bay of Islands ophiolite (Canada), where such structures are thought to have formed very close to the expanding mid-ocean ridge axis, marking a flow component normal to the ridge.

Pegmatoid clinopyroxenite dykes and/or pockets are of metric dimension (Fig. 11), where the thickest portions exhibit internal differentiation, both in mineralogy (originating wehrlitic portions) and grain-size. The contact between dyke and harzburgite is marked by a dunite depletion halo. Dunite and harzburgite xenoliths commonly occur within the dykes.

Wehrlite dykes are strongly serpentinized, showing medium grain size and harzburgite xenoliths, occurring locally in the complex. A folded pegmatoid orthopyroxenite sill has undeformed orthopyroxene grains and random orientation in the fold hinge, showing that its intrusion took place simultaneously with the formation and folding of the harzburgite mantle foliation. Effects of this folding can also be seen in the websteritic banding and are associated with melt migration processes.

Olivine gabbro dykes are narrow (3–15 cm thick) and characteristically present weak to strong propylitic alteration. Some of them do not display chilled margins, indicating that the temperature of the dyke was very close to that

Fig. 10. Folded pegmatitic orthopyroxenite sill hosted in harzburgite. Orthopyroxene crystals in the hinge are not deformed, indicating that crystallization occurred concomitantly with folding. Pen is 12.5 cm long.

of the harzburgite host, while others display compositional and textural banding (layering) parallel to sub-parallel to the contacts with the host-rock.

Diabase dykes vary somewhat in thickness (2 cm–1.5 m) and cut all lithotypes, including chromitite pods. Sometimes they form boudins in serpentine schist as a result of deformation along local shear zones within harzburgite. In this instance its original mineralogy is transformed to an actinolite, chlorite and epidote-rich assemblage.

Hence, based on textural features and field relationships between the pyroxenitic and gabbroic

Fig. 11. Wedged-shaped pegmatitic clinopyroxenite dyke (dark grey) cutting harzburgite (pen on harzburgite is 12.5 cm long.).

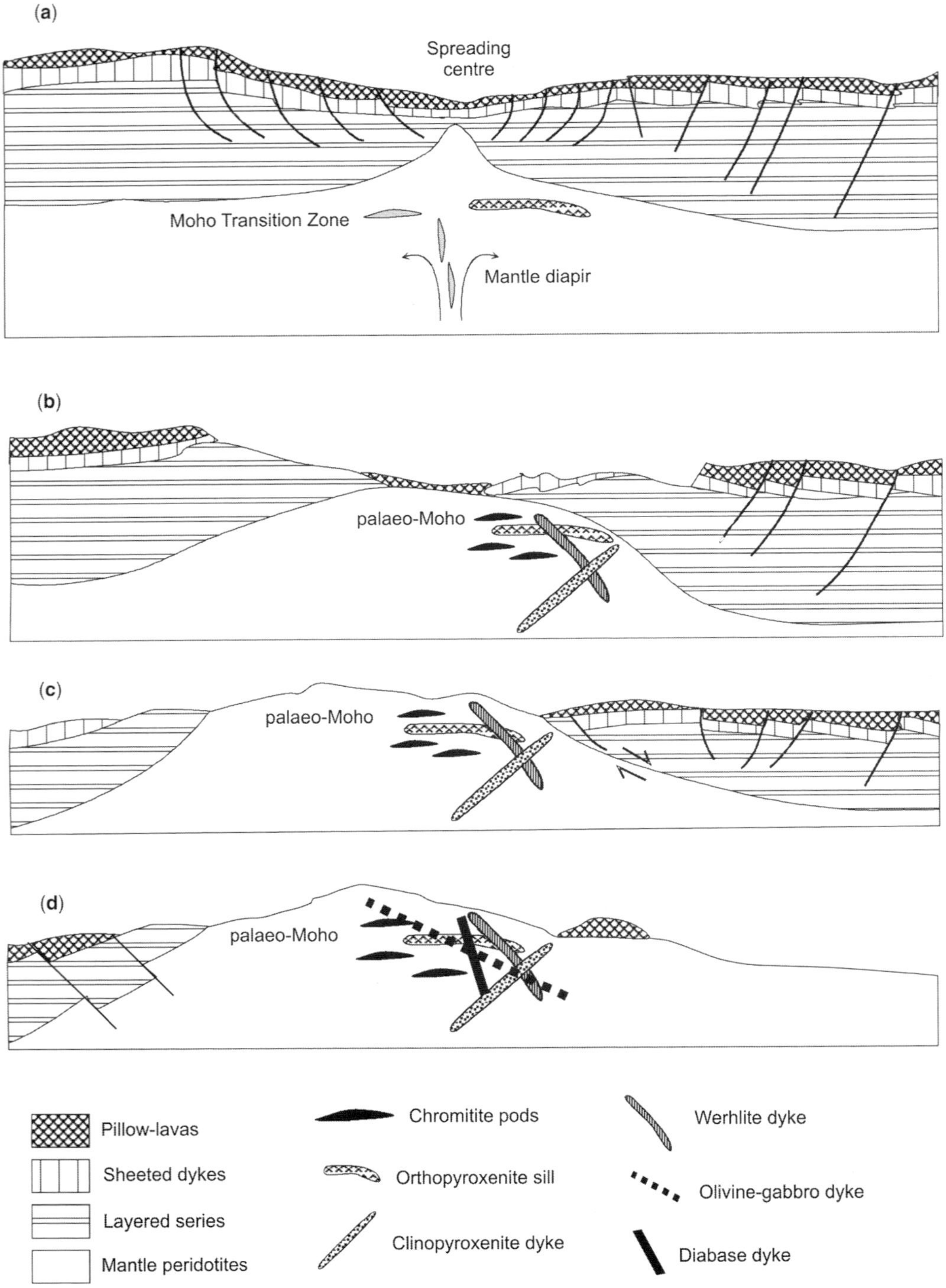

Fig. 12. Schematic model for seafloor spreading and diapiric uprise of the Quatipuru ophiolite (A). Adapted from Dilek & Thy (1998).

dykes, we suggest that the Quatipuru ophiolite peridotites record several stages of uplift in the oceanic environment, beginning in asthenospheric conditions (ductile deformation below the ocean-spreading zone) to shallower levels in an oceanic crustal environment. In the next stage they became detached and were subsequently obducted onto continental crust. The evolution of the mantle peridotite diapir, as well as the dyke formation process, is similar to those described in the Semail ophiolite (Ceuleneer *et al.* 1996). In this model, mantle ductile flow conditions were maintained until the injection of the clinopyroxenite sills (Fig. 12). Subsequently crustal conditions were marked by the intrusion of wehrlite and olivine gabbro dykes and, at still shallower conditions, by the injection of dolerite dykes (Fig. 12). The preservation of the mesh texture and of structural elements of mantle origin in the peridotites indicates that serpentinization took place under static conditions, as suggested by, for example, for serpentinized peridotites of the Hess Deep region of the East Pacific Ocean (Früh-Green *et al.* 1996).

The deformational history recorded in the Quatipuru ophiolite rocks is divided into high temperature and low temperature structures, like as proposed by Nicolas *et al.* (1999). High temperature structures are represented by imperfect, diffuse foliation in harzburgite, associated websterite banding and the litho-structural arrangement of dunite lenses within harzburgite. The latter probably originated from the combination of deformational and melt migration processes (cf. Kelemen *et al.* 1995). Low temperature structures correspond to ductile shear zones, represented by zones of serpentine-schist with anastomosing main foliation, intrafolial westward-verging microfolds and harzburgite and olivine gabbro dyke boudins, the latter exhibiting the chlorite–actinolite association and recrystallised plagioclase indicating shear zone formation under greenschist facies conditions.

The Morro do Agostinho ophiolite

The Morro do Agostinho ophiolite is located in the vicinity of the city of Araguacema (Tocantins State), and is about 3 km long in the NW–SE direction (Fig. 3). It consists of an association of harzburgite and basalt with pillow structures. Such rocks form mega-lenses in the general NNE direction, isolated and tectonically emplaced in metasedimentary country rocks. The harzburgite outcrops show alternating preserved and strongly deformed portions, with a characteristic silexite envelope. In contrast, basalt outcrops are largely undeformed.

The harzburgite exhibits proto-granular texture with local sheared bands of serpentine schist. It is cut by pegmatoid websterite dykes with dunite and harzburgite xenoliths. These relationships, together with the presence of ferruginous silexite associated with serpentinized peridotites, are similar to those found in the Quatipuru Complex. The basalts exhibit pillow structures, pillow breccia fragments and hyaloclastite breccia (Fig. 13). Individual pillows show variolitic texture along the contact of the unaltered basalt with the highly spillitized external parts of the pillows.

Fig. 13. Pillow lavas (rounded limits indicated by traced line), Morro do Agostinho Complex.

The pillow breccia fragments are mostly sub-rounded and subordinately sub-angular and vary from 0.5 to 10 cm in diameter. The geometric arrangement with the groundmass that supports such fragments indicates movement, though restricted, during lava flow. The hyaloclastite breccia exhibits a fragmentary texture (Cas & Wright 1988), characterized by abundance of lithic fragments of variable size; the lapilli fraction predominates in relation to larger vitreous groundmass supported fragments (blocks).

Late hydrothermal activity associated with these basalts is represented by metrical parts of epidotized breccia, commonly having associated carbonate and quartz fine-veins; sulphide is characteristically absent.

Structures found in basaltic rocks from Morro do Agostinho, are similar, for example, to those described by Busby-Spera (1987) for different facies of basaltic lava flows in the back-arc basin of Cedros Island, Mexico.

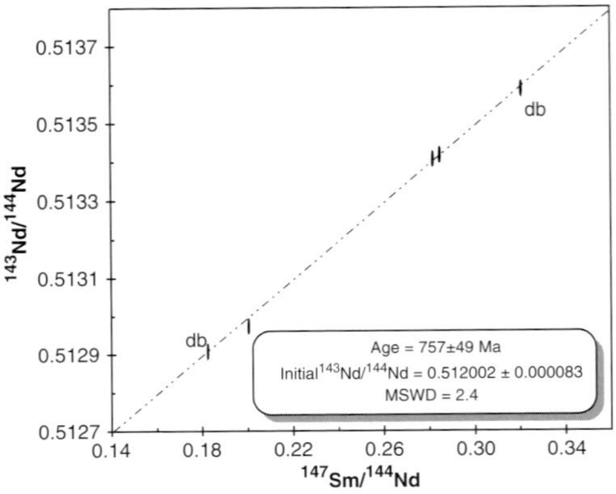

Fig. 14. Sm–Nd isochron diagram for dykes from the Quatipuru Complex. Letters db represent dolerite dyke samples, while the other three points represent olivine gabbro samples.

Isotope geology of the Quatipuru ophiolite and associated rocks

Geochronological and isotopic geochemical analyses were carried out in the Geochronology Laboratory of the University of Brasilia, according to procedures described by Gióia & Pimentel (2000).

Age of the Quatipuru ophiolite

The major peridotite rocks of the Quatipuru ophiolite represent residual mantle (harzburgite) and consequently are not appropriate for dating the ocean crust formation stage in the Araguaia Belt. For this reason, we selected five dyke rocks cutting the mantle peridotites of the Quatipuru ophiolite for Sm–Nd analysis (three olivine gabbros and two dolerite samples; Table 1). Such rocks correspond to the latest and most differentiated magmatic crystallization products in the process of the oceanic lithosphere formation.

The Sm–Nd data yield a whole-rock isochron age of 757 ± 49 Ma (Fig. 14), with a calculated initial $^{143}Nd/^{144}Nd$ ratio of 0.512002 ($\varepsilon Nd_{(t)} = +6.6$). The positive $\varepsilon Nd_{(t)}$ values from all analysed rocks are indicative of a depleted mantle source (MORB). We may conclude that contamination by continental crust material is not evident for these rocks. Thus, we interpret our result as the crystallization age of magmatic products related to the construction of oceanic lithosphere of the Araguaia Belt in Neoproterozoic times.

Isotopic composition of basaltic and country rocks

Sm–Nd analyses of some representative samples were also obtained for the Morro do Agostinho ophiolite and metasedimentary rocks of Estrondo and Tocantins groups (Table 2).

Positive $\varepsilon Nd_{(t)}$ values of the Morro do Agostinho basalt shows typical MORB signature,

Table 1. *Sm–Nd isotopic data for Quatipuru ophiolite dyke rocks*

Sample	Rock type	Sm (ppm)	Nd (ppm)	$^{147}Sm/^{144}Nd$	$^{143}Nd/^{144}Nd$ ($\pm 2\sigma$)	$\varepsilon Nd_{(757)}$
QT-55.B	Diabase	1.173	2.207	0.3213	0.513590 (31)	+6.55
QT-36.5B	Diabase	3.0799	10.228	0.1820	0.512915 (11)	+6.88
QT-53.C	Olivine gabbro	0.495	1.062	0.2820	0.513409 (15)	+6.65
QT-48.B	Olivine gabbro	1.038	3.136	0.2001	0.512980 (25)	+6.39
QT-47.N	Olivine gabbro	0.3193	0.6773	0.2850	0.513420 (33)	+6.76

$^{143}Nd/^{144}Nd$ normalized to $^{146}Nd/^{144}Nd = 0.71290$. Model ages ($T_{DM}$) calculated according to the single-stage depleted mantle model of DePaolo (1981); the primary age used for $\varepsilon Nd_{(t)}$ is based on the Sm–Nd isochron obtained in this paper, assuming 2σ errors of 0.1 % for $^{147}Sm/^{144}Nd$ and 0.003 % for $^{143}Nd/^{144}Nd$.

Table 2. *Sm–Nd data for basalt from the Morro do Agostinho ophiolite and metasedimentary rocks from the Tocantins and Estrondo groups.*

Sample	Rock type	Sm (ppm)	Nd (ppm)	$^{147}Sm/^{144}Nd$	$^{143}Nd/^{144}Nd$ ($\pm 2\sigma$)	$\varepsilon Nd_{(t)}$	T_{DM}(Ga)
Morro Agostinho Complex							
MA-01	Spillitized basalt	3.61	10.38	0.2103	0.512979 (33)	+5.39	–
MA-01.F	Unaltered basalt	3.69	11.0	0.2028	0.512907 (17)	+4.71	–
Estrondo Group							
QTR-01	Amphibole schist	3.99	18.35	0.1316	0.512056 (28)	−6.18	1.81
QTR-02	Amphibolite	8.41	41.92	0.1213	0.511998 (07)	−6.43	1.70
Tocantins Group							
QTR-05.B	Slate	10.27	51.04	0.1216	0.511709 (19)	− 12.09	2.19
QT-04	Phyllite	11.02	49.25	0.1375	0.511920 (14)	− 9.25	2.23
NO-03	Greywacke	1.75	8.895	0.1200	0.511975 (08)	− 6.77	1.72

which together with the values of mafic dykes from the Quatipuru Complex, clearly points to a depleted mantle derivation for the ophiolitic rocks of the Araguaia basin.

Negative $\varepsilon Nd_{(t)}$ for the metasediments, considering an age of deposition around 600–630 Ma, point to reworked continental crust as the main source rock. Samples QT-04 and QTR-05.B belong to the Tocantins Group and their model ages, when interpreted as a rough estimate of crustal residence, indicate a Palaeoproterozoic source age. Sample NO-3, from near the western border of the Quatipuru ophiolite, is an epiclastic country rock that was previously described by Osborne (2001). The 1.72 Ga T_{DM} model age may be considered as an average for the crustal residence of the metasediment, but U–Pb data for detrital zircons obtained by Osborne (2001) indicate that these meta-greywackes show contributions of both Palaeoproterozoic and Brasiliano source rocks. Samples QTR-02 and QTR-01 from the Estrondo Group have comparable T_{DM} model ages. Although only a few samples have been analyzed for a provenance study, all show a contribution from Palaeoproterozoic sources, whereas sedimentary provenance studies in the Baixo Araguaia Supergroup rocks show contribution from younger terranes (Meso–Neoproterozoic ages) and suggest a more complicated scenario for the evolution of this belt (Moura *et al.* 2005, 2008).

Discussion and correlations

The possible continuity of the Araguaia–Paraguay belts into the Mauritanide–Bassaride–Rokelide belt or Dahomey Belt in north of West Africa is based on similarities between several geological features: the age and nature (lithotypes) of the basement rocks, the lithostratigraphic record, ophiolite type and age, suture zones, extrusive and intrusive calc-alkaline suite rocks, glacial deposits, etc. (Table 3). Some geological features of the Araguaia Belt are also similar to those found in the Dahomey Belt: (1) a suture zone, identified by high gravity anomalies, sometimes associated with magnetic anomalies related to mafic–ultramafic bodies, (2) mafic–ultramafic bodies, such as the Amalaoulaou and Timetrine complexes, are interpreted as ophiolites (Black *et al.* 1979) and (3) the age of the Timetrine Complex is about 800 ± 50 Ma (Caby 1987).

In the Araguaia and Mauritanide belts the basement rocks show similar ages and sometimes crop out as inliers bordered by supracrustal rocks; in the former these inliers are spatially associated with high gravity anomalies in the Bouguer profiles, suggesting that the structural framework is caused by the uplift of portions of the upper mantle.

From the stratigraphical point of view the Araguaia and Paraguay belts have similar platform cover sequences represented by banded iron formation and carbonates with fossils of Vendian age (Alvarenga *et al.* 2000; Osborne 2001). Together with their marginal location relative to the Amazon Craton, this indicates that these belts can be treated as a single Brasiliano fold belt (Almeida 1974). In addition, the Araguaia and Paraguay belts show a clear tectonic–metamorphic polarity, with increasing metamorphic grade and deformation intensity from west to east, exemplified by amphibolite-facies of the Estrondo Group on the eastern part of the Araguaia Belt compared to non-metamorphic conditions in the external zone and cratonic covering of the Paraguay Belt (e.g., the Puga and Diamantino formations, the former characterized by the presence of glacial sediments).

The Mauritanide Belt is characterized by a foreland succession, represented by a sequence of very low-grade, immature, flyschoid metasedimentary rocks of the upper Proterozoic Tichilit el Beïda

Table 3. *Comparative aspects of the Mauritanide and Araguaia belts*

	Mauritanide Belt	Araguaia Belt
Basement age	Liberian (2.7 Ga) Eburnian (2.0–1.8 Ga)	2.85 Ga Colméia Complex 1.85 Ga Cantão Gneiss
Basement lithologies	Gneisses, granite and volcano-sedimentary rocks	TTG-type gneisses
Basement inliers	Eastern Senegal and Kayes	Colméia, Xambioá, Lontra and Grota Rica
Metamorphism	Low to absent close to cratonic region, reaching amphibolite facies in the internal zone	Low to absent close to cratonic region, reaching amphibolite facies in the internal zone
Calc-alkaline suites	Intrusive (Kelbé Complex) and extrusive (El Hneïkât Unit)	Extrusive (624 Ma)
Ophiolitic material	El Aoueïja Unit and Oued Amoûr Unit; Termesse and Guinguan groups Farkâka Association	Quatipuru, Morro do Agostinho, Serra do Tapa complexes
Age of ophiolitic material	650 to 700 Ma	757 ± 49 Ma
Glacial deposits	Tichilit el Beïda Group; Rokel River Group (Rokelide Belt)	Documented only in the southern portion (Paraguay Belt–Puga Formation)
Geophysical patterns	High gravity	High gravity associated with high magnetic anomalies

Group. Early Cambrian microfossils (*Aldanella*) were identified in a similar stratigraphic level in the Rokelide orogen. The basal portion is characterized by a distinctive Neoproterozoic tillite associated with baritic carbonate, chert and stromatolitic dolostone (Dallmeyer & Lécorché 1989).

In the Mauritanide–Bassaride–Rokelide belt the stage of oceanic lithosphere formation is indicated by the presence of a fragmented ophiolite sequence represented by the El Aoueija and Oued Amour units and tholeiitic volcanic rocks of the Guinguan Group and Farkaka Association, related to plutono-volcanic calc-alkaline suites whose ages range between 650 and 700 Ma (Dallmeyer & Villeneuve 1987; Dallmeyer & Lécorché 1989). Island-arc associated terranes, such as the Kouloun-tou branch, are characterized by calc-alkaline volcanic and plutonic rocks with ages varying from 683 to 673 Ma (Dallmeyer & Villeneuve 1987).

The beginning of the crustal rifting stage in the Araguaia Belt is marked by alkaline rocks approximately 1 Ga old (Moura *et al.* 2000). The oceanization stage of the Araguaia Belt is dated as 757 ± 49 Ma ago, represented by the rectilinear ophiolitic assemblage, forming an expressive lineament in the Araguaia Belt. This lineament borders the Amazon Craton, extending over 500 km from the Araguacema (Quatipuru and Morro do Agostinho ophiolites) as far as the Tucuruí Formation (Trouw *et al.* 1976) (Fig. 2). Another important ophiolite occurrence is located halfway between these two extremes, represented by basaltic pillow-lavas and serpentinites of the Serra do Tapa ophiolite. It has been interpreted as representing a proto-oceanic basin, similar to the north and central portions of the modern Red Sea, or similar to poorly evolved Alpine–Apennine oceanic basins (Kotschoubey *et al.* 2005).

Ophiolitic bodies have not been identified in the Paraguay Belt. However, the eastern portion of the Cuiabá Group makes contact with volcano-sedimentary sequences of the Goiás magmatic arc terrane. This terrane is represented in this area by the Bom Jardim de Goiás and Arenópolis–Piranhas sequences (Seer 1985; Pimentel & Fuck 1992), which correspond to island-arc basins with juvenile isotopic signatures that could have associated ophiolites. We suggest that the age of the ophiolite could be used as a diagnostic correlation element for the Neoproterozoic fold belts during western Gondwana assembly, especially with reference to central Brazil and Africa. Figure 15 illustrates a proposed model for this Neoproterozoic scenario.

In contrast, the oceanic lithosphere of the Brasília Belt was formed around 800 Ma ago, as determined in Araxá Group amphibolites in the Bonfinópolis region (Piuzana *et al.* 2003). Diachronous subduction occurred between 0.9 and 0.85 Ga, leading to accretion of the Mara Rosa intra-oceanic arc over the continental fragment of the Goiás

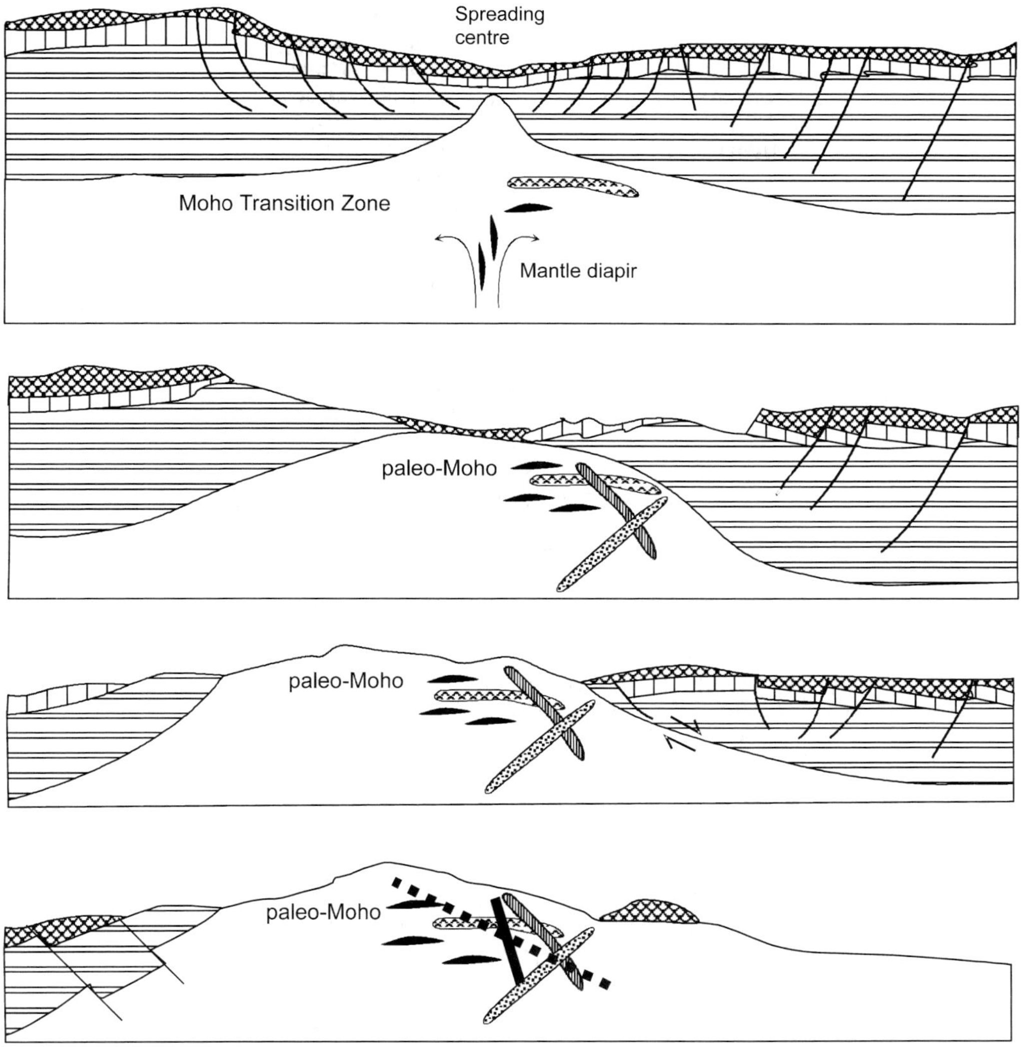

Fig. 15. Schematic model proposed for Araguaia belt and Quatipuru ophiolite formation. (**a**) initial break-up of continental crust (*c*. 1 Ga); (**b**) development of an island-arc due to intra-oceanic subduction (900–800 Ma); (**c**) formation of a supra-subduction zone with an oceanic spreading centre in a back-arc setting, with associated transform faults (Quatipuru original site), *c*. 700 Ma; (**d**) inversion and ocean closure *c*. 650–600 Ma; (**e**) final disposition of the Araguaia Belt with late granite intrusions *c*. 550–500 Ma.

Archaean block around 0.79 Ga (Pimentel *et al.* 1997). Two events of young intra-oceanic crust formation occur between 860 and 630 Ma in the Brasília Belt (Viana *et al.* 1995; Laux *et al.* 2004). The age of 634 Ma obtained by Osborne (2001) from a rhyolitic tuff was interpreted as not representing the upper sequence of ophiolitic body, but instead as corresponding either to calc- alkaline magmatism associated with inversion of the Araguaia basin, or to arc magmatism as identified in Mauritanide–Bassaride–Rokelide and Brasilia belts (Dallmeyer & Villeneuve 1987; Pimentel *et al.* 1997).

The centrifugal structural vergences in the Araguaia and Mauritanides–Rokelides belts are compatible with a model of a closing ocean, with mass escape and tectonic transport towards the cratonic blocks, respectively the Amazon and

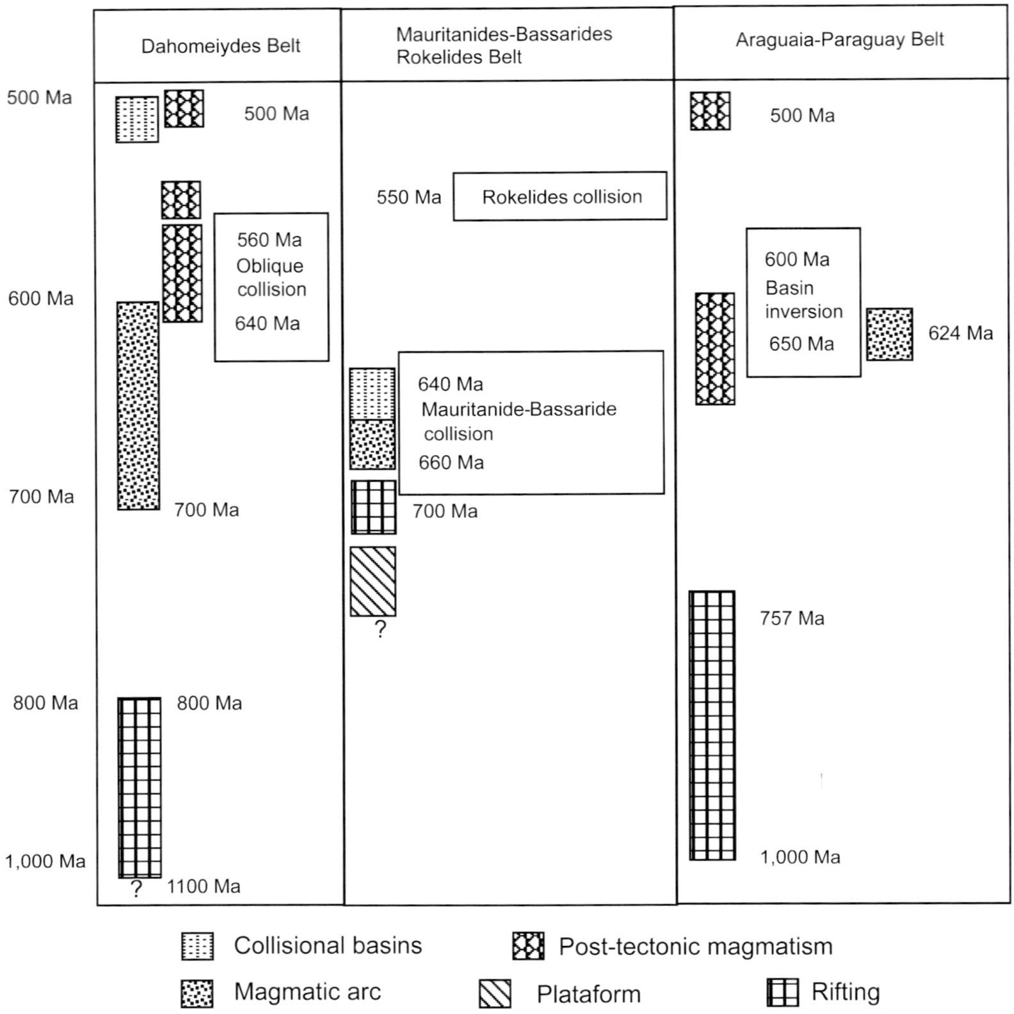

Fig. 16. Principal events for correlation between Brasiliano and Pan-African fold belts.

West Africa cratons. This is corroborated by gravimetric profiles of both belts, where high gravity anomalies are located in the respective internal zones of these belts, indicating that portions of upper mantle were involved in the tectonic evolution and suggesting the development of suture zones between these belts (Dallmeyer & Villeneuve 1987; Ussami & Molina 1999). Ritz & Robineau (1988), based on geological and geophysical data (geo-electrical models and gravity), proposed that the Mauritanide Belt is the result of a continental collision, with a west-dipping suture zone characterized by the presence of tholeiitic volcanic and volcaniclastic rocks (Dallmeyer & Villeneuve 1987).

The final framework of the Tocantins Province (Araguaia, Paraguay and Brasília belts) is marked by the stabilization of the Brasiliano orogenesis and the intrusion of post-orogenic granites between 550 and 500 Ma (Moura *et al.* 2000). In contrast, during the Cambrian, deformation and metamorphism in the Bassaride and perhaps in the central Mauritanide belts are interpreted as distal effects of a continent–continent collision that led to formation of the Rokelide orogen.

In a general way we can outline some principal events that occurred between the Araguaia–Paraguay belt and the Mauritanide–Bassaride–Rokelide belt (Fig. 16). Such events allied to geological characteristics shown in Table 3 demonstrate a strong correlation between these areas—'ties that bind'—in West Gondwana.

Conclusions

The following points are stressed as conclusions of this study.

(1) The Araguaia Belt borders the eastern part of the Amazon Craton, and its geometry can be related to the original boundaries of the craton. The Quatipuru ophiolite and its correlatives in Africa are the best markers of the suture zones, marked by high gravimetric anomalies, sometimes associated with magnetic anomalies, and linking the Araguaia–Paraguay and Mauritanide–Bassaride–Rokelide belts.

(2) These belts present similarities in the basement age and lithology, stratigraphic record, glacial deposits, metamorphic polarity, pattern of Bouguer anomaly and age of ophiolites, demonstrating that magmatic and tectonic processes or events were operating at the same time and/or in the same region and indicating that a branch of a large Neoproterozoic ocean surrounded the West African palaeocontinent.

(3) The largest ophiolitic slices in the Araguaia Belt indicate a straight planar limit bordering the Amazon Craton, the best being represented by the elongate Quatipuru ophiolite (40 km long) and the smaller Morro do Agostinho ophiolite.

(4) The Quatipuru ophiolite is characterized by associations typical of the base of the Moho Transition Zone, where the record of partial melting process and magma impregnation is represented by refractory harzburgites (residual mantle), dunite bodies, a suite of scattered narrow mafic and ultramafic sills and dykes, and chromitite pods, indicating a high rate of magma supply and consequently an environment of fast spreading mid-ocean ridge (Nicolas 1989; Hekinian *et al.* 1992; Constantin 1999). However, field evidence points to restricted outpouring of magma, forming pillowed basalts. Morro do Agostinho is characterized by a distinct association of serpentinized harzburgite and basaltic pillow-lavas similar to that found in the modern ocean floor (e.g., the Garret Transform Fault in the Pacific Ocean), showing that the outcrop level of ophiolitic slices and fragments in the Araguaia Belt could reflect the original structure of its oceanic site.

(5) Olivine gabbros and diabase dykes signify shallow conditions during magma uprise and yield a Sm–Nd isochron age of 757 ± 49 Ma for the oceanization stage of the Araguaia Belt. We also suggest that the age of the Quatipuru ophiolite can be used as a major correlation element between the Neoproterozoic fold belts of central Brazil and NW Africa.

(6) The final position of the ophiolitic bodies of the Araguaia Belt was controlled by thrusting towards the WNW, resulting from obduction and accretion of this mobile belt to the Amazon Craton border. The resulting deformation imprinted on the peridotite is represented by serpentine schist (mylonite) bands with microfolds showing vergence towards the Amazon Craton.

(7) A model of ocean closure due to block collisions is characterized by centrifugal structural vergences between the Araguaia and Rokelide belts, with mass escape and tectonic transport towards the cratonic blocks, respectively the Amazonian and West Africa cratons. Slip-line features identified in the Araguaia Belt show that the Amazon Craton (a rigid indenter) had a straight limit in the Quatipuru region. However, this geometry could have been different in the region of the Paraguay Belt, thus contributing to differences in the lithostratigraphical record of the two belts.

Some issues are not yet solved, such as: (1) when did the ophiolite obduction occur? (2) when exactly did the Brasiliano collision take place? and (3) did another block, besides the Amazon and West African cratons, participate in the collision process (intra-oceanic arc)?

M. A. P. Paixão is grateful to CNPQ (Proc. N. 146034/1999-6) for a Doctorate scholarship and to Simone Gióia for assistance with the isotopic analyses. We are grateful to R. Van Schmus and A. C. Pedrosa-Soares for their reviews of the paper and R. A. J. Trouw and R.J. Pankhurst for editorial reviews. Special gratitude is dedicated to B. B. Brito Neves for his enthusiastic incentive.

References

ABREU, F. A. M. 1978. O Supergrupo Baixo Araguaia. *In*: *Congresso Brasileiro de Geologia 30, Recife, Anais*, **2**, Sociedade Brasileira de Geologia, 539–545.

ABREU, F. A. M., GORAYEB, P. S. S. & HASUI, Y. 1994. Tectônica e inversão metamórfica no Cinturão Araguaia. *In*: *4th Simpósio de Geologia da Amazônia, Boletim de Resumos*, 1–4.

ALMEIDA, F. F. M. 1974. Sistema tectônico marginal do Cráton do Guaporé. *In*: *Congresso Brasileiro de Geologia 28, Anais*, **4**. Sociedade Brasileira de Geologia, 9–17.

ALMEIDA, F. F. M., FERREIRA, F. J. F., CARNEIRO, C. D. R. & STEFANI, F. L. 1986. Aspectos evolutivos da Geossutura Tocantins–Araguaia. *In*: *Congresso Brasileiro de Geologia 34, Goiânia, Anais*, **3**. Sociedade Brasileira de Geologia, 1269–1285.

ALVARENGA, C. J. S., MOURA, C. A. V., GORAYEB, P. S. S. & ABREU, F. A. M. 2000. Paraguay and Araguaia belts. *In*: CORDANI, U. G., MILANI, E. J., THOMAZ FILHO, A. & CAMPOS, D. A. (eds) *Tectonic evolution of South America, 31st International Geological Congress, Rio de Janeiro*, 183–193.

BLACK, R., CABY, R. *ET AL.* 1979. Evidence for late Precambrian plate tectonics in West Africa. *Nature*, **278**, 223–227.

BOUDIER, F. & NICOLAS, A. 1995. Nature of the Moho transition zone in the Oman ophiolite. *Journal of Petrology*, **36**, 777–796.

BRITO NEVES, B. B., CAMPOS NETO, M. C. & FUCK, R. A. 1999. From Rodinia to Western Gondwana: an approach to the Brasiliano-Pan African cycle and orogenic collage. *Episodes*, **22**, 155–166.

BUSBY-SPERA, C. 1987. Lithofacies of deep marine basalts emplaced on a Jurassic backarc apron, Baja California, Mexico. *Journal of Geology*, **95**, 671–686.

CABY, R., 1987. The Pan African belt of West Africa from the Sahara Desert to the Gulf of Benin. *In*: SCHAER, J. P. & RODGERS, J. (eds) *The Anatomy of Mountain Ranges*. Princeton University Press, 129–170.

CAS, R. A. F. & WRIGHT, J. V. 1988. *Volcanic successions: modern and ancient: a geological approach to processes, products and successions*. Allen & Unwin Ltd, London.

CASTAING, C., FEYBESSE, J. L., THIÉBLEMONT, D., TRIBOULET, C. & CHEVREMONT, P. 1994. Paleogeographical reconstructions of the Pan-African/Brasiliano orogen: closure of an oceanic domain or intracontinental convergence between major blocks? *Precambrian Research*, **69**, 327–344.

CEULENEER, G., MONNEREAU, M. & AMRI, I. 1996. Thermal structure of a fossil mantle diapir inferred from the distribution of mafic cumulates. *Nature*, **379**, 149–153.

CONSTANTIN, M. 1999. Gabbroic intrusions and magmatic metasomatism in harzburgite from the Garrett transform fault: implications for the nature of the mantle-crust transition at fast-spreading ridges. *Contributions to Mineralogy and Petrology*, **136**, 111–130.

CORDANI, U. G., NEVES, B. B. B. & D'AGRELLA, M. S. 2003. From Rodinia to Gondwana: A review of the available evidence from South America. *Gondwana Research*, **6**, 275–284.

CORDEIRO, A. A. C. & MCCANDLESS, G. 1976. Maciço ultramáfico do Quatipuru. *In*: *Congresso Brasileiro de Geologia 29, Anais*, **3**. Sociedade Brasileira de Geologia, 9–15.

COSTA, J. B. S. & HASUI, Y. 1997. Evolução geológica da Amazônia. *In*: COSTA, M. L. & ANGÉLICA, R. S. (eds) *Contribuições à geologia da Amazônia*. Sociedade Brasileira de Geologia, Núcleo Norte, Belém, **2**, 15–90.

COSTA, J. B. S., ARAÚJO, O. J. B., HASUI, Y. & BORGES, M. S. 1998. O Bloco Araguacema. *In*: *Congresso Brasileiro de Geologia 40, Belo Horizonte, Anais*. Sociedade Brasileira de Geologia, 65.

DALĹAGNOL, R., BETTENCOURT, J. S., JORGE-JOÃO, X. S., MEDEIROS, H., COSTI, H. T. & MACAMBIRA, M. J. B. 1987. Granitogenesis in northern Brazilian region: a review. *Revista Brasileira de Geociências*, **17**, 382–403.

DALLMEYER, R. D. & VILLENEUVE, M. 1987. $^{40}Ar/^{39}Ar$ polyorogenic mineral age record of a poliphase tectonothermal evolution in the Southern Mauritanide orogen, southeastern Senegal. *Geological Society of America, Bulletin*, **98**, 602–611.

DALLMEYER, R. D. & LÉCORCHÉ, J. P. 1989. $^{40}Ar/^{39}Ar$ polyorogenic mineral age record within the central Mauritanide orogen, West Africa. *Geological Society of America Bulletin*, **101**, 55–70.

DEPAOLO, D. J. 1981. A neodymium and strontium isotopic study of the Mesozoic calc-alkaline granitic batholiths of Sierra Nevada and Peninsular Ranges, Califórnia. *Journal of Geophysical Research*, **86**, 10470–10488.

DILEK, Y. & THY, P. 1998. Structure, petrology and seafloor spreading tectonics of the Kizildag ophiolite, Turkey. *In*: MILLS, R. A. & HARRISON, K. (eds) *Modern Ocean floor processes and geological record*. Geological Society, London, Special Publications, **148**, 43–69.

FONSECA, M. A., EVANGELISTA, H. J. & OLIVEIRA, C. G. 1999. Estruturas transcorrentes na Faixa Araguaia e seu significado no contexto da interação dos crátons Amazônico–São Francisco/Congo e oeste Africano. *In*: *6th Simpósio de Geologia da Amazônia, Boletim de Resumos Expandidos*, 251–254.

FONSECA, M., DE OLIVEIRA, C. & EVANGELISTA, H. 2004. The Araguaia Belt, Brazil: part of a Neoproterozoic continental-scale strike-slip fault system. *In*: WEINBERG, R., TROUW, R., FUCK, R. & HACKSPACHER, P. (eds) *The 750–550 Ma Brasiliano Event of South America. Journal of the Virtual Explorer, Electronic Edition*, **17**, Paper 6.

FRÜH-GREEN, G. L., PLAS, A. & DELL' ANGELO, L. N. 1996. Mineralogic and stable isotope record of polyphase alteration of upper crustal gabbros of the East Pacific Rise (Hess Deep, Site 984). *In*: MÉVEL, C., GILLIS, K. M., ALLAN, J. F. & MEYER, P. S. (eds) *Proceedings of the Ocean Drilling Program, Scientific Results*, **147**, 235–254.

GIÓIA, S. M. & PIMENTEL, M. M. 2000. The Sm–Nd isotopic method in the geochronology laboratory of the University of Braslia. *Anais da Academia Brasileira de Ciências*, **72**, 219–245.

GORAYEB, P. S. S. 1989. Corpos serpentiníticos da Faixa Araguaia na região de Araguacema-Pequizeiro-Conceição do Araguaia (Goiás-Pará). *Revista Brasileira de Geociências*, **19**, 51–62.

GORAYEB, P. S. S., MOURA, C. A. V. & BARROS, G. R. 2000. Pb–Pb zircon ages of the Porto Nacional high-grade metamorphic terrain, northern portion of the Goiás massif, Central Brazil. *Revista Brasileira de Geociências*, **30**, 190–194.

HASUI, Y. & COSTA, J. B. S. 1990. O Cinturão Araguaia: um novo enfoque estrutural-estratigráfico. *In*: *Congresso Brasileiro de Geologia 36, Natal, Anais*, **6**, Sociedade Brasileira de Geologia, 2535–2549.

HASUI, Y., ABREU, F. A. M. & SILVA, J. M. R. 1977. Estratigrafia da Faixa de Dobramentos Paraguaia-Araguaia na região centro-norte do Brasil. *Boletim Instituto de Geociência-USP*, **8**, 107–108.

HASUI, Y., COSTA, J. B. S. & ABREU, F. A. M. 1984. Província Tocantins: Setor setentrional. *In*: ALMEIDA, F. F. M. & HASUI, Y. (eds) *O Pré-Cambriano do Brasil*. Editora Edgard Blücher, São Paulo, 137–204.

HEFFERAN, K. P., ADMOU, H., KARSON, J. A. & SAQUAQUE, A. 2000. Anti-Atlas (Morocco) role in Neoproterozoic Western Gondwana reconstruction. *Precambrian Research*, **103**, 89–96.

HEKINIAN, R., BIDEAU, D., CANNAT, M., FRANCHETEAU, J. & HÉBERT, R. 1992. Volcanic activity and crust-mantle exposure on the ultrafast Garrett transform fault near 13°28' S in the Pacific. *Earth and Planetary Science Letters*, **108**, 259–275.

JACKSON, E. D. 1969. Chemical variation in coexisting chromite and olivine in chromitite zones of the Stillwater Complex. *In*: WILSON, H. B. D. (ed.) *Magmatic Ore Deposits, a Symposium*. Economic Geology Monographs, **4**, 61–75.

KELEMEN, P. B., SHIMIZU, N. & SALTERS, V. J. M. 1995. Extraction of mid-ocean ridge basalt from the upwelling mantle by focused flow of melt in dunite channels. *Nature*, **375**, 747–753.

KOTSCHOUBEY, B., HIERONYMUS, B., RODRIGUES, O. B. & AMARAL, R. T. 1996. Basaltos e serpentinitos da Serra do Tapa (PA). Prováveis testemunhos de um complexo ofiolítico pouco evoluído edesmembrado. *In*: *Congresso Brasileiro de Geologia 39, Salvador, Anais*, **6**. Sociedade Brasileira de Geologia, 25–28.

KOTSCHOUBEY, B., HIERONYMUS, B. & ALBUQUERQUE, C. A. R. 2005. Disrupted peridotites and basalts from the Neoproterozoic Araguaia Belt (northern Brazil): remnants of a poorly evolved oceanic crust? *Journal of South American Earth Sciences*, **20**, 211–230.

KRÖNER, A. & CORDANI, U. 2003. African, southern Indian and South American cratons were not part of the Rodinia supercontinent: evidence from field relationships and geochronology. *Tectonophysics*, **375**, 325–352.

LAUX, J. H., PIMENTEL, M. M., DANTAS, E. L., ARMSTRONG, R., ARMELE, A. & NISON, A. A. 2004. Mafic magmatism associated with the Goiás magmatic arc in the Anicuns region, Goiás, central Brazil: Sm–Nd isotopes and new ID-TIMS and SHRIMP U–Pb data. *Journal of South American Earth Sciences*, **16**, 599–614.

LÉCORCHÉ, J. P., DALLMEYER, R. D. & VILLENEUVE, M. 1989. Definition of tectonostratigraphic terranes in the Mauritanide, Bassaride, and Rokelide orogens, west Africa. *In*: DALLMEYER, R. D. (ed.) *Terranes in the Circum-Atlantic Paleozoic Orogens*. Geological Society of America, Special Papers, **230**, 131–143.

MERCIER, J. C. & NICOLAS, A. 1975. Textures and fabrics of upper mantle peridotites as illustrated by xenoliths from basalts. *Journal of Petrology*, **16**, 454–487.

MOLNAR, P. & TAPPONIER, P. 1975. Cenozoic tectonics of Asia: effects of a continental collision. *Science*, **189**, 419–426.

MOURA, C. A. V. & GAUDETTE, H. E. 1993. Evidence of Brasiliano/Panafricano deformation in the Araguaia belt: implication for Gondwana evolution. *Revista Brasileira de Geociências*, **23**, 117–123.

MOURA, C. A. V., GORAYEB, P. S. S. & ABREU, F. A. M. 2000. The Araguaia Belt revisited. *In*: *31st International Geological Congress, Rio de Janeiro, Abstracts volume*. Sociedade Brasileira de Geologia, (CD-ROM).

MOURA, C. A. V., PINHEIRO, B. L. S., NOGUERIRA, A. C. R, GORAYEB, P. S. S. & GALARZA, M. A. 2005. Sedimentary provenance in paleoenvironment of, the Baixo Araguaia Supergroup: constraints on the paleogeography evolution of the Araguaia Belt in the assembly of west Gondwana. *In*: PANKHURST, R. J. & VEIGA, G. D. (eds) *Gondwana 12: Geological and Biological Heritage of Gondwana*. Academia Nacional de Ciências, Córdoba, Argentina, 260–261.

MOURA, C. A. V., PINHEIRO, B. L. S., NOGUEIRA, A. C. R., GORAYEB, P. S. S. & GALARZA, M. A. 2008. Sedimentary provenance and palaeoenvironment of the Baixo Araguaia Supergroup: constraints on the palaeogeographical evolution of the Araguaia Belt and assembly of West Gondwana. *In*: PANKHURST, R. J., TROUW, R. A. J., BRITO NEVES, B. B. & DE WIT, M. J. (eds) *West Gondwana: Pre-Cenozoic Correlations Across the South Atlantic Region*. Geological Society, London, Special Publications, **294**, 173–196.

NICOLAS, A. 1989. *Structures of ophiolites and dynamics of oceanic lithosphere*. Kluwer Academic Publishers, Dordrecht, The Netherlands, 367 p.

NICOLAS, A., BOUDIER, F. & MESHI, A. 1999. Slow spreading accretion and mantle denudation in the Mirdita ophiolite (Albania). *Journal of Geophysical Research*, **104** (B7), 15155–15167.

NILSON, A. A. 1984. O atual estágio de conhecimento dos complexos máfico-ultramáficos pré-cambrianos do Brasil – uma avaliação preliminar. *In*: *Congresso Brasileiro de Geologia 33, Rio de Janeiro, Anais*, **4**, Sociedade Brasileira de Geologia, 4166–4203.

OSBORNE, G. C. 2001. Geotectonics and Mineralization in the Ophilite Assemblages of the Araguaia Marginal Basin, Pará and Tocantins States, Central Brazil. *In*: *7 Simpósio de Geologia da Amazônia, Resumo Expandidos* (CD-ROM).

PAIXÃO, M. A. P. & NILSON, A. A. 2001*a*. Characterization of the Moho Transition Zone in the Quatipuru Complex - a slab of oceanic lithosphere in southern Pará, Brazil - and implications for the tectonic evolution of the Araguaia Belt. *In*: *International Conference: Geology of Oman, Abstracts*, 68.

PAIXÃO, M. A. P. & NILSON, A. A. 2001*b*. Basaltos almofadados e harzburgitos do Morro do Agostinho (Araguacema, Tocantins): caracterização geológica e petrográfica de fragmento ofiolítico na Faixa Araguaia. *In*: *7 Simpósio de Geologia da Amazônia, Resumos Expandidos* (CD-ROM).

PAIXÃO, M. A. P. & NILSON, A. A. 2002. Fragmentos ofiolíticos da Faixa Araguaia: caracterização geológica e implicações tectônicas. *Contribuições à Geologia da Amazônia*, **3**, 85–103.

PIMENTEL, M. M. & FUCK, R. A. 1992. Neoproterozoic crustal accretion in Central Brazil. *Geology*, **20**, 375–379.

PIMENTEL, M. M., WHITEHOUSE, M. J., VIANA, M. G., FUCK, R. A. & MACHADO, N. 1997. The Mara Rosa arc in the Tocantins Province: further evidence for Neoproterozoic crustal accretion in Central Brazil. *Precambrian Research*, **81**, 299–310.

PIMENTEL, M. M., FUCK, R. A. & GIOIA, M. C. L. 2000. The Neoproterozoic Goiás magmatic arc, Central Brazil: a review and new Sm-Nd isotopic data. *Revista Brasileira de Geociências*, **30**, 35–39.

PIUZANA, D., PIMENTEL, M. M., FUCK, R. A. & AMSTRONG, R. 2003. SHRIMP U–Pb and Sm–Nd data for the Araxá Group and associated magmatic rocks: Constraints for the age of sedimentation and geodynamic context of the southern Brasília Belt, central Brazil. *Precambrian Research*, **125**, 139–160.

PORADA, H. 1989. Pan African rifting and orogenesis in southern to equatorial Africa and eastern Brazil. *Precambrian Research*, **44**, 103–106.

RITZ, M. & ROBINEAU, B. 1988. Tectonic interpretation of electrical structures beneath the West African Craton edge in eastern Senegal. *American Journal of Science*, **288**, 756–776.

SEER, H. J. 1985. *Geologia, deformação e mineralização de cobre no complexo vulcano-sedimentar de Bom Jardim de Goiás*. MSc thesis, University of Brasília.

SOUZA, J. O., MORETON, L. C. & CAMARGO, M. A. 1995. Geologia das seqüências metavulcanossedimentares da Serra do Tapa (SE do Pará) e Xambica (NW do Tocantins): ocorrência de metabasaltos com pillow-lavas. *Boletim de Geociências do Centro-oeste*, **18**, 20–31.

SUHR, G. 1992. Upper mantle peridotites in the Bay of Islands ophiolite, Newfoundland: formation during the final stages of a spreading centre? *Tectonophysics*, **206**, 31–53.

THAYER, T. P. 1970. Chromite segregations as petrogenetic indicators. *In*: Geological Society of South Africa, Special Publications, **1**, 380–390.

TOHVER, E., VAN DER PLUIJM, B. A., VAN DER VOO, R., RIZZOTTO, G. & SCANDOLÂRA, J. E. 2002. Paleogeography of the Amazon craton at 1.2 Ga: Early Grenvillian collision with the Llano segment of Laurentia. *Earth and Planetary Science Letters*, **199**, 185–200.

TROMPETTE, R. 1994. *Geology of Western Gondwana (2000–500 Ma). Pan-Anfrican-Brasiliano Agregation of South America and Africa*. A.A. Balkema, Rotterdam.

TROMPETTE, R. 1997. Neoproterozoic (~600 Ma) aggregation of Western Gondwana: a tentative scenario. *Precambrian Research*, **82**, 101–112.

TROUW, R. A. J., VAZ, L. F., SLONGO, T. T. & NAKASATO, N. 1976. Geologia da região de Tucuruí, Baixo Tocantins, Pará. *In*: *Congresso Brasileiro de Geologia 29, Anais*, **2**, Sociedade Brasileira de Geologia, 137–148.

USSAMI, N. & MOLINA, E. C. 1999. Flexural modeling of the neoproterozoic Araguaia belt, central Brazil. *Journal of South American Earth Sciences*, **12**, 87–98.

VIANA, M. G., PIMENTEL, M. M., WHITEHOUSE, M. J., FUCK, R. A. & MACHADO, N. 1995. O Arco Magmático de Mara Rosa, Goiás: geoquímica e geocronologia e suas implicações regionais. *Revista Brasileira de Geociências*, **25**, 111–123.

VILLENEUVE, M. & CORNÉE, J. J. 1994. Structure, evolution and palaeogeography of the West African craton and bordering belts during the Neoproterozoic. *Precambrian Research*, **69**, 307–326.

VILLENEUVE, M. & DALLMEYER, R. D. 1987. Geodynamic evolution of the Mauritanide, Bassaride and Rokelide orogens (West Africa). *Precambrian Research*, **37**, 19–28.

WALL, S. A. 1975. *The mineralogy, chemistry and certain aspects of reactivity of chromite from the Bushveld igneous complex*. National Institute of Metallurgy, Johanesburg, Report, **1709**.

Correlations between the classic Paraná and Cape–Karoo sequences of South America and southern Africa and their basin infills flanking the Gondwanides: du Toit revisited

E. J. MILANI[1] & M. J. DE WIT[2]

[1]*Petrobras Research Center, Exploration R&D, Tectonics Group, 950 Horácio Macedo Av., Cidade Universitária, 21941.915, Rio de Janeiro, RJ, Brazil (e-mail: ejmilani@petrobras.com.br)*

[2]*AEON and Department of Geological Sciences, University of Cape Town, Rondebosch 7701, South Africa.*

Abstract: Early during the twentieth Century, pioneering correlations between the Palaeozoic–Mesozoic basins of South America and southern Africa were used by Alexander du Toit to support the initial concepts of continental drift and the proposal of a united Gondwana continent. New stratigraphic tools and data can now be used to further tease out similarities and differences to reconstruct the detailed histories of these, the Paraná and Cape–Karoo basins. In turn this knowledge can be used also to increase our understanding of the origin and evolution of Gondwana. Recent advances in tectonics and stratigraphy showed that both basins evolved together along a common early Palaeozoic Gondwana margin facing the Panthalassa. Thereafter, this margin was transformed into a series of linked foreland basins coupled to the evolution of the Gondwanides. In detail, the foreland successions differ considerably due to spatial and temporal differences in tectonic histories along the Gondwanides. Only towards the end of the Palaeozoic did both basins evolve and merge into a single continental-scale, and truly intracratonic, terrestrial Gondwana basin that persisted until the early Cretaceous. This shared history was once again disrupted in the Early Cretaceous during the opening of the South Atlantic Ocean.

The two largest, long-lived and once contemporaneous Phanerozoic sedimentary basins of Gondwana occur in South America and southern Africa, and are known as the Paraná and Cape–Karoo basins, respectively. These basins now flank opposite margins of the South Atlantic Ocean. As early as 1916, Juan Keidel recognized their geological similarities and, in 1927, Alexander du Toit published the first stratigraphic account of these similarities following his lengthy sojourn through South America, in a publication 'A geological comparison of South America with South Africa', sponsored by the Carnegie Institution of Washington. Du Toit's mandate was to test Alfred Wegener's then highly controversial concept of continental drift (Wegener 1912) rooted in a low-resolution correlation between these two basins. Du Toit found the bio- and litho-stratigraphy of the South American rock sequences of the Paraná Basin in Brazil and the distant flanking mountain of the Sierra de la Ventana in Argentina to be remarkably similar to those that he had himself mapped out so carefully for many years in the Cape–Karoo Basin and its peripheral Cape Fold Belt mountains of southern Africa (Fig. 1). Many geologists have remarked on the similarities ever since, often incrementally improving on the correlations that du Toit synthesized in his famous 1937 book 'Our Wandering Continents'. By the 1990s modern stratigraphic tools further confirmed many of these similarities, but also started to pinpoint some substantial differences. Here we explore some of these recent results in detail. In order to fully appreciate the correlations and the shared evolution of the two basins, we also briefly highlight stratigraphic sequences of basins adjacent to the Paraná Basin that are traditionally treated separately, and yet clearly have a common history with parts of the Paraná Basin and the larger framework of basin evolution along Gondwana's Panthalassan margin.

The present Paraná Basin (Fig. 2), with a surface area of about 1.1×10^6 km^2, is the remnant of a vast sedimentary basin of central-eastern South America that preserves a Phanerozoic stratigraphic record of almost 400 million years, ranging from Late Ordovician to Late Cretaceous times, with a maximum cumulative thickness, including Mesozoic igneous rocks, of about 7 km. Six supersequences (major unconformity-bounded, second-order allostratigraphic units, in the sense of the cratonic sequences

From: PANKHURST, R. J., TROUW, R. A. J., BRITO NEVES, B. B. & DE WIT, M. J. (eds) *West Gondwana: Pre-Cenozoic Correlations Across the South Atlantic Region.* Geological Society, London, Special Publications, **294**, 319–342. DOI: 10.1144/SP294.17 0305-8719/08/$15.00

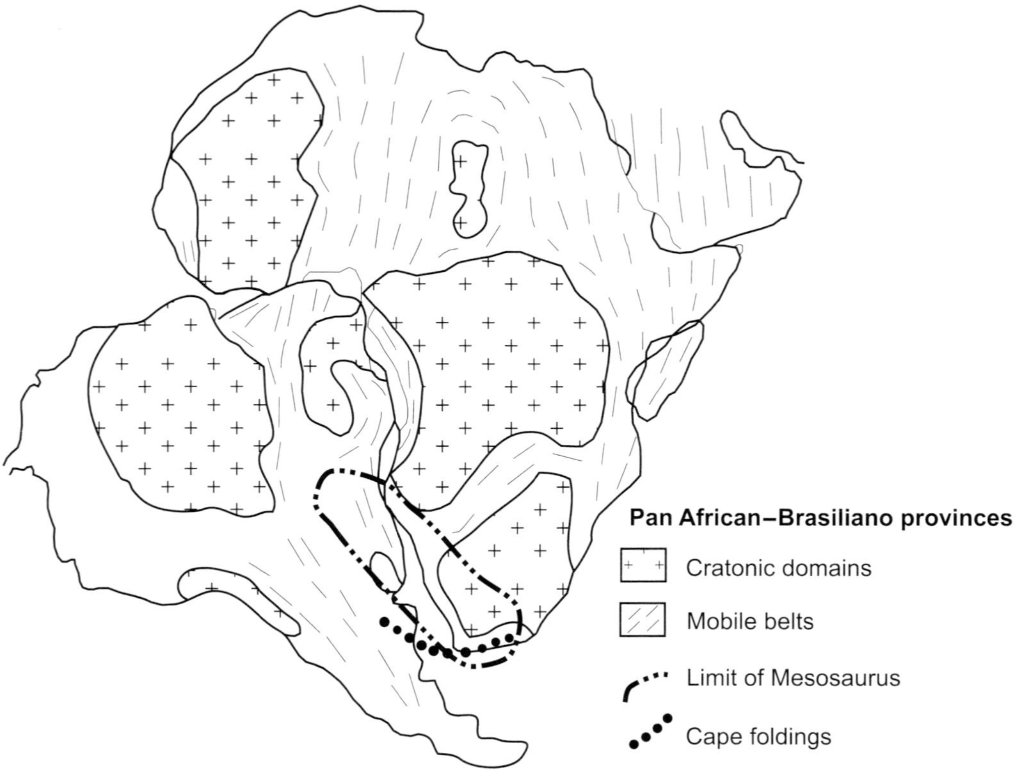

Fig. 1. Present-day knowledge of the West Gondwana pre-break up configuration, showing Pan-African–Brasiliano structures and provinces (modified after Powell 1993). The '*Limit of Mesosaurus*' (found in the Permian Irati and Whitehill beds) and the '*Cape foldings*' (Permian–Triassic Sierra de la Ventana and Cape fold belts) supported du Toit's regional correlations between the Cape–Karoo and Paraná basins and were his main criteria for a '*Suggested Continental Restoration under the Displacement Hypothesis*', first published in the pioneering '*A geological comparison of South America with South Africa*' (1927).

of Sloss 1963) record successive phases of sediment accumulation that alternated with substantial times of erosion in the Paraná Basin during the Phanerozoic (Milani 1997). The two lower supersequences, Ordovician–Silurian and Devonian in age, document two Early Palaeozoic transgressive–regressive marine cycles. A substantial period of erosion (up to 50 Ma long) preceded the deposition of the third supersequence, which spans the range Carboniferous to Lower Triassic. The three upper supersequences are Mesozoic continental sedimentary packages associated with abundant igneous rocks.

The lowermost sediments are represented by the Ordovician–Silurian Rio Ivaí Supersequence. This comprises poorly preserved relics, up to 300 m thick, of a marine package of siliciclastic rocks (Alto Garças Formation in Brazil, Caacupé Group in Paraguay) deposited in a series of regional SW–NE orientated troughs and overlain by diamictites (Iapó Formation) that record in South America the widespread Ashgill glaciation of Gondwana (Milani *et al.* 1996). These basal units are overlain by early Silurian (Llandovery) shales of the Vila Maria Formation (Brazil) and the equivalent Vargas Peña Formation (Paraguay), which represent second-order maximum flooding conditions for the Ordovician–Silurian cycle.

The second supersequence in the Paraná Basin is Devonian in age. It is represented by the continental to shallow-marine Furnas Formation, an extensive blanket of white, kaolinitic sandstones up to 250 m thick, in turn overlain from Pragian times onwards by neritic shales of the Ponta Grossa Formation. In the central Paraná Basin this Devonian package is up to 850 m thick, but this increases westward to reach a few kilometres in thickness in Devonian depocentres in Argentina and Bolivia (Gohrbandt 1993).

The third supersequence of the Paraná Basin represents the classic Carboniferous–Permian Gondwana section. The sequence is up to 2.5 km in thickness, including a maximum of 1.5 km of

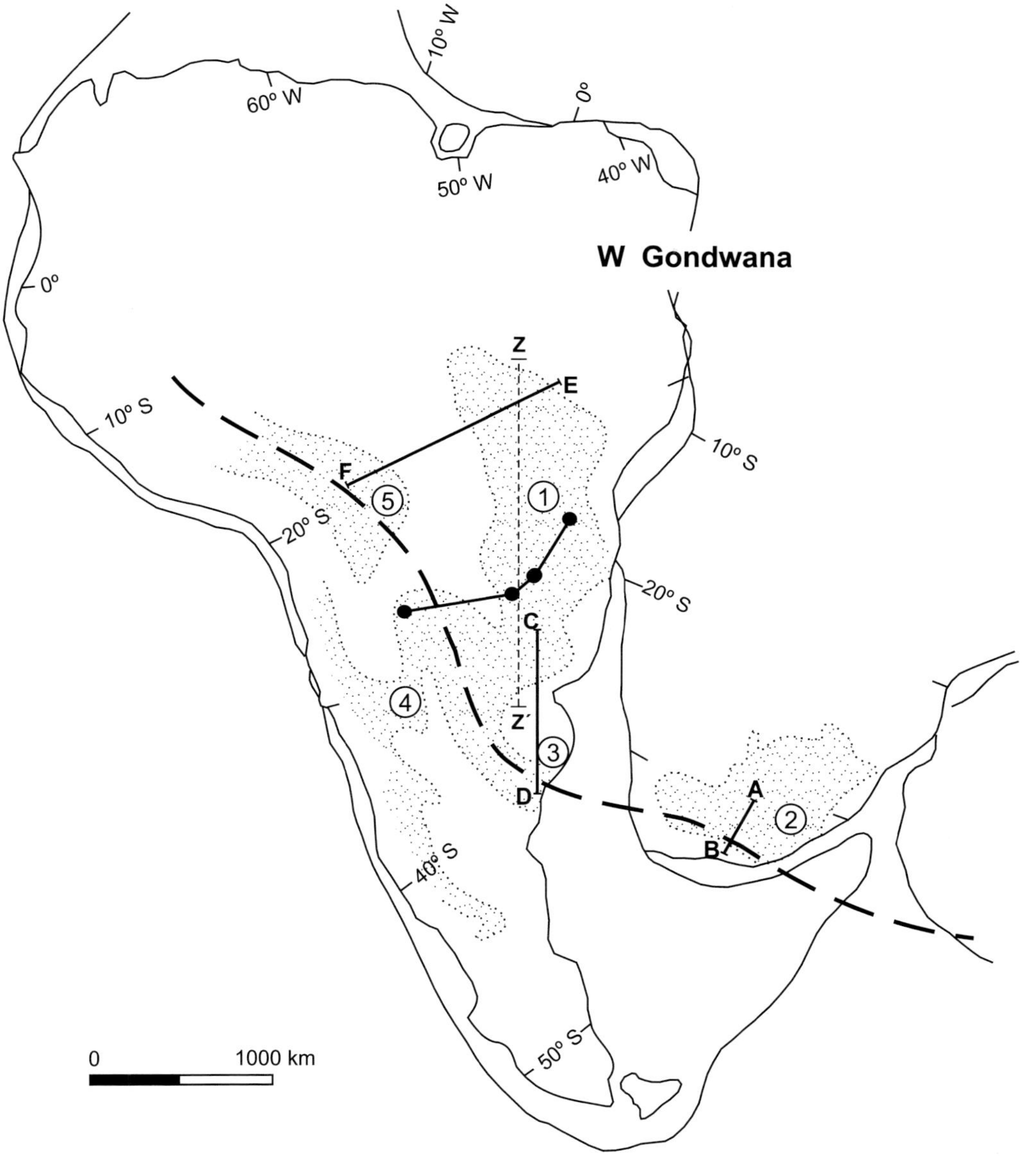

Fig. 2. Palaeo-geological reconstruction of Carboniferous–Permian basins of SW Gondwana, drawn over the present-day geography of Africa and South America (modified after Veevers *et al.* 1994; López-Gamundí *et al.* 1994; de Wit *et al.* 1988). Basins: 1, Paraná; 2, Karoo; 3, Sauce Grande; 4, Precordillera–Paganzo; 5, Tarija. Bold dashed line is the northern limit of the Gondwanides, as represented by de Wit *et al.* (1988). A–B, C–D and E–F are the geological cross-sections shown in Figure 3. Bold dots mark the section of Figure 6. Z–Z′ is the position of the cross-section presented in Figure 9.

glacial diamictites, sandstones and shales of the Itararé Group. The latter define three major cycles of de-glaciation and sedimentation patterns that record a retreat of the ice cap towards the south while Gondwana was moving to the north (França & Potter 1988). By Artinskian–Kungurian times the climate had changed to allow deposition of extensive coal beds (Rio Bonito Formation), followed by a significant thickness of Kazanian siltstones and shales: the Palermo Formation, which represents the maximum palaeobathymetric conditions of the entire Carboniferous–Permian package. Overlying this is a package of black shales with very high organic carbon content, which includes the *Mesosaurus* fauna-bearing Irati Formation, deposited during increasingly arid conditions (Araújo *et al.* 2001). This manifests the beginning of the drying-up of the 'Paraná sea', a cycle that ended in the continental red beds of the Rio do Rasto Formation, and was followed by terrestrial sedimentation until the Cretaceous.

Towards the south, about 1000 km apart from the main present-day Paraná Basin in Brazil, is the Sauce Grande Basin in Argentina (Fig. 2), adjacent to the Sierra de la Ventana fold belt (López-Gamundí & Rossello 1998). Representing a stratigraphical record ranging from Ordovician to Permian (including pre- and syn-orogenic cycles) and exhibiting an outstanding expression of the Late Permian southern Gondwana orogenesis, both the basin and the fold belt are key elements for inter-regional correlation and for the understanding of this sector of SW Gondwana palaeomargin, in spite of the uncertain geodynamic relationship between the Paraná and the Sauce Grande basins due to the lack of present-day geological continuity between the two areas (Fig. 3).

The Cape–Karoo Basin (Fig. 2) stretches across much of South Africa, and northwards intermittently across into Namibia and Zimbabwe, with an original area in excess of 1×10^6 km^2. The basin was once almost certainly more extensive, with local remnants preserved as far as central Africa and Madagascar. In this regional context the rocks span a history of almost 400 Ma. The main depocentres are, however, confined to South Africa where the total thicknesses can reach over 10 km, and the preserved stratigraphic record spans just over 300 Ma. The stratigraphic record of the basin is classically divided into two supergroups, traditionally treated as having formed in two separate successor basins, in turn subdivided into a number of groups. These supergroups are, in fact, first-order allostratigraphic units that include a series of supersequences, in a stratigraphical framework and hierarchy similar of that described for the Paraná Basin.

The lowermost of these is the Cape Supergroup, which ranges in age from late Mid Cambrian (*c.* 500 Ma) to Late Devonian (*c.* 360 Ma) and comprises a number of second-order units with well defined marine transgression–regression sequences (Broquet 1992), devoid of any volcanic material. The overlying supergroup is known as the Karoo Supergroup. This starts with an extensive section of glacial sediments, with up to seven major ice advance–retreat episodes that represent *c.* 50 Ma of the predominantly Carboniferous–Early Permian 'Dwyka' Gondwana glaciation (Opdyke *et al.* 2001). In the southern part of the basin, the lowermost diamictite contact is gradational with the uppermost units of the underlying Cape Supergroup. The Dwyka is relatively abruptly terminated by a thin and marine transgressive unit of black

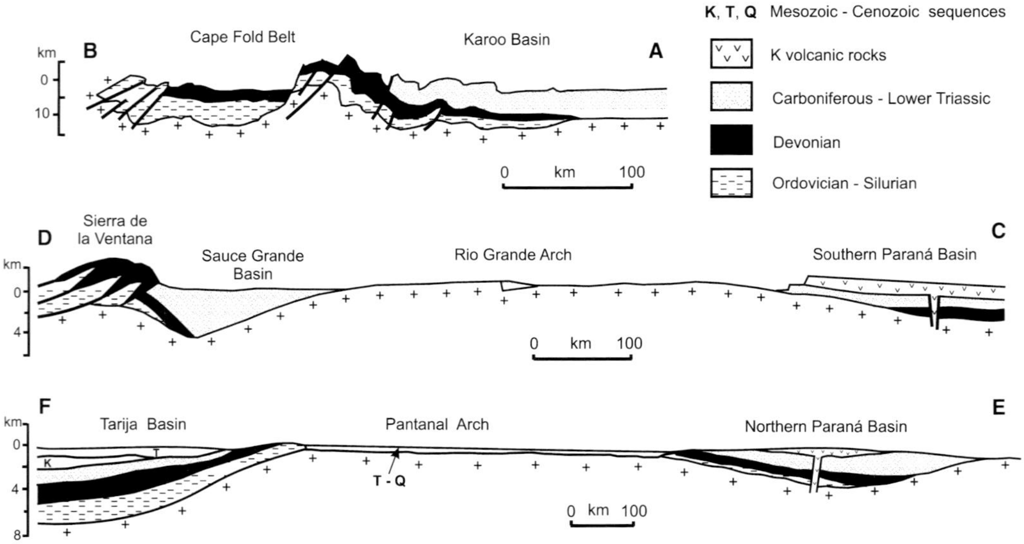

Fig. 3. Schematic, geological cross-sections showing the tectonic and stratigraphic configuration of three sites along SW Gondwana margin. Note distribution of major supersequences: in sections A–B and C–D Ordovician–Silurian and Devonian packages thicken towards the palaeo-margin of the continent, whereas the Carboniferous–Lower Triassic section is clearly controlled by the evolving orogen. Source of data: A–B modified after Duane & Brown (1992); C–D and E–F modified after França *et al.* (1995).

shales and sporadic carbonates (Prince Albert Formation), followed by organic carbon and uranium-rich pyritiferous lacustrine deposits (Whitehill Formation), in turn transitionally overlain by a thick package of upward-coarsening turbidite deposits (Wickens 1992). The turbidites grade into a complex diachronous delta–shoreline section that records a change in depositional style between the lower and upper Karoo Basin (Rubidge *et al.* 2000). The uppermost Karoo rocks represent a major terrestrial cycle of sedimentation formed under increasingly arid conditions (Decker & de Wit 2005), culminating with extensive dune-sands that in turn are overlain abruptly by the Early Jurassic Drakensberg lavas of the Karoo continental flood basalts (*c.* 182 Ma, Duncan *et al.* 1997), best preserved in Lesotho. Farther north, in Namibia and southern Angola, aeolian sedimentation continued up until the early Cretaceous before being covered by the *c.* 134 Ma Etendeka basalts and rhyolites, a small African remnant outlier of the great Paraná continental flood basalt province that virtually terminated sedimentation in the Paraná Basin of South America.

Thus in both Gondwana basins sedimentation was terminated by the same large igneous event that heralded the initiation and progressive opening of the South Atlantic Ocean, whose proto-continental margins south of the Walvis Ridge became the foci of extensive seaward-dipping basalt sequences (Stern & de Wit 2004). During this transition phase the former intracontinental volcanic and sediment depocentres of the contemporaneous Paraná and Karoo basins shifted to their respective continental margins flanking the South Atlantic (Turner *et al.* 1994), ending their long-lived shared Palaeozoic–early Mesozoic history.

Classically, the Paraná and Karoo basins were considered typical intracratonic basins in which all stratigraphic features represent the interplay between tectonics (subsidence-uplift) and sediment accumulation rates (e.g., Rust 1975; Tankard *et al.* 1982; Zalán *et al.* 1990; Cole 1992) to produce an apparent framework of sequences similar to those described by Sloss (1963) for the North American intracratonic basins. Alternatively, sequence stratigraphies of the Gondwana basins have been analysed using the Vail *et al.* (1977) model of eustatic sea-level changes (e.g., Broquet 1992; Pereira *et al.* 2005). None of these models can fully account for the observed stratigraphy in the Paraná and the Cape–Karoo basins (e.g., Cloetingh *et al.* 1992; Milani 1997; Catuneanu *et al.* 1998; Milani & Ramos 1998).

The evolution of the two Gondwana basins under scrutiny can be summarized in its simplest form as follows. First, throughout the Palaeozoic, outward growth of continental lithosphere within the Gondwanides along the southern margin of Gondwana was an important process through which the Paraná and Cape–Karoo basins became progressively isolated from their once conterminous 'south'-facing continental-margin position, open to the Panthalassa Ocean throughout much of pre-Carboniferous Palaeozoic times (Zalán *et al.* 1990; Milani 1992; de Wit & Ransome 1992; Catuneanu *et al.* 1998). This evolving tectonic framework resulted in progressive hinterland migration of the depocentres, to finally become trapped as an intracratonic basin within the heartland of Gondwana by Late Permian times.

In this broad geodynamic framework, we emphasize correlations in order to highlight similarities and differences between subsidence and sedimentation histories of the Paraná and Cape–Karoo basins on a Gondwana scale. Subsidence analysis of both basins reveals the existence of notably synchronous episodes of accelerated subsidence in both the foreland and intracratonic settings, suggesting that these areas have had a common evolutionary history sharing not only regional sedimentary environments but also mechanisms of subsidence. Following the opening of the South Atlantic Ocean, terrestrial sedimentation continued to dominate the interior of both continents. Whilst the most continuous subsequent stratigraphic record is preserved in the dominantly marine sections around the margins of both continents, here we restrict ourselves to the onshore stratigraphic record of both basins up until about the time of first rifting in the South Atlantic. The later is best signalled onshore by the Serra Geral–Etendeka Large Igneous Province.

Regional tectonic framework

The tectonic evolution of the Paraná and Cape–Karoo basins, and their respective similarities and differences, are influenced on a first order basis by two major lithospheric domains (Fig. 4).

The Gondwana shield

This corresponds to the core of the palaeo-continent, being a complex collage of Proterozoic lithospheric blocks and Archaean cratons welded together along Neoproterozoic–Lower Palaeozoic orogens that are commonly referred to as Pan-African and Brasiliano in Africa and South America, respectively (see Introduction to this volume).

The central portion of the Paraná Basin contains the greatest thicknesses of almost all of the Phanerozoic supersequences, with a maximum cumulative thickness of about 7 km. Both the

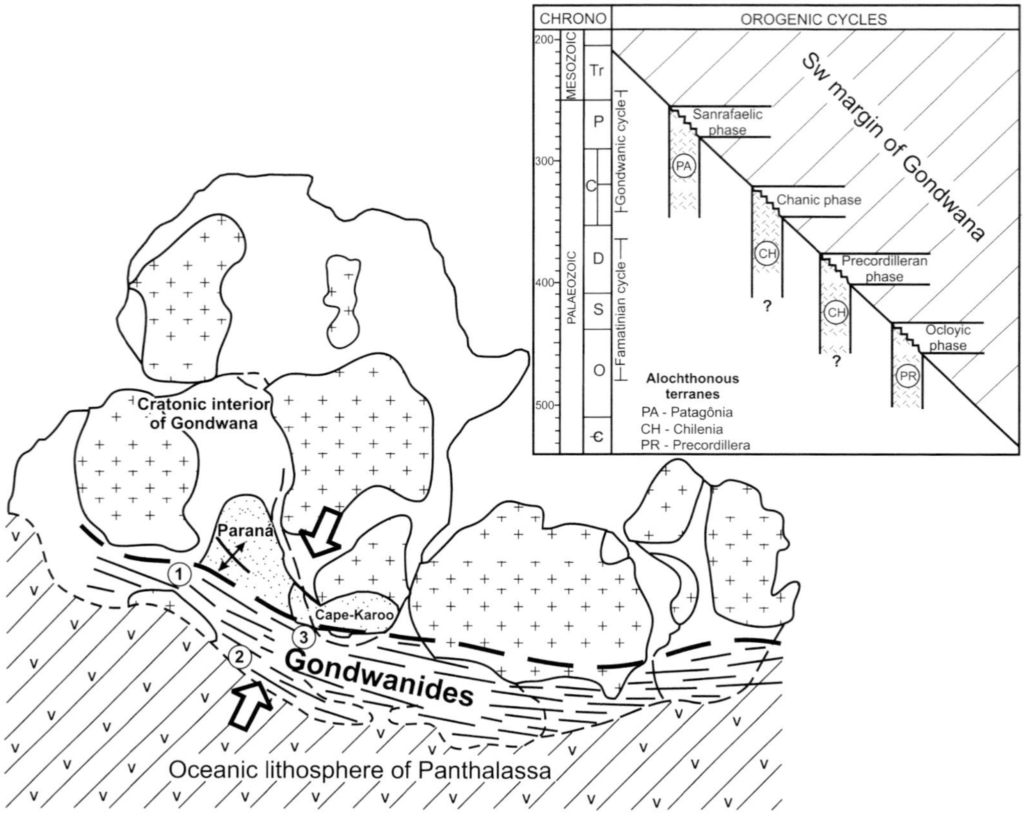

Fig. 4. Regional tectonic setting of the southern margin of Gondwana during the Phanerozoic. A long-lived zone of convergence and recurrent collisional tectonics developed (the Gondwanides) due to the interaction between the palaeocontinent and the oceanic floor of Panthalassa (arrows). Compiled from Powell (1993) and de Wit *et al.* (1988), after the concept of du Toit (1927). Inset shows a summary of the classical Palaeozoic orogenic periods recognized in the South American segment of SW Gondwana (after Ramos 1988). Numbers 1 to 3 correspond to some areas inside the Gonwanides used as references in the regional subsidence and/or stratigraphic analysis, respectively: Bolivian Tarija basin, Precordillera–Paganzo and Sauce Grande–Sierras Australes of Argentina.

Ordovician–Silurian and Devonian packages thicken towards the west, but the Carboniferous–Permian sections attain maximum thickness in the central-northern region of the basin and thin to the south. About 80 deep exploration boreholes have delineated an underlying shield with cratonic nuclei, traversed by Brasiliano belts (Cordani *et al.* 1984; Zalán *et al.* 1990; Basei *et al.* 2000). Milani (1997) and Milani & Ramos (1998) interpreted new gravity and magnetic data together with borehole information on basement rocks and their ages and emphasized the role of reactivation of the Brasiliano belts during the Palaeozoic tectonic and sedimentation history of the Paraná Basin. Across the central Paraná Basin, geophysical data reveal a persistent SW–NE trend of basement anomalies (Marques *et al.* 1993 – unpublished Petrobras internal report cited by Milani 1997; Milani & Ramos 1998). Seismic and well data confirm this to be a central rift filled with Ordovician–Silurian sedimentary rocks (Milani 2004). A deep borehole in this rift penetrated a lower magmatic unit (the Três Lagoas basalt, Mizusaki 1989 – unpublished Petrobras internal report cited by Milani 1997) and associated volcaniclastic rocks intercalated with the sediments. The occurrence in this basin of Early Palaeozoic igneous material (443 ± 10 Ma; Ar–Ar, Milani 2004) suggests a phase of significant tectonic extension during the initial subsidence history of this basin.

In southern Africa, the Cape Basin package is thickest (*c.* 6 km) just north of the tectonic front of the Cape Fold Belt; and from there, the sequences of the Cape Basin thicken southwards farther still to a maximum of about 8 km (Broquet 1992). Towards the north, the Cape sequences thin rapidly, and are overlapped by the Karoo sequences (Fig. 3). The Dwyka glacial deposits also thin towards the north; this has been attributed to the original topography and basin shape at the time of their deposition (du Toit 1937). About half of the basement underlying the main Cape–Karoo Basin is comprised by the Archaean Kaapvaal Craton in the north, the rest of the basin to the south being underlain by the Mesoproterozoic Natal–Namaqua metamorphic complex (Pitts *et al.* 1992; Eglinton & Armstrong 2001). The crustal and lithospheric thicknesses and the chemistry of these two basement terranes are distinctly different (Fouch *et al.* 2004), but subsidence rates and sediment thickness of the overlying basin do not vary significantly across this basement boundary (Cloetingh *et al.* 1992; Catuneanu *et al.* 1998). Recent magnetotelluric and seismic reflection sections in the southern half of the Karoo Basin confirm that the internal stratigraphy is undisturbed across this boundary (Branch *et al.* 2007; Weckmann *et al.* 2007). The Mesoproterozoic basement below the region overlapping the tectonic front of the Cape Fold Belt contains the largest continental magnetic anomaly in the world (the Beattie anomaly, e.g., Pitts *et al.* 1992; Harvey *et al.* 2001; Lindeque *et al.* 2007; Stankiewicz *et al.* 2007; Weckman *et al.* 2007), and a deeply buried Mesoproterozoic palaeosuture has been invoked for this east–west trending anomaly. A possible extension of the anomaly has recently been recognised just north of the Sierra de la Ventana in Argentina (Villar *et al.* 2005).

To the north and west, the main Karoo Basin is separated from the areas in Namibia with comparable stratigraphic units below the Etendeka flood basalts by a major Neoproterozoic province, the Damara Belt. Along the west coast of southern Africa, to the west of the north–south branch of the Cape Fold Belt, the Cape–Karoo Basin overlies parts of the southern arm of the Damara Belt in the form of the Neoproterozoic Gariep–Malmesbury Belt and its flanking remobilised Mesoproterozoic basement (Frimmel & Frank 1998). To the south and east of the Cape syntaxis, Mesoproterozoic basement probably extends under the entire east–west trending section of the Cape Fold Belt. Previously it had been assumed that the basement here was Neoproterozoic, but many of these rock sequences are now known to be isolated sediments with mafic igneous rocks that were deposited in local rifts below the lowermost thermal subsidence sequences (e.g., the Table Mountain Group) of the Cape Supergroup (Barnett *et al.* 1997), indicating that, as in the early Paraná Basin, here too significant tectonic extension occurred during the initial subsidence of the basin.

The orogenic 'Gondwanides'

This name is used to describe an extensive belt of contemporaneous Phanerozoic orogens and related basins (Fig. 4) flanking the southern border of Gondwana (Keidel 1916; du Toit 1927, 1937) also referred to as 'Samfrau geosyncline' by du Toit (1937) and clearly displayed on the Geological Map of Gondwana by de Wit *et al.* (1988).

It is now known that the basement to the Gondwanides comprises a mosaic of smaller crustal domains (or blocks) with distinct geological history and geophysical–geochemical characteristics (e.g., Vaughan *et al.* 2005). It is not the objective of this paper to describe these variations in detail. Suffice it to say that the lithosphere of the Gondwanides is chemically and physically distinct from that of the interior of the Gondwana shield.

A second order difference between the outboard edge of the Paraná and the Cape–Karoo basins is related to regional, along-strike contrasts in Palaeozoic tectonothermal evolution of the Gondwanides. Some of this has been attributed to an inherited 'orocline' along the continental margin of Gondwana. This oroclinal bend may have 'shielded' southern Africa from direct convergent tectonism, in contrast to the margin of the South American sector of Gondwana (e.g., Johnston 2000). This difference in tectonothermal history along the Gondwanides reveals itself in the different early foreland subsidence histories of the respective foreland basins, especially their Palaeozoic stratigraphic infills, as described below.

Throughout most of the Palaeozoic the southern margin of Gondwana, particularly the sector that now corresponds to the Andean border of South America was the focus of active convergence between the Gondwana shield and the oceanic lithosphere of Panthalassa (Bahlburg & Breitkreuz 1991; Gohrbandt 1993; Vaughan *et al.* 2005). In more detail, a succession of orogenic cycles marks the Palaeozoic history of the southwestern edge of the Gondwana shield in South America (Fig. 4). These include three major tectono-sedimentary/magmatic cycles: the Pampean (early Cambrian), the Famatinian (Ordovician to Devonian) and the Gondwanic (or Gondwanian: Carboniferous to early Triassic) cycles (e.g., Ramos 1988, 1990; Pankhurst & Rapela 1998; Rapela 2000). The Pampean cycle is related to the final assembly of Gondwana, with sediment provenance from the Kalahari and Natal-Namaqua areas and with tectonics dominated by transpression (Rapela *et al.* 2008). The Famatinian cycle encompasses two pulses of compressional deformation and associated phenomena, referred to as the Ocloyic and the Precordilleran orogenies. The Gondwanic cycle includes the Chanic and the Sanrafaelic (Sierra de la Ventana) orogenies. These important orogenic episodes have been related to accretion of a series of exotic terranes to the margin of Gondwana during the Palaeozoic (Ramos 1988, 1990, 2005; Pankhurst *et al.* 2006); they are well documented in South America but not all found in southern Africa. The mid-Ordovician orogenic peak of the Ocloyic orogeny has been ascribed to a more extensive Laurentia–Gondwana continent–continent collision (Dalla Salda 2005), but again there is no evidence for this in southern Africa.

Farther outboard from the present South African coast, accretion tectonics along the Gondwanides was probably not as important, although it is less easy to verify, because this area has been severely disrupted during subsequent Gondwana break-up into a number of microplates that are now dispersed between southern Africa and Antarctica (e.g., Vaughan & Livermore 2005). The precise reconstruction of these relatively small microplates has not yet reached a consensus, and most of the pertinent Palaeozoic tectonic history is preserved only in the poorly exposed stratigraphy of the Antarctic Peninsula. Nevertheless, within the simple reconstruction framework that we use for our analyses here, correlations can be established with reasonable confidence. In southern Africa only the equivalent of the Sanrafaelic orogeny is well developed as the Cape orogeny that was active during the timespan from Late Permian to Early Triassic (Hälbich 1992). This tectonism also affected the interior of the Gondwana shield by reactivating old structures in the basement below the Paraná Basin and surrounding areas (Zalán *et al.* 1990; Daly *et al.* 1991; Milani 1997; Trouw & de Wit 1999).

One first order similarity of the Palaeozoic sedimentary geology along southwestern Gondwana (Fig. 5) is the vast volume of mature orthoquartzitic siliciclastic rocks (de Wit & Ransome 1992; França *et al.* 1995). It is said that upon viewing these siliciclastic rocks in South Africa, the legendary F. J. Pettijohn commented: '*this is the biggest pile of sand I have ever seen*' (Arthur Fuller, pers. comm. 1990). A significant exception to this is found in the Argentine Precordillera. There, a thick succession of carbonates, bearing typical species of the Cambrian *Olenellus* fauna (Borrello 1965) is exposed in association with Mesoproterozoic basement (*c.* 1 Ga, Grenvillian age), separated from adjacent regions by major tectonic sutures (Ramos *et al.* 1986; Astini *et al.* 1996). This has led to the interpretation that the Precordillera was an exotic terrane that originated in Laurentia, rifted and drifted away from it, and finally docked against Gondwana (Ramos *et al.* 1986; Astini *et al.* 1995; Astini 1996; Astini *et al.* 2005; Dalla Salda 2005; Thomas & Astini 2005), and that the collision during Middle to Late Ordovician times created the structures of the Ocloyic orogeny in Argentina and Bolivia. Nothing similar to this carbonate sequence occurs inside cratonic Gondwana; this is definitively a phenomenon related to the history of the palaeo-border of the continent.

A second, major cycle of deformation is recognized in the Precordillera region as the Precordilleran orogeny (Furque 1965; Astini 1996). This episode induced important deepening in the foreland basin and accumulation of up to 2200 m of Early to Middle Devonian turbidites known as the Punta Negra Formation. Astini (1996) ascribes this Precordilleran orogeny to the collision of a sialic block known as Chilenia (Ramos *et al.* 1984). At this time a considerable amount of thermal subsidence is recorded in the lower clastic sequence of the Cape–Karoo Basin, including the first marine sediments of the Cape Supergroup (the Bokkeveld Group, Broquet 1992). This heralds the first significant marine incursion along the South African sector of the Gondwana continental margin and a possible response to far field plate boundary stresses in South Africa at that time (e.g., Cloetingh *et al.* 1992). Throughout the Devonian, the source for these sediments was in the north. Thus, although these deposits of the Cape Supergroup represent the lateral equivalent of the deeper water Punta Negra Formation foreland basin turbidites in South America, their source regions lay in opposite directions. In the Paraná Basin, the remarkably rapid deepening in palaeo-environmental conditions experienced by the Devonian sea during Pragian times can be attributed to craton-ward propagation of a Gondwanide-related flexural down-warp (Fig. 6; Milani & Ramos 1998).

The precise timing of collision of Chilenia against the complex and segmented margin of Gondwana is still debated. Previously it was linked to the Late Devonian–Early Carboniferous Chanic orogeny (Ramos *et al.* 1984; Ramos 1988). Adjacent to the Chanic deformation zone, Early Carboniferous tectonism produced transtensional subsidence along older sutures in the foreland domain (Fernández-Seveso & Tankard 1995) to accommodate the lower section in the Paganzo, Río Blanco and Calingasta–Uspallata basins of Western Argentina. It is not known how far the Chanic-induced subsidence reached inside SW Gondwana. No corresponding Lower Carboniferous sedimentary section could have been recorded in the Paraná Basin due to the lack of depositional space; at that time important ice caps were located right over the basin (França & Potter 1988).

The Carboniferous–Permian subsidence cycle terminated during the Sanrafaelic orogeny, and the subsequent unconformably overlying upper sequences represent the climax of are-volcanism

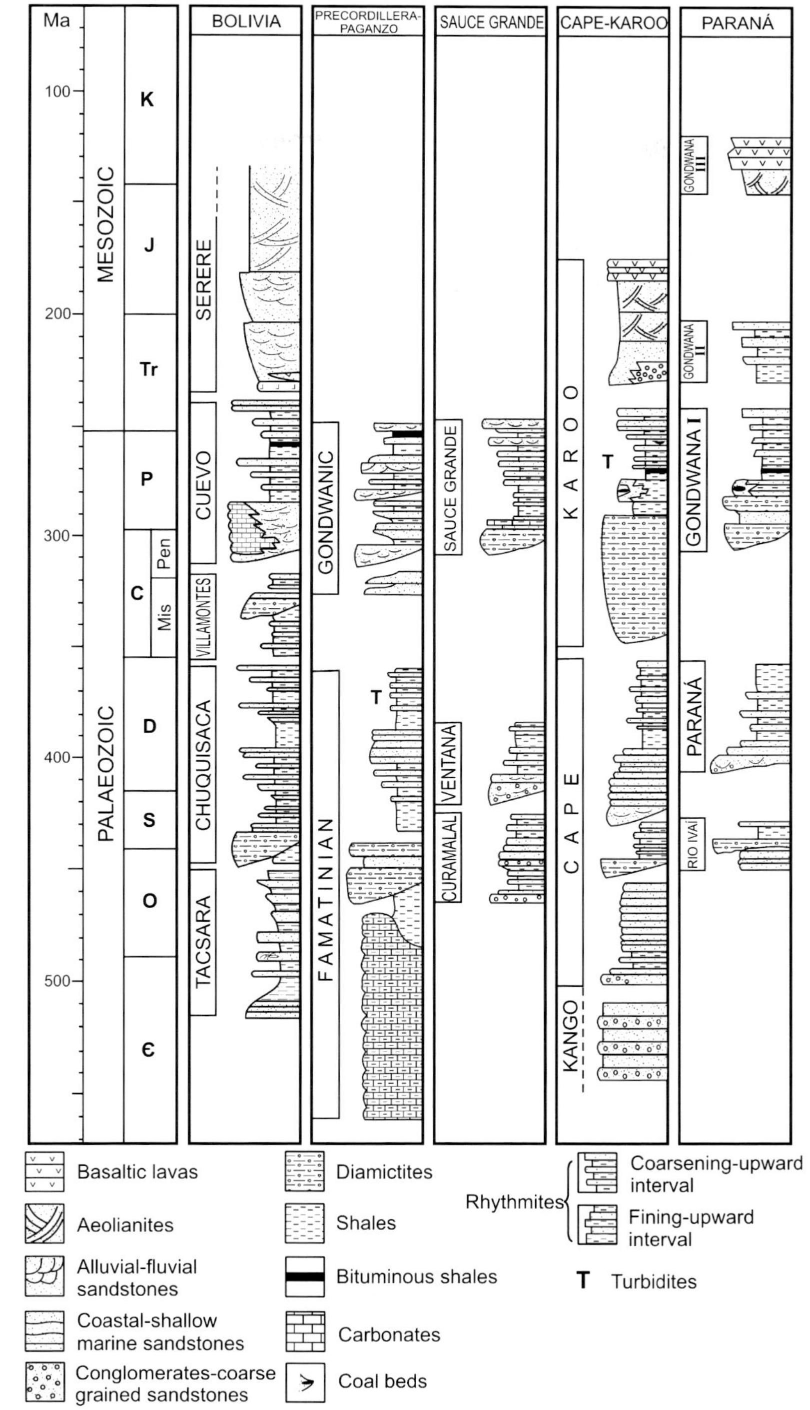

Fig. 5. Summary of sedimentological and stratigraphic aspects of basins discussed in this paper (for location see Fig. 2; see Legend for lithologies and facies). Main source of data for Bolivia, Sempere (1995); Precordillera–Paganzo, Ramos (1990), Fernández-Seveso & Tankard (1995), and Kokogian *et al.* (1993); Sauce Grande, López-Gamundí *et al.* (1994, 1995) and Iñiguez *et al.* (1989); Cape–Karoo: Veevers *et al.* (1994) and Cole (1992), and Paraná, Milani (1997). Time-scale after Gradstein *et al.* (2004).

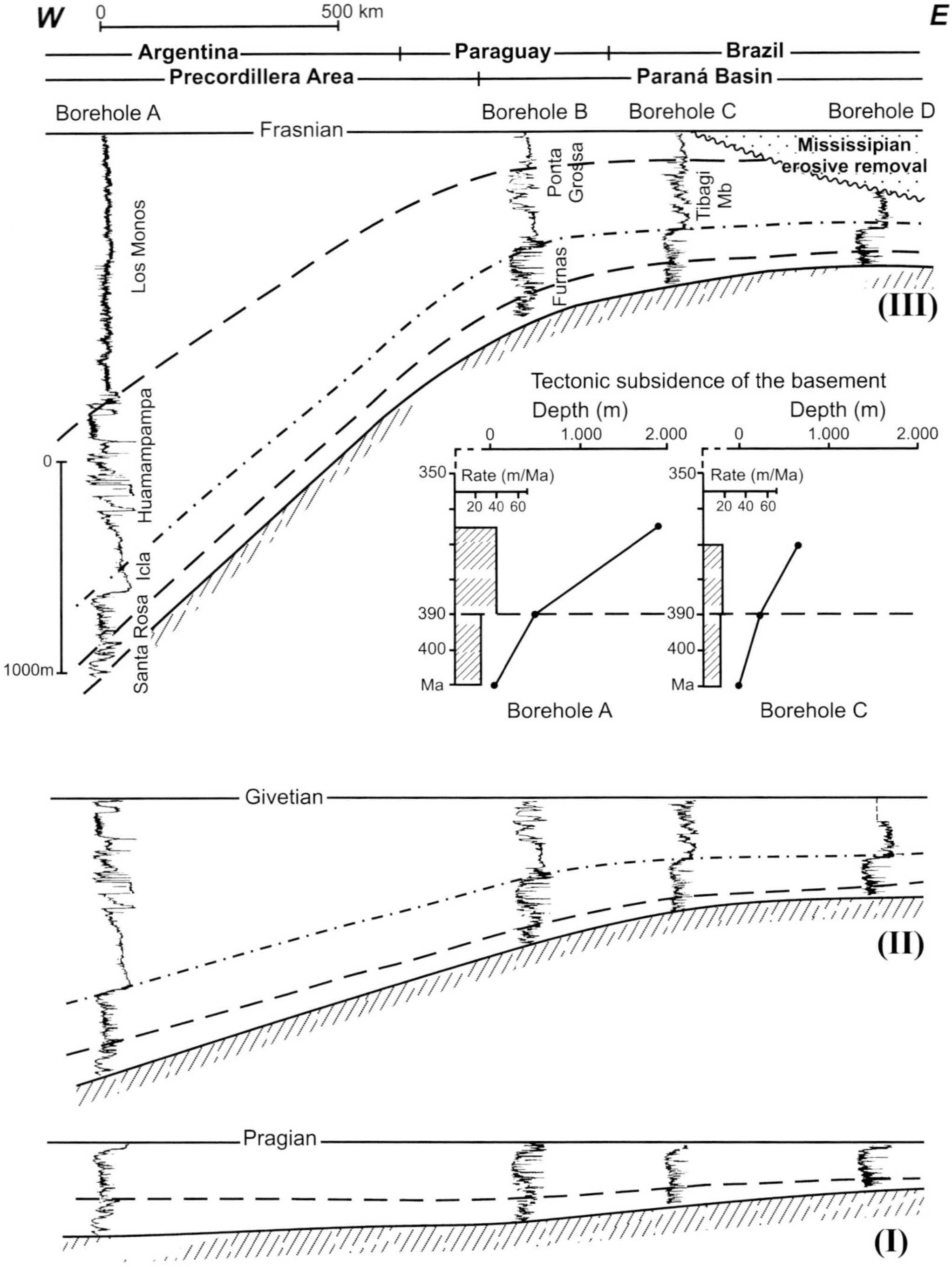

Fig. 6. Regional correlation of gamma ray borehole data showing the distribution of Devonian strata in southern South America (Milani 1997). Ages for the Argentinian units are from Barrett & Isaacson (1988), and for the Paraná Basin from Melo (1988). (I) During Lockovian to Pragian times, a stable substratum led to the development of a wide blanket-like, continental to shallow marine sandy platform (Santa Rosa–Furnas formations). Pragian times marked the beginning of an accelerated cycle of subsidence (II), leading to a major Devonian flooding. The subsidence plots show an important break both in the foreland and in the cratonic domains during Pragian times, and this event of accelerated subsidence is likely to have caused Early Devonian simultaneous drowning of the entire area. (III) The pattern of higher rates of subsidence continued up to Frasnian times. For location see Figure 2.

and continental volcaniclastic sedimentation (the Choyoi) between 275–250 Ma (López-Gamundí *et al.* 1994). Pankhurst *et al.* (2006) ascribe much of the intense silicic magmatism, in the south at least, to slab tear-off and lower crustal melting subsequent to Mid Carboniferous collision of a southern Patagonian continental block. During this same time period in South Africa and Namibia, many thin rhyolitic–andesitic air-fall tuffs are represented as bentonite layers throughout the Karoo Basin (e.g., Cole 1992). U–Pb dating on zircons from tuffs in the Collingham Formation shows them to be especially common at around 270–280 Ma, but their total age range is 250–300 Ma (Bangert *et al.* 1999). Several workers have linked these Karoo tuffs to volcanic activity along the active convergent margin of the Gondwanides in Argentina (e.g., Cole 1992). In South Africa, these ash-fall deposits predate and/or overlap with the first episodes of deformation phases of the Cape Fold Belt, and record the onset of the foreland basin deposition in southern Africa, with sediments dominantly sourced for the first time from the south and east (Cole 1992; Cloetingh *et al.* 1992; Catuneaunu *et al.* 2002). Similar bentonitic layers (air-fall tuffs) in the Permian sequences of the Paraná Basin have also been dated to range between 278 and 299 Ma (Guerra-Sommer *et al.* 2005; Santos *et al.* 2006). The bentonites offer a robust basis for potential detailed chronostratigraphic correlation between the Palaeozoic sequences of the two basins.

Stratigraphical record and subsidence analysis

The early foreland domains: Palaeozoic depocentres along the Gondwanides

Three selected areas (Fig. 4) highlight various aspects of the early Palaeozoic foreland basin evolution marginal to the cratonic Gondwana: the Tarija Basin of Bolivia, the Precordillera–Paganzo basin and the Sauce Grande Basin flanking the Sierras Australes (Sierra de la Ventana) Fold Belt of Argentina. We compare these to the Palaeozoic sections of the Cape–Karoo Basin and the Cape Fold Belt in southern Africa, where true foreland basin deposition did not occur until the late Palaeozoic (Fig. 5).

The Palaeozoic successions in Bolivia are thick and widespread. They comprise five major supersequences known as the Tacsara, Chuquisaca, Villamontes, Cuevo and Serere allostratigraphic units that include dominantly latest Cambrian to the Early Triassic terrigenous clastic sediments (Sempere 1995). Pennsylvanian to Lower Permian Copacabana limestones (Díaz-Martínez 1995) in the Cuevo Supersequence, are among the scarce occurrences of carbonate rocks within the Gondwanides of South America. Another is found in the Precordillera region of Argentina, where a surprisingly thick preserved Cambro-Ordovician platform occurs. As discussed above, these carbonates represent deposition on an exotic fragment of Laurentia that was welded to Gondwana during Middle to Late Ordovician times. In the Precordillera, these exotic carbonates are overlain by a Silurian to Devonian siliciclastic package bearing faunas and floras typically endemic to Gondwana. Inboard of the Precordillera, Cambrian–Ordovician and Silurian–Devonian packages constitute a set of autochthonous siliciclastic sequences derived from erosion of the Gondwana shield to the north.

The Sauce Grande Basin adjacent to the Sierra de la Ventana (or Sierras Australes) Fold Belt of Argentina and the Cape–Karoo Basin in the Cape Fold Belt of southern Africa share a close Gondwanide geological history (Keidel 1916; du Toit 1927, 1937; López-Gamundí & Rossello 1998; Rapela *et al.* 2003). Both regions are underlain by a thick package of shield-derived and relatively mature clastic sediments, including thick quartzite sections of Early Palaeozoic age (uncertain mid-Cambrian to Devonian, Andreis *et al.* 1989; Johnson 1991; Broquet 1992; Armstrong *et al.* 1998). An angular unconformity separates these deposits from Meso- to Neo-Proterozoic basement intruded by Cambrian A-type granites. U–Pb dating and geochemistry has shown that the granites and associated rhyolitic extrusive rocks below the famous Cape unconformity, which separates them from the overlying siliciclastic rocks of the Cambrian–Ordovician Table Mountain Group, overlap in age and chemical composition with similar rock types below the unconformity of the thick clastic sediments (Curamalal Group) in the Sierra de la Ventana (Rapela *et al.* 2003). This is the most dramatic follow-up correlation work yet vindicating the earliest lithostratigraphic correlations between Africa and South America made almost a century ago by Keidel (1916). The absence of body fossils in both sequences on these opposite sides of the Atlantic is conspicuous; yet the same trace fossils (*Cruziana*) have been described in both sequences (Broquet 1992; Rapela *et al.* 2003). In both the western Cape and in the Sierra de la Ventana, basal conglomerates with locally derived felsic igneous clasts occur sporadically (Barnett *et al.* 1997; Rapela *et al.* 2003). Throughout the Cape Basin, the thick siliciclastic rocks of the lower Table Mountain Group are abruptly overlain, and locally scoured into, by a thin sequence of diamictites (the Pakhuis Formation) that represent the short-lived latest Ordovician glaciation. The related palaeo-pole and extensive polar ice cap were centred near what is today southern Chad in central Africa. There is no record of related glacial deposits in the Sierra de la Ventana, but since this is a globally recognised glacial event, and because these glacial deposits are present in the internal basal sections of the Paraná Basin (see below), we may assume its non-preservation in the Sierra de la Ventana.

The subsidence history of the southwestern Gondwana Palaeozoic foreland basin (Fig. 7) is summarized by using data from the reference areas above. A subsidence plot was calculated for the Bolivian case using the back-stripping method of Steckler & Watts (1978). Previously published curves were used for the other areas. All plots show the existence of periods of accelerated subsidence and suggest a lithospheric flexural loading mechanism for the subsidence (e.g., Williams 1995). In all regions there appears to be a strong correlation between the time when these basins first experienced accelerated subsidence and the age of onset of the classical orogenic episodes recognized in the geology of the Gondwanides, as discussed above.

The internal domain: correlating the Paraná and Karoo Basin fills

Six supersequences, each one comprising a geological record of some tens of millions of years,

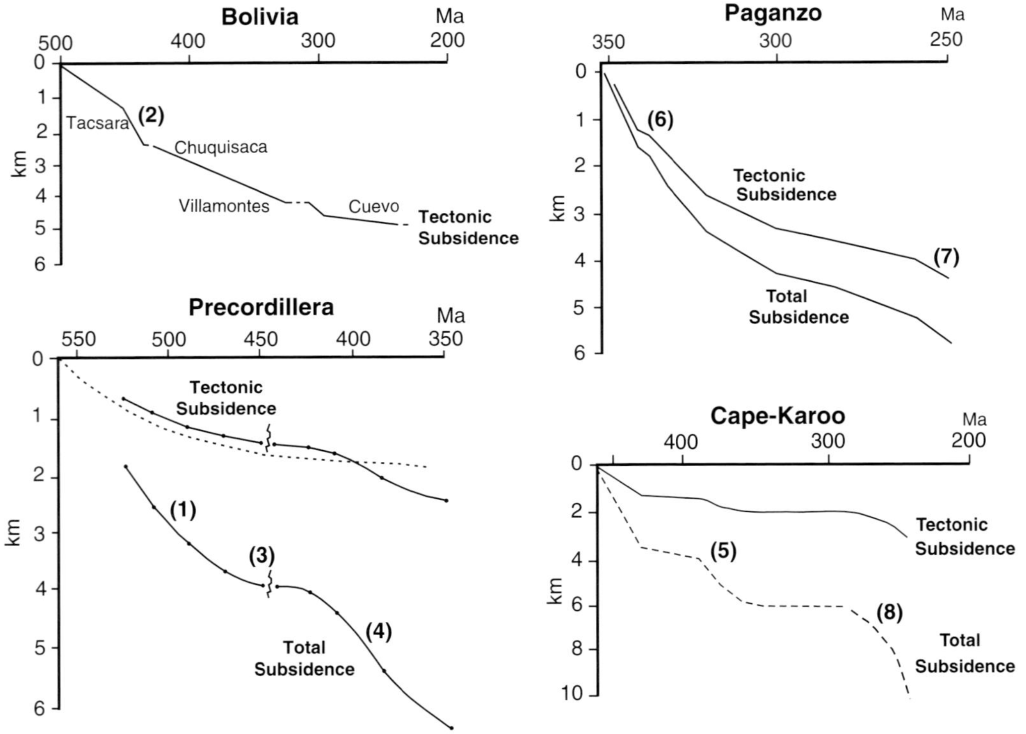

Fig. 7. Subsidence plots for some areas along the Gondwanides, as shown in Figure 2. The curve for Bolivia was calculated according to stratigraphic data from Gohrbandt (1993); Precordillera–Paganzo—curve for the Famatinian cycle by Ramos (1993), and that for the Gondwanic cycle by Fernández-Seveso *et al.* (1993). The stratigraphy of the Sauce Grande basin is poorly constrained in age and does not allow the calculation of a reliable subsidence curve. Cape–Karoo curve after Cloetingh *et al.* (1992). Relevant features are: (1) the passive margin response of the Cambrian–Early Ordovician carbonate platform of the Precordillera; (2) the accelerated period of subsidence in the Bolivia curve starting at *c.* 450 Ma, which can be correlated with the collision and docking of the Precordillera terrane against Gondwana; (3) the resultant Ocloyic orogeny and foreland subsidence. A second period of accelerated subsidence in the Precordillera curve was during Early Devonian, with its highest rates at *c.* 400–390 Ma (4); a similar signal was also detected in the Cape diagram (5), but, as yet, no cause-and-effect relationship can be established. The high rates of subsidence observed in the Paganzo curve during the interval 350–320 Ma (6) were interpreted by Fernández-Seveso (1993) as related to transtensional subsidence, most likely derived from the stresses of the Chanic orogeny. The Late Permian period of accelerated subsidence in the Paganzo curve (7) and in the Karoo curve (8) is the classic flexural response for the Sanrafaelic (Cape–Sierra de la Ventana) orogeny.

constitute the stratigraphic framework of the interior Paraná Basin (Milani 1997). Much of the time is represented by a series of lacunae that separate the supersequences (Figs 5 and 8). The Rio Ivaí (Caradoc–Llandovery), Paraná (Lochkovian–Frasnian) and Gondwana I (Pennsylvanian–Scythian) supersequences represent major Palaeozoic transgressive–regressive cycles, whereas Gondwana II (Anisian–Norian), Gondwana III (Upper Jurassic–Berriasian) and Bauru (Senonian) are fully continental packages, the latter being an up to 250 m thick post-volcanic section accumulated in the flexural depression created by the load of the Gondwana III lava pile.

The Rio Ivaí Supersequence comprises the oldest sedimentary rocks of the Paraná Basin. This Ordovician–Silurian package is widespread, but its thickness varies considerably with some thick elongated depocentres striking SW–NE. There is also a general trend of thickening to the west, with the package reaching about 1000 m in the Paraguayan portion of the basin. A regional, fluvial palaeocurrent pattern from NE to SW is evident (Milani *et al.* 1996) in the lowermost section of the Rio Ivaí package. Seismic data (Marques *et al.* 1993 *in* Milani 1997 (unpublished Petrobras internal report) show that the thickest Rio Ivaí occurrence is confined to a 600 km long, SW–NE orientated rift system that runs from Paraguay to the northeastern portion of the basin. The sequence includes basal conglomerates and sandstones (Alto Garças Formation) with a section of basalts (the Três Lagoas basalt) that suggests significant extension and rifting during the inception of the formation of the Paraná Basin. The Balcarce Formation (Rapela *et al.* 2003; Zimmermann & Spaletti 2005) of the Tandilia System, adjacent to the Sierra de la Ventana, is the most likely correlative unit of the Rio Ivaí Supersequence in that area. The latest Ordovician glacial diamictites (Iapó Formation) are also preserved here and, together with overlying fossiliferous shales and siltstones (Vila Maria Formation), span Caradoc–Llandovery times. The shales record maximum flooding of the Ordovician–Silurian cycle, following rapid de-glaciation, as they do in the Cape Basin (where the Cedarberg Shales contain 'giant', cold-water conodonts). The top of the Rio Ivaí Supersequence is cut by a peneplain covered by the sheet-like Devonian Paraná Supersequence. The latter represents a complete transgressive–regressive cycle of sedimentation starting with continental to neritic Lower Devonian sandy rocks (Furnas Formation) followed by marine shaly sediments (Ponta Grossa Formation) that span the Pragian to Frasnian stages. The Pragian shales have sedimentological and stratigraphic characteristics indicating maximum Devonian flooding and rapid drowning of the shallow Furnas platform. A second basin-scale unconformity surface marks the upper limit of the Devonian package. This sub-Pennsylvanian unconformity is a benchmark surface in the Paraná Basin, and represents a lacuna of almost 55 Ma. This unconformity is distinctly angular in the Gondwanides affected by the Chanic orogeny, such as in the western Argentine basins and the Sauce Grande foreland basin (López-Gamundí & Rossello 1993).

The subsequent accumulation of the Gondwana I Supersequence of the Paraná Basin occurred during de-glaciation episodes and reflected increased sedimentary influx from those areas freed from ice cover. Sediments were dominated by mass flows and re-sedimentation, defining a singular depositional style for the Pennsylvanian to Lower Permian interval throughout the Paraná Basin. The 1500 m thick de-glaciation related section (named Itararé Group in the southern portion of the Paraná Basin and the Aquidauana Formation in the north) is composed of diamictites intercalated with sandstone and shale packages (França & Potter 1988), of both glacio-terrestrial (minor) and glacio-marine environments. The interpretation of marine affinities of the Itararé sedimentation is supported by the overall presence of *Tasmanites* and acritarchs (Daemon & Quadros 1970). The glacial deposits onlap the sub-Pennsylvanian unconformity from north to south and extend over progressively wider areas. In the Early Permian, onlapping sedimentation reached the southernmost portion of the basin, presently located in Uruguay. Basin-wide correlations of the glacial record in the Paraná Basin (and for the rest of the sedimentary section as well) is almost exclusively constructed from extensive biostratigraphy (Daemon & Quadros 1970; several other references for local sections). Calibrated to European palinozones, the resulting chronostratigraphy (particularly for the Carboniferous–Permian package) is imprecise and uncertain. This has recently been improved through U–Pb dating of zircons from intercalated bentonites (ash-fall tuff beds, as mentioned above, Coutinho *et al.* 1991; Guerra-Sommer *et al.* 2005; Santos *et al.* 2006). Through this work, as in the Karoo Basin, pre-existing stratigraphic schemes are now known to be substantially incorrect by up to tens of millions of years, highlighting the necessity for intensive geochronology-based stratigraphy that is only just starting. In any case, the glacial record in the Paraná Basin has been traditionally ascribed to the range Westphalian–Artinskian, a time-span of about 35 Ma. A Mississippian record is lacking in the Paraná Basin, but is present in western depocentres like the Paganzo Basin of Argentina, where depositional space was created in a time when the Paraná Basin domain was experiencing peak glacial conditions.

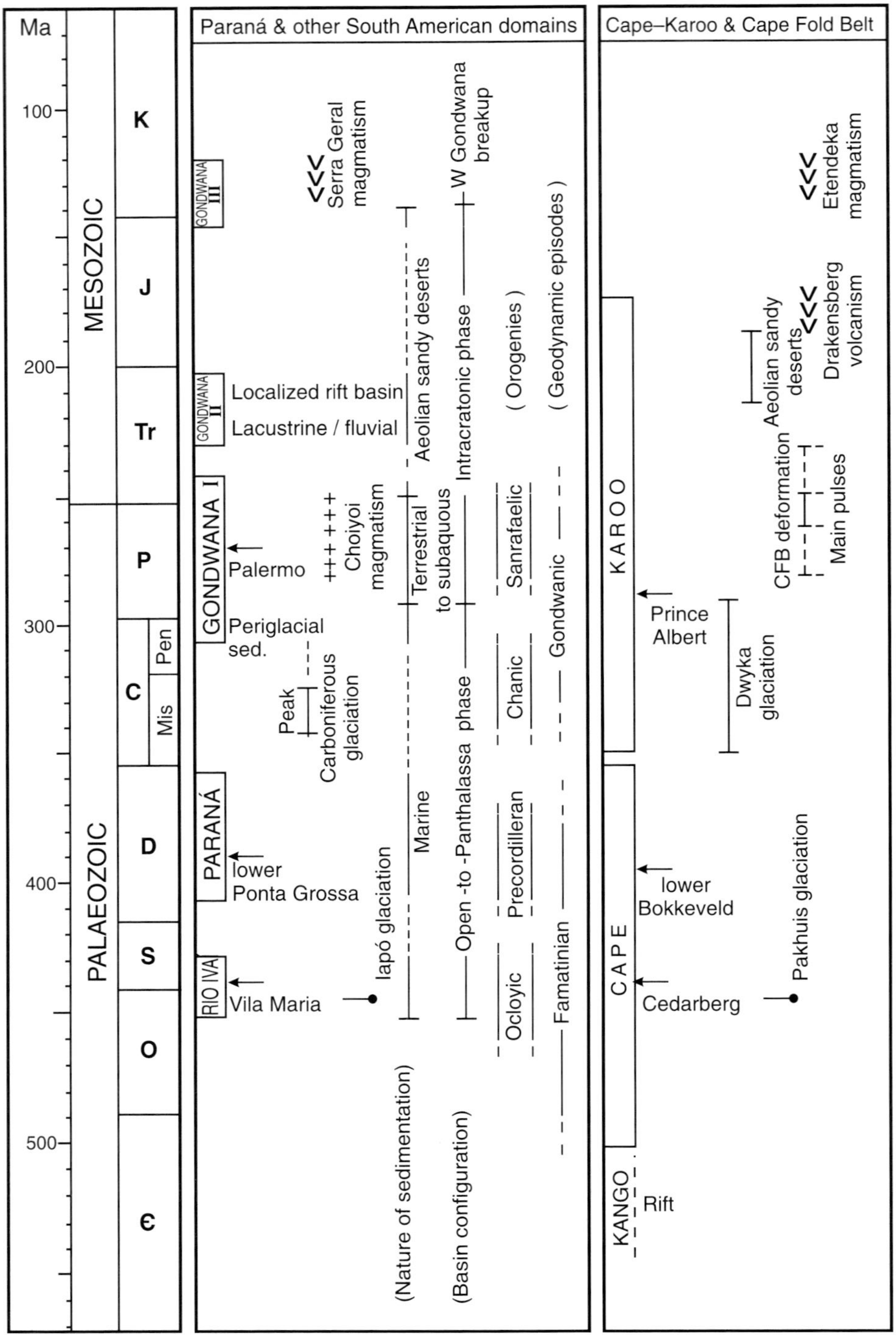

Fig. 8. Major stratigraphic units of the Paraná and Cape–Karoo basins, and correlation in time with some important tectonic, magmatic and climatic events that took place in SW Gondwana during the Phanerozoic. Arrows indicate major events of flooding. Note the conspicuous trend of drying in these basins, which were marine in their origin (open towards Panthalassa) and became intracratonic, trapped inside Gondwana, by Permian times. Main source of data for Paraná: Milani (1997) and Milani & Ramos (1998); Cape–Karoo: Cole (1992) and Broquet (1992); other South American domains: Milani (1997) and references therein. Time scale after Gradstein *et al.* (2004).

By contrast, in southern Africa there is no sign of a major hiatus before the onset of the mid-Palaeozoic glaciation (Figs 5 and 8). In the Karoo Basin the glacial period is represented by sediments of the Dwyka Group (Visser 1997). Sedimentation in the Cape Basin terminated during the Lower Carboniferous period (Visean) after deposition of the tidal flat to lacustrine sandstones of the Waaipoort Formation of the Witteberg Group. The age of the overlying Dwyka deposits has been a matter of long debate. It was previously believed that a hiatus of approximately 30 Ma duration may have occurred before deposition of the Karoo Sequence commenced (Visser 1990, 1997). However, palaeontology has shown that the top of the Waaipoort Formation is Visean in age (Streel & Theron 1999) and lacustrine (Evans 1999). Dropstones and soft sediment deformation features believed to be glacial in origin are present in the uppermost beds of the Witteberg Group (Streel & Theron 1999). Therefore, glaciation was in progress during deposition of the uppermost Witteberg sediments, and it appears that there is in fact a minimal hiatus before glaciers advanced across the Witteberg surface.

Glaciogenic sedimentation and erosion dominated until the earliest Permian (upper Asselian–lowermost Sakmarian) accumulating up to 800 m of thick diamictites and rhythmites in the southern part of the Karoo Basin, but much less in the vicinity of the classical striated pavements along the northern fringes of the basin (Visser 1989, 1997; Cole 1992). Ice-flow directions indicate sources to the north (Cargonian Highlands), east (Eastern Highlands, now in East Antarctica), and southwest (Southern Highlands, now in West Antarctica), which may represent major ice-spreading centres at that time (Visser 1989, 1997). Visser (1997) estimated an ice sheet cover of the basin floor of more than 4 km, similar to that in much of present-day East Antarctica. In the Carboniferous, at the time of maximum glaciation of Gondwana, the south pole was located on the Antarctic shield (Opdyke *et al.* 2001) then situated less than 1000 km to the east of the Karoo Basin (e.g., Rakotosolofo *et al.* 1999; Scotese 2000; Reeves *et al.* 2002). In Africa, the northern margins of the ice sheets reached at least as far as southern Madagascar (Rakotosolofo *et al.* 1999), the Central African Republic, northern Angola and southern Sudan.

Sections of up to 800 m of sediment in the form of lodgement till, rain-out and sub-aqueous and sub-glacial melt-water sands, suspended mud, and turbidity-current sands and silts, were deposited in at least seven upward-fining cycles related to advance and retreat of the ice cap, each starting with thick coarse tillites/diamictites and terminating in thin shaly horizons and rhythmites/varves. Although some investigators infer a marine setting for the deposition of part of these diamictites and related varves (Cole 1992; Johnson 1997; Visser 1997; Catuneanu *et al.* 1998; Rubidge *et al.* 2000), there is no unequivocal evidence for this. It is possible therefore that the bulk of the Dwyka sequence of the Karoo Basin was deposited in a terrestrial setting (du Toit 1926). Marine fossils occur at the top of the Dwyka along the western margins of the basin, particularly in southern Namibia and along the deep glacial valley through the Vryburg–Prieska region along the northern extents of the Karoo Basin. These are witnesses of a short marine transgression related to eustatic sea-level rise following rapid global de-glaciation (du Toit 1954; McLachlan & Anderson 1973; Visser 1987, 1989). Elsewhere, these sediments indicate non-marine conditions, including those of the overlying carbonaceous mudstones of the Prince Albert, Whitehill and Collingham formations, when the Karoo Basin had become a gigantic inland lake (Faure & Cole 1999). These observations imply that the water-lain diamictites, previously modelled as deposited in a wide marine-shelf environment (Visser 1989, 1997), may represent glaciogenic lake sediments deposited beneath and peripheral to the major continental ice sheet that covered much of Gondwana at that time.

Two occurrences of rhyolitic–andesitic volcanic tuff are present in the Dwyka Group of southern Africa (Bangert *et al.* 1999). These were most likely derived from a magmatic arc to the SW of the Paraná Basin (Cole 1992; Bangert *et al.* 1999). U–Pb dating of zircons from tuffs near Laingsburg about 400 m above the base of the Dwyka yielded a date of 297 $\pm$ 1.8 Ma (Bangert *et al.* 1999). The age of the top of the Dwyka has been similarly derived using zircon dating from tuffs in the lowermost beds of the overlying Prince Albert Formation (288 $\pm$ 3 Ma and 289 $\pm$ 3.8 Ma). Thus, the Dwyka Group in southern Africa spans about 50 Ma, with the upper 200 m of the Dwyka Group representing less than 10 Ma or at most 20% of the glacial period. Although the mid-Palaeozoic glacial deposits of the Dwyka and the Itararé Groups have been correlated in general terms since the time of du Toit (1927), it is clear that there are some differences between them in terms of environmental conditions.

The Karoo Basin evolved further as a flexural foreland basin around the Permo-Triassic boundary, linked to the emerging Cape Fold Belt (Hälbich 1992; Cloetingh *et al.* 1992; Catuneanu *et al.* 1998, 2002). Many workers have related the Upper Palaeozoic Karoo sequences to tectonic processes associated with subduction of the palaeo-Pacific plate beneath the Gondwana plate during the Permo-Carboniferous (Lock 1980; Cole 1992;

Smith 1995; Johnson 1997; Catuneanu *et al.* 1998, 2002). However, the above-mentioned models are based on lithostratigraphic analysis complemented by only limited biostratigraphy due to a scarcity of zonal fossils with precise time resolution. This has hampered detailed tectonic modelling and basin analysis, and is the cause of significant controversy about the age and position of lithostratigraphic boundaries (e.g., Zawada & Cadle 1988; Rubidge *et al.* 2000) and sequence stratigraphy (e.g., Catuneanu *et al.* 1998; Turner 1999). There is simply not enough reliable chronostratigraphy to verify detailed correlations, but available data show that some of the basin evolution models are extrapolations based on low resolution biostratigraphy. For example, Catuneanu *et al.* (1998) place the lithostratigraphic top of the Dwyka sequence at *c.* 268 Ma and 265 Ma in the southern and northern parts of the Karoo Basin, respectively, 30–35 Ma younger than the lithostratigraphy based on more recent U–Pb zircon chronostratigraphic data. Similarly, their dating of the Collingham Formation is at least 12 Ma younger than the U–Pb dates derived from zircons in the widespread air-fall tuffs from this formation, as will be mentioned below.

Considerable debate remains also whether the Karoo Basin at this time is entirely terrestrial or if there are substantial marine sequences preserved (e.g., Faure & Cole 1999; Visser 1992*a*). Rapid melting of the ice sheets in the Early Permian was followed by slow deposition of black suspended muds (Prince Albert Formation, lower Ecca Group; Sakmarian–Kungurian in age). Geochemistry (Zawada & Cadle 1988; Faure & Cole 1999) indicates that most of these deposits are fresh- to brackish-water lake deposits and not marine as is often inferred (e.g., Cole 1992; Visser 1997; Catuneanu *et al.* 1998, 2002; Rubidge *et al.* 2000). Thereafter, post-glacial isostatic rebound of the northern and eastern provenance areas resulted in an influx of fluvial deltaic sands and onset of extensive coal deposition in the northeastern part of the Karoo Basin.

At the end of the Early Permian, a distinctive black pyrite-rich mud horizon, the Whitehill Formation, was deposited throughout a gigantic and highly reducing basin, which by then extended across into the Paraná Basin of South America, where the coeval Irati Formation was deposited. Both these formations have a characteristic fauna of the fresh-water *Mesosaurus* and dragonflies (du Toit 1927, 1937; Milani 1992; Visser 1992*b*; Araújo *et al.* 2001). This distinctive geochemical marker is a true 'time line' for correlation; and it has recently been shown that it can be easily traced in subsurface using relatively fast and simple magnetotelluric sounding (Branch *et al.* 2007). Both the Whitehill and Irati shales have a very high total organic carbon content (up to 24% in condensed sections of the latter), a sulphur content as high as 8%, and elevated uranium concentrations. In South Africa, the Whitehill Formation is abruptly overlain by the upward-coarsening sequence of turbidites of the Collingham and Laingsburg formations.

In turn, distal turbidites of the Collingham Formation are overlain by coarser turbidites of the Vischkuil and Laingsburg formations and then the deltaic sequences of the Fort Brown Formation. These deposits are succeeded by a package of shoreline and braided-river deposits of the Beaufort Group (Upper Permian, Rubidge *et al.* 2000); these contain the terrestrial mammal-like reptile *Dicynodon*, described in detail by King (1990). A tuff horizon in the lowermost Beaufort, just below the *Cistecephalus–Dicynodon* boundary (Rubidge, pers. comm. 2001) yields a U–Pb zircon date of 258 ± 1 Ma (uppermost Guadalupian). These deposits were abruptly followed by meandering river deposits of the upper Beaufort (Triassic) across the Permo-Triassic boundary at *c.* 251 Ma (Smith 1995; MacLeod *et al.* 2000; Ward *et al.* 2000). Thereafter, a new Mesozoic fauna, characterized by *Lystrosaurus*, assumed the landscape, and terrestrial sedimentary processes continued to dominate with a progressive desertification, interrupted at times by wet–dry cycles straddling the Triassic–Jurassic boundary (Decker & de Wit 2005). This environment was terminated abruptly by the Drakensberg flood basalts of the Karoo Large Igneous Province, at 182 ± 1 Ma (Duncan *et al.* 1997; Turner 1999). At the contact, basalt flows can be seen to fill a landscape of dune deposits.

Comparative subsidence histories

In South America, an important phase of structural rearrangement of the Paraná Basin geometry also followed the Carboniferous–Permian glaciation. Accompanying the Late Permian deformation along southwestern Gondwana (the evolving Sierra de la Ventana Fold Belt, von Gosen *et al.* 1991; Cobbold *et al.* 1992), subsidence and sediment accumulation in the Paraná Basin followed a period of accelerated flexural subsidence until earliest Triassic times, accommodating about 1400 m of mostly terrestrial sedimentary rocks of the Teresina and Rio do Rasto formations. López-Gamundí *et al.* (1995) pointed out the syntectonic character of sedimentation in the foreland basin adjacent to the Sierra de la Ventana Fold Belt during the late Permian. In the Paraná these are basin-centred discontinuous aeolian fields and shallow lake deposits framed by huge sandy deserts (Pirambóia Formation in the northern portion; Sanga do Cabral and Buena Vista formations in the southern area; Cabacuá Formation in Paraguay). A progressive and irreversible continentalization of the depositional systems of the Paraná Basin occurred from the Late Permian to the late Triassic and can be seen from the sedimentary record of the upper portion of the Gondwana I Supersequence (Rio do Rasto Formation) onwards. By then, sandy deserts covered the entire basin until the latest Jurassic (Botucatu Formation), followed by the Early Cretaceous basaltic lavas and intrusives of the Serra Geral Formation, and its equivalents in the Etendeka region of Namibia, whose lava sequences have been correlated both geochemically and isotopically (Marsh *et al.* 2001).

The presence of thick and extensive Ecca Permian turbidites in the Karoo, therefore, defines an important difference between the two basins. This appears to signify a greater degree of flexural response of the foreland basin to the Cape orogeny than is the case in South America during the Sanrafaelic orogeny. In South America, the regional sense of onlap of sedimentary beds was suddenly inverted (Fig. 9), as shown by the northward-wedging retrograding sequence of the Guatá Group (Milani & Ramos 1998). Maximum palaeo-bathymetric conditions for the Gondwana I Supersequence (external neritic at most) are documented in the Palermo Formation of earliest Late Permian age (Daemon & Quadros 1970). The overlying package, accommodated by a renewed cycle of flexural subsidence of the basement, is a regressive section up to 1400 m thick (Passa Dois Group) that culminates in Early Triassic aeolian sandstones (Sanga do Cabral, Pirambóia and Buena Vista formations). Subsequently, the two interior basins of Gondwana merged into a giant sand desert. This was first interrupted in the Cape–Karoo Basin only by basalt flows of the Karoo igneous province during the late Jurassic; then in both basins by outpourings of basalts during the second large (Paraná) igneous province, focussed predominantly along the west coast of southern Africa but best preserved in South America.

To summarise, the subsidence history of the interior domain of the Paraná and Cape–Karoo basins are compared using their respective subsidence curves (Fig. 10). In South America, the Ordovician–Silurian phase cannot be adequately evaluated quantitatively due to poor chronostratigraphic resolution within the dominantly sandy section of the Rio Ivaí Supersequence. The geometry of this unit, defining narrow, elongated depocentres along SW–NE weakness zones of the basement, suggests some kind of rifting or transtension as the initial tectonic mechanism responsible for inception of the basin. This is also seen in the subsidence curves of the Cape Basin of southern Africa, where extensional processes were responsible for beta factors ranging between 1.2 and 2.2 (Cloetingh *et al.* 1992).

In the Paraná Basin, Devonian subsidence started with low rates related to a flat and stable substratum, in accordance with the overall characteristics of its basal section, the Furnas Formation. This sandy unit exhibits a blanket-like geometry with remarkably constant thickness and sedimentological characteristics across the basin. From

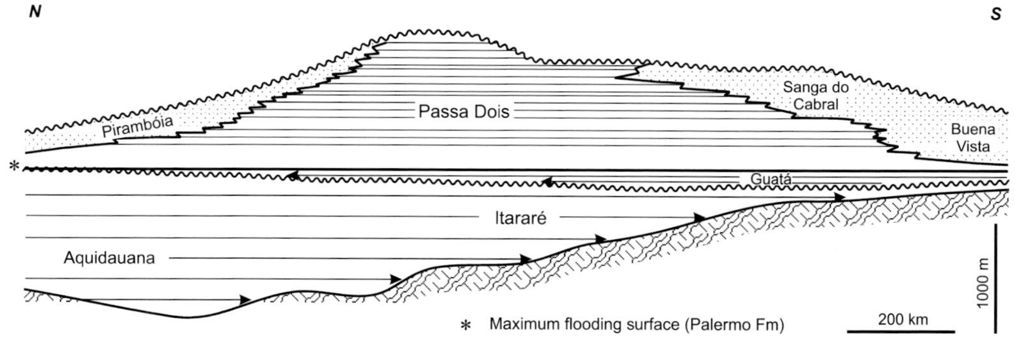

Fig. 9. Gondwana I Supersequence (Carboniferous to Lower Triassic) basin-scale correlation of stratigraphic data (Milani 1997) using information of 40 deep boreholes (not shown). The stratigraphic record of this supersequence of the Paraná Basin documents an abrupt change in the sense of onlap: accompanying the southward-retreating ice cap it developed from north to south during the accumulation of the basal, glacially-influenced package of the Itararé Group and Aquidauana Formation, whereas the overlying post-glacial transgressive package of the Guatá Group onlaps to the north, with subaerial exposure and the development of a regional unconformity in the northern portion of the basin. Such an important change in basin configuration is attributed to intracratonic response to plate margin tectonics. The end of Palaeozoic history of the Paraná Basin was marked by the advance of continental depositional systems (Pirambóia, Sanga do Cabral and Buena Vista formations) towards the remnant central water body (Rio do Rasto Formation). For location see Figure 2.

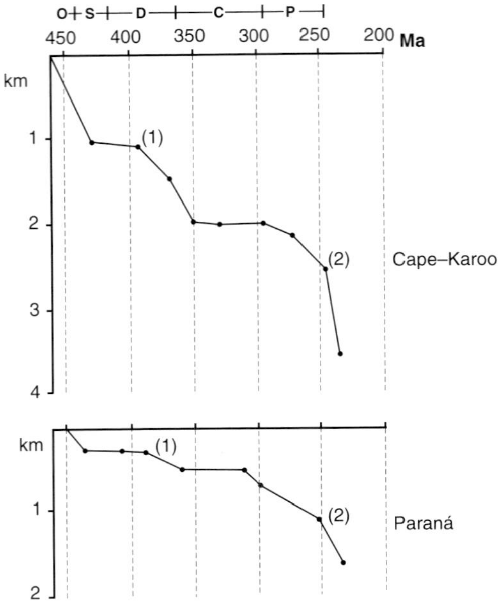

Fig. 10. Backstripped, tectonic subsidence of basement of Cape–Karoo (Cloetingh *et al.* 1992) and Paraná basins (Milani 1997). Note (1) the important Devonian cycle of accelerated subsidence; (2) the second cycle of accelerated subsidence during Late Permian–Early Triassic. The similarity of tectonic histories implicit in these diagrams, in spite of the attenuated rates of Paraná relative to Karoo, is very significant considering the position of each basin and the differential distance to Palaeozoic plate margins. This suggests a coupled foreland–intracratonic flexural mechanism of subsidence.

Pragian time onwards, a pattern of increasing subsidence rates was established. This period of accelerated subsidence led to the maximum flooding by the Devonian sea, recorded by the laminated black shales of the basal Ponta Grossa Formation. Shaly sedimentation under highstand marine conditions, punctuated by deltaic prograding sandy bodies, proceeded up to Frasnian time. Similar accelerated subsidence is observed at *c.* 400 Ma in the Cape subsidence plot (Fig. 10). Both areas presumably responded rapidly to lithosphere flexures generated far away in the Gondwanides. The general response of sandy shallow marine to coastal deposits (Furnas and Nardouw) changing to shaly neritic environments with second-order maximum flooding conditions (basal Ponta Grossa and basal Ceres) and then a highstand setting with Devonian deltas (middle/upper Ponta Grossa and upper Bokkeveld/Witteberg) are the sedimentological responses that we infer to trace a linked subsidence history of these basins.

The Late Permian–Early Triassic subsidence evolution in the history of the Paraná and Karoo basins is also remarkably similar (Fig. 10), in spite of very different lithostratigraphy at those times: the Paraná Basin was definitively trapped inside Gondwana, experiencing the final drying-up of its sea and the accumulation of extensive continental deposits, surprisingly thick (1.4 km), whereas in the Karoo a *c.* 1-km thick pile of distal turbidites, deposited during low subsidence rates, was succeeded by several kilometres of more rapidly accumulated coarse sediments derived from the emerging Cape mountains during latest Permian, in a true foreland domain. Despite these differences, the subsidence plots delineate distinct accelerated subsidence at around 250 Ma from the stratigraphic records of both basins. The 'classic' response of foreland development in the Paraná Basin is an unusual and still poorly understood mechanism of intracratonic flexure. Nevertheless, it appears that the general model of a Late Permian Gondwanide 'foredeep', which included the Karoo and Sauce Grande basins, and a contemporaneous Gondwanide 'foreland', where the Paraná Basin developed, was an outstanding insight of du Toit's genius.

Concluding remarks

The data set at hand today allows a more confident correlation between the Gondwana basins of South America and Southern Africa than was possible in du Toit's time. First, it is now more evident that a close relationship existed between the development of the Paraná Basin in the continental interior and the Palaeozoic tectonic regime along the flanking Gondwanides. By treating the Cape and Karoo basins as part of a continuum, such a relationship also emerges more clearly. The lithosphere of southwestern Gondwana reacted by flexure under the stresses generated along the Gondwanides, and this provided an effective mechanism to create 'intracratonic' depositional space well into the Gondwana shield. In both South America and southern Africa, this mechanism of subsidence seems to reflect the craton-ward propagation of the flexural bending of the lithosphere that characterizes the foreland domain, so that the Paraná Basin experienced phases of accelerated subsidence that correlate well with those in the adjacent foreland. Yet, in detail the lithostratigraphies of the foreland basin sequences do not correlate well between the Cape–Karoo and Paraná basins because the Palaeozoic collisional tectonic history along the Gondwanides was different and/or diachronous (Fig. 11). Only by Permian–Early Triassic times did plate boundary processes operate in

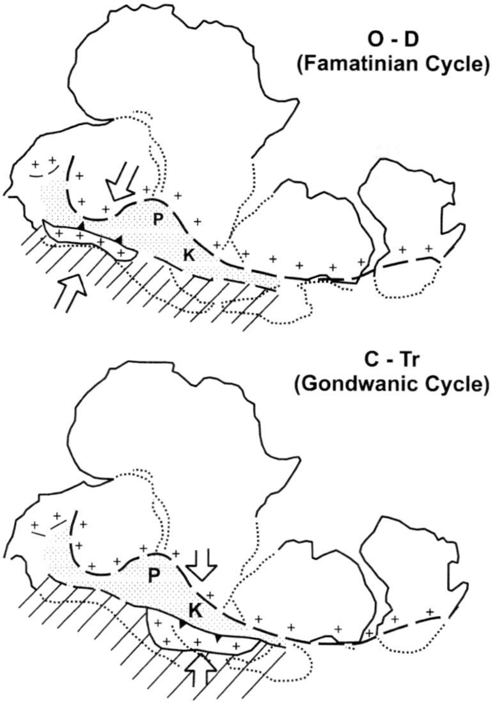

Fig. 11. Schematic tectonic setting of SW Gondwana margin during the two main Palaeozoic geodynamic cycles that affected the region. The main area of collision and orogenic deformation migrated from NW (during the Famatinian cycle) to SE (during the Gondwanic cycle). Paraná (P) and Cape–Karoo (K) basins were affected to different degrees by the various tectonic episodes active along the margin of the palaeo-continent.

unison to yield similar coupled foreland basin sequences. Even then distinct stratigraphic successions are evident, related to differences in local subsidence rates, and presumably to spatial and temporal variations in degrees of tectonic loading. These differences still need to be fully understood. During the early Mesozoic the respective foreland basins transformed and merged into a truly trans-Gondwana terrestrial basin. But even at this stage the Karoo Basin history differed in that the terrestrial desert conditions were interrupted by the widespread rapid outpourings of the short-lived (<1 Ma) Karoo LIP. However, at the very end of their common history, towards the time when the two continents finally separated, the Serra Geral–Etendeka basalts and their associated seaward-dipping volcanic sequences were erupted all along both margins of the South Atlantic.

The evolution of both the Paraná and Cape–Karoo basins as described herein is an integrated product of a particular local history of subsidence (and the variable elastic thickness of their basement), and the variable rates of transformation into intracratonic basins during the concomitant, but diachronous Palaeozoic tectonics within the Gondwanides. Thickness variations across both basins are therefore controlled to a much greater degree by loading effects during the evolution of the Gondwanides than by global eustasy, and this provides a fundamental key towards improving understanding of the geological history of central, continental Gondwana.

We both acknowledge discussions with many Gondwana colleagues over a long time, especially at the Gondwana conferences. E.J.M acknowledges Petrobras for providing the means for his research into southern South American geology. M.d.W particularly thanks Arthur Fuller, John Rogers, Ingo Hälbich and Peter Booth for introducing him to the lure of the Cape Fold Belt and the Cape–Karoo Basin. His research over the years was supported through the National Research Foundation of South Africa. Pat Eriksson and Stephen Flint provided great inputs towards the improvement of the manuscript. This is AEON contribution no. 29.

References

Andreis, R. R., Iñiguez, A. M., Lluch, J. L. & Rodríguez, S. 1989. Cuenca paleozoica de Ventania, Sierras Australes, Provincia de Buenos Aires. *In*: Chebli, G. A. & Spalletti, L. A. (eds) *Cuencas sedimentarias argentinas*. Universidad Nacional, Tucuman, Geological Correlation Series, **6**, 265–298.

Araújo, L. M., Rodrigues, R. & Scherer, C. M. S. 2001. Expressão estratigráfica do carbono orgânico, resíduo insolúvel, enxofre, parâmetros de pirólise e elementos-traço nas seqüências deposicionais Irati, Bacia do Paraná. *In*: *7 Congresso Brasileiro de Geoquímica, Curitiba, abstracts* 207.

Armstrong, RA., de Wit, M. J., Reid, D., York, D. & Zartman, R. 1998. Cape Town's Table Mountain reveals rapid Pan-African uplift of its basement rock. *Journal of African Earth Sciences*, **27–1A**, 10–11.

Astini, R. A. 1996. Las fases diastróficas del Paleozoico Medio en la Precordillera del oeste argentino—evidencias estratigráficas. *In*: *13 Congreso Geologico Argentino y 3 Congreso de Exploración de Hidrocarburos, Buenos Aires, Actas*, **5**, 509–526.

Astini, R. A., Benedetto, J. L. & Vaccari, N. E. 1995. The early Paleozoic evolution of the Argentine Precordillera as a Laurentian rifted, drifted, and collided terrane—a geodynamic model. *Geological Society of America Bulletin*, **107**, 253–273.

Astini, R. A., Ramos, V. A., Benedetto, J. L., Vaccari, N. E. & Cañas, F. L. 1996. La Precordillera: un terreno exótico a Gondwana. *In*: *13 Congreso Geologico Argentino y 3 Congreso de Exploración de Hidrocarburos, Buenos Aires, Actas*, **5**, 293–324.

Astini, R. A., Thomas, W. A. & McClelland, W. C. 2005. Provenance of Middle Ordovician detritus in the Argentine Precordillera. *In*: Pankhurst, R. J. & Veiga, G. D. (eds) *Gondwana 12: Geological and*

Biological Heritage of Gondwana, Abstracts. Academia Nacional de Ciencias, Córdoba, tArgentina, 51.

BAHLBURG, H. & BREITKREUZ, C. 1991. Paleozoic evolution of active margin basins in the Southern Central Andes (Northwestern Argentina and Northern Chile). *Journal of South American Earth Sciences*, **4**, 171–188.

BANGERT, B., STOLLHOFEN, H., LORENZ, V. & ARMSTRONG, R. 1999. The geochronology and significance of ashfall tuffs in the glaciogenic Carboniferous–Permian Dwyka Group of Namibia and South Africa. *Journal of African Earth Sciences*, **29**, 33–50.

BARNETT, W., ARMSTRONG, R. & DE WIT, M. 1997. Stratigraphy of the upper Neoproterozoic Kango and lower Paleozoic Table Mountain Groups of the Cape Fold Belt, revisited. *South African Journal of Geology*, **100**, 237–250.

BARRETT, S. F. & ISAACSON, P. E. 1988. Devonian paleogeography of South America. *In*: MCMILLAN, N. J., EMBRY, A. F. & GLASS, D. J. (eds) *Devonian of the world*. Canadian Society of Petroleum Geologists, Memoirs, **14**, 655–667.

BASEI, M. A. S., SIGA, O., JR., MASQUELIN, H., HARARA, O. M., REIS NETO, J. M. & PRECIOZZI, F. P. 2000. The Dom Feliciano belt of Brazil and Uruguay and its foreland domain, the Rio de la Plata craton: framework, tectonic evolution and correlation with similar provinces of southwestern Africa. *In*: CORDANI, U. G., MILANI, E. J., THOMAZ FILHO, A. & CAMPOS, D. A. (eds) *Tectonic evolution of South America. 31st International Geological Congress, Rio de Janeiro*, 311–334.

BORRELLO, A. V. 1965. Sobre la presencia del Cámbrico inferior olenellidiano en la Sierra de Zonda, Precordillera de San Juan. *Ameghiniana*, **10**, 313–318.

BRANCH, T., WECKMANN, U. & RITTER, O. 2007. The Whitehill Formation—a highly conductive marker in the Karoo Basin. *South African Journal of Geology*, **110**, 465–476.

BROQUET, C. A. M. 1992. The sedimentary record of the Cape Supergroup: a review. *In*: DE WIT, M. J. & RANSOME, I. D. (eds) *Inversion tectonics of the Cape Fold Belt, Karoo and Cretaceous basins of Southern Africa*. A. A. Balkema, Rotterdam, 159–184.

CATUNEANU, O., HANCOX, P. J. & RUBIDGE, B. S. 1998. Reciprocal flexural behaviour and contrasting stratigraphies: a new basin development model for the Karoo retroarc foreland system, South Africa. *Basin Research*, **10**, 417–439.

CATUNEANU, O., HANCOX, P. J., CAIRNCROSS, B. & RUBIDGE, B. S. 2002. Foredeep submarine fans and forebulge deltas: orogenic off-loading in the underfilled Karoo Basin. *Journal of South African Earth Sciences*, **35**, 489–502.

CLOETINGH, S., LANKREIJER, A., DE WIT, M. J. & MARTINEZ, I. 1992. Subsidence history analyses and forward modelling of the Cape and Karoo Supergroups. *In*: DE WIT, M. J. & RANSOME, I. D. (eds) *Inversion tectonics of the Cape Fold Belt, Karoo and Cretaceous basins of Southern Africa*. A. A. Balkema, Rotterdam, 239–248.

COBBOLD, P. R., GAPAIS, D., ROSSELLO, E. A., MILANI, E. J. & SZATMARI, P. 1992. Permo-Triassic intracontinental deformation in SW Gondwana. *In*: DE WIT, M. J. & RANSOME, I. D. (eds) *Inversion tectonics of the Cape Fold Belt, Karoo and Cretaceous basins of Southern Africa*. A. A. Balkema, Rotterdam, 23–26.

COLE, D. I. 1992. Evolution and development of the Karoo Basin. *In*: DE WIT, M. J. & RANSOME, I. D. (eds) *Inversion tectonics of the Cape Fold Belt, Karoo and Cretaceous basins of Southern Africa*. A. A. Balkema, Rotterdam, 87–100.

CORDANI, U. G., BRITO-NEVES, B. B., FUCK, R. A., PORTO, R., THOMAZ FILHO, A. & CUNHA, F. M. B. 1984. *Estudo preliminar de integração do Pré-Cambriano com os eventos tectônicos das bacias sedimentares brasileiras*. Petrobras, Rio de Janeiro, Ciência-Técnica-Petróleo Series, **15**.

COUTINHO, J. M. V., HACHIRO, J., COIMBRA, A. M. & SANTOS, P. R. 1991. Ash-fall derived vitroclastic tuffaceous sediments in the Permian of the Paraná Basin and their provenance. *In*: ROCHA-CAMPOS, A. C. (ed.) *Gondwana Seven Proceedings*. University of São Paulo, Brazil, 147–160.

DAEMON, R. F. & QUADROS, L. P. 1970. Bioestratigrafia do Neopaleozóico da Bacia do Paraná. Brasília. *In*: *Congresso Brasileiro de Geologia* 24, *Anais*. Sociedade Brasileira de Geoloiga, 359–412.

DALLA SALDA, L. A. 2005. The Appalachian-Famatinian orogenic belt. *In*: PANKHURST, R. J. & VEIGA, G. D. (eds) *Gondwana 12: Geological and Biological Heritage of Gondwana, Abstracts*. Academia Nacional de Ciencias, Córdoba, Argentina, 127.

DALY, M. C., LAWRENCE, S. P., KIMUN'A, D. & BINGA, M. 1991. Late Palaeozoic deformation in central Africa: a result of distant collision? *Nature*, **350**, 605–607.

DECKER, JE. & DE WIT, M. J. 2005. Carbon isotope evidence for Crassulacean Acid Metabolism in the Mesozoic. *Terra Nova*, **18**, 9–17.

DE WIT, M. J. & RANSOME, I. D. 1992. Regional inversion tectonics along the southern margin of Gondwana. *In*: DE WIT, M. J. & RANSOME, I. D. (eds) *Inversion tectonics of the Cape Fold Belt, Karoo and Cretaceous basins of Southern Africa*. A. A. Balkema, Rotterdam, 15–22.

DE WIT, M. J., JEFFERY, M., BERGH, H. & NICOLAYSEN, L. 1988. *Geological map of sectors of Gondwana, reconstructed to their disposition c. 150 Ma. 1:10,000,000*. American Association of Petroleum Geologists/University of Witwatersrand.

DÍAZ-MARTÍNEZ, E. 1995. *Devonico Superior y Carbonifero del Altiplano de Bolivia—estratigrafia, sedimentologia y evolución palaeogeográfica*. Informes de ORSTOM Bolívia, La Paz, **46**.

DUANE, M. J. & BROWN, R. 1992. Geochemical open-system behaviour related to fluid-flow and metamorphism in the Karoo Basin. *In*: DE WIT, M. J. & RANSOME, I. D. (eds) *Inversion tectonics of the Cape Fold Belt, Karoo and Cretaceous basins of Southern Africa*. A. A. Balkema, Rotterdam, 127–140.

DU TOIT, A. L. 1926. *The Geology of South Africa*. Oliver and Boyd, Edinburgh.

DU TOIT, A. L. 1927. *A geological comparison of South America with South Africa*. The Carnegie Institution, Washington, Publication **381**.

DU TOIT, A. L. 1937. *Our wandering continents*. Oliver and Boyd, Edinburgh.

DU TOIT, A. L. 1954. *The Geology of South Africa*. Haughton and Boyd, London.

DUNCAN, R. A, HOOPER, P. R., REHACEK, J., MARSH, J. S. & DUNCAN, A. R. 1997. The timing and duration of the Karoo igneous event, southern Gondwana. *Journal of Geophysical Research*, **102**, 18127–18138.

EGLINTON, B. M. & ARMSTRONG, R. A. 2001. Geochronological and isotope constraints on the Mesoproterozoic Natal-Namaqua Belt: evidence from deep borehole intersections in South Africa. *Precambrian Research*, **125**, 179–189.

EVANS, F. J. 1999. Paleobiology of Early Carboniferous biota of the Waaipoort Formation (Witteberg Group), South Africa. *Palaeontologia Africanus*, **35**, 1–6.

FAURE, K. & COLE, D. 1999. Geochemical evidence for lacustrine microbial blooms in the vast Permian Main Karoo, Paraná, Falkland Islands and Haub basins of southwestern Gondwana. *Palaeogeography Palaeoclimatology Palaeoecology*, **152**, 189–213.

FERNÁNDEZ-SEVESO, F. & TANKARD, A. J. 1995. Tectonics and stratigraphy of the Late Palaeozoic Paganzo Basin of Western Argentina and its regional implications. *In*: TANKARD, A. J., SUÁREZ SORUCO, R. & WELSINK, H. J. (eds) *Petroleum basins of South America*. American Association of Petroleum Geologists, Memoirs, **62**, 285–301.

FERNÁNDEZ-SEVESO, F., PEREZ, M. A., BRISSON, I. E. & ALVAREZ, L. 1993. Sequence stratigraphy and tectonic analysis of the Paganzo Basin, West Argentina. *In*: *International Conference on Carboniferous and Permian, Comptes Rendus, Buenos Aires*, **2**, 223–260.

FOUCH, M. J., JAMES, D. E., VANDECAR, J. C. & VAN DER LEE, S. KAAPVAAL SEISMIC GROUP, 2004. Mantle seismic structure beneath the Kaapvaal and Zimbabwe cratons. *South African Journal of Geology*, **107**, 33–44.

FRANÇA, A. B. & POTTER, P. E. 1988. Estratigrafia, ambiente deposicional e análise de reservatório do Grupo Itararé (Permocarbonífero), Bacia do Paraná (Parte 1). *Boletim de Geociências da Petrobras*, **2**, 147–191.

FRANÇA, A. B., MILANI, E. J. *ET AL*. 1995. Phanerozoic correlation in Southern South America. *In*: TANKARD, A. J., SUÁREZ SORUCO, R. & WELSINK, H. J. (eds) *Petroleum basins of South America*. American Association of Petroleum Geologists, Memoirs, **62**, 129–161.

FRIMMEL, H. E. & FRANK, W. 1998. Neoproterozoic tectono-thermal evolution of the Gariep Belt and its basement, Namibia and South Africa. *Precambrian Research*, **90**, 1–28.

FURQUE, G. 1965. Geologia de la región del Cerro la Bolsa (Provincia de La Rioja). *In*: *Segundas Jornadas Geológicas Argentinas, Buenos Aires, Actas*, 181–215.

GOHRBANDT, K. H. A. 1993. Palaeozoic palaeogeographic and depositional developments on the central proto-Pacific margin of Gondwana: their importance to hydrocarbon accumulation. *Journal of South American Earth Sciences*, **6**, 267–287.

GRADSTEIN, F., OGG, J. & SMITH, A. 2004. *A Geologic Time Scale 2004*. Cambridge University Press.

GUERRA-SOMMER, M., CAZZULO-KLEPZIG, M., FORMOSO, M. L., MENEGAT, R. & BASEI, M. A. S. 2005. New radiometric data from ash fall rocks in Candiota coal-bearing strata and the palynostratigraphic framework in southern Paraná Basin (Brazil). *In*: PANKHURST, R. J. & VEIGA, G. D. (eds) *Gondwana 12: Geological and Biological Heritage of Gondwana*. Abstracts, Academia Nacional de Ciencias, Córdoba, Argentina, 189.

HÄLBICH, I. W. 1992. The Cape Fold Belt orogeny: state of the art 1970s–1980s. *In*: DE WIT, M. J. & RANSOME, I. D. (eds) *Inversion tectonics of the Cape Fold Belt, Karoo and Cretaceous basins of Southern Africa*. A. A. Balkema, Rotterdam, 141–158.

HARVEY, J. D., DE WIT, M. J., STANKIEWICZ, J. & DOUCOURE, C. M. 2001. Structural variations of the crust in the southwestern Cape deduced from seismic receiver functions. *South African Journal of Geology*, **104**, 231–242.

IÑIGUEZ, A. M., VALLE, A., POIRÉ, D. G., SPALLETTI, L. A. & ZALBA, P. E. 1989. Cuenca precambrica/palaeozoica inferior de Tandilia, Provincia de Buenos Aires. *In*: CHEBLI, G. A. & SPALLETTI, L. A. (eds) *Cuencas sedimentarias argentinas*. Universidad Nacional, Tucuman, Geological Correlation Series, **6**, 245–264.

JOHNSON, M. R. 1991. Sandstone petrography, provenance and plate tectonic setting in Gondwana context of the southeastern Cape-Karoo Basin. *South African Journal of Geology*, **94**, 137–154.

JOHNSON, M. R. 1997. The foreland Karoo Basin, South Africa. *In*: SELLEY, R. C. (ed.) *African basins. Sedimentary basins of the World*. Elsevier, Amsterdam, 269–317.

JOHNSTON, S. T. 2000. The Cape Fold Belt and Syntaxis, and the rotated Falkland Island: dextral transpression tectonics along the southwest margin of Gondwana. *Journal of African Earth Sciences*, **31**, 51–63.

KEIDEL, J. 1916. La geología de las sierras de la Província de Buenos Aires y sus relaciones con las montañas de Sud Africa y los Andes. *Anales del Ministerio de Agricultura de la Nación, Sección Geología, Mineralogía y Minería*, Buenos Aires, **3**, 1–78.

KING, G. 1990. *The dicynodonts*. Chapman and Hall, London.

KOKOGIAN, D. A., FERNÁNDEZ-SEVESO, F. & MOSQUERA, A. 1993. Las secuencias sedimentarias triasicas. *In*: RAMOS, V. A. (ed.) *Geologia y recursos naturales de Mendoza*. Asociación Geológica Argentina—Instituto Argentino de Petróleo, Buenos Aires, 65–78.

LINDEQUE, A. S., RYBERG, T., STANKIEWICZ, J., WEBER, M. H. & DE WIT, M. J. 2007. Seismic imaging of the crust and Moho in a section across the Beattie Magnetic Anomaly, South Africa. *South African Journal of Geology*, **110**, 419–438.

LOCK, B. E. 1980. Flat-plate subduction and the Cape Fold Belt of South Africa. *Geology*, **8**, 35–39.

LÓPEZ-GAMUNDÍ, O. & ROSSELLO, E. A. 1993. Devonian-Carboniferous unconformity in Argentina and its relation to Eo-Hercynian orogeny in southern South America. *Geologische Rundschau*, **82**, 136–147.

LÓPEZ-GAMUNDÍ, O. & ROSSELLO, E. A. 1998. Basin fill evolution and palaeotectonic patterns along the Samfrau geosyncline: the Sauce Grande basin–Ventana foldbelt (Argentina) and Karoo basin–Cape foldbelt (South Africa). *Geologische Rundschau*, **86**, 819–834.

LÓPEZ-GAMUNDÍ, O. R., ESPEJO, I. S., CONAGHAN, P. J. & POWELL, C. M. A. 1994. Southern South America. *In*: VEEVERS, J. J. & POWELL, C. M. A. (eds) *Permian-Triassic Pangean basins and foldbelts along the Panthalassan margin of Gondwanaland*. Geological Society of America, Memoirs, **184**, 281–329.

LÓPEZ-GAMUNDÍ, O. R., CONAGHAN, P. J., ROSSELLO, E. A. & COBBOLD, P. R. 1995. The Tunas Formation (Permian) in the Sierras Australes Foldbelt, east central Argentina—evidence for syntectonic sedimentation in a foreland basin. *Journal of South American Earth Sciences*, **8**, 129–142.

MACLEOD, K. G., SMITH, R. M. H., KOCH, P. I. & WARD, P. 2000. Timing of mammal-like reptile extinction across the Permian-Triassic boundary in South Africa. *Geology*, **28**, 227–230.

MARSH, J. S., EWART, A., MILNER, S. C., DUNCAN, A. R. & MILLER, R. MCG. 2001. The Etendeka Igneous Province: magma types and their stratigrapic distribution with implications for the evolution of the Paraná- Etendeka flood basalt province. *Bulletin of Volcanology*, **62**, 464–486.

MCLACHLAN, I. R. & ANDERSON, A. 1973. A review of the evidence for marine conditions in southern Africa during Dwyka times. *Palaeontologia Africanus*, **15**, 37–64.

MELO, J. H. G. 1988. The Malvinokaffric realm in the Devonian of Brazil. *In*: MCMILLAN, N. J., EMBRY, A. F. & GLASS, D. J. (eds) *Devonian of the world*. Canadian Society of Petroleum Geologists, Memoirs, **14**, 669–704.

MILANI, E. J. 1992. Intraplate tectonics and the evolution of the Paraná Basin, Southern Brazil. *In*: DE WIT, M. J. & RANSOME, I. D. (eds) *Inversion tectonics of the Cape Fold Belt, Karoo and Cretaceous basins of Southern Africa*. A. A. Balkema, Rotterdam, 101–108.

MILANI, E. J. 1997. *Evolução tectono-estratigráfica da Bacia do Paraná e seu relacionamento com a geodinâmica fanerozóica do Gondwana sul-ocidental*. DSc thesis, Universidade Federal do Rio Grande do Sul.

MILANI, E. J. 2004. Comentários sobre a origem e evolução tectônica da Bacia do Paraná. *In*: MANTESSO-NETO, V., BARTORELLI, A., CARNEIRO, C. D. R. & BRITO-NEVES, B. B. (eds) *Geologia do continente Sul-Americano: evolução da obra de Fernando Flávio Marques de Almeida*. Beca Editora, São Paulo, 265–279.

MILANI, E. J. & RAMOS, V. A. 1998. Orogenias paleozóicas no domínio sul-ocidental do Gondwana e os ciclos de subsidência da Bacia do Paraná. *Revista Brasileira de Geociências*, **28**, 473–484.

MILANI, E. J., ASSINE, M. L., SOARES, P. C. & DAEMON, R. F. 1996. A Seqüência Ordovício-Siluriana da Bacia do Paraná. *Boletim de Geociências da Petrobras*, **9**, 301–320.

OPDYKE, N. D., MUSHAYANDEBUN, M. & DE WIT, M. J. 2001. A new palaeomagnetic pole for the Dwyka System and correlative sediments in sub-Saharan Africa. *Journal of African Earth Sciences*, **33**, 143–154.

PEREIRA, E., RODRIGUES, R. & BERGAMASCHI, S. 2005. Stratigraphical, bioestratigraphical and geochemical characterization of Silurian and Devonian world-wide transgressive events in the Paraná Basin: a high-resolution approach. *In*: PANKHURST, R. J. & VEIGA, G. D. (eds) *Gondwana 12: Geological and Biological Heritage of Gondwana, Abstracts*. Academia Nacional de Ciencias, Córdoba, Argentina, 290.

PANKHURST, R. J. & RAPELA, C. W. 1998. Introduction. *In*: PANKHURST, R. J. & RAPELA, C. W. (eds) *The Proto-Andean Margin of Gondwana*. Geological Society, London, Special Publications, **142**, 1–9.

PANKHURST, R. J., RAPELA, C. W. Fanning C. M. & MÁRQUEZ, M. 2006. Gondwanide continental collision and the origin of Patagonia. *Earth-Science Reviews*, **76**, 235–257.

PITTS, B. E., MAHER, M. J., DE BEER, J. H. & GOUGH, D. I. 1992. Interpretation of magnetic, gravity and magnetotelluric data across the Cape Fold Belt and Karoo basin. *In*: DE WIT, M. J. & RANSOME, I. D. (eds) *Inversion tectonics of the Cape Fold Belt, Karoo and Cretaceous basins of Southern Africa*. A. A. Balkema, Rotterdam, 27–32.

POWELL, C. M. A. 1993. Assembly of Gondwanaland—open forum. *In*: *Proceedings of Gondwana Eight*. A. A. Balkema, Rotterdam, 219–237.

RAKOTOSOLOFO, N. A., TORSVIK, T. H., ASHWAL, L. D., EIDE, E. A. & DE WIT, M. J. 1999. The Karoo supergroup revisited and Madagascar-Africa fits. *Journal of African Earth Sciences*, **29**, 135–151.

RAMOS, V. A. 1988. Late Proterozoic — Early Palaeozoic of South America—a collisional history. *Episodes*, **11**, 168–174.

RAMOS, V. A. 1990. Field guide to geology of the Central Andes (31°–33° SL). Buenos Aires, Universidad de Buenos Aires, Central Andes Field Seminar.

RAMOS, V. A. 1993. Interpretación tectónica. *In*: RAMOS, V. A. (ed.) *Geologia y recursos naturales de Mendoza*. Asociación Geológica Argentina—Instituto Argentino de Petróleo, Buenos Aires, 257–268.

RAMOS, V. A. 2005. The Proterozoic-Early Palaeozoic margin of western Gondwana. *In*: PANKHURST, R. J. & VEIGA, G. D. (eds) *Gondwana 12: Geological and Biological Heritage of Gondwana, Abstracts*. Academia Nacional de Ciencias, Córdoba, Argentina, 304.

RAMOS, V. A., JORDAN, T. E., ALLMENDINGER, R. W., KAY, S. M., CORTES, J. M. & PALMA, M. A. 1984. Chilenia: un terreno aloctono en la evolucion palaeozoica de los Andes centrales. *In*: *IX Congreso Geologico Argentino, Buenos Aires, Actas*, **2**, 84–106.

RAMOS, V. A., JORDAN, T. E. *ET AL.* 1986. Palaeozoic terranes of the central Argentine-Chilean Andes. *Tectonics*, **5**, 855–880.

RAPELA, C. W. 2000. The Sierras Pampeanas of Argentina: Paleozoic building of the southern proto-Andes. *In*: CORDANI, U. G., MILANI, E. J., THOMAZ FILHO, A. & CAMPOS, D. A. (eds) *Tectonic evolution of South America. 31st International Geological Congress, Rio de Janeiro*, 381–387.

RAPELA, C. W., PANKHURST, R. J., FANNING, C. M. & GRECCO, L. E. 2003. Basement evolution of the

Sierra de la Ventana Fold Belt: new evidence for Cambrian continental rifting along the southern Margin of Gondwana. *Journal of the Geological Society, London*, **160**, 613–628.

RAPELA, C. W., CASQUET, C. *ET AL.* 2008. The Río de la Plata craton and the assembly of SW Gondwana. *Earth-Science Reviews*, **83**, 49–82.

REEVES, C. V., SAHU, B. K. & DE WIT, M. J. 2002. A re-examination of the palaeo-positions of Africa's eastern neighbours in Gondwana. *Journal of African Earth Sciences*, **34**, 101–108.

RUBIDGE, B. S., HANCOX, P. J. & CATUNEANU, O. 2000. Sequence analysis of the Ecca-Beaufort contact in the southern Karoo of South Africa. *South African Journal of Geology*, **103**, 81–96.

RUST, I. C. 1975. Tectonic and sedimentary framework of Gondwana basins in southern Africa. *In*: *Third Gondwana Symposium, Australia, Actas*, 554–564.

SANTOS, R. V., SOUZA, P. A. *ET AL.* 2006. Shrimp U–Pb zircon dating and palynology of bentonitic layers from the Permian Irati Formation, Paraná Basin, Brazil. *Gondwana Research*, **9**, 456–463.

SCOTESE, C. R. 2000. *Paloemap Project: Earth History (palaeogeographic maps)*. Department of Geology, University of Texas, Arlington.

SEMPERE, T. 1995. Phanerozoic evolution of Bolivia and adjacent regions. *In*: TANKARD, A. J., SUÁREZ SORUCO, R. & WELSINK, H. J. (eds) *Petroleum basins of South America*. American Association of Petroleum Geologists, Memoirs, **62**, 207–230.

SLOSS, L. L. 1963. Sequences in the cratonic interior of North America. *Geological Society of America Bulletin*, **74**, 93–114.

SMITH, R. M. H. 1995. Fluvial environments across the Permian-Triassic boundary in the Karoo Basin, South Africa and possible causes of the tetrapod extinctions. *Palaeogeography Palaeoclimatology Palaeoecology*, **117**, 84–104.

STANKIEWICZ, J., RYBERG, T., SCHULZE, A., LINDEQUE, A. S., WEBER, M. H. & DE WIT, M. J. 2007. Results from the wide-angle seismic refraction lines in the Southern Cape. *South African Journal of Geology*, **110**, 407–418.

STECKLER, M. & WATTS, A. B. 1978. Subsidence of the Atlantic-type continental margin off New York. *Earth and Planetary Science Letters*, **41**, 1–13.

STERN, C. R. & DE WIT, M. J. 2004. Rocas Verdes ophiolite, southernmost South America: remnants of progressive stages of development of oceanic type crust in a continental back arc basin. *In*: DILEK, Y. & ROBINSON, P. T. (eds) *Ophiolites in Earth History*. Geological Society, London, Special Publications, **218**, 665–683.

STREEL, M. & THERON, J. N. 1999. The Devonian-Carboniferous boundary in South Africa and the age of the earliest episode of the Dwyka glaciation: New palynological result. *Episodes*, **22**, 41–44.

TANKARD, A. J., JACKSON, M. P. A., ERIKSSON, K. A., HOBDAY, D. K., HUNTER, D. R. & MINTER, W. E. L. 1982. *Crustal evolution of southern Africa*. Springer-Verlag, Berlin.

THOMAS, W. A. & ASTINI, R. A. 2005. Vestiges of an Ordovician Ocloyic west-vergent, thin-skinned thrust belt in the Argentine Precordillera, southern central Andes. *In*: PANKHURST, R. J. & VEIGA, G. D. (eds) *Gondwana 12: Geological and Biological Heritage of Gondwana, Abstracts*. Academia Nacional de Ciencias, Córdoba, Argentina, 350.

TROUW, R. A. J. & DE WIT, M. J. 1999. Relation between the Gondwanide Orogen and contemporaneous intracratonic deformation. *Journal of African Earth Sciences*, **28**, 203–213.

TURNER, B. R. 1999. Tectonostratigraphical development of the upper Karoo foreland basin: orogenic unloading versus thermally-induced Gondwana rifting. *Journal of African Earth Sciences*, **28**, 215–238.

TURNER, S., REGELOUS, M., KELLEY, S., HAWKESWORTH, C. & MANTOVANI, M. 1994. Magmatism and continental break-up in the South Atlantic: high precision Ar–Ar geochronology. *Earth and Planetary Science Letters*, **121**, 333–348.

VAIL, P. R., MITCHUM, R. M. & THOMPSON, S. 1977. Seismic stratigraphy and global changes of sea level, part 3: relative changes of sea level from coastal onlap. *In*: PAYTON, C. E. (ed.) *Seismic stratigraphy—applications to hydrocarbon exploration*. American Association of Petroleum Geologists, Memoirs, **26**, 63–81.

VAUGHAN, A. P. M. & LIVERMORE, R. A. 2005. Episodicity of Mesozoic terrane accretion along the Pacific margin of Gondwana: implications fro superplume–plate interactions. *In*: VAUGHAN, A. P. M., LEAT, P. T. & PANKHURST, R. J. (eds) *Terrane processes at the margins of Gondwana*. Geological Society, London, Special Publications, **246**, 143–178.

VAUGHAN, A. P. M., LEAT, P. T. & PANKHURST, R. J. (eds) 2005. *Terrane processes at the margins of Gondwana*. Geological Society, London, Special Publications, **246**.

VEEVERS, J. J., COLE, D. I. & COWAN, E. J. 1994. Southern Africa: Karoo Basin and Cape Fold Belt. *In*: VEEVERS, J. J. & POWELL, C. M. A. (eds) *Permian-Triassic pangean basins and foldbelts along the panthalassan margin of Gondwanaland*. Geological Society of America, Memoirs, **184**, 223–280.

VILLAR, L. M., CHERNICOFF, C. J. & ZAPPETTINI, E. 2005. Evidence of a Famatinian continental magmatic arc at Paso del Bote, La Pampa province, Argentina. *In*: PANKHURST, R. J. & VEIGA, G. D. (eds) *Gondwana 12: Geological and Biological Heritage of Gondwana, Abstracts*. Academia Nacional de Ciencias, Córdoba, Argentina, 365.

VISSER, J. N. J. 1987. The palaeogeography of part of southwestern Gondwana during the Permian-Carboniferous glaciation. *Palaeogeography Palaeoclimatology Palaeoecology*, **61**, 205–219.

VISSER, J. N. J. 1989. The Permo-carboniferous Dwyka Formation of southern Africa: deposition by a predominantly marine ice-sheet. *Palaeogeography Palaeoclimatology Palaeoecology*, **70**, 377–391.

VISSER, J. N. J. 1990. The age of the late Palaeozoic glaciogene deposits in southern Africa. *South African Journal of Geology*, **93**, 366–375.

VISSER, J. N. J. 1992*a*. Sea-level changes in a back-arc-foreland transition: the late Carboniferous-Permian Karoo Basin of South Africa. *Sedimentary Geology*, **83**, 115–131.

VISSER, J. N. J. 1992*b*. Deposition of the early to late Permian Whitehill Formation during a sealevel high stand in a juvenile foreland basin. *South African Journal of Geology*, **95**, 181–193.

VISSER, J. N. J. 1997. Deglaciation sequences in the Permo-Carboniferous Karoo and Kalahari basins of southern Africa: a tool in the analysis of cyclic glacio-marine basin fills. *Sedimentology*, **44**, 507–521.

VON GOSEN, W., BUGGISCH, W. & KRUMM, S. 1991. Metamorphism and deformation mechanisms in the Sierras Australes fold and thrust belt (Buenos Aires Province, Argentina). *Tectonophysics*, **185**, 335–356.

WARD, P. D., MONTGOMERY, D. R. & SMITH, R. 2000. Altered river morphology in South Africa related to the Permian-Traissic extinction. *Science*, **289**, 1740–1743.

WECKMANN, U., RITTER, O., JUNG, A., BRANCH, T. & DE WIT, M. J. 2007. Magnetotelluric measurements across the Beattie magnetic anomaly and the Southern Cape Conductive Belt, South Africa. *Journal of Geophysical Research*, **112**, B5, B05416.

WEGENER, A. 1912. Die Entstelhung der Konitinente. *Geologische Rundschau*, **3**, 267–292.

WICKENS, H. D. V. 1992. Submarine fans of the Permian Ecca Group in the SW Karoo Basin: Their origin and reflection on the tectonic evolution of the basin and its source areas. *In*: DE WIT, M. J. & RANSOME, I. D. (eds) *Inversion tectonics of the Cape Fold Belt, Karoo and Cretaceous basins of Southern Africa*. A. A. Balkema, Rotterdam, 117–126.

WILLIAMS, K. E. 1995. Tectonic subsidence analysis and Palaeozoic palaeogeography of Gondwana. *In*: TANKARD, A. J., SUÁREZ SORUCO, R. & WELSINK, H. J. (eds) *Petroleum basins of South America*. American Association of Petroleum Geologists, Memoirs, **62**, 79–100.

ZALÁN, P. V., WOLFF, S. *ET AL*. 1990. The Paraná Basin, Brazil. *In*: LEIGHTON, M. W., KOLATA, D. R., OLTZ, D. F. & EIDEL, J. J. (eds) *Interior cratonic basins*. American Association of Petroleum Geologists, Memoirs, **51**, 681–708.

ZAWADA, P. K. & CADLE, A. B. 1988. Position of the Ecca-Beaufort boundary contact in the southwestern Free State: an evaluation of four possible alternatives. *South African Journal of Geology*, **91**, 49–56.

ZIMMERMANN, U. & SPALLETTI, L. A. 2005. Provenance of the Balcarce Formation: an indicator for Early Palaeozoic volcanism in Eastern Argentina. *In*: PANKHURST, R. J. & VEIGA, G. D. (eds) *Gondwana 12: Geological and Biological Heritage of Gondwana, Abstracts*. Academia Nacional de Ciencias, Córdoba, Argentina, 377.

The Neoproterozoic glacial record in the Río de la Plata Craton: a critical reappraisal

P. J. PAZOS[1], L. S. BETTUCCI[2] & J. LOUREIRO[2]

[1]*CONICET-UBA, Dpto. de Ciencias Geológicas, Fac. Cs. Ex. y Nat., Ciudad Universitaria Pabellón II (1428), Ciudad Autónoma de Buenos Aires, Argentina (e-mail: pazos@gl.fcen.uba.ar)*

[2]*Dpto. de Geología, Facultad de Ciencias, Universidad de la República, Iguá 4225 (11400), Montevideo, Uruguay*

Abstract: Neoproterozoic glacial successions have been described in South America, but the glacial deposits of the Río de la Plata Craton have been neglected in previous studies addressing the global distribution of glacially influenced successions. The Río de la Plata Craton contains Neoproterozoic glacial deposits in the Sierra del Volcán Formation (Tandilia System, Argentina), glacial influenced deposits in the Playa Hermosa and Zanja del Tigre formations (Dom Feliciano Belt, Uruguay) and suspected glacially influenced deposits in Passo da Areia (São Gabriel block, Brazil). The Tandilia System glacial record includes diamictites, dropstones and rhythmites deposited in glaciomarine conditions in a tectonically stable depositional setting. The Dom Feliciano Belt includes a thin section with ice-rafted clasts in carbonates and a thicker section containing diamictites, rhythmites, outsized clasts and deformed beds in a volcano-sedimentary succession. The São Gabriel block occurrence deserves more attention to confirm any glacial influence in the fine-grained part of the succession. Glaciation is considered to be contemporaneous with the Gaskiers glaciation (580 Ma), with the exception of the carbonates with dropstones that may represent a previous event correlative with one of the glaciations described in the Kalahari Craton, prior to Kalahari–Río de la Plata assembly in the proto-western Gondwana margin.

The Precambrian is one of the most enigmatic times in Earth history, involving drastic changes in the biosphere, mostly documented in the last part of the Neoproterozoic. This was a time when dramatic fluctuations in climate, sea-water composition, continent assembly, interaction between tectonism and climate, and life diversification took place (e.g., Brasier & Shields 2000; Butterfield 2000; Evans 2000; Hurtgen *et al.* 2002; Bowring *et al.* 2003; Eyles & Januszczak 2004; Allen & Hoffman 2005; Halverson *et al.* 2005). Sedimentary successions archive crucial information for the evaluation of the magnitude and pattern of such transformations. For instance, the Neoproterozoic rock record contains significant diamictite intervals that have traditionally been interpreted as evidence for two severe global glaciations, informally known as Sturtian and Marinoan (Hambrey & Harland 1985; Hoffman *et al.* 1998), but a younger and probably more geographically restricted glaciation termed Gaskiers has also been well established (see Halverson *et al.* 2005 for references). However, the nature, timing and correlation of these 'global' glaciations are still the subject of intense debate, and other Neoproterozoic glaciations cannot be completely ruled out. Some interpretations suggest general synchronicity and glaciation extending into low latitudes, offering a picture of the Earth's surface that was totally frozen with complete shut-down of the hydrological cycle and life crisis (the Snowball Earth hypothesis); the aftermath was a rapid change to greenhouse conditions (Hoffman *et al.* 1998). Criticism of the hypothesis suggests that global correlation of glaciations requires more detailed age constraints to corroborate the fundamental idea of a quasi-synchronous record. For instance, new Re–Os ages from Australia indicate a long-lived Sturtian glaciation (711–643 Ma), or diachronism not much different from that of the Pleistocene glaciation (Kendall *et al.* 2006). Other papers question the true nature of the assumed glacial deposits. Some of the best known glacial diamictites have been interpreted as meteoritic impact eject (e.g., Rampino 1994). Alternatively, subaqueous debris flow, sometimes not related to glacial influence at all but triggered by tectonic instability in synchronous mega-rift 'zipper-rift' basins, has been suggested for deposits previously interpreted as tillites (e.g., Eyles & Januszczak 2004), although such synchronous rift evolution has been proven mistaken by Fanning & Link (2004). Iceberg circulation (e.g., Condon *et al.* 2002) and multiple glacial advances and retreats indicate incomplete

From: PANKHURST, R. J., TROUW, R. A. J., BRITO NEVES, B. B. & DE WIT, M. J. (eds) *West Gondwana: Pre-Cenozoic Correlations Across the South Atlantic Region.* Geological Society, London, Special Publications, **294**, 343–364. DOI: 10.1144/SP294.18 0305-8719/08/$15.00

shut-down in the hydrological cycle (e.g., Leather *et al.* 2002), contrary to the Snowball Earth hypothesis. Interestingly, a low-latitude glacial record is rarely questioned (see Evans 2000) and has been confirmed for Dobrzinski *et al.* (2005) for the Sturtian and Marinoan deposits in the Yangtze platform of South China. Such uncertainty about climate evolution during the Neoproterozoic explains the re-invigorated investigation of sedimentological aspects of diamictites with a supposed glacial origin (e.g., Arnaud & Eyles 2002, 2006; Arnaud 2004), the palaeomagnetism of glaciated intervals (e.g., Li *et al.* 2004; Macouin *et al.* 2004), U–Pb dating (e.g., Brasier *et al.* 2000; Lund *et al.* 2003; Fanning & Link 2004; Hoffmann *et al.* 2004; Condon *et al.* 2005), Re–Os dating (Kendall *et al.* 2006), geochemistry and chemostratigraphy (e.g., Kennedy *et al.* 1998; Gorjan *et al.* 2000; Alvarenga *et al.* 2004; Gaucher *et al.* 2005*a*, *b*; Halverson *et al.* 2005), tectonic and palaeogeographic correlations (Veevers 2004, 2005) and palaeobiological approaches (Corsetti *et al.* 2006), among many others.

South America contributes to the database of the Neoproterozoic glacial record with undisputed glacial or glacially influenced successions, mainly in Brazil. The best known and studied glacial successions are located in the São Francisco Craton (Fig. 1). Their sedimentary record includes diamictites and striated pavements assigned to the Sturtian glacial event (Martins Neto *et al.* 2001; Pedrosa-Soares *et al.* 2001), but other glacial deposits are also distributed in tectonically deformed belts between cratonic blocks (e.g., Cukrov *et al.* 2005). The younger glacial deposits were assigned to the Marinoan glaciation and restricted to the southern part of the Amazon Craton and Paraguay Belt (Alvarenga & Trompette 1992), but have been recently re-interpreted and assigned to the younger (580 Ma) Gaskiers glaciation by Nogueira *et al.* (2003). The Río de la Plata Craton is traditionally considered to be devoid of equivalent glacial deposits, even though glaciomarine deposits are documented at its northern border (Corumbã Basin, Alvarenga & Trompette 1992). But the glacial deposits of the Corumbã Basin, those documented in the Tandilia System by Spalletti & del Valle (1984) and in southern Brazil by Eerola (1995) were not considered by Eyles & Januszczak (2004) and the glacially influenced origin of some Neoproterozoic units in the Río de la Plata Craton has been not mentioned in discussions of the tectono-sedimentary and climatic evolution of the eastern margin of this craton during the Neoproterozoic (e.g., Gaucher *et al.* 2003, 2005*a*). In consequence a 'climatic window' between Namibian and Brazilian records has been suggested to explain the absence of glacial record in the Río de la Plata Craton. However, in the last decade further glacially influenced deposits have been documented (e.g., Pazos *et al.* 2003), confirming that this craton, like the neighbouring ones, was at least partially glaciated during the Neoproterozoic. Different palaeomagnetic reconstructions for the Neoproterozoic locate the Río de la Plata Craton in either mid-high or lower latitudes (see Meert & Torsvik 2004). Sánchez Bettucci & Rapalini (2002) suggested a low to intermediate palaeolatitude and a congruent polar wander path at 600 Ma based on preliminary results from the Neoproterozoic Playa Hermosa Formation in southeastern Uruguay. Uncertainties in the stratigraphic position and sedimentary features of glacially related deposits within the sedimentary cover of the Río de la Plata Craton, as well as their correlation with other documented glacial deposits in West Gondwana, are the focus of this paper.

Río de la Plata Craton

The assembly of the western margin of Gondwana was a progressive process that included several collisions between large blocks (cratons), but also small and complex micro-blocks, during the closure of ancient oceanic basins. However, the exact extent of each craton is not well established, e.g., the Río de la Plata Craton includes igneous and metamorphic rocks with Palaeoproterozoic ages (2.2–1.6 Ga), with no Grenvillian ages (Cordani *et al.* 2000), distributed in outcrops in Uruguay and Argentina. Grenvillian ages between 1.0 and 0.9 Ga were obtained in the Punta del Este terrane by U–Pb (ID-TIMS, Preciozzi *et al.* 1999, 2003), in rocks that were strongly reworked during the Brasiliano orogeny. According to Rapela *et al.* (2007), the western boundary of the Río de la Plata Craton is against the Pampean (Cambrian) rocks of the Eastern Sierras Pampeanas. Suggested results of collision of the Pampean Belt with the Río de la Plata Craton include magmatism and metamorphism of latest Neoproterozoic to Early Cambrian age (Escayola *et al.* 2007). The northern extent of the Río de la Plata Craton is speculative: Alkmim *et al.* (2001) interpreted the Paraguay fold belt as an ancient basin closed during collision of the Río de la Plata Craton with Amazonia in the late Precambrian–Early Cambrian. However, a recent proposal by Trindade *et al.* (2006) is that the main collisional interval was Cambrian (530–525 Ma). The eastern margin is also a matter for debate, particularly given the complex tectonic interpretations for the geological evolution of southeastern Brazil and Uruguay. A complex nomenclature of tectono-stratigraphic domains divides the Precambrian geology of Uruguay and southern Brazil into a puzzle of

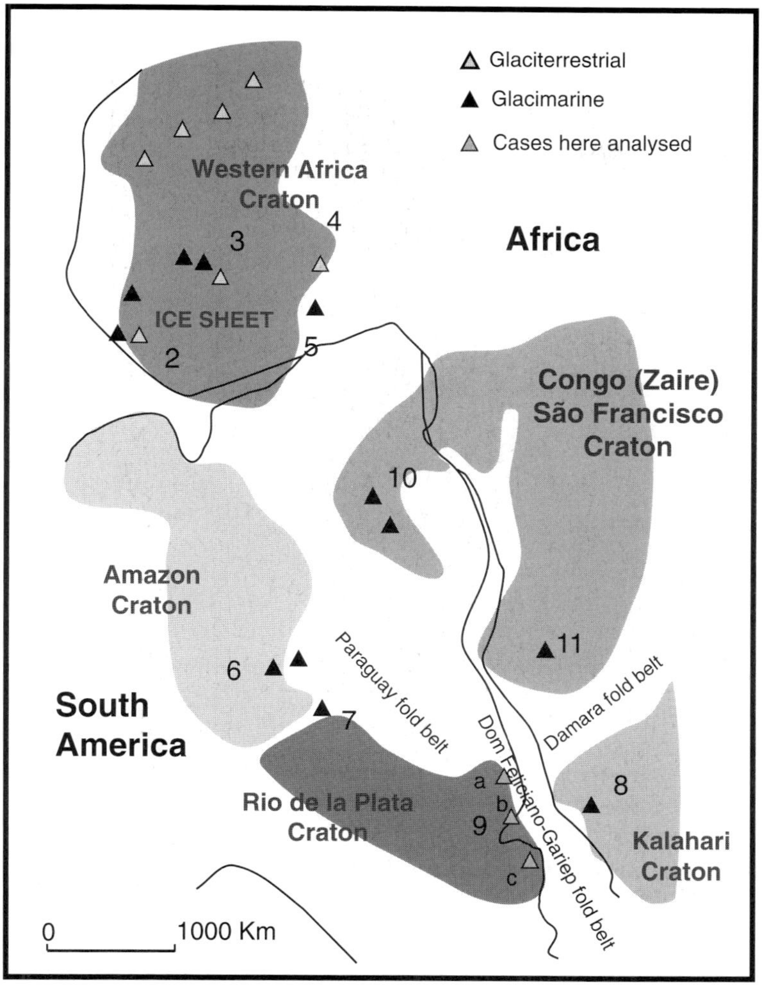

Fig. 1. Distribution of Neoproterozoic glacial deposits, cratons and orogenic belts of West Gondwana (modified from Alvarenga & Trompette 1992 and Pazos *et al.* 2003): 1, tillites (*s.s*) of the northern part of the Taudeni Basin including those of the Adrar and Mauritania; 2, glacial rocks of Mali and Senegal boundary (Kayes area) and the Mali and Guinea boundary; 3, glacio-marine deposits of southwestern Mali; 4, tillites (*s.s*) of the northern Volta Basin; 5, glacial deposits of the Pan-African Dahomeyides; 6–7, glacio-marine deposits and diamictites of the Brasiliano–Paraguay Belt and southern border of the Amazon Craton; 8, Ghaub Formation (diamictites and dolomite) and Nama Group; 9, diamictites of (a) Sierra del Volcán Formation, (b) Playa Hermosa Formation, and (c) Passo da Areia sequence; 10, Jequitaí Formation, glacio-marine deposits (São Francisco Craton); 11, Congo Craton, Otavi Group.

blocks, termed 'terranes' that collided diachronously. In Uruguay, at least four terranes are recognized (Fig. 2). One of them, the Dom Feliciano Belt located to the east of Nico Pérez terrane, contains the glacial sections discussed in this paper. Bossi & Gaucher (2004) speculated that the Nico Pérez terrane was a block attached to a northern extension of the Río de la Plata Craton (the Piedra Alta terrane) in the Mesoproterozoic. Another exotic block (Punta del Este terrane) is situated to the east of Sierra Ballena shear zone (Fig. 2) and, in fact, is part of the Kalahari Craton (Preciozzi *et al.* 2003), which collided with the Río de la Plata Craton during the latest Precambrian–Early Cambrian (Frimmel & Basei 2006). The Alférez-Cordillera shear zone, with a NE-SW to east–west trend, is interpreted as the expression of the suture of that collision (Basei *et al.* 2005). The sinistral Sierra Ballena mega-shear zone (Fig. 2) has a NE–SW orientation and was correlated with the Purros shear zone of the Kaoko Belt of Namibia by Oyhantçabal (2005). In southern Brazil other exotic blocks (e.g., São Gabriel block, Luis Alves) were accreted during the

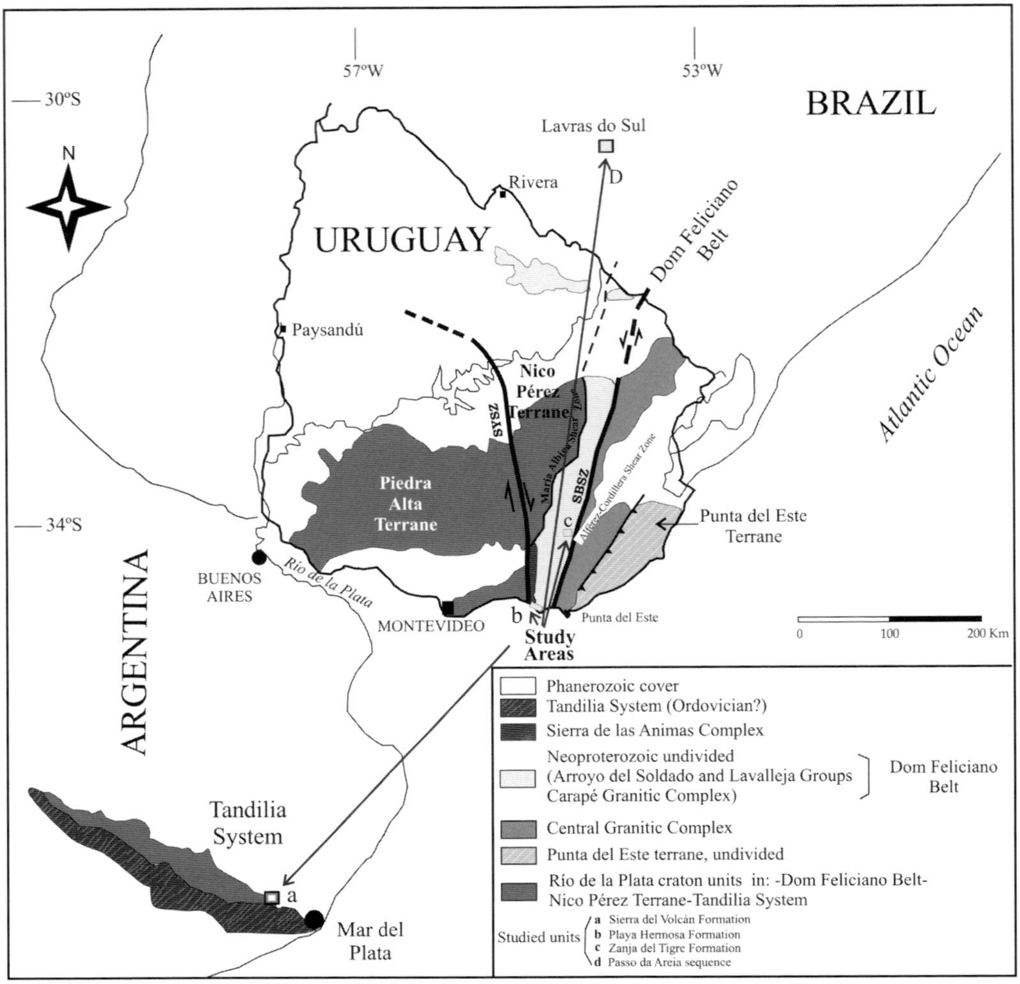

Fig. 2. Geological sketch of the eastern margin of the Río de la Plata Craton and location of the study areas. SYSZ, Sarandí del Yí Shear Zone; SBSZ, Sierra Ballena Shear Zone. Studied units indicated: a, Sierra del Volcán Fm; b, Playa Hermosa Fm; c, Zanja del Tigre Fm; d, Paso de Areia sequence.

Cryogenian and Ediacaran (Chemale 2000; Campanha *et al.* 2005), but the complete tectonic evolution of the craton is still far from being completely understood.

Magmatism and deformation involved in the different stages of amalgamation correspond to the Brasiliano/Pan-African cycle (650–530 Ma). According to Campos Neto & Figueredo (1995), this comprises an older or classical Brasiliano magmatic event (650–600 Ma) and a younger, or 'Río Doce' event (590–530). The tectonic evolution and assembly envisaged by Alkmim *et al.* (2001) implies an earlier collision between the Río de la Plata and São Francisco cratons at 0.75 Ga. Later, movement between these cratons generated the Dom Feliciano Belt, with subsequent collision and incorporation of minor blocks. The accretion of Pampia marks the last pre-Ordovician collision event in the western margin; other terranes, such as Cuyania and Chilenia, were annexed during Ordovician and Devonian–Carboniferous, representing the final accretionary phase of the western margin of Gondwana (Astini 2003).

The glacial record

Tandilia System

A synthesis of the sedimentology and palaeontology of the Tandilia System (Fig. 3a) has recently been presented by Poiré *et al.* (2003). The

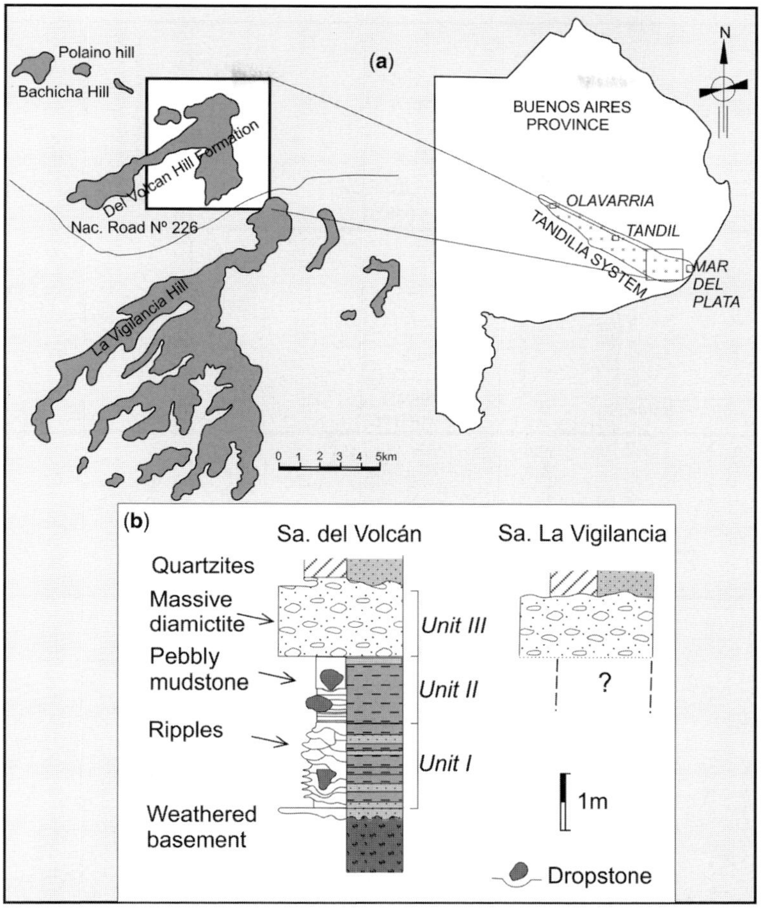

Fig. 3. (**a**) Location of the studied area. (**b**) Logged sections at the Del Volcán and La Vigilancia hills. Taken from Spalletti & del Valle (1984).

sedimentary succession includes limestones, quartzites, shales, abundant stromatolites and banded-iron levels that comprise the lithostratigraphic units of the Sierras Bayas Group. This group is 175 m thick and contains at the base a dolomitic succession (Villa Mónica Formation) attributed to the Cryogenian (700–900 Ma) on the basis of diagenetic evolution and stromatolites (see Gómez Peral *et al.* 2003; Poiré & Spalletti 2005). It is unconformably overlain by other units that have less diagenetic overprint, suggesting a considerable time lapse in the unconformity. The Sierras Bayas Group is unconformably overlain by the Cerro Negro Formation, which includes tide-dominated deposits containing acritarchs and possible *Cloudina* shells that confirm an Ediacaran age (Poiré & Spalletti 2005). However, according to Gaucher *et al.* (2005*b*), the Sierras Bayas Group is Ediacaran based on palynomorphs, presence of *Cloudina* and C and Sr-chemostratigraphy. These authors suggest correlation of the Cerro Largo, Loma Negra and Cerro Negro formations (Fig. 4) with the lower part of the Arroyo del Soldado Group, but also suggest an Ediacaran age for the Villa Mónica Formation based only on the low diversity of acritarchs. Although low diversity could be an artefact of preservation or a response to palaeoenvironmental controls in tidal and marginal marine facies. This new interpretation contradicts previous isotopic ages of 800 Ma for pelites of the Villa Mónica Formation (e.g., Rb–Sr, Cingolani & Bonhomme 1982) and is difficult to reconcile with the observed diagenetic differences and the stromatolite stratigraphy that suggest 800–900 Ma (Poiré 2002). The Cambro-Ordovician Balcarce Formation (Poiré *et al.* 2003; Poiré & Spalletti 2005) (pre-Late Ordovician according to the age of the dykes that cut the unit, Rapela *et al.* 1974)

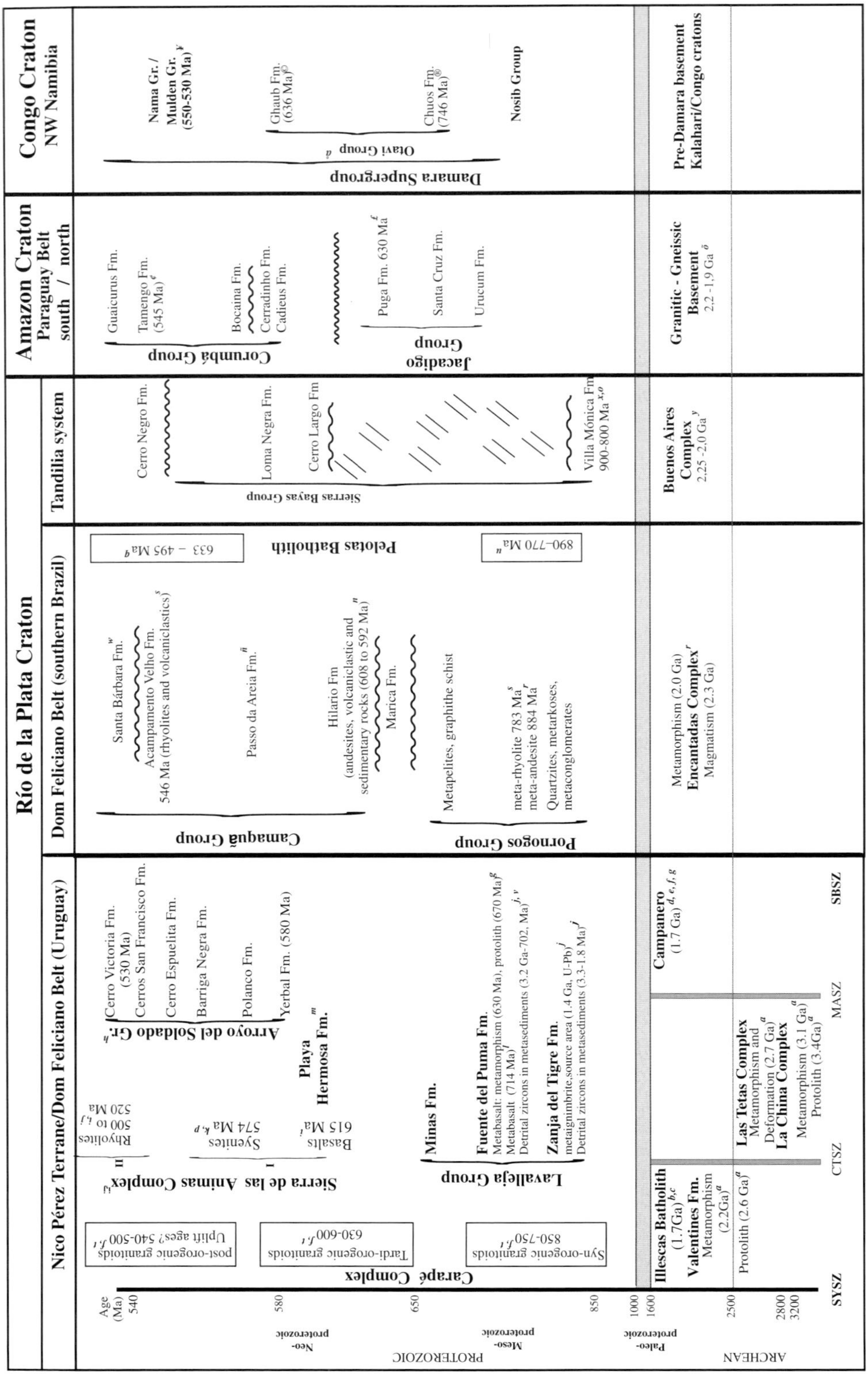
Río de la Plata Craton
Nico Pérez Terrane/Dom Feliciano Belt (Uruguay)
Dom Feliciano Belt (southern Brazil)
Tandilia system
Amazon Craton
Paraguay Belt
south / north
Congo Craton
NW Namibia
Age (Ma)
540
580
650
850
1000
1600
2500
2800
3200
PROTEROZOIC
Neo-proterozoic
Meso-proterozoic
Paleo-proterozoic
ARCHEAN
Carapé Complex
post-orogenic granitoids Uplift ages? 540-500
Tardi-orogenic granitoids 630-600
Syn-orogenic granitoids 850-750
Sierra de las Animas Complex
Rhyolites 500 to 520 Ma
Syenites 574 Ma
Basalts 615 Ma
Arroyo del Soldado Gr.
Cerro Victoria Fm. (530 Ma)
Cerros San Francisco Fm.
Cerro Espuelita Fm.
Barriga Negra Fm.
Polanco Fm.
Yerbal Fm. (580 Ma)
Playa Hermosa Fm.
Lavalleja Group
Minas Fm.
Fuente del Puma Fm.
Metabasalt: metamorphism (630 Ma), protolith (670 Ma)
Metabasalt (714 Ma)
Detrital zircons in metasediments (3.2 Ga-702, Ma)
Zanja del Tigre Fm.
metaignimbrite,source area (1.4 Ga, U-Pb)
Detrital zircons in metasediments (3.3-1.8 Ma)
Illescas Batholith (1.7Ga)
Valentines Fm. Metamorphism (2.2Ga)
Protolith (2.6 Ga)
Campanero (1.7 Ga)
Las Tetas Complex
Metamorphism and Deformation (2.7 Ga)
La China Complex
Metamorphism (3.1 Ga)
Protolith (3.4Ga)
SYSZ
CTSZ
MASZ
SBSZ
Camaquã Group
Santa Bárbara Fm.
Acampamento Velho Fm.
546 Ma (rhyolites and volcaniclastics)
Passo da Areia Fm.
Hilario Fm (andesites, volcaniclastic and sedimentary rocks (608 to 592 Ma)
Marica Fm.
Porongos Group
Metapelites, graphithe schist
meta-rhyolite 783 Ma
meta-andesite 884 Ma
Quartzites, metarkoses, metaconglomerates
Pelotas Batholith
633 – 495 Ma
890–770 Ma
Metamorphism (2.0 Ga)
Encantadas Complex
Magmatism (2.3 Ga)
Sierras Bayas Group
Cerro Negro Fm.
Loma Negra Fm.
Cerro Largo Fm
Villa Mónica Fm
900-800 Ma
Buenos Aires Complex
2,25 -2,0 Ga
Corumbá Group
Guaicurus Fm.
Tamengo Fm. (545 Ma)
Bocaina Fm.
Cerradinho Fm.
Cadieus Fm.
Jacadigo Group
Puga Fm. 630 Ma
Santa Cruz Fm.
Urucum Fm.
Granitic - Gneissic Basement
2,2 -1,9 Ga
Damara Supergroup
Otavi Group
Nama Gr. / Mulden Gr. (550-530 Ma)
Ghaub Fm. (636 Ma)
Chuos Fm. (746 Ma)
Nosib Group
Pre-Damara basement Kalahari/Congo cratons

unconformably overlies the Sierras Bayas Group and other units which do not form part of that group, such as the diamictitic Sierra del Volcán Formation and the subsurface Punta Mogotes Formation. The glacial-related Sierra del Volcán Formation (Fig. 3b) itself unconformably overlies the Palaeoproterozoic to Mesoproterozoic igneous–metamorphic Buenos Aires Complex (Spalletti & del Valle 1984), but by lithostratigraphic correlation it has been considered younger (Ediacaran–Cambrian) than the Sierras Bayas Group; unfortunately, these units are never exposed in the same section, making interpretation uncertain. One of the elements taken into account by Spalletti & del Valle (1984) for suggesting a post-Sierras Bayas Group age included the different grade and type of alteration of the basement compared with the saprolitization that preceded deposition of the Sierras Bayas Group, implying at least two stages of palaeo-weathering prior to deposition of the Sierra del Volcán diamictites. These features developed under different climate conditions preclude any correlation with the basal unit of the group. The other feature mentioned by Spalletti & del Valle (1984) is the presence of unmetamorphosed quartzitic, outsized blocks in the diamictites, seemingly precluding provenance from the Sierras Bayas Group. However, Gómez Peral *et al.* (2007) concluded that the Sierras Bayas Group is unmetamorphosed, so that the exotic blocks could indeed be sourced from the quartzites, as in the new interpretation of Poiré & Spalletti (2005). The Sierras Bayas Group includes tidal and wave deposits, karstic surfaces and internal unconformities, evidencing shallow marine environmental deposition through several sedimentary cycles (see Andreis *et al.* 1992). The Arroyo del Soldado Group is a succession more than 3000 m thick that, according to Gaucher *et al.* (2003), spanned 580–530 Ma based on palynomorphs, skeletal fossils and chemostratigraphy. The passive margin basin suggested by Gaucher *et al.* (2003) seems incompatible with the rate of accommodation space typical of passive margins and with the immature 1500 m thick conglomerates (Barriga Negra Formation) interpreted by these authors as deposits related to a drop in sea level. Generation of accommodation space and deposition of thick coarse-grained rocks is more compatible with a tectonic control during sedimentation.

Nico Pérez Terrane – Dom Feliciano Belt

The Nico Pérez terrane (see Bossi & Campal 1992) is located between the Sarandí del Yí–Solís and María Albina shear zone (Fig. 2) and forms part of the Río de la Plata Craton (Brito Neves & Alkmim 1993); traditionally, the Sierra Ballena shear zone was its the eastern limit (Bossi & Campal 1992, among others). Hartmann *et al.* (2001) have separated the Nico Pérez terrane into lithotectonic units (Arroyo del Soldado and Lavalleja groups, Valentines, La China and Las Tetas complexes). The age of the units is not accurately constrained (see Bossi & Gaucher 2004) but the Arroyo del Soldado Group is stratigraphically assigned to 580–530 Ma (Gaucher *et al.* 2003; 2005*b*) and correlated with the postglacial (post-580 Ma) units of the Corumbã Basin (Nogueira *et al.* 2003).

The Dom Feliciano Belt (Fragoso Cesar 1980), named by Preciozzi *et al.* (1999) the Cuchilla de Dionisio Belt, is represented in Uruguay by sedimentary and volcanic rocks metamorphosed and deformed during the Brasiliano orogenic cycle (*sensu* Almeida *et al.* 1973) and also by coeval igneous rocks (Figs 2 & 4). The Campanero Unit (Sánchez Bettucci 1998; Sánchez Bettucci *et al.* 2003*a*) is the Palaeoproterozoic basement to Neoproterozoic supracrustal rocks (Lavalleja Group). This unit is represented by granitoids with variable deformation grade. The U–Pb ages yield a 1735 +32/−17 Ma upper intercept and a 723 +240/−210 Ma lower intercept (Sánchez Bettucci *et al.* 2003*b*); the latter suggesting deformation at high temperature. Mallman *et al.* (2003) obtained a similar U–Pb age to the upper intercept (1754 ± 6.8 Ma) using the Sensitive High Resolution Ion Micro-Probe (SHRIMP) method. The Brasiliano magmatism is recorded in the Carapé Complex as three major tectono-magmatic events at 850–750 Ma, *c.* 600 and 540–500 Ma (Sánchez Bettucci *et al.* 2003*a*). This granitic complex is

Fig. 4. Distribution of the main lithostratigraphic units. References: a, Hartman *et al.* (2001); b, Bossi *et al.* (1998); c, Heaman *in* Campal & Schipilov (1995); d, Mallmann *et al.* (2003); e, Preciozzi *et al.* (2005); f, Sánchez Bettucci *et al.* (2003*b*); g, Sánchez Bettucci *et al.* (2003*a*); h, Gaucher *et al.* (2005*b*); i, Sánchez Bettucci & Linares (1996); j, Oyhantçabal *et al.* (2005*a*); k, Oyhantçabal *et al.* (2007); l, Sánchez Bettucci & Ramos (1999); m, Pazos *et al.* (2003); n, Sommer *et al.* (2006); ñ, Eerola 1995; o, Cingolani & Bonhomme (1982); p, Oyhantçabal *et al.* (2005*b*); q, Philipp *et al.* (2002); r, Hartman *et al.* (2000); s, Chemale (2000); t, Sánchez Bettucci (1998); u, Fragoso César (1991); v, Basei *et al.* (2006); w, Borba & Mizusaki (2002); x, Gómez Peral *et al.* (2003); y, Cingolani *et al.* (2002); õ, Colombo *et al.* (1999); £, Alvarenga & Trompette (1992); ¥, Wood *et al.* (2002); ©, Hoffmann *et al.* (2004); ®, Hoffmann *et al.* (1996); ¢, Boggiani *et al.* (2005). SYSZ, Sarandí del Yí shear zone; CTSZ, Cueva del Tigre shear zone; MASZ, María Albina shear zone; SBSZ, Sierra Ballena shear zone.

correlated with the Pelotas batholith of southern Brazil (Fig. 4).

The Lavalleja Group displays a complex stratigraphic organization (Bossi & Navarro 1991; Sánchez Bettucci 1998; Sánchez Bettucci & Ramos 1999; Sánchez Bettucci *et al.* 2001). This group is represented by a metamorphic association of very low to medium grade, occurring in narrow bands of volcanic and sedimentary rocks (Ryoke–Abukuma type) and it was separated (Sánchez Bettucci 1998; Sánchez-Bettucci & Ramos 1999) into three formations: Zanja del Tigre, Fuente del Puma and Minas. This group is affected by folding, thrusting, and transcurrent faults. Regardless of the tectonic setting, the Lavalleja Group was affected by metamorphism and deformation during the Brasiliano orogenic event. The two lower formations attained middle amphibolite facies to greenschist tacies conditions, whereas the upper formation only reached greenschist facies to anchimetamorphism. The Lavalleja Group was correlated with other tectono-stratigraphic units of southern Brazil (Fig. 4), such as the Porongos Group (Fragoso Cesar *et al.* 1982*a*, *b*, 1987, 1994; Fragoso Cesar & Machado 1997; Trompette 1994; Sánchez Bettucci *et al.* 2001; Saalmann *et al.* 2006; among others). The Lavalleja Group would be equivalent to the Brusque Complex in Santa Catarina (Jost & Bitencourt 1980; Almeida *et al.* 2000). The Porongos Group and Brusque Complex together form the Tijucas Belt.

The Zanja del Tigre Formation (Fig. 5) is in tectonic contact with the Campanero Unit (basement) and the Fuente del Puma Formation. It is constituted by quartzites with muscovite and/or fuchsite, andalusite, biotite and muscovite schists, marbles, meta-gabbros, carbonates, banded iron formation and amphibolites. Oyhantçabal *et al.* (2005*b*) present U–Pb ages (both ID-TIMS and SHRIMP) ranging from 1.42 to 1.49 Ga for re-worked meta-ignimbrites and metasedimentary rocks, which is indicative of a Mesoproterozoic source area.

The Fuente del Puma Formation (Fig. 5) appears in tectonic contact with the underlying unit and is overlain by the Minas Formation. It is affected by lower (chlorite zone) to upper (garnet zone) greenschist facies metamorphism. This formation is represented by a volcano-sedimentary sequence and was separated by Sánchez Bettucci *et al.* (2001) into three members: volcanic (basic and acidic), sedimentary (carbonates, siliciclastic and volcaniclastic rocks), and igneous (gabbros). U–Pb data for rutile from metabasalts of this formation yield ages of *c.* 670 Ma (crystallization) and 643 Ma (metamorphism; Sánchez Bettucci *et al.* 2003*b*, 2004). U–Pb SHRIMP analyses of detrital zircons from metasediments has yielded ages between 3197 and 702 Ma (Preciozzi *et al.* 2005), the youngest confirming a Neoproterozoic source.

The Minas Formation is exposed in the neighbourhood of the city of Minas (Fig. 5) and comprises metasedimentary rocks with very low grade metamorphism and deformation. The most representative lithologies correspond to grey calcareous siltstones and dolomites with pelites, quartzites and brown-coloured carbonates with psammitic intercalations (Sánchez Bettucci *et al.* 2001).

The Brasiliano magmatism in this area is represented by intrusive rocks emplaced in the Campanero basement and the Lavalleja Group. It is characterized by several late- to post-orogenic metaluminous to peraluminous granitic suites named the Carapé Granitic Complex by Sánchez Bettucci *et al.* (2003), and can be correlated with the Dom Feliciano granitic suite (Fragoso Cesar 1980, Figueiredo *et al.* 1990; Gastal *et al.* 2005, among others).

The Sierra de Las Animas Complex is constituted by bimodal magmatism representing two events (Sánchez Bettucci & Rapalini 2002), the first one with Neoproterozoic ages of *c.* 615–570 Ma (Sánchez Bettucci & Linares 1996; Oyhantçabal *et al.* 2007), and the second one ranging from 520 to 500 Ma (Bossi *et al.* 1993; Sánchez Bettucci & Linares 1996; Sánchez Bettucci 1998). This post-orogenic magmatism was a consequence of an extensional relaxation episode related to the Brasiliano and Rio Doce orogenic cycles. The volcanic, sub-volcanic and plutonic Sierra de Las Animas Complex is composed of basalts, trachytes, syenites, rhyolites and volcanic breccias with alkaline and subalkaline affinities, interstratified with sedimentary deposits (Sánchez Bettucci 1997).

A sedimentary succession with a major volcano-sedimentary composition at the top constitutes the Playa Hermosa Formation, exposed in the southwestern extreme of the Nico Pérez terrane, on the coast of the Río de la Plata estuary (Sánchez Bettucci & Pazos 1996; Pazos *et al.* 1998). This unit is affected by tectonic tilt, sometimes related to magmatic intrusions, but internal deformation or metamorphism is generally absent. Although precise dates are lacking, the lower member of Playa Hermosa Formation is coetaneous with the first bimodal volcanic effusions of the Sierra de Las Animas Complex (Loureiro *et al.* 2006), suggesting a Neoproterozoic age. The presence of peperites and vesicular basalts suggests shallow water deposition. The absence of regional metamorphism in Playa Hermosa Formation is consistent with a younger age than the Lavalleja Group (Pazos *et al.* 2003).

Sánchez Bettucci & Rapalini (2002) obtained a preliminary virtual geomagnetic pole (VGP) for

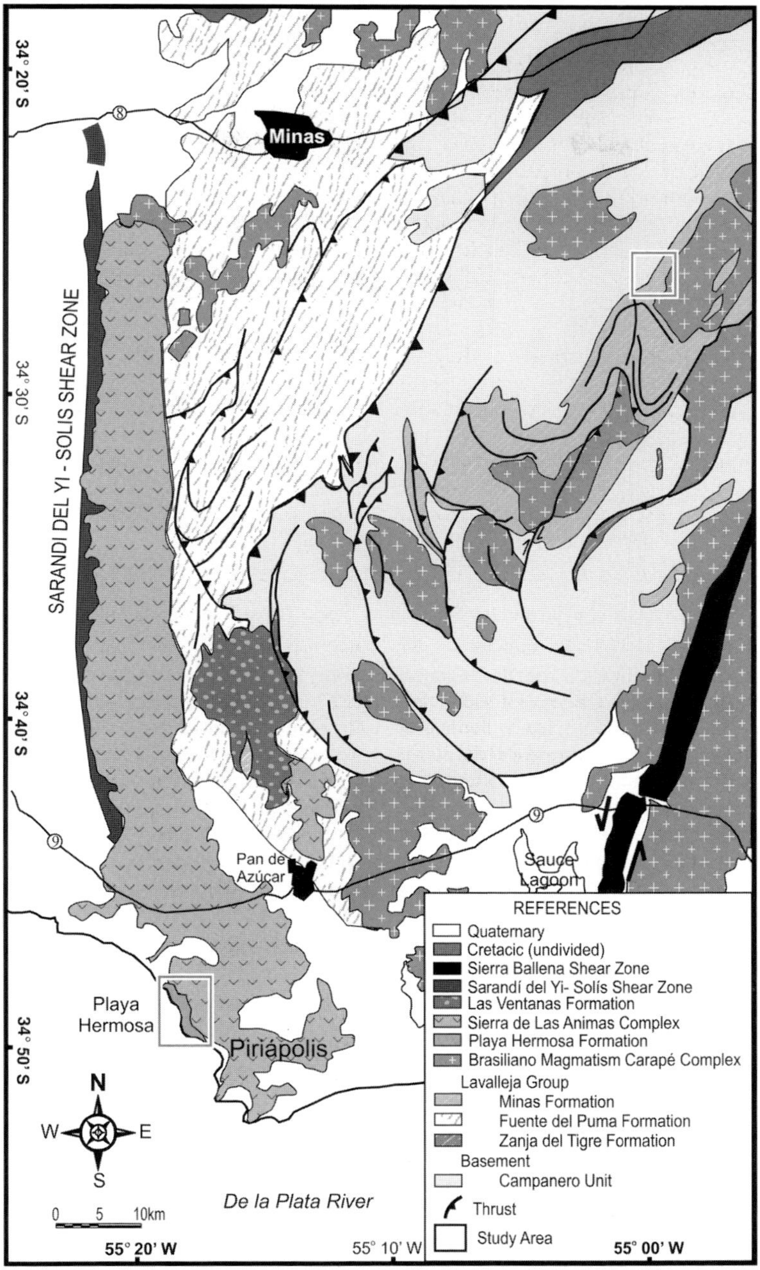

Fig. 5. Geology of the southern area of Uruguay with the location of the glacial sections.

c. 600 Ma from the lower member of Playa Hermosa Formation that agrees with the Campo Alegre (Tohver *et al.* 2006) pole of similar age. The result suggests that the glacially influenced deposits could have been produced at low to intermediate latitudes.

São Gabriel block

The São Gabriel block is a terrane located in southern Brazil, close to the Brazil–Uruguay boundary (Fig. 6), immediately to the west of the Dom Feliciano Belt and to the east of the Río de

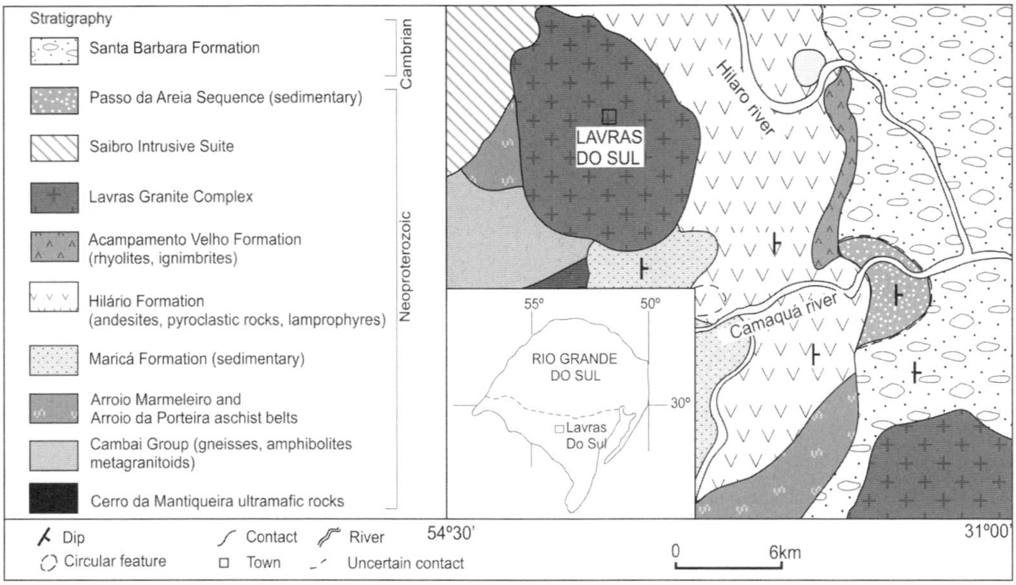

Fig. 6. Geology of the Lavras do Sul area in the São Gabriel block. Taken from Eerola (2001).

la Plata Craton. The geology of the São Gabriel block has been analysed by Chemale (2000), but new Sm–Nd isotopic data from volcano-sedimentary units evidence juvenile Neoproterozoic oceanic crust to the east of the Río de la Plata Craton (Saalmann *et al.* 2005) and give more information about the NE border of the craton. The basement is composed of gneisses of Cryogenian age (750–700 Ma, U–Pb), covered by a metamorphosed volcano-sedimentary succession. The undeformed succession that unconformably overlies the volcano-sedimentary sequence comprises the units of the Camaquã Basin, intruded by late tectonic granites of 595 Ma (see Saalmann *et al.* 2005). The suspected glacial deposits in the São Gabriel block were mentioned for the first time by Carvalho & Pinto (1938). In the area of Lavras do Sul (Fig. 6), Eerola (1995, 2001) suggested glacial influence in deposits belonging to the Camaquã Basin. The stratigraphy of this basin was extensively described by Paim *et al.* (1995, 2000) who defined five allogroups: I Maricá (620–600 Ma), II Bom Jardim (592–580 Ma) III Cerro do Bugio (573–560 Ma), IV Santa Barbara (559–540 Ma) and V Guaritas (470 Ma). Eerola (1995) named a succession with glacial influence as the Passo da Areia sequence and mentioned that it is covered by the 'Santa Barbara units'. For this reason, the glacially influenced facies should have been deposited contemporaneously with alkaline volcanism of the Cerro do Bugio allogroup. Sommer *et al.* (2006) suggest that the shoshonitic volcanism (Hilário Formation, Bom Jardim Group) represents the oldest (608–592 Ma), whereas the magmatic unit in the Camaquã Basin and the bimodal magmatism of Acampamento Velho (Cerro do Bugio Group) and Campo Alegre yield ages between 602 and 549 Ma. It is possible to correlate the younger (Acampamento Velho) volcanism with the Sierra de Las Animas Complex (II event).

Palaeoclimatic evidence

Tandilia System

The Sierra del Volcán Formation is a very thin unit (Fig. 3b), less than 8 m thick (Spalletti & del Valle 1984). The succession overlies a weathered (kaolinite-rich) basement and begins with rhythmic, fine-grained sandstones and pelites containing dropstones of varied size and composition (unit 1) exhibiting roughing and bending structures (e.g., Thomas & Connell 1985) and syn-sedimentary folding (Spalletti & del Valle 1984). Outsized clasts (Fig. 3b) present different sizes (mainly pebbles) and roundness; they are usually faceted with glacially related bullet shapes. Compositionally they include quartz, quartzites, migmatites and other igneous–metamorphic lithologies, sometimes with similar weathering grade to the basement, providing good evidence of pre-depositional alteration of the basement (Spalletti & del Valle 1984). Sandstones and carbonates are absent. The host rocks show undulatory lamination indicative of incipient ripple lamination. Pelites are the result of settling from a suspension in a shallow marine environment (Spalletti & del Valle 1984). The succession continues with 1.20 m of fine-grained siltstones and claystones containing drop-stones (unit 2) with a dominance of quartzite fragments. This level passes upward to pebbly sandstone (unit 3), and presents normal grading and faint parallel stratification to the top. The Balcarce Formation is deposited over the Sierra del Volcán Formation with sharp angular unconformity.

The diamictites were revisited by van Staden *et al.* (2005) to analyse the provenance and depositional framework of the 'tillites'. More than 15 samples distributed in the intervals previously described by Spalletti & del Valle (1984) were analysed geochemically. They also studied grain surfaces using electron microscope techniques, yielding evidence of transport under glacial conditions (e.g., striations, conchoidal breakage on larger grains, flat cleavage fractures and faceted grains). The surface studies indicate subaqueous transport and the large number of impact craters indicates a highly energetic beach environment (van Staden *et al.* 2005). Geochemically, major and trace elements analysis did not help to decipher the grade of alteration of samples, Nb/Y and Zr/Ti ratios being homogeneous throughout the entire 'tillite'. Rare earth element data indicate an upper crustal composition and some samples revealed differences in the light rare earth element (LREE) concentrations. Overall the results allowed van Staden *et al.* (2005) to suggest an evolution from oxic to anoxic conditions from base to top. Spalletti & del Valle (1984) pointed out variable domains of kaolinite or illite in the analysed samples, but the authors emphasized the dominance of kaolinite in the altered basement. The geochemical composition suggests a source area composed of both igneous-metamorphic basement and sedimentary units (van Staden *et al.* 2005). More recently Zimmermann (*pers. comm.*), having in mind that kaolinite is the most abundant clay present in the diamictite matrix (see Spalletti & del Valle 1984) and is also extremely abundant in the overlying Balcarce Formation (Zimmermann & Spalletti 2005), speculated that glacial deposits of the Sierra del Volcán Formation could be Ordovician, like the overlying Balcarce Formation. This would imply that the glacial record is late Ordovician (Ashgillian?). This speculation is not followed in this paper since the angular unconformity necessarily places the Balcarce Formation in the Silurian, although radiometric ages support an Ordovician age (e.g., Rapela *et al.* 1974, 1998, 2007). The kaolinitization of the basement preceded deposition of the Sierra del Volcán Formation and explains the abundance of kaolinite in the diamictites. Interestingly, van Staden *et al.* (2005) did not discuss the angular unconformity or comment about the geometrical disposition, which could be the result of pinching out of facies or local tilting. However, a Precambrian age for the glacial deposits was suggested. The contrasting change from anoxic conditions (upper diamictite) to tidal dominated deposits, well oxygenated substrates, compositionally mature sandstones, and a *Cruziana*-type ichnofauna that characterizes the Balcarce Formation (see Poiré & Spalletti 2005) confirm the contrasting depositional setting between these units and confirm a sequence boundary between them. In our view, geological evidence (angular unconformity) supports the traditional differentiation of two independent stratigraphic units and favours the previously suggested Precambrian age for the Sierra del Volcán Formation (e.g., Spalletti & del Valle 1984; Poiré & Spalletti 2005; van Staden *et al.* 2005). The stratigraphic relationship of the unit with respect to the Sierras Bayas Group is still dubious, but a latest Neoprotoerozoic age seems probable. The glacial origin is not questionable and the succession represents the first confirmed glacial deposits in southern South America during the Precambrian. Sedimentologically, it is necessary to point out that the term 'tillite' used by van Staden *et al.* (2005) is inappropriate, for the following reasons: the section includes three units rather than a single 'tillite'; the lower and middle ones contains evidence of rain-out and can be termed rain-out diamictites rather than tillite; the host deposits are heterolithic rhythmites deposited by weak currents (ripples) and suspension and fall-out, and were not directly deposited from a glacier. The upper unit is regarded as a 'diamictite' rather than a 'tillite' following Spalletti & del Valle (1984) because it does not contain evidence of glacio-tectonic deformation or erosive features indicative of glacial erosion and lodgement deposition, but also because normal grading and faint lamination to the top suggest a cohesion-less debris flow deposited in a probably unstable depositional setting below wave base level.

Dom Feliciano Belt

(a) Playa Hermosa Formation. A complete analysis of the sedimentary facies of the Playa Hermosa Formation is still lacking. However, the lower section has been particularly targeted in investigations regarding the possible glacial origin of the succession. The lower section is approximately 50 m thick (Fig. 7) and comprises two facies associations according to Pazos *et al.* (2003). Facies Association I (Fig. 7) includes breccias, conglomerates, sandstones and mudstones, and records subaqueous deposition on unstable slopes adjacent to glacial

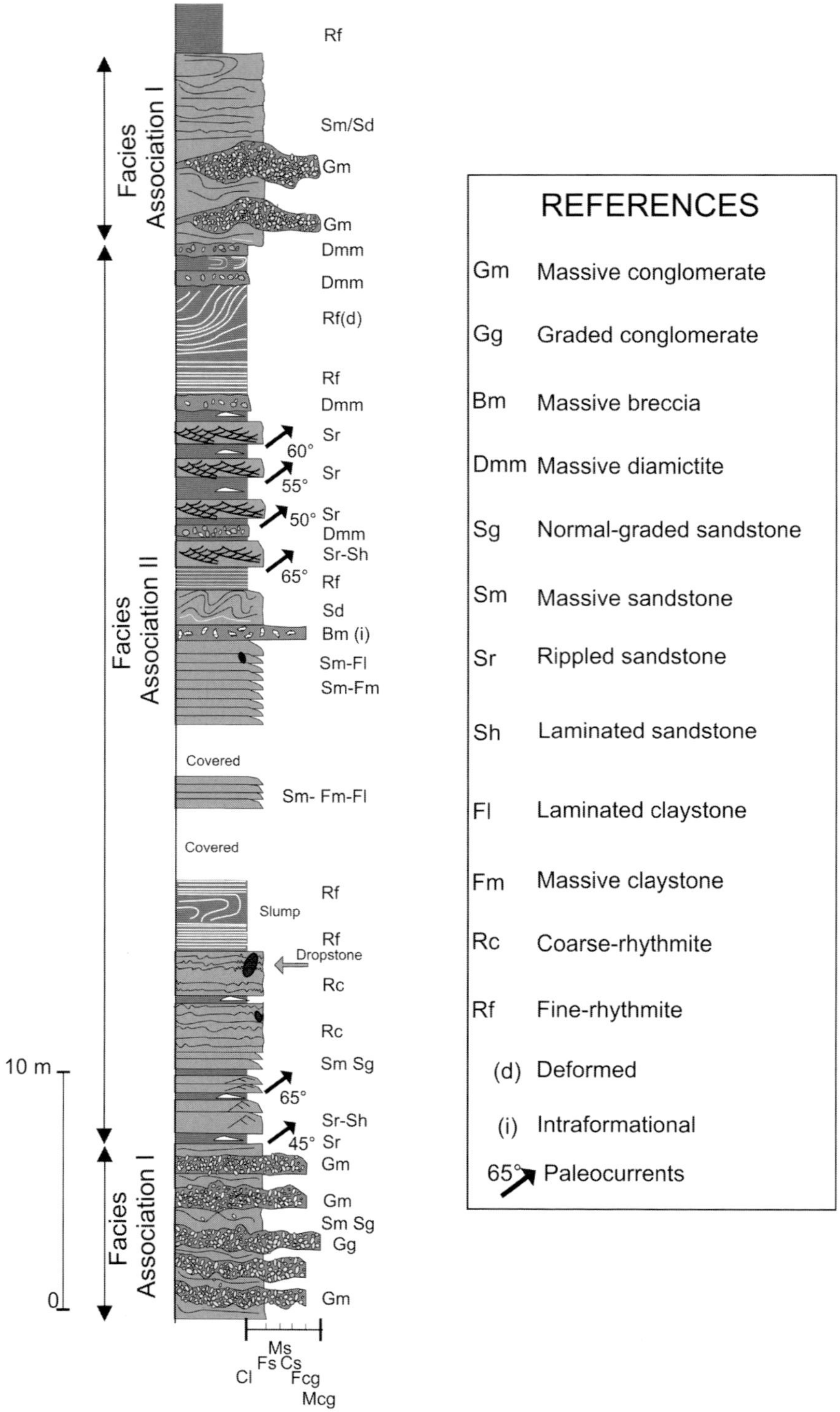

Fig. 7. Logged section in the lower part of the Playa Hermosa Formation, modified from Pazos *et al.* (2003).

centres and probable active faults (Pazos *et al.* 2003). The combined features of interbedded breccias, conglomerates, sandstones and minor mudstones indicate a wide availability of rock detritus but the intense soft-sediment deformation and complex geometry of psefitic deposits suggest synsedimentary remobilization, probably due to high sedimentation rate or seismically induced slumping. Facies Association II includes diamictites, sandstones and mudstones. This facies association presents clear climate indicators of at least seasonal freezing of the body water and melting of icebergs. The diamictites represent tabular deposits containing angular clasts of different sizes, composed of granites, quartz, quartzites, feldspar and sometimes include deformed sedimentary clasts of rhythmite-type (Fig. 8a). Clasts may be randomly distributed or form patches. These deposits do not represent 'tillites' according to Pazos *et al.* (2003, p. 69) but should be regarded as good examples of resedimented till, deposited like gravity flows with a variable grade of homogenization in a pro-glacial and unstable depositional setting. Sandstones are massive, graded or, more commonly, contain drift ripple-cross lamination, show strongly asymmetrical sections, were affected by down-slope deformation prior to lithification, and indicate palaeo-flow and palaeo-slopes to the NE (Figs 7 & 8b). Some ripples present high climbing angles indicative of rapid aggradation. These ripples were interpreted as 'micro-hummocky' by Fambrini *et al.* (2003) and used as strong evidence of a marine depositional setting. The storm origin is incorrect and any inference about the marine or lacustrine depositional setting may be based on drift-cross lamination. However, the fine-grained section of the Las Ventanas Formation, which may be correlated with the fine-grained section of the Playa Hermosa Formation (Pazos *et al.* 2005), contains palynomorphs and indicates marine influence at least for the fine-grained intervals (see Blanco & Gaucher 2004).

The clearest evidence of seasonal freezing or rain-out processes from icebergs is a large quartzite dropstone (block size), with bullet shape and polished surface, that disrupts the underlying deposits (Fig. 8c). No clast in the host deposits is larger than a small pebble. Moreover, the host deposits constitute crudely stratified rhythmites with a thickness of less than 5 cm. The impact structures are coincident with the examples typified by Thomas & Connell (1995). Rhythmites, resembling varves, (Fig. 8d) also occur stratigraphically higher and contain diamictite intervals with isolated blocks, fine-grained sandstones with granite clasts and soft-sediment deformation structures. These deposits are intercalated with pelites, starved rippled deposits and fine-grained turbidites (Fig. 5d in Pazos *et al.* 2003). The association indicates a more distal and possible deeper depositional setting than Facies Association I.

Evidence from the analysed section is not conclusive about a direct glacial origin but the combination of diamictites, rhythmites with varve-like deposits and dropstones is a strong evidence to support glacially influenced deposition (Pazos *et al.* 2003). The palaeocurrent pattern indicates a palaeo-high area situated to the south-west, possibly in the vicinity of the Tandilia System.

(b) Zanja del Tigre Formation. The metamorphosed and deformed limestones of this unit include a stratigraphically thin (2 m) interval containing outsized clasts (Pazos *et al.* 2005) that vary from pebbles to blocks and include gabbros, quartz-quartzites (Fig. 8e) and granites (Fig. 8f). They are isolated, disrupting the lamination in different form to that resulting from tectonic deformation (rotational deformation). Some are very similar to the examples illustrated by Condon *et al.* (2002) in the compilation of dropstones intervals from Neoproterozoic successions around the world.

São Gabriel Block

Passo da Areia sequence. Eerola (1995) described and illustrated a sedimentary succession that presents features that may be ascribed as 'glacially influenced' deposits: outsized clasts isolated in fine-grained, rhythmically laminated layers (Figs 9 and 10 in Eerola 1995,), diamictites with clasts (pebbly sandstones and mudstones), and rhythmically laminated shales. Eerola (1995) also mentions other areas of suspect glacial deposits but, unfortunately, did not mention the composition of the outsized blocks. The local abundance of volcanic and pyroclastic deposits introduces more uncertainty about the origin of this sequence.

Discussion

The differentiation between tillites and diamictites of non-glacial origin was emphasized by Eyles & Januszczak (2004). However, in tectonically active basins or in topographically irregular depositional scenarios the most abundant deposits are the result of gravity flows and reworking that converted them into diamictites. This is not to negate a primary glacial origin. Tectonism was active during deposition of some of the units discussed in this paper, related to the Brasiliano cycle, and indicates that extension, probably related to strike-slip basins, was prominent in the eastern part of the Río de la Plata Craton. Volcanism is documented in the tectonically active areas of the craton (Eerola 2001; Pazos *et al.* 2003; Saalmann *et al.*

Fig. 8. Sedimentary features of Playa Hermosa Formation: (**a**) diamictites with mudstone intra-clasts, (**b**) rippled-sandstones with soft-deformation overprinted (taken from Pazos *et al.* 2003), (**c**) dropstone deforming coarse rhythmites and detail of the top view, (**d**) varve-like rhythmites (taken from Pazos *et al.* 2003). Sedimentary features of the Zanja del Tigre Formation: (**e**) quartzite clasts of IRD (ice rafted diamictites) origin, (**f**) granite clast immersed in carbonates.

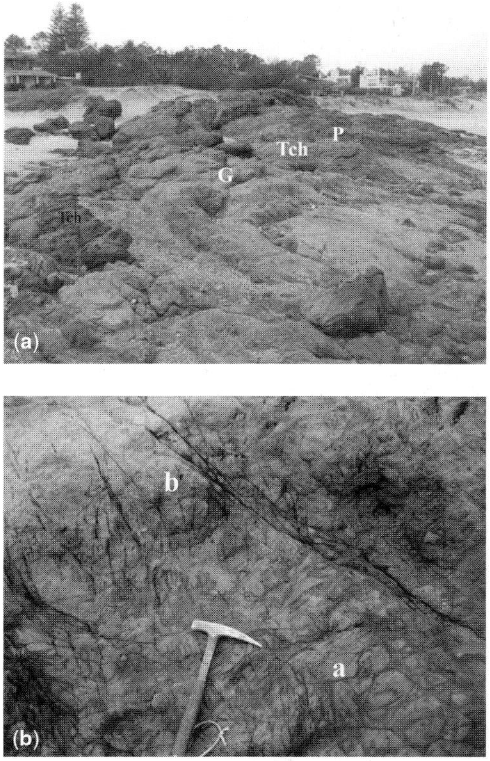

Fig. 9. Interaction between magma and wet unconsolidated sediments (peperites) at Playa Hermosa beach. (**a**) Outcrop view of magma–sediment interaction; the bedding has been partially destroyed (Tch, trachyte; P, pelite; G, conglomerate). (**b**) Fluidal peperitic texture related to trachyte. The host sediment is pelite. a, Irregular (amoeboid) contact; b, juvenile clast is trachyte.

2005) and is also coetaneous (Fig. 9a, b) with the Playa Hermosa Formation (Loureiro *et al.* 2006), suggesting that seismic shock during high-rate debris production, a common feature in glacial depositional settings, may be the trigger of slumping and folding in the Playa Hermosa Formation (see Pazos *et al.* 2003). However re-sedimentation and deformation controlled by high sedimentation rates in a retreating glacial environment is here suggested, taking into account that the fine-grained deposits contain intra-basin clasts and that soft-deformation and scarce entombed angular granite pebbles suggest rain-out processes. The palaeogeographic and palaeotopographic framework is not well known, but in an active tectonic basin, relatively high topography and glaciers advancing from mountains into the coast (fjords?), as in southern Chile, are not ruled out. Such cases permit explanation of glacial deposits under conditions that are not severe and also not in high-latitude locations. Recently, Gaucher *et al.* (2005*b*) reported palynomorphs from the fine-grained intervals of the Las Ventanas Formation that forms part of the same depositional cycle as the Playa Hermosa Formation (Blanco & Gaucher 2004; Pazos *et al.* 2005), evidencing marine influence. The combination of psefites, diamictites, rhythmites, slumping and dropstone-levels is similar to glacial–postglacial transitions in Carboniferous successions of Argentina where a fjord environment has been suggested (e.g., Kneller *et al.* 2004; Pazos *et al.* 2007). Such environments are characterized by high rates of sedimentations, rapid facies shifting and high discharge from melt-water (e.g., Hansen 2004). This could equally explain the absence of marine palynomorphs in the fine-grained intervals of the type locality of the Playa Hermosa Formation compared with the fertile Las Ventanas Formation. Thus the absence of fossils may be explained by palaeo-ecological and taphonomical controls and is a weak criterion to support the conclusion of Gaucher *et al.* (2005*b*) of different ages for the two units. Gaucher *et al.* (2003, 2005*b*) suggested than the relatively diverse association of palynomorphs present in the platform limestones of the Arroyo del Soldado Group and now in the siliciclastic rocks of the Las Ventanas Formation, is strong evidence of the inconsistency with the Snowball Earth hypothesis, which assumes cut-off of marine life. Interestingly, Pazos *et al.* (2003) pointed out that the Playa Hermosa Formation marks an earlier stage of the basin infill, previous to the carbonate platform deposition superbly recorded in the Arroyo del Soldado Group after glaciation disappeared (Pazos *et al.* 2003). Corsetti *et al.* (2006) compared the biota contained in Neoproterozoic glacial deposits with frozen sea areas of Antarctica and concluded that most of the forms that appeared previous to the glaciation are not affected by climate deterioration, and the terminal Proterozoic diversification in the biota is not connected with the glacial aftermath because it occurred million of years after the last glacial vestige disappeared.

The dropstone levels documented in the metamorphosed Zanja del Tigre Formation may be regarded as distal glacially influenced deposits or the result of a Heinrich-type event. The Heinrich layers occur at times of relative cold within the glacial period, related to increased iceberg production (Heinrich 1988). There are numerous glacially related successions that comprise distal glaciomarine deposits where only rain-out debris marks the glacial event (see Condon *et al.* 2002). The distribution of icebergs is not random and marine currents and winds control their trajectory, and the size and quantity of the released debris during melting is dependant on such control (Prins *et al.* 2002).

West Gondwana correlations

The glacially influenced deposits in the centre of the Río de la Plata Craton (Tandilia System) are confirmed by macroscopic and microscopic evidence. The stratigraphic age is not well constrained and absolute dating is lacking. However, differences in the grade of diagenesis, higher in the oldest unit of the Sierras Bayas Group (the Villa Mónica Formation) and lower in the glacially related Sierra del Volcán Formation and the rest of the Sierras Bayas Group, suggest a younger age for the latter. As mentioned, the age of the Villa Monica Formation is matter of debate but it was traditionally considered to be pre-Ediacaran (700–800 Ma) until the proposal of an Ediacaran age from palynomorph content (Gaucher *et al.* 2005*a*). The gap involved in the unconformity between Villa Mónica and Cerro Largo formations may be of more than 200 Ma in the traditional stratigraphic interpretation, or less, if the Gaucher *et al.* (2005*a*) proposal is accepted. Thus the glacial deposits of the Sierra del Volcán Formation could be correlated with any of the widely accepted glacial events. However, clasts composition in the diamictites suggest provenance from unmetamorphosed quartzites (Spalletti & del Valle 1984), possibly a quartzite interval in the Cerro Largo Formation (see Poiré & Spalletti 2005), and in consequence the age of the glacial deposits would need to be Ediacaran. The absence of carbonates clasts in the diamictites may suggest that they are younger than the glacial deposits. This scenario is congruent with the stratigraphic scheme of the Corumbã Basin, and strongly suggests a correlation between both glacial sections. Poiré & Spalletti (2005, Fig. 2) suggested that the diamictites of the Sierra del Volcán are Cambrian like the Cerro Negro Formation, but without any explanation for this stratigraphic position. Similarly, the glacially influenced deposits of the Playa Hermosa Formation were originally correlated with the Varanger glaciation by Pazos *et al.* (2003) based on preliminary and not very robust palaeomagnetic data that exhibit a polar wonder path congruent with the *c.* 600 Ma palaeo-pole (see Sánchez Bettucci & Rapalini 2002). Interestingly, volcanic intrusions with micro-syenites confirm syn-sedimentary volcanism in the lower part of the succession. This effusive event was related to the early stages of the Sierra de Animas Complex (Sánchez Bettucci *et al.* 2006). The syenites in the area were dated at 579 ± 1 Ma (Ar–Ar) by Oyhantçabal *et al.* (2007). In consequence, the glacially related deposits are definitively not Marinoan, but more probably Gaskiers or even younger. In the Paraguay fold belt, glacial deposits of the Puga Formation have been correlated with the Marinoan glacial event taking into account a 627 Ma (Pb–Pb) age for postglacial carbonates (Babinsky *et al.* 2006), but other diamictite intervals interpreted as glacial in origin were described in the Serra Azul Formation (Figueiredo *et al.* 2006) and correlated with the Gaskiers glaciation. The absence of cap-carbonates seems to be a common feature in Gaskiers-age deposits, while Marinoan or Sturtian deposits usually have them (see Halverson *et al.* 2005). The glacial influence suggested for the Zanja del Tigre Formation by Pazos *et al.* (2005), where banded iron formation deposits are common, may be connected with the earlier postglacial stages of the Marinoan glaciation and can be partially correlated with the Puga Formation in Brazil and the Ghaub Formation in Namibia. In this context, and accepting the new ideas that suggest that Punta del Este terrane is a part of the Kalahari Craton, the dropstone-rich interval could be connected distally with glacial centres located in Namibia. The Marinoan glacial record in Namibia exhibits sedimentation in a slope and outer platform environment (Domack & Hoffman 2003). Thus in Uruguay this event would be a very distal glaciomarine ice-rafted interval, during periods of extensive iceberg production, similar to the Heinrich events in the Quaternary (see Hesse *et al.* 2004).

Surprisingly, the Gaskiers glaciation is probably more widespread in southern South America than previously imagined. This event has not been described in Africa and is almost absent in West Gondwana with the exception of Tasmania, where Calver *et al.* (2004) suggested a Gaskiers record for diamictites. Assumed isochronism and lithostratigraphic correlations for the Sturtian glaciation have been challenged recently in Australia by Kendall *et al.* (2006), suggesting a long term glaciation encompassing the classical Sturtian and Marinoan records, with multistage advances and retreats as in Phanerozoic counterparts. These authors also questioned a glacial origin for the Gaskiers-age glacial deposits analysed by Calver *et al.* (2004). The idea of diachronism and multiple advances and retreats is a very important and indicates that Neoproterozoic glaciations were severe but comparable with other well-understood Palaeozoic and Cenozoic glaciations. Diachronism is widely accepted in the Late Palaeozoic glaciation where at least three stages between Lower Carboniferous and Lower Permian have been defined (see Isbell *et al.* 2003). For instance, glacial centres were located in western South America in the Lower Carboniferous but migrated to the east and Africa in the Late Carboniferous (see Pazos 2002). This scenario may explain the absence of Gaskiers-age glacial deposits in Africa, taking into account that the glacial centres were in southern South America. However, sea level drops described by Saylor (2003) in Late Ediacaran successions in Namibia may be connected with a glacio-eustatic control.

Conclusions

The salient conclusions of this paper include stratigraphical, sedimentological and palaeogeographical aspects.

(1) Glacially influenced deposits of Precambrian age are confirmed in the Río de la Plata Craton and most probably represent a latest Neoproterozoic record. It suggests that Gaskiers-age glacial deposits were more widespread in the craton than previously envisaged.

(2) A tectono-sedimentary framework indicates tectonic activity (mainly extensional) and magmatism during coetaneous glacial activity in the eastern border of the craton and stability with limited accommodation space around the centre (Tandilia System).

(3) An older (Marinoan-age?) distal glaciomarine record is documented in the Zanja del Tigre Formation and may be correlative with the glacial deposits of the Ghaub Formation and equivalents in Africa. These predate the collision and amalgamation of the Kalahari (Punta del Este terrane) and Río de la Plata cratons.

(4) In the Río de la Plata Craton there is no evidence of direct or primary glacial deposition and most of the record is composed of diamictites deposited in an unstable setting with rain-out intervals. However, a glacially related origin is undisputed in the Sierra del Volcán diamictites. The Playa Hermosa Formation exhibits glacial influence under active tectonism, while glacial influence in the Passo da Areia section must be a matter of debate until new detailed sedimentological information provides information on the composition of the outsized clasts and the nature of the facies deposition.

This work received financial support by FCE (8255), Comisión Sectorial de Investigación Científica, Universidad de la República, Uruguay. We thank Santiago Stareczek and Gonzalo Sánchez for providing polished rock slides. This paper is a contribution to the IGCP 512. Finally, we express our gratitude to Bob Pankhurst for the invitation to contribute to this special volume.

References

ALKMIM, F. F., MARSHAK, S. & FONSECA, M. 2001. Assembling west Gondwana in the Neoproterozoic: Clues from the São Francisco craton region, Brazil. *Geology*, **29**, 319–322.

ALLEN, P. A. & HOFFMAN, P. F. 2005. Extreme winds and waves in the aftermath of a Neoproterozoic galciation. *Nature*, **433**, 123–127.

ALMEIDA, F., AMARAL, G., CORDANI, U. & KAWASHITA, K. 1973. The Precambrian evolution of the South American cratonic margin south of the Amazon river. *In*: NAIRM, A. & STEHLI, F. (eds) *The Ocean Basin and Margins*. Plenum Press, New York, 441–446.

ALMEIDA, F. F. M., BRITO NEVES, & CARNEIRO, C. D. R. 2000. The origin and evolution of the South American Platform. *Earth-Science Reviews*, **50**, 77–111.

ALVARENGA, C. J. S. & TROMPETTE, R. 1992. Glacially influenced sedimentation in the Later Proterozoic of the Paraguay belt (Mato Grosso, Brazil). *Palaegeography, Palaeoclimatology, Palaeoecology*, **92**, 85–105.

ALVARENGA, C. J. S., SANTOS, R. V. & DANTAS, E. L. 2004. C–O–Sr isotopic stratigraphy of cap carbonates overlying Marinoan-age glacial diamictites in the Paraguay Belt, Brazil. *Precambrian Research*, **131**, 1–21.

ANDREIS, R. R., ZALBA, P. E. & IÑIGUEZ RODRÍGUEZ, A. M. 1992. Paleosuperficies y Sistemas Depositacionales en el Proterozoico Superior de Sierras Bayas, Sierras Septentrionales, Pcia. Buenos Aires, Argentina. *In*: *IV Reunión Argentina de Sedimentología, La Plata, Argentina, Proceedings*, **1**, 283–290.

ARNAUD, E. 2004. Giant cross-beds in the Neoproterozoic Port Askaig Formation, Scotland: implications for snowball Earth. *Sedimentary Geology*, **165**, 155–174.

ARNAUD, E. & EYLES, C. H. 2002. Glacial influence on Neoproterozoic sedimentation: The Smalfjord Formation, northern Norway. *Sedimentology*, **49**, 765–788.

ARNAUD, E. & EYLES, C. H. 2006. Neoproterozoic environmental change recorded in the Port Askaig Formation, Scotland: climatic and tectonic controls on sedimentation. *Sedimentary Geology*, **183**, 99–124.

ASTINI, R. A. 2003. The Ordovician Proto-Andean Basins. *In*: BENEDETTO, J. L. (ed.) *Ordovician Fossils of Argentina*. Secretaría de Ciencia y Tecnología, Universidad de Cordoba, 1–74.

BABINSKI, M., TRINDADE, R. I. F., ALVARENGA, C. J. S., BOGGIANI, P. C., LIU, D. & SANTOS, R. V. 2006. Geochronological constraits on the Neoproterozoic glaciations in Brazil. *In*: *Snowball Earth Conference, Ascona, Switzerland, Proceedings*, **1**, 19–20.

BASEI, M. A. S., FRIMMEL, H. E., NUTMAN, A. & PRECIOZZI, F. 2006. Provenance and depositional age of the Dom Feliciano Belt supracrustal units, Brazil-Uruguay: Correlations with SW Africa. *In*: *V South American Symposium on Isotope Geology, Punta del Este, Uruguay, Short Papers*, **1**, 45–47.

BASEI, M. A. S., FRIMMEL, H. E., NUTMAN, A. P., PRECIOZZI, F. & JACOB, J. 2005. A connection between the Neoproterozoic Dom Feliciano (Brazil/Uruguay) and Gariep (Namibia/South Africa) orogenic belts - evidence from a reconnaissance provenance study. *Precambrian Research*, **139**, 195–221.

BLANCO, G. & GAUCHER, C. 2004. Stratigraphy, paleontology and age of Las Ventanas Formation (Neoproterozoic, Uruguay). *In*: *1st Symposium on Neoproterozoic–Early Paleozoic Events in SW-Gondwana, Extended Abstracts*, IGCP Project 478, Second Meeting, São Paulo, Brazil, 8–9.

BOGGIANI, P. C., BABINSKI, M. *ET AL*. 2005. U–Pb SHRIMP investigation of ash beds in the Corumbã

Group (Ediacaran), Paraguay Belt, Brazil. *In*: *2nd Symposium on Neoproterozoic–Early Palaeozoic Events in southwestern Gondwana. Windhoek, Namibia, Extended Abstracts*. IGCP 478, 8–9.

Borba, A. W. & Mizusaki, A. M. P. 2002. Significado tectonico de leques aluviais da Formaçao Santa Bárbara (Eo-Paleozoico) na regiao de Caçapava do Sul (RS, Brasil). *Pesquisas em Geociencias*, **29**, 31–41.

Bossi, J. & Campal, N. 1992. Magmatismo y tectónica transcurrente durante el Paleozoico inferior del Uruguay. *In*: Gutiérrez, J., Saavedra, J. & Rábano, I. (eds) *Paleozoico Inferior de Ibero-América*. Universidad de Extremadura, Extremadura, Spain, 343–356.

Bossi, J. & Gaucher, C. 2004. The Cuchilla Dionisio Terrane, Uruguay: an allochthonous block accreted in the Cambrian to SW-Gondwana. *Gondwana Research*, **7**, 661–674.

Bossi, J. & Navarro, R. 1991. *Geología del Uruguay I*. Departamento de Publicaciones de la Universidad de la República, Montevideo.

Bossi, J., Cingolani, C., Llambias, E., Varela, R. & Campal, N. 1993. Características del magmatismo post-orogénico finibrasiliano en el Uruguay: Formaciones Sierra de Ríos y Sierra de Animas. *Revista Brasilera de Geociencias*, **23**, 282–288.

Bossi, J., Ferrando, L. *et al*. 1998. *Carta Geológica del Uruguay, escala 1:500.000*. Geoeditores., Montevideo.

Bowring, S. A., Myrow, P. M., Landing, E. & Ramezani, J. 2003. Geochronological constraints on Neoproterozoics events and the rise of metazoans. *Astrobiology*, **2**, 112.

Brasier, M. & Shields, G. 2000. Neoproterozoic chemostratigraphy and correlation of the Port Askaig glaciation, Dalradian Supergroup of Scotland. *Journal of the Geological Society, London*, **157**, 909–914.

Brasier, M. D., McCarron, G., Tucker, R., Eláter, J., Allen, P. & Shields, G. 2000. New U–Pb zircon dates for the Neoproterozoic Grubrah glaciation and for the top of the Huqf Supergroup, Oman. *Geology*, **28**, 175–178.

Brito Neves, B. B. & Alkmim, F. F. 1993. Cráton: evolução de um conceito. *In*: Domingues, J. M. L. & Misi, A. (eds) *Livro sobre o II Simpósio do Cráton do São Francisco*. **1**. Núcleo Bahia, Sergipe, Sociedade Brasilera de Geologia–Sociedade Mineira de Engenheiros, 1–10.

Butterfield, N. J. 2000. Bangiomorpha pubescens n. gen, n. sp.; implications for the evolution of sex, multicellularity, and the Mesoproterozoic/Neoproterozoic radiation of eukaryotes. *Paleobiology*, **26**, 386–404.

Calver, C. R., Black, L. P., Everand, J. L. & Seymour, D. B. 2004. U–Pb zircon age constraints on the late Neoproterozoic glaciation in Tasmania. *Geology*, **32**, 893–896.

Campal, N. & Schipilov, A. 1995. The Illescas bluish quartz rapakivi granite (Uruguay–South America): some geological features. *In*: *Symposium on Rapakivi Granites and Related Rocks, Belem, Brazil, Proceedings*, **18**.

Campanha, G. A. C., Basei, M. A. S., Tassinari, C., Nutman, A. P. & Faleiros, F. M. 2005. U–Pb SHRIMP age determinations for the Iporanga formation, SE Brazil: implications for the tectonic evolution of the Ribeira Belt. *In*: Pankhurst, R. J. & Veiga, G. D. (eds) *Gondwana 12: Geological and Biological Heritage of Gondwana, Abstracts*. Academia Nacional de Ciencias, Córdoba, Argentina, **82**.

Campos Neto, M. C. & Figueiredo, M. C. H. 1995. The Río Doce Orogeny, Southeastern Brazil. *Journal of South American Earth Sciences*, **8**, 143–62.

Carvalho, P. F. & Pinto, E. A. 1938. Reconhecimento geológico no Estado de Santa Catarina. *Boletim do Serviço Geológico e Mineralógico do Brasil*, **92**, 1–30.

Chemale, F., Jr. 2000. Evoluçao geológica do escudo Sul-rio-grandense. *In*: Holz, M. & De Ros, L. F. (eds) *Geologia do Rio Grande do Sul*. Centro de Investigação do Gondwana/Universidade Federal do Rio Grande do Sul, Porto Alegre, 13–52.

Cingolani, C. & Bonhomme, M. G. 1982. Geochronology of La Tinta Upper Proterozoic Sedimentary Rocks, Argentina. *Precambrian Research*, **18**, 119–132.

Cingolani, C. A., Hartmann, L. A., Santos, J. O. S. & McNaughton, N. J. 2002. U–Pb SHRIMP dating of zircons from the Buenos Aires Complex of the Tandilia Belt, Río de la Plata Craton, Argentina. *In*: *XV Congreso Geológico Argentino, Santa Cruz, Argentina*, CD-ROM.

Colombo, C., Tassinari, G. & Macambira, M. J. B. 1999. Geochronological provinces of the Amazonian Craton. *Episodes*, **22**, 174–182.

Condon, D., Prave, A. R. & Benn, D. I. 2002. Neoproterozoic glacial-rainout intervals: Observations and implications. *Geology*, **30**, 35–38.

Condon, D., Zhu, M., Bowring, S., Wang, W., Yang, A. & Jim, Y. 2005. U–Pb ages from the Neoproterozoic Doushantuo Formation, China. *Science*, **308**, 95–98.

Cordani, U. G., Sato, K., Teixeira, W., Tassinari, C. C. G. & Basei, M. A. S. 2000. Crustal evolution of the South American Platform. The Amazonian Craton. *In*: Cordani, U. G., Milani, E. J., Thomáz Filho, A. & Campos, D. A. (eds) *Tectonic evolution of South America, 31st International Geological Congress, Rio de Janeiro*, 19–40.

Corsetti, F. A., Olcott, A. N. & Bakermans, C. 2006. The biotic response to Neoproterozoic snowball Earth. *Palaeogeography, Palaeoclimatology, Palaeoecology*, **232**, 114–130.

Cukrov, N., Alvarenga, C. J. S. & Uhlein, A. 2005. Litofácies da Glaciação Neoproterozóica nas Porções Sul do Cráton do São Francisco: Exemplos de Jequitaí (MG) e Cristalina (GO). *Revista Brasileira de Geociencias*, **35**, 69–76.

Dobrzinski, N., Bahlburg, H. & Strauss, H. 2005. Palaeoenvironmental implications of sedimentological records combined with geochemical data: a Cryogenian succession on the Yangtze platform in South China. *In*: *IAS International Conference on Glacial Sedimentary Processes and Products, Aberystwyth*, poster.

DOMACK, E. & HOFFMAN, P. 2003. Stratigraphic transition into and out of a snowball glacial: Evidence from the Otavi platform and Fransfontein slope, Namibia: Eos Transactions of the American Geophysical Union, *Fall Meeting, San Francisco*, C11B-0819, 84–86.

EEROLA, T. 1995. From ophiolites to glaciers? Review on geology of the Neoproterozoic - Cambrian Lavras Do Sul region, southern Brazil. *In*: AUTIO, S. (ed.) *Geological Survey of Finland, Current Research: 1993–1994*. Geological Survey of Finland, Special Papers, **20**, 5–16.

EEROLA, T. 2001. Climate change at the Neoproterozoic-Cambrian transition. *In*: ZHURAVLEV, A. Y. & RIDING, R. (eds) *The ecology of the Cambrian radiation New York: Perspectives in Paleobiology and Earth History*, Columbia University Press, 90–106.

ESCAYOLA, M., PIMENTEL, M. M. & ARMSTRONG, R. A. 2007. Neoproterozoic Back-Arc Basin: SHRIMP U–Pb and SM–Nd isotopic evidence from the Eastern Pampean Ranges, Argentina. *Geology*, **35**, 495–498.

EVANS, D. A. D. 2000. Stratigraphic, geochronological, and paleomagnetic constraints upon the Neoproterozoic climatic paradox. *American Journal of Science*, **300**, 347–433.

EYLES, N. & JANUSZCZAK, N. 2004. 'Zipper-rift': a tectonic model for Neoproterozoic glaciations during the break up of Rodinia after 750 Ma. *Earth-Science Reviews*, **65**, 1–73.

FAMBRINI, G. L., PAES-DE-ALMEIDA, R., RICCOMINI, C. & FRAGOSO-CESAR, A. R. S. 2003. Tempestitos com Influência Glacial da Formação Playa Hermosa (Neoproterozóico), Piriápolis, Uruguai. *Revista Brasileira de Geociências*, **33**, 1–12.

FANNING, C. M. & LINK, P. K. 2004. U–Pb SHRIMP ages of Neoproterozoic (Sturtian) glaciogenic Pocatello Formation, southeastern Idaho. *Geology*, **32**, 881–884.

FIGUEIREDO, M. C. H., FRAGOSO CÉSAR, A. R. S. & KRONBERG, B. L. 1990. Litogeoquímica das principais unidades do Batólito Pelotas no Rio Grande do Sul. *In*: *Congresso Brasileiro de Geologia 36, Natal, Anais*, Sociedade Brasileira de Geologia, **4**. 1623–1738.

FIGUEIREDO, M. F., BABINSKI, M., ALVARENGA, C. J. S., PINHO, F. E. C. & SIMON, C. M. 2006. Chemostratigraphy (C, O and Sr) of Ediacaran post-glacial carbonates of Paraguay-Belt, Matto Grosso State, Brazil. *In*: *V South American Symposium on Isotope Geology, Punta del Este, Uruguay*, 141–144.

FRAGOSO CESAR, A. R. S. 1980. O Cráton do rio de la Plata e o Cinturão Dom Feliciano no Escudo Uruguaio-Sul-Riograndense. *In*: *Congresso Brasileiro de Geologia 31, Balneário Camboriú, Anais*, **5**. Sociedade Brasileira de Geologia, 2879–2892.

FRAGOSO CÉSAR, A. R. S. 1991. *Tectonica de placas no ciclo brasiliano: as orogenias dos cinturoes Dom Feliciano e Ribeira no Rio Grande do sul*. PhD thesis. Univerisdade de São Paulo.

FRAGOSO CESAR, A. R. S. & MACHADO, R. 1997. Neoproterozoic terranes of the Gaucho Shield (Southern Brazil and Uruguay). *In*: *South American Symposium on Isotope Geology, São Paulo, Brazil*, 65–67.

FRAGOSO CESAR, A. R. S., WERNICK, E. & SOLIANI, E. 1982*a*. Evoluçao Geotectónica do cinturão Dom Feliciano - uma contribuçao através da aplicaçao do modelo de tectônica de Placas. *In*: *Congresso Brasileiro de Geologia 32, Salvador, Anais*, **1**. Sociedade Brasileira de Geologia, 13–22.

FRAGOSO CESAR, A. R. S., WERNICK, E. & SOLIANI, E. 1982*b*. Associaçoes petrotectónicas do cinturao Dom Feliciano (SE da Plataforma sul-americana). *In*: *Congresso Brasileiro de Geologia 32, Salvador, Anais*, **1**. Sociedade Brasileira de Geologia, 1–12.

FRAGOSO CESAR, A. R. S., MACHADO, R. & GÓMEZ RIFAS, C. 1987. Observaçoes sobre o cinturao Dom Feliciano no Escudo Uruguaio e correlaçòes com o escudo do R.G do Sul. *In*: *III Simposium Sul-Brasileiro de Geologia, Curitiba, Brazil*, **2**, 791–809.

FRAGOSO CESAR, A. R. S., MACHADO, R., MELLO, F. M. D. E., SAYEG, H. S., SILVA FILHO, W. F. & FAMBRINI, G. L. 1994. Acresçao Neoproterozóica No Sul Do Brasil E Uruguai: A placa Rio de la Plata e os terrenos Piratini e Rio Vacacaí. *In*: *Congresso Brasileiro de Geologia 38, Balneário Camboriú, Anais*, **1**. Sociedade Brasileira de Geologia, 206–207.

FRIMMEL, H. E. & BASEI, M. A. S. 2006. Tracking down the Neoproterozoic connection between Southern Africa and South America - a revised geodynamic model for SW-Gondwana amalgamation. *In*: *V South American Symposium on Isotope Geology, Punta del Este, Uruguay*, 94–97.

GASTAL, M. C. P., LAFON, J. M., HARTMANN, L. A. & KOESTER, E. 2005. Sm–Nd isotopic compositions as a proxy for magmatic processes during the Neoproterozoic of the southern Brazilian Shield. *Journal of South American Earth Sciences*, **18**, 255–276.

GAUCHER, C., BOGGIANI, P. C., SPRECHMANN, P., SIAL, A. N. & FAIRCHILD, T. 2003. Integrated correlation of the Vendian to Cambrian Arroyo del Soldado and Corumbã Groups (Uruguay and Brazil) -palaeogeographic, palaeoclimatic and palaeobiologic implications. *Precambrian Research*, **120**, 241–278.

GAUCHER, C., FRIMMEL, H. E. & GERMS, G. J. B. 2005*a*. Organic walled microfossils and biostratigraphy of the upper Port Nolloth Group (Namibia): implications for the latest Neoproterozoic glaciations. *Geological Magazine*, **142**, 539–559.

GAUCHER, C., POIRÉ, D., PERAL, L. & CHIGLINO, L. 2005*b*. Litoestratigrafía, Bioestratigrafía y correlaciones de las sucesiones sedimentarias del Neoproterozoico–Cambrico del Cratón del Río de la Plata (Uruguay y Argentina). *Latin American Journal of Sedimentology and Basin Analysis*, **12**, 145–160.

GÓMEZ PERAL, L., POIRÉ, D. G., STRAUSS, H. & ZIMMERMANN, U. 2007. Chemostratigraphy and diagenetic constraints on Neoproterozoic carbonate successions from the Sierras Bayas Group, Tandilia System, Argentina. *Chemical Geology*, **237**, 109–128.

GORJAN, P., VEEVERS, J. J. & WALTER, M. R. 2000. Neoproterozoic sulfur-isotope variation in Australia and global implications. *Precambrian Research*, **100**, 151–179

HALVERSON, G. P., HOFFMAN, P. F., SCHRAG, D. P., MALOOF, A. C. & RICE, A. H. N. 2005. Toward a Neoproterozoic composite carbon-isotope record.

Geological Society of American Bulletin, **117**, 1181–1207.

HAMBREY, M. J. & HARLAND, W. B. 1985. The Late Proterozoic Glacial Era. *Palaegeography Palaeoclimatology, Palaeoecology*, **51**, 255–272.

HANSEN, L. 2004. Deltaic infill of a deglaciated artic fjord east Greenland: sedimentary facies and sequence stratigraphy. *Journal of Sedimentary Research*, **74**, 422–437.

HARTMANN, L. A., PORCHER, C. C. & REMUS, M. V. D. 2000. Evolucão das rochas metamórficas do Rio Grande do Sul. *In*: HOLZ, M. & DE ROS, L. F. (eds) *Geologia do Rio Grande do Sul*. Centro de Investigação do Gondwana, Universidade Federal do Rio Grande do Sul, Porto Alegre, 79–118.

HARTMANN, L. A., CAMPAL, N. *ET AL*. 2001. Archean crust in the Rio de la Plata craton, Uruguay-SHRIMP U–Pb zircon reconnaissance geochronology. *Journal of South American Earth Sciences*, **14**, 557–570.

HEINRICH, H. 1988. Origin and consequences of cyclic ice rafting in the northeast Atlantic ocean during the past 130,000 years. *Quaternary Research*, **29**, 142–152.

HESSE, R., RASHID, H. & KHODABAKHSH, S. 2004. Fine-grained sediment lofting from meltwater-generated turbidity currents during Heinrich events. *Geology*, **32**, 449–452.

HOFFMAN, P. F., HAWKINS, D. P., ISACHSEN, C. E. & BOWRING, S. A. 1996. Precise U–Pb zircon ages for early Damaran magmatism in the Summas Mountains and Welwitschia inlier, northern Damara belt, Namibia. *Geological Survey of Namibia Communications*, **11**, 47–52.

HOFFMAN, P. F., KAUFMAN, A. J., HALVERSON, G. P. & SCHRAG, D. P. 1998. A Neoproterozoic Snowball Earth. *Science*, **281**, 1342–1346.

HOFFMANN, K. H., CONDON, D. J., BOWRING, S. A. & CROWLEY, J. L. 2004. U–Pb zircon date from the Neoproterozoic Ghaub Formation, Namibia: constraints on Marinoan glaciation. *Geology*, **32**, 817–820.

HURTGEN, M. T., ARTHUR, M. A., SUITS, N. S. & KAUFMAN, A. J. 2002. The sulfur isotopic composition of Neoproterozoic seawater sulfate; implications for a snowball Earth? *Earth and Planetary Science Letters*, **203**, 413–429.

ISBELL, J. L., MILLER, M. F., WOLFE, K. L. & LENAKER, P. A. 2003. Timing of the Late Paleozoic Glaciation in Gondwana: was Glaciation responsible for the development of northern hemisphere Cyclothem. *In*: CHAN, M. A. & ARCHER BINDING, A. W. (eds) *Extreme depositional environments: Mega end members in geologic time*. Geological Society of America, Special Papers, **370**, 1–20.

JOST, H. & BITENCOURT, M. F. 1980. Estratigrafia e tectônica de uma fração da faixa de Dobramentos de Tijucas no Rio Grande do Sul. *Acta Geologica Leopoldensia*, **4**, 27–60.

KENDALL, B., CREASER, R. A. & SELBY, D. 2006. Re–Os geocronology of postglacial black shales in Australia: Constraints on the timing of 'sturtian' glaciation. *Geology*, **34**, 729–732.

KENNEDY, M. J., RUNNEGAR, B., PRAVE, A. R., HOFFMANN, K. H. & ARTHUR, M. A. 1998. Two or four Neoproterozoic glaciations? *Geology*, **26**, 1059–1063.

KNELLER, B., MILANA, J. P., BUCKEE, C. & AL JA'AIDI, O. 2004. A depositional record of deglaciation in a paleofjord (Late Carboniferous [Pennsylvanian] of San Juan Province, Argentina): The role of catastrophic sedimentation. *Geological Society of America Bulletin*, **116**, 348–367.

LEATHER, J., ALLEN, P. A., BRASIER, M. D. & COZZI, A. 2002. Neoproterozoic snowball Earth under scrutiny: evidence from the Fiq glaciation of Oman. *Geology*, **30**, 891–894.

LI, Z. X., EVANS, D. A. D. & ZHANG, S. A. 2004. 90° spin on Rodinia: possible causal links between the Neoproterozoic supercontinent, superplume, true polar wander and low-latitude glaciation. *Earth and Planetary Science Letters*, **220**, 409–421.

LOUREIRO, J., SÁNCHEZ BETTUCCI, L., PAZOS, P., SÁNCHEZ, G. & STARECZEK, S. 2006. Hyaloclastic rocks at the Playa Hermosa Formation. *In*: *V South American Symposium on Isotope Geology, Punta del Este, Uruguay*, 397–398.

LUND, K., ALEINIKOFF, J. N., EVANS, K. V. & FANNING, C. M. 2003. SHRIMP U–Pb geochronology of Neoproterozoic Windermere Supergroup, central Idaho: Implications for rifting of western Laurentia and synchroneity of Sturtian glacial deposits. *Geological Society of America Bulletin*, **115**, 349–372.

MACOUIN, M., BESSE, J., ADER, M., GILDER, S., YANG, Z., SUN, Z. & AGRINIER, P. 2004. Combined paleomagnetic and isotopic data from the Doushantuo carbonates, South China: implications for the 'snowball Earth' hypothesis. *Earth and Planetary Science Letters*, **224**, 387–398.

MALLMANN, G., CHEMALE, F., JR., ARMSTRONG, R. & KAWASHITA, K. 2003. Sm–Nd and U–Pb Shrimp zircon of the Nico Pérez Terrane, Reworked Río de la Plata Craton, *In*: *Uruguay. IV South American Symposium on Isotope Geology, Salvador, Brazil*, 207–209.

MARTINS NETO, M. A., PEDROSA-SOARES, A. C. & LIMA, S. A. A. 2001. Tectono-sedimentary evolution of sedimentary basins from late Paleoproterozoic to late Neoproterozoic in the São Francisco craton and Araçuai fold belt, eastern Brazil. *Sedimentary Geology*, **141**, 343–370.

MEERT, J. G. & TORSVIK, T. H. 2004. Palaeomagnetic constraints on Neoproterozoic 'Snowball Earth' continental reconstructions. *In*: JENKINS, G., MCMENAMIN, M., SOHL, L. & MCKAY, C. (eds) *The Extreme Proterozoic: Geology, Geochemistry, and Climate*. American Geophysical Union, Geophysical Monographs, **146**, 5–12.

NOGUEIRA, A. C. R., RICCOMINI, C., SIAL, A. N., MOURA, C. & FAIRCHILD, T. 2003. Soft-sediment deformation at the base of the Neoproterozoic Puga cap carbonate (southwestern Amazon Craton, Brazil): Confirmation of rapid icehouse-greenhouse transition in snowball earth. *Geology*, **31**, 613–616.

OYHANTÇABAL, P. 2005. The Sierra Ballena Shear zone: kinematics, timing and its significance for the geotectonic evolution of southeast Uruguay. PhD thesis Georg-August-Universität Göttingen.

OYHANTÇABAL, P., SIEGESMUND, S, WEMMER, K. & SÁNCHEZ-BETTUCCI, L. 2005*a*. The evolution of Sierra Ballena Shear Zone (Dom Feliciano Belt) and its correlation with the Purros Shear Zone (Kaoko Belt). *In*: PANKHURST, R. J. & VEIGA, G. (eds) *Gondwana 12: Geological and Biological Heritage of Gondwana, Abstracts*. Academia Nacional de Ciencias, Córdoba, Argentina, 97.

OYHANTÇABAL, P., SÁNCHEZ BETTUCCI, L. *ET AL*. 2005*b*. Nueva Propuesta Estratigrafica para las Supracorticales del Cinturón Dom Feliciano (Proterozoico, Uruguay). *In*: *XII Congreso Latinoamericano de Geología, Ecuador*, CD-ROM.

OYHANTÇABAL, P., SIEGESMUND, S., WEMMER, K., FREI, R. & LAYER, P. 2007. Post-collisional transition from calc-alkaline to alkaline magmatism during transcurrent deformation in the southernmost Dom Feliciano Belt (Braziliano–Pan-African, Uruguay). *Lithos*, doi: 10.1016/j.lithos.2007.03.001.

PAIM, P. S. G., CHEMALE, F., JR. & LOPES, R. C. 1995. Aloestratigrafia, sistemas deposicionais e evolução paleogeográfica da Bacia do Camaquã – Vendiano superior/Ordoviciano inferior do RS. 6° *In*: *Simpósio Sul-Brasileiro de Geologia, Porto Alegre, Brazil*, **4**, 39–50.

PAIM, P. S. G., CHEMALE, F., JR. & LOPES, R. C. 2000. A Bacia do Camaquã. *In*: HOLZ, M. & DE ROS, L. F. (eds.) *Geologia do Rio Grande do Sul*. Centro de Investigação do Gondwana, Universidade Federal do Rio Grande do Sul, Porto Alegre, 231–274.

PAZOS, P. J. 2002. Palaeoenvironmental framework of the Late Palaeozoic glacial to postglacial transition in the Paganzo-Calingasta Basin (South America) and Karoo-Kalahari Basins (southern Africa): Ichnological implications. *Gondwana Research*, **5**, 619–640.

PAZOS, P. J., TÓFALO, R. & SÁNCHEZ BETTUCCI, L. 1998. Procesos sedimentarios e indicadores paleoclimáticos en la sección inferior de la Formación Playa Hermosa, Cuenca Playa Verde, Piriápolis, Uruguay. *In*: *II Congreso Uruguayo de Geología, Punta del Este, Uruguay*, 64–69.

PAZOS, P. J., SÁNCHEZ BETTUCCI, L. & TÓFALO, R. O. 2003. The record of the Varanger glaciation at the Río de la Plata craton, Vendian-Cambrian of Uruguay. *Gondwana Research*, **6**, 65–78.

PAZOS, P. J., SÁNCHEZ BETTUCCI, L., PEÇOITS, E. & AUBET, N. 2005. Carbonates with dropstones in Uruguay: a possible linking lithology with the Namibian and Brasilian Neoproterozoic glacial record. *In*: *IAS International Conference on Glacial Sedimentary Processes and Products, Aberystwyth*, 23–27.

PAZOS, P. J., DI PASQUO, M. R. & AMENABAR, C. 2007. Trace Fossils of the glacial to post-glacial transition in the El Imperial Formation (upper Carboniferous), San Rafael basin, Argentina. *In*: BROMLEY, R., BUATOIS, L. A., GENISE, J., MÁNGANO, M. G. & MELCHOR, R. (eds) *Ichnology at the crossroads: A multidimensional approach to the science of organism-substrate interactions*. Society of Sedimentary Geology (SEPM), Special Publication, in press.

PEDROSA-SOARES, A. C., NOCE, C. M., WIEDEMANN, C. & PINTO, C. P. 2001. The Araçuai-West-Congo Orogen in Brazil: an overview of a confined orogen formed during Gondwanaland assembly. *Precambrian Research*, **110**, 307–323.

PHILIPP, R. P., MACHADO, R., NARDI, R. V. S. & LAFON, J. M. 2002. O magmatismo granítico Neoproterozoico do Batolito Pelotas no sul do Brasil: novos dados e revisao da geocronología regional. *Revista Brasileira de Geociencias*, **32**, 277–290.

POIRÉ, D. G. 2002. The Precambrian/Lower Paleozoic sedimentary cover of Tandilia System, Argentina. *In*: GAUCHER, C. & POIRÉ, D. G. (eds) Field trip Guide, *In*: *II International Colloquium on the Vendian–Cambrian of W-Gondwana*, 55–66.

POIRÉ, D. G. & SPALLETTI, L. A. 2005. La cubierta sedimentaria precámbrica/paleozoica inferior del Sistema de Tandilia,. *In*: DE BARRIO, R. E., ETCHEVERRY, R. O., CABALLÉ, M. F. & LLAMBÍAS, E. J. (eds) *Geología y Recursos Minerales de la provincial de Buenos Aires. Relatorio del XXVI Congreso Geológico Argentino*, 51–68.

POIRÉ, D., SPALLETTI, L. & DEL VALLE, A. 2003. The Cambrian-Ordovician siliciclastic platform of the Balcarce Formation (Tandilia System, Argentina): facies, trace fossils, paleoenvironments and sequence stratigraphy, *Geologica Acta*, **1**, 41–60.

PRECIOZZI, F., BASEI, M. A. S. & MASQUELIN, H. 1999. New geochronological data from the Piedra Alta Terrane (Rio de la Plata Craton). *In*: *II South American Symposium on Isotope Geology, Villa Carlos Paz, Argentina*, 341–343.

PRECIOZZI, F., BASEI, M. A. S., PEEL, E., SÁNCHEZ BETTUCCI, L., CORDANI, U. & OYANTÇABAL, P. 2003. Punta del Este Terrane: Mesoproterozoic basement and Neoproterozoic cover. *In*: *IV South American Symposium on Isotope Geology, Salvador, Brazil*, 338–340.

PRECIOZZI, F., PEEL, E., SÁNCHEZ BETTUCCI, L., BASEI, M. A. S., LEDESMA, J. & MUZIO, R. 2005. *El Cinturón Dom Feliciano y sus relaciones con el Craton del Río de la Plata y el Terreno Punta del Este - Geología y Geocronología*. Proyecto Fondo Clemente Estable 6009 (unpublished internal report), Uruguay.

PRINS, M., BOUWER, L. M. *ET AL*. 2002. Ocean circulation and iceberg discharge in the glacial North Atlatantic: Inferences from unmixing of sediment size distributions. *Geology*, **30**, 555–558.

RAMOS, V. 2005. The proterozoic-Early Paleozoic margin of western Gondwana. *In*: PANKHURST, R. J. & VEIGA, G. (eds) *Gondwana 12: Geological and Biological Heritage of Gondwana. Abstracts*. Academia Nacional de Ciencias, Córdoba, Argentina, 304.

RAMPINO, M. R. 1994. Tillites, Diamictites, and Ballistic ejecta of large impacts. *Journal of Geology*, **102**, 439–456.

RAPELA, C., DALLA SALDA, L. & CINGOLANI, C. 1974. Un intrusivo básico Ordovícico en la Formación La Tinta (Sierra de los Barrientos), Provincia de Buenos Aires. *Revista de la Asociación Geológica Argentina*, **29**, 319–331.

RAPELA, C., PANKHURST, R. J. *ET AL*. 1998. The Pampean orogeny of the southern proto-Andes: Cambrian continental collision in the Sierras Pampeanas. *In*: PANKHURST, R. J. & RAPELA, C. W. (eds) *The Proto-Andean Margin of Gondwana*. Geological Society, London, Special Publications, **142**, 181–217.

RAPELA, C. W., CASQUET, C. *ET AL.* 2008. The Río de la Plata craton and the assembly of SW Gondwana. *Earth-Science Reviews*, **83**, 49–82.

SAALMANN, K., HARTMANN, L. A., REMUS, M. V. D., KOESTER, E. & CONCEIC, A. R. V. 2005. Sm–Nd isotope geochemistry of metamorphic volcanosedimentary successions in the São Gabriel Block, southernmost Brazil: evidence for the existence of juvenile Neoproterozoic oceanic crust to the east of the Río de la Plata craton. *Precambrian Research*, **136**, 159–75.

SAALMANN, K., REMUS, M. V. D. & HARTMANN, L. A. 2006. Structural evolution and tectonic setting of the Porongos belt, southern Brazil. *Geological Magazine*, **143**, 59–88.

SÁNCHEZ BETTUCCI, L. 1997. Los Basaltos postorogénicos de la Región Piriápolis - Pan de Azúcar, República Oriental del Uruguay. *Revista de la Asociación Geológica Argentina*, **52**, 3–16.

SÁNCHEZ BETTUCCI, L. 1998. *Evolución tectónica del Cinturón Dom Feliciano en la región Minas - Piriápolis, Uruguay*. PhD thesis, Universidad de Buenos Aires.

SÁNCHEZ BETTUCCI, L. & LINARES, E. 1996. Primeras edades en basaltos del Complejo Sierra de Ánimas, Uruguay. *In*: *XIII Congreso Geológico Argentino, Buenos Aires, Argentina, Actas*, **I**, 399–404.

SÁNCHEZ BETTUCCI, L. & PAZOS, P. 1996. Análisis Paleoambiental y Marco tectónico en la Cuenca Playa Verde, Piriápolis, Uruguay. *In*: *XIII Congreso Geológico Argentino, Buenos Aires, Argentina, Actas*, **I**, 405–412.

SÁNCHEZ BETTUCCI, L. & RAMOS, V. A. 1999. Aspectos Geológicos de las rocas metavolcánicas y metasedimentarias del Grupo Lavalleja, Sudeste de Uruguay. *Revista Brasilera de Geociencias*, **29**, 557–570.

SÁNCHEZ BETTUCCI, L. & RAPALINI, A. E. 2002. Paleomagnetism of the Sierra de Las Animas Complex, Southern Uruguay: Its Implications in the Assembly of Western Gondwana *Precambrian Research*, **118**, 243–265.

SÁNCHEZ BETTUCCI, L., COSARINSKY, M. & RAMOS, V. A. 2001. Tectonic setting of the Late Proterozoic Lavalleja Group (Dom Feliciano Belt), Uruguay. *Gondwana Research*, **4**, 395–407.

SÁNCHEZ BETTUCCI, L., OYHANTÇABAL, P., PAGE, S. & RAMOS, V. A. 2003*a*. Petrography and Geochemistry of the Carapé Complex, (Southeastern Uruguay). *Gondwana Research*, **6**, 89–105.

SÁNCHEZ BETTUCCI, L., PRECIOZZI, F., BASEI, M. A. S., OYHANTÇABAL, P., PEEL, E. & LOUREIRO, J. 2003*b*. Campanero Unit: A Probable Paleoproterozoic Basement and its Correlation to Other Units of Southeastern Uruguay. *In*: *IV South American Symposium on Isotope Geology, Salvador, Brazil*, 673–674.

SÁNCHEZ BETTUCCI, L., OYHANTÇABAL, P., PRECIOZZI, F., LOUREIRO, J., RAMOS, V. A. & BASEI, M. A. S. 2004. Mineralizations of The Lavalleja Group (Uruguay), A Neoproterozoic Volcano - Sedimentary Sequence. *Gondwana Research*, **7**, 745–751.

SAYLOR, B. Z. 2003. Sequence stratigraphy and carbonate–siliciclastic mixing in a terminal Proterozoic foreland basin, Urusis Formation, Nama Group, Namibia. *Journal of Sedimentary Research*, **73**, 264–279.

SOMMER, C. A., LIMA, E. F., NARDI, L. J. D. & WAICHEL, B. L. 2006. The evolution of Neoproterozoic magmatism in Southernmost Brazil: shoshonitic, high-K tholeiitic and silica saturated, sodic alkaline volcanism in post-collisional basins. *Anais da Academia Brasileira de Ciencias*, **78**, 573–589.

SPALLETTI, L. & DEL VALLE, A. 1984. Las diamictitas del sector oriental de Tandilia: caracteres sedimentológicos y origen. *Revista de la Asociación Geológica Argentina*, **39**, 188–206.

THOMAS, G. S. P. & CONNELL, R. J. 1985. Iceberg drop, dump and grounding structures from Pleistocene glacio-lacustrine sediments, Scotland. *Journal of Sedimentary Petrology*, **55**, 243–249.

TOHVER, E., D'AGRELLA-FILHO, M. S. & TRINDADE, R. I. F. 2006. Paleomagnetic record of Africa and South America for the 1200–500 Ma interval, and evaluation of Rodinia and Gondwana assemblies. *Precambrian Research*, **147**, 193–222.

TRINDADE, R. I. F., D'AGRELLA-FILHO, M. S., EPOF, I. & BRITO NEVES, B. B. 2006. Paleomagnetism of Early Cambrian Itabaiana mafic dikes (NE Brazil) and the final assembly of Gondwana. *Earth and Planetary Science Letters*, **244**, 361–377.

TROMPETTE, R. 1994. *Geology of Western Gondwana (2000–500 Ma)*. A.A. Balkema, Rotterdam.

VAN STADEN, A., ZIMMERMANN, U. & GERMS, G. J. 2005. Provenance and depositional study on tillites from the Volcan Hill, Tandilia System in east Argentina: Preliminary Results. *In*: *XXVI Congreso Geológico Argentino, Actas*, **I**, 239–246.

VEEVERS, J. J. 2004. Gondwanaland from 650–500 Ma assembly through 320 Ma merger in Pangea to 185–100 Ma breakup: supercontinental tectonics via stratigraphy and radiometric dating. *Earth-Science Reviews*, **68**, 1–132.

VEEVERS, J. J. 2005. Edge tectonics (trench rollback, terrane export) of Gondwanaland-Pangea synchronized by supercontinental heat. *Gondwana Research*, **8**, 449–456.

WOOD, R. A., GROTZINGER, J. P. & DICKSON, J. A. D. 2002. Proterozoic Modular Biomineralized Metazoan from the Nama Group, Namibia. *Science*, **296**, 2383–2386.

ZIMMERMANN, U. & SPALLETTI, L. 2005. Provenance of the Balcarce Formation: an indicator for early Palaeozoic volcanism in eastern Argentina? *In*: PANKHURST, R. J. & VEIGA, G.(eds) *Gondwana 12: Geological and Biological Heritage of Gondwana, Abstracts*. Academia Nacional de Ciencias, Córdoba, Argentina, 377.

South Atlantic divergent margin evolution: rift-border uplift and salt tectonics in the basins of SE Brazil

W. MOHRIAK[1], M. NEMČOK[2] & G. ENCISO[3]

[1]*Petroleo Brasileiro S.A.–UN-EXP–GP-SSE, Avda. Chile, 65–s. 1302 20.035-900 Rio de Janeiro, RJ, Brazil (email: webmohr@petrobras.com.br)*

[2]*Energy & Geoscience Institute, 423 Wakara Way, Suite 300, Salt Lake City, UT 84108, USA*

[3]*Spinnaker Exploration, 1200 Smith Street, Suite 800, Houston, TX 77002, USA*

Abstract: The South Atlantic Ocean evolved after rupture of the São Francisco–Congo–Rio de la Plata–Kalahari cratonic landmass and the Late Proterozoic fold belts. Break-up in the South Atlantic realm developed diachronously: rifting started in the south (Argentina) during the Jurassic and progressed towards the equatorial segment. The central portion was controlled by a rift-resistant cratonic nucleus (the São Francisco–Congo craton) and as a result underwent development of narrow basins; parts controlled by Neoproterozoic fold belts developed wide basins. The final break-up of western Gondwana and the onset of plate divergence were marked by thick wedges of seaward-dipping reflectors, located near the incipient ocean-ridge spreading centre that had already been formed by the time Aptian evaporites were deposited. Subsequently, a few episodes of intraplate tectonic and magmatic activity affected the Santos, Campos and Espírito Santo basins.

Post-break up development of the offshore basins was affected by gravity gliding over the Aptian evaporites. Continental uplift may be invoked as the main cause of salt mobilization, generating prograding clastic wedges that thickened basin-wards and produced a loading effect on the salt basin. Coupled with onshore erosional unloading and the effects of the gravity gliding, this probably resulted in further flexural uplift of the continental margin.

We discuss the development of the sedimentary basins located on the eastern Brazilian margin after Gondwana break-up and South Atlantic opening. We focus on the most important tectono-magmatic episodes during the formation of the rift basins, their relationship to salt tectonics, and the impact of continental margin uplift on the development of massive clastic progradation wedges in the Santos Basin. The rift basins represent two main phases of evolution. An early syn-rift sequence, characterized by rotated blocks observed in seismic data, is associated with continental lacustrine sediments. A late syn-rift sequence, characterized by a thick sedimentary succession that is less controlled by basement faults, forms a sag basin with sub-horizontal layers below an evaporitic, transitional sequence. The sag basin is characterized by regional distribution of sediments that thin both landwards and basin-wards, forming depocentres in the deep-water region. Locally these sediments onlap volcanic highs and wedges of seaward-dipping reflectors (Mohriak 2001).

A palinspastic reconstruction of Gondwana (Fig. 1), assembling the South American, African, Australian, Indian and Antarctic continents, shows the distribution of Precambrian cratons and Neoproterozoic mobile belts. The main Archaean cratonic regions (São Francisco and Congo) and Late Precambrian to Early Palaeozoic mobile belts (Borborema, Rio Doce and Ribeira in eastern Brazil, West Congo and Kaoko in West Africa) underwent Mesozoic continental rupture that evolved into the South Atlantic Ocean. Stable cratonic areas such as the Amazon, São Francisco and Rio de la Plata blocks rifted apart from the West Africa Congo and Kalahari cratons. The extensional stresses that led to continental rifting were followed by the development of the active spreading centre that later evolved into the present-day Mid-Atlantic Ridge.

The break-up of Western Gondwana and the opening of the South Atlantic Ocean started in Late Jurassic/Early Cretaceous times in the southernmost parts of the South American continent. It advanced towards the northeastern Brazilian margin (Rabinowitz & LaBrecque 1979; Chang *et al.* 1992; Heilbron *et al.* 2000). The link between the South and the Central–North Atlantic was

From: PANKHURST, R. J., TROUW, R. A. J., BRITO NEVES, B. B. & DE WIT, M. J. (eds) *West Gondwana: Pre-Cenozoic Correlations Across the South Atlantic Region.* Geological Society, London, Special Publications, **294**, 365–398. DOI: 10.1144/SP294.19 0305-8719/08/$15.00 © The Geological Society of London 2008.

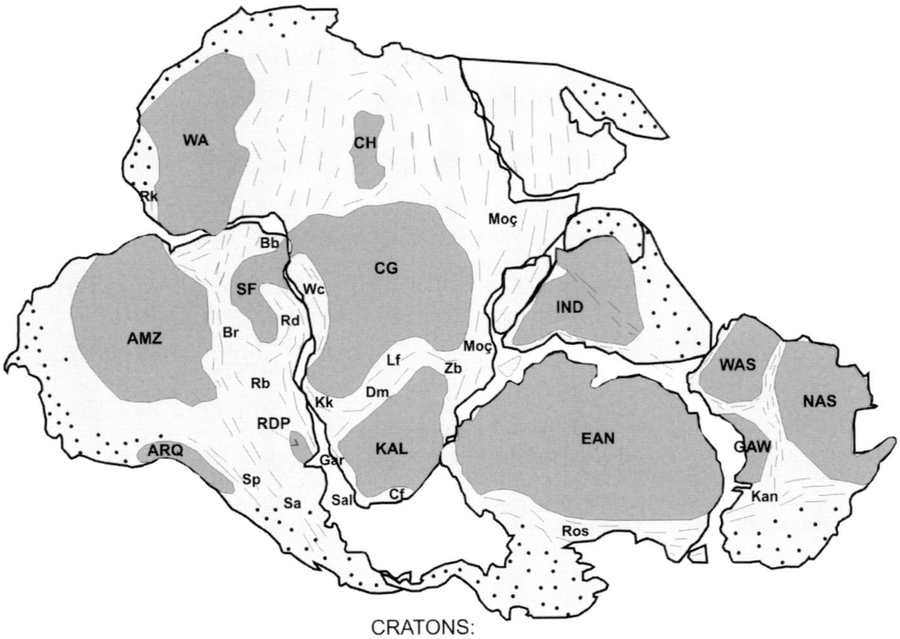

CRATONS:
ARQ, Arequipa; AMZ, Amazonas; RDP, Rio de La Plata; SF, São Francisco;
WA, West African; CG, Congo; CH, Chad; KAL, Kalahari;
EAN, East Antarctica; IND,Indian; WAS, West Australian; GAW, Gawler; NAS, North Australian.

Mobile belts:
Sp, Sierras Pampeanas; Sa, Sierras Australes; Br, Brasília; Rb, Ribeira; Rd, Rio Doce; Bb, Borborema;
Rk, Rokellides; Wc, West Congo; Kk, Kaoko; Gar, Gariep; Sal, Saldania; Cf, Cape Fold;
Dm, Damara; Lf, Lufilian; Zb, Zambezi; Moç, Moçambique; Ros, Ross; Kan, Kanmatoo.

Modified from Schmitt et al., 2004.

Fig. 1. Pre-break-up reconstruction of Gondwana with Precambrian cratons and Neoproterozoic fold belts.

established by Albian–Turonian times (Koutsoukos & Dias-Brito 1987; Koutsoukos *et al.* 1991).

The São Francisco–Congo craton connection trends almost perpendicular to the present-day coastline of NE Brazil. The development of the sedimentary basins that are underlain by Archaean granulites in this region is clearly controlled by the crystalline basement, which is more resistant to rifting than the metamorphosed sedimentary successions and igneous intrusions of the Brasiliano/Pan-African fold belts that are observed further south in basins such as the Espírito Santo, Campos and Santos on the Brazilian margin, as well as in the conjugate West African margin (offshore Angola and Gabon). These fold belts, named Ribeira and Rio Doce on the Brazilian side (Heilbron *et al.* 2000; Schmitt *et al.* 2004), control rift structures and salt tectonic styles that are markedly different from equivalent structures and styles in the region underlain by the São Francisco Craton (Mohriak 2005; Rosendahl *et al.* 2005).

The schematic distribution of the Aptian salt provinces along both the Brazilian and West African margins (Fig. 2) is marked by a main salt province along the divergent margin (Sergipe–Gabon to Santos–Namibe basins) and by small salt pods in the equatorial margin (Ceará Basin). The Brazilian margin is characterized by presence of two different evaporitic assemblages (Davison 2005) that co-exist in the Sergipe–Alagoas basin (Mohriak *et al.* 2000). The first evaporitic sequence accumulated in syn-rift troughs of the Sergipe Basin during arid episodes in the Early Aptian (*c.* 120–124 Ma). This assemblage includes thick halite layers and is known as the Paripueira sequence (Ojeda 1982). The second evaporitic assemblage was deposited from the Sergipe to the Santos basins as massive halite with intercalations of less soluble salts (anhydrite) and more soluble K-rich salts, such as sylvite, carnalite, and tachydrite. These evaporites are dated as Late Aptian (*c.* 112–115 Ma), and correspond to the massive salt layer that forms huge salt diapirs and salt walls in the deep-water region of the Espírito Santo, Campos and Santos basins (Fig. 2). Davison (2005) suggests that the massive salt layers of the

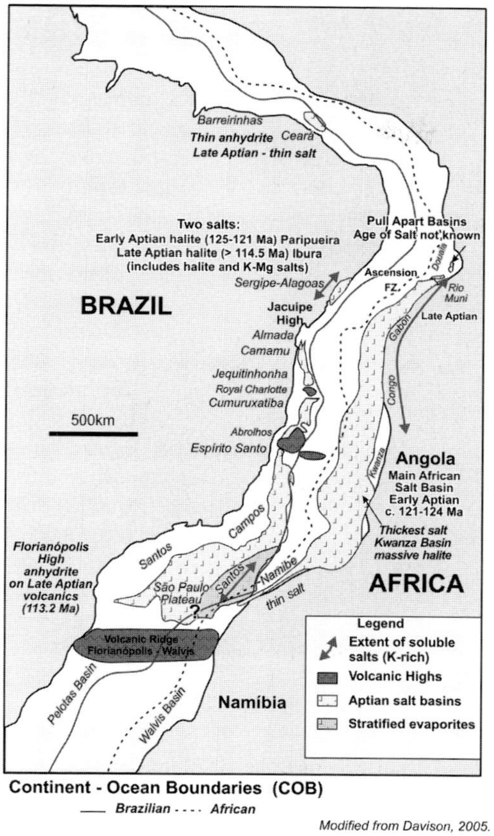

Fig. 2. General distribution of salt basins along the eastern Brazilian and West African continental margins, with the ages of evaporite deposition and major volcanic ridges.

Angolan Kwanza Basin correspond to the first evaporitic assemblage, suggesting a marked asymmetry of the salt basin in this northern part of the South Atlantic (Fig. 2).

The western border of the Brazilian salt basin, as well as the southernmost segment, near the Florianópolis High (Fig. 2) are characterized by a reduced proportion of halite (Meisling *et al.* 2001). Some boreholes have drilled only anhydrite above volcanic rocks dated as Late Aptian (*c.* 113 Ma) to Early Cretaceous (*c.* 130 Ma). There are no evaporites to the south of the Florianópolis Fracture Zone–Walvis Ridge. This ridge corrresponds to a conspicuous volcanic zone that extends from the deep oceanic crust towards the Florianópolis platform, where it is expressed as a volcanic high. South of the Florianópolis–Walvis Ridge, the Aptian stratigraphic sequence is associated with open marine environments, indicating that oceanic crust was already formed to the south (Pelotas and Walvis basins, Fig. 2). At this time the environment of deposition along the Santos and Namibe–Kwanza basins corresponded to a shallow-water gulf, transitional from the Barremian lacustrine environments to the Aptian evaporite assemblage. The final opening the South Atlantic Ocean culminated with a marine transgression from Late Aptian to Early Albian. This resulted in widespread open marine, shallow-water carbonate platforms identified along the Brazilian and West African margins, which were subsequently drowned by Cenomanian–Turonian times.

The amount of salt in the SE Brazilian margin (in the Santos Basin, in particular) is considerably larger than the amount of salt in the conjugate Namibe Basin, located north of the Walvis Ridge (Fig. 2). This indicates that the continent–ocean boundary is located much closer to the African (dotted line in Fig. 2) than to the Brazilian margin (continuous line Fig. 2), suggesting a dextral shifting of the ocean spreading centre (Mohriak 2001; Mohriak & Paula 2005). Previous interpretations (Kowsmann *et al.* 1977; Dias 1993) have suggested that this boundary extends towards the Florianópolis Fracture Zone, and the uncertainty is indicated by a question mark in Figure 2 at the southernmost extremity of the São Paulo Plateau. In contrast, the North Angola–Congo–Gabon Atlantic segment (Fig. 2) of the African rift basin is characterized by a salt section much wider and thicker than the corresponding elements on the Brazilian conjugate margin (Mohriak *et al.* 2000; Rosendahl *et al.* 2005).

The bathymetry along the margin of SE Brazil is characterized by a prominent platform narrowing in the NE direction (Figs 2 and 3a, b). The marked widening of the platform to the north of the Vitória–Trindade Lineament is associated with a Late Cretaceous–Early Tertiary magmatic episode that is widespread in southeast Brazil (Mizusaki *et al.* 2002; Mizusaki & Thomaz Filho 2004). The expanded platform in the Abrolhos region is associated with siliciclastic and carbonate rocks overlying volcanic rocks that locally form huge antiformal structures (Mohriak 2004).

Several areas have shallow bathymetry platforms, the most prominent of them lying along the Santos Basin (light colours in Fig. 3b), which is affected by curvilinear irregularities that are concave basin-wards, indicating listric faults affecting the present-day sea-floor. The shallow bathymetry of the São Paulo Plateau is partly caused by salt tectonics, due to the increased thickness of the Aptian evaporite layer in the deep-water region. Gravity gliding from platform to deep basin has inflated the thickness of the original salt layer and uplifted the post-salt sedimentary successions (Mohriak *et al.* 1995). The deep-water province of the São Paulo Plateau is characterized by several groups of bathymetric highs (Fig. 3b) surrounding

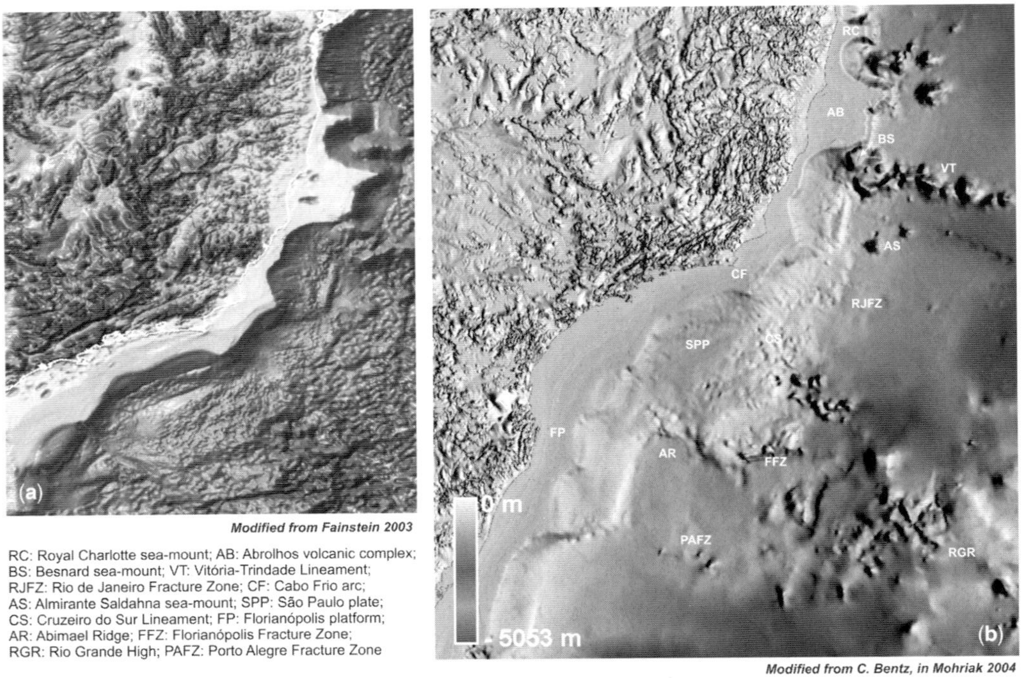

Fig. 3. (**a**) Topographic and bathymetric map of the eastern Brazilian region; (**b**) geomorphological map with location of regional structures and major tectonic features such as the Royal Charlotte volcanic complex (RC), the Abrolhos Volcanic Complex (AVC), the Besnard Seamount (BSM), the Cabo Frio High (CFH), the São Paulo Plateau (SPP), the Florianópolis Fracture Zone (FFZ) and Florianópolis Platform (FP), the Abimael Ridge (AR), and the Rio Grande Rise (RGR).

local lows which correspond to mini-basins filled with Late Cretaceous to Tertiary sediments. Another cause of the shallow bathymetry in the Santos Basin is the extensive clastic progradation that affected the platform from the Late Cretaceous until the Late Tertiary, particularly in the area between the Cabo Frio High and the Florianópolis Platform (Fig. 3b). The Rio Grande Rise corresponds to an elevated plateau or aseismic ridge on the oceanic crust adjacent to the Santos Basin (Fig. 3b). This shallow-water ridge contains volcanic rocks that formed a high on the basin floor during the Late Cretaceous (Kowsmann *et al.* 1977; Thiede 1977; Barker *et al.* 1983).

Based on a regional integration of geological and geophsical data we present a more detailed structural–tectonic map with simplified bathymetric contours and major fracture zones along the SE Brazilian margin, the boundaries of the Aptian salt basin, and the continent–ocean boundary marking the transition to volcanic crust (Fig. 4). Several onshore intrusive bodies, dated as Late Cretaceous to Early Tertiary, form the Cabo Frio igneous lineament, which extends from the Mantiqueira and Serra do Mar ranges to the Cabo Frio Arch (Almeida 1976; Mohriak *et al.* 1995; Mohriak 2004). The continental region adjacent to the Santos, Campos and Espírito Santos basins is characterized by high topographic relief, with elevations reaching up to 2000 m (Figs 3b and 4). The Mantiqueira and Serra do Mar are mountain ranges located less than 200 km from the Cretaceous hinge zone. The drainage divide and the mountain front run sub-parallel to the present-day coastline (Fig. 4). The continental margin in northeastern Brazil, north of the Abrolhos platform, does not exhibit such high elevations near the coastline (Fig. 3a).

The Campos and Santos basins lie predominantly on thinned continental crust. The position of the continent–ocean boundary was determined from gravity data and extrapolated oceanic fracture zones (Fig. 4). The distance between the Cretaceous hinge zone and the ridge of the Serra do Mar and Mantiqueira mountain ranges corresponds to the retreat of the uplifted margin from Cretaceous until present (Fig. 4) and roughly corresponds to the width of the part of the basin affected by gravity gliding of the salt layer. Isopach maps between the top and base of the salt horizon in SE Brazil (Demercian *et al.* 1993) indicate that the deep-water region is characterized by elongated

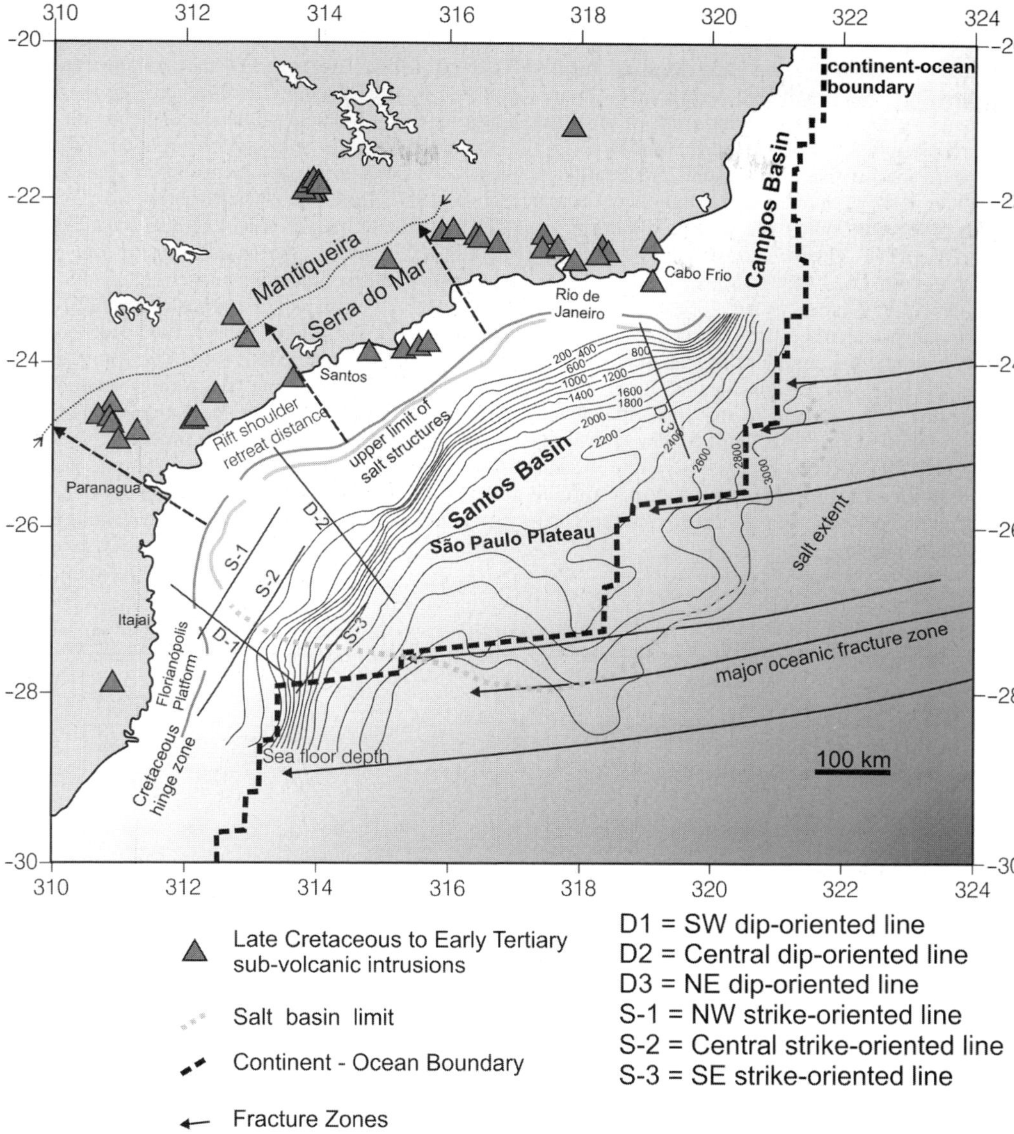

Fig. 4. Simplified bathymetric map of the eastern Brazilian margin with major structural elements offshore (Cretaceous hinge zone, salt boundaries, continent–ocean limit, oceanic fracture zones). Sea-floor bottom contours indicate a SE-facing double-horseshoe shape of the Santos Basin. The region adjacent to the rift border is characterized by volcanic massifs and intrusive plugs; the mountain front and drainage divide run parallel to the coastline. Boundaries of Aptian salt basins and the continental–oceanic boundary are also shown on the map. This map combines features from Almeida (1976), Ojeda (1982), Pereira *et al.* (1986), Pereira & Macedo (1990), Demercian *et al.* (1993), Gallagher *et al.* (1994), Karner (2000) and Mohriak (2001).

salt walls and very thick salt masses that may reach up to several kilometres in thickness. The uppermost portion of the salt horizon is very thin and the lowermost portion is significantly thickened by salt flow, inflating the original layer by several kilometres. The distance between the continent–ocean boundary and the salt extent approximately documents the distance travelled by salt during gravity gliding, ranging from 56 to 162 km at different locations. Such transport is supported by the existence of allochthonous salt tongues that have been identified along the margin as having advanced onto the oceanic crust, as will be discussed below. However, there are also several regions with local evidence for autochthonous salt deposited on a volcanic basement (Mohriak 2001). Volcanic plugs and igneous intrusions are locally onlapped by Aptian salt and sedimentary layers, and these locally cover the landward edge of the seaward-dipping reflectors (Mohriak 2001).

Several studies discuss the extent and location of the volcanic substratum and the continent–ocean boundary for the Santos Basin (e.g., Chang *et al.* 1992; Dias 1993; Gladczenko *et al.* 1997; Karner 2000; Cobbold *et al.* 2001; Meisling *et al.* 2001; Mohriak 2001; Gomes *et al.* 2002). We suggest that the salt basin in the deep-water province may extend beyond the continent–ocean boundary in some areas (Fig. 4). We consider the main syn-rift depocentres to be located to the west of volcanic rocks identified as seaward-dipping wedges (Mohriak 2001). However, this does not eliminate the possibility of post-salt source rocks formed during anoxic periods such as occurred in the Turonian–Cenomanian.

Some authors (Macedo 1990; Dias 1993) have suggested that the pre-salt syn-rift sequence may extend as far as towards the oceanic fracture zones (e.g., Florianópolis Fracture Zone in the Santos Basin); others advocate that the salt mass advanced towards a basement high composed of extrusive and intrusive igneous rocks (Gladzenko *et al.* 1997; Mohriak 2001; Mohriak *et al.* 2002). These two alternatives for the Santos Basin are presented in Figure 5. A regional crustal transect by Macedo (1990) indicates syn-rift sediments overlying the salt basin as far as the Florianópolis Fracture Zone (São Paulo Ridge, Fig. 5a); a more detailed geoseismic profile (Fig. 5b) from the platform towards the oceanic ridge (corresponding to the SE portion of profile 5a) suggests a thicker salt mass pinching-out towards an extrusive volcanic complex. The latter interpretation suggests that some salt may have been originally deposited on a volcanic substratum instead of overlying siliciclastic syn-rift sediments. A present-day analogue

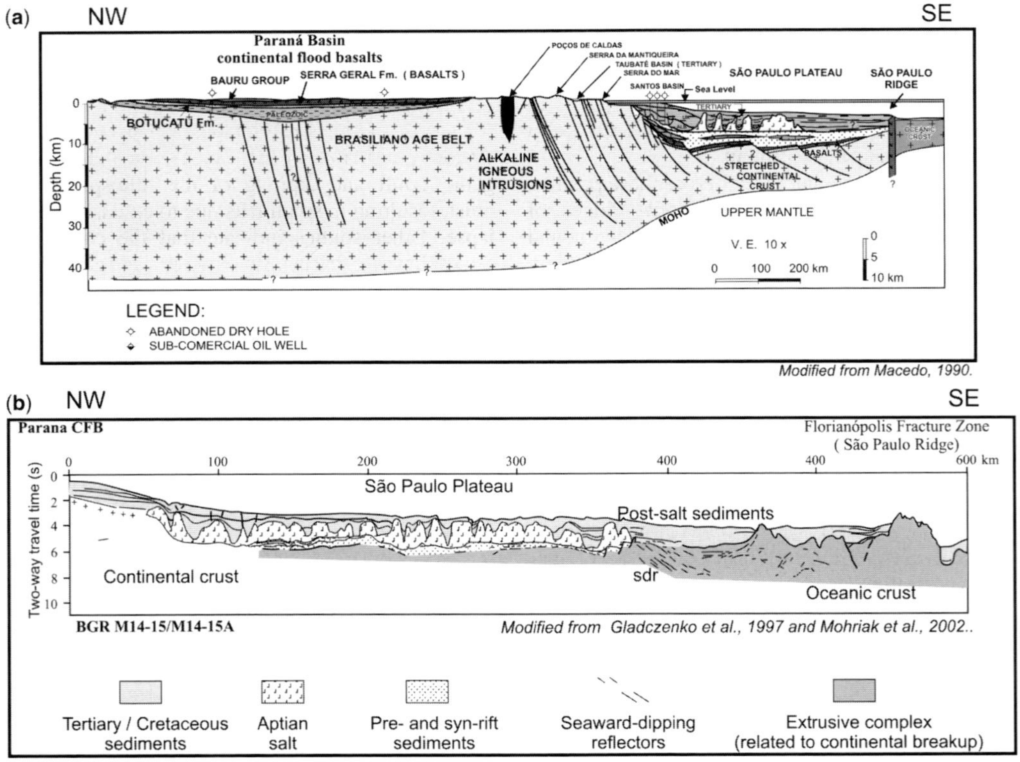

Fig. 5. Alternative interpretations of rift structures and salt tectonics in the Santos Basin. (**a**) Schematic geological transect from the onshore Paraná Basin towards the oceanic crust, crossing the Florianópolis Fracture Zone. (**b**) Schematic geoseismic section across the Santos Basin, from the platform toward the continental–oceanic crust limit, showing alternative interpretations for the rift sequence in the ultra-deep water region. The Paraná Basin continental flood basalts occur to the west of the profile, and the transition to oceanic crust is characterized by wedges of seaward-dipping reflectors (sdr).

for this situation would be the Assal Lake in the Afar region (Jackson *et al.* 2000; Mohriak 2001). In this area, halite is directly deposited on a volcanic substratum formed during the advance of the Gulf of Aden oceanic propagator towards the onshore region.

Demercian *et al.* (1993) suggested that volcanic rocks in the southern Santos Basin might correspond to continental basalts or igneous intrusions in the continental crust (e.g., the Avedis Chain). Based on the regional interpretation of seismic and potential field data, Mohriak (2001) identified a positive anomaly associated with a bathymetric low extending towards the Pelotas Basin (e.g., the Abimael ridge in Fig. 6). This structure could be a sea-floor spreading centre that propagated from the Pelotas Basin towards the Santos Basin but was abandoned when a new spreading centre formed to the east of the ridge. Mohriak (2001) further suggested that plugs described by Demercian *et al.* (1993) as continental volcanic rocks might correspond to igneous intrusions derived from mantle material, related to newly-formed oceanic crust. These ridges are characterized by positive gravity anomalies. We place the continent–ocean boundary landwards and laterally from these abandoned sea-floor spreading centres, rather than along their axes as proposed by Karner (2000) on the basis of a crustal Bouguer anomaly map.

Another possible interpretation of these positive anomalies is that they are peridotite ridges of unroofed upper mantle belonging to the continental lithosphere, as documented in the Iberian margin (Boillot *et al.* 1989; Manatschal 2004 and references therein). Alternatively, Modica & Brush (2004) interpreted the large gravity and magnetic anomalies in the ultra-deep water region of the southernmost segment of the Santos Basin as Precambrian highs. Seismic refraction data and exploratory drilling should help elucidate these questions in the near future.

Interpretation of the continent–ocean boundary is further hampered by the relationship between the seaward-dipping reflector (SDR) wedges and abandoned sea-floor spreading ridges and/or ultramafic bodies. A number of regional deep seismic profiles along the Brazilian margin can be interpreted as imaging SDR-wedges corresponding to sub-aerial basalts and other volcanic rocks (Mohriak *et al.* 2002). The volcanic rocks can be linked to the ultramafic intrusive bodies that would subside more rapidly, together with surrounding oceanic crust, due to their high density. However, the landward limit of the SDR-wedges apparently overlies continental crust, because the pre-salt layers seem to be affected by normal faults. The age of the seaward-dipping reflector wedges and igneous intrusions associated with abandoned ridges seems to be very close to that of salt deposition, since salt sometimes overrides volcanic layers and sometimes it is rifted apart by abandoned ridges (Mohriak 2001).

Following continental break-up, the continental margin did not behave as a typical passive margin. Both extensional and compressional stress events operated during the post-break up development of several small rift basins near the continental margin of São Paulo and Rio de Janeiro states (Fig. 4), particularly along the Serra da Mantiqueira and Serra do Mar ranges (Cobbold *et al.* 2001). The largest is the NE–SW-trending Taubaté Basin, which is about 200 km long and about 20–30 km wide (Melo *et al.* 1985). Its stratigraphy is known only roughly because of the lack of deep exploratory boreholes. The basin fill is less than 1000 m thick (Marques 1990; Mohriak 2003), and the basin architecture is characterized by a series of half-grabens separated by accommodation zones at locations where the rift border faults change their polarity. The interpreted seismic profiles through the basin indicate that its controlling faults were active up to late Tertiary time (Mohriak 2003). The rift grabens are located along the southern border of the uplifted Serra da Mantiqueira. This has led some researchers (Chang *et al.* 1989; Fernandes & Chang 1992) to attribute formation of the Taubaté Basin to epeirogenic uplift and arching of the lithosphere and crust, resulting in small rifts due to extensional collapse. Other studies indicate that the controlling faults have a transpressional component. Zalán (1986) associated the linear configuration of the Taubaté Basin and the high-angle geometry of its border faults to strike-slip processes. More recently, Zalán & Oliveira (2005) suggest that the onshore basins are part of a Tertiary regional mega-rift system that extends from the São Paulo towards the Rio de Janeiro states, bordering the Santos and Campos basins.

The Upper Cretaceous–Lower Tertiary igneous rocks along the borders of the Santos and Campos basins reveal a complex post-break up tectonic development along the eastern Brazilian margin. Present-day activity is indicated by several earthquake epicentres, located both onshore and offshore. The 1955 earthquake in the Espírito Santo Basin (offshore Abrolhos region) exceeded magnitude 6 on the Richter scale (Assumpção *et al.* 1980).

There are several competing interpretations for the post-break-up tectonic events along the Brazilian continental margin. Physical models (Szatmari & Mohriak 1995) indicate that the pattern of post-break up tectonics may be controlled by compression driven by the mid-Atlantic ridge push combined with Nazca plate subduction along the

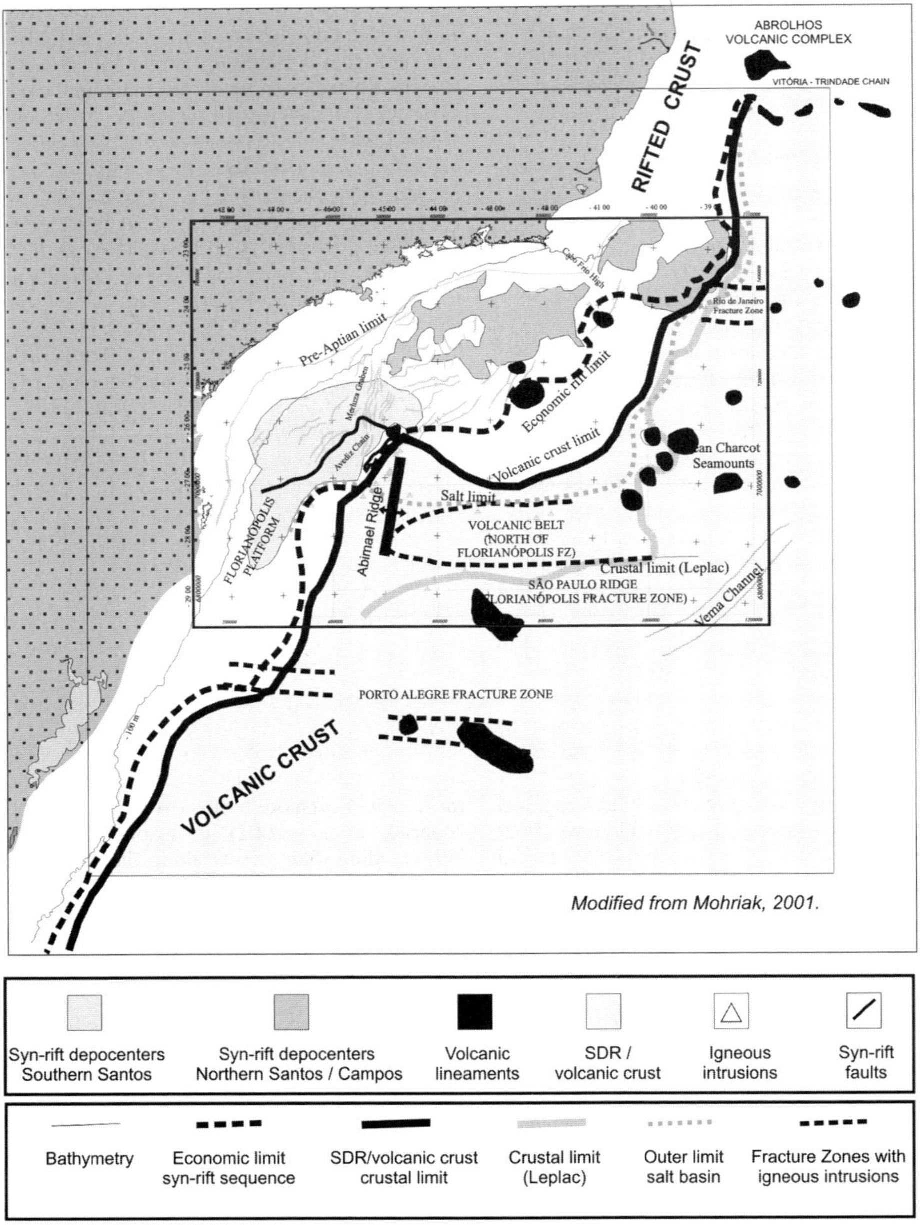

Fig. 6. Tectonic map of the eastern Brazilian margin showing distribution of syn-rift depocentres, volcanic crust limit, and the interpretation of probable propagators advancing from the Pelotas Basin towards the Santos Basin (Mohriak 2001).

Pacific margin of South America. In fact, several coeval events are recognized in the South Atlantic margin and the Andean–Pacific continental margin. The tectonic and magmatic events of the 100–90 Ma old Mirano orogeny in the Andes are coeval with tectono-magmatic episodes in the Santos and Campos basins (Mizusaki & Mohriak 1992; Szatmari & Mohriak 1995). The widespread Eocene magmatism that affected the Eastern Brazilian margin was synchronous with the Incaic orogeny in the Andes (Bussell 1983; Szatmari & Mohriak 1995). These periods of Late Cretaceous–Early Tertiary tectono-magmatic activity correspond to intervals of sand-rich turbidite sedimentation in the deep basin, indicating rejuvenation of the sediment source area along the Brazilian margin.

Tectonic evolution of sedimentary basins along the South Atlantic divergent margin

The tectonic evolution of the South Atlantic divergent margins is discussed on the basis of regional seismic and potential field data, integrated with analysis of several exploratory boreholes drilled along the eastern Brazilian sedimentary basins.

A simplified tectono-stratigraphic column of the Santos and Campos basins (Fig. 7) starts with pre- to syn-rift volcanic rocks that are related to the Camboriú and Cabiúnas formations, respectively. These Neocomian tholeiitic basalts (135–130 Ma, Mizusaki *et al.* 1992, 1998, 2002; Mizusaki & Thomaz Filho 2004) can be found as far north as the Espírito Santo Basin, but are absent along the northeastern Brazilian margin, where the syn-rift sedimentary successions are underlain by Mesozoic and Palaeozoic pre-rift siliciclastic and carbonate sequences (Mohriak *et al.* 1998; Milani & Thomaz Filho 2000).

A detailed topographic–bathymetric map of the Eastern Brazilian margin (Fig. 8) with the continent–ocean boundary defined by seismic, gravity and magnetic data interpreted to yield regional seismic profiles (Leplac Project: Gomes 1992) will be discussed in conjunction with potential field maps and regional deep seismic profiles. As indicated by the Bouguer anomaly map (Fig. 9), the continent–ocean boundary in the Santos, Campos and Espírito Santo basins is located farther away from the coastline than in the Camamu, Jacuípe and Sergipe-Alagoas basins of NE Brazil (Karner 2000; Mohriak *et al.* 2000). This is in accordance with the observation that the rift basins underlain by Neoproterozoic fold belts are wider than those underlain by the Archaean–Palaeoproterozoic São Francisco Craton (Rosendahl *et al.* 2005). The width and thickness of the Aptian salt in the Santos Basin is also much larger than in the northeastern basins (Fig. 2). Figures 8 and 9 also show the location of regional seismic profiles and regional geological sections discussed below, illustrating the seismic signature of the deep crustal structures along the margin.

A geodynamical model for the development of the sedimentary basins along the Eastern Brazilian margin (Fig. 10) is characterized by five tectonic phases (Cainelli & Mohriak 1999; Mohriak *et al.* 2002; Mohriak 2004):

The first phase (Fig. 10-I) represents the onset of lithospheric extension that led to the separation of the South America and Africa in the Late Jurassic–Early Cretaceous. The model for this phase assumes a small degree of asthenospheric uplift and a regionally distributed thinning of the continental crust and upper mantle, with incipient faults in the upper crust controlling local depocentres. The depocentres are associated with widespread, thin sedimentary sequences that are often included in the pre-rift sequence of a large sag basin designated the Afro-Brazilian depression (Cainelli & Mohriak 1999).

The second phase (Fig. 10-II) is characterized by increased lithospheric stretching and asthenospheric uplift (McKenzie 1978; White & McKenzie 1988). This represents the main episode of intracontinental rifting in this part of the South Atlantic, which was associated with effusion of Late Jurassic–Early Cretaceous tholeiitic basalts both onshore and along the incipient continental margins. The northern rift segment, which lacks this magmatism, includes both onshore rifts such as the Recôncavo–Tucano–Jatobá basins and offshore rifts represented by Almada, Camamu, Jacuipe and Sergipe–Alagoas basins (Fig. 2).

The basalt volcanism of the onshore Paraná Basin and offshore Pelotas–Espírito Santo basins is interpreted as precursory to the main rift phase in SE Brazil (White & McKenzie 1989; Mohriak *et al.* 2002). The widespread extrusion of continental flood basalts (e.g., the Serra Geral volcanic rocks in the Paraná Basin) and related tholeiitic basalts in the southern offshore basins (Mizusaki *et al.* 1998; Mohriak *et al.* 2002) is associated with the NW-trending dike swarm along the Ponta Grossa Arch (Milani & Thomaz-Filho 2000). Along the Santos and Campos basin margin, a number of NE-trending Late Jurassic–Early Cretaceous dykes and igneous intrusions have also been found (Cobbold *et al.* 2001; Meisling *et al.* 2001; Mizusaki *et al.* 2002). Ar–Ar dating of these rocks has given an age of *c.* 133 Ma (Renne *et al.* 1992; Turner *et al.* 1994). This magmatic event was followed by normal faulting that affected the whole continental crust and formed half-grabens located to the east of the Ribeira fold belt. This extensional episode did not significantly affect the onshore Serra Geral basalts which cover the Palaeozoic Paraná Basin, although there is local evidence of extensional faults controlling volcaniclastic successions (Riccomini *et al.* 2005).

Extensional stresses of the second phase were concentrated along the present-day continental margin, along intra-continental rifts that developed as a series of coast-parallel elongated and deep lakes, filled with Neocomian to Barremian volcanic and siliciclastic rocks. These rifts evolved to the present-day sedimentary basins of the eastern Brazilian margin. They have their counterpart on the Angolan margin (Chang *et al.* 1992; Mohriak

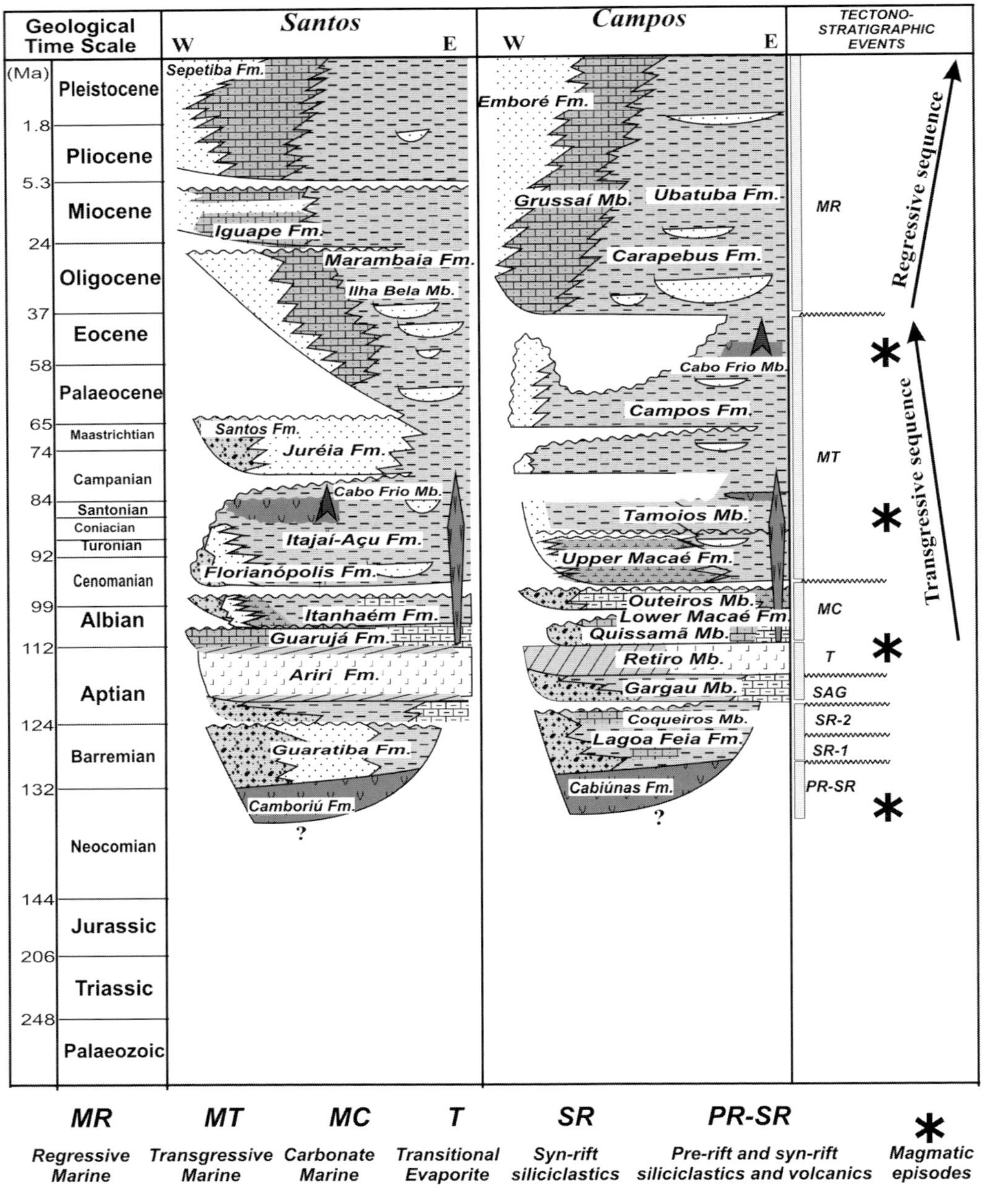

Modified from Mohriak, 2004.

Fig. 7. Lithostratigraphy of the Santos and Campos basins (modified from Pereira *et al.* 1986; Pereira 1990; Pereira & Macedo 1990; Pereira & Feijó 1994; Cainelli & Mohriak 1999; Mohriak 2004). Syn-rift sequences with lithostraphic formations: Camboriú (Santos Basin) and Cabiúnas (Campos Basin), basalts with minor siliciclastic units; Guaratiba and Lagoa Feia, alluvial and lacustrine sediments. Transitional sequences with lithostratigraphic formations: Ariri and Retiro, shallow marine evaporites, predominantly halite and anhydrite. Post-rift marine sequences with lithostratigraphic formations: Guarujá and Macaé, transgressive carbonate platform rocks; Florianópolis and Quissamã, regressive wedge of siliclastic rocks; Itajaí and Tamoios, shaly carbonate rocks with interbedded turbidites; Santos and Campos, regressive wedges with siliciclastic rocks; Sepetiba–Iguape–Marambaia, sandstone–carbonate–shale prograding wedges in the Santos Basin; Emborê–Grussaí–Ubatuba, sandstone–carbonate–shale prograding wedges in the Campos Basin.

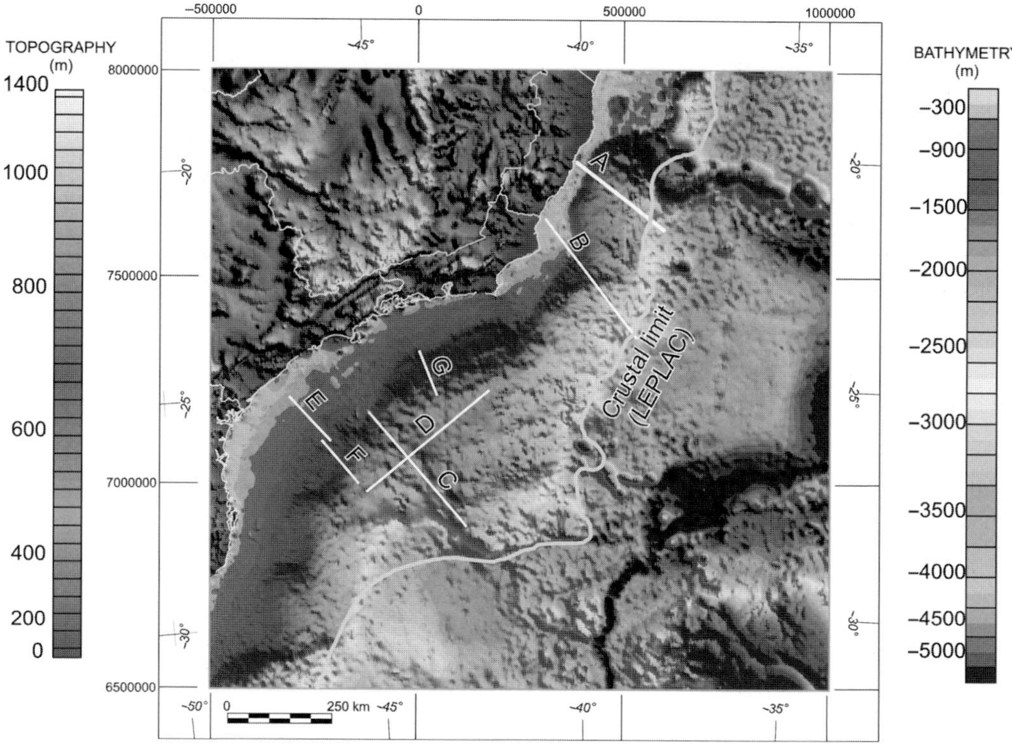

Fig. 8. Topography and bathymetry along the Eastern Brazilian region with location of regional seismic profiles and geological sections discussed in this work. The continent–ocean boundary according to the Leplac Project interpretation is plotted as a grey line.

et al. 2004; Rosendahl *et al.* 2005), forming thick sedimentary troughs associated with the Lukunga, Falcão, Águia and Cuvo formations (Karner *et al.* 2003).

The third phase (Barremian to Aptian, Fig. 10-III) represents the end of syn-rift extension, and is characterized by diminishing activity of large faults that rotated the rift blocks with their sedimentary cover. This phase is locally associated with volcanism, large-fault reactivation, and rift-block erosion, resulting in a regional unconformity that levels the topography, frequently referred as the Break-up Unconformity (Falvey 1974). This unconformity separates continental lacustrine sediments from sediments of transitional to marine environments. Above this unconformity but below the evaporite transitional sequence, some sedimentary basins contain a substantial thickness of Aptian siliciclastic and carbonate rocks, interpreted as sag basin fill (Henry & Brumbaugh 1995). Biostratigraphic and geochemical evidence suggests that this sequence heralds the first marine incursions into the eastern Brazilian–West African gulf, and locally it contains hydrocarbon source rocks (Henry *et al.* 1996).

By Late Aptian times, the arid climate and episodic marine-water influx resulted in an elongated salt basin extending between the West African and eastern Brazilian margins, locally separated by igneous intrusions and volcanic massifs. The gulf was more than 1000 km long with variable width, down to a minimum of *c.* 30 km in the northeastern Brazilian margin, offshore the Jacuípe High (Fig. 2). Its widest segment (*c.* 300 km) was located in the Santos Basin where there is a thick evaporite assemblage characterized by a peculiarly stratified seismic image in the central deep basin. The section contains several evaporite cycles, indicating maximum aridity concentrations capable of depositing highly soluble K-rich salts (Fig. 2).

The fourth phase (Fig. 10-IV) is characterized by development of a mid-Atlantic ridge and focussed spreading processes (Chang *et al.* 1992; Harry & Sawyer 1992; Mohriak 2001; Mohriak *et al.* 2002). The timing of ridge initiation can be

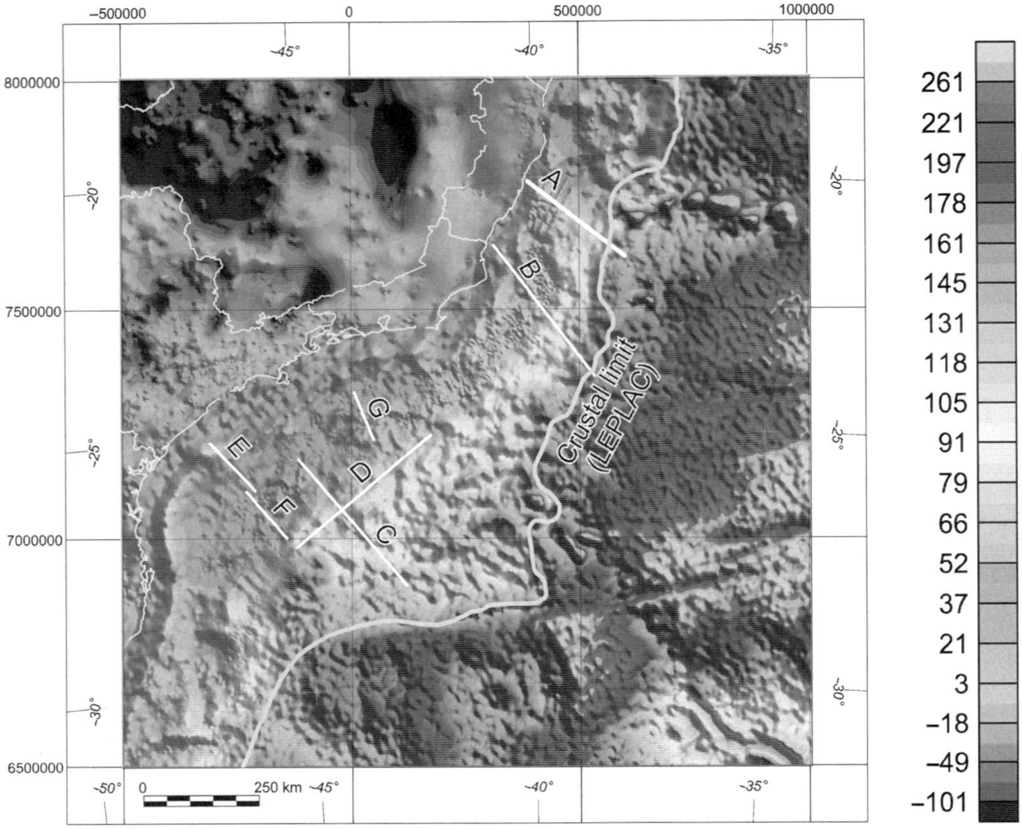

Fig. 9. Bouguer anomaly map (derived from Geosat and Petrobras data) of the eastern Brazilian region with location of regional seismic profiles and geoseismic transects. The continent–ocean boundary according to the Leplac Project interpretation is plotted as a grey line.

determined from the SDRs that lie on top of the outermost rift blocks of the Santos and Campos basins (Souza *et al.* 1993; Gladczenko *et al.* 1997). Here the wedges have the appearance of sub-aerial volcanic rocks formed in the incipient oceanic spreading ridge (Hinz 1981; Mutter *et al.* 1982). They are coeval with development of the break-up unconformity and sedimentation in the sag basin depocentres, and thus predate the evaporites (Jackson *et al.* 2000).

Following Early Cretaceous continental break-up, most of the tectonic activity in the eastern Brazilian sedimentary basins was related to the evolution of the South Atlantic Ocean, with thermal subsidence typical of passive continental margins. Sedimentary successions post-dating the Aptian salt accumulation became predominantly carbonatic by the Albian, suggesting a shallow-water environment that progressively deepened with time.

The fifth phase (Fig. 10-V) extends from Albian to Recent and represents a progressive increase in the bathymetry, which resulted in a marked deepening of the depositional environment by the Late Albian and the demise of the shallow-water carbonates by Cenomanian time. Post-Albian time is characterized by widespread salt tectonics, which apparently controlled most of the structures present on the eastern Brazilian margin. Gravity gliding on top of the evaporites was apparently triggered by extensional tectonics (Jackson & Vendeville 1994) and episodic progradation of clastic wedges related to the uplift of the continental margin (Mohriak *et al.* 1995).

The Late Cretaceous–Early Tertiary period of this phase is characterized by tectono-magmatic events along the southeastern margin. These events include alkaline intrusions, dated at 90 to about 50 Ma (Mizusaki & Mohriak 1992), and tholeiitic extrusions in the Abrolhos volcanic complex of the Espírito Santo Basin that are sometimes associated with compressional structures imaged on seismic data (Lima 2003; Mohriak *et al.* 2003; Mohriak 2004).

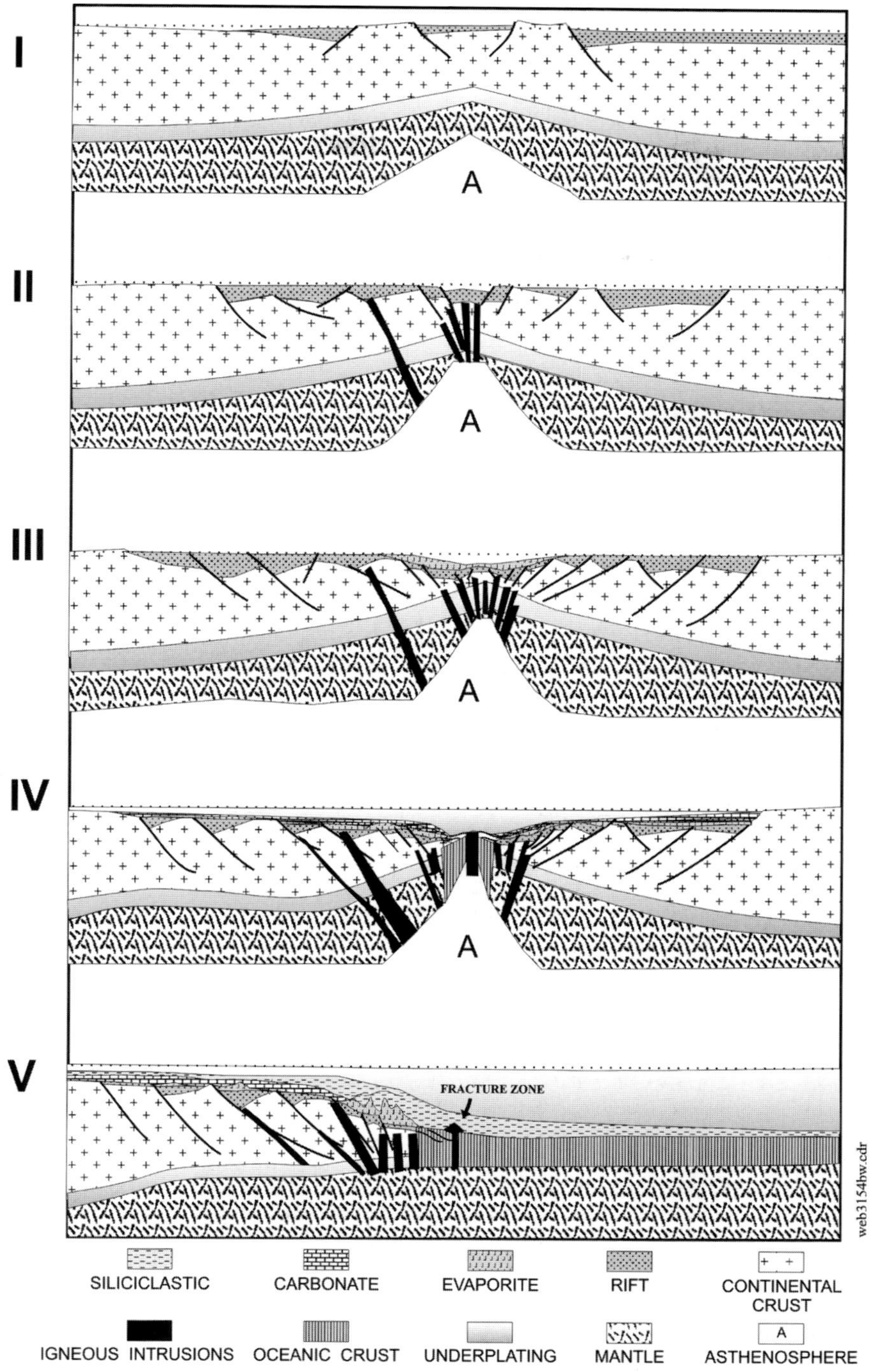

Fig. 10. Schematic diagram showing sequence of events during the break-up of Gondwana and formation of the passive margin sedimentary basins in the eastern Brazilian margin. Dotted line represents lake and sea-level at different stages of basin evolution.

The tectonic evolution of the eastern Brazilian and West African margins can be best evaluated using robust geological and geophysical constraints from the deep basin. Only a few regional deep seismic profiles have been published from this region, and modern refraction data is available only along the West African sedimentary basins (Moulin *et al.* 2005; Rosendahl *et al.* 2005). Most of the regional seismic profiles along the Brazilian margin are located in the Campos (Mohriak *et al.* 1990; Fainstein *et al.* 2001; Fainstein 2003) and Sergipe (Mohriak *et al.* 2000) basins. Below we discuss several new examples of deep crustal architecture using seismic images from the Espírito Santo Basin, to complement equivalent data along the Angolan margin (e.g., Moulin *et al.* 2005).

Figure 11 shows an approximately 270 km long seismic transect through the Espírito Santo Basin, extending from the platform to the deep basin (see location of Transect A in Figs 8 and 9). This

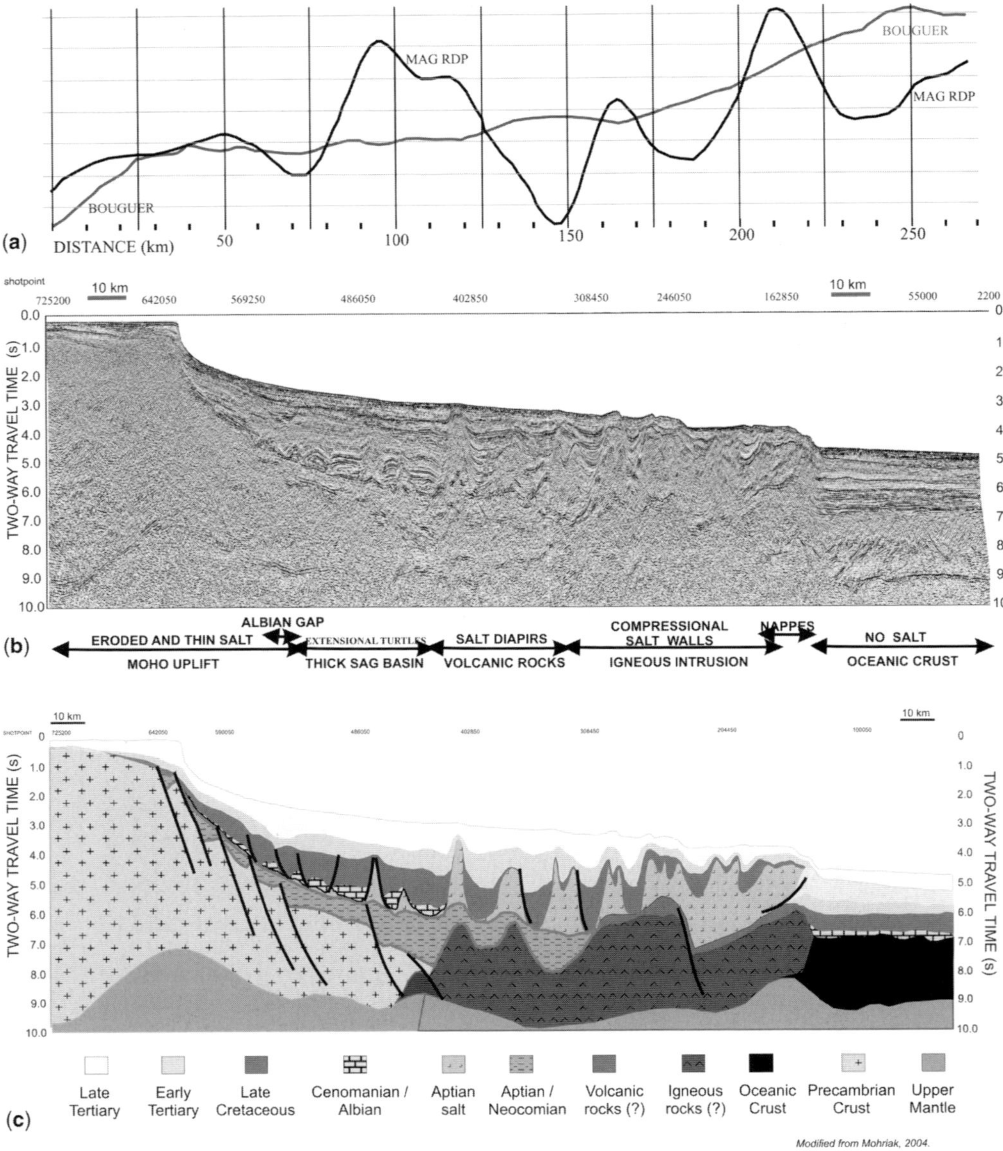

Fig. 11. (**a**) Plot of gravimetric and magnetic anomalies along the regional seismic profile along the Espírito Santo Basin, extending from the platform towards the oceanic crust (Transect A on Fig. 8); (**b**) regional deep seismic profile for the transect A; (**c**) interpretation of major structural and tectonic elements along the deep seismic profile A.

seismic transect is combined with the Bouguer gravity and total magnetic field profiles (Fig. 11a). The shallow-water portion of the profile shows a narrow syn-rift trough filled with Neocomian to Barremian siliciclastic rocks, penetrated by several wells (Cainelli & Mohriak 1999). Near the shelf break, an abrupt up-doming of the Moho is imaged in the seismic profile (10 to 7 s TWT; see Fig. 11b, c for a schematic interpretation), associated with crustal thinning from more than 30 km to less than 20 km over a distance of about 30 km. A thick and broad sag basin is imaged below the thin salt layer, between the shelf break and the first salt diapirs in the deep basin. The profile shows a large hiatus referred to as the Albian gap related to a major listric fault about 25 km beyond the shelf break (Fig. 11b, 60–70 km from the origin at the NW extremity of the profile). The transect reveals evidence of the extensional tectonics that affected the Albian and younger Cretaceous sediments, forming Albian rafts and large antiformal turtleback structures controlled by synthetic and antithetic listric normal faults.

The extensional domain of salt tectonics in this area (Fig. 11b, 50–100 km from the origin) is marked by gravity sliding of Albian rafts above a thin layer of salt and anhydrite. The pre-salt succession is characterized by a very thick sag basin that does not seem to be affected by basement-involved faults. The thickest part of the sag basin is characterized by a very large, positive magnetic anomaly. Salt diapirs, compressional faults and folds typical of a contractional domain are observed above the thickest part of the sag basin. The evaporite layer seems to be highly inflated by salt flow and imbrication towards the oceanic limit of the basin (Fig. 11b, 150–200 km from the origin). The outermost part of the basin in the transect contains an allochthonous salt tongue advancing towards the oceanic crust (Mohriak *et al.* 2004). In front of the salt tongue, the oceanic crust is covered by sedimentary successions with mostly sub-horizontal layering, similar to equivalent features observed along the Angolan margin (Fainstein & Krueger 2005). The oceanic crust (220–270 km from the origin) is characterized by a transparent upper layer, an intermediate layer containing sets of seaward- and landward-dipping reflectors, and a transparent lower layer whose base is marked by a strong reflector at about 9 s TWT, which most likely corresponds to the Moho.

Comparison between the seismic and the potential field data (Fig. 11b) shows that the oceanic crust is characterized by flattening of the Bouguer gravity anomalies, and the transition from continental to oceanic crust is associated with increased complexity of salt tectonics. Large magnetic anomalies indicate the presence of highly magnetized intrusive bodies underlying the salt diapir province (Fig. 11b, 150–210 km from the origin).

The transition between the sag basin and the oceanic crust corresponds to a salt deformation front in a contractional domain that is not underlain by any typical rift structures. An outer high, associated with strong magnetic anomalies, probably corresponds to an intrusive complex located near the continent–ocean boundary that was overridden by salt nappes (Fig. 11b, c).

Seismic data from this basin have been interpreted as showing allochthonous salt tongues similar to those in the Gulf of Mexico basin (e.g., Demercian *et al.* 1993; Mohriak *et al.* 2004). One such salt tongue is imaged in the deep seismic profile of Figure 11b (205–215 km from the origin). A simple depth-conversion of the time data (Fig. 12) places the base of the oceanic crust (Moho) at a depth of about 13–15 km.

A schematic interpretation of the regional deep-seismic profile through the Campos Basin (Fig. 13, see Transect B in Figs 8 and 9 for location) extends from the platform towards the oceanic crust. It illustrates the crustal architecture and the gravity-glide domains on this continental margin (Mohriak *et al.* 1998; Mohriak *et al.* 2004). The gravity glide contains five domains: (1) Domain I displays extensional tectonics and small amounts of salt; (2) Domain II displays extensional tectonics and larger volumes of salt, associated with salt flow and gravity gliding, and gravity spreading of the overburden; (3) Domain III displays very large salt diapirs associated with extensional to compressional tectonics; (4) Domain IV displays large volumes of autochthonous and allochthonous salt associated with compressional tectonics; (5) Domain V is characterized by oceanic crust without salt, in which the volcanic basement is overlain by relatively thin layers of post-salt sediments that are locally intruded by igneous bodies.

A dip-oriented regional seismic profile though the Santos Basin (Fig. 14, see Transect C in Figs 8 and 9 for location) extending from the plaftorm towards the oceanic crust across the Florianópolis Fracture Zone, illustrates the main tectono-sedimentary compartments of the continent–ocean transition. The profile documents that the platform in the Santos Basin is much larger and wider than the platform in the Espírito Santo Basin. In contrast with the Campos and Espírito Santo basins, the platform is covered by a much thicker Cretaceous sedimentary succession that progrades basin-wards. The profile shows that the Cretaceous sedimentary layers are strongly controlled by salt tectonics. Part of the profile extends towards the ultra-deep water region, which is marked by a very thick evaporite sequence with a peculiarly stratified seismic signature and an enigmatic reflector that extends

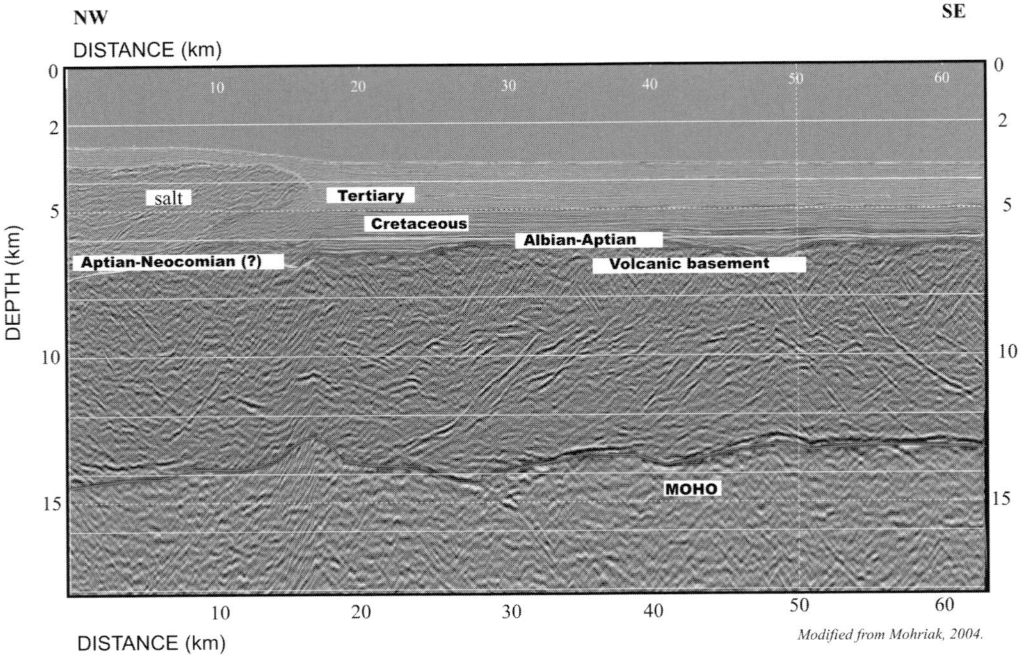

Fig. 12. Depth-converted segment of deep seismic profile showing allochthonous salt tongue advancing towards the oceanic crust in the deep water region of the Espírito Santo Basin. This image results from depth-conversion of the time data using simple velocity tables.

from the syn-rift depocentres towards the volcanic body located north of the Florianópolis Fracture Zone and thus overlies a volcanic basement region.

The stratified evaporite succession is also imaged in a strike-oriented regional seismic profile through the Santos Basin (Fig. 15, see Transect D in Figs 8 and 9 for location). This profile extends from the thick salt basin in the deep-water region of the Santos Basin towards the Pelotas Basin. The SW end of the profile is characterized by absence of salt, in an area characterized by a bathymetric low interpreted as underlain by

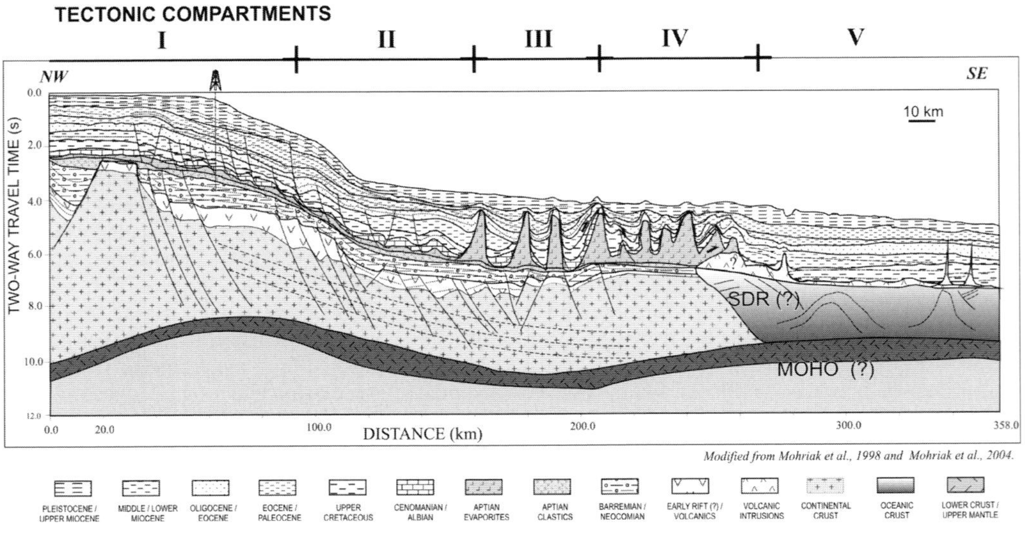

Fig. 13. Salt tectonics domains along the continental margin (Transect B, Campos Basin; see location on Fig. 8).

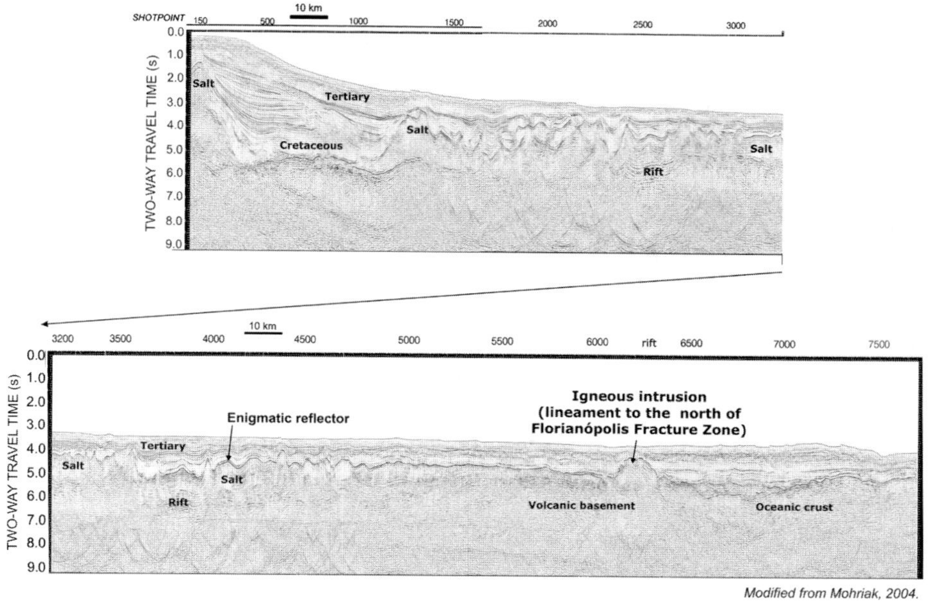

Fig. 14. Seismic profile in the Santos Basin (Transect C, Fig. 8) extending from the platform towards the oceanic crust.

oceanic basement (Fig. 15b). The low-bathymetry region has a triangular geometry in gravity and magnetic maps, which is apparently controlled by aborted spreading ridges that propagated northwards (Mohriak 2001). The very thick sedimentary cover in transect D includes Cretaceous and Tertiary sequences slightly affected by folding. The lowermost part of the sedimentary succession is strongly affected by igneous intrusions, which are onlapped by post-salt Cretaceous layers; the Late Cretaceous and Early Tertiary layers are only slightly affected by these intrusions.

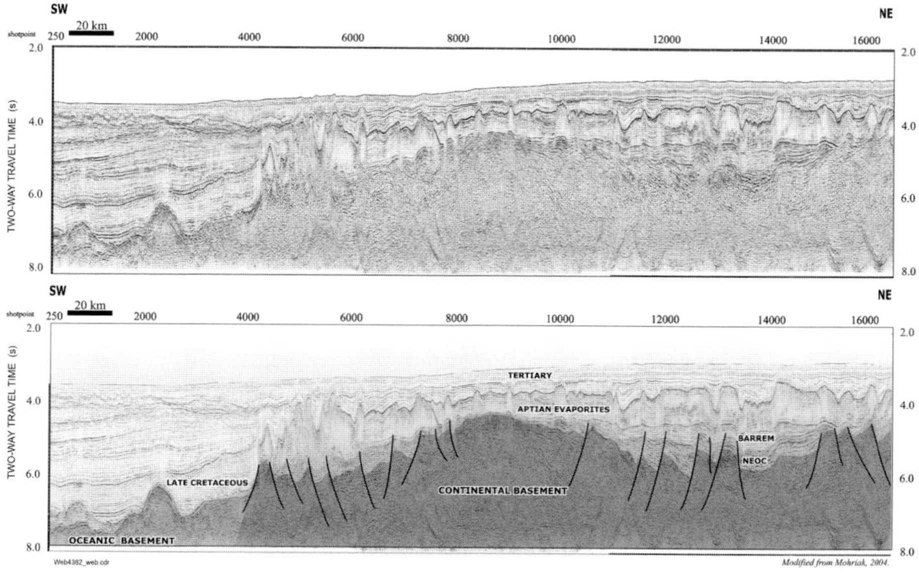

Fig. 15. Seismic profile in the Santos Basin (Transect D, Fig. 8) along a strike direction, crossing the Southeastern High and advancing towards oceanic crust in the Pelotas Basin, beyond the salt limit.

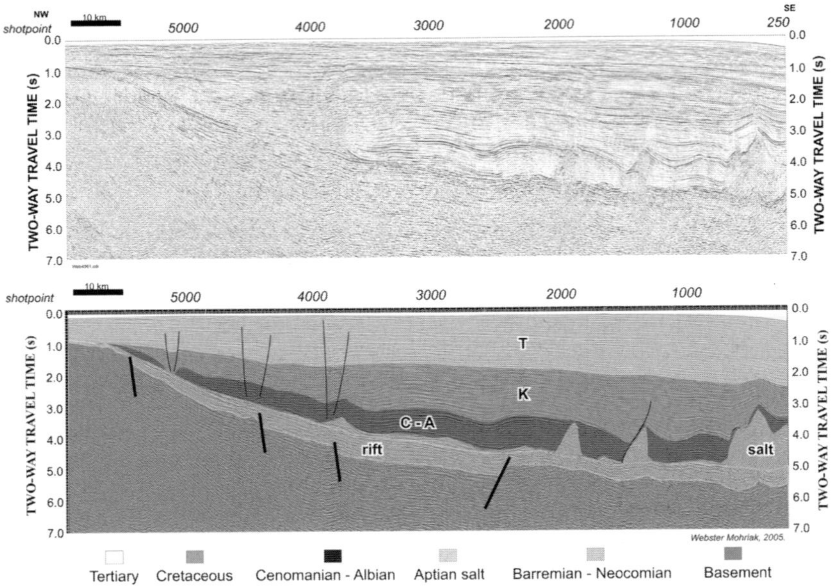

Fig. 16. Seismic profile (Transect E, Fig. 8) along a dip direction in the Santos Basin. Top: uninterpreted seismic profile; below, schematic interpretation with main stratigraphic sequences.

A more proximal profile running through the southern Santos Basin from platform to shelf-break (Fig. 16, see Figs 8 and 9 for location of Transect E) illustrates the pinch-out of the syn-rift and pre-Tertiary post-rift sequences. Although not well imaged below the salt, Neocomian–Barremian syn-rift strata (probably including volcanic layers) are inferred along most of the profile, except in the first 10 km, where the sequence seems to be largely eroded. The proximal, shallow-water region in the profile (*c.* 30 km in width) is characterized by residual evaporites resulting from salt flow and dissolution. Salt pillows and diapirs are imaged near the shelf break. The extensional domain contains gentle anticlines and turtle structures.

A regional seismic profile through the southern part of the Santos Basin (Fig. 17, see Transect F in Figs 8 and 9) illustrates the control of salt tectonics on the geometry of the post-rift sedimentary sequences. The syn-rift sequence is observed towards the western segment of the profile, but is not unequivocally interpreted in the deep-water region, where the salt layer seems to be underlain by volcanic rocks assumed to be basement. This profile illustrates the development of very tall salt diapirs and a series of antithetic faults that resulted from the salt flow towards the deep basin. These extensional faults are related to episodes of massive clastic progradation on the platform (Mohriak *et al.* 1995).

The development of the Cabo Frio fault zone and the Albian gap (Fig. 18) is related to salt tectonics and magmatism. The northernmost extremity of the fault zone coincides with the Cabo Frio High, which is punctured by several intrusive plugs dated from Late Cretaceous to Early Tertiary. Alkaline intrusive bodies also occur onshore, apparently forming a trend that becomes younger from the Poços de Caldas intrusion towards the Cabo Frio intrusion near the coastline (Figs 4 and 18). The location of the Albian gap, between the Cabo Frio High and central portion of the Santos Basin, is strikingly parallel to the area with the elevated Serra do Mar and Serra da Mantiqueira ranges (Fig. 2) and sub-parallel to the Tertiary rifts (São Paulo, Taubaté and Resende basins, Fig. 18).

A regional seismic profile through the Cabo Frio fault zone in the central part of the Santos Basin (Fig. 19, see Fig. 18 for location) illustrates the main features related to salt tectonics controlling post-rift sedimentary sequences. Figure 19b shows the interpretation of massive clastic progradation wedges expelling salt from the basin and forcing it to move basin-wards, while a major counter-regional listric fault (*sensu* Rowan *et al.* 1999) was formed and forced down by the weight of sediments accumulated on the salt layer. This mechanism also created a broad zone where Albian carbonates are absent due to their basin-ward gliding: the Albian gap (Mohriak *et al.* 1995).

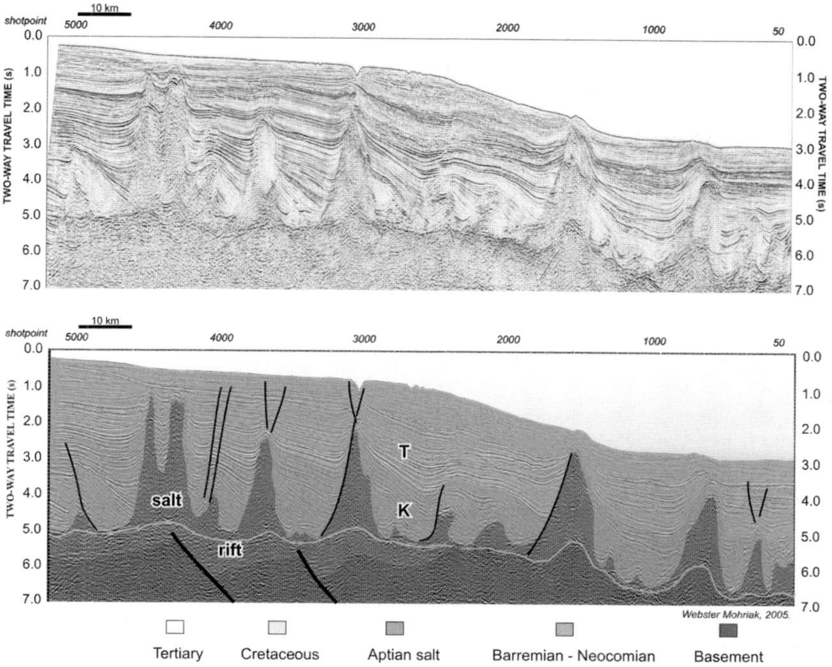

Fig. 17. Seismic profile (Transect F, Fig. 8) along a dip direction in the Santos Basin. Top: uninterpreted seismic profile; below, schematic interpretation with main stratigraphic sequences.

Several alternative hypotheses have been proposed for the origin of the Cabo Frio fault zone. Some authors suggest salt withdrawal with negligible extension as the main mechanism (Ge *et al.* 1997), whereas others suggest salt mobilization with major extension forming carbonate rafts that moved basin-wards (Szatmari *et al.* 1996; Mohriak & Szatmari 2001).

Uplifted flanks of the Eastern Brazilian margin rift basins

The sedimentary basins along the Eastern Brazilian margin are characterized by a mountainous region onshore, suggestive of uplifted rift flanks. A regional satellite image depicting the topography of continental SE Brazil (Fig. 20a), tilted with vertical exaggeration to emphasize the geomorphological details of the elevated regions bordering the Santos Basin northwards of the pre-Aptian limit of the basin (Fig. 18). The onshore Taubaté and Resende basins, located between the Serra da Mantiqueira (to the north) and Serra do Mar (to the south) are bordered to the north by several igneous intrusions that show a general east–west trend. The topographic relief from the highlands towards the coastline (Fig. 20a and b) is marked by depressions which correspond to the Tertiary rifts. Figure 20b also shows the location of a topographic profile from the highlands towards the coastline. This profile starts from high plateau elevations of almost 2000 m in the Mantiqueira region, crosses topographic depressions of the Taubaté Basin and the Guanabara rift (which are divided by high elevations of the Serra do Mar, near the inflection point B), and terminates beyond the coastline, crossing an elevated offshore region at Ilha Grande (Fig. 20c). The eastern extremity of the profile approaches the rift-border fault that limits the Early Cretaceous volcanic rocks (point C). This profile corresponds to an uplifted, highly eroded, rift shoulder, segmented and modified by Tertiary rifts (Zalán & Oliveira 2005). A simple restoration to its pre-Tertiary state results in a cross section of residual elevations rising smoothly from NW to SE. This provides a post-erosional picture of the uplifted rift flank with the Serra da Mantiqueira and the Serra do Mar ranges that make up a single mountain chain (Fig. 20c, d). This restoration shows an uplifted shoulder, commonly observed in other rift regions worldwide, such as the North Sea, Gulf of Suez or the East African rifts (e.g., Roberts & Yielding 1991; Kusznir & Ziegler 1992; Hendrie *et al.* 1994). The rift shoulder in the Santos Basin is

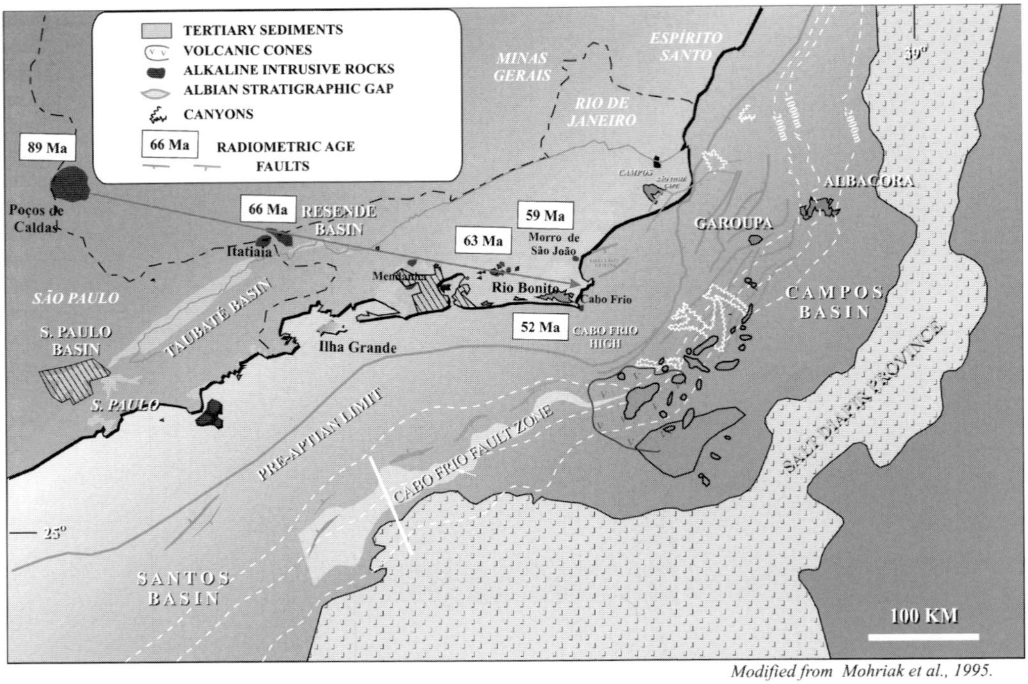

Fig. 18. Regional tectonic map of the eastern Brazilian margin showing the east-west trending Cabo Frio lineament with igneous intrusions and the location of the Cabo Frio fault zone in the Santos Basin. Transect D is a regional seismic profile along a dip direction crossing the Albian gap. The Taubaté Basin extends across the eastern part of São Paulo State, and the Resende Basin is located in Rio de Janeiro State. Ilha Grande is located in a major bay near the border of these states, landwards of the pre-Aptian hinge line, which shows a major inflection from a NE to an east-west trend towards the Cabo Frio High.

typically asymmetric, with a steep scarp facing the sea and a gentle slope facing the land (Fig. 20c). A conceptual model balancing erosion in such uplifted shoulder regions with deposition of sediments in adjacent depressions is shown in Figure 20d. The boundary between the rift and uplifted shoulder is usually marked by a master fault (e.g., the Camamu–Jequitinhonha margin in Eastern Brazil) or a complex fault system (e.g., the South Gabon margin, Rosendahl *et al.* 2005).

The development of rift shoulders have been explained by models based on both transient and permanent uplift mechanisms. Transient uplift models are related to the thermal effects of rifting and include depth-dependent extension (Royden & Keen 1980; Hellinger & Sclater 1983; Watts & Thorne 1984; Morgan *et al.* 1985), lateral heat flow (Steckler 1981; Cochran 1983; Alvarez *et al.* 1984; Buck *et al.* 1988) and secondary convection under rift shoulders (Keen 1985; Steckler 1985; Buck 1986). These mechanisms, which can create 500–1500 m shoulder elevation, operate only during the elevated thermal regime of the lithosphere. Such positive topography will decay over a time period equivalent to the thermal time constant of the lithosphere, which is roughly 60 Ma (McKenzie 1984). Permanent uplift models are related to magmatic underplating (Cox 1980; Ewart *et al.* 1980; White & McKenzie 1988) and lithospheric unloading and/or plastic necking (Zuber & Parmentier 1986; Parmentier 1987; Braun & Beaumont 1989; Issler *et al.* 1989; Chery *et al.* 1992; Weissel & Karner 1994).

Based on fission-track geochronology, processes of erosion and deposition have recently begun to be incorporated into existing models of rift shoulder uplift (e.g., van Balen *et al.* 1995; van der Beek *et al.* 1995; Burov & Cloetingh 1997). Initial lithospheric stretching brings hot mantle material closer to the surface, resulting in an accentuated thermal regime in the crust beneath the basin. This results in a decrease in the effective elastic thickness values beneath both the basin depocentre and the rift shoulders. In response, ductile material of the lower crust flows from the basin centre towards the rift shoulders, which are placed farther from

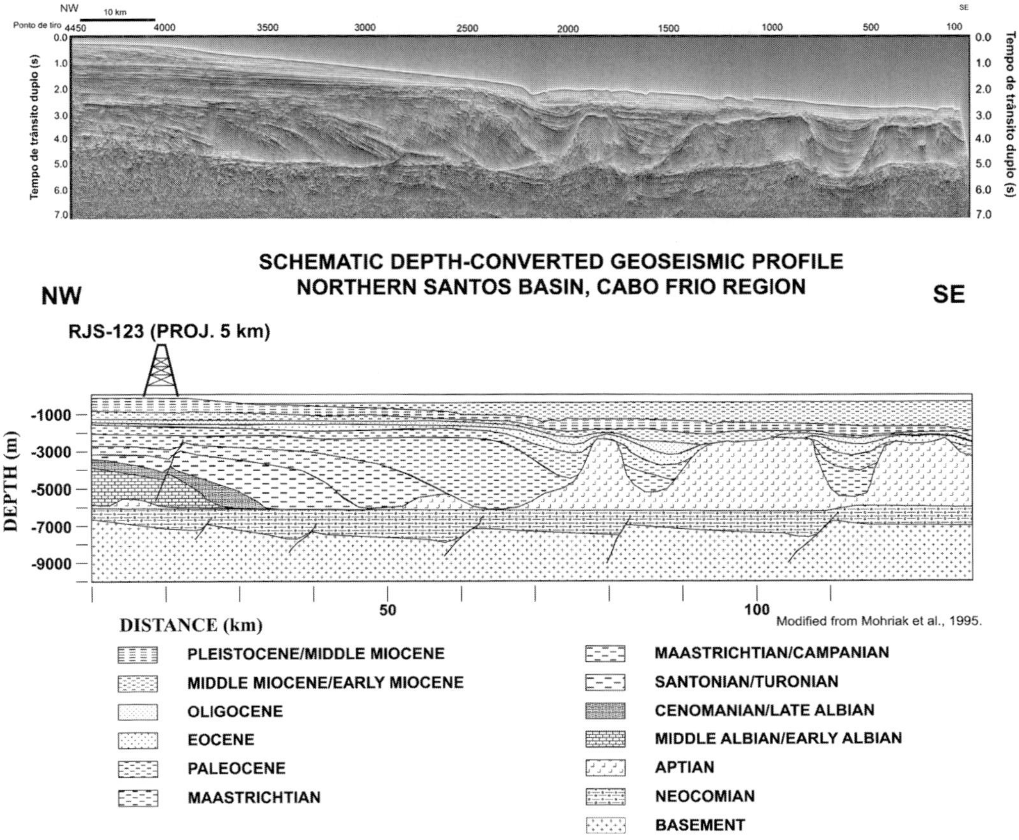

Fig. 19. Top: regional seismic profile corresponding to Transect D (Fig. 18). This time section was plotted with special processing to enhance the salt bodies and the prograding siliciclastic wedges; below, geological depth-converted section with schematic interpretation of the stratigraphic sequences.

the initial extensional value, facilitating their uplift. Erosion of rift shoulders also enhances their uplift through isostasy.

The Santos Basin (Fig. 4), chosen as an example, cannot have its rift shoulder morphology explained by transient uplift models, since it reaches elevations of up to 2 km today, and would have been as high as 2.8–3.3 km prior to the Tertiary rifting that formed Taubaté and Guanabara basins (Fig. 20c). Onshore apatite fission-track data (Gallagher *et al.* 1994) and offshore sedimentological data (e.g., Almeida 1976; Pereira *et al.* 1986; Pereira 1989; Pereira & Macedo 1990; Pereira & Feijó 1994; Cainelli & Mohriak 1999) indicate that rift-shoulder uplift took place roughly from Late Cretaceous to Quaternary time. The most significant erosional remnant of the rift shoulder, the Serra do Mar and Mantiqueira mountain fronts, are located 172–247 km landward from its assumed initial Late Cretaceous position at the pre-Aptian hinge line (Fig. 4). If we place the elevated topography at Ilha Grande (Fig. 20c, d) as the most external remnant, the erosional retreat of the shoulder decreases to 112–187 km.

Regional geoseismic transects in the Santos Basin

The analysis of regional transects in the Santos Basins indicates that the erosion/deposition patterns observed in seismic data are not in accordance with results of flexural models of passive margin development.

Stratigraphic data obtained from the integration of hundreds of exploratory boreholes do not conform to the results of flexural models coupled with erosion and deposition. The characteristic offlap pattern observed in the seismic profiles along the basin (Figs 19 & 22) do not follow the site of Neocomian–Aptian rifting. But the offlap pattern formed during the Senonian, not during

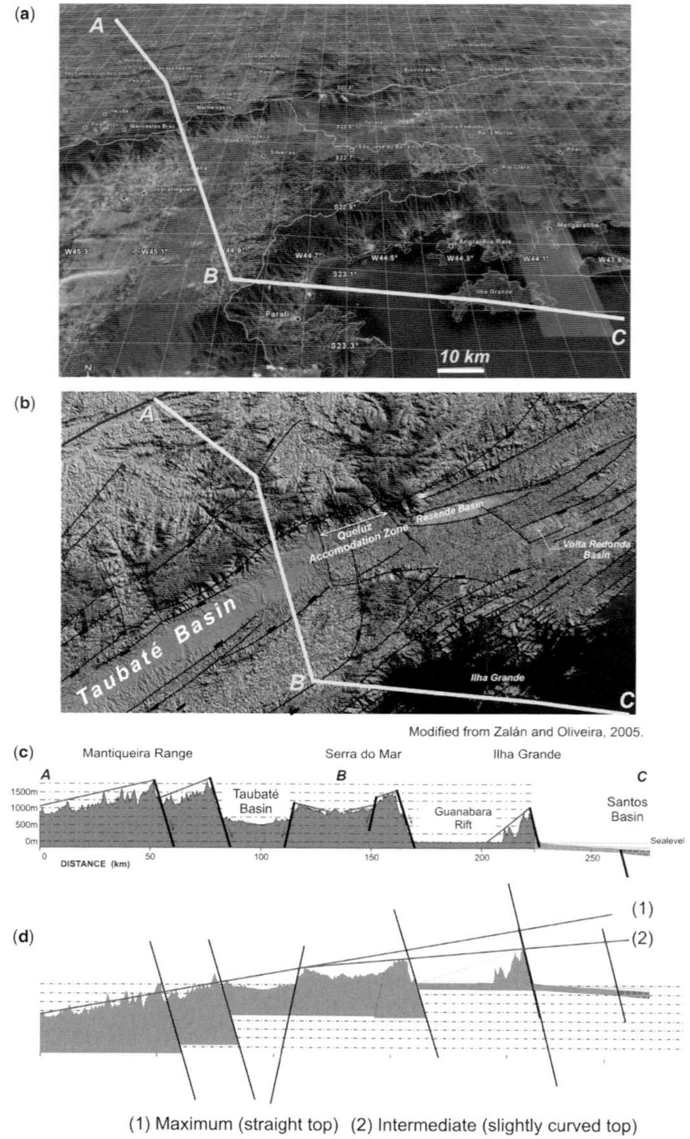

Fig. 20. (**a**) Earth satellite image of southeastern Brazil with 3× vertical exaggeration to illustrate geomorphological features. A topographic profile A–B–C crosses the elevated regions of Serra da Mantiqueira and Serra do Mar, and reaches the coastline near Ilha Grande. (**b**) Satellite image of southeastern Brazil showing topography and the approximate location of the topographic profile A–B–C from Zalán & Oliveira (2005). (**c**) Topographic profile A–B–C from the Mantiqueira Range towards the Santos Basin. Straight lines above the mountain ranges are interpreted by Zalán & Oliveira (2005) to correspond to residual topography resulting from faulted flanks along the depressions (Taubaté Basin and Guanabara rift). Basin-wards of Ilha Grande, geological maps identify the pre-Aptian hinge line, which limits the Neocomian syn-rift rocks and the Aptian Ariri Formation evaporites. On the platform, Quaternary–Tertiary sediments cover the Precambrian basement and tholeiitic lava flows associated with the pre-salt Camboriú Formation. (**d**) Schematic restoration of topographic profile A–B–C previous to Tertiary rifting.

the Albian, with a delay of about 24 Ma indicating that the timing of rift shoulder uplift must have been affected by subsurface crustal processes and another surface process in addition to erosion and deposition. One of the goals of the following theoretical discussion is to assess the seismic evidence for tectonic and sedimentary patterns that may lead to a refined definition of this process.

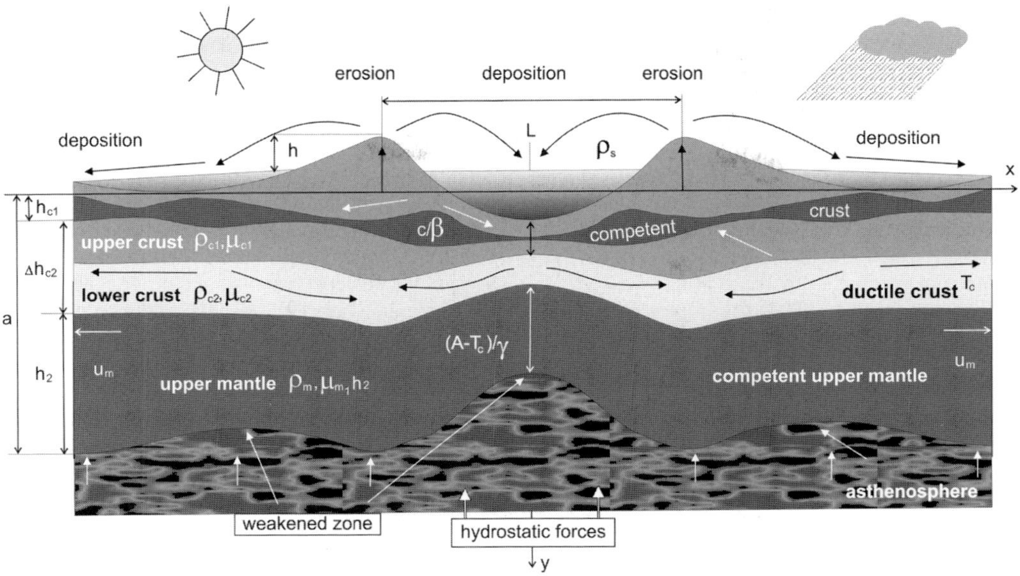

Fig. 21. Rift-related basin subsidence and rift-shoulder uplift, after van Balen *et al.* (1995) and Burov & Cloetingh (1997). Black arrows and ρ_s show sediments eroded from rift shoulders that in turn exert a sedimentary load on basin; white arrows indicate relatively strong parts of the crust and mantle lithosphere, which flex and weaken.

Flexural models of the rift shoulder uplift incorporating erosion and deposition indicate that positive or negative topography of the rift shoulder is affected by the depth of processes involved in rift necking and rates of surface denudation (e.g., Burov & Cloetingh 1997; Fig. 21). Subsidence of the deep-seated necking and uplift of the rift shoulders create pressure gradients that can drive

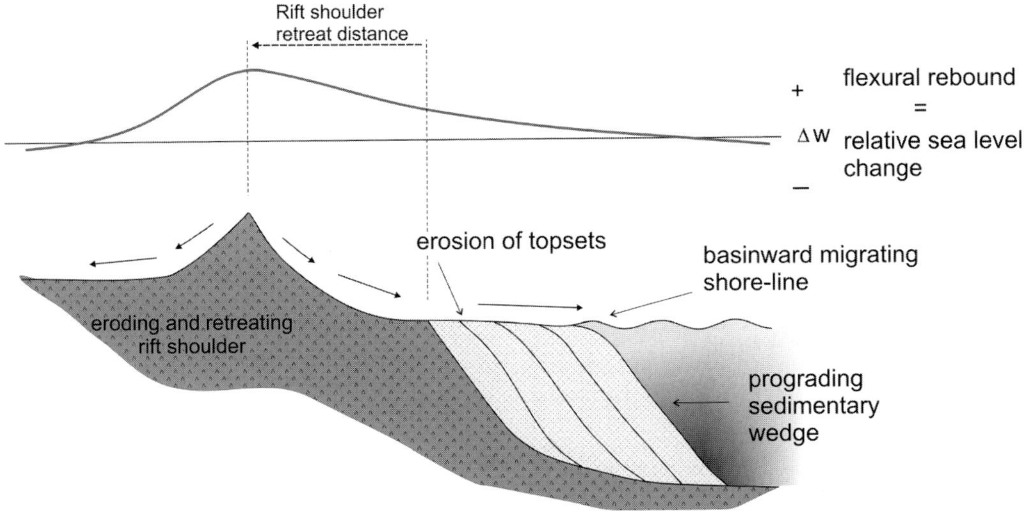

Fig. 22. Conceptual model showing the initial post-rifting coastal offlap and the rift shoulder retreat at passive margin, after van Balen *et al.* (1995) and Burov & Cloetingh (1997). The flexural rebound related to erosion of the rift shoulder causes uplift, which affects the adjacent portion of the basin, causing erosional truncation of topsets of the sedimentary wedge. Coastal onlap can occur when the uplift rate is lower than basin subsidence, which is controlled by thermal contraction and the flexural response to the increase of the lithospheric rigidity due to cooling and sedimentary loading. $+\Delta w$ indicates the instantaneous flexural uplift and $-\Delta w$ the subsidence.

ductile flow within the lower crust, resulting in additional basin subsidence and shoulder uplift. This mechanism can be further enhanced by rapid erosion of the shoulders or rapid deposition in the basin (Fig. 22). On the other hand, shallow-seated necking, slow erosion of shoulders or slow deposition in the basin can retard or even reverse ductile flow, reducing uplift of shoulders and basin subsidence. Erosion of topography distributes the load areally. Thus, surface processes may enhance or counteract the deep processes.

The sedimentary fill of the Santos Basin indicates that deep-seated necking is characterized by Senonian and Eocene–Recent offlap patterns extending from the margin towards the centre. Slow erosion of the basin margin is indicated by Cenomanian–Turonian and Palaeocene sediments that onlap the margin, although the effects of global sea-level variations cannot be ruled out.

The seismic stratigraphic patterns along the offshore region indicate neither post-rift offlap nor prominent post-rift onlap immediately following the deposition of evaporites (Figs 16, 17 & 19). Figures 23 and 24 indicate these relationships in dip- and strike-oriented regional geoseismic transects across the basin. Syn-rift Neocomian lacustrine sediments accumulated in grabens formed by extensional tectonics that affected the Early Cretaceous basalts (Figs 16, 17 & 19). Barremian–Aptian alluvial fan/fan deltas and shallow-water sediments were deposited in a sag basin only slightly affected by normal faults. These sediments, in turn, were overlain by shallow marine Late Aptian evaporites, forming a salt basin during the early phases of post-rift thermal subsidence (Jackson *et al.* 2000). The evaporites were overlain by Albian platform carbonates with shallow-water facies in the lower successions (Modica & Brush 2004). The seismic interpretation suggests that marine transgression formed a short-lived onlap pattern only during later phases of thermal subsidence, particularly during the Cenomanian–Turonian interval. This sequence is marked by deep-water anoxic environments and deposition of rich source rocks (Modica & Brush 2004).

This marine transgression was soon followed by a Senonian regression associated with the large prograding clastic wedge of the Santos Formation (Figs 16, 17, 23 & 24). Deposition was coeval with erosion of the rift shoulder, as indicated by apatite fission-track analysis (Gallagher *et al.* 1994). The prograding wedge of the Santos Formation was overlain by Palaeocene onlapping sediments and several younger regressive sedimentary packages (Fig. 19).

In summary, offshore stratigraphic patterns indicate no major onlap or offlap immediately after the end of Neocomian–Aptian rifting (as interpreted in Figs 23 and 24); onlap was delayed by *c.* 13 Ma. Sedimentary layer areas (from Figs 23 and 24) divided by deposition intervals indicate slow deposition rates during Albian–Turonian in comparison with younger periods. After *c.* 11 Ma, this was followed by an offlap pattern that lasted *c.* 21 Ma. Our calculations of the sediment area deposited during one year indicate an increase in the deposition rate during this time period: 1.9–4.8 times faster than the Albian–Turonian rate. This in turn was followed by a Palaeocene onlap and progradation until the Neogene (Fig. 19). All this evidence points to erosion of uplifted regions affecting the post-rift development of the continental margin as a feedback between surface and subsurface processes. Deep-seated processes modified the initial post-rift setting and caused long-delayed but distinct erosion of the rift shoulder during Senonian and Eocene–Quaternary.

Flexural models of basin subsidence and uplift indicate that the combination of subsurface and surface mechanisms should have triggered shoulder uplift much earlier than the Santonian. The observed prograding clastic wedges developed after a period of insignificant erosion and deposition. This requires an additional surface mechanism that progressively increased in effect from Albian to Maastrichtian time, and was as important as erosion/deposition.

Influence of salt tectonics in the post-rift development of the Santos Basin

Widespread Aptian evaporites have been mapped regionally by several researchers in the Santos Basin (Ariri Formation), as well as in the other basins of the eastern Brazilian margin (e.g., Demercian *et al.* 1992; Cobbold *et al.* 1995; Mohriak *et al.* 1995; Cobbold *et al.* 2001). The salt layer acts as a detachment zone between the syn-rift and sag-basin sediments and the post-salt Albian carbonates. The sediments above the detachment are controlled by extensional tectonics and are deformed by gravity gliding and gravity spreading (e.g., Petrobras 1983; Pereira & Macedo 1990; Cobbold & Szatmari 1991; Duval *et al.* 1992; Cainelli & Mohriak 1999). These thin-skinned structures are present in both the Brazilian and West African margins and follow the tenets of extensional salt tectonics (Jackson & Vendeville 1994), with listric normal faults detaching along the evaporite ductile zone (Fig. 16).

The upper limit of the Aptian salt basin lies basin-ward from the Cretaceous rift margin. This can be determined by the presence of the first remnant salt body in a down-dip direction from the hinge zone (Figs 16 & 23). The lower or distal limit of the salt basin lies on the transition from

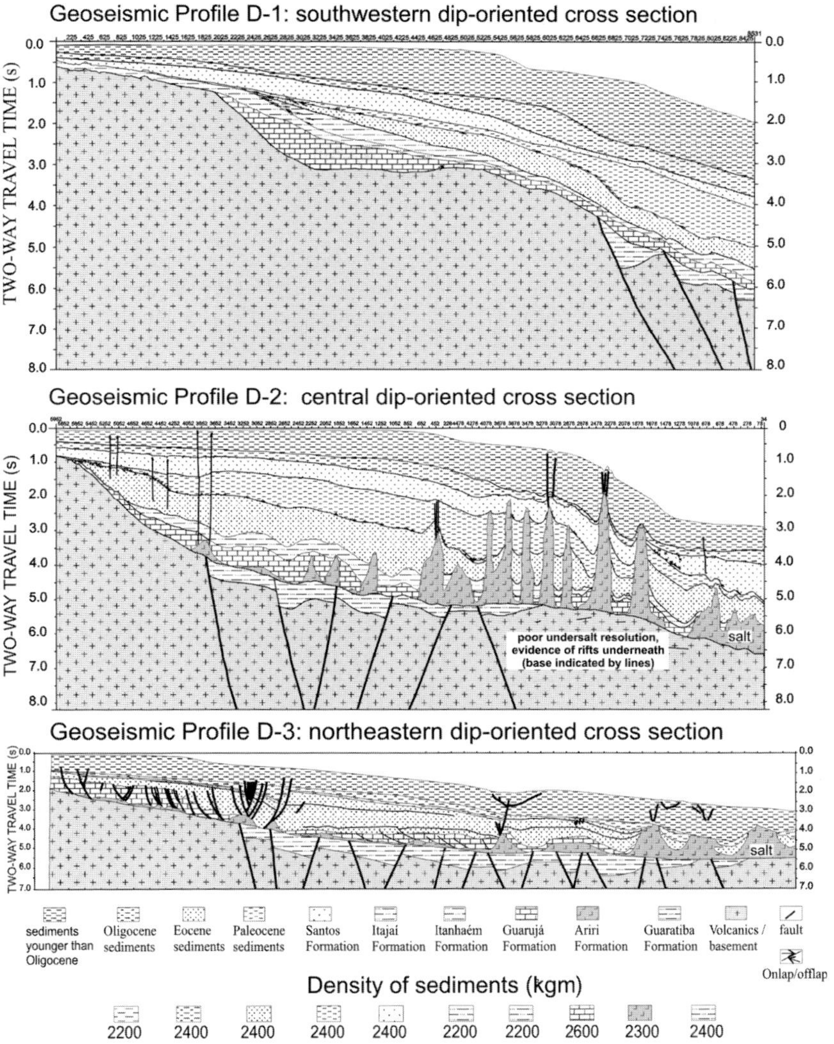

Fig. 23. Geologic cross sections interpreted along three dip-oriented (NW–SE) reflection seismic sections through the southwestern, central and northeastern portions of the Santos Basin (D-1, D-2 and D-3; see location on Figure 4). The southwestern section is located outside of the region affected by gravity gliding on top of the evaporite layer; the central section is in the centre of the gliding body; and the northeastern section is in the inside margin of the gliding body. The central and northeastern profiles are in the basin portion located in front of the Serra do Mar Mountains. The southwestern one is to the SW of this portion. Note that the southwestern profile has the Santos Formation relatively thin and it does not form any prominent offlap pattern. The Santos Formation in remaining two profiles is 2–4 times thicker and forms a prominent offlap pattern in its NW portion. NE profile shows that upper portion of the gliding sediments is affected by extensional tectonic while the lower portion is affected by contractional tectonic. Average densities of sediments in all three profiles are listed in the legend. Cross-sectional areas of sediments in profiles (in km^2) are: SW profile—Neogene 66.24, Santos/Palaeocene/Eocene/Oligocene 385.10, Itajaí 61.84, Guarujá 69.67, Guaratiba 74.07; Central profile—Neogene 138.48, Oligocene 212.75, Eocene 64.16, Palaeocene 129.05, Santos 138.68, Itajaí 215.13, Guarujá 61.95, evaporites 194.87; NE profile—Neogene 207.23, Oligocene 126.62, Eocene 49.83, Palaeocene 19.59, Santos 137.63, Guarujá 49.40, evaporites 56.77. Note that Oligocene and younger sediments are not separated in the northeastern profile.

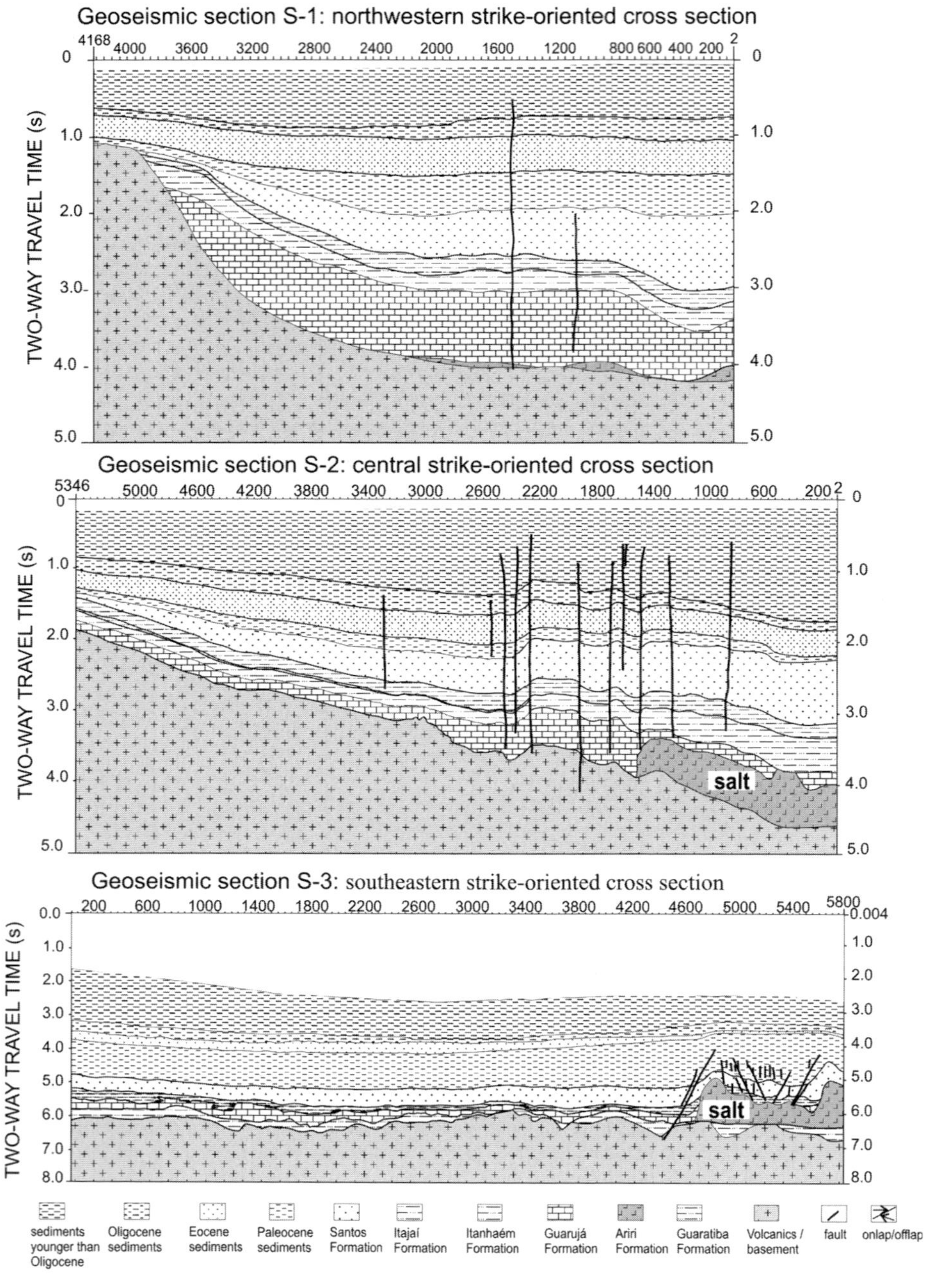

Fig. 24. Geological cross sections along three strike-oriented (SW–NE) reflection seismic sections (S-1, S-2 and S-3; see location on Fig. 4) through the northwestern, central and southeastern portions of the southwestern boundary of the region affected by gravity gliding on the top of the evaporitic horizon. NW profile shows the uppermost part of sediments gliding on top of the evaporites. Only two dextral strike-slip faults are needed to accommodate deformation between sediments that do not glide and sediments that glide. The central profile indicates that a lot of dextral strike-slip faults are needed for this accommodation. The SE profile shows the lowermost part of the gliding sediments. It contains several accommodation strike-slip and oblique-slip faults.

continental to oceanic crust. Salt may also advance towards the oceanic crust as allochthonous salt sheets and salt tongues (Fig. 12) or as salt masses thrusted over volcanic highs (Mohriak 2001).

The SSW boundary of the Aptian salt basin is formed by a system of strike-slip faults separating the province affected by gravity gliding from stable sediments (Figs 4 & 24). Figure 24 documents how the amount of displacement along the strike-slip system first increases in a down-dip direction from the hinge zone, and then decreases approaching the front of the gliding province.

Large horizontal strains related to gravity gliding have been determined in the evaporitic unit and overlying sediments. They are represented by both extensional and compressional structures. The extensional structures, including salt rollers of the normal-fault footwalls, salt walls along conjugate normal faults and turtleback anticlines, and rafts that were displaced basin-wards, are mainly located on the inclined margin of the basin (Fig. 23). They accommodate a significant amount of the horizontal down-dip extension of the post-salt overburden, which is detached from the more rigid basement by the Aptian salt layer. Demercian *et al.* (1993) have shown that the salt tectonics extensional domain in the adjacent Campos Basin, having similar down-dip length of 150–200 km, accommodated about 100 km of extension of the post-salt overburden. The compressional structures, including growth folds, asymmetric salt walls above reverse faults and salt tongues, are located on the flat basin floor (Fig. 23). They accommodate about 100 km of horizontal contraction in the adjacent Campos Basin (Demercian *et al.* 1993). By analogy we infer approximately the same amount for the Santos Basin. The horizontal contraction in the deep-water region apparently balances the down-dip extension of the post-salt overburden observed along the sedimentary basin margin, particularly in the platform and slope. Apart from gravity gliding accommodated by internal deformation of the evaporites and overlying sediments, the gliding resulted in the down-dip distance travelled by the salt front from its original location on the thinned continental crust. This can be determined from the location of the evaporite distal edge on the oceanic crust, which is younger than the evaporite deposition (Figs 2 & 4).

The boundary of the continental and oceanic crusts (Fig. 4), was partially determined from the extent of oceanic fracture zones and gravity data by Karner (2000), indicating several discrepancies when compared with the Leplac Project continent–ocean boundary. This limit was overlain by the salt province map (Ojeda 1982; Pereira & Macedo 1990; Demercian *et al.* 1993), showing the advance of the salt mass towards major oceanic fracture zones, thus overlying volcanic crust in the SE segment of the Santos Basin. The overlap indicates that the salt, together with overlying sediments, glided about 56–162 km across the oceanic crust. Gliding roughly followed the maximum concave curvature of the basin slope in a direction away from the Mantiqueira and Serra do Mar mountain ranges (Fig. 4).

Assessment of load changes caused by salt tectonics and massive clastic progradation episodes

The timing of gliding on the evaporites can be determined from the ages of syntectonic sediment wedges thickening towards listric normal faults detached along the evaporitic horizon. These wedges indicate normal faulting during the deposition of the older Albian Guarujá Formation (Pereira *et al.* 1986; Pereira & Macedo 1990; Pereira & Feijó 1994), the younger Albian Itanhaém Formation, the Cenomanian–Turonian Itajaí Formation, the Senonian Santos Formation (and its lateral equivalents), and the Tertiary sediments (Figs 7, 16, 17, 19 and 23).

The described gravity gliding is responsible for the salt withdrawal and escape of the overlying sediments from the margin of the Santos Basin and their accumulation in its deeper parts. This transport mechanism unloads the basin margin and loads the basin centre in a seesaw fashion.

Line balancing along the northeastern profile (Fig. 23) for Palaeocene, Upper Cretaceous and Albian stratigraphic tops indicates that the amount of gravity gliding down the continental slope along the salt horizon was 26.8 and 48.2 km between Upper Cretaceous and top Paleocene (Santos Fm), and between top Albian and top Upper Cretaceous, respectively. Calculated gliding rates of 1.38 mm a^{-1} and 2.68 mm a^{-1} between top Albian–top Santos and top Santos–top Palaeocene, respectively, indicate a progressively faster gliding rate from Albian to Palaeocene, which was most probably caused by the progressively steeper dip of the continental slope.

The calculated gliding rates indicate how fast gravity gliding removes a sedimentary section from the basin margin and adds it to the basin floor. This mechanism unloads the basin margin and loads its floor. The importance of this mechanism is best evaluated when we compare its efficiency with efficiency of erosion/deposition. Average vertical denudation of dry and marine cliffs, measured in numerous present-day cases, ranges from 0.1 to 1 mm a^{-1} (Saunders & Young 1983). Their average lateral denudation amounts to 2.5–16,000 mm a^{-1}. Removal of the whole sedimentary section, from several hundred to several thousand metres thick, by gravity gliding from the basin margin at rates 1.38–2.68 mm a^{-1} results in several orders of magnitude greater unloading of the margin than that caused by erosion.

Average deposition and lateral advance rates of the largest world deltas are 0.88–400 mm a^{-1} and 6000–268,000 mm a^{-1} (Kukal 1990). Average deposition rates decrease basin-ward, through 1–5 mm a^{-1} for pro-delta slope, 0.2–2 mm a^{-1} for upper fan, and 0.05–0.5 mm a^{-1} for lower fan, to 0.002–0.01 mm a^{-1} for siliceous ooze (Kukal 1990; Einsele 1992). Addition of a kilometre-scale thick sedimentary section to the basin floor at rates 1.38–2.68 mm a^{-1} through gravity gliding results in loading several orders of magnitude greater than loading caused by hemipelagic deposition. This indicates that the load removed from the margin and added to the basin through salt tectonics is much greater than values imposed by the fast erosion coupled with fast deposition.

Recent numeric models (e.g., Burov & Cloetingh 1997) indicate that focused erosional removal of sediment from elevated basin margin and its deposition in the adjacent basin creates additional vertical loads on the lithosphere, capable of generating bending stresses of several hundreds of MPa. This can be 3–5 times larger than the excess normal pressure caused by the sedimentary load in the basin, which is in the order of 100 MPa.

We can evaluate whether gliding redistributes normal pressures of comparable magnitude. Figure 4 shows that sediments above the evaporites glided on average 109 km towards the centre of the basin. To simplify the calculation, the cross-sectional areas of sedimentary layers from Figure 23 were calculated and reshaped into rectangles with average thicknesses. We used the average basin width of 600 km and densities listed in Figure 23. The calculations in the central profile indicate that the estimated 109 km glide removes normal pressure of 99 MPa from the basin margin and adds it to the basin centre. In the case of the northeastern profile, a value of the redistributed normal pressure of 78 MPa was calculated. We can conclude, therefore, that excess normal pressures redistributed by the gravity gliding compare approximately with excess normal pressures associated with erosion/deposition.

Surface load redistribution by gravity gliding could have a temporary self-enhancing character since progressive load redistribution would cause lithospheric flexure, which increases dip of the basin slope, which in turn accelerates the gliding rate. Our sedimentary data from seismic sections through the basin together with apatite fission-track data from the basin margin (Gallagher *et al.* 1994) indicate that the load redistribution by gravity gliding could have triggered the rift-shoulder uplift during Senonian.

The deposition rate of the Senonian Santos Formation (*c.* 2–5 times higher) indicates yet another self-enhancing mechanism of subsidence: uplift, represented by accelerated erosion reacting to the rift-shoulder uplift and associated deposition of the thick siliciclastic wedges of the Santos Formation.

During the Palaeogene, gravity collapse of the upper slope above the salt horizon stopped when sufficient friction developed along the detachment in places where salt was completely evacuated (Fig. 23). After evaporite extraction the slope became stable and resisted further collapse, even during the deposition of large Tertiary prograding wedges caused by progressively accelerated erosion of the rift shoulder. These wedges eventually started to load the slope again, above the starved basin, progressively slowing rift-shoulder uplift and erosion. In extreme cases such a starved basin can react by forming localized uplifts, as has been simulated by numeric modelling (Burov & Cloetingh 1997) and documented by our observation of the post-Palaeocene uplift of the São Paulo Plateau (Figs 3 & 4). The Late Tertiary uplift of the sedimentary successions in the deep-water region might be due to inflation of the evaporite layer by basin-ward flow of salt. This compressional thickening of the salt layer is characterized by intra-salt reflectors that might correspond to thrust faults with a basin-ward vergence (Fig. 12).

A very thick sedimentary layer is present in the central and southeastern portions of the Santos Basin, reaching a thickness of more than 2 km (Cobbold *et al.* 1995). These layers have a distinct and peculiar seismic signature (Figs 14 & 15), corresponding to stratified evaporites with a strong reflector located on top of the sequence, which is locally deformed by compressional structures. Before drilling, this reflector was known as 'the enigmatic reflector'. Exploratory boreholes indicate that the stratified assemblage corresponds to different evaporite cycles, ranging from sulphate (anhydrite) to K-bearing salts (carnalite, sylvite), and the strong reflector corresponds to top of Aptian salt. The seismic images also suggest different rates of flow for different lithologies. A rather transparent zone of halite is particularly mobile and is locally associated with canopies and recumbent folds.

The critical evidence for the rift-shoulder activity being related to salt tectonics is their spatial association. Figures 4 and 18 show that the rift shoulder is located mainly in front of the portion of the Santos Basin, with a peculiar style of salt tectonics characterized by antithetic faults

(Mohriak *et al.* 1995). The development of the Albian gap (Fig. 19) is controlled by the very thick sedimentary wedge that loaded the salt basin and evacuated and expelled the mobile evaporites from the platform, forming thick salt diapirs and salt walls in the deep-water province (Fig. 19).

In the Santos Basin, the central and northeastern dip-orientated cross-sections (Fig. 23, see Fig. 4 for location) document a thick wedge composed of the Senonian Santos Formation, fed by the uprising rift shoulder. Marginal parts of this wedge indicate a prominent offlap pattern, resulting from the uplift of the basin margin adjacent to the rift shoulder. The southwestern cross-section in Fig. 23 (D-1, located outside of the main depocentre of the salt province) is marked by a relatively stable post-salt overburden. The profile is located behind the strike-slip fault system dividing the gravitationally stable and unstable parts of the Santos Basin slope. It is characterized by a thin wedge of the Santos Formation, which was not fed by eroding a distinct rift shoulder. The prominent offlap pattern for the sedimentary successions immediately following deposition of the salt, which is a typical feature observed in the other profiles, is also missing in this section. And finally, the syn-rift Neocomian to Barremian successions seem to be missing in the platform and slope along the dip-orientated central cross-section (D-2 in Fig. 23) suggesting a more elevated region during the formation of the Aptian salt basin, with evaporites locally being deposited on a volcanic basement.

Conclusions

The development of intra-continental rifts filled with lacustrine and fluvial sediments was controlled by the type of substratum to the sedimentary successions (cratonic vs. fold belt basement). These differences are clearly observed in the offshore sedimentary basins along the eastern Brazilian margin. The Santos, Campos and Espírito Santo basins, located on fold-belt basement, have much wider syn-rift and salt deposition regions than the basins that are located on cratonic basement at the northeastern margin. The substratum of the Almada, Camamu and Jacuípe basins corresponds to granulites from the São Francisco Craton; these basins are narrower and have a very thick syn-rift trough with smaller thickness of sedimentary successions associated with subsequent thermal subsidence. They are also characterized by reduced amounts of evaporite deposits.

The emplacement of Early Cretaceous intrusive and extrusive rocks was related to extensional stresses responsible for the development of syn-rift lacustrine depocentres. Huge volumes of igneous rocks were also formed during the continental break-up and formation of active spreading centres, which resulted in thick wedges of seaward-dipping reflectors. These volcanic rocks loaded the divergent margin which rapidly subsided during the Aptian, and the depressions were filled with lacustrine to marine sediments, culminating in widespread evaporite deposition by Late Aptian time. Salt tectonics was one of the main controls behind the major changes in deposition patterns along the continental margin, forming several extensional and compressional domains in the area between the rift border and the continent–ocean boundary in the deep basin.

Shallow-marine conditions began with platform carbonates and siliciclastic rocks in the Albian, and a transgressive marine environment progressively resulted in deep-water sedimentation by the Late Cretaceous. Late Cretaceous to Early Tertiary magmatic episodes are also registered along the eastern Brazilian margin, both onshore and offshore. Compressional events are registered in some provinces, such as the Abrolhos platform. Particularly in the onshore region adjacent to the Mantiqueira and Serra do Mar ranges, extensional stresses resulted in the development of small intracontinental rift basins. These Tertiary onshore basins are characterized by a very thin sediment thickness, indicating a negligible amount of extension and uplift of their borders in contrast to the adjacent offshore basins along the continental margin; these are characterized by relatively thick syn-rift troughs and very thick post-rift sedimentary successions, particularly along the Santos, Campos and Espírito Santo basins. The original thickness of the Aptian salt layer, as well as the siliciclastic and carbonatic Cretaceous sequences, may exceed several kilometres in the deep water region of these basins.

The break-up of western Gondwana advanced from the southern segment (Argentina) towards the equatorial margin. In the Brazilian margin, the southern basins (Santos, Campos and Espírito Santo) are bordered by mountain ranges with high elevations (Mantiqueira Range) whose uplift in post-Albian times was responsible for episodes of massive clastic progradation in the Santos Basin. This Late Cretaceous progradation resulted in loading of the basin and renewed border uplift. It also triggered a peculiar style of salt tectonics characterized by listric, detached counter-regional normal faults that were formed during the expulsion of the mobile salt by sediment loading. The progression of (1) erosional unloading of the margin; (2) depositional loading of the basin; (3) removal of the salt and sediment load from the margin to the centre of the basin by gravity gliding was a self-enhancing process. The gradual increase of the flexure of the continental margin eventually resulted in flank uplift, the erosion of which in turn forced the massive progradation events.

The rift-shoulder uplift was coeval with strong subsidence in the Santos Basin during post-Albian times. The uplift occurred *c.* 24 Ma after Neocomian–Aptian rifting, indicating that rifting processes were not the exclusive control. Rift-shoulder uplift was also controlled by gravity gliding, as documented by the shoulder development in front of the basin portion strongly affected by salt tectonics. Removal of the whole sedimentary section above the evaporites, and evaporite withdrawal from the basin margin, led to unloading several orders of magnitude faster than simple erosional unloading. Equally, addition of the glided sedimentary section and accumulated evaporites to the basin floor was a loading process several orders of magnitude faster than depositional loading of the basin by hemipelagic sediments. The excess normal pressure of about 100 MPa redistributed by gravity gliding was comparable with maximum excess pressures that could be caused by erosion coupled with deposition.

Progressive load redistribution caused by gliding resulted in a lithospheric flexure, which increased dip of the basin slope, which in turn accelerated the gliding rate. Gliding rates increased from Albian–Senonian values of 1.38 mm a^{-1} to Palaeocene values of 2.68 mm a^{-1}. The lithospheric flexure was further accelerated by rift-shoulder erosion coupled with increased deposition resulting from the shoulder uplift. Depositional rates became approximately 2–5 times faster than Albian–Turonian rates.

The African margin is also characterized by a shoulder uplift in the region bordering the Angolan and Namibian offshore basins. However, these basins are characterized by much narrower rift and salt basins. The asymmetric opening of this portion of the South Atlantic was associated with oceanic propagators stepping right-laterally, towards the African side. A more symmetric rift, with a balanced distribution of evaporites, is observed in the Campos–Kwanza conjugate basins. The northern segment of the South Atlantic rift and salt basins is also characterized by asymmetry, with the Brazilian margin characterized by narrow and deep rifts with much less salt than the Congo–Gabon conjugate margin.

We thank B. B. Brito Neves for his tireless disposition and encouragement to elaborate this contribution to the Gondwana volume. We also acknowledge the patience of editors, their cooperation and dedication to have this publication prepared on time. We thank the management of Petroleo Brasileiro, particularly P. M. M. Mendonça, L. A. Reis and E. Porsche for permission to publish this work. We are grateful to many institutions and geoscientists who provided geological and geophysical data, figures and illuminating suggestions, which have been synthesized in this paper. Part of this work is a follow-up after a previous regional basin analysis project developed at University of Utah which was supported by Ocean Energy Corporation. Geoseismic cross sections in the Santos Basin are not precisely located for confidentiality reasons. We thank the referees Soenke Neben and Magdalena Scheck-Wenderoth for their helpful comments and suggestions that greatly improved the manuscript, and M. de Wit for his careful revision and corrections in the final version of this work. We also appreciate the critical reading and suggestions by Cláudia Queiroz (Petrobras–E&P) in the edited version.

References

ALVAREZ, F., VIRIEUX, J. & LE PICHON, X. 1984. Thermal consequences of lithosphere extension: The initial stretching phase. *Geophysical Journal of the Royal Astronomic Society*, **78**, 389–411.

ALMEIDA, F. F. M. 1976. The system of continental rifts bordering the Santos Basin, Brazil. *Anais da Academia Brasileira de Ciências*, **48**, 15–26.

ASSUMPÇÃO, M., DIAS NETO, C. M., BERROCAL, J., ARTEZAMA, R., FRANÇA, H. & ORTEGA, R. 1980. Sismicidade do sudeste do Brasil. *In*: *Congresso Brasileiro de Geologia 31, Camboriú, Anais*, **2**. Sociedade Brasileira de Geologia, 1075–1092.

BARKER, P. F., BUFFLER, R. T. & GAMBOA, L. A. 1983. A seismic reflection study of the Rio Grande Rise. *In*: BARKER, P. F., CARLSON, R. L., JOHNSON, D. A. ET AL. (eds) *Initial Reports of the Deep Sea Drilling Project*, **72**, 499–517.

BOILLOT, G., MOUGENOT, D, GIRARDEAU, J. & WINTERER, E. L. 1989. Rifting processes on the West Galicia margin, Spain. *In*: TANKARD, A. J. & BALKWILL, H. R. (eds) *Extensional Tectonics and Stratigraphy of the North Atlantic Margins*. American Association of Petroleum Geologists Memoirs, **46**, 363–377.

BRAUN, J. & BEAUMONT, C. 1989. A physical explanation of the relation between flank uplifts and the breakup unconformity at rifted margins. *Geology*, **17**, 760–764.

BUCK, W. R. 1986. Small-scale convection induced by passive rifting: The cause for uplift of rift shoulders. *Earth and Planetary Science Letters*, **77**, 362–372.

BUCK, W. R., MARTINEZ, F., STECKLER, M. S. & COCHRAN, J. R. 1988. Thermal consequences of lithospheric extension: pure and simple. *Tectonics*, **7**, 213–234.

BUROV, E. & CLOETINGH, S. 1997. Erosion and rift dynamics; new thermomechanical aspects of post-rift evolution of extensional basins. *Earth and Planetary Science Letters*, **150**, 7–26.

BUSSELL, M. A. 1983. Timing of tectonic and magmatic events in the Central Andes of Peru. *Journal of the Geological Society, London*, **140**, 279–286.

CAINELLI, C. & MOHRIAK, W. U. 1999. Some remarks on the evolution of sedimentary basins along the

Eastern Brazilian continental margin. *Episodes*, **22**, 206–216.

CHANG, H. K., APPI, C. J., RICCOMINI, C., CASTRO, J. C., ARAI, M., FREITAS, E. L. & NETO, E. V. S. 1989. Geologia da Bacia de Taubaté. *In*: *Primeiro Simpósio de Geologia do Sudeste, Rio de Janeiro, Boletim de Resumos*. Sociedade Brasileira de Geologia, 10.

CHANG, H. K., KOWSMANN, R. O., FIGUEIREDO, A. M. F. & BENDER, A. 1992. Tectonics and stratigraphy of the East Brazil Rift system: an overview. *Tectonophysics*, **213**, 97–138.

CHERY, J., LUCAZEAU, M., DAIGNIERES, M. & VILOTTE, J. P. 1992. Large uplift of rift flanks: A genetic link with lithospheric rigidity? *Earth Science and Planetary Letters*, **112**, 195–211.

COBBOLD, P. R. & SZATMARI, P. 1991. Radial gravitational gliding on passive margins. *Tectonophysics*, **188**, 249–289.

COBBOLD, P. R., SZATMARI, P., DEMERCIAN, L. S., COELHO, D. & ROSSELLO, E. A. 1995. Seismic experimental evidence for thin-skinned horizontal shortening by convergent radial gliding on evaporites, deep-water Santos Basin. *In*: JACKSON, M. P. A., ROBERTS, D. G. & SNELSON, S. (eds) *Salt tectonics: a global perspective*. American Association of Petroleum Geologists Memoirs, **65**, 305–321.

COBBOLD, P. R., MEISLING, K. E. & MOUNT, V. S. 2001. Reactivation of an obliquely rifted margin, Campos and Santos basins, southeastern Brazil. *American Association of Petroleum Geologists, Bulletin*, **11**, 1925–1944.

COCHRAN, J. R. 1983. Effects of finite extension times on the development of sedimentary basins. *Earth and Planetary Science Letters*, **66**, 289–302.

COX, K. G. 1980. A model for flood basalt volcanism. *Journal of Petrology*, **21**, 629–650.

DAVISON, I. 2005. Tectonics of the South Atlantic Brazilian Salt Basin. *In*: *GCSSEPM 25th Annual Bob F. Perkins Research Conference: Petroleum Systems of Divergent Continental Margin Basins, Abstracts CD*, 468–480.

DEMERCIAN, L. S., SZATMARI, P., COBBOLD, P. R. & COELHO, D. F. 1992. Halocinese na Bacia de Campos. *In*: *Congresso Brasileiro de Geologia 37, São Paulo, Anais*, **1**. Sociedade Brasileira de Geologia, 560–561.

DEMERCIAN, S., SZATMARI, P. & COBBOLD, P. R. 1993. Style and pattern of salt diapirs due to thin-skinned gravitational gliding, Campos and Santos basins, offshore Brazil. *Tectonophysics*, **228**, 393–433.

DIAS, J. L. 1993. Evolução da fase rift e a transição rift/drift nas bacias das margens leste e sudeste do Brasil. *In*: *3rd International Congress of the Brazilian Geophysical Society, Rio de Janeiro, Expanded Abstracts*, **2**, 1328–1332.

DUVAL, B., CRAMEZ, C. & JACKSON, M. P. A. 1992. Raft tectonics in the Kwanza Basin, Angola. *Marine and Petroleum Geology*, **9**, 389–404.

EINSELE, G. 1992. *Sedimentary basins-evolution, facies, and sediment budget*. Springer-Verlag, New York.

EWART, A., BAXTER, K. & ROSS, J. A. 1980. The petrology and petrogenesis of the Tertiary anorogenic mafic lavas of southern and central Queensland, Australia-Possible implications for crustal thickening: *Contributions to Mineralogy and Petrology*, **75**, 129–152.

FAINSTEIN, R. 2003. A geophysical tour in the South Atlantic. *In*: MOHRIAK, W. U. (ed.) *Rifted sedimentary basins of the South Atlantic: turbidite reservoirs; sedimentation and tectonic processes*. Petrobras-Universidade Estatal de Rio de Janeiro, Rio de Janeiro, Brazil, CD-ROM.

FAINSTEIN, R. & KRUEGER, A. 2005. Salt tectonics comparisons near there continent-ocean boundary escarpments. *In*: POST, P. J., ROSEN, N. C., OLSON, D. L., PALMES, S. L., LYONS, K. T. & NEWTON, G. B. (eds) *GCSSEPM 25th Annual Bob F. Perkins Research Conference: Petroleum Systems of Divergent Continental Margin Basins, Abstracts CD*, 510–540.

FAINSTEIN, R., JAMIESON, G., HANNAN, A., EILES, N., KRUEGER, A. & SCHELANDER, D. 2001. Offshore Brazil Santos Basin exploration potential from recently acquired seismic data. *In*: *7th International Congress of the Brazilian Geophysical Society, Abstracts* CD. Sociedade Brasileira de Geofísica, Salvador, 52–55.

FALVEY, D. A. 1974. The development of continental margins in plate tectonic theory. *The APEA Journal*, **14**, 95–106.

FERNANDES, F. L. & CHANG, H. K. 1992. Compartimentação e evolução tectônica da Bacia de Taubaté. *In*: *Congresso Brasileiro de Geologia 37, São Paulo, Anais*, **1**. Sociedade Brasileira de Geologia, 570–572.

GALLAGHER, K., HAWKESWORTH, C. J. & MANTOVANI, M. S. M. 1994. The denudation history of the onshore continental margin of SE Brazil inferred from apatite fission track data. *Journal of Geophysical Research*, **99**, 18117–18145.

GE, H., JACKSON, M. P. A. & VENDEVILLE, B. C. 1997. Kinematics and dynamics of salt tectonics driven by progradation. *American Association of Petroleum Geologists Bulletin*, **81**, 393–423.

GLADCZENKO, T. P., HINZ, K., ELDHOLM, O., MEYER, H., NEBEN, S. & SKOGSEID, J. 1997. South Atlantic volcanic margins. *Journal of the Geological Society, London*, **154**, 465–470.

GOMES, B. S. 1992. Integração preliminar dos dados gravimétricos marítimos da Petrobras e do Projeto Leplac: Bacias de Campos, Santos e Pelotas. *In*: *Congresso Brasileiro de Geologia 37, São Paulo, Anais*, **1**. Sociedade Brasileira de Geologia, 559–560.

GOMES, P. O., PARRY, J. & MARTINS, W. 2002. The Outer High of the Santos Basin, Southern São Paulo Plateau, Brazil: Tectonic Setting, Relation to Volcanic Events and Some Comments on Hydrocarbon Potential. *American Association of Petroleum Geologists Search and Discovery Article #90022 (2003)*, 1–5.

HARRY, D. L. & SAWYER, D. S. 1992. Basaltic volcanism, mantle plumes, and the mechanics of rifting: the Paraná flood basalt province of South America. *Geology*, **20**, 207–210.

HEILBRON, M., MOHRIAK, W. U., VALERIANO, C. M., MILANI, E. J., ALMEIDA, J. & TUPINAMBÁ, M. 2000. From collision to extension: the roots of the southeastern continental margin of Brazil. *In*: MOHRIAK, W. U. & TALWANI, M. (eds) *Atlantic rifts and continental margins*, American Geophysical Union, Geophysical Monographs, **115**, 1–32.

HELLINGER, S. J. & SCLATER, J. G. 1983. Some remarks on two-layer extensional models or the evolution of sedimentary basins. *Journal of Geophysical Research*, **88**, 8251–8269.

HENDRIE, D. B., KUSZNIR, N., MORLEY, C. K. & EBINGER, C. J. 1994. Cenozoic extension in northern Kenya; a quantitative model of rift basin development in the Turkana region. *Tectonophysics*, **236**, 409–438.

HENRY, S. G. & BRUMBAUGH, W. 1995. Pre-salt rock development on Brazil's conjugate margin: West African examples. *In*: *4th International Congress of the Brazilian Geophysical Society, Rio de Janeiro, Expanded Abstracts*, **I**, 68–70.

HENRY, S. G., MOHRIAK, W. U. & MELLO, M. R. 1996. South Atlantic sag basins: new petroleum system components. *American Association of Petroleum Geologists Bulletin*, **80**, 1300.

HINZ, K. 1981. A hypothesis on terrestrial catastrophes: wedges of very thick oceanward dipping layers beneath passive continental margins. *Geologisches Jahrbuch*, **E-22**, 3–28.

ISSLER, D. H., MCQUEEN, H. & BEAUMONT, C. 1989. Thermal and isostatic consequences of simple shear extension of the continental lithosphere. *Earth and Planetary Science Letters*, **91**, 341–358.

JACKSON, M. P. A. & VENDEVILLE, B. C. 1994. Regional extension as a geologic trigger for diapirism. *Geological Society of America Bulletin*, **106**, 57–73.

JACKSON, M. P. A., CRAMEZ, C. & FONCK, J. M. 2000. Role of subaerial volcanic rocks and mantle plumes in creation of South Atlantic margins: implications for salt tectonics and source rocks. *Marine and Petroleum Geology*, **17**, 477–498.

KARNER, G. D. 2000. Rifts of the Campos and Santos Basins, Southeastern Brazil: Distribution and timing. *In*: MELLO, M. R. & KATZ, B. J. (eds) *Petroleum systems of South Atlantic margins*. American Association of Petroleum Geologists Memoirs, **73**, 301–315.

KARNER, G. D., DRISCOLL, N. W. & BARKER, D. H. N. 2003. Syn-rift regional subsidence across the West African continental margin: the role of lower plate ductile extension. *In*: ARTHUR, T. J., MACGREGOR, D. S. & CAMERON, N. R. (eds) *Petroleum Geology of Africa: New themes and developing technologies*. Geological Society, London, Special Publications, **207**, 105–129.

KEEN, C. E. 1985. The dynamics of rifting: Deformation of the lithosphere by active and passive driving mechanisms. *Royal Astronomical Society Geophysical Journal*, **80**, 95–120.

KOUTSOUKOS, E. A. M. & DIAS-BRITO, D. 1987. Paleobatimetria da margem continental do Brasil durante o Albiano. *Revista Brasileira de Geociências*, **17**, 86–91.

KOUTSOUKOS, E. A. M., MELLO, M. R., AZAMBUJA FILLHO, N. C., HART, M. B. & MAXWELL, J. R. 1991. The Upper Aptian-Albian succession of the Sergipe Basin, Brazil: an integrated paleoenvironmental assessment. *American Association of Petroleum Geologists, Bulletin*, **75**, 479–498.

KOWSMANN, R. O., LEYDEN, R. & FRANCISCONI, O. 1977. Marine seismic investigations, Southern Brazil margin. *American Association of Petroleum Geologists, Bulletin*, **61**, 546–557.

KUKAL, Z. 1990. The rate of geological processes. *Earth-Science Reviews*, **28**, 1–284.

KUSZNIR, N. & ZIEGLER, P. 1992. The mechanics of continental extension and sedimentary basin formation; a simple-shear/pure-shear flexural cantilever model. *Tectonophysics*, **215**, 117–131.

LIMA, C. C. 2003. Ongoing compression across South American plate: observations, numerical modeling and some implications for petroleum geology. *In*: AMEEN, M. (ed.) *Fracture and in-situ stress characterization of hydrocarbon reservoirs*. Geological Society, London, Special Publications, **209**, 87–100.

MCKENZIE, D. 1978. Some remarks on the development of sedimentary basins. *Earth and Planetary Science Letters*, **40**, 25–32.

MCKENZIE, D. P. 1984. A possible mechanism for epeirogenetic uplift. *Nature*, **307**, 616–619.

MACEDO, J. M. 1990. Evolução tectônica da Bacia de Santos e áreas continentais adjacentes. *In*: GABAGLIA, G. P. R. & MILANI, E. J. (eds) *Origem e evolução de bacias sedimentares*. Petrobrás, Rio de Janeiro, 361–376.

MANATSCHAL, G. 2004. New models for evolution of magma-poor rifted margins based on a review of data and concepts from West Iberia and the Alps. *International Journal of Earth Sciences*, **93**, 432–466.

MARQUES, A. 1990. Evolução tectono-sedimentar e perspectivas exploratórias da Bacia de Taubaté, São Paulo, Brasil. *Revista Brasileira de Geociências*, **4**, 253–262.

MEISLING, K. E., COBBOLD, P. R. & MOUNT, V. S. 2001. Segmentation of an obliquely rifted margin, Campos and Santos basins, southeastern Brazil. *American Association of Petroleum Geologists, Bulletin*, **11**, 1903–1924.

MELO, M. S., RICCOMINI, C., HASUI, Y., ALMEIDA, F. F. M. & COIMBRA, A. M. 1985. Geologia e evolução do sistema de bacias tafrogênicas continentais do sudeste do Brasil. *Revista Brasileira de Geociências*, **15**, 193–201.

MILANI, E. J. & THOMAZ FILHO, A. 2000. Sedimentary basins of South America. *In*: CORDANI, U. G., MILAN, E. J., THOMAZ FILHO, A. & CAMPOS, D. A. (eds) *Tectonic evolution of South America, 31st International Geological Congress, Rio de Janeiro*, 389–449.

MIZUSAKI, A. M. P. & MOHRIAK, W. U. 1992. Sequências vulcano-sedimentares na região da plataforma continental de Cabo Frio, RJ. *In*: *Congresso Brasileiro de Geologia 37, São Paulo, Anais*, **2**. Sociedade Brasileira de Geologia, 468–469.

MIZUSAKI, A. M. P., THOMAZ FILHO, A. & CESERO, P. 1998. Ages of the magmatism and the opening of the South Atlantic Ocean. *Universidade Federal do Rio Grande do Sul, Instituto de Geociências, Pesquisas*, **25-2**, 47–57.

MIZUSAKI, A. M. P. & THOMAZ FILHO, A. 2004. O magmatismo pós-Paleozóico no Brasil. *In*: MANTESSO-NETO, V., BARTORELLI, A., CARNEIRO, C. D. R. & BRITO-NEVES, B. B. (eds) *Geologia do continente sul-americano: evolução da obra de Fernando Flávio Marques de Almeida*. Beca Produções Culturais Ltda., São Paulo, 281–291.

MIZUSAKI, A. M. P., PETRINI, R. *ET AL.* 1992. Basalt magmatism along the passive continental margin of

SE Brazil (Campos Basin). *Contributions to Mineralogy and Petrology*, **111**, 143–160.

MIZUSAKI, A. M. P., THOMAZ FILHO, A. & CESERO, P. 1998. Ages of the magmatism and the opening of the South Atlantic Ocean. *Universidade Federal do Rio Grande do Sul, Instituto de Geociências, Pesquisas*, **25**, 47–57.

MIZUSAKI, A. M. P., THOMAZ-FILHO, A., MILANI, E. J. & CÉSERO, P. 2002. Mesozoic and Cenozoic igneous activity and its tectonic control in northeastern Brazil. *Journal of South American Earth Sciences*, **15**, 183–198.

MODICA, C. J. & BRUSH, E. R. 2004. Postrift sequence stratigraphy, paleogeography, and fill history of the deep-water Santos Basin, offshore southeast Brazil. *American Association of Petroleum Geologists Bulletin*, **88**, 923–946.

MOHRIAK, W. U. 2001. Salt tectonics, volcanic centers, fracture zones and their relationship with the origin and evolution of the South Atlantic Ocean: geophysical evidence in the Brazilian and West African margins. *In*: *7th International Congress of the Brazilian Geophysical Society, Salvador, Brazil, Expanded Abstracts*, 1594.

MOHRIAK, W. U. 2003. Bacias Sedimentares da Margem Continental Brasileira. *In*: BIZZI, L. A., SCHOBBENHAUS, C., VIDOTTI, R. M. & GONÇALVES, J. H. (eds) *Geologia, Tectônica e Recursos Minerais do Brasil*. Companhia de Pesquisas de Recursos Minerais, Brasília, 87–165.

MOHRIAK, W. U. 2004. Recursos energéticos associados à ativação tectônica mesozóico-cenozóica da América do Sul. *In*: MANTESSO-NETO, V., BARTORELLI, A., CARNEIRO, C. D. R. & BRITO-NEVES, B. B. (eds) *Geologia do continente sul-americano: evolução da obra de Fernando Flávio Marques de Almeida*. Beca Produções Culturais Ltda., São Paulo, 293–318.

MOHRIAK, W. U. 2005. Interpretação geológico-geofísica do craton do São Francisco e da plataforma continental da Bahia: limite crustal e estruturas oceânicas. *In*: ROSA, M. L. S., KOSIN, M., RIOS, D. C., PINTO, M. S. & TEIXEIRA, J. B. G. (eds) *III Simpósio sobre o Cráton do São Francisco, Salvador, Anais*, 13–16.

MOHRIAK, W. U. & PAULA, O. B. 2005. Major tectonic features in the southeastern Brazilian margin. *In*: *9th International Congress of the Brazilian Geophysical Society, Salvador, Brazil, Expanded Abstracts*. Sociedade Brasileira de Geofísica, SBGf174.

MOHRIAK, W. U. & SZATMARI, P. 2001. Salt tectonics and sedimentation along Atlantic margins: insights from seismic interpretation and physical models. *In*: KOYI, H. A. & MANCKTELOW, N. S. (eds) *Tectonic Modeling: A volume in Honor of Hans Ramberg*. Geological Society of America, Memoirs, **193**, 131–151.

MOHRIAK, W. U., HOBBS, R. & DEWEY, J. F. 1990. Basin-forming processes and the deep structure of the Campos Basin, offshore Brazil. *Marine and Petroleum Geology*, **7**, 94–122.

MOHRIAK, W. U., MACEDO, J. M. *ET AL*. 1995. Salt tectonics and structural styles in the deep-water province of the Cabo Frio region, Rio de Janeiro, Brazil. *In*: JACKSON, M. P. A., ROBERTS, D. G. & SNELSON, S. (eds) *Salt tectonics: a global perspective*. American Association of Petroleum Geologists, Memoirs, **65**, 273–304.

MOHRIAK, W. U., BASSETTO, M. & VIEIRA, I. S. 1998. Crustal architecture and tectonic evolution of the Sergipe-Alagoas and Jacuípe basins, offshore northeastern Brazil. *Tectonophysics*, **288**, 199–220.

MOHRIAK, W. U., MELLO, M. R., BASSETTO, M., VIEIRA, I. S. & KOUTSOUKOS, E. A. M. 2000. Crustal architecture, sedimentation, and petroleum systems in the Sergipe-Alagoas Basin, Northeastern Brazil. *In*: MELLO, M. R. & KATZ, B. J. (eds) *Petroleum systems of South Atlantic margins*. American Association of Petroleum Geologists, Memoirs, **73**, 273–300.

MOHRIAK, W. U., ROSENDAHL, B. R., TURNER, J. P. & VALENTE, S. C. 2002. Crustal architecture of South Atlantic volcanic margins. *In*: MENZIES, M. A., KLEMPERER, S. L., EBINGER, C. J. & BAKER, J. (eds) *Volcanic Rifted Margins*. Geological Society of America, Special Papers, **362**, 159–202.

MOHRIAK, W. U., PAULA, O. *ET AL*. 2003. Volcanic provinces in the Eastern Brazilian margin: geophysical models and alternative geodynamic interpretations. *In*: *8th International Congress of The Brazilian Geophysical Society, CDROM with abstracts (workshop Geodynamic Evolution of the South Atlantic and adjacent margins)*.

MOHRIAK, W. U., FERNANDEZ, B. & BIASSUSSI, A. S. 2004. Salt tectonics domains and structural provinces: analogies between the South Atlantic and the Gulf of Mexico. *In*: POST, P. J., OLSON, D. L., LYONS, K. T., PALMES, S. L., HARRISON, P. F. & ROSEN, N. C. (eds) *Salt–sediment interactions and hydrocarbon prospectivity: concepts, applications, and case studies for the 21st century. 24th Annual GCSSEPM Foundation, Bob F. Perkins Research Conference, Houston, Texas, USA, CD ROM*, 551–587.

MORGAN, P., BOULOS, F. K. *ET AL*. 1985. Heat flow in eastern Egypt: The thermal signature of a continental breakup. *Journal of Geodynamics*, **4**, 107–131.

MOULIN, M., ASLANIAN, D. *ET AL*. 2005. Geological constraints on the evolution of the Angolan margin based on reflection and refraction data (ZaiAngo project). *Geophysical Journal International*, **162**, 793–810.

MUTTER, J. C., TALWANI, M. & STOFFA, P. L. 1982. Origin of seaward-dipping reflectors in oceanic crust off the Norwegian margin by "subaerial sea-floor spreading". *Geology*, **10**, 353–357.

OJEDA, H. A. O. 1982. Structural framework, stratigraphy and evolution of Brazilian marginal basins. *American Association of Petroleum Geologists, Bulletin*, **66**, 732–749.

PARMENTIER, E. M. 1987. Dynamic topography in rift zones: Implications for lithospheric heating. *Royal Society of London Philosophical Transactions*, **A321**, 23–25.

PEREIRA, M. J. 1989. Santos Basin: perspectives of the new petroleum province on the SE Brazilian platform. *In*: *Petrobras, Seminário de Interpretação Exploratória, Rio de Janeiro*, 157–167.

PEREIRA, M. J. & MACEDO, J. M. 1990. A Bacia de Santos: perspectivas de uma nova província petrolífera na plataforma continental sudeste brasileira. *Boletim de Geociências da Petrobras*, **4**, 3–11.

PEREIRA, M. J. & FEIJÓ, F. J. 1994. Bacia de Santos. *Boletim de Geociências da Petrobrás*, **8**, 219–234.

PEREIRA, M. J., BARBOSA, C. M. *ET AL.* 1986. Stratigraphy of the Santos Basin: sequence analysis, depositional systems and revision of lithostratigraphy (in Portuguese). *In*: *Congresso Brasileiro de Geologia 34, Goiânia, Anais*, **1**, Sociedade Brasileira de Geologia, 65–71.

PETROBRAS, 1983. Campos and Espirito Santo Basins, offshore Brazil. *In*: BALLY, A. W. (ed.) *Seismic expression of structural styles-a picture and work atlas*. American Association of Petroleum Geologists, Studies in Geology, **15**, 2351–2358.

RABINOWITZ, P. D. & LABRECQUE, J. 1979. The Mesozoic South Atlantic Ocean and evolution of its continental margins. *Journal of Geophysical Research*, **84**, B11, 5973–6002.

RENNE, P., ERNESTO, M. *ET AL.* 1992. The age of Paraná flood volcanism, rift of Gondwanaland, and the Jurassic-Cretaceous boundary. *Science*, **258**, 975–979.

RICCOMINI, C., SANT'ANNA, L. G. & FAMBRINI, G. L. 2005. Jacuí: nova bacia sedimentar eocretácica no sul do Brasil. *In*: *X SNET-Simpósio Nacional de Estudos Tectônicos, Curitiba, Boletim de Resumos Expandidos*, 48–50.

ROBERTS, A. M. & YIELDING, G. 1991. Deformation around basin-margin faults in the North Sea/mid Norway rift. *In*: ROBERTS, A. M., YIELDING, G. & FREEMAN, B. (eds) *The geometry of normal faults*. Geological Society of London, Special Publications, **56**, 61–78.

ROSENDAHL, B. R., MOHRIAK, W. U., NEMČOK, M., ODEGARD, M. E., TURNER, J. P. & DICKSON, W. G. 2005. West African and Brazilian conjugate margins: Crustal types, architecture, and plate configurations. *In*: *Proceedings of the 4th HGS/PESGB International Conference on African E&P, Houston, USA, Abstracts*, 3.

ROWAN, M. G., JACKSON, M. P. A. & TRUDGILL, B. 1999. Salt-related fault families and fault welds in the northern Gulf of Mexico. *American Association of Petroleum Geologists Bulletin*, **83**, 1454–1484.

ROYDEN, L. H. & KEEN, C. E. 1980. Rifting processes and thermal evolution of the continental margin of eastern Canada determined from subsidence curves. *Earth and Planetary Science Letters*, **51**, 343–361.

SAUNDERS, I. & YOUNG, A. 1983. Rates of surface processes on slopes, slope retreat and denudation. *Earth Surface Processes and Landforms*, **8**, 473–501.

SCHMITT, R. S., TROUW, R. A. J., SCHMUS, W. R. V. & PIMENTEL, M. M. 2004. Late amalgamation in the central part of West Gondwana: new geochronological data and the characterization of a Cambrian collisional orogeny in the Ribeira Belt (SE Brazil). *Precambrian Research*, **133**, 29–61.

SOUZA, K. G., FONTANA, R. L., MASCLE, J., MACEDO, J. M., MOHRIAK, W. U. & HINZ, K. 1993. The southern Brazilian margin: an example of a South Atlantic volcanic margin. *In*: *3rd International Congress of the Brazilian Geophysical Society, Rio de Janeiro*, **2**, 1336–1341.

STECKLER, M. S. 1981. *The thermal and mechanical evolution of Atlantic-type continental margins* PhD. thesis, Columbia University, New York.

STECKLER, M. S. 1985. Uplift and extension of the Gulf of Suez: Indications of induced mantle convection. *Nature*, **317**, 135–139.

SZATMARI, P. & MOHRIAK, W. U. 1995. Plate model of post-breakup tectono-magmatic activity in SE Brazil and the adjacent Atlantic. *In*: *V Simpósio Nacional de Estudos Tectônicos, Gramado*, 213–214.

SZATMARI, P., GUERRA, M. C. M. & PEQUENO, M. A. 1996. Genesis of large counter-regional normal fault by flow of Cretaceous salt in the South Atlantic, Santos Basin, Brazil. *In*: ALSOP, G. I., BLUNDELL, D. J. & DAVISON, I. (eds) *Salt Tectonics*. Geological Society, London, Special Publications, **100**, 259–264.

THIEDE, J. 1977. Subsidence of aseismic ridges: evidence from sediments on Rio Grande Rise (Southwest Atlantic Ocean). *American Association of Petroleum Geologists, Bulletin*, **61**, 929–940.

TURNER, S., REGELOUS, M., KELLEY, S., HAWKESWORTH, C. & MANTOVANI, M. 1994. Magmatism and continental break-up in the South Atlantic: high precision Ar/Ar geochronology. *Earth and Science Planetary Science Letters*, **121**, 333–348.

VAN BALEN, R. T., VAN DER BEEK, P. A. & CLOETINGH, S. A. P. L. 1995. The effect of rift shoulder erosion on stratal patterns at passive margins; implications for sequence stratigraphy. *Earth and Planetary Science Letters*, **134**, 527–544.

VAN DER BEEK, P., ANDRIESSEN, P. & CLOETINGH, S. 1995. Morphotectonic evolution of rifted continental margins: Inferences from a coupled tectonic-surface processes model and fission track thermochronology. *Tectonics*, **14**, 406–421.

WATTS, A. B. & THORNE, J. A. 1984. Tectonics, global changes in sea level and their relationship to stratigraphic sequences at the U. S. Atlantic continental margin. *Marine and Petroleum Geology*, **1**, 319–339.

WEISSEL, J. K. & KARNER, G. D. 1994. Flexural uplift of rift flanks due to mechanical unloading of the lithosphere during extension. *Journal of Geophysical Research*, **94**, 13919–13950.

WHITE, N. & MCKENZIE, D. 1988. Formation of the "steer's head" geometry of sedimentary basins by differential stretching of the crust and mantle. *Geology*, **16**, 250–253.

WHITE, R. S. & MCKENZIE, D. P. 1989. Magmatism at rift zones: the generation of volcanic continental margins and flood basalts. *Journal of Geophysical Research*, **94**, 7685–7730.

ZALÁN, P. V. 1986. A tectônica transcorrente na exploração de petróleo: uma revisão. *Revista Brasileira de Geociências*, **16**, 245–257.

ZALÁN, P. V. & OLIVEIRA, J. A. B. 2005. Origem e evolução estrutural do Sistema de Riftes Cenozóicos do Sudeste do Brasil. *Boletim de Geociências da Petrobras*, **13**, 269–300.

ZUBER, M. T. & PARMENTIER, E. M., 1986. Lithospheric necking: a dynamic model for rift morphology. *Earth and Planetary Science Letters*, **77**, 373–383.

Restoring Pan-African–Brasiliano connections: more Gondwana control, less Trans-Atlantic corruption

M. J. DE WIT[1,2], J. STANKIEWICZ[2] & C. REEVES[3]

[1]*AEON-Africa Earth Observatory Network and Department of Geological Sciences, University of Cape Town, Rondebosch 7700, South Africa (e-mail: maarten.dewit@uct.ac.za)*

[2]*GFZ–GeoForschungsZentrum Potsdam, Telegraphenberg, Potsdam 14473, Germany*

[3]*Earthworks BV, Achterom 41A, 2611 PL, Delft, The Netherlands*

Abstract: The concept of South America and Africa as rigid continents during the formation, growth and motion of their respective plates has frustrated reconstruction of a tight, geologically economic fit between these two fragments in their Gondwana framework. We recognize that (1) internal strains released during and following Gondwana break-up have distorted their actual shapes within Gondwana and (2) these two continents comprise mosaics of smaller microblocks, or platelets, of relatively undistorted Precambrian terrains that experienced modest, episodic relative motions along rift zones that separate them. This permits a fresh approach to quantitative reconstructions of palaeo-continents. Former geological ties forged at the time of Gondwana amalgamation, now exposed at the continental margins of the South Atlantic as piercing points, provide robust anchors for new paleo-cartographic experiments. We present two new tectonic maps of the Brasiliano and Pan-African structures of West Gondwana on which we identify ten piercing points that, if re-joined simultaneously, could facilitate quantification of a well-substantiated Gondwana fit and help retrace the evolution of its continental margins with greater accuracy than has been achieved until now. This has significant bearing on understanding the origin and evolution of passive continental margins, and the geodynamics of Gondwana break-up.

Relative rapid amalgamation of a number of continental fragments during the Late Neoproterozoic resulted in a united Gondwana supercontinent (e.g., de Wit *et al.* 1988; Unrug 1992, 1997; Trompette 1994; Brito Neves *et al.* 1999; Bizzi *et al.* 2003). Whilst accretion of smaller blocks continued along the peripheries of this supercontinent throughout the Palaeozoic (e.g., de Wit & Ransome 1992; Nance & Murphy 1996; Linnemann *et al.* 2004; Vaughan *et al.* 2005), central Gondwana became peneplained and remained relatively undisturbed until a series of Mesozoic intra-continental rifts fractured this stability (e.g., Burke & Dewey 1976; de Wit *et al.* 1988; Daly *et al.* 1989; Fairhead & Brink 1991; Reeves 1999). The precise history of Gondwana's subsequent break-up and early separation into the present day continents of the southern hemisphere is recorded in the rock record along the continental margins of South America, Africa, Antarctica, Australia and India (e.g., Tankard *et al.* 1996; Hinz *et al.* 1999; Mohriak *et al.* 2007, 2008). This history has yet to be deciphered to a degree that can differentiate cause and effect between convecting mantle and lithosphere processes, since there are significant gaps in our knowledge about the kinematics of the break-up of Gondwana (e.g., Storey 1995; Hawkesworth *et al.* 1999; Krabbendam & Barr 2000; Stern & de Wit 2004). How and why this supercontinent formed and came apart lies at the heart of understanding the geodynamics, episodic growth and break up of continents in general. A first step towards unravelling this interactive history requires a more accurate pre-break up fit for Gondwana than is presently available (e.g., Reeves 1999; Ghosh *et al.* 2004). Surprisingly, a robust fit still has to be achieved between the two continents flanking the South Atlantic Ocean, which have featured most prominently in all models of Gondwana reconstructions. This quest is unfulfilled because the internal deformations of Africa and South America since Gondwana break-up are not yet well-quantified (Fig. 1; Burke & Dewey 1976; de Wit *et al.* 1988; Fairhead & Brink 1991; Reeves 1999; Eagles 2007). Reeves (1999) partially solved this problem by recognising the need to subdivide the continents, particularly Africa, into a number of Precambrian sub-domains that were semi-independent during early Mesozoic rifting and fragmentation of Gondwana (Fig. 1c), and modelling the fit between Africa and South America using modest Euler rotations of these rigid microplates (Reeves *et al.* 2004). This model yields satisfactory fits between the fragments of former Gondwana, and deserves rigorous testing. Testing requires identification of *vertical* litho- and tectonic-markers on a number of different

From: PANKHURST, R. J., TROUW, R. A. J., BRITO NEVES, B. B. & DE WIT, M. J. (eds) *West Gondwana: Pre-Cenozoic Correlations Across the South Atlantic Region.* Geological Society, London, Special Publications, **294**, 399–412. DOI: 10.1144/SP294.20 0305-8719/08/$15.00

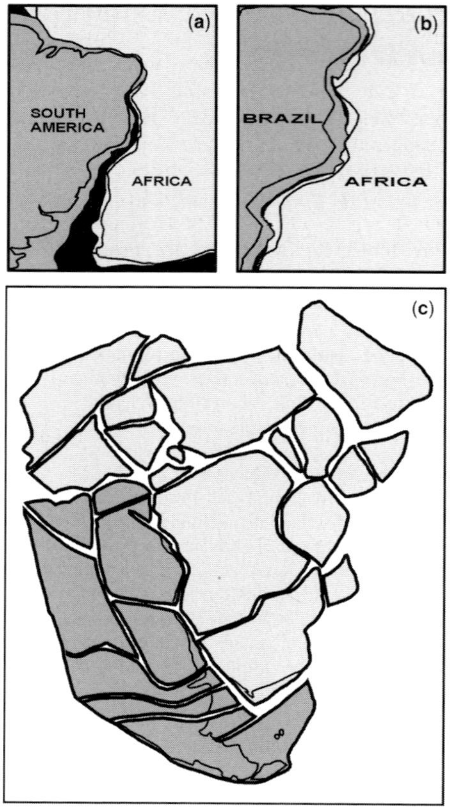

Fig. 1. Schematic representation of continental fits between South America (dark grey) and Africa (pale grey). (**a**) *c.* 120–150 Ma fit using rigid-plate rotations based on marine magnetic anomalies and transform faults (modified after Rabinowitz & La Breque 1979; de Wit *et al.* 1988). Black represents oceanic depths. (**b**) Possible tight fit of the South Atlantic continents between Nigeria and Namibia, with *c.* 60 km separation between the edges of the Precambrian blocks in Africa and South America during the earliest stages of rifting, analogous to that between Precambrian blocks along the present East African Rift with its rift-confined lakes (after Reeves *et al.* 2004). Black represents lakes. (**c**) *c.* 150–200 Ma fit of West Gondwana, schematically represented as a mosaic of Precambrian blocks separated by narrow (<50 km) rift zones. Small internal adjustments and rotations of Precambrian blocks (platelets) of central and North Africa and Brazil, to undo known Mesozoic–Cenozoic rifts and slip along major Pan-African, Brasiliano, Cretaceous and Miocene fault zones and composite tectonic lineaments (see Reeves 1999; Reeves *et al.* 2004), create a good fit along the central South Atlantic. The southernmost Atlantic is left with notable gaps because the Palaeozoic (Karoo) rift systems of southern Africa (e.g., Daly *et al.* 1989, 1992) and the Mesozoic–Cenozoic rifts of southern South America that strike at near right angles to the continental margin of South America (e.g., Tankard *et al.* 1996; Hinz *et al.* 1999), have not yet been adequately restored (e.g., Reeves *et al.* 2004; Eagles 2007).

Precambrian blocks along both continental margins of the South Atlantic that can be matched and reconnected with confidence. Many former contiguous Neoproterozoic faults and shear zones that now cut continental margins on both sides of the ocean provide 'piercing points' that could fulfil this requirement (e.g., Reeves & de Wit 2000). Vertical or subvertical structures should be insensitive to present-day erosion level.

We present a west South America–Africa map with a compilation of Neoproterozoic structural and tectonic data from the *c.* 200 million year period during which the continental fragments that comprise west Gondwana amalgamated along five major orogenic systems, each comprising a number of tectonic belts (Fig. 2). On this map, we identify ten key piercing points along the circum-South Atlantic that need improved aeromagnetic characterization and accurate dating, so that they can be reconnected simultaneously to yield a yet more satisfactory finite fit.

The time period in question spans the late Neoproterozoic, at the very end of the Precambrian, between about 750 and 550 Ma, albeit with some events that 'spill-over' into the early Palaeozoic. The geological activity in that time period within South America is generally referred to as 'Brasiliano' (e.g., Hurley *et al.* 1967; Trompette 1994; Brito Neves *et al.* 1999; Bizzi *et al.* 2003; Van Schmus *et al.* 2007) and in Africa as 'Pan-African' (e.g., Kennedy 1964; Clifford 1970; Cahen & Snelling 1984; de Wit *et al.* 1988, 2001; Black & Liégeois 1993; Stern 1994; Trompette 1994; Meert 2003; Johnson & Woldehaimanot 2003; Collins & Pisarevsky 2005). Although we restrict ourselves to regions that flank both sides of the South Atlantic Ocean, we extend our tectonic analyses eastward to include a larger part of west Gondwana, especially in Africa, to fully appreciate the extent that late Neoproterozoic geodynamics affected this region, and how Neoproterozoic tectonic inheritance influenced the shaping of the South Atlantic during subsequent Phanerozoic events.

Previous work

The first quantitative reconstruction between South America and Africa was the early computer model of Bullard *et al.* (1965) that economized overlap of continental crust whilst maintaining an optimally aesthetic fit. More reliable plate tectonic rotations using marine magnetic anomalies and fracture zones (e.g., Rabinowitz & LaBrecque 1979; de Wit *et al.* 1988; Lawver *et al.* 1999; Scotese 2000; Eagles & König 2007) and satellite-derived ocean floor topography (Smith & Sandwell 1997) followed. This arsenal of data has not, however,

ever achieved a complete satisfactory closure of the South Atlantic, and has continued, therefore, to call for greater geological and geophysical understanding and input from the surrounding continental margins (e.g., Reeves 1999; Bauer *et al.* 2000; Eagles 2007; Fig. 1).

One issue is exactly *which* conjugate continental outlines should be matched. Bullard *et al.* (1965) matched the 500 fathom (*c.* 1000 m) isobaths as a rough estimate of the likely limits of the (extended) continental crust. Coastlines are ephemeral geologically but easily recognizable in illustrations. Yet they mostly fall within the transition zone from true continental to true oceanic crust. For much of Gondwana, including the South Atlantic margins, we favour using the limit of the extant Precambrian geology. This can sometimes be seen in outcrop or inferred (where below shallow cover) from aeromagnetic surveys. Inboard of this limit we assume the Precambrian rocks within each platelet are essentially undistorted and unextended by Gondwana break-up processes. Outboard, as far as the boundary with true oceanic crust, we expect to find laterally extended (remnants of) crust that existed at the end of the Precambrian (Pan-African/Brasiliano). We favour assembling these Precambrian 'outcrop' limits with their margins essentially parallel. The optimum separation in the reconstructions or, in other words, the width of Precambrian crust that has been extended into the transitional crust of the two continental margins, is unknown. Numerous experimental geometrical reassemblies suggest that only 50–100 km (divided between the two margins) has been lost in this way (Fig. 1; Reeves *et al.* 2002).

Testing different reconstruction models for independent verification over the last 40 years has relied mostly on geochronology of onshore geology along both sides of the Atlantic. This approach was first pioneered using K–Ar and Rb–Sr dating techniques in NE Brazil and West Africa, by Hurley and co-workers (Hurley *et al.* 1967; Hurley 1972, 1974), and other parts of Africa and Brazil using, for example, the extensive data sets of Cahen & Snelling (1966, 1984) and Cordani (see Cordani *et al.* 1988 for a review). Although correlations verified the concepts of continental drift/plate tectonics, they lacked the accuracy required to convincingly correlate between individual stratigraphic marker horizons, intrusions or tectonic structures on either side of the Atlantic. It was not until the 1990s when precise U–Pb dates became widely available that attempts to test correlations and to quantify the fits and kinematic history of Gondwana break-up was attempted using, for example, well-dated continental shear zones that were active only during the formation of Gondwana (e.g., of Pan-African and Brasiliano age; Daly *et al.* 1989; de Wit *et al.* 2001). The combination of this analysis with data from magnetic anomalies and transform faults in the Indian Ocean to reconnect subvertical shear zones along the margins of India, Madagascar and Africa was attempted by Reeves, and a refined fit between East and West Gondwana to accuracies within tens of kilometres was derived (e.g., Reeves 1999; Reeves & de Wit 2000; Reeves *et al.* 2002, 2004). A prerequisite to ensure the accuracy in this technique is to join only conjugate points along opposite continental edges defined by intersection of *subvertical* shear-zones, faults, and dykes. This approach avoids unknown differential uplift and erosion on the separate fragments subsequent to their separation. Using piercing points along the continental margins of once-conterminous *shallow* dipping thrusts, suture zones or sills, creates uncertainties of several hundreds of kilometres when rejoining them along continental margins that have been eroded to different crustal levels since their separation. In the latter case additional detailed thermochronology is needed to resolve precise 3D correlation between such piercing points (e.g., Reeves & de Wit 2000), a required approach that is still frequently violated (e.g., Collins *et al.* 2007).

A Brasiliano/Pan-African tectonic map

Our new tectonic map is presented in Figure 3 (fold-out) and its simplified geological counterpart in Figure 4. The map shows the major shear zones and suture zones of Pan-African age, taken from the Geological Map of Gondwana (de Wit *et al.* 1988), converted to GIS format in 1992, and updated where appropriate, using reports from the peer-refereed geological literature and from published geological maps. Where such structural data are not available, other information is sometimes substituted, for example, the trend of foliations in parts of Libya, Sudan, Chad, CAR, DRC and Uganda. Local maps and compilations are used for this alternative data (e.g., Gsell & Sonet 1960; Sonet 1963; Vail 1976; El-Makhrouf 1978; UNESCO 2000). Fault, shear zone and geochronological data used for the purpose of identifying 'piercing points' are summarized in specific sections below. The Brasiliano data are originally also from the Geological Map of Gondwana, but are extensively edited using the new digital compilations published by the geological survey of Brazil (Bizzi *et al.* 2003).

Our maps are compiled using ESRI GIS and Mapping Software (http://www.esri.com/software/arcview/) that allows a variety of displays. Figures from publications in journals and books had to be

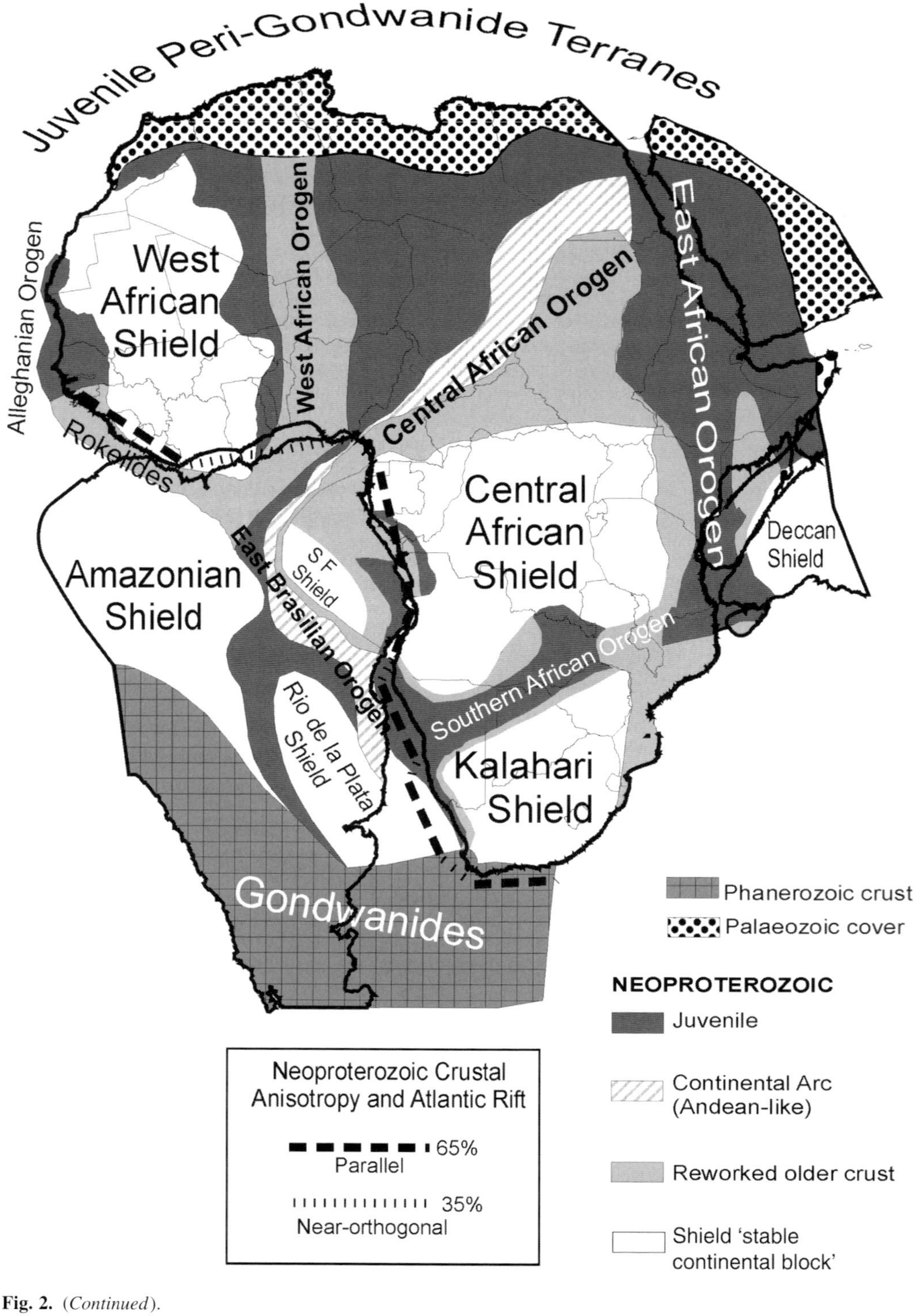

Fig. 2. (*Continued*).

scanned and imported into our GIS database. In this process keeping cartographic accuracy is essential. When co-ordinates and projections were provided in a published map, it was easy to geo-reference the figure to ensure placing it at the correct location. Often, however, the projection was not given and co-ordinates were 'eyeballed in' on the map. We have also come across several publications where the co-ordinates provided were incorrect. Fortunately, virtually all figures included an accurate map of the rivers draining the area—these constitute an essential reference point when doing field work in remote parts of Africa. With the AEON Africa River Database at our disposal (which includes a detailed digitized drainage map of Africa, see Stankiewicz & de Wit 2006; de Wit & Stankiewicz 2006, for details), judicious trial-and-error usually allows accurate positioning of even the poorest maps.

Ten selected Neoproterozoic 'piercing points' to close the South Atlantic

The following piercing points (see Fig. 4 and Table 1 for their locations) are suggested as prime targets for further geochronology and geophysics to aid the reconstruction of the continental margins flanking the South Atlantic, and to facilitate a tighter fit between South America and Africa. This is by no means an exhaustive list, but serves to illustrate the approach and the requirements of the Pan-African/Brasiliano features needed to complete manifold simultaneous rotations.

(1) The Rokelide front of subvertical Neoproterozoic transpressive shear zones that delineate the southern tectonic margin of the Archaean–Eburnian age West African Shield (MacFarlane 1980) and the equivalent southern margin of the Archaean–Trans-Amazonian age São Luís fragment in northern Brazil (Bizzi *et al.* 2003; Klein & Moura 2008).

(2) The Dahomide thrust front/suture zone (including ophiolites) and the vertical Kandi Shear System that separate the West African shield from the Neoproterozoic West Saharan orogen (Castaing *et al.* 1993; Attoh *et al.* 1996; Agbossoumondé *et al.* 2004), and the equivalent tectonic boundary along the eastern margin of the São Luis fragment in NE Brazil (Bizzi *et al.* 2003). In both locations, this tectonic zone hosts a number of subvertical shear zones, and separates this Pan-African zone from Palaeoproterozoic basement (*c.* 2.0 Ga) to the west that has not been affected by Pan-African tectonothermal overprinting (e.g., the Birimian of the West African Shield and the Trans-Amazonian of São Luís). This is in contrast with all Archaean–Palaeoproterozoic basement east of the West African Shield and São Luis fragment (the Nigerian shield and the Ceará terrane of the Borborema province, respectively), which has been thoroughly remobilized by Pan-African thermotectonic processes, often densely invaded by late Pan-African and Brasiliano granites, and interleaved with minor Neoproterozoic sequences (e.g., Caby 1989; Black & Liégeois 1993; Castaing *et al.* 1993; Trompette 1994; Ekwueme & Attoh 1996; Fetter *et al.* 2003; see Fig. 4).

(3) Major subvertical crustal lineament(s) within the West Sahara orogen extending from the Tuareg shield through the Nigerian shield, flanking the Latea block (Caby 1989, 2003; Castaing *et al.* 1993; Black *et al.* 1994; Trompette 1994; Bournas *et al.* 2003; Caby & Monié 2003; Liegéois *et al.* 2003; Ouzegane *et al.* 2003; Peucat *et al.* 2003), into the Trans-Brasiliano Lineament (TBL) and associated shear zones of the Ceará terrane of NE Brazil; and the junction between the Araguaia and the western sections of the Brasília Belts (Pimental & Fuck 1992; Bizzi *et al.* 2003; Fetter *et al.* 2003; Piuzana *et al.* 2003; Neves 2003; Van Schmus 2003, 2008).

(4) Major subvertical shear zones flanking the fold belt along the eastern margin of the Tuareg shield (Caby 1989, 2003; Trompette 1994; Ferre *et al.* 2002; the most easterly is known as the Rangane shear zone, RSZ) that strike into the north–south trending Seridó Fold Belt of the Borborema province, NE Brazil (Bizzi *et al.* 2003; Van Schmus *et al.* 2003, 2008).

Fig. 2. (*Continued*) Simplified map of major Neoproterozoic–Palaeozoic orogenic systems and associated tectonic belts of Gondwana. Also shown are the extents of relatively stable continental lithosphere in the Neoproterozoic (white represents Neoproterozoic shields) and those areas where parts of these shields and other older fragments were significantly reworked during the Neoproterozoic (pale grey). The Pan-Gondwana Andean-like belt from Egypt to Argentina (*c.* 7500 km) flanks the Neoproterozoic Central African Shield (which includes its extension into Brazil as the largely remobilised São Francisco Shield). The poor fit across the southernmost Atlantic Ocean does not yet permit extending this Andean-like arc farther south beyond the coast of Argentina. Those sections of Neoproterozoic orogenic belts that trend approximately parallel to the early South Atlantic rift, are shown in dashed lines: here, rifting took place near-parallel to the Pan-African/Brasiliano tectonic fabric (*c.* 65% of the total South Atlantic rift length); sections where the Atlantic rifts cross-cuts such Pan-African/Brasiliano tectonic trends at a significant angle (*c.* 35%) are shown in dotted lines. Such conterminous ties may play important roles in creating the *en echelon* geometries and overlap of early continental rifts (e.g., Fig. 1b).

(5) The Central African Shear System (the Adamawa, Tchollire–Banyo, Sanaga, Foumban shear zones) and the Bohorema Shear System (the Pernambuco and Patos shear zones in Brazil). These essentially vertical major mylonitic systems stretch from NE Brazil into Central Africa (e.g., Poidovin 1985; de Wit 1988; Daly *et al.* 1989; Neves 2003; Van Schmus *et al.* 2008), through Cameroon, northern Gabon to the Central African Republic and Chad (Poidovin 1985; Rolin 1995; Toteu *et al.* 2001, 2004; Ngako *et al.* 2003). There is no consensus as to how precisely the individual shears link up, despite the long history of study (compare, for example, Poidovin 1985; de Wit *et al.* 1988; Daly *et al.* 1989; Castaing *et al.* 1993; Trompette 1994; Neves 2003; Van Schmus *et al.* 2008, and Fig. 3). They are well-defined positionally in NE Brazil from geological mapping and aeromagnetic surveys; detailed aeromagnetic mapping of Nigeria currently underway may throw new light on the precise location of their continuation into Africa.

(6) The Sergipano–Oubengides fold belt of central Africa (Fig. 4; often not shown on tectonic maps of Africa, even in relatively recent works, despite its detailed description by Poidovin 1985). There are two important components of this belt that should be considered for correlations across the Atlantic. The first concerns the near east–west striking and south-verging tectonic front with thrusts and vertical strike-slip systems that cut and remobilize the northern boundary of the Central African Shield. This shield represents a large Neoproterozoic continental fragment that stretches into South America, and comprises a mosaic of Archaean cratons (Sao Francisco (São Miguel do Aleixo system), Ntem (Yaoundé system), Congo (Nola–Bangui system) in Brazil, Gabon, and CAR/DRC) and their Palaeoproterozoic outboard margins (reworked in the Neoproterozoic—Eburnian in Africa and Trans-Amazonian in South America; Poidovin 1985; Rolin 1995; Toteu *et al.* 2001, 2004; Bizzi *et al.* 2003; Van Schmus *et al.* 2003, 2008; see Fig. 4). The second concerns the recognition that a long-lived Neoproterozoic magmatic belt, apparently of Andean-like proportions, with associated vertical shear zones, flanks the western edge of the Central African Shield (de Wit *et al.* 2005), extending southwestwards across the Sahara, from Egypt, through Darfur, Chad, Cameroon (Schurmann 1974; Kusnir 1993; Kushnir & Moutaya 1998; Tagne-Kamga 2003; Toteu *et al.* 2004; de Wit *et al.* 2005) into NE Brazil (marked by the Alto Pajeú domain; Bizzi *et al.* 2003; Van Schmus *et al.* 2008). Flanking the São Francisco craton (Riacho do Pontal domain), it continues through the Brasília Belt (Pimental & Fuck 1992; Piuzana *et al.* 2003, 2005), and possibly farther south still along the edge of the Río de la Plata craton (Dalla Salda *et al.* 1988; Bizzi *et al.* 2003), a total distance of >7500 km. This is a fundamental Neoproterozoic tectono-magmatic belt of central Gondwana. Because it thins considerably (due to Neoproterozoic tectonism) as it approaches the Atlantic coast in Cameroon (Mayo Kebi domain), and across in NE Brazil (Alto Pajeú domain), it defines a piercing point of 'golden-spike' proportions (Fig. 4). Batholiths of this belt in north-central Africa and NE Brazil are some of the only known associations with extensive *c.* 1.0 Ga Mesoproterozoic granitoids in northern Gondwana.

(7) The southern margin of the West Congo Fold Belt is enigmatic. In Angola, the (low-grade) West Congo Fold Belt is abruptly cut of by a series of steep transcurrent faults and an east–west striking granulite terrain that cut the Angolan coast at high angle (e.g., de Wit *et al.* 1988; Tack *et al.* 2001; Fig. 4). This obliquity has been interpreted by some workers to indicate a significant transcurrent fault that offsets the fold belt front eastward to join with the Neoproterozoic Lufilian arc of the Katangan fold belt in Zambia/DRC (e.g., Daly *et al.* 1989, 1992; Collins & Pisarevsky 2005). Alternatively this transfer structure may form part of a deep crustal triple junction to the west joining the subvertical coast-parallel high-grade shear belts of the Ribeira-Araçuaí Belt of Brazil (e.g., de Wit *et al.* 1988; Bizzi *et al.* 2003; Heilbron & Machado 2003; Schmitt *et al.* 2004, 2005) and in particular with the shear systems in the Cabo Frio terrane (Heilbron & Machado 2003; M. Heilbron, pers. comn. 2007).

(8) The subvertical shear zones of the Kaoko Belt (Schmitt *et al.* 2005; Goscombe *et al.* 2005; Gray *et al.* 2006; Goscombe & Gray 2007) in northern Namibia and southern Angola transgress obliquely across the South Atlantic margins towards the subvertical shear zones of the Brasiliano-age Ribeira Belt (Schmitt *et al.* 2004, 2005; Gray *et al.* 2006). Because the shear zones of these belts are strike-oblique with a low angle to the present coastlines, and because they are of similar age, identifying their respective 'partners' on each side of the South Atlantic may prove difficult.

(9) The northern extensions of vertical shear zones associated with the Alferez–Cordillera–Punta del Este shear zone (ACPESZ) in the Braziliano-age Dom Feliciano Belt (e.g., Dalla Salda *et al.* 1988; Bizzi *et al.* 2003; Basei *et al.* 2005) strike with low obliquity northwards across the South Atlantic margin to apparently project into the westernmost vertical strike-slip shear zones that cut the Pan-African age Western Kaoko batholith and, at their southernmost extremity, strike southwestwards into the South Atlantic.

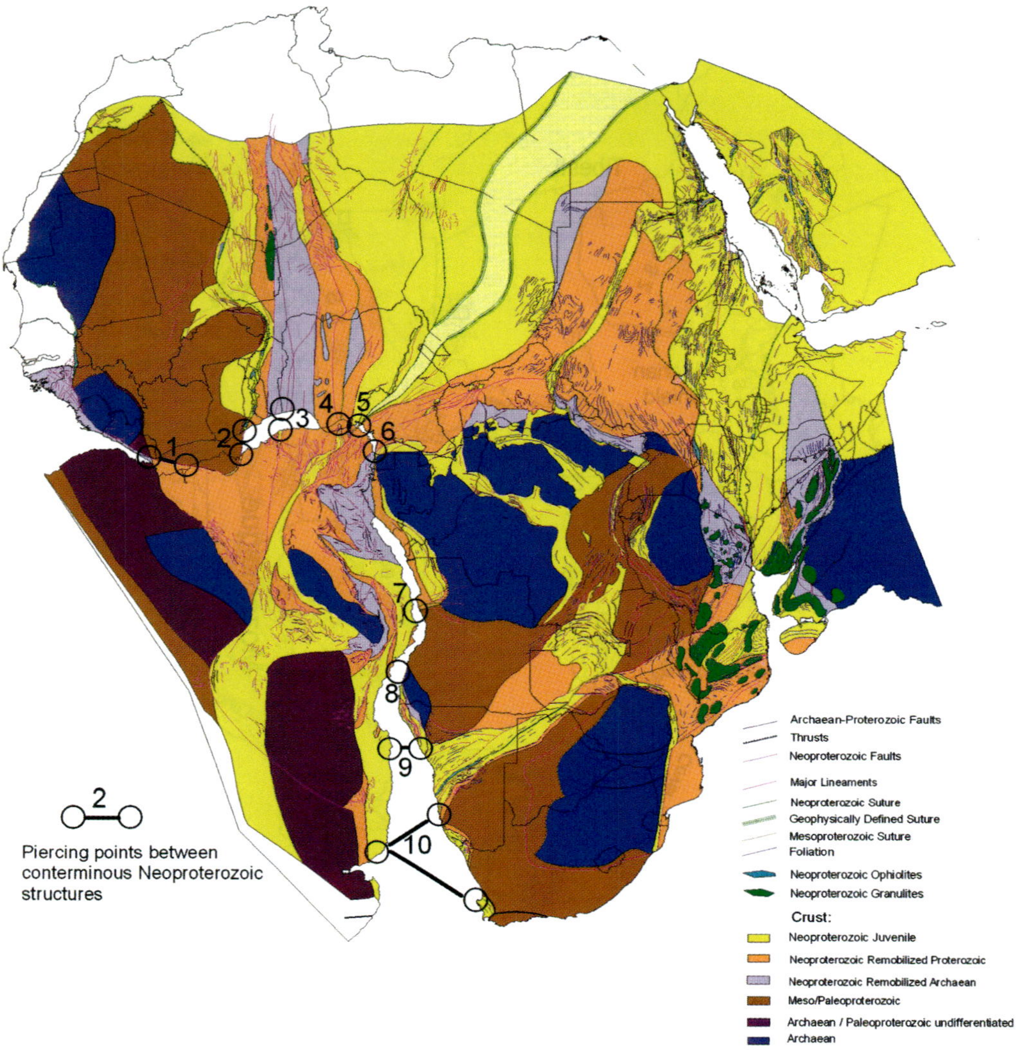

Fig. 4. A simplified geological map of West and Central Gondwana showing the major Neoproterozoic Shields with their embedded Archaean cratons, and their margins that were remobilized in the Pan-African/Brasiliano orogens. Major remobilized Archaean–Proterozoic fragments within the Neoproterozoic orogenic systems are also shown. Nine Neoproterozoic 'piercing' points between once conterminous Pan-Gondwana sub-vertical lineaments are identified along opposite sides of the South Atlantic. In addition, the mid-Phanerozoic (*c.* 250 Ma) 'piercing points' associated with the Cape Fold Belt and the Sierra de la Ventana, are also shown. The precise locations and details of these 'piercing points', which might facilitate restoring the South Atlantic margin to derive a more reliable Gondwana fit, are given in Table 1.

angle across the South Atlantic margins (Figs 3 & 4). These belts have long been correlated on lithostratigraphic grounds (e.g., Keidel 1916; du Toit 1927, 1937) and the correlation has been robustly vindicated by recent detailed geochronology (Rapela *et al.* 2003). Because the tectonic features of parts of these belts are steeply dipping, and because they formed well before Gondwana break-up, they qualify as the most southerly tectonic 'piercing points' in Africa with counterparts in South America. However, a word of caution is warranted here: the extreme western section of the Cape Fold Belt bifurcates into SW- and NW-striking tectonic branches, flanking the easternmost outcrops of the Saldanian basement of the Western Cape mentioned above (e.g., de Wit & Ransome 1992). Which branch

Table 1. *Piercing points for Transatlantic reconstruction of West Gondwana*

Point	Africa long.	Africa lat.	African country	America long.	America lat.	Today's distance (km)
1	8.0°W	4.5°N	Liberia	48.0°W	0.5°S	4500
2	0.5°W	5.5°N	Ghana	43.5°W	2.5°S	4900
3	3.0°E	6.5°N	Benin/Nigeria	40.0°W	3.0°S	4900
4	7.0°E	4.5°N	Nigeria	35.5°W	5.0°S	4800
5	8.5°E	4.5°N	Nigeria/Cameroon	35.0°W	6.0°S	5000
6	10.0°E	2.5°N	Cameroon/Eq. Guinea	35.5°W	9.0°S	5200
7	13.0°E	10.0°S	Angola	40.5°W	20.5°S	5800
8	12.0°E	14.5°S	Angola	45.0°W	23.5°S	6000
9	13.5°E	21.0°S	Namibia	49.0°W	28.5°S	6300
10a	15.0°E	26.0°S	Namibia	55.5°W	35.0°S	6700
10b	18.0°E	33.0°S	South Africa			6600

correlates directly with the Sierra de la Ventana, if either, will have to await detailed geophysical work along both continental margins.

Discussion

The recognition that major continents have not always responded as single rigid blocks during plate motions has led to the qualitative understanding that internal continental distortions must be better understood and quantified if continental margins around the South Atlantic (and elsewhere) are to be refitted to their correct Gondwana positions. Matching pre-drift subvertical shear zones on both sides of the South Atlantic that were conterminous just prior to rifting facilitates such an approach. There are, however, significant uncertainties in precisely correlating such structures across the Atlantic. We note, for example, that some correlations of suitable piercing points derived from sub-vertical structures differ substantially in the current literature (e.g., 2 and 3 in Fig. 4), in part because the structures have yet to be accurately dated. This underscores the point that the shear zones associated with the piercing points that we have identified require rigorous examination, geophysically and isotopically.

In addition, the internal Precambrian microplates and the edges of most cratons/shields have not yet been clearly differentiated because, in most cases, Pan-African and Brasiliano tectonothermal overprints and remobilizations have distorted their pre-Gondwana geometries (e.g., Fig. 4). Indeed there is still vigorous debate whether or not large tracts of central Africa affected by Pan-African thermotectonism constitute juvenile or remobilized earlier crusts, such as the NE extension of the Central African Shield (sometimes referred to as the Sahara metacraton or ghost craton), Tibesti and surroundings, parts of the Nigerian metacraton, and so forth (e.g., Kennedy 1964; Clifford 1970; Ghuma & Rogers 1976; Nagy *et al.* 1976; Schurmann 1976; Vail 1976; El-Makhrouf 1978; Cahen & Snelling 1984; Harris *et al.* 1984; Pelgram *et al.* 1987; Sultan *et al.* 1990, 2000; Harms *et al.* 1996; Krabbendam & Barr 2000; Kuster & Liégeois 2001; Abdelsalam *et al.* 2002, 2003). This will require a lot more careful isotopic and geophysical analysis on both sides of the Atlantic to resolve.

The fact that Neoproterozoic tectonic piercing points exists at all along the circum-continental margins of the South Atlantic corroborates findings that continental rifting does not always follow the inherent lithospheric anisotropy of the separating fragments. Indeed our geological map (Fig. 4) emphasizes that the east–west striking central South Atlantic margins cut across major Pan-African/Brasiliano tectonic belts as well as older tectonic trends on Archaean cratons and Palaeoproterozoic shields. There has been analysis in more detail for the entire Gondwana supercontinent elsewhere (Krabbendam & Barr 2000): on average about 50% of Gondwana's 25 000 km rifted margins are parallel to pre-existing structures or craton margins. Although we find this to be (significantly?) different for the circum-Atlantic margins (*c.* 65%, Fig. 2), we concur with Krabbendam & Barr (2000) that it remains to be resolved why Gondwana broke up along only one out of five of its major Neoproterozoic Gondwana orogens that cut across Central and West Gondwana (Fig. 2). Resolving this enigma will also help guide our as yet limited understanding the fundamental geodynamics of the demise of supercontinents, and the role of inherited ties between their fragments.

Conclusions

Obtaining a 'tight' reconstruction for Gondwana will remain an elusive goal unless better integration between marine geophysics and on-land geology is achieved. Reversing the oceanic crust between continental margins as they have grown apart is the relatively easy part. Recognition that the major Gondwana fragments have acted as a mosaic of platelets for part of their pre- and post-drift histories, and characterizing these smaller Precambrian fragments to aid in reconstructing their initial (and mostly aborted) rifting and rotations, is now a more challenging task. Aeromagnetic anomalies over the continents may map the boundaries between these rigid continental blocks separated by small rotations, degrees of extension and/or amounts of strike-slip motions, and can identify rift shoulders where the fossil boundaries are deeply buried (e.g., Reeves 1999; Reeves *et al.* 2004). Fitting these blocks into their pre-drift positions requires matching of geological structures that were once conterminous. One way to realize this is to re-join Neoproterozoic 'piercing points' on the conjugate margin of continental blocks that are well-dated and otherwise geophysically and geologically characterized. We have argued that for South America and Africa this might be achievable in the case of at least ten well-defined Neoproterozoic ties around the margins of the South Atlantic Ocean, and we presented a detailed map of the Brasiliano/Pan-African structures of Gondwana to facilitate the planning of such experiments. Playing back the continental motions whilst simultaneously pinning these original ties together, should result in a more robust and accurate fit from which to re-track interactive mantle–lithosphere break-up mechanisms and help quantify internal Gondwana strains that led to the break-up and separation in the first place. Involving a greater degree of such geological and geophysical control to reconstruct the Gondwana break-up history will help to correct for early Trans-Atlantic rift distortions and thus to better reconstruct the evolution of continental margins of the South Atlantic; and this in turn will improve our general understanding of the evolution of continents.

We would like to thank Margie Jeffery who started the Pan Gondwana Neoproterozoic Tectonic Map project with M.d.W. in the late 1980s. This led to a draft version by 1992 that was widely circulated amongst interested colleagues and informally reviewed by the late Robert Shackleton. Their comments were instrumental in persevering with this project. The present version of the map has benefited greatly from these comments and from the large number of U–Pb dates that have become available over the last 15 years. We are grateful to many colleagues for sharing their data, preprints and reprints throughout this period, and in particular to Richard Armstrong and Sam Bowring for a number of joint projects that benefited from their expertise in geochronology. We have enjoyed many discussions with colleagues over the years, especially at Gondwana conferences (particularly Gondwana 9 to 12). The map has benefited also from the emergence and maturing of GIS technology. We thank members of the CIGCES at the University of Cape Town that were instrumental in helping to master this. Most of the work was funded for many years through sustained NRF (National Research Foundation of South Africa) support, and some through the South African exploration industry. In the final year the GFZ-Potsdam, Germany has made it possible to finish this phase of the project. We are grateful in particular to Rolf Emmermann, Jörg Erzinger and Michael Weber to facilitate this; and to Ariane Siebert for help with Figure 1. We thank Brendan Murphy, Brian Storey, Ian Dalziel and Bob Pankurst for constructive reviews. This is AEON contribution number 32.

References

ABDELSALAM, M. G., LIÉGEOIS, J.-P. & STERN, R. J. 2002. The Saharan metacraton. *Journal of African Earth Sciences*, **34**, 119–136.

ABDELSALAM, M. G., ABDEL-RAHMAN, El.-S. M. *ET AL.* 2003. Neoproterozoic deformation in the northeastern part of the Saharan metacraton, northern Sudan. *Precambrian Research*, **123**, 203–221.

AGBOSSOUMONDÉ, Y., GUILLOT, S. & MENOT, R.-P. 2004. Pan-African subduction-collision event evidenced by high-P coronas in metanorites from the Agou massif (southern Togo). *Precambrian Research*, **135**, 1–21.

ATTOH, K., DALLMEYER, R. D. & AFFATON, P. 1996. Chronology of nappe assembly in the Pan-African Dahomeyide orogen: evidence from $^{40}Ar/^{39}Ar$ mineral ages. *Precambrian Research*, **82**, 153–171.

BASEI, M. A. S., FRIMMEL, H. E., NUTMAN, A. P., PRECIOZZI, F. & JACOB, J. 2005. A connection between Neoproterozoic Dom Feliciano (Brazil/Uruguay) and Gariep (Namibia/South Africa) orogenic belts – evidence from a reconnaissance provenance study. *Precambrian Research*, **139**, 195–221.

BAUER, K., NEBEN, S. *ET AL.* 2000. Deep structure of the Namibia continental margin as derived from integrated geophysical studies. *Journal of Geophysical Research*, **105**, 25 829–25 853.

BLACK, R. & LIÉGEOIS, J. P. 1993. Cratons, mobile belts, alkaline rocks and continental lithospheric mantle: the Pan-African testimony. *Journal of the Geological Society, London*, **150**, 89–98.

BLACK, R., LATOUCHE, L., LIÉGEOIS, J. P., CABY, R. & BERTRAND, J. M. 1994. Pan-African displaced terranes in the Tuareg Shield (central Sahara). *Geology*, **22**, 641–644.

BIZZI, L. A., SCHOBBENHAUS, C., VIDOTTI, R. M. & CONÇALVES, J. H. (eds) 2003. *Geology, tectonics and mineral resources of Brazil.* CPRM - Serviço Geológico do Brasil, Brasília.

BOURNAS, N., GALDEANO, A., HAMOUDI, M. & BAKER, M. 2003. Interpretation of the aeromagnetic map of Eastern Hoggar (Algeria) using the Euler deconvolution, analytic signal and local wave-number methods. *Journal of African Earth Sciences*, **37**, 191–206.

BRITO NEVES, B. B., CAMPUS NETO, M. C. & FUCK, R. A. 1999. From Rodinia to western Gondwana: an approach to the Brasiliano–Pan-African cycle and orogenic collapse. *Episodes*, **22**, 155–166.

BULLARD, E., EVERETT, J. E. & SMITH, A. G. 1965. The fit of the continents around the Atlantic. A symposium on continental drift. *Philosophical Transactions of the Royal Society, London*, **258**, 41–51.

BURKE, K. C. & DEWEY, J. F. 1976. Two plates in Africa during the Cretaceous? *Nature*, **249**, 313–317.

CABY, R. 1989. Precambrian terranes of Benin–Nigeria and northeast Brazil and the late Proterozoic south Atlantic fit. *In*: DALLMEYER, R. D. (ed.) *Terranes in the Circum-Atlantic Paleozoic Orogens*. Geological Society of America, Special Papers, **230**, 145–158.

CABY, R. 2003. Terrane assembly and geodynamic evolution of central-western Hoggar: a synthesis. *Journal of African Earth Sciences*, **37**, 133–159.

CABY, R. & MONIÉ, P. 2003. Neoproterozoic subduction and differential exhumation of western Hoggar (south west Algeria): new structural, petrological and geochronological evidence. *Journal of African Earth Sciences*, **37**, 269–294.

CAHEN, L. & SNELLING, N. J. 1966. *The geochronology of Equatorial Africa*. North Holland Publishing Co., Amsterdam.

CAHEN, L. & SNELLING, N. J. 1984. *The geochronology and evolution of Africa*. Clarendon Press, Oxford.

CASTAING, C., TRIBOULET, C., FEYBESSE, J. L. & CHÈVREMONT, P. 1993. Tectonometamorphic evolution of Ghana, Togo and Benin in the light of the Pan African/Brasiliano orogeny. *Tectonophysics*, **218**, 323–342.

CLIFFORD, T. N. 1970. The structural framework of Africa. *In*: CLIFFORD, T. N. & GASS, I. G. (eds) *African Magmatism and Tectonics*. Oliver and Boyd, Edinburgh, 1–26.

COLLINS, A. S. & PISAREVSKY, S. A. 2005. Amalgamating eastern Gondwana: the evolution of the circum-Indian orogens. *Earth-Science Reviews*, **71**, 229–270.

COLLINS, A. S., CLARK, C., SAJEEV, K., SANTOSH, M., KELSEY, D. E. & HAND, M. 2007. Passage through India, The Mozambique suture, high pressure granulites & the Palghat- Chauvery shear zone. *Terra Nova*, in press.

CORDANI, U. G., TEIXEIRA, W., TASSINARI, C. G. G., KAWASHITA, K. & SATO, K. 1988. The growth of the Brazilian Shield. *Episodes*, **11**, 163–167.

DALLA SALDA, L., BOSSI, J. & CINGOLANI, C. 1988. The Río de la Plata cratonic region of southwestern Gondwanaland. *Episodes*, **11**, 263–269.

DALY, M. C., CHOROWICZ, J. & FAIRHEAD, J. D. 1989. Rift basin evolution in Africa: the influence of reactivated step basement shear zones. *In*: COOPER, M. A. & WILLIAMS, G. D. (eds) *Inversion Tectonics*. Geological Society, London, Special Publications, **44**, 309–334.

DALY, M. C., LAWRENCE, S. R., DIENU-TSHIBAND, K. & MATOUANA, B. 1992. Tectonic evolution of the Cuvette Centrale, Zaire. *Journal of the Geological Society*, London, **149**, 539–546.

DE WIT, M. J. & RANSOME, I. G. D. 1992. Regional inversion tectonics along the southern margin of Gondwana. *In*: DE WIT, M. J. & RANSOME, I. D. G. (eds) *Inversion tectonics of the Cape Fold Belt, Karoo and Cretaceous basins of southern Africa*. A.A. Balkema, Rotterdam, 15–22.

DE WIT, M. J. & STANKIEWICZ, J. 2006. Changes in surface water supply across Africa with predicted climate change. *Science*, **311**, 1917–1921. doi: 10.1126/science.1119929.

DE WIT, M. J., JEFFERY, M., BERGH, H. & NICOLAYSON, L. O. 1988. *Geological Map of sectors of Gondwana reconstructed to their position ~150 Ma (with explanatory notes). Scale 1: 1 000 000*. American Association of Petroleum Geologists, Tulsa, USA.

DE WIT, M. J., BOWRING, S. A., ASHWAL, L. D., RANDRAINASOLO, L. G., MOREL, V. P. I. & RAMBELOSON, R. A. 2001. Age and tectonic evolution of Neoproterozoic ductile shear zones in southwestern Madagascar, with implications for Gondwana studies. *Tectonics*, **20**, 1–45.

DE WIT, M. J., BOWRING, S. A. *ET AL.* 2005. The great Neoproterozoic central Saharan arc and the amalgamation of the North African Shield. *In*: *GAG-MAC-CSPG-CSSS Conference, Halifax, 2005, Abstracts with Programs*, 43.

DU TOIT, A. L. 1927. *A geological comparison of South America with South Africa*. Carnegie Institution of Washington, Publication **381**.

DU TOIT, A. L. 1937. *Our wandering continents*. Edinburgh, Oliver and Boyd.

EAGLES, G. 2007. New angles on South Atlantic opening. *Geophysical Journal International*, **166**, 353–361.

EAGLES, G. & KÖNIG, M. 2007. Jurassic and early Cretaceous plate kinematics of central Gondwana. *Geophysical Journal International*, in press.

EL-MAKHROUF, A. A. 1978. Tectonic interpretation of Jebal Eghei area and its regional application to Tibesti orogenic belt, south central Libya (SPLAJ). *Journal of African Earth Sciences*, **7**, 945–967.

EKWUEME, B. N. & ATTOH, K. 1996. The West African Shield. *In*: DE WIT, M. J. & ASWAL, L. D. (eds) *Greenstone Belts*. Oxford University Press, 517–528.

FAIRHEAD, J. D. & BRINK, R. M. 1991. Differential opening of the Central and South Atlantic Oceans and the opening of the West African rift systems. *Tectonophysics*, **187**, 191–203.

FERRE, E., GLEIZES, G. & CABY, R. 2002. Oblique convergent tectonics and granite emplacement in the Trans-Saharan belt of eastern Nigeria: a synthesis. *Precambrian Research*, **114**, 199–219.

FETTER, A. H., SARAIVA DOS SANTOS, T. J. *ET AL.* 2003. Evidence for Neoproterozoic continental arc magmatism in the Santa Quiteria batholith of Ceara State, NW Borborema province, NE Brazil: implications for the assembly of West Gondwana. *Gondwana Research*, **6**, 265–273.

FRIMMEL, H. E. & FRANK, W. 1998. Neoproterozoic tectonothermal evolution of the Gariep Belt and its basement, Namibia and South Africa. *Precambrian Research*, **90**, 1–28.

GHOSH, J. G., DE WIT, M. J. & ZARTMAN, R. E. 2004. Age and tectonic evolution of Neoproterozoic ductile shear zones in the southern granulite terrain of India, with implications for Gondwana studies. *Tectonics*, **23**, 1–38, doi:101029/2002TC001444.

GHUMA, M. A. & ROGERS, J. J. W. 1976. Geology, geochemistry & tectonic setting of the Ben Ghnema batholith, Tibesti massif, southern Libya. *Geological Society of America Bulletin*, **89**, 1351–1358.

GOSCOMBE, B., GRAY, D., ARMSTRONG, R., FOSTER, D. A. & VOGL, J. 2005. Event geochronology of the Pan-African Kaoko Belt, Namibia. *Precambrian Research*, **140**, 103–131.

GOSCOMBE, B. & GRAY, D. 2007. The coastal terrane of the Kaoko Belt, Namibia: outboard arc-terrane and tectonic significance. *Precambrian Research*, **155**, 139–158.

GRAY, D. R., FOSTER, D. A., GOSCOMBE, B., PASSHIER, C. W. & TROUW, R. A. J. 2006. $^{40}Ar/^{39}Ar$ thermochronology of the Pan-African Damara orogen, Namibia, with implications for tectonothermal and geodynamic evolution. *Precambrian Research*, **150**, 49–72.

GSELL, J. & SONET, J. 1960. *Notice explicative sur la feuille Adré. Carte Geologique de reconnaissance a l'échelle du 1/500,000 Republique du Tchad*. Institut Equatorial de Reserches et d'Etudes Geologiques et Minieres, Paris.

HARMS, U., SCHANDELMEIER, H. & DARBYSHIRE, D. P. F. 1996. Pan African reworked early-middle Proterozoic crust in NE Africa west of the Nile: Sr and Nd isotopic evidence. *Journal of the Geological Society, London*, **147**, 859–872.

HARRIS, N. B. W., HAWKESWORTH, C. J. & REIS, A. C. 1984. Crustal evolution in NE and E Africa from model Nd ages. *Nature*, **309**, 773–776.

HAWKESWORTH, C. J., KELLEY, S., TURNER, S., LE ROEX, A. P. & STOREY, B. C. 1999. Mantle processes during Gondwana break-up. *Journal African Earth Sciences*, **28**, 239–261.

HEILBRON, M. & MACHADO, N. 2003. Timing of terrane accretion in the Neoproterozoic-Eopaleozoic Ribeira orogen (SE Brazil). *Precambrian Research*, **125**, 87–112.

HINZ, K., NEBEN, S. *ET AL.* 1999. The Argentine continental margin north of 48°S: sedimentary successions, volcanic activity during breakup. *Marine and Petroleum Geology*, **16**, 1–25.

HURLEY, P. M. 1972. Can the subduction process of mountain building be extended to Pan-African and similar orogenic belts? *Earth and Planetary Science Letters*, **15**, 305–314.

HURLEY, P. M. 1974. Pangeaic Orogenic System. *Geology*, **2**, 373–376.

HURLEY, P. M., ALMEIDA, F. F. *ET AL.* 1967. Test of drift by comparison of radiometric ages. *Science*, **157**, 495–500.

JOHNSON, P. R. & WOLDEHAIMANOT, B. 2003. Development of the Arabian-Nubian Shield: perspectives on accretion and deformation in the northern East African Orogen and the assembly of Gondwana. *In*: YOSHIDA, M., WINDLEY, B. F. & DASGUPTA, S. (eds) *Proterozoic East Gondwana Supercontinent Assembly and Break-up*. Geological Society, London, Special Publications, **206**, 289–325.

KENNEDY, W. Q. 1964. The structural differentiation of Africa in the Pan African (500 m.y.) tectonic episode. *Annual Reports of the Institute of African Geology*, **8**, session 1962–1963, Leeds University, UK, 48–49.

KEIDEL, J. 1916. La geología de las sierras de la Província de Buenos Aires y sus relaciones con las montañas de Sud Africa y los Andes. Buenos Aires. *Anales del Ministerio de Agricultura de la Nación, Sección Geología, Mineralogía y Minería*, IX, **3**, 1–78.

KLEIN, E. L. & MOURA, C. A. V. 2008. The São Luís craton and the Gurupi Belt (Brazil): possible links with the West African craton and surrounding Pan-African belts. *In*: PANKHURST, R. J., TROUW, R. J., BRITO NEVES, B. B. & DE WIT, M. J. (eds) *West Gondwana: Pre-Cenozoic Correlations Across the South Atlantic Region*. Geological Society, London, Special Publications, **294**, 000–000.

KRABBENDAM, M. & BARR, T. D. 2000. Proterozoic orogens and the break up of Gondwana: why did some orogens not rift? *Journal African Earth Sciences*, **31**, 35–49.

KUSNIR, I. 1993. *Geologie, resources minerals et resources en eau du Tchad*. Connaissance du Tchad 1, Centre National d'Appui a la Recherche, Ministrie des Mines de l'Énergy et des Resources en Eau, N'Djamena.

KUSNIR, I. & MOUTAYA, H. A. 1998. Resources minerals du Tchad: une revue. *Journal of African Earth Sciences*, **24**, 549–562.

KUSTER, D. & LIÉGEOIS, J.-P. 2001. Sr, Nd isotopes and geochemistry of the Bayuda Desert high grade metamorphic basement (Sudan): an early Pan African oceanic convergent margin, not the edge of the east Saharan ghost craton? *Precambrian Research*, **109**, 1–23.

LAWVER, L. A., GAHAGAN, L. M. & DALZIEL, I. W. D. 1999. A tight fit–Early Gondwana, a plate reconstruction perspective. *In*: MOTOYOSHI, Y. & SHIRAISHI, K. (eds) *Origin and evolution of the continents*. Memoirs, National Institute Polar Research, Tokyo, Japan, Special Issue, **53**, 214–229.

LIÉGEOIS, J. P., LATOUCHE, L., BOUGHRARA, M., NAVEZ, J. & GUIRAUD, M. 2003. The LATEA metacraton (central Hoggar, Tuareg Shield, Algeria): behaviour of an old passive margin during Pan-African orogeny. *Journal of African Earth Sciences*, **37**, 161–190.

LINNEMANN, U., MCNAUGTHON, N. J., ROMER, R. L., GEHMLICH, M., DROST, K. & TONK, C. 2004. West African provenance for Saxo-Thuringa (Bohemian massif): Did Amorica ever leave pre-Pangean Gondwana?—U/Pb-SHRIMP zircon evidence and the Nd-isotope record. *International Journal of Earth Science (Geologische Rundschau)*, **93**, 683–705.

MACFARLANE, A. (ed.) 1980. *Geology and mineral deposits of Sierra Leone*. British Geological Survey, Overseas and Technical Report, OM07.

MEERT, J. G. 2003. A synopsis of events related to the assembly of eastern Gondwana. *Tectonophysics*, **362**, 1–40.

MOHRIAK, W., NEMČOK, M. & ENCISO, G. 2008. South Atlantic divergent margin evolution: Rift border uplift and salt tectonics along the southeastern Brazilian Basins. *In*: PANKHURST, R. J., TROUW, R. J., BRITO NEVES, B. B. & DE WIT, M. J. (eds) *West Gondwana: Pre-Cenozoic Correlations Across the*

South Atlantic Region. Geological Society, London, Special Publications, **294**, 000–000.

NANCE, R. D. & MURPHY, J. B. 1996. Basement isotopic signatures and Neoproterozoic paleogeography of Avalonian-Cadomian and related terranes in the circum-North Atalntic. *In*: NANCE, R. D. & THOMPSON, M. D. (eds) *Avalonian and related peri-Gondwanan terranes of the circum-North Atlantic*. Geological Society of America, Special Papers, **304**, 333–346.

NAGY, R. M., GHUMA, M. A. & ROGERS, J. J. W. 1976. A crustal suture and lineament in North Africa. *Tectonophysics*, **31**, 67–72.

NGAKO, V., AFFATON, P., NNANGE, J. M. & NJANKO, Th. 2003. Pan-African tectonic evolution in central and southern Cameroon: transpression and transtension during sinistral shear movement. *Journal of African Earth Sciences*, **36**, 207–214.

NEVES, S. P. 2003. Proterozoic history of the Borhorema Province (NE Brazil): Correlations with neighboring cratons and Pan-African belts and implications for the evolution of western Gondwana. *Tectonics*, **22**, 1031, doi:10.1029/2001TC001352.

OUZEGANE, K., KIENAST, J.-R., BENDAOUD, A. & DRARENI, A. 2003. A review of Archean and Paleoproterozoic evolution of the Ouzzal granulitic terrane (western Hoggar, Algeria). *Journal of African Earth Sciences*, **37**, 207–228.

PELGRAM, W. J., REGISTER, J. K. & FULLAGAR, P. D. 1987. Pan-African ages from Tibesti Massif batholith, southern Libya. *Earth and Planetary Science Letters*, **30**, 123–128.

PEUCAT, J. J., DRARENI, A., LATOUCHE, L., DELOULE, E. & VIDAL, P. 2003. U/Pb zircon (TIMS and SIMS) and Sm-Nd whole rock geochronology of the Gour Oumelalen granulite basement, Hoggar massif, Tuareg shield, Algeria. *Journal of African Earth Sciences*, **37**, 229–240.

PIMENTAL, M. M. & FUCK, R. A. 1992. Neoproterozoic accretion in central Brazil. *Geology*, **20**, 375–379.

PIUZANA, D., PIMENTEL, M. M. & FUCK, R. A. 2003. Neoproterozoic granulite facies metamorphism and coeval granitic magmatism in the Brasília Belt, central Brazil: regional implications of new SHRIMP U-Pb and Sm-Nd data. *Precambrian Research*, **125**, 245–273.

PIUZANA, D., PIMENTEL, M. M., FUCK, R. A. & ARMSTRONG, R. 2005. SHRIMP U-Pb and Sm-Nd data for the Araxá Group and associated magmatic rocks: constraints for the age of sedimentation and geodynamic context of the southern Brasília Belt, central Brazil. *Precambrian Research*, **125**, 139–160.

POIDOVIN, J.-L. 1985. *Le Proterozoique Superior de la Republique Centrafricaine*. Annals of Royal Museum for Central Africa, Tervuren, serie 8, number 91.

RABINOWITZ, P. D. & LABRECQUE, J. 1979. The Mesozoic South Atlantic Ocean and evolution of its continental margins. *Journal of Geophysical Research*, **84**, B11, 5973–6002, 1979.

RAPELA, C. W., PANKHURST, R. J., FANNING, C. M. & GRECCO, L. E. 2003. Basement evolution of the Sierra de la Ventana fold belt: new evidence for Cambrian continental rifting along the southern margin of Gondwana. *Journal of the Geological Society*, London, **160**, 613–628.

REEVES, C. V. 1999. Aeromagnetic and gravity features of Gondwana and their relation to continental break-up: more pieces, less puzzle. *Journal of African Earth Sciences*, **28**, 263–277.

REEVES, C. & DE WIT, M. J. 2000. Making ends meet in Gondwana: retracing the transforms of the Indian Ocean and reconnecting continental shear zones. *Terra Nova*, **12**, 272–280.

REEVES, C. V., SAHU, B. K. & DE WIT, M. J. 2002. A re-examination of the paleo-positions of Africa's eastern neighbours in Gondwana. *Journal of African Earth Sciences*, **34**, 101–108.

REEVES, C., DE WIT, M. J. & SAHU, B. K. 2004. Tight reassembly of Gondwana exposes Phanerozoic shears in Africa as global players. *Gondwana Research*, **7**, 7–19.

ROLIN, P. 1995. *Carte Tectoniqie de la Republique Centrafricaine, 1: 1 Million*. BRGM Orleans, France.

ROZENDAAL, A., GRESSE, P. G., SCHEEPERS, R. & LE ROUX, J. P. 1999. Neoproterozoic to early Cambrian crustal evolution of the PanAfrican Saldanian Belt, South Africa. *Precambrian Research*, **97**, 303–323.

SCHMITT, R. S., TROUW, R. A. J., VAN SCHMUS, W. R. & PIMENTEL, M. M. 2004. Late amalgamation in the central part of West Gondwana: new geochronological data and the characterization of a Cambrian collisional orogeny in the Ribeira Belt (SE Brazil). *Precambrian Research*, **133**, 29–61.

SCHMITT, R., TROUW, R. A. J. & PASSCHIER, C. W. 2005. Cambrian tectonic evolution of Ribeira Belt (SE-Brazil) and Kaoko Belt (NW Namibia); ties that bind underwater. *In*: PANKHURST, R. J. & VEIGA, G. D. (eds) *Gondwana 12: Geological and Biological Heritage of Gondwana Abstracts*. Academia Nacional de Ciencias, Córdoba, Argentina, 327.

SCHURMANN, H. M. E. 1974. *The Precambrian in North Africa*. E.J. Brill Publishing, Leiden, Netherlands.

SCOTESE, C. R. 2000. *Palaeomap Project: Earth History (palaeogeographic maps)*. Department of Geology, University of Texas, Arlington.

SMITH, W. & SANDWELL, D. 1997. *Measured and estimated seafloor topography (version 4.2)*. World Data Centre for Marine Geology and Geophysical research. Publication RP-1 (poster, 34″ × 53″). National Geophysical Data Center, Boulder, CO USA.

SONET, J. 1963. *Notice explicative sur la feuille Niére. Carte Geologique de reconnaissance a l'échelle du 1/500,000 Republique du Tchad*. Institut Equatorial de Reserches et d'Etudes Geologiques et Minieres, Paris.

STANKIEWICZ, J. & DE WIT, M. J. 2006. A proposed drainage evolution model for Central Africa—Did the Congo flow east? *Journal of African Earth Sciences*, **44**, 75–84.

STERN, R. J. 1994. Arc assembly and continental collision in the Neoproterozoic East African orogen: implications for the consolidation of Gondwana(land). *Annual Reviews of Earth and Planetary Sciences*, **22**, 319–351.

STERN, C. R. & DE WIT, M. J. 2004. Rocas Verdes ophiolite, southernmost South America: remnants of progressive stages of development of oceanic type crust in a continental back arc basin. *In*: DILEK, Y. & ROBINSON, P. T. (eds) *Ophiolites in Earth History*. Geological Society, London, Special Publications, **218**, 665–683.

STOREY, B. C. 1995. The role of mantle plumes in continental break up: case histories from Gondwanaland. *Nature*, **377**, 301–308.

SULTAN, M., CHAMBERLAIN, K. R., BOWRING, S. A., ARVIDSON, R. E., ABUZIED, H. & EL KALIOUBY, B. 1990. Geochronology and isotope evidence for involvement of pre-Pan African crust in the Nubian Shield. *Geology*, **18**, 761–764.

SULTAN, M., TUCKER, R. D., EL ALFY, Z., ATTIA, R. & RAGAB, A. G. 2000. U/Pb (zircon) ages for the gneissic terrane west of the Nile, southern Egypt. *Geologische Rundschau*, **83**, 514–522.

TACK, L., WINGATE, M. T. D., LIÉGEOIS, J.-P., RERNANDEZ-ALONSO, M. & DEBLONDE, A. 2001. Early Neoproterozoic magmatism (1000–910 Ma) of the Zadinian and Mayumbian Groups (Bas-Congo): onset of Rodinia rifting at the western edge of the Congo craton. *Precambrian Research*, **110**, 277–306.

TAGNE-KAMGA, G. 2003. Petrogenesis of the Neoproterozoic Ngondo plutonic complex (Cameroon, west-central Africa): a case of late collisional ferro-potassic magmatism. *Journal of African Earth Sciences*, **36**, 149–172.

TANKARD, A. J., SUAREZ, R. & WELSINK, H. J. (eds) 1996. *Petroleum Basins of South America*. American Association of Petroleum Geologists, Memoirs, **62**.

TOTEU, S. F., VAN SCHMUS, W. R., PENAYE, J. & MICHARD, A. 2001. New U–Pb and Sm–Nd data from north-central Cameroon and its bearing on the pre-Pan African history of central Africa. *Precambrian Research*, **108**, 45–73.

TOTEU, S. F., PENAYE, J. & DJOMANI, Y. P. 2004. Geodynamic evolution of the Pan African belt in Central Africa with special reference to Cameroon. *Canadian Journal of Earth Sciences*, **41**, 73–85, 2004.

TROMPETTE, R. 1994. *Geology of Western Gondwana (2000–500 Ma). Pan African-Brasiliano aggregation of South America and Africa*. A.A. Balkema, Amsterdam.

UNESCO 2000. *Geological Map of Africa at a scale of 1:5 million*. Geological Commission of the Geological Map of the World, Paris.

UNRUG, R. 1992. The supercontinental cycle and Gondwanaland assembly: component cratons and the timing of suturing events. *Journal of Geodynamics*, **16**, 215–240.

UNRUG, R. 1997. Rodinia to Gondwana. The geodynamic map of Gondwana supercontinent assembly. *GSA Today*, **2**, 2–6.

VAN SCHMUS, W. R., BRITO NEVES, B. B. *ET AL*. 2003. The Serido Group of NE Brazil, a late Neoproterozoic pre- to syn-collisional basin in the West Gondwana: insights from SHRIMP U-Pb detrital zircon ages and Sm-Nd crustal residence (Tdm) ages. *Precambrian Research*, **127**, 287–327.

VAN SCHMUS, W. R., OLIVEIRA, E. P., DA SILVA FILHO, A. F., TOTEU, S. F., PENAYE, J. & GUIMARÃES, I. P. 2008. Proterozoic links between the Borborema Province, NE Brazil, and the Central African Fold Belt. *In*: PANKHURST, R. J., TROUW, R. J., BRITO NEVES, B. B. & DE WIT, M. J. (eds) *West Gondwana: Pre-Cenozoic Correlations Across the South Atlantic Region*. Geological Society, London, Special Publications, **294**, 69–99.

VAIL, J. R. 1976. Outline of the geochronology and tectonic units of the basement complex of northeast Africa. *Proceedings of the Royal Society, London*, **A350**, 127–141.

VAUGHAN, A. P. M., LEAT, P. T. & PANKHURST, R. J. (eds) 2005. *Terrane processes at the Margins of Gondwana*. Geological Society, London, Special Publications, **246**.

Index

Page numbers in *italic* denote figures. Page numbers in **bold** denote tables.